Gaurav Bhalla

Biomimetic Oxidations Catalyzed by Transition Metal Complexes

Biomimetic Oxidations Catalyzed by Transition Metal Complexes

Editor

Bernard Meunier

Imperial College Press

Published by

Imperial College Press
57 Shelton Street
Covent Garden
London WC2H 9HE

Distributed by

World Scientific Publishing Co. Pte. Ltd.
P O Box 128, Farrer Road, Singapore 912805
USA office: Suite 1B, 1060 Main Street, River Edge, NJ 07661
UK office: 57 Shelton Street, Covent Garden, London WC2H 9HE

Library of Congress Cataloging-in-Publication Data
Biomimetic oxidations catalyzed by transition metal complexes /
editor, Bernard Meunier.
p. cm.
Includes bibliographical references and index.
ISBN 1-86094-098-6 (alk. paper)
1. Metalloenzymes. 2. Biomimetics. 3. Metal catalysts.
4. Oxidases. 5. Oxidation. I. Meunier, Bernard.
QP601.7.B57 1999
572'.51--dc21 99-36713
CIP

British Library Cataloguing-in-Publication Data
A catalogue record for this book is available from the British Library.

Printed in Singapore.

Acknowledgments

This book has been possible because of the enthusiastic participation of all contributors. I would like to express my deep gratitude for their engagement in the writing of this collective book. I also would like to thank all the talented researchers, colleagues and collaborators who have co-signed all my contributions in the field of biomimetic oxidations since 1979, when I moved from organometallic chemistry and crystallography to the fascinating world of oxygen atom and electron transfers catalyzed by transition metal complexes.

Toulouse, June 1998
Bernard Meunier

Contents

List of Contributors

Gary W. BRUDVIG
Department of Chemistry, Yale University,
P.O. Box 208107, New Haven, CT 06520–8107, USA

Diane E. CABELLI
Department of Chemistry, Brookhaven National Laboratory,
Upton, NY 11973, USA

Robert H. CRABTREE
Department of Chemistry, Yale University,
P.O. Box 208107, New Haven, CT 06520–8107, USA

Sergiu M. GORUN
Department of Chemistry, Brown University,
Providence, Rhode Island 02912, USA

John T. GROVES
Department of Chemistry, Princeton University,
New Jersey 08544, USA

Zhengbo HU
Department of Chemistry, Brown University,
Providence, Rhode Island 02912, USA

Keith U. INGOLD
Steacie Institute for Molecular Sciences,
National Research Council of Canada,
Ottawa, Ontario, Canada KIA OR6

Kenneth D. KARLIN
Department of Chemistry, The Johns Hopkins University, Baltimore, MD 21218, USA

Naruyoshi KOMIYA
Osaka University, Department of Chemistry, Faculty of Engineering Science, Toyonaka, Osaka 560–8531, Japan

Marie-Aude KOPF
Department of Chemistry, The Johns Hopkins University, Baltimore, MD 21218, USA

Hans-Jörg KRÜGER
Institut für Anorganische und Angewandte Chemie der Universität Hamburg, Martin-Luther-King-Platz 6, D-20146 Hamburg, Germany

Neil A. LAW
Department of Chemistry, The Willard H. Dow Laboratories, University of Michigan, Ann Arbor, MI 48109–1055, USA

Jinbo LEE
Department of Chemistry, Princeton University, New Jersey 08544, USA

Julian LIMBURG
Department of Chemistry, Yale University, P.O. Box 208107, New Haven, CT 06520–8107, USA

Philip A. MACFAUL
Steacie Institute for Molecular Sciences, National Research Council of Canada, Ottawa, Ontario, Canada KIA OR6

James M. MAYER
Department of Chemistry, Box 351700, University of Washington, Seattle, Washington 98195–1700, USA

Jennifer L. McLAIN
Department of Chemistry, Princeton University,
New Jersey 08544, USA

Bernard MEUNIER
Laboratoire de Chimie de Coordination du CNRS,
205 route de Narbonne, 31077 Toulouse cedex 4, France

Shun-Ichi MURAHASI
Osaka University, Department of Chemistry,
Faculty of Engineering Science, Toyonaka,
Osaka 560–8531, Japan

Vincent L. PECORARO
Department of Chemistry, The Willard H. Dow Laboratories,
University of Michigan, Ann Arbor, MI 48109–1055, USA

Dennis RILEY
Monsanto Co., 800 N. Lindbergh Blvd.,
St. Louis, MO 63167, USA

Anne ROBERT
Laboratoire de Chimie de Coordination du CNRS,
205 route de Narbonne, 31077 Toulouse cedex 4, France

Jorge A. RODRIGUEZ
Department of Chemistry and Biochemistry, UCLA,
Los Angeles, CA 90095–1569, USA

Roger A. SHELDON
Delft University of Technology, Faculty of Chemical Technology and Materials Science, Laboratory for Organic Chemistry and Catalysis, Julianalaan 136, 2628 BL Delft, The Netherlands

Carla SLEBODNICK
Department of Chemistry, The Willard H. Dow Laboratories,
University of Michigan, Ann Arbor, MI 48109–1055, USA

Joan S. VALENTINE
Department of Chemistry and Biochemistry, UCLA, Los Angeles, CA 90095–1569, USA

Charles G. YOUNG
School of Chemistry, University of Melbourne, Parkville, Vic. 3052, Australia

Haining ZHU
Department of Chemistry and Biochemistry, UCLA, Los Angeles, CA 90095–1569, USA

Introduction

Selective oxidation of raw materials is an important area for the chemical industry, due to the necessity to produce oxygen-containing chemicals from fossil hydrocarbons without complete conversion to carbon dioxide (such an easy reaction!). Selective oxidation is often a difficulty during the preparation of fine chemicals, since it is still challenging to cleanly introduce an alcohol function at the desired position of a drug precursor in the same manner as enzymes.

The first aspect of the field of metal-catalyzed oxidations is still concerned with large scale industrial processes using the old-fashioned, but efficient, autoxidation of alkanes, involving the initiation of a free radical chain reaction by transition metal salts (intermediate organic radicals react with the triplet state of molecular oxygen to produce alkoxyl radicals which are also able to abstract hydrogen atoms from the starting alkane). Researchers have tried very hard for the last two decades to mimic metalloenzyme-catalyzed oxidations by using rather simple metal complexes associated with different oxidants. The primary aim of this book is to present the efforts which have been produced in this field of "biomimetic oxidations." The choice of the oxidant is crucial to success in this challenging area. There are now several well-known traps which can be avoided in the modeling of enzyme-catalyzed oxidations:

* Nature is able to efficiently perform monooxygenation reactions by introducing one oxygen atom of molecular oxygen within a defined substrate, the second oxygen atom of O_2 being eliminated as a water molecule with the associated consumption of two electrons, provided by NAD(P)H, and two protons according to Eq. (1):

$$R\text{–}H + O_2 + 2e + 2H^+ \xrightarrow[\text{(e.g. cytochrome P450)}]{\text{monooxygenase}} R\text{–}OH + H_2O \qquad (1)$$

This electron transfer to monooxygenases is carefully controlled by reductases able to provide the necessary two electrons, and only two, to avoid an over-consumption of reducing equivalents. Chemists have difficulties in mimicking such elegant processes performed by multi-enzymatic systems. Many systems based on molecular oxygen and an electron source have been reported to catalyze hydroxylation and epoxidation reactions, but also the formation of water via the 2e-reduction of the intermediate metal-oxo species, this undesired reaction being the main reaction.[1]

* One way to "shunt" this difficulty is to use single oxygen atom donors like iodosylbenzene PhIO, hypochlorite NaOCl or monopersulfate $KHSO_5$ (PhIO was first used to demonstrate the "peroxide shunt" with cytochrome P450[2] and with synthetic metalloporphyrins[3,4]). Other oxidants like hydrogen peroxide H_2O_2 or alkylperoxides ROOH are obvious candidates as oxygen atom donors, in particular hydrogen peroxide is the suitable "green" oxidant since the only side product after an oxidation reaction is water. However, both peroxides have a weak and rather symmetrical O-O bond. The dissociation energy of the homolytic cleavage of the peroxidic O-O bond is 50 ± 2 kcal/mol, making its spontaneous homolysis very easy. This homolytic cleavage is even more facile when an electron is transferred into the antibonding orbital of the O-O bond leading to formation of hydroxyl radicals HO• (this is the well-known Fenton reaction when the electron is provided to H_2O_2 by an iron(II) salt).

* Another problem with hydrogen peroxide is to avoid the dismutation of hydrogen peroxide to dioxygen and water which is catalyzed by many different transition metal complexes (modeling of the activity of catalase).

* Alkylhydroperoxides ROOH can be easily cleaved by transition metal complexes and the resulting alkoxyl radicals RO• are able to efficiently abstract hydrogen atoms from alkanes leading to a mixture of alcohols and ketones as in autoxidation reactions. These alkane oxidations with ROOH have often been claimed as being biomimetic hydroxylations involving metal-oxo species, but both product distribution and absence of selective epoxidation activity indicate that the metal center is not involved in the control of the oxygen atom transfer as expected for metalloenzyme models.[4,5]

Effective monooxygenase models are able to perform both hydroxylation of alkanes and epoxidation of olefins, since the active species is an electrophilic high-valent metal-oxo entity like in the corresponding metalloenzymes. The metal center is able to activate the oxidant and then to control the oxygen atom transfer from the metal-oxo species to the substrate, even if some fugace radical intermediates are generated during the catalytic oxygenation reaction (the oxygen rebound mechanism involves an organic radical intermediate controlled by the metal center, without formation of free radicals).

* All the different studies on the modeling of oxygenases clearly confirms that the coordination of olefins to the metal-oxo center is not necessary to produce epoxides. The concept of olefin activation by low-valent metal complexes in catalytic hydrogenation does not apply to biomimetic catalysts which involve high-valent metal species. As a matter of fact, the order of reactivity of olefins in metal-catalyzed epoxidations is strictly the inverse of that observed for metal-catalyzed hydrogenations (in this latter case, primary olefins are more easily hydrogenated than tetrasubstituted olefins. In the case of metal-catalyzed epoxidations, terminal olefins are always less reactive than electron rich di-, tri- or tetrasubstituted olefins).

Finally, all these biomimetic oxidations have largely contributed to development of methods for the characterization of high-valent metal-oxo species [6,7] and to the understanding of their chemical reactivity with respect to organic substrates.[8–12] The incorporation of oxygen atoms from water into products during epoxidations or hydroxylations performed in water solutions has been useful to understand the exchange of the oxygen atom of high-valent metal-oxo species with water via the axial *trans* hydroxo ligand ("redox tautomerism" or more precisely "oxo-hydroxo tautomerism").[8,11] The lifetime of a manganese(V)-oxo porphyrin complex is below one second in water.[10] Apart from these electrophilic metal-oxo species, metal(III)-hydroperoxo porphyrin or phthalocyanine complexes M^{III}-OOH are also involved as active entities in the oxidation of electron-deficient double bonds. The nucleophilic properties of these hydroxo complexes has been evidenced in the epoxidation of quinones[13] and the degradation of chlorophenols.[14,15]

Modeling of non-heme oxygenases and oxidases has also provided useful information on some intermediates involved in the catalytic cycles of these enzymes (many of them having an active site with two metallic centers like methane monooxygenase or ribonucleotide reductase). The activation of molecular oxygen by a dicopper complex has recently provided key data on the reversible cleavage and formation of the dioxygen O-O bond by a bimetallic system.[16] As for oxo-metalloporphyrin catalysts, the rate–determining step in the oxidation of hydrocarbons by a μ-oxo manganese dimer is the abstraction of a hydrogen atom.[17]

All these efforts on the elaboration of biomimetic oxidation catalysts have provided useful proposals on the catalytically active species for monooxygenases and peroxidases before the recent determinations of tri-dimensional structures of monooxygenases (cytochrome P450,[18,19] methane monooxygenase[20]) or peroxidases (ligninase,[21] chloroperoxidase[22]).

After Sheldon and Kochi's[23] milestone book — "Metal-Catalyzed Oxidations of Organic Compounds" and also Mijs and De Jonge's[24] "Organic Syntheses by Oxidation with Metal Compounds," this present book will provide views on the different state-of-the-art aspects of biomimetic oxidations catalyzed by transition metal complexes (e.g. models of flavin-based oxygenases not included) distributed in 14 different chapters:

- Chapter 1: *Thermodynamic influences of C-H bond hydroxylation* by James M. MAYER.
- Chapter 2: *Distinguishing biomimetic oxidations from oxidations mediated by freely diffusing radicals* by Keith U. INGOLD and Philip MacFAUL.
- Chapter 3: *Biomimetic oxygenations related to cytochrome P450. Metal-oxo and metal-peroxo intermediates* by Jennifer L. McLAIN, Jinbo LEE and John T. GROVES.
- Chapter 4: *Models of heme peroxidases and catalases* by Bernard MEUNIER.
- Chapter 5: *Non-heme peroxidases and catalases: mechanistic implications from the studies of manganese and vanadium model compounds* by Carla SLEBODNICK, Neil A. LAW and Vincent L. PECORARO.
- Chapter 6: *Methane monooxygenase models* by Zhengbo HU and Sergiu M. GORUN.

- Chapter 7: *Models of copper enzymes and heme-copper oxidases* by Marie-Aude KOPF and Kenneth D. KARLIN.
- Chapter 8: *Models of catechol dioxygenases* by Hans J. KRÜGER.
- Chapter 9: *Biomimetic chemistry of molybdenum* by Charles G. YOUNG.
- Chapter 10: *Models of superoxide dismutases* by Joan S. VALENTINE, Diane CABELLI and Dennis RILEY.
- Chapter 11: *Modeling of the oxygen-evolving complex in photosystem II* by Julia LIMBURG, Gary W. BRUDVIG and Robert H. CRABTREE.
- Chapter 12: *Asymmetric biomimetic oxidations* by Anne ROBERT and Bernard MEUNIER.
- Chapter 13: *Bioinspired oxidations catalyzed by ruthenium complexes* by Shun I. MURAHASHI and Naruyoshi KOMIYA.
- Chapter 14: *Biocatalytic and biomimetic oxidations from an industrial perspective* by Roger A. SHELDON.

This book is directed towards chemists (both in academic and industrial institutions) engaged in organic synthesis, coordination chemistry and chemical or enzymatic catalysis. One of our primary concerns is to develop the understanding of catalytic oxidations and we hope that students will enjoy the reading of this book after an initial academic background of 2–3 years.

The principal literature has been covered through the middle of 1998.

References

1. Tabushi, I., Kodera, M. and Yokoyama, M. *J. Am. Chem. Soc.* 1985, **107:** 4466–4473.
2. Lichtenberger, F., Nastainczyk, W. and Ullrich, V. *Biochem. Biophys. Res. Commun.* 1976, **70:** 939–946.
3. Groves, J.T., Nemo, T.E. and Myers, R.S. *J. Am. Chem. Soc.* 1979, **101:** 1032–1033.
4. Meunier, B. *Chem. Rev.* 1992, **92:** 1411–1456.
5. (a) MacFaul, P.A., Arends, I.W.C.E., Ingold, K.U. and Wayner, D.D.M. *J. Chem. Soc. Perkin Trans. 2*, 1997, 135–145; (b) MacFaul, P.A., Ingold, K.U., Wayner, D.D.M and Que, L. *J. Am. Chem. Soc.* 1997, **119:**

10594–10598; (c) MacFaul, P.A., Wayner, D.D.M. and Ingold, K.U. *Acc. Chem. Res.* 1998, **31:** 159–162.

6. Groves, J.T., Haushalter, R.C., Nakamura, M., Nemo, T.E. and Evans, B. *J. Am. Chem. Soc.* 1981, **103:** 2884–2886.
7. Ayougou, K., Mandon, D., Fisher, J., Weiss, R., Müther, M., Schünemann, V., Trautwein, A.X., Bill, E., Terner, J., Jayaraj, K., Gold, A. and Austin, R.N. *Chem. Eur. J.* 1996, **2:** 1159–1163.
8. (a) Bernadou, J., Fabiano, A.S., Robert, A. and Meunier, B. *J. Am. Chem. Soc.* 1994, **116:** 9375–9376; (b) Bernadou, J. and Meunier, B. *Chem. Commun.* 1998, 2167–2173.
9. Lee, K.A. and Nam, W. *J. Am Chem. Soc.* 1997, **119:** 1916–1922
10. Groves, J.T., Lee, J. and Maria, S.S. *J. Am. Chem. Soc.* 1997, **119:** 6269–6273.
11. Song, R., Sorokin, A., Bernadou, J. and Meunier, B. *J. Org. Chem.* 1997, **62:** 673–678.
12. Shaik, S., Filatov, M., Schröder, D. and Schwarz, H. *Chem. Eur. J.* 1998, **4:** 193–199.
13. Selke, M. and Valentine, J.S. *J. Am. Chem. Soc.* 1998, **120:** 2652–2653.
14. Sorokin, A., Séris, J.L. and Meunier, B. *Science* 1995, **268:** 1163–1166.
15. Meunier, B. and Sorokin, A. *Acc. Chem. Res.* 1997, **30:** 470–476.
16. Halfen, J.A., Mahapatra, S., Wilkinson, E.C., Kaderli, S., Young, V.G., Que, L., Zuberbühler, A.D. and Tolman, W.B. *Science* 1996, **271:** 1397–1400.
17. Wang, K. and Mayer, J.M. *J. Am. Chem. Soc.* 1997, **119:** 1470–1471.
18. Poulos, T.L., Finzel, B.C. and Howard, A.J. *J. Mol. Biol.* 1987, **195:** 687–700.
19. Poulos, T.L., Cupp-Vickery, J. and Huiying, L. in *Cytochrome P450, Structure, Mechanism, and Biochemistry* (ed.) Ortiz de Montellano P.R. Plenum Press, New York, 1995, Chap. 4, 125–150.
20. Rosenzweig, A.C., Nordlund, P., Takahara, P.M., Frederick, C.A. and Lippard, S.J. *Chem. & Biol.* 1995, **2:** 409–418.
21. Poulos, T.L., Edwards, S.L., Wariishi, H. and Gold, M.H., *J. Biol. Chem.* 1993, **268:** 4429–4440.
22. Sundaramoorthy, M., Terner, J. and Poulos, T.L. *Structure* 1995, **3:** 1367–1377.
23. Sheldon, R.A. and Kochi, J.K., *Metal-Catalyzed Oxidations of Organic Compounds.* Academic Press, New York, 1981.
24. *Organic Syntheses by Oxidation with Metal Compounds* (eds.) Mijs, W.J and De Jonge, C.R.H.I. Plenum Press, New York, 1986.

Chapter 1

Thermodynamic Influences on C–H Bond Oxidation

James M. Mayer

Department of Chemistry, Box 351700,
University of Washington, Seattle,
Washington 98195–1700, USA

1.1 Introduction

Selective partial oxidations of organic molecules are important in chemistry and biology, and are of substantial commercial significance. A wide range of biochemical systems accomplish such transformations under mild conditions and with high selectivity, as discussed throughout this volume. These systems provide guideposts for the discovery of chemical oxidation systems, termed biomimetic systems. The following chapters examine different classes of reactions and different groups of redox enzymes. The goal of this chapter is to provide a framework for these various reactions rooted in their thermochemistry. We begin with the thermochemistry of the overall transformations, for instance methane to methanol, then move to discussions of mechanisms and the thermodynamics of individual reaction steps. A key issue is how the kinetics of a reaction step are affected by its thermodynamic driving force. The primary focus will be on reactions that involve transfer of a hydrogen atom.

1.2 Thermochemical Analysis of Oxidation Reactions

The oxidation of hydrocarbons and other organic molecules is not difficult — consider a room filled with methane, air, and a match. This is because of the large overall exothermicity of combustion. The real challenge lies in partial oxidation i.e. stopping the oxidation at an intermediate product such as methanol. This is difficult as the initial hydroxylation is the least exothermic of the four two-electron steps (Eqs. 1.1–1.4; all species gas phase).[1] These are written as reactions with an oxygen ("O") atom to facilitate the comparison with oxygen atom donor reagents below; analogous reactions with $\frac{1}{2}O_2$ are less exothermic by 59.6 kcal/mol per oxygen atom. Hydroxylations of higher alkanes are more favorable than that for methane, by up to 12 kcal/mol (Table 1.1). Alkene epoxidation is slightly less favorable: ΔH° for ethylene + O → ethylene oxide is –84.7 kcal/mol, and epoxidation of higher olefins are up to ca. 7 kcal/mol more favorable.[2]

$$CH_4 + O \rightarrow CH_3OH \qquad \Delta H^\circ = -89.8 \text{ kcal/mol} \qquad (1.1)$$

$$CH_3OH + O \rightarrow H_2C{=}O + H_2O \qquad \Delta H^\circ = -95.3 \text{ kcal/mol} \qquad (1.2)$$

$$H_2C{=}O + O \rightarrow HCOOH \qquad \Delta H^\circ = -124.1 \text{ kcal/mol} \qquad (1.3)$$

$$HCOOH + O \rightarrow CO_2 + H_2O \qquad \Delta H^\circ = -120.9 \text{ kcal/mol} \qquad (1.4)$$

To convert methane to methanol, an oxidant must be able to donate an oxygen atom for less than about 90 kcal/mol, so that the reaction will be exothermic. The thermodynamic oxygen donor abilities of common oxygen atom donors exhibit an enormous range, from 40 kcal/mol for N_2O to

Table 1.1 C–H bond strengths and heats of hydroxylation for simple hydrocarbons[a]

R–H	R–H → R• + H•	R–H + O → ROH
CH_3–H	104.9 ± 0.1	–89.8
CH_3CH_2–H	101.1± 0.4	–95.6
$(CH_3)_2CH$–H	98.6 ± 0.4	–99.6
$(CH_3)_3C$–H	96.5 ± 0.4	–102.0
C_6H_5–H	111.2 ± 0.8	–102.4
$C_6H_5CH_2$–H	89.9 ± 0.6	–95.5

[a]ΔH° values for gas-phase species, in kcal/mol, from Ref. 1.

Table 1.2 Thermochemistry of selected oxygen atom transfer reactions[a]

XO / X	$XO \rightarrow X + O$
O_3 / O_2	25
H_2O_2 / H_2O	34
N_2O / N_2	40
$\frac{1}{2}$ O_2 / –[b]	60
C_5H_5NO / C_5H_5N	72
epoxide / alkene	85 – ~92
Me_2SO / Me_2S	87
ROH / RH	90–102[c]
CO_2 / CO	127
Ph_3PO / Ph_3P	135[d]

[a]In kcal/mol; all species gas phase. Adapted from Ref. 4. Data from Refs. 1, 2 (epoxides), 3 and 5 (pyO). [b]$\Delta H°$ for 1/2 $O_2(g) \rightarrow O(g)$. [c]Table 1.1. [d]See Ref. 6.

135 kcal/mol for Ph_3PO (Table 1.2). The ~100 kcal/mol range corresponds to a difference in equilibrium constants at ambient temperatures of ca. 10^{70}. Entropic effects are neglected in this discussion because they are expected to be small for $XO + Y \rightarrow X + OY$ reactions. (A large ΔS of ± 34 e.u. corresponds to ± 10 kcal/mol in free energy at 25°C.) This thermochemical approach to oxygen atom transfer follows that of Holm.[3] Me_2SO (DMSO) is a good midpoint of this scale,[4] with D[S–O] = 87 kcal/mol. Reagents with weaker X–O bonds are good oxygen atom donors and have the thermodynamic ability to hydroxylate alkanes and epoxidize most olefins. The chemistry of compounds with stronger X–O bonds is typically X acting as an oxygen atom acceptor to form XO.

Having the thermodynamic driving force for a reaction is a necessary but not sufficient condition for reactivity. The oxidation of methane to methanol by O_2 is 30 kcal/mol downhill yet does not occur at a significant rate under normal conditions. This is why metal complexes (and metal or metal-oxide surfaces) are widely used as catalysts. In one common mechanism, the terminal oxidant generates a metal-oxo complex which in turn oxidizes the substrate. This requires M–O bond strengths of about 60 ± 20 kcal/mol. High-valent iron-, manganese-, and ruthenium-oxo complexes such as discussed in Chaps. 2, 3, and 13 appear to fall in this range. The molybdenum-oxo bonds that are the focus of Chap. 9 are clearly stronger, as these enzymes deoxygenate DMSO. In general, the earlier and heavier transition metals tend to make stronger bonds to oxygen, for instance, in vanadium(V), rhenium(V) and tungsten(IV) complexes. The W≡O bond in $W(O)Cl_2(PR_3)_3$ has been estimated to be ≥ 140 kcal/mol, and this provides the driving force for the deoxygenation of alcohols to alkanes and epoxides to alkenes.[4,7] The tungsten-oxygen bond is so strong that tungsten alkoxide complexes can simply homolyze to give the W≡O and R•.[7] Compounds with more than one oxo ligand tend to have weaker metal-oxo bonds. The solution $(\eta^5\text{-}C_5Me_5)(O)_2Re{=}O$ bond strength of 117 ± 1 kcal/mol is one of very few for which direct experimental data is available.[8] The remarkable range of metal-oxo bond strengths, perhaps 80 kcal/mol or more between tungstyl and ferryl complexes, may be without peer in transition metal chemistry.

1.3 Oxidation Mechanisms and the Energetics of Individual Steps

1.3.1 Mechanisms of C–H bond oxidation

Many different mechanisms have been established or suggested for the activation of C–H bonds by transition metal complexes. This chapter is focused on pathways thought to be relevant to biomimetic oxidations; more complete discussions and references can be found elsewhere.[9] Concerted insertion of an oxygen atom from a metal-oxo or peroxo complex into a C–H bond (Eq. 1.5) — analogous to carbene insertions[10] — has been much discussed though a paradigmatic example is lacking. The enthalpy of this step is the sum of the M=O bond strength and the $\Delta H°$ for hydroxylation of RH (Table 1.1).

$$L_nM{=}O + R{-}H \rightarrow L_nM + ROH \qquad (1.5)$$

Other mechanisms for C–H bond oxidation involve intermediates, and typically formation of the first intermediate is the rate-determining step. Examples are illustrated in Eqs. 1.6–1.10 for a metal-oxo complex. The energetics of such steps can provide mechanistic and kinetic insight in favorable circumstances, enabling prediction of rate constants (see below). Outer sphere electron transfer (Eq. 1.6) and hydrogen atom transfer (Eq. 1.7) are frequently invoked in biomimetic oxidations. Hydride transfer (Eq. 1.8) has received less attention but is strongly implicated in permanganate- and ruthenium-oxo oxidations of alcohols/alkoxides, formate, and other good hydride donors.[11] Initial proton transfer (Eq. 1.9) is relevant primarily for acidic substrates such as alcohols, phenols, and a few unusual hydrocarbons. This most often occurs as a pre-equilibrium prior to the rate-limiting redox step. [2+2] addition of a C–H bond across a metal-ligand π bond (Eq. 1.10), has been observed for oxo, imido, and alkylidene complexes.[9a] This is an organometallic mechanism since a metal-carbon bond is formed. The other organometallic pathways of oxidative addition, σ-bond metathesis, and deprotonation of a σ-complex appear to be less relevant to biomimetic oxidations. They are of much interest because of their unusual selectivity.[9]

$$L_nM{=}O + R{-}H \rightarrow L_nM{=}O^- + RH^+ \tag{1.6}$$

$$L_nM{=}O + R{-}H \rightarrow L_nM{-}O{-}H + R\bullet \tag{1.7}$$

$$L_nM{=}O + R{-}H \rightarrow L_nM{-}O{-}H^- + R^+ \tag{1.8}$$

$$L_nM{=}O + R{-}H \rightarrow L_nM{-}O{-}H^+ + R^- \tag{1.9}$$

$$L_nM{=}O + R{-}H \rightarrow L_nM(R)OH \tag{1.10}$$

1.3.2 Energetics of electron transfer

Electron transfer is one of the most fundamental chemical and biochemical reactions and it is the paradigm of a reaction understood in great detail. Marcus theory and its extensions indicate what features determine electron transfer rates, from vibrations within a reacting center to the influence of the solvent. A recent report successfully predicted electron transfer self-exchange rates solely based on parameters from optical spectra.[12] Bimolecular reaction of an electron donor D with an acceptor A occurs by initial formation of a precursor complex D||A in which electron transfer takes place (Eq. 1.11). The successor complex $D^+||A^-$ then separates to products.

$$D + A \rightleftarrows D\|A \rightarrow D^+\|A^- \rightarrow D^+ + A^- \tag{1.11}$$

The free energy of the overall electron transfer is given by the redox potentials $D^{0/+}$ and $A^{0/-}$, which can often be measured electrochemically. Within a solvent cage, the effective $E°(\Delta G°)$ should be corrected for electrostatic effects, giving $E°(\Delta G°)$, a correction which can be quite large in solvents of low dielectric constant.[92] The potentials and the reaction chemistry should, if possible, be studied under the same conditions: the same solvent, ionic strength, temperature, etc. This applies to any use of thermodynamic quantities. In cases where the relevant values are not available for reaction conditions, arguments should be made with caution. For example, applying the gas-phase heats of reaction discussed above to solution reactions requires assumptions about the constancy of heats of solvation — often reasonable for relatively nonpolar organic molecules. When charged species are involved, however, the heats of solution are much larger, and gas phase values cannot be used.

Electron transfer reactions in solvents of low polarity, such as methylene chloride, are complicated by ion pairing effects.[13] Ion pairing energies can be as high as 10 kcal/mol (K_{eq} ~ 10^7 M^{-1}). As long as the chemical and electrochemical reactions are examined at the same ionic strength and with similarly sized ions, the electrochemical potentials should closely indicate the free energy of formation of $D^+||A^-$. But there may be additional effects if the mechanism requires subsequent dissociation of the ion pair.

1.3.3 Thermochemistry of hydrogen atom transfer

Hydrogen atom transfer is a frequent step in chemical and biological reactions. Examples include combustion, industrial partial oxidation processes,[14] trapping of oxygen radicals by antioxidants,[15] and vitamin B-12 reactions.[16] The industrial processes include both homogeneous and heterogeneous oxidations, for instance *p*-xylene to terephthalic acid and propylene to acrylonitrile, and account for more than 10^{10} lb/yr of U.S. production alone.[17]

ΔH° for a hydrogen atom transfer reaction (Eq. 1.12) is simply the difference in the two bond dissociation energies, D(H–Y) – D(H–X). Such

$$X + H{-}Y \rightarrow X{-}H + Y \qquad (1.12)$$

bond dissociation energies, or bond strengths, are available for many small molecules (c.f. Table 1.1). The determination of X–H bond strengths for many larger molecules is now possible using a thermochemical cycle developed by Bordwell and others (Scheme 1.1).[18] Conceptually, a hydrogen atom is composed of a proton and an electron so the affinity of X for H• is directly related to its affinity for e^- (a redox potential, $E°$) and H^+ (a pK_a). The X–H bond strength is derived from $E°$ and pK_a values by adding a constant C, which depends on the solvent and reference electrode

Scheme 1.1

$$\begin{array}{rcll} X + e^- & \rightarrow & X^- & \Delta G^\circ = -nFE^\circ = -23.06E^\circ\,(\mathrm{kcal/mol}) \\ X^- + H^+ & \rightarrow & X{-}H & \Delta G^\circ = RT\ln K_a = -1.37\mathrm{p}K_a\,(\mathrm{kcal/mol}) \\ H\bullet & \rightarrow & e^- + H^+ & -C\ (\text{a constant}) \\ \hline X + H\bullet & \rightarrow & X{-}H & \Delta H^\circ = -23.06E^\circ - 1.37\mathrm{p}K_a - C\,(\mathrm{kcal/mol}) \end{array} \qquad (1.13)$$

(Table 1.3). To convert from free energy measurements ($E°$ and pK_a) to an enthalpy, D(H–X), it is necessary to assume that the entropies of X and XH are equal,[18] which seems reasonable for large molecules that differ only by a hydrogen atom (but note the counterexample in Sec. 1.6.2 below). In

Table 1.3 Values of C (kcal/mol)

Solvent	Electrochemical Reference	C	Reference
DMSO	$Cp_2Fe^{+/o}$	73.3	18a
MeCN	$Cp_2Fe^{+/o}$	59.5	18g,h
H_2O	H^+/H_2 (NHE)	57 ± 2	19

the formalism of Scheme 1.1, C includes both the free energy of oxidation of H• and the entropy of H•. Bordwell and co-workers empirically determined C for DMSO solvent with $Cp_2Fe^{+/o}$ as the reference using E° and pK_a data for species with known bond strengths.[18a] The same constant was obtained for a number of different substrates, indicating that the method would be widely applicable. A similar cycle has been used by Tilset and Parker to calculate metal-hydride bond strengths in acetonitrile.[18e–g]

We have adapted this approach to determine the H• affinity of permanganate (MnO_4^-), in other words the O–H bond strength in monoprotonated manganate, $HMnO_4^-$ (Scheme 1.2).[19] Following Tilset and Parker, the normal hydrogen reference electrode (NHE) is explicitly included in our Scheme so that the constant can be expressed in terms of other physical quantities (Eq. 1.14). These quantities can be obtained from standard sources except for the enthalpy of solution of H•, which was taken as equal to the solvation of H_2 (1 kcal/mol[20]) as has been previously suggested.[21] Our value of C for aqueous solutions with potentials referenced to NHE is 57 ± 2 kcal/mol.[19]

As a check on our value of C, the $[O_3MnO\text{–}H]^-$ bond strength was calculated with an alternative cycle involving H• transfer from another species, XH (Eq. 1.15). ΔG° for reaction 1.15 is given by the difference between the E° and pK_a values (Eq. 1.16) and equal to ΔH° with the assumption that $\Delta S \cong 0$. ΔH° is the difference between the $H\text{–}OMnO_3^-$ and H–X bond strengths. The relevant E° and K_a values for phenol and aniline have recently been reported,[24] along with the solution bond strengths. These bond strengths are different from the gas phase values because of differential solvation of XH and X•; the $\Delta\Delta G$ for

Scheme 1.2

$$\left.\begin{array}{r} MnO_4^-(aq) + e^- \rightarrow MnO_4^{2-}(aq) \\ 1/2H_2(g) \rightarrow H^+(aq) + e^- \end{array}\right\} E^\circ = 0.564\ \text{V}\ (\equiv 13.0\ \text{kcal/mol})^{22}$$

$$MnO_4^{2-}(aq) + H^+(aq) \rightarrow HMnO_4^-(aq) \quad pK_a = 7.4\ (\equiv 10.1\ \text{kcal/mol})^{23}$$

$$H\bullet(aq) \rightarrow 1/2H_2(g) \quad C = 57\ \text{kcal/mol}$$

$$MnO_4^-(aq) + H\bullet(aq) \rightarrow HMnO_4^-(aq) \quad \Delta H^\circ = -80 \pm 3\ \text{kcal/mol}$$

$$C = \Delta H^\circ\left\{H\bullet(g) \rightarrow \frac{1}{2}H_2(g)\right\} - \Delta H^\circ\{H\bullet(g) \rightarrow H\bullet(aq)\} - \frac{1}{2}TS^\circ\{H_2(g)\} \tag{1.14}$$

solvation of PhOH vs. PhO• in water was estimated to be 3 kcal/mol. All the calculations — Scheme 1.2 and Eqs. 1.15 and 1.16 — are in agreement with $D(H\text{–}OMnO_3^-) = 80 \pm 3$ kcal/mol.[19]

$$MnO_4^- + X\text{–}H \leftrightarrows HMnO_4^- + X\bullet \tag{1. 15}$$

$$\Delta G = 23.06[E(X\bullet) - E(MnO_4^-)] + 1.37[pK_a(XH) - pK_a(HMnO_4^-)]$$

$$\cong \Delta H^\circ = BDE(X\text{–}H) - BDE(H\text{–}OMnO_3^-) \tag{1.16}$$

As the discussion above indicates, there are a number of equivalent ways to apply this thermochemical approach. Instead of adding an electron and then a proton, H^+ can be added first and then e^-, so that the cycle uses the pK_a of AH^+ and the E° of $AH^{+/o}$ (versus the pK_a of AH and the E° of A in Scheme 1.1). Aqueous electrochemical measurements may reflect proton-coupled electron transfer as one step. For instance, the observed reduction potential for $[(bpy)_2(py)Ru{=}O]^{2+}$ corresponds to its addition of an electron and a proton (Eq. 1.17; *E* versus NHE).[25] Evidence for this includes a dependence of *E* on pH with a slope of 59 mV per pH unit. When such Nernstian behavior is followed, the bond strength can be determined from the redox potential extrapolated to pH 0, according to $\Delta H^\circ = 23.06E^\circ + 57$ kcal/mol (Eq. 1.18). The various approaches are equivalent because the bond strength is a thermodynamic quantity that is path-independent, as illustrated by the square Scheme 1.3.[26]

Scheme 1.3

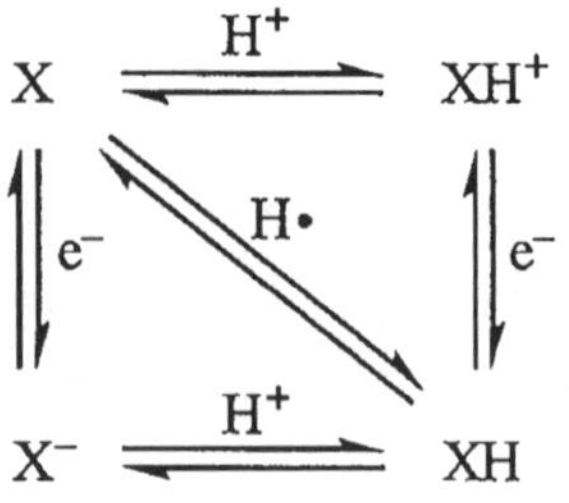

$$[(bpy)_2(py)Ru{=}O]^{2+} + H^+ + e^- \rightarrow [(bpy)_2(py)RuOH]^{2+} \; E = 0.77 \text{ V @ pH7} \tag{1.17}$$

$$[(bpy)_2(py)Ru{=}O]^{2+} + H\bullet \rightarrow [(bpy)_2(py)RuOH]^{2+} \; \Delta H^\circ = 84 \text{ kcal/mol} \tag{1.18}$$

These cycles are conceptually simple but present some pitfalls, some of which we have fallen into. Our original report erroneously included $\Delta H_{solv}(H_2)$ in the definition of C,[27a,b] which is not needed because the normal hydrogen electrode uses H_2 (g). This has taught us to always show the phase and/or solvent for each of the components in the cycle. We also used a slightly different formulation of Scheme 1.2 with an incorrect reference state for $\Delta G_{solv}(H_2)$.[27b] The $D(O_3MnO{-}H^-)$ reported here and in Ref. 19 is also different from our earlier reports in using the recently re-determined pK_a of $HMnO_4^-$ from pulse radiolysis studies.[23] Even these values are not ideal because the pulse radiolysis conditions are not identical to the electrochemical ones, for instance in ionic strength (which has a small effect on the $MnO_4^{-/2-}$ potential[28]). Given the potential problems, it would be valuable for reports of such calculations to describe the approach used and the values of E°, pK_a, and C.

The thermochemical cycles can be configured to give solution or gas phase bond strengths. The strict definition of bond strengths requires all gas phase species. Gas phase bond strengths are routinely used in solution arguments and are often assumed to be phase and solvent independent. Bordwell derives gas-phase bond strengths from the solution E° and pK_a values by assuming equal heats of solvation for X and XH. This is reasonable for nonpolar materials but there is debate for species such as phenol when the added hydrogen can engage in a hydrogen bond.[18c,i,24] If A and AH have similar heats of solvation, solution bond strengths should differ only because of the heat of solution of H•, which is a small number [~1 kcal/mol in H_2O (see above)]. We have chosen to calculate solution bond strengths so that assumptions do not have to be made about solvation of (potentially charged) metal complexes L_nMX and L_nMXH. In addition, we aim to understand solution reactivity. It remains to be determined whether bond strengths involving metal complexes are as solvent independent as the organic bond strengths are usually assumed to be.

1.3.4 *Thermochemistry of other C–H activation steps*

The free energy for proton transfer (Eq. 1.9) is given by the difference in the two pK_a values, as long as attention is paid to the caveats above about similar conditions. Tables of pK_a values of organic compounds are available.[29] The abilities of organic molecules to act as hydride donors, as in Eq. 1.8, have been analyzed by Parker and co-workers using a thermochemical cycle that includes one pK_a and two redox potentials.[30] Such values, like the bond strengths discussed above, are available only in experimentally favorable cases. It is even more difficult to determine the thermochemistry for other kinds of mechanisms of C–H activation, such as the organometallic [2+2] path of Eq. 1.10 without the good fortune of detecting the intermediate and measuring an equilibrium constant.

1.4 Kinetic / Thermodynamic Relationships[31]

There is no general relationship between kinetic parameters (e.g. $\Delta G^{\ddagger}$, $\Delta H^{\ddagger}$, $\Delta S^{\ddagger}$) and the ground state thermochemistry of a reaction (ΔG, ΔH, ΔS). However, there are a few classes of reactions, most notably electron transfer, for which there is good understanding of how the driving force affects the rate constant. And most classes of reactions, over a sufficiently narrow range of reactants and energy differences, show a linear correlation between relative rate constants ($\Delta\Delta G^{\ddagger}$) and differences in free energy ($\Delta\Delta G$). This is a principle underlying Hammett σ-ρ analyses, Brønsted plots for proton transfer reactions, Evans-Polanyi relations for hydrogen atom transfer, and other linear free energy relations.[31] One can mathematically expand $\Delta\Delta G^{\ddagger}$ as a power series in $\Delta\Delta G$ (Eq. 1.19); for sufficiently small values of the latter only the first two terms are important.

$$\Delta\Delta G^{\ddagger} = \zeta + \alpha(\Delta\Delta G) + \beta(\Delta\Delta G)^2 + \ldots \qquad (1.19)$$

In this report, we will also correlate $\Delta G^{\ddagger}$ and $\Delta H^{\ddagger}$ with enthalpies of reaction ($\Delta\Delta H$), which appear to be essentially equally valid.[31a] Such correlations require that the entropy changes $\Delta\Delta S$ be small or proportional to enthalpy and free energy changes, as is often observed.

The discussion here will utilize transition state theory (Fig. 1.1), as implied by the symbol ‡. It should be noted, however, that recent work suggests that more complex treatments are needed to understand some biochemical C–H bond oxidations. Klinman and co-workers have emphasized the importance of quantum mechanical tunneling, in which not only the height but also the width of the barrier is important.[32] A clear example is the rate-limiting C–H bond cleavage step in the linoleic acid oxidation by soybean lipoxygenase, in which tunneling is evidenced by a temperature-independent k_H/k_D of 56 and unusually low apparent activation energies (1–2 kcal/mol). In most hydrogen transfer reactions (although perhaps not this one), tunneling is unlikely to contribute more than an order of magnitude or perhaps two to the overall rate constant, even when it dominates the isotope effects. Tunneling and other quantum effects are particularly important in electron transfer reactions.[33]

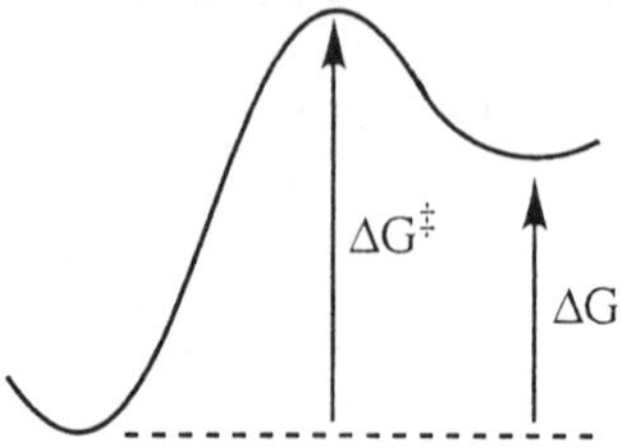

Fig. 1.1

Transition state theory may also be insufficient for reactions that occur on very short time scales of a few vibrational periods, when dynamical effects could become important.[34] This has come up in discussions of the so-called rebound step in the common mechanism for C–H oxidation by cytochrome P450 enzymes (the second step in Eq. 1.20, in which *P* stands for the porphyrin ligand).[34b] Newcomb *et al.* have used very rapid radical clocks to suggest that, in some cases, the putative radical species would have an extremely short lifetime.[35] Chapter 2 presents a more complete discussion on the oxygen rebound mechanism (see also Ref. 37e).

$$(P^{\bullet+})\text{Fe=O} + \text{R–H} \rightarrow (P^{\bullet+})\text{FeOH} + \text{R}\bullet \rightarrow (P)\text{Fe(ROH)} \qquad (1.20)$$

1.4.1 Kinetic constraints from transition state theory

Kinetic insight can be derived from the transition state theory requirement that $\Delta G^{\ddagger}$ be larger than or equal to ΔG (Fig. 1.1). It is also found that $\Delta H^{\ddagger}$ is larger than or within a few kcal/mol of ΔH (though this is not strictly required). $\Delta H^{\ddagger}$ is smaller than ΔH only when the enthalpic barrier to the reverse reaction is negative, which is rare and only small negative values are known. Such constraints can rule out possible mechanisms. For instance, Meyer and co-workers showed that the oxidation of hydroquinone by $[(bpy)_2(py)Ru(O)]^{2+}$ ($k = 10^6\ M^{-1}s^{-1}$, $\Delta G^{\ddagger} = 9.3$ kcal/mol) cannot occur by initial electron transfer ($\Delta G^{\circ\prime} > 16.1$ kcal/mol) or initial proton transfer ($\Delta G^{\circ\prime} > 13.4$ kcal/mol) using the known $E^{\circ\prime}$ and pK_a values.[36] It should be noted that the thermochemical constraint for an observed intermediate, $\Delta G \lesssim 0$ kcal/mol, is much stricter than for an unobserved intermediate ($\Delta G < \Delta G^{\ddagger}$)

As an illustration of the value and potential difficulties of this approach, consider the oxidation of alkenes by ferryl porphyrins. Among the many mechanistic proposals for these reactions, discussed in more detail elsewhere,[37] electron transfer has been suggested as the initial step. To pick one example, the reactions of $[(TMP^{\bullet+})Fe^{IV}(O)(OClO_3)]^+$ with styrene and cyclooctene in CH_2Cl_2 quantitatively form an intermediate, which it was suggested could be the result of electron transfer (Eq. 1.21; TMP = tetramesityl porphyrin).[38] Such a proposal can be considered in light of the relevant redox potentials. However, the potentials are not available for the reaction conditions so these arguments should be applied with caution. The

$$(TMP^{\bullet+})Fe^{IV}O(OClO_3) + PhCH{=}CH_2 \xrightarrow{?} [(TMP^{\bullet+})Fe^{III}(O)(OClO_3)^-][PhCHCH_2^{\bullet+}] \rightarrow\rightarrow \quad (1.21)$$

irreversible potential at which styrene is oxidized in acetonitrile has been reported as +1.90 or +1.65 V versus SCE.[39] Potentials for ferryl-porphyrin complexes are more difficult to obtain because they are of limited stability; estimates range from 1.2 to 1.4 V versus SCE in various solvents.[40] With the assumption that these values apply to the CH_2Cl_2 solution of reaction 1.21, the electron transfer from styrene to iron is uphill by >0.2 V ($\Delta G° > +5$ kcal/mol; $K_{eq} < 10^{-3}$ M^{-1}). This would appear to be inconsistent with the quantitatively formed intermediate being the product of outer sphere electron transfer.[41] The conclusion is stronger for aliphatic alkenes, which are harder to oxidize than styrene.[42] It seems likely that bond formation accompanies the redox step in order to provide the necessary driving force. For instance, C–O bond formation would give $(TMP^{\bullet+})FeO\text{-}CH_2\dot{C}HPh$. One can *very* crudely estimate that this reaction should be mildly exothermic, as formation of the C–O bond (typically ~90 kcal/mol) is balanced against cleavage of the alkene π bond (~60 kcal/mol) and conversion of an iron-oxygen double bond to a single bond (maybe 20 kcal/mol?). A small exothermicity would be consistent with the report that added reagents can reverse formation of the intermediate.[38] A more detailed thermochemical analysis of C–O bond homolysis has been presented for a tungsten benzyloxide.[7]

Thermochemical arguments can also shed light on the mechanism of methane oxidation by soluble methane monooxygenase enzymes. The reactive di-iron center (compound Q) reacts very rapidly with methane[43] suggesting that this step is not substantially endothermic. A mechanism that has been considered is hydrogen atom transfer from CH_4 to the active site. This would require that the active site make a ca. 100 kcal/mol bond to H•, since the methane C–H bond is 105 kcal/mol (Table 1.1). Following Eq. 1.13 (Scheme 1.1), one can calculate the redox potential and pK_a values that would be required in the active site (Fig. 1.2). A pK_a of 7 at the active site, for instance, would imply a potential of 1.45 V. Since this value is at the edge of what is reasonable for an enzyme active site, I am moved to favor mechanisms other than simple hydrogen atom transfer.

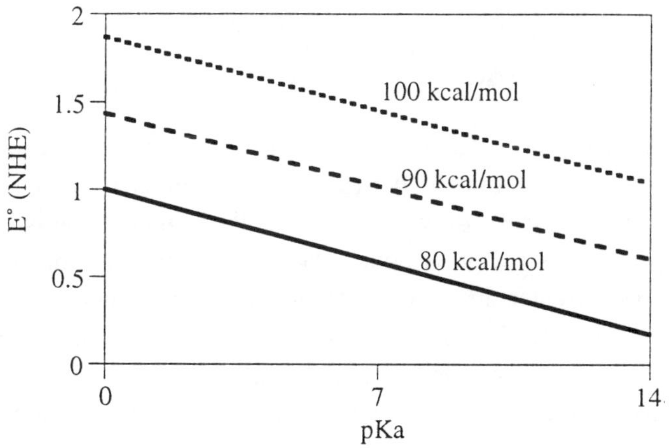

Fig. 1.2 Possible aqueous $E°$ and pK_a values for active sites that make 100, 90 and 80 kcal/mol bonds to H•.

1.4.2 Kinetic barriers and driving force for electron transfer

Marcus theory of electron transfer gives an explicit relationship between kinetic barriers and the net free energy of reaction (Eq. 1.22),[33] perhaps the only such well-established relationship in solution chemistry. This equation is equivalent (Eq. 1.23) to the first three terms of the power series expansion in Eq. 1.19 and gives explicit meaning to the constants: w^r, the work required to bring the reactants together, and the reorganization energy λ. The predicted quadratic dependence of $\Delta G^{\ddagger}$ on ΔG has been confirmed in a number of systems, most notably in the observation of the "inverted region."[33] Over a narrow range of driving force, this expression can be approximated by a linear relationship with a slope α (Eq. 1.24).

$$\Delta G^{\ddagger} = w^{r} + \frac{\lambda}{4}\left(1 + \frac{\Delta G}{\lambda}\right)^{2} \tag{1.22}$$

$$\Delta G^{\ddagger} = \left(w^{r} + \frac{\lambda}{4}\right) + \frac{\Delta G}{2} + \frac{\Delta G^{2}}{4\lambda} \tag{1.23}$$

$$\frac{d(\Delta G^{\ddagger})}{d(\Delta G)} = \alpha = \frac{1}{2} + \frac{\Delta G}{2\lambda} \tag{1.24}$$

Equation 1.22 can be derived from a simple model of two intersecting parabolic free energy surfaces, with the crossing point being the transition state. In the symmetrical case when $\Delta G = 0$, the height of the barrier within the reacting complex is $\lambda/4$, which is therefore termed the intrinsic barrier. $\lambda/4$ is typically divided into two components. Inner sphere reorganization accounts for the distortion in vibrational coordinates of the redox partners needed to adopt the geometry of the transition state and outer sphere reorganization reflects changes in the solvent or other surrounding medium. When no chemical bonds are made or broken, the reactions are termed outer sphere electron transfer and there is relatively little interaction of the two surfaces. Then the inner sphere reorganization energy can be divided into the reorganization occurring at reactant A and that occurring at reactant B, which leads to the Marcus cross relation in terms of self-exchange rates.

1.4.3 Kinetic barriers and driving force for hydrogen atom transfer

Evans and Polanyi in the 1930s proposed that, for a series of related hydrogen atom transfer reactions, the activation energies E_a are proportional to the driving force ΔH (Eq. 1.25).[44] This is similar to the Brønsted relation in which rates of proton transfer (log k or $\Delta G^{\ddagger}$) are linearly related to the driving force (log K_{eq} or ΔG). These expressions can be seen as the simplest case of the power series expansion (Eq. 1.19) or as a simplification of a Marcus treatment. (Marcus theory has been shown to be valuable in studies of proton transfer, methyl transfer, and other processes.[45]) The Polanyi corrrelation of E_a and ΔH works well as long as similar reactions are being compared (Fig. 1.3[46]).

$$E_a = \zeta + \alpha(\Delta H) \quad (1.25)$$

Figure 1.3 also shows that over a large enough range of driving force, the actual dependence is nonlinear and is not too far from the quadratic dependence expected from a Marcus theory-like treatment. For very endothermic reactions, the transition state resembles the products (the Hammond postulate) and the slope α approaches 1. In other words, adding one kcal/mol to the endothermicity adds almost one kcal/mol to the activation energy. In a Marcus treatment, this corresponds to ΔG approaching λ (Eq. 1.24). When the reactions are very exothermic there is a very early transition state which is insensitive to the driving force so $\alpha \rightarrow 0$ (the $\Delta G = -\lambda$ case). In this interpretation, α indicates the position of the transition state along the reaction coordinate (although this has been shown to be an oversimplification[47]).

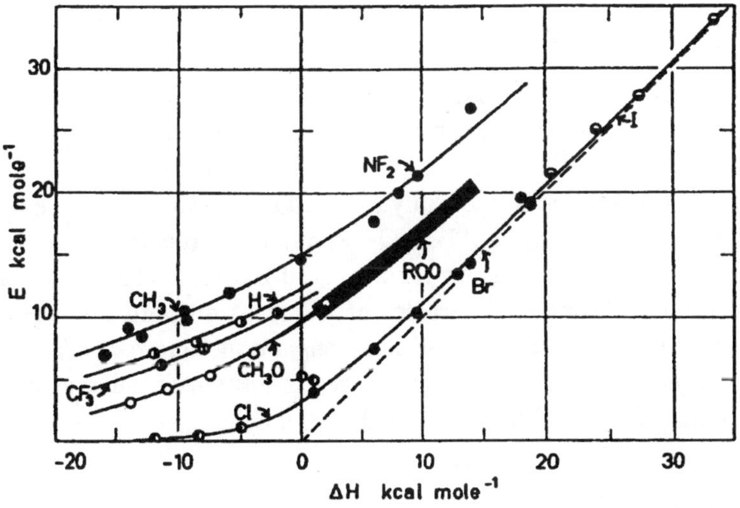

Fig. 1.3 Dependence of activation energy, E_a on heat of reaction, ΔH, for hydrogen atom transfer reactions of various radicals. The shaded area represents the supposed distribution of E and ΔH for ROO• radicals. Reprinted from Ref. 46.

It should be emphasized that the Polanyi relation (Eq. 1.25) is applicable only to hydrogen atom transfer reactions and only when comparing similar reactions, for instance oxygen radicals with C–H bonds. As can be seen in Fig. 1.3 there are substantial differences between radicals even at the same driving force. The origin of this apparent variation in "intrinsic barriers" is discussed below.

1.5 Hydrogen Atom Abstraction by Metal-Oxo Complexes

The classic agents to abstract hydrogen atoms from organic molecules are main group radicals such as chlorine atoms and hydroxyl, alkoxyl, and peroxyl radicals. In recent years, it has become increasingly clear that metal-containing active sites can also abstract hydrogen atoms. This is true on the surface of heterogeneous catalysts,[48] in reagents used in organic chemistry,[49] and at the reactive centers of metalloenzymes. Metal-mediated hydrogen atom abstraction has been implicated, for instance, in the oxidation of fatty acids by the iron center in lipoxygenases,[50] in the copper-mediated oxidations by dopamine β-hydroxylase,[51] and the processing of metabolites and xenobiotics by cytochrome P450 enzymes.[52] Many other examples appear throughout this volume.

It has often been assumed that these metal sites must have some "radical character" in order to accomplish hydrogen atom abstraction, usually radical character at a coordinated ligand. For instance, the radical-like reactivity of ferryl (Fe=O) species is frequently ascribed to radical character at oxygen, as has been indicated by an "Fe—O•" resonance structure. Spin densities have been calculated, in part as a surrogate for radical character[53] but our work suggests that radical character has little to do with hydrogen atom abstracting reactivity. Rather these reactions are in large part controlled by the exo- or endothermicity of the hydrogen atom transfer step, as predicted by the Polanyi relation (Eq. 1.25 and Fig. 1.3).

1.5.1 Oxidations of C–H bonds by CrO_2Cl_2 and MnO_4^-

To develop an understanding of C–H bond oxidations by metal-oxo complexes, we have studied oxidations of alkanes and arylalkanes by chromyl chloride (CrO_2Cl_2)[27,54] and permanganate (MnO_4^-).[19,55] This work has been reviewed in greater depth elsewhere.[56] CrO_2Cl_2 and MnO_4^- reactions have been studied since the 19th century and appear in many organic texts but, surprisingly, their mechanisms were still a matter of debate.[57,58] We began with CrO_2Cl_2 in part because it is available in high purity and because it is miscible with alkanes, thereby avoiding complications from a potentially reactive solvent. CrO_2Cl_2 can be handled conveniently on a vacuum line, though it must be treated with care as it is light-sensitive, highly corrosive, and a serious carcinogen.

CrO_2Cl_2 oxidizes neat cyclohexane over a period of hours at 80 °C to give chlorocyclohexane, a trace of cyclohexene, and a dark precipitate (Eq. 1.26).[27] The precipitate includes oxygenated products, present as ligands, and it contains all of the chromium, with an average Cr oxidation state of ~3.5. The yields in Eq. 1.26 are per mole of CrO_2Cl_2 which is a little misleading because chlorocyclohexanone, for instance, is a six-electron oxidation of cyclohexane and therefore requires ~3 equivalents of CrO_2Cl_2. Still, the observed products only account for 26% of the chromium oxidative equivalents consumed. This difficulty in obtaining a complete mass balance or stoichiometry is common in oxidation reactions. The complete mass balance is especially important when deriving selectivities or kinetic isotope effects from product ratios. In the CrO_2Cl_2 + C_6H_{12} reaction, indirect experiments indicate that the "missing" oxidative equivalents are used in large part to form adipate ($^-O_2C(CH_2)_4CO_2^-$) which remains bound to the

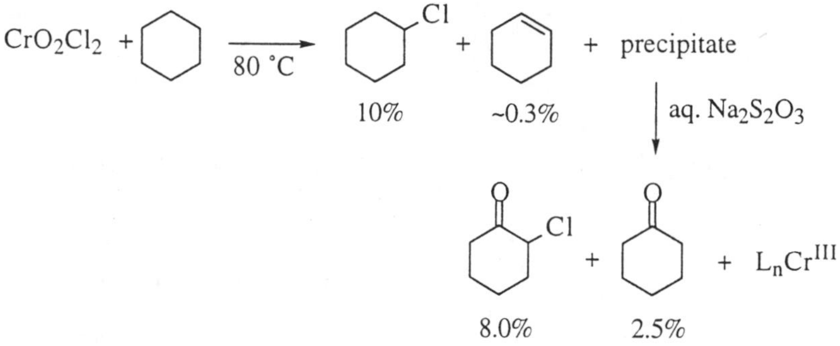

(1.26)

Cr^{III} product. The oxidation of toluene gives benzaldehyde (32%), benzyl chloride (18%) and benzyl alcohol (5%), accounting for 84% of the consumed CrO_2Cl_2 oxidative equivalents.

All of the data support a mechanism of initial hydrogen atom transfer from the substrate to CrO_2Cl_2, followed by rapid trapping of the organic radical[27,54] (illustrated for cyclohexane in Scheme 1.4). The conclusion of radical intermediates follows that of Wiberg and co-workers in the 1960s,[59] though carbocation and concerted pathways have also been suggested.[60] The site of attack is at the weakest C–H bond, so benzylic and *t*-butyl radicals are formed from toluene and isobutane, respectively. The reactions exhibit simple second order kinetics without induction periods or any other evidence for a chain mechanism. Primary isotope effects were observed. Cyclohexyl radicals were trapped by $CBrCl_3$, following its use by Groves and Nemo in an iron porphyrin system.[61] The dependence of the bromocyclohexane yield on the initial chromium concentration supports the formation and trapping of $C_6H_{11}\bullet$ by CrO_2Cl_2.

CrO_2Cl_2 oxidation of isopropylcyclopropane proceeds by hydrogen abstraction to give the dimethyl cyclopropyl carbinyl radical,[54] a well-studied radical clock[62] (Eq. 1.27). Ring-opening occurs competitively with trapping

Scheme 1.4

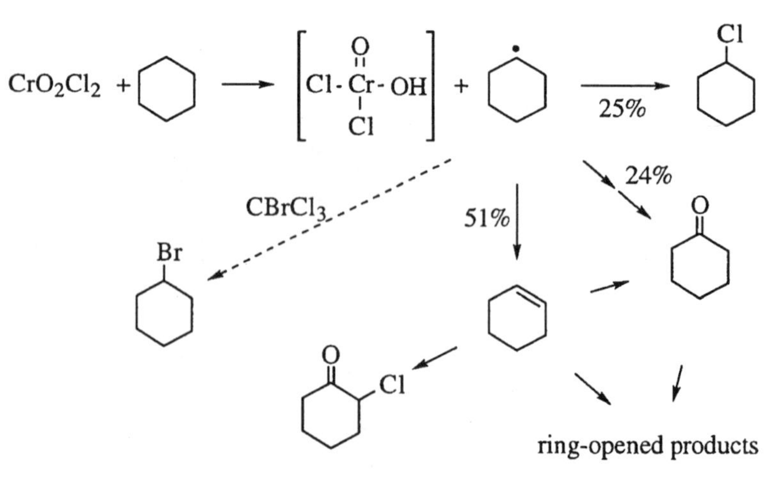

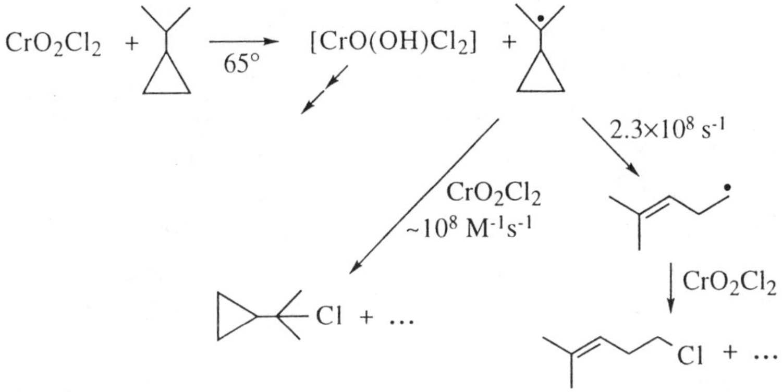

(1.27)

by CrO_2Cl_2 under our conditions, yielding a myriad of products because CrO_2Cl_2 reacts by multiple paths with both the alkene and radical functionalities. The CrO_2Cl_2 mechanism is quite similar to the "oxygen rebound" pathway proposed for cytochrome P450 and other enzymatic oxidations (see Chap. 2; the rebound mechanism for P450 has recently been questioned based on studies with very rapid radical clock substrates[63]). A rebound-type mechanism had been previously proposed for chromium(VI) oxidations by Wiberg[59] and by Stephenson[64] to account for the predominant retention of stereochemistry on C–H bond oxidation.

Radical trapping by CrO_2Cl_2 is very rapid, with $k(C_6H_{11}\bullet + CrO_2Cl_2) \cong 3 \times 10^9\ M^{-1}s^{-1}$, determined by competition with $CBrCl_3$.[27a] This is consistent with pulse radiolysis studies showing almost diffusion-limited radical trapping by permanganate and dichromate.[65] Rapid radical trapping is not surprising from a thermochemical perspective. CrO_2Cl_2 must form a strong O–H bond in order to act as an H• abstracting agent, so it must also form a strong C–O bond (ca. 50 kcal/mol). The exothermic nature of the trapping explains the lack of selectivity: $C_6H_{11}\bullet + CrO_2Cl_2$ produces roughly equal amounts of chlorinated, oxygenated, and olefin products (Scheme 1.4). These products are reminiscent of chloroperoxidase, oxygenase, and desaturase reactions.

The ability of CrO_2Cl_2, a stable red liquid, to abstract hydrogen atoms from hydrocarbons contrasts with the situation in most enzymatic and biomimetic oxidations. In these cases (examples appear throughout this volume) the reactive oxidant is generated *in situ* from a stable metal complex or enzyme resting state. Dioxygen, peroxides, iodosyl benzene, hypochlorite, and other oxidants have been used as terminal oxidants. In enzymatic processes, there are often mechanisms to prevent formation of the active oxidant until the substrate is present.[52] While this approach is efficient for catalysis, it makes study of the C–H activation step difficult as the active oxidant is typically not observed. The work described here is focused on reactions where the reactive oxidant is isolable and therefore its properties can be studied.

Permanganate is another example of a well-known and isolable oxidant that accomplishes C–H bond oxidation.[58] $^{n}Bu_4NMnO_4$, an organic soluble permanganate salt, has sufficient solubility in toluene (~0.3 mM) for mechanistic studies (*caution*: it is reported to be explosive).[66] Toluene is oxidized to benzoic acid with a small amount of benzaldehyde under our conditions.[19] As found for CrO_2Cl_2, complete mass balances were not achieved (possibly due to analytical limitations). The permanganate is converted to colloidal MnO_2. For toluene, ethylbenzene, diphenylmethane, triphenylmethane, dihydroanthracene, fluorene and xanthene, the initial rates are first order in oxidant and first order in substrate, without induction periods. Reaction with toluene-d_8 is 6 (± 1) times slower than with C_7H_8 at 45 °C. Reactions in *o*-dichlorobenzene solvent had similar rate constants, although the reproducibility was not good.

All of the data are consistent with the same mechanism of initial hydrogen atom abstraction followed by radical trapping (Eq. 1.28).[19] Alternative pathways involving electron transfer or hydride transfer could be ruled out by the lack of substantial solvent or substituent effects. Thus the electron deficient 4-methylbenzophenone reacts faster than toluene and *p*-xylene reacts only four times faster than toluene despite being ~0.4 V easier to oxidize[67] (0.4 V corresponds to a difference

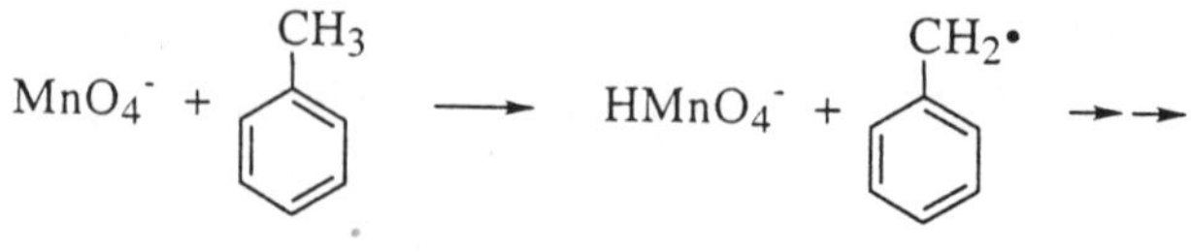

(1.28)

of 10^7 in K_{eq}). Outer sphere electron transfer was also ruled out by estimating the redox potentials, indicating that formation of MnO_4^{2-} + $C_7H_8^+$ would be at least 1.9 V (44 kcal/mol) uphill, much larger than the observed $\Delta G^\ddagger$.

1.5.2 Correlation of rates with C–H bond strengths

The strongest evidence that $^nBu_4NMnO_4$ reacts by initial hydrogen atom transfer is that the rate constants follow a Polanyi correlation with ΔH°. Rate constants for the hydrogen atom transfer step are calculated from the measured rate constants for disappearance of $^nBu_4NMnO_4$ by multiplying by a stoichiometric factor to account for the number of permanganate equivalents consumed per radical generated. The kinetic results reported per reactive substrate hydrogen, are given in Table 1.4. The rate constants and activation enthalpies both correlate well with the ΔH° for hydrogen atom transfer (Figs. 1.4 and 1.5).[68] This ΔH° is simply the difference between the strength of the C–H bond being broken and that of the O–H bond being formed, calculated for $HMnO_4^-$ in Scheme 1.2 above. A similar correlation is observed for the four CrO_2Cl_2 reactions studied, plotting $\Delta H^\ddagger$ and $\Delta G^\ddagger$ against the C–H bond strengths (Fig. 1.6).[68]

Table 1.4 Rate constants and activation parameters for permanganate oxidations.[a]

substrate	k ($M^{-1}s^{-1}$)	$\Delta G^\ddagger$	$\Delta H^\ddagger$	$\Delta S^\ddagger$	D(C–H)
$PhCH_3$	1.4×10^{-7}	26 (± 1)	21.0 (± 1.0)	−20 (± 3)	89.9
$PhCH_2CH_3$	3.3×10^{-6}	25 (± 1)	20.8 (± 0.6)	−19 (± 3)	84.7
Ph_2CH_2	6.5×10^{-5}	23 (± 1)	17.1 (± 0.5)	−21 (± 3)	82
Ph_3CH	2.7×10^{-3}	21 (± 2)	16.5 (± 0.8)	−15 (± 5)	81
DHA[b]	1.2×10^{-1}	19 (± 2)	13.8 (± 1.0)	−17 (± 7)	78
xanthene	5.6×10^{-1}	18 (± 3)	11.7 (± 1.2)	−20 (± 8)	75.5

[a]Values in kcal/mol (e.u. for $\Delta S^\ddagger$) at 298 K for the hydrogen atom transfer step, per reactive hydrogen, from Refs. 19 and 1f [D($PhCH_2$–H)]. [b]DHA = 9,10-dihydroanthracene.

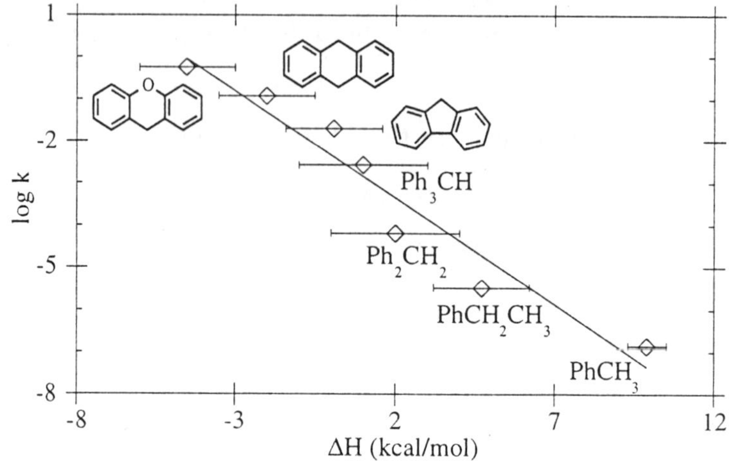

Fig. 1.4 log k versus ΔH° for the hydrogen atom transfer step in $^{n}Bu_4NMnO_4$ reactions.[68]

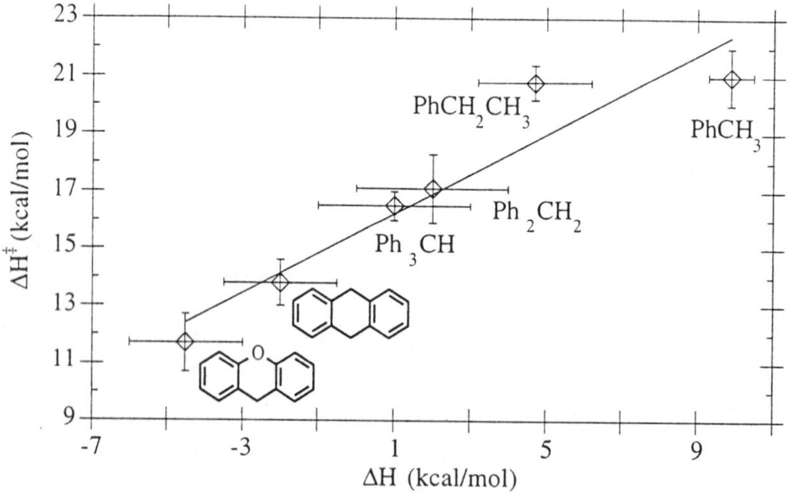

Fig. 1.5 ΔH‡ versus ΔH° for the hydrogen atom transfer step in $^{n}Bu_4NMnO_4$ reactions.[68]

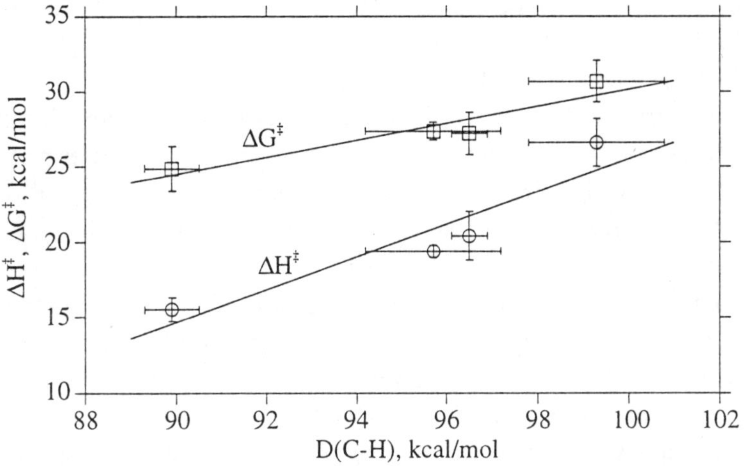

Fig. 1.6 $\Delta H^{\ddagger}$ and $\Delta G^{\ddagger}$ versus C–H bond strength for the hydrogen atom transfer step in CrO_2Cl_2 reactions.[68]

The correlations are equally linear with $\Delta H^{\ddagger}$ or with $\Delta G^{\ddagger}$ because $\Delta S^{\ddagger}$ is roughly constant ($^{n}Bu_4NMnO_4$) or proportional to $\Delta H^{\ddagger}$ (CrO_2Cl_2). It is most appropriate to plot $\Delta H^{\ddagger}$ versus $\Delta H°$ because both are enthalpies. The correlation of log *k* (proportional to $\Delta G^{\ddagger}$) with $\Delta H°$ has the advantages that the rate constants are more accurately determined, and (as shown below) that this facilitates comparisons with main group radicals. The unitless slopes of the $\Delta H^{\ddagger}$ versus $\Delta H°$ plots — α in the Polanyi relation (Eq. 1.25) — are 0.7 for $^{n}Bu_4NMnO_4$ and 1.1 for CrO_2Cl_2 (Figs. 1.5 and 1.6). Over the > 10 kcal/mol range of these plots, some nonlinearity might be predicted (c.f. Fig. 1.3) but it is not evident within the scatter of the data.

A comment should be made about the C–H bond strengths used in these correlations. While accurate values are available for small organic molecules,[1a] for larger molecules the uncertainties are substantial (as indicated by the horizontal error bars in Figs. 1.4–1.6). In addition, there is no up-to-date and self-consistent compilation. Many C–H bond strengths have been revised upward in the last decade to account for systematic measurement errors,[1a] but others await necessary revision. Our struggles to compile internally consistent C–H bond strengths are detailed in footnotes in Refs. 19 and 27. This report uses the newly revised (slightly larger) benzylic C–H bond strength for toluene.[1f]

These correlations are mechanistically very informative, despite the uncertainties involved. Figures 1.4–1.6 rule out electron transfer or hydride transfer as rate-limiting steps because these would not follow the same pattern. In the permanganate series, xanthene is a much better electron donor and forms a much more stable carbocation than its hydrocarbon counterparts, yet it is oxidized at a rate predicted by its homolytic C–H bond strength. In the CrO_2Cl_2 oxidations, toluene would not correlate with alkanes if e^- or H^- transfer were rate-limiting. The correlations also rule out rate-limiting formation of intermediates with metal-carbon bonds (as has been suggested for related reactions[69]). There are few stable metal complexes with tertiary alkyl ligands, and steric factors would certainly displace Ph_3CH from the lines in Figs. 1.4 and 1.5. Figure 1.6 similarly includes reactions of primary, secondary, and tertiary C–H bonds, but steric effects are present only to the extent that they change D(C–H). [2+2] addition of C–H bonds across early transition metal imido and alkylidene linkages has been documented for methane and for primary and secondary C–H bonds, but not for tertiary ones.[9a] The same is true for σ-bond metathesis and related mechanisms.[9] Barriers to [2+2] addition correlate *inversely* with C–H bond strength, opposite to the trend observed here. Apparently these reactions are dominated by the strength of the metal-carbon bond formed.

1.5.3 Correlations of metal-oxo reactivity with oxygen radicals

Figures 1.4–1.6 are classic implementations of the Polanyi correlation: one oxidant reacting with a series of substrates. It is also informative to correlate reactions of various oxidants with one substrate. Figure 1.7 is such a plot for the oxidation of toluene by $^nBu_4NMnO_4$ and oxygen radicals. Permanganate fits right on the line extrapolated from the oxygen radicals. In other words, permanganate reacts at essentially the rate predicted for an oxygen radical that makes an 80 kcal/mol O–H bond, even though

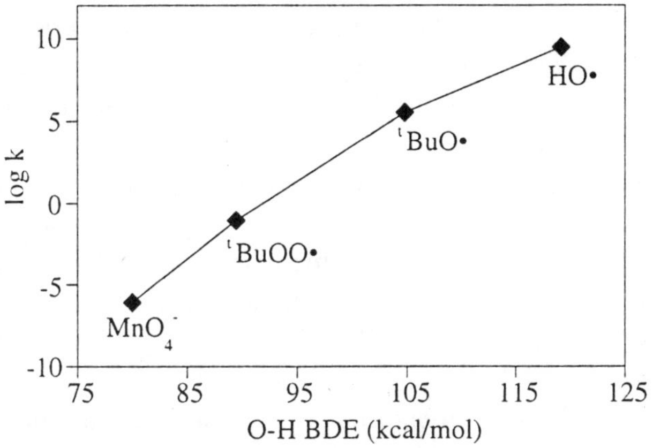

Fig. 1.7 log *k* versus O–H bond strength for reactions of toluene with $^{n}Bu_4NMnO_4$ and oxygen radicals.[72,73]

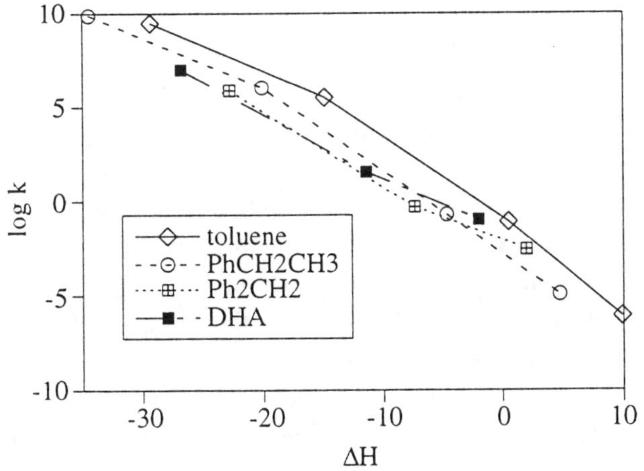

Fig. 1.8 log *k* versus ΔH° for reactions of arylalkanes with OH•, ᵗBuO•, ᵗBuOO•, and $^{n}Bu_4NMnO_4$ (DHA = 9,10-dihydroanthracene).[74]

permanganate is not a radical. Permanganate, like CrO_2Cl_2, is d^0 and closed-shell. The lowest lying triplet states in these molecules are higher in energy than the activation barriers $\Delta G^{\ddagger}$,[70] so they cannot be intermediates in the reactions. The key feature of permanganate that enables it to abstract a hydrogen atom is the 80 kcal/mol O–H bond it can make, not any radical character.

Plots similar to Fig. 1.7 can be made for reactions of ethylbenzene, diphenylmethane, and 9,10-dihydroanthracene (DHA), with deviations from a smooth curve of not much more than an order of magnitude.[19] This can be seen in a combined plot (Fig. 1.8), including the four substrates and four oxidants (OH•, ^{t}BuO•, ^{t}BuOO•, and $^nBu_4NMnO_4$). It is striking that permanganate and oxygen radicals form such a unified data set. Figures 1.7 and 1.8 cover an enormous range of rate constants ($>10^{15}$) and driving force (39 kcal/mol). Equally good correlations are observed for CrO_2Cl_2 if an 83 kcal/mol bond strength is assumed.[27] The curvature in the figures is due in part to the hydroxyl radical rates approaching the diffusion limit, and, by analogy with Marcus theory (Eq. 1.22), the dependence of rate on driving force may be nonlinear. Figure 1.8 also illustrates the limited accuracy of the Polanyi relation when comparing a range of reagents (the 30-year-old Fig. 1.3 may be somewhat idealized).

There are many reasons why the permanganate rate constants would not correlate with those of oxygen radicals. The plots implicitly assume that the aqueous permanganate bond strength (Scheme 1.2) applies to ion-paired toluene solutions of $^nBu_4NMnO_4$, and rate constants in different solvents are being compared. Hydrogen atom abstraction by alkoxy radicals has been shown to be essentially solvent independent,[71] but it is not clear that this will hold for the other oxidants. At least the $^nBu_4NMnO_4$ oxidations are similar in toluene and *o*-dichlorobenzene.[19]

We have compared permanganate with oxygen radicals because they all form O–H bonds and because the correlation works. But it is not clear that permanganate *should* correlate with XO• since the Polanyi equation holds only for similar radicals (Fig. 1.3). The classic deviation from the Polanyi relation[44] is the very slow gas-phase reaction of CH_3• with CH_4 (21 cm^3 $mole^{-1}s^{-1}$) versus the very rapid reaction of Cl• + CH_4 (6.0×10^{10} cm^3 $mole^{-1}s^{-1}$) at essentially the same driving force

(ΔH° = 0, +2 kcal/mol, respectively).[75] The standard explanation — termed polar effects[44] — is essentially a frontier orbital argument. This is focused on the energy of interaction between the C–H σ and $\sigma*$ orbitals and the radical SOMO (semi-occupied molecular orbital).[76,77] An electron-deficient radical, such as an oxygen or chlorine radical, interacts in the transition state more strongly with an electron rich hydrogen than with an electron poor one (Scheme 1.5a). Thus ^{t}BuO• abstracts from the methoxy group in methyl acetate rather than from the acetate, even though the latter has the weaker C–H bonds. Similarly, CH_3CH_2COOH is preferentially attacked by CH_3• at the weaker α-C–H (8:1), but by electrophilic Cl• at the more electron rich β position (0.03:1).[44] But polar effects are not the whole answer to why different classes of radical reactions fall on different lines in Fig. 1.3. For instance, if polarity and SOMO energy are the key features, Cl• and I• should not lie on the same line. Furthermore, polar effects are unable to explain the very facile reaction of oxygen radicals with phenols (see Sec. 1.6.3 below). Extending these arguments to metal-oxo complexes is particularly complicated because the HOMO and LUMO of the oxidant are involved instead of a SOMO (Scheme 1.5b, with the energies drawn roughly to scale).

An alternative approach to understand why different radicals fall on different lines in Fig. 1.3 starts from a Marcus-type intersecting parabola model (Fig. 1.9).[78] Only the inner sphere component of the reorganization

Scheme 1.5[27c]

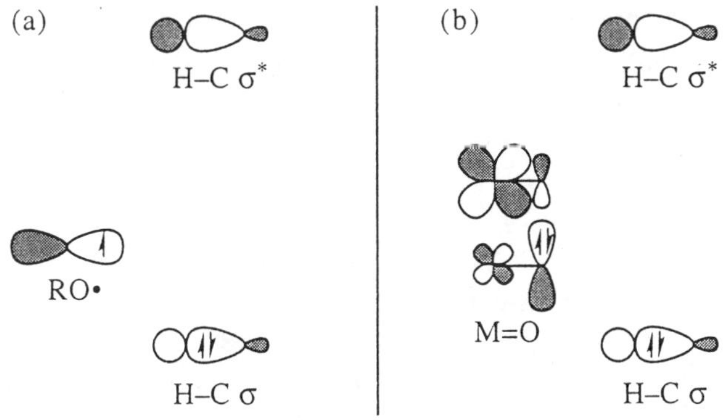

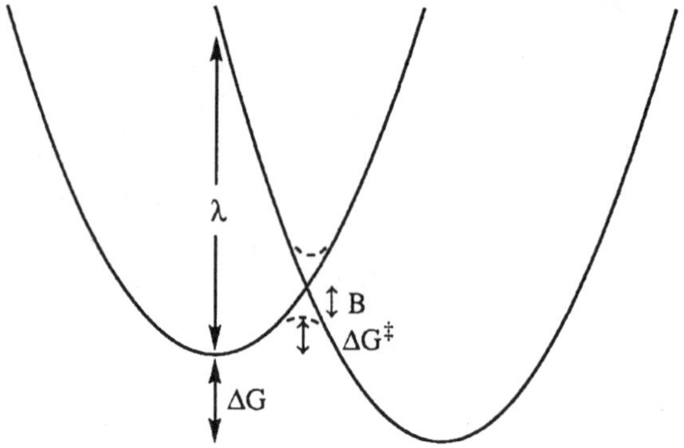

Fig. 1.9 Intersecting parabola model for chemical reactions.

energy λ need be considered, since the solvent reorganization energy is small. This conclusion is based on the lack of a solvent dependence for hydrogen atom transfer reactions.[71] This is a result of very little charge being transferred.[71] As an example of an inner sphere reorganization, it seems likely that pyramidalization of the methyl group contributes to the relatively high barriers for CH_3• reactions. This is a totally different explanation than polar effects because it is a property of one reacting partner (CH_3•), while the frontier orbital picture involves both reactants and is therefore unique to the cross reaction. In the intersecting parabola model, the frontier orbital interaction appears primarily in the interaction between the reactant and product surfaces near the transition state (B in Fig. 1.9). Most electron transfer reactions are in the weak coupling regime of $B \ll \Delta G^{\ddagger}$, but this is not likely to be the case for H• transfers. This discussion is closely related to the curve crossing model developed by Shaik and Pross[78b] which seems a likely framework on which to build a more complete understanding of these reactions.

1.6 Hydrogen Atom Abstraction by Metal Coordination Complexes

The thermochemical approach to hydrogen atom abstraction should apply to complexes other than metal-oxo's. The reason permanganate and CrO_2Cl_2

can abstract hydrogen lies in the strength of the bond they make to hydrogen, not in any peculiarity of the metal-oxo functional group. This argument is used in organic radical abstractions: Chlorine atoms are very reactive while I• is essentially inert because the Cl• reactions are *32 kcal/mol* more exothermic. ^{t}BuO• is much more reactive than ^{t}BuOO• because it forms a 16 kcal/mol stronger O–H bond (Table 1.5). In principle, hydrogen atom abstraction should be observed for any active site with affinity for H•, in other words an affinity for both an electron and a proton. This should include coordination complexes, surface sites on heterogeneous catalysts, and active sites in metalloenzymes. Our ongoing efforts to test these ideas are briefly summarized below.

Table 1.5 X–H bond strengths[a]

X	D(X–H)
OH•	119
RO•	105
Cl•	103
ROO•	89
Br•	88
$[MnO_4^-]$	80
I•	71

[a]in kcal/mol, from Refs. 1 and 19.

1.6.1 Manganese μ-oxo dimers

$[Mn_2(\mu\text{-}O)_2(phen)_4]^{3+}$ (phen = 1,10–phenanthroline) has been reported to undergo two reductions by proton-coupled electron transfer.[79] This and related compounds have long been of interest as models for the oxygen evolving complex in photosystem II, whose catalytic cycle has recently been suggested to involve hydrogen atom transfer from the manganese cluster to a tyrosyl radical.[80] A more detailed discussion of such complexes can be found in Chaps. 4, 5 and 11. Our redox potential and pK_a measurements[81] imply the O–H bond strengths in Eqs. 1.30 and 1.31 (L = phen), where H• is added to a bridging oxygen. Hydrogen atom abstracting reactivity was therefore predicted.

$$[L_2Mn(\mu{-}O)_2MnL_2]^{3+} + H\bullet \xrightarrow[\text{MeCN}]{-79\ \text{kcal/mol}} [L_2Mn(\mu{-}O)(\mu{-}OH)MnL_2]^{3+} \tag{1.30}$$

$$[L_2Mn(\mu{-}O)(\mu{-}OH)MnL_2]^{3+} + H\bullet \xrightarrow[\text{MeCN}]{-75\ \text{kcal/mol}} [L_2Mn(\mu{-}OH)_2MnL_2]^{3+} \tag{1.31}$$

$[Mn_2(\mu\text{-}O)_2(phen)_4]^{3+}$ oxidizes dihydroanthracene (DHA) over 11 h at 55 °C to anthracene and traces of anthrone and anthraquinone (Eq. 1.32).[81] Fluorene is similarly oxidized to bifluorenyl and 9-fluorenone, supporting the predicted mechanism of hydrogen atom abstraction to give a carbon radical. Both the di-μ-oxo dimer and the μ-oxo-μ-hydroxo dimer are found to abstract H• from DHA, as suggested by Eqs. 1.30 and 1.31. The rate constants for both oxidants are within a factor of 10^2 of the values predicted by linear extrapolation of the rate constants for ^{t}BuO• and ^{sec}BuOO• (Fig. 1.10). However, the two dimer rate constants do not fit a Polanyi correlation very well, as the difference in driving force of 4 kcal/mol (a difference in K_{eq} of ca. 10^3) would predict a difference in rate constants much larger than the observed factor of four. Furthermore, the predictions do not hold for reactions of $Mn_2(\mu\text{-}O)_2(X\text{-salpn})_2$ complexes (generously provided by Baldwin and Pecoraro).[82,83] Oxidation of DHA is not observed under comparable conditions, yet they form O–H bonds of comparable strength to $[Mn_2(\mu\text{-}O)_2(phen)_4]^{3+}$. The origin of this difference is not yet understood.

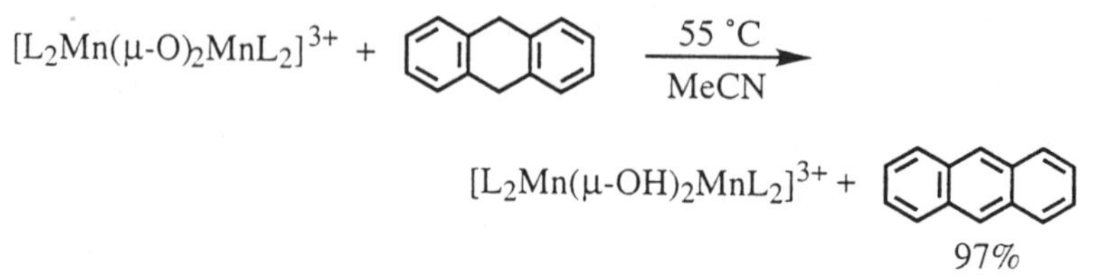

(1.32)

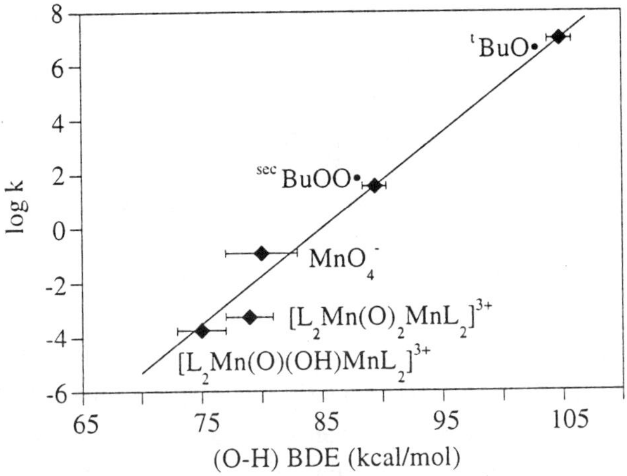

Fig. 1.10 log *k* versus O–H bond strength for reactions of 9,10-dihydroanthracene with oxygen radicals, ^{n}Bu$_4$NMnO$_4$, and manganese phenanthroline dimers.[81]

1.6.2 Copper(III) and iron(III) complexes

Oxidizing copper(III) and iron(II) complexes are being examined as potential models for the C–H activation step in dopamine β-hydroxylase[51] (see Chap. 7) and lipoxygenase.[50] Application of literature $E°$ and pK_a values for Cu(H$_{-3}$Aib$_3$) (below left)[84] and Cu(Pre)$^+$ (below center)[85] suggest bond strengths that predict hydrogen atom abstraction reactivity. Consistent with the prediction, Cu(Pre)$^+$ oxidizes DHA to anthracene in ~50% yield.[86] In contrast, Cu(H$_{-3}$Aib$_3$) is inert to hydrocarbons with weak C–H bonds.[86]

These systems are not amenable to straightforward application of the bond strength calculation or the Polanyi relation. Two solvent molecules add to the copper center on reduction of square planar d^8 Cu^{III} to d^9 Cu^{II}, so the bond strengths (and rates) should be quite solvent-dependent. There is a significant entropy change on reduction,[87] contrary to the assumption made in Sec. 1.3.3 above. The reduced form of Cu(Pre)$^+$, Cu(PreH)$^+$, has a hydrogen bond between the oximes [and it is not clear where protonation occurs in the reduced form of Cu(H$_{-3}$Aib$_3$)]. These effects complicate the reaction coordinate and make it less likely that the $\Delta G^‡$ is simply related to the energies of the ground state reactants and products.

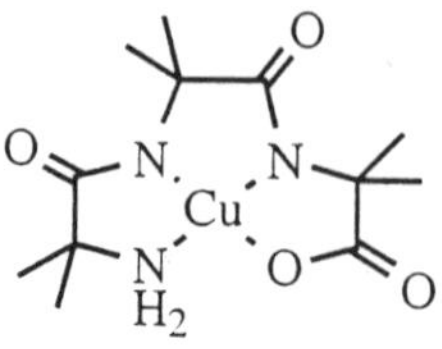

$Cu(H_{-3}Aib_3)$

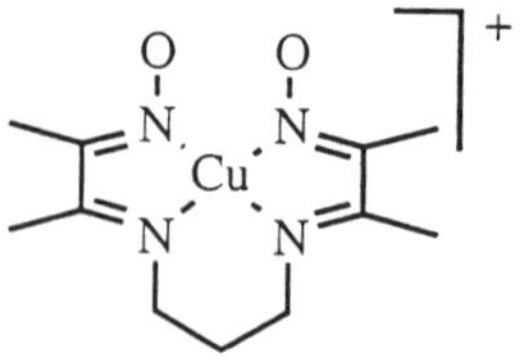

$Cu(Pre)^+$

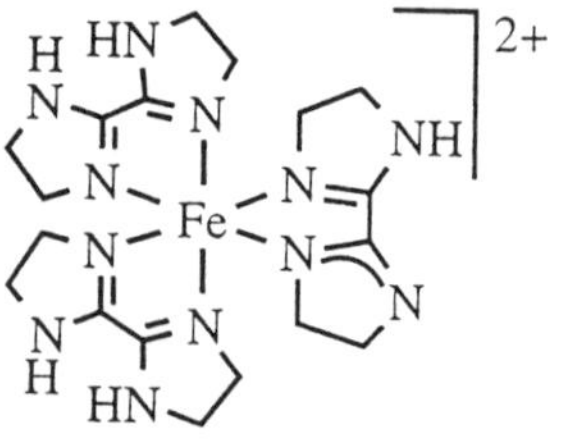

$[Fe(Hbim)(H_2bim)_2]^{2+}$

$[Fe^{III}(Hbim)(H_2bim)_2]^{2+}$ (above right)[88] also oxidizes DHA to anthracene.[89] The iron product is predominantly the fully protonated $Fe^{II}(H_2bim)_3{}^{2+}$. A hydrogen atom abstraction mechanism is indicated, with the hydrogen being transferred to the deprotonated nitrogen atom that is three bonds removed from the redox active iron center. Jonas and Stack have reported similar reactivity for an iron(III)-methoxide complex, oxidizing substrates with weak C–H bonds and forming the corresponding iron(II) methanol compound.[50b] The redox potential and pK_a values for both systems are consistent with the ability of the complexes to abstract H•, and the rate constants correlate with substrate C–H bond strengths. The rate constant for $[Fe^{III}(Hbim)(H_2bim)_2]^{2+}$ oxidation of DHA correlates with those for oxygen radicals.[89]

1.6.3 Oxidations of phenols

In all the cases we have examined, oxidation of a phenol is much faster than oxidation of a C–H bond whose bond strength is equal to or weaker than the phenolic O–H bond strength. $Cu(H_{-3}Aib_3)$ does not oxidize 1-fluorene carboxylic acid but it does oxidize 2,4-di-^{t}butylphenol to its dimer. $Mn_2(\mu\text{-}O)_2(X\text{-}salpn)_2$ complexes oxidize this phenol but are unreactive with 9,10-dihydroanthracene. $[Mn_2(\mu\text{-}O)_2(phen)_4]^{3+}$ reacts with phenols rapidly. Oxygen radicals exhibit the same selectivity, reacting much faster with phenols than with weaker C–H bonds (four comparisons are shown in Table 1.6). Thus Polanyi correlations of C–H bond reactions cannot be applied to oxidation of O–H bonds. The origin of this kinetic preference

Table 1.6 Comparisons of H• abstractions from O–H versus C–H bonds[a]

XO•	X–H [D(X–H)]	k
^{t}BuO• +	PhOH [87]	3×10^8
	$PhCH_2CH_3$ [85]	5×10^5
^{t}BuO• +	BHT [81]	1×10^8
	DHA [78]	1×10^6
^{t}BuOO• +	PhOH [87]	3×10^3
	$PhCH_2CH_3$ [85]	1×10^{-1}
^{t}BuOO•+	BHT [81]	10^4
^{s}BuOO• +	DHA [78]	35

[a]Bond strengths [D(X–H)] in kcal/mol and rate constants per hydrogen ($M^{-1}s^{-1}$ at ~300 K) from Refs. 1g, h, 19, 73. BHT = 4-Me-2,6-^{t}Bu-C_6H_2OH; DHA = 9,10-dihydroanthracene.

has been discussed but is not established.[90] Mechanistic studies are in progress to determine whether the metal reactions proceed by simple hydrogen atom transfer, whether the phenol coordinates prior to oxidation, or if there is some other mechanism. Formation of a hydrogen bond between the phenol and the oxidant could enhance or inhibit hydrogen atom transfer. In reactions of oxygen radicals with phenols, it has been shown that abstraction occurs only from a free phenol, so formation of a hydrogen bond inhibits H• transfer.[91]

1.7 Concluding Comments

Knowledge of the relevant thermochemistry, especially ΔG and ΔH, provides insight into a chemical process. This holds for both the thermodynamics of the overall transformation and for the rate-limiting step. Useful values can often be derived from published data and/or from pK_a and electrochemical measurements. Such data inform mechanistic discussions, for instance using the requirement that $\Delta G^\ddagger \geq \Delta G$. For electron transfer

(Marcus theory), hydrogen atom transfer (Polanyi relation), and a few other classes of reactions, the activation parameters $\Delta G^{\ddagger}$ and $\Delta H^{\ddagger}$ correlate closely with ΔG and ΔH under certain conditions. The presence or lack of such correlations can be valuable mechanistic evidence, particularly if a range of reactions is examined. For example, the correlation between rates of permanganate oxidations of alkylaromatics and the substrate C–H bond strengths (Figs. 1.4 and 1.5) is strong evidence for a hydrogen atom transfer pathway. The lack of correlation with redox potentials, hydride donor abilities, or substrate sterics rule out alternative mechanisms.

The Polanyi correlation of rate constants with ΔH provides a means to understand hydrogen atom transfer reactions of metal-containing active sites. While these are often thought to require radical character at the abstracting atom, studies of CrO_2Cl_2 and MnO_4^- reactions show that unpaired spin density is not required for hydrogen atom abstraction. Instead, a high thermodynamic affinity of L_nMX for H• (a large L_nMX–H bond strength) is necessary. A high affinity for H• requires an oxidizing metal center and the reduced form must have significant basicity. Rates of hydrogen atom abstraction by $^nBu_4NMnO_4$ and apparently CrO_2Cl_2 correlate reasonably well with rates of abstraction by oxygen radicals, based on the Polanyi relation. This enabled *predictions* of rate constants for hydrogen atom abstraction by $[Mn_2(\mu\text{-}O)_2(phen)_4]^{3+}$ and $[Mn_2(\mu\text{-}O)(\mu\text{-}OH)(phen)_4]^{3+}$, predictions that proved to be reasonably accurate. In other systems, including related manganese dimers, the reactivity is less than that predicted. It is not clear what other factors play a role in the higher barriers of these reactions. But even here, the thermochemistry is the key starting point. More studies are needed to test the validity and generality of this approach.

1.8 Acknowledgments

I would very much like to thank all who have been involved in this collaborative effort: Dr. Jerry Cook, Dr. Kim Gardner, Dr. Kun Wang, Dr. Becky Conry, Teresa Blubaugh, Andrea Collier, Dr. Tom Crevier, Justine Roth, Dr. Mark Lockwood, Linda Kuehnert, Dr. Ester Spaltenstein, Dr. David Thorsell, Julie Adams, Jason Adkins, Audrey Nakamura, Matthew Farrow, and Jasmine Bryant. I am also very grateful to those who have provided funding: the National Institutes of Health, the donors of the Petroleum Research Fund administered by the American Chemical Society, the National Science Foundation, Union Carbide, Chevron Research, BP America, DuPont, and the Exxon Education Foundation.

References

1. Thermochemical data from: (a) Berkowitz, J., Ellison, G.B. and Gutman, D. *J. Phys. Chem.* 1994, **98:** 2744–2765; (b) *Thermodynamic Tables.* Thermodynamic Research Center, Texas A&M University, College Station, Texas (1986–1991); (c) *JANAF Thermochemical Tables*, 3rd Ed. *J. Phys. Chem. Ref. Data* 1985, **14:** Supplement #1; (d) Lias, S.G., Bartmess, J.E., Liebman, J.F., Holmes, J.L. and Levin, R.D. *Gas-Phase Ion and Neutral Thermochemistry. J. Phys. Chem. Ref. Data* 1988, **17:** Supplement #1; (e) Colussi, A.J. in *Chemical Kinetics of Small Organic Radicals* (ed.) Alfassi, Z.B., CRC Press Boca Raton, FL, 1988, p. 33; (f) D($PhCH_2$–H): Bierbaum, V., DePuy, C., Davico, G., and Ellison, B. *Int. J. Mass Spectro. Ion Physics* 1996, **156:** 109–131; (g) D(PhO–H): Wright, J.S., Carpenter, D.J., McKay, D.J. and Ingold, K.U. *J. Am. Chem. Soc.* 1997, **119:** 4245–4252; (h) D(ArO–H): Lucarini, M., Pedrielli, P., Pedulli, G.F., Cabiddu, S. and Fattuoni C. *J. Org. Chem.* 1996, **61:** 9259–9263.
2. Atagi, L.M., Over, D.E., McAlister, D.R. and Mayer, J.M. *J. Am. Chem. Soc.* 1991, **113:** 870–874.
3. Holm, R H. *Chem. Rev.* 1987, **87:** 1401–1449. Holm, R.H. and Donahue, J.P. *Polyhedron* 1993, **12:**571–589.
4. Mayer, J.M. *Adv. Trans. Met. Coord. Chem.* 1996, **1:** 105–157.
5. Shaofeng, L. and Pilcher, G. *J. Chem. Thermodyn.* 1988, **20:** 463–465.
6. Kirklin, D.R. and Domalski, E.S. *J. Chem. Thermodyn.* 1988, **20:** 743–754. A value of 124 ± 4 kcal/mol is given in an earlier report: Watt, G.D. McDonald, J.W. and Newton, W.E. *J. Less Common Metals* 1977, **54:** 415–423.
7. Crevier, T.J. and Mayer, J.M. *J. Am. Chem. Soc.* 1997, **119:** 8485–8491.
8. Gable, K.P., Juliette, J.J.J., Li, C. and Nolan, S.P. *Organometallics* 1996, **15:** 5250–5251.
9. (a) Bennett, J.L. and Wolczanski, P.T. *J. Am. Chem. Soc.* 1997, **119:** 10696–10719 and references therein; (b) Shilov, A.E. and Shul'pin, G.B. *Chem. Rev.* 1997, **97:** 2879–2932.

10. March, J. *Advanced Organic Chemistry,* 4th Ed. Wiley, New York, 1992, pp. 603–605.
11. (a) Stewart, R. *Oxidation Mechanisms*. Benjamin, New York, 1964; (b) Roeker, L. Meyer, T.J. *J. Am. Chem. Soc.* 1987, **109:** 746–754; *ibid.* 1986, **108:** 4066–4073, Thompson, M.S. and Meyer, T.J. *ibid.* 1982, **104:** 5070–5076.
12. Nelson, S.F., Ismagilov, R.F. and Trieber, D.A. *Science* 1997, **278:** 846–849.
13. (a) Wherland, S. *Coord. Chem. Rev.* 1993, **128:** 169–199; (b) Wherland, S., personal communication.
14. Olah, G.A. and Molnár, Á. *Hydrocarbon Chemistry*. Wiley, New York, 1995.
15. Valgimigli, L. Ingold, K.U. and Lusztyk, J. *J. Am. Chem. Soc.* 1996, **118:** 3545–3549 and references therein.
16. Dolphin, D, **B12**. Wiley, New York, 1981.
17. *Chemistry and Engineering News.* June 24, 1996, pp. 40ff.
18. Leading references: (a) Bordwell, F.G. *et al. J. Am. Chem. Soc.* 1991, **113:** 9790; (b) *ibid.* 1996, **118:** 8777; (c) *ibid.* 10819–10823; (d) Parker, V.D. *ibid.* 1992, **114:** 7458 and 1993, **115:** 1201; (e) Tilset, M. and Parker, V.D. *ibid.* 1989, **111:** 6711 and *ibid.* 1990, **112:** 2843; (f) Parker, V.D., Handoo, K.L., Roness, F. and Tilset, M. *ibid.* 1991, **113:** 7493; (g) Skagestad, V. and Tilset, M. *ibid.* 1993, **115:** 5077–5083; (h) Tilset, M. personal communication, 1997; (i) Wayner, D.D.M., Lusztyk, E., Pagé, D., Ingold, K.U., Mulder, P., Laarhoven, L.J.J. and Aldrich, H.S. *ibid.* 1995, **117:** 8737.
19. Gardner, K.A., Kuehnert, L.L. and Mayer, J.M. *Inorg. Chem.* 1997, **36:** 2069.
20. Young, C.L. (ed.) *Solubility Data Series.* Pergamon, New York, 1981, Vol. 5/6, p. 2.
21. Following Ref. 18d (including correction), the solvation of H• is modeled by the solvation of H_2, giving $\Delta H°\{H•(g) \rightarrow H•(aq)\} \cong \Delta H°\{H_2(g) \rightarrow H_2(aq)\}$.
22. Latimer, W.M. *The Oxidation States of the Elements and Their Potentials in Aqueous Solution*, 2nd Ed. Prentice Hall, New York, 1952.
23. Rush, J.D. and Bielski, B.H. *Inorg. Chem.* 1995, **34:** 5832–5838.
24. (a) For phenol: Lind, J., Shen, X., Eriksen, T.E. and Merényi, G. *J. Am. Chem. Soc.* 1990, **112:** 479–482 give $\Delta\Delta G_{solv} = \Delta G_{solv}(PhO•) - \Delta G_{g\text{-}aq}(PhOH)] \cong 3 \pm 1$ kcal/mol and $D(PhO\text{-}H)_g = 87 \pm 2$ kcal/mol (see also Refs. 1h,18i), yielding $D(H\text{–}OMnO_3^-) = 81 \pm 3$ kcal/mol. (b) For aniline, a slightly different thermochemical cycle is used involving $E°(PhNH_2)$ and $pK_a(PhNH_2^+)$ as described in: Jonsson, M., Lind, J., Eriksen, T.E. and Merényi, G. *J. Am. Chem. Soc.* 1994, **116:** 1423–1427. $[\Delta\Delta G_{g\text{-}aq} = \Delta G_{g\text{-}aq}(PhNH•) - \Delta G_{g\text{-}aq}(PhNH_2)] \cong 0.5$ kcal/mol and $\Delta(PhNH\text{–}H)_g = 89$ kcal/mol yield $\Delta(H\text{–}OMnO_3^-) = 80 \pm 2$ kcal/mol.

25. Moyer, B.A. and Meyer, T.J. *Inorg. Chem.* 1981, **20:** 436–444. Dovletoglou, A. Adeyemi, A.S. and Meyer, T.J. *Inorg. Chem.* 1996, **35:** 4120–4127.
26. c.f. Caudle, M.T. and Pecoraro, V.L. *J. Am. Chem. Soc.* 1997, **119:** 3415–3416.
27. (a) Cook, G.K. and Mayer, J.M. *J. Am. Chem. Soc.* 1994, **116:** 1855–1868. (b) *ibid.* 8859. (c) Cook, G.K. and Mayer, J.M. *ibid.* 1995, **117:** 7139–7156.
28. Carrington, A. and Symons, M.C.R. *J. Chem. Soc.* 1956, 3373.
29. Izutsu, K. *Acid-Base Dissociation Constants in Dipolar Aprotic Solvents.* IUPAC Chemical Data Series No. 35. Blackwell Scientific Publications, Boston, MA, 1990. Bordwell, F.G. *Acc. Chem. Res.* 1988, **21:** 456–463. Ref. 10, pp. 250–253.
30. Cheng, J.P. Handoo, K.L. and Parker, V.D. *J. Am. Chem. Soc.* 1993, **115:** 2655–2660. See also, *ibid.* 5067–5072.
31. (a) Exner, O. in *Correlation Analysis in Chemistry* (eds.) Chapman, N.B. and Shorter, J. Plenum, New York, 1978, pp. 439–540; (b) Lowry, R.H. and Richardson, K.S. *Mechanism and Theory in Organic Chemistry*, 3rd Ed. Harper & Row New York, 1987; (c) Shaik, S.S., Schlegel, H.B. and Wolfe, S. *Theoretical Aspects of Physical Organic Chemistry: The S_N2 Mechanism.* Wiley, New York, 1992.
32. Bahnson, B.J. and Klinman, J.P. *Methods Enzymol.* 1995, **249:** 373–397. See also Moiseyev, N., Rucker, J. and Glickman, M.H. *J. Am. Chem. Soc.* 1997, **119:** 3853–3860.
33. (a) Barbara, P.G., Meyer, T.J. and Ratner, M.A. *J. Phys. Chem.* 1996, **100:** 13148–13168; (b) Sutin, N. *Prog. Inorg. Chem.* 1983, **30:** 441; (c) Bolton, J. R. and Archer, M.D. in *Electron Transfer in Inorganic, Organic, and Biological Systems* (eds.) Bolton J.R., Mataga N. and McLendon G., Amer. Chem. Soc., Washington D.C., 1991 (*Adv. Chem. Series* #228), pp. 7–23.
34. For example, see (a) Carpenter, B.K. *J. Am. Chem. Soc.* 1995, **117:** 6336–6344 and (b) Carpenter, B.K. *American Scientist* 1997, **85:** 138.
35. Newcomb, M. La Tadic, M.-H. Chestney, D.L. Roberts, E.S. and Hollenberg, P. F *J. Am. Chem. Soc.* 1995, **117:** 12085–12091.
36. Binstead, R.A., McGuire, M.E., Dovletoglou, A., Seok, W.K., Roeker, L.E. and Meyer, T.J. *J. Am. Chem. Soc.* 1992, **114:** 173–186.
37. (a) Meunier, B. *Chem. Rev.* 1992, **92:** 1411–1456; (b) Ostović, D. and Bruice, T.C. *Acc. Chem. Res.* 1992, **25:** 314–320; (c) Traylor, T.G., Nakano, T., Dunlap, B.E., Traylor, P.S. and Dolphin, D. *J. Am. Chem. Soc.* 1986, **108:** 2784; (d) Selke, M., Sisemore, M.F. and Valentine, J.S. *ibid.* 1996, **118:** 2008–2012; (e) McMurry, T.J. and Groves, J.T. in *Cytochrome P-450: Structure,*

Mechanism, and Biochemistry, lst Ed. (ed.) Ortiz de Montellano, P.R. Plenum, New York, 1985, pp. 1–28.

38. (a) Gross, Z., Nimri, S., Barzilay, C.M. and Simkhovich, L. *J. Bioinorganic Chem.* 1997, **2:** 492–506; (b) Gross, Z. and Nimri, S. *J. Am. Chem. Soc.* 1995, **117:** 8021–8022; (c) Groves, J.T. and Watanabe, Y. *J. Am. Chem. Soc.* 1986, **108:** 507–508.
39. (a) 1.90 V: *Encyclopedia of Electrochemistry of the Elements* (ed.) Bard A.J., Marcel Dekker, New York, 1978, Vol. 11, pp. 48–49; (b) 1.65 V: Garrison, J.M. Ostović, D. and Bruice, T.C. *J. Am. Chem. Soc.* 1989, **111:** 4960–4966.
40. (a) Jeon, S. and Bruice, T.C. *Inorg. Chem.* 1992, **31:** 4843–4848; (b) Groves, J.T., Gross, Z. and Stern, M.K. *ibid.* 1994, **33:** 5065–5072; (c) Fuji, H. *J. Am. Chem. Soc.* 1993, **115:** 4641–4648; (d) Balch, A.L., Latos-Grazynski, L. and Renner, M.W. *ibid.* 1985, **107:** 2983–2985; (e) Gross, Z., personal communication.
41. (a) See also Mirafzal, G.A., Kim, T., Liu, J. and Bauld, N.L. *J. Am. Chem. Soc.* 1992, **114:** 10968–10969; (b) Electron transfer is likely to be less favorable in the less polar solvent CH_2Cl_2. The solvent dependence of E° should outweigh the electrostatic stabilization of the ion pair.
42. (a) Kochi, J..K., in *Comprehensive Organic Synthesis*, Vol. 7 (Oxidation) (ed.) Trost, B.M., Pergamon, New York, 1991, pp. 849–855; (b) Ross, S.D. Finkelstein, M. and Rudd, E.J. *Anodic Oxidation.* Academic Press, New York, 1975, p. 117.
43. Wallar, B.J. and Lipscomb, J.D. *Chem. Rev.* 1996, **96:** 2625–2657.
44. (a) Kochi, J.K. (ed.) *Free Radicals,* Wiley, New York, 1973, especially; (b) Ingold, K.U., Vol. 1, Chap. 2, pp. 69ff and; (c) Russell, G.A. Vol. 1, Chap. 7, pp. 283–293; (d) Tedder, J.M. *Ang. Chem., Int. Ed. Engl.* 1982, **21:** 401–410; (e) Korzekwa, K.R., Jones, J.P. and Gillette, J.R. *J. Am. Chem. Soc.* 1990, **112:** 7042–7046.
45. For recent work and leading references: Kristjánsdóttir, S.S. and Norton, J.R. *J. Am. Chem. Soc.* 1991, **113:** 4366–4367; Creutz, C. and Sutin, N. *J. Am. Chem. Soc.* 1988, **110:** 2418–2427; Albery, W.J. *Annu. Rev. Phys. Chem.* 1980, **31:** 227–263; Saveant, J.-M. *Adv. Phys. Org. Chem.* 1990, **6:** 1; Choi, M.-G. and Brown, T.L. *Inorg. Chim. Acta* 1992, **198–200:** 823. Schwarz, C.L. and Endicott, J.F. *Inorg. Chem.* 1995, **34:** 4572–4580; Schwarz, C.L., Bullock, R.M. and Creutz, C. *J. Am. Chem. Soc.* 1991, **113:** 1225–1236; Lee, I.-S.H., Jeoung, E.H. and Kreevoy, M.M. *J. Am. Chem. Soc.* 1997, **119:** 2722–2728. Ref.31c.
46. Knox, J.H. in *Oxidation of Organic Compounds,* Vol. 2 (ed.) Mayo F.R. Amer. Chem. Soc. (*Adv. Chem. Series*) 1968, **76:** 11–22.

47. (a) Reference 31b, pp. 671–676; (b) Pross, A. *Theoretical and Physical Principles of Organic Reactivity.* Wiley, New York, 1995 pp. 175–182; (c) Pross, A., Yamatoaka, H. and Nagase, S. *J. Phys. Org. Chem.* 1991, **4:** 135.
48. For instance, Lunsford, J.H. *Ang. Chem., Int. Ed. Engl.* 1995, **34:** 970–980.
49. For example "$Mn(OAc)_3$": Snider, B.B. *Chem. Rev.* 1996, **96:** 339.
50. (a) For a recent review, see Solomon, E.I., Zhou, J., Neese, F. and Pavel, E.G. *Chem. Biol.* 1997, **4:** 795–808; (b) Jonas, R.T. and Stack, T.D.P. *J. Am. Chem. Soc.* 1997, **119:** 8566–8567 and references therein.
51. Klinman, J.P. *Chem. Rev.* 1996, **96:** 2541–2561.
52. *Cytochrome P-450: Structure, Mechanism, and Biochemistry.* (ed.) Ortiz de Montellano, P.R. Plenum, New York; (a) 2nd Ed., 1995; (b) 1st Ed., 1985; (c) Groves, J.T. Krishnan, S. Avaria, G.E. and Nemo, T.E. in *Biomimetic Chemistry, ACS Advances in Chemistry Series* 1980, **191:** 277; (d) Chapter 3 of this volume.
53. For example, Loew, G.H. and Herman, Z.S. *J. Am. Chem. Soc.* 1980, **102:** 6173–6174.
54. Wang, K. and Mayer, J.M. *J. Org. Chem.* 1997, **62:** 4248.
55. Gardner, K.A. and Mayer, J.M. *Science* 1995, **269:** 1849.
56. Mayer, J.M. *Acc. Chem. Res.* 1998, **31:** 441–450.
57. Reviews of CrO_2Cl_2 and Cr^{VI} reactions: (a) Wiberg, K.B. in *Oxidation in Organic Chemistry* Part A (ed.) Wiberg K.B. Academic, New York, 1965, pp. 69–184; (b) Freeman, F. in *Organic Syntheses by Oxidation with Metal Compounds* (eds.) Mijs, W.J. and de Jonge, C.R.H.I. Plenum, New York, 1986, Chap. 2.
58. Permanganate reviews: Ref. 11a and (a) Waters, W.A. *Quart. Rev. (London)* 1958, 277; (b) Fatiadi, A.J. *Synthesis (Stuttgart)* 1987, 85.
59. (a) Reference 57a; (b) Wiberg, K.B. and Foster, G. *J. Am. Chem. Soc.* 1961, **83:** 423; (c) Wiberg, K.B. and Eisenthal, R. *Tetrahedron* 1964, **20:** 1151.
60. (a) Nenitzescu, C.D. *Bull. Soc. Chim. Fr.* 1968, **4:** 1349; (b) Rappé, A.K. and Goddard, W.A., III. *J. Am. Chem. Soc.* 1982, **104:** 3287.
61. Groves, J.T. and Nemo, T.E. *J. Am. Chem. Soc.* 1983, **105:** 6243.
62. Bowry, V.W., Lusztyk, J. and Ingold, K.U. *J. Am. Chem. Soc.* 1991, **113:** 5687.
63. Newcomb, M., La Tadic, M.-H., Chestney, D.L., Roberts, E.S. and Hollenberg, P. F. *J. Am. Chem. Soc.* 1995, **117:** 12085.
64. Stephenson, L.M., Egnatchik, J. and Speth, D.R. *J. Org. Chem.* 1979, **44:** 346.
65. (a) MnO_4^-: Steenken, S. and Neta, P. *J. Am. Chem. Soc.* 1982, **104:** 1244; (b) Cr^{VI}: Al-Sheikhly, M. and McLaughlin, W.L. *Radiat. Phys. Chem.* 1991, **38:** 203.

66. Karaman, H., Barton, R.J., Robertson, B.E. and Lee, D.G. *J. Org. Chem.* 1984, **49:** 4509.
67. (a) E°_{ox} = 1.54 V for *p*-xylene versus 1.96 V for toluene (in MeCN, versus $Ag^+/AgNO_3$):[67b]; (b) Kochi, J.K. in *Comprehensive Organic Synthesis*, Vol. 7 (Oxidation) (ed.) Trost, B.M., Pergamon, New York, 1991, pp. 849–855. Slightly different values appear in Ref. 39; (c) The conversion of $Ag/AgNO_3/MeCN$ to aqueous NHE is ~+0.5 V: Sawyer, D.T.; Sobkowiak, A. Robers, J.L., Jr. *Electrochemistry for Chemists,* 2nd Ed. Wiley, 1995, pp. 192, 201.
68. Figures adapted from Ref. 56.
69. Lee, D.G. and Chen, T. *J. Am. Chem. Soc.* 1993, **115:** 11231–11236. See also Ref. 60b.
70. The lowest lying permanganate triplet state lies roughly 51 kcal/mol (18 000 cm^{-1}, 2.2 eV) above the ground state: Blasse, G. *Structure and Bonding* (*Berlin*) 1980, **42:** 1–42, especially p. 11. The value for CrO_2Cl_2 is similar: Cielak-Golonka, M. *Coord. Chem. Rev.* 1991, **109:** 223–249, especially 233–237. Miller, R.M.; Tinti, D.S. and Case, D.A. *Inorg. Chem.* 1989, **28:** 2738–2743.
71. Avila, D.V. Brown, C.E. Ingold, K.U. Lusztyk, J. *J. Am. Chem. Soc.* 1993, **115:** 466–470.
72. Adapted from Ref. 55. Bond strengths and radical rate constants from refs. 1 and 73.
73. (a) Howard, J.A. and Scaiano, J.C. Oxyl-, peroxyl- and related Radicals. In *Radical Reaction Rates in Liquids,* Landolt-Börnstein New Series (ed.) Fischer, H. Springer-Verlag, New York, 1984, Vol. 13, subvol. d; (b) ^{t}BuO• + DHA: Mulder, P., Arends, I.W.C.E., Clark, K.B. and Wayner, D.D.M. *J. Phys. Chem.* 1995, **99:** 8182–8189; (c) ^{s}BuOO• + DHA: Howard, J. A. and Ingold, K.U. *Can. J. Chem.* 1968, **46:** 2661.
74. Data from Refs. 19 and 73.
75. Baulch, D.L., Duxbury, J., Grant, S.J. and Montague, D.C. Evaluated kinetic data for high temperature reactions, Vol. 4. *J. Phys. Chem. Ref. Data* 1981, **10:** Supplement #1. Kerr, J.A. and Moss, S.J. *CRC Handbook of Bimolecular and Termolecular Gas Reactions,* Vol. 1, CRC Press, Boca Raton, Florida, 1981.
76. Fleming, I. *Frontier Orbitals and Organic Chemical Reactions.* Wiley, London, 1976, pp. 182–188.
77. An alternative to polar effects is presented in Zavitsas, A.A. and Chatgilialoglu, C. *J. Am. Chem. Soc.* 1990, **112:** 10645.
78. (a) References 33 and 45; (b) References 31c and 47.
79. Manchanda, R., Thorp, H.H., Brudvig, G.W. and Crabtree, R.H. *Inorg. Chem.* 1992, **31:** 4040–4041.

80. Hoganson, C.W. and Babcock, G.T. *Science* 1997, **277:** 1953.
81. Wang, K. and Mayer, J.M. *J. Am. Chem. Soc.* 1997, **119:** 1470.
82. Adams, J.C. and Mayer, J.M. unpublished results.
83. Baldwin, M.J. and Pecoraro, V.L. *J. Am. Chem. Soc.* 1996, **118:** 11325 and references therein. salpn = *N*, *N'*-bis(salisylidene)-1,3-propanediamine; X = phenolate substituent.
84. Diaddario, L.L., Robinson, W.R. and Margerum, D.W. *Inorg. Chem.* 1983, **22:** 1021–1025. Aib_3 is the tripeptide of α-amino isobutyric acid.
85. Sulfab, Y. and al-Shatti, N.I. *Inorg. Chim. Acta* 1984, **87:** L23.
86. Lockwood, M.A., Blubaugh, T.J., Collier, A.M., Lovell, S. and Mayer, J.M. *Angew. Chem. Int. Ed.* 1999, **38:** 225–227.
87. Youngblood, M.P. and Margerum, D.W. *Inorg. Chem.* 1980, **19:** 3068.
88. Burnett, M.G. McKee, V. and Nelson, S.M. *J. Chem. Soc., Chem. Comm.* 1980, 599–601 and *J. Chem. Soc., Dalton Trans.* 1981: 1492–1497.
89. Roth, J.P. and Mayer, J.M. *Inorg. Chem.* 1999, **38:** 2760–2761.
90. Folti, M. Ingold, K.U. and Lusztyk, J. *J. Am. Chem. Soc.* 1994, **116:** 9440–9447. For an alternative view, see Ref. 77.
91. Valgimigli, L., Banks, J.T., Ingold, K.U. and Lusztyk, J. *J. Am. Chem. Soc.* 1995, **117:** 9966–9971; MacFaul, P.A., Ingold, K.U. and Lusztyk, J. *J. Org. Chem.* 1996, **61:** 1316–1321.
92. Eberson, L. *Electron Transfer Reactions in Organic Chemistry*. Springer-Verlag, New York, 1987, pp. 27–28.

Chapter 2

Distinguishing Biomimetic Oxidations from Oxidations Mediated by Freely Diffusing Radicals

K. U. Ingold* & Philip A. MacFaul

Steacie Institute for Molecular Sciences,
National Research Council of Canada,
Ottawa, Ontario, Canada K1A 0R6

A large number of enzymatic oxidation processes and current views regarding their mechanisms are described elsewhere in this book. Scientists have attempted to understand these enzymes at the molecular level by mimicking their chemical behavior and by checking proposed mechanisms by reacting the enzymes with substrates (generally non-natural) which can serve as mechanistic probes. In this chapter, our concern will be restricted to biomimetic oxygenation systems and will specifically address the question: *How can a researcher determine whether their new "biomimetic oxidation system" occurs via what is effectively an "insertion" of an oxygen atom into a chemical bond or via* ***free*** *radical chemistry*? The printing of free in bold face is intended to emphasize that the radicals involved in the overall chemistry can diffuse freely throughout the solution or local environment. "Insertion" does *not* necessarily mean that the chemistry does not occur by radicals but, simply, that if radicals are involved they do not escape from the molecular environment in which they were created. As the reader of this book will appreciate, it appears likely that at least some of the monooxygenases "insert" an oxygen atom into unactivated C–H bonds by a process that may involve the transient or partial formation of carbon-centered radicals but these radicals do not escape from the enzymes' active site to become **free** radicals.

Procedures for Distinguishing Between Biomimetic and Free Radical Oxidations

Today, the quantitative aspects of the kinetics of **free** radical reactions in solution are very extensive (in terms of reactions, solvent effects, etc.) and very firmly established (many thousands of reliable absolute rate constants for elementary reaction steps).[1] It is therefore a *relatively* simple task to distinguish unequivocally between radical-free biomimetic oxidative chemistry and **free** radical-mediated oxidative chemistry *provided* the researcher has a reasonably comprehensive and quantitative understanding of **free** radical kinetics and reaction mechanisms. With such knowledge, firm predictions can be made regarding the reaction products and the relative and absolute overall rates of reaction which will be caused by a change in the reagents and their concentrations, by a change in the solvent, by the addition of a new compound at a particular concentration to the reaction mixture, and so on. If, however, a researcher lacks a clear and quantitative understanding of **free** radical kinetics, incorrect mechanistic conclusions are all too likely to be drawn.

The involvement of **free** radicals in purported non-radical biomimetic oxidations has generally been demonstrated by product analyses in experiments designed to produce *predictable* changes in the nature of the products, or their relative, or absolute yields if **free** radicals are involved. Most of the main procedures are itemized and described briefly below. The wise researcher should always use more than one of these procedures before drawing any mechanistic conclusions.

1. *Free* Radical Clocks[2]

(i) Choice of substrate to be oxidized

If the new, potentially biomimetic, catalytic system has been demonstrated to convert certain hydrocarbons to corresponding alcohols, it should also be used to oxidize similar types of hydrocarbons, i.e. alkanes, alkyl aromatics or alkenes, which *can* yield different products from those of a true "oxygen atom insertion" if **free** carbon-centered radicals are intermediates. This *general* procedure can often be quantified. For example, suppose experiments under an atmosphere of $^{18}O_2$ have shown that the oxygenated products (alcohol, ketone, etc.) have incorporated their oxygen atom(s) from dioxygen (from air or generated *in situ*). The experiment should be repeated (under $^{16}O_2$) using a hydrocarbon, **UH**, which, *if* **free** radicals are involved, will yield a radical, **U•**, which undergoes rearrangement to a new radical, **R•**, with a known rate constant, k_r, under the experimental conditions. The **free** radical mechanism of oxidation is:

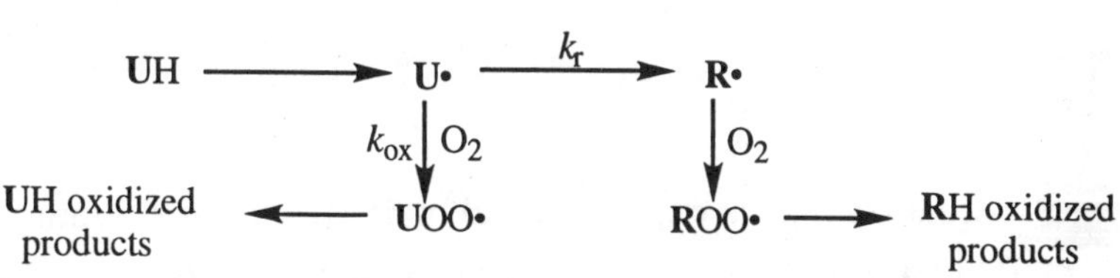

Provided the rearrangement of **U•** is sufficiently rapid, some or all of the oxidation products will be derived from **R•**. This situation will be obtained if $k_r > k_{ox}[O_2]$. The observed products will then provide strong presumptive evidence for a **free** radical mechanism and against an "oxygen insertion" mechanism. However, if $k_r << k_{ox}[O_2]$, all the products will be derived from the un-rearranged radical, **U•**. Thus, *an inappropriate choice of the **free** radical clock will lead to the incorrect mechanistic conclusion that this particular oxidation system is "biomimetic" and not **free** radical.*

Table 2.1 lists a few of the hundreds of **U•**⟶ **R•** reactions for which k_r is known.[1–3] These few reactions have been chosen to illustrate the enormous range of k_r values available to the researcher. To return to the general reaction discussed above: for nearly all carbon-centered radicals $k_{ox} \sim k_{diff} \sim 10^{10}$ $M^{-1}s^{-1}$ in solvents of normal viscosity and, under 1 atm of air, $[O_2] \sim 2 \times 10^{-3}$ M in most organic solvents. It is therefore pointless to use *tert*-butyl benzene as a mechanistic probe since there will be no rearranged products, [i.e. $k_{ox}[O_2]$ ($\sim 2 \times 10^7$ s^{-1}) >> k_r ($= 9 \times 10^2$ s^{-1}). However, it would be eminently sensible to employ the hypersensitive probe for **free** radicals, *trans*-2-phenyl-1-methylcyclopropane ($k_r = 3 \times 10^{11}$ s^{-1}) which would give *only* rearranged products and thus would provide strong presumptive evidence for a **free** radical mechanism. The stereospecific hydroxylation of *cis*- and *trans*-1,2 dimethylcyclohexane is also a useful probe for "oxygen insertion" into alkanes because oxidation via a carbon-centered **free** radical would yield a mixture of the epimeric alcohols, e.g.

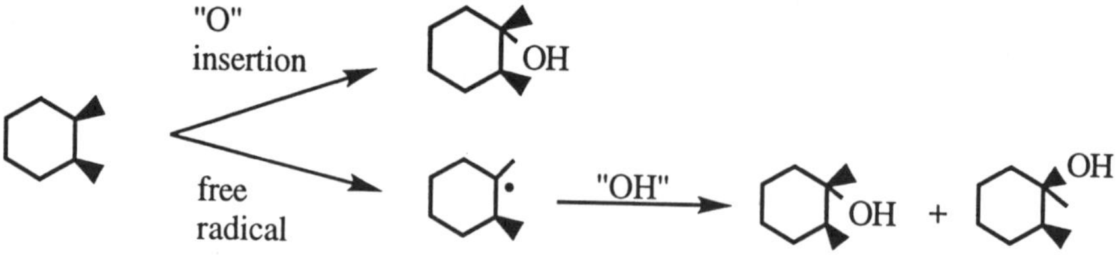

However, oxidation via a **caged** radical pair mechanism would yield stereospecific hydroxylation products if cage collapse (oxygen rebound) was faster than rotation of the dimethylcyclohexyl radical with respect to the oxygen atom donor, ~ 10^{-10} s.

Table 2.1 Rate constants for selected free alkyl radical clocks at 298 K[3].

U•		R•	k_r
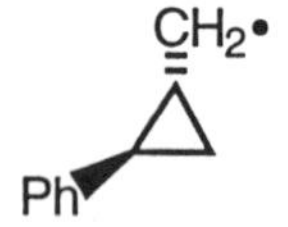	→	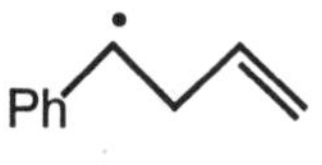	3×10^{11} s^{-1}
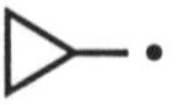	→	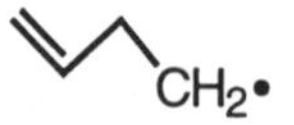	9.4×10^{7} s^{-1}
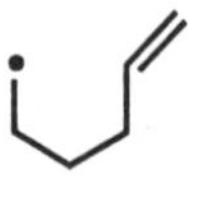	→	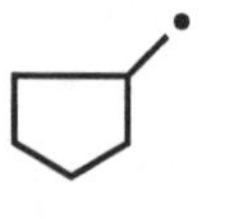	2.3×10^{5} s^{-1}
$PhCMe_2\dot{C}H_2$	→	$PhCH_2\dot{C}Me_2$	9×10^{2} s^{-1}

For alkene oxidations to alcohols the **U•** ⟶ **R•** rearrangement probe can be replaced by an allyl radical probe, e.g.

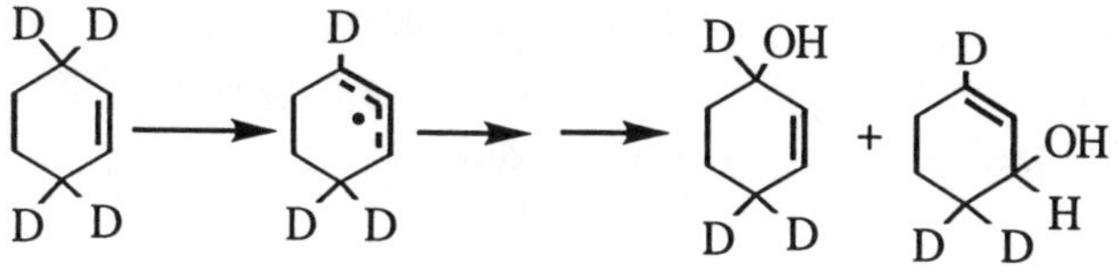

(2.1)

$$PhCH_2CH{=}CH_2 \longrightarrow Ph\dot{C}HCHCH_2 \longrightarrow PhCH{=}CHCH_2OH \qquad (2.2)$$

(2.3)

For alkene oxidations to epoxides, a useful probe is *cis*-stilbene which yields *cis*-stilbene oxide if oxygenation is a genuine two-electron process, i.e.[5]

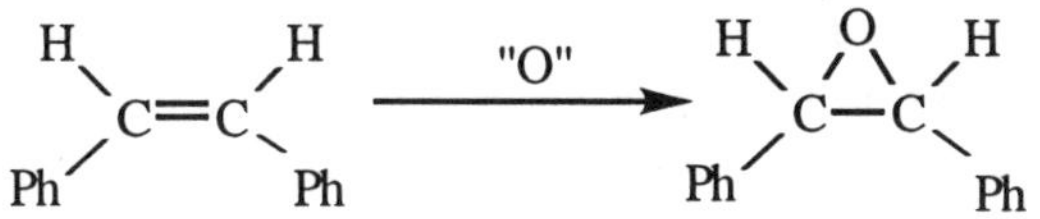

(2.4)

but *trans*-stilbene oxide via one-electron, **free** radical processes, i.e.

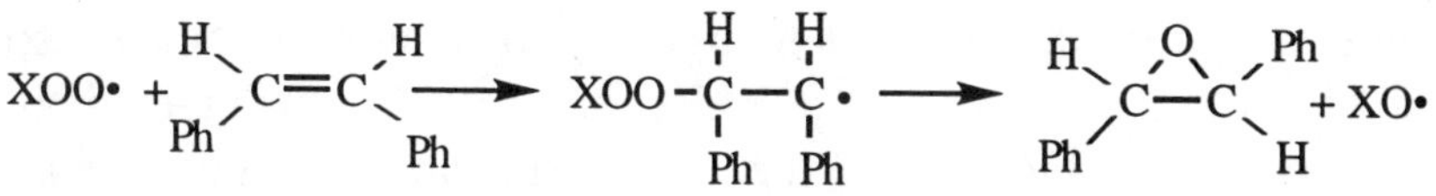

(2.5)

Benzaldehyde is also a characteristic **free** radical oxidation product of *cis*-stilbene:

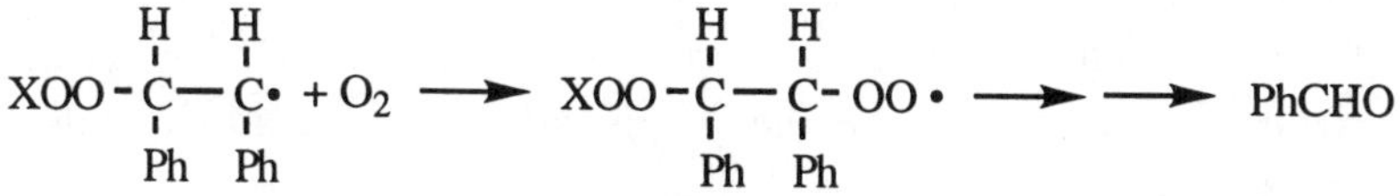

(2.6)

(ii) Choice of "activated oxygen" source

As their name implies, monooxygenases oxidize their substrates by "inserting" a single oxygen atom into the substrate molecule. This oxygen atom is derived from dioxygen. The second oxygen atom of the dioxygen is reduced to water via a sacrificial reductant as part of the catalytic cycle, e.g.

$$RH + O_2 + 2\,e^- + 2\,H^+ \longrightarrow ROH + H_2O \qquad (2.7)$$

Probably a majority of attempts to mimic the chemistry of monooxygenases have made use of some two-electron reduced form of dioxygen, viz. hydrogen peroxide or an organic hydroperoxide, i.e.

$$RH + XOOH \longrightarrow ROH + XOH \qquad (2.8)$$

The advantage of this strategy is that it eliminates the need for a sacrificial reductant. Two commercially available hydroperoxides have been most favored, hydrogen peroxide and *tert*-butyl hydroperoxide (TBHP). The latter has the advantage of excellent solubility in many organic solvents.

Claims for non-**free** radical biomimetic oxidations accomplished using a hydroperoxide and a transition metal catalyst (which is commonly an Fe^{II} or Fe^{III} chelate added initially or formed *in situ*), rest on the *assumption* that oxidation involves a reactive, high-valent, iron-oxo species, generally formalized as $[Fe^{V}{=}O]$ to avoid the need to specify ligands. However, it seems doubtful that any such high-valent, iron-oxo species exist. (Note that even in cytochrome P450, the fifth oxidizing equivalent is believed to be located on the porphyrin ring and hence P450's active oxidizing agent is probably best formalized as $[^{\cdot+}PFe^{IV}{=}O]$.) In the absence of a porphyrin ring or a chelator analogous to porphyrin, the highest oxidation state of iron can probably best be formulated as $[Fe^{IV}{=}O]$. The chemistry involving the putative $[Fe^{V}{=}O]$ species is speculated by different researchers to take different pathways (sometimes even involving Fe^{V} with an iron-carbon bond to the substrate, $Fe^{V}(OH)R$). However, in a generalized way (and ignoring the chemically non-relevant ligands), the catalytic cycle is usually formulated as:

$$YFe^{III} + XOOH \rightleftharpoons Fe^{III}OOX + HY$$

$$Fe^{III}OOX \xrightarrow{\text{heterolysis}} Fe^{V}{=}O + XO^{-} \xrightarrow{H^{+}} XOH$$

$$Fe^{V}{=}O + RH \longrightarrow Fe^{III} + ROH$$

The **free** radical mechanism for oxidation involves *homolysis* of the peroxy-iron species rather than heterolysis. This yields a hydroxyl radical from H_2O_2 and an alkoxyl radical from an alkyl hydroperoxide. These highly reactive radicals attack the substrate and initiate its oxidation by producing (from most substrates) a carbon-centered radical, R•. For alkanes this attack occurs by hydrogen atom abstraction, but for unsaturated hydrocarbons (including alkyl aromatics in the case of HO• radicals) the carbon-centered radicals may also be formed by addition. The generalized **free** radical mechanism for Fe^{II} and Fe^{III} can be represented by:

$$XOOFe^{III\,(II)} \xrightarrow{\text{homolysis}} Fe^{IV(III)} = O + XO\bullet$$

$$XO\bullet + RH \longrightarrow XOH + R\bullet \text{ (or } XORH\bullet)$$

$$R\bullet \text{ (or } XORH\bullet) + O_2 \longrightarrow \text{oxidized products}$$

The radical clock methodology can be employed when the claimed biomimetic oxidation process has utilized TBHP. The TBHP is replaced by a *tert*-alkyl hydroperoxide which, **if** alkoxyl radicals are produced via homolysis of the peroxyl iron species, will have a much shorter intrinsic lifetime than *tert*-butoxyl. The reduced intrinsic lifetime could be due to a more facile β-scission than *tert*-butoxyl or to a rearrangement not available to *tert*-butoxyl. Both mechanisms would convert the reactive *tert*-alkoxyl radical into a relatively unreactive carbon-centered radical. Some unimolecular *tert*-alkoxyl radical reactions (which determine their intrinsic lifetimes) are given in Table 2.2 together with the room temperature rate constants for the processes shown. It should be noted that while the unimolecular rate constants for intramolecular 1,5 hydrogen atom abstractions and 5-exo cyclizations are unlikely to be solvent-dependent, the rate constants for β-scission are accelerated (sometimes dramatically) by an increase in the polarity of the solvent.[6] *Tert*-alkoxyl radicals abstract hydrogen atoms from most hydrocarbons with solvent-independent rate constants in the range $(1\text{–}10) \times 10^6$ $M^{-1}s^{-1}$, [1,6,7] e.g. from cyclohexane

$k = 1.2 \times 10^6\ M^{-1}s^{-1}$, from ethylbenzene $k = 1.05 \times 10^6\ M^{-1}s^{-1}$ and from cyclohexene $k = 5.7 \times 10^6\ M^{-1}s^{-1}$. Therefore, even in the presence of 1 M hydrocarbon substrate, a *tert*-alkoxyl radical which undergoes a β-scission or rearrangement with a rate constant of ca. $10^7\ s^{-1}$ will give much smaller yields of RH oxidized products than *tert*-butoxyl radicals derived from TBHP, and if the *tert*-alkoxyl radical decomposes with a rate constant $> 10^8\ s^{-1}$, there will be little or no hydrogen atom abstraction from RH *by this radical.*

To date, the most useful *tert*-alkoxyl radical clock for distinguishing between radical-free biomimetic and **free** radical hydrocarbon oxidations has proven to be that derived from 2-methyl-1-phenylprop-2-yl hydroperoxide, MPPH,[8] because it undergoes extremely rapid β-scission:

$$\underset{\text{MPPH}}{PhCH_2C(CH_3)_2OOH} \longrightarrow \underset{\text{MPPO}}{PhCH_2C(CH_3)_2O\bullet} \xrightarrow{\beta} PhCH_2\bullet + (CH_3)_2CO \qquad (2.9)$$

The corresponding β-scission of *tert*-butoxyl radicals is much slower:

$$\underset{\text{TBHP}}{(CH_3)_3COOH} \longrightarrow \underset{\text{TBO}}{(CH_3)_3CO\bullet} \xrightarrow{\beta} CH_3\bullet + (CH_3)_2CO \qquad (2.10)$$

Thus, many purported examples of alkane biomimetic oxidations in which TBHP was used as the oxidizing agent have subsequently been demonstrated to occur via **free** radical mechanisms simply by replacing the TBHP by MPPH. The products are no longer derived from the alkane but instead show the clear signature of the benzyl radical, e.g. benzyl alcohol, benzaldehyde, bibenzyl, etc. *However, great care is required when interpreting the results obtained by replacing TBHP by MPPH when the oxidizable substrates are alkyl aromatics or alkenes.* Such hydrocarbons are likely to yield their characteristic oxidation products both with TBHP *and* with MPPH whether the reaction is radical-free or occurs by normal **free** radical chemistry. This is because hydrogen atom abstraction from alkyl aromatics and alkenes does not (in contrast to alkanes) require a highly reactive alkoxyl radical but can also be achieved at a significant rate by the much less reactive alkylperoxyl radicals. Thus, both TBHP via the TBO radical and MPPH via the benzylperoxyl radical can initiate classical autoxidation chain reactions with these substrates, e.g.

$$PhCH_2\bullet + O_2 \longrightarrow \left.\begin{matrix} Me_3CO\bullet \\ PhCH_2OO\bullet \end{matrix}\right\} + RH \longrightarrow R\bullet + \left\{\begin{matrix} Me_3COH \\ PhCH_2OOH \end{matrix}\right.$$

$$R\bullet + O_2 \longrightarrow ROO\bullet \xrightarrow{RH} ROOH + R\bullet$$

Table 2.2 Rate constants for unimolecular reactions for selected alkoxyl radicals in CH_3CN at 298 K.

Reaction			*k* **(Ref.)**
$PhCH_2CMe_2O\bullet$	$\longrightarrow$	$PhCH_2\bullet + Me_2CO$	~ $2 \times 10^8\ s^{-1}$ (8)
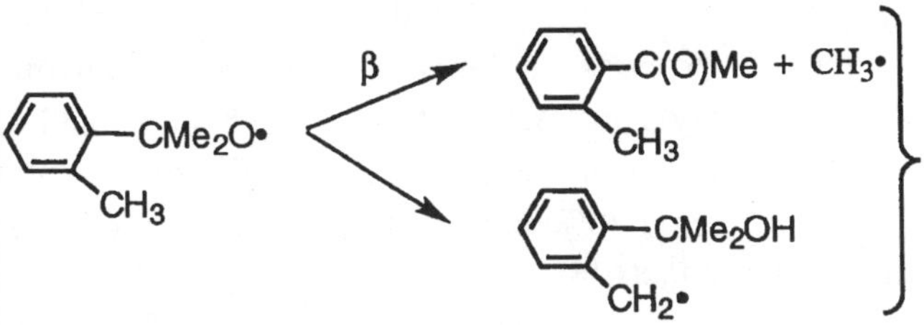			$2.5 \times 10^6\ s^{-1}$ (9)
			$6.3 \times 10^5\ s^{-1}$ (6)
$Me_3CO\bullet$	$\longrightarrow$	$Me_2CO + Me\bullet$	$8.3 \times 10^4\ s^{-1}$ (1)
			$5 \times 10^8\ s^{-1}$ (10)
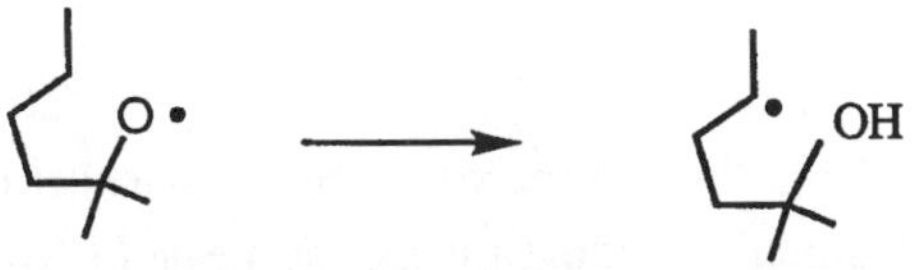			~ $5 \times 10^6\ s^{-1}$ (1)

It is therefore essential to check any apparent radical-free MPPH result with one or more of the other critical tests indicated in this chapter (as well as checking the products for benzyl alcohol, etc.). This "autoxidation problem" can only arise with substrates containing activated C–H bonds. Hydrogen atom abstraction from alkanes by alkylperoxyl radicals is simply too slow (e.g. $k^{ROO\bullet}$ (c–C_6H_{12}) = 0.04 $M^{-1}s^{-1}$)[11] to compete with their bimolecular self-reactions (i.e. chain termination).

2. Ambiguous Alkene Substrates

The epoxidation of an alkene to its epoxide provides a very nice probe for high-valent metal-oxo species which are known to be capable of effecting such direct (possibly concerted) two-electron oxidations. If epoxidation with a purported high-valent metal-oxo species is not achieved (as is true of many iron/hydroperoxide based systems) this is strong evidence that these species are absent or, at least, are not chemically competent intermediates. Unfortunately, some alkenes give very respectable yields of epoxides via **free** radical autoxidation processes. Thus, in the autoxidation of cyclopentene, 88% of the peroxyls abstract allylic hydrogen atoms and 12% add to the double bond.[12] However, in the autoxidation of cyclooctene, the corresponding figures are 30% and 70% and yields of epoxide can approach 50% or more of the total products.[12]

High yields of epoxides can also be produced in the **free** radical autoxidation of 1-alkenes. In this case, the "culprit" is aldehyde formation. As aldehydes accumulate in the reaction mixture, the co-oxidation of the aldehyde with the alkene results in epoxide formation by two additional routes, i.e. additional to the alkylperoxyl addition, cyclization route, Reaction 2.11.[13] One route is analogous to Reaction 2.11 but involves the

$$\mathrm{ROO^{\bullet} + H_2C{=}CHR' \longrightarrow ROOCH_2\dot{C}HR' \longrightarrow RO^{\bullet} + \overset{\frown O \frown}{CH_2{-}CHR'}} \qquad (2.11)$$

much better oxygen atom donor, the acylperoxyl radical (Reaction 2.12), and the other route is a non-radical epoxidation by the peracid (formed via oxidation of the aldehyde).

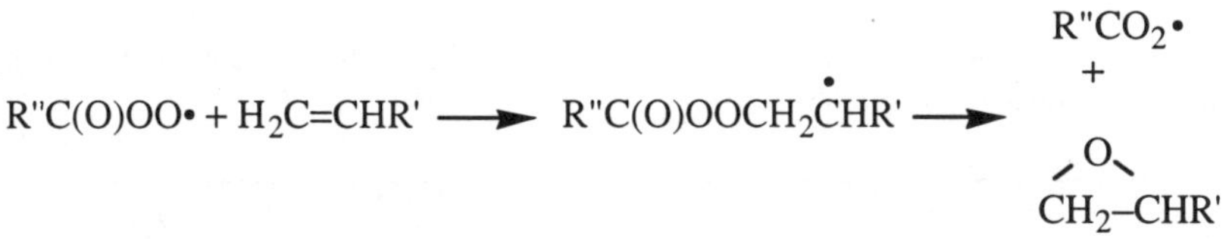

(2.12)

Thus, high yields of epoxide do not, in general, prove that alkene oxidation is occurring by a non-radical process. The formation of products (including epoxides) by autoxidation of the substrate can be readily checked by addition of *antioxidants* (*vide infra*).

3. Effects of Additives

The mechanism of a purported biomimetic system can be examined by the addition of certain compounds to the oxidation reactions. The presence of a high-valent iron-oxo species has been probed in the past by the addition of sulfides with the idea being that if an (Fe^V=O) active species is formed then it will be diverted from its normal fate and instead will oxidize the sulfide to the sulfoxide:[14]

$$(Fe^{V}{=}O) + R'SR'' \longrightarrow (Fe^{III}) + R'S(O)R'' \qquad (2.13)$$

However, sulfoxides can also be formed in the presence of hydroperoxides under free radical conditions[9,15], hence the use of sulfides as traps for high-valent iron-oxo species in metal/hydroperoxide systems is not recommended. Nevertheless, there are a variety of well-characterized additives that can be used to probe the presence of both oxygen-centered and carbon-centered radicals.

(i) Traps for oxygen-centered radicals

Phenols and aromatic amines are used (both commercially and by nature) as antioxidants because they react much more rapidly with alkylperoxyl and alkoxyl radicals than these radicals react with hydrocarbons. A truly vast quantity of kinetic data is available[1] _ a very small fraction of which is given in Table 2.3.

Table 2.3 Selected rate constants for reactions of oxygen-centered radicals with aromatic phenols and amines in CH_3CN at 298 K.

Reactants		Products	k (Ref.)
$PhOH + PhCMe_2O\bullet$	$\longrightarrow$	$PhO\bullet + PhCMe_2OH$	$5.8 \times 10^6\ M^{-1}s^{-1}$ (16)
$2,6-{}^tBu_2\text{-}4\text{-}MeC_6H_4OH + Me_3CO\bullet$	$\longrightarrow$	$2,6-{}^tBu_2\text{-}4\text{-}MeC_6H_4O\bullet + Me_3COH$	$2.0 \times 10^7\ M^{-1}s^{-1}$ (8)
$Ph_2NH + Me_3CO\bullet$	$\longrightarrow$	$Ph_2N\bullet + Me_3COH$	$3.0 \times 10^8\ M^{-1}s^{-1}$ (17)
$2,6-{}^tBu\text{-}4\text{-}MeC_6H_4OH + ROO\bullet$	$\longrightarrow$	$2,6-{}^tBu\text{-}4\text{-}MeC_6H_4O\bullet + ROOH$	$1.4 \times 10^4\ M^{-1}s^{-1}$ (18)
$Ph_2NH + ROO\bullet$	$\longrightarrow$	$Ph_2N\bullet + ROOH$	$\sim 4 \times 10^4\ M^{-1}s^{-1}$ (19)

To illustrate the use of the data in Table 2.3, consider the rate constants for hydrogen atom abstraction from cyclohexane and cyclohexene by alkylperoxyl radicals and alkoxyl radicals. For an added antioxidant, AH, to provide 99% protection to 1 M hydrocarbon, RH, against attack by peroxyl radicals requires k_{AH}^{ROO} [AH] = $100k_{RH}^{ROO}$ and by alkoxyl radicals requires k_{AH}^{RO} [AH] = $100k_{RH}^{RO}$. Thus, if 2,6-di-*tert*-butyl-4-methylphenol, BHT, was chosen as the antioxidant, the BHT concentrations required to provide 99% protection to cyclohexane against attack by peroxyl radicals and alkoxyl radicals would be 2.9×10^{-4} M and an impossible 6 M, respectively, while for cyclohexene the corresponding concentrations would be 4.4×10^{-2} M and a quite ridiculous 28.5 M. Thus, BHT can provide excellent protection to hydrocarbons against peroxyl radicals even at low concentrations but it provides little or no protection against attack by alkoxyl radicals.

A careful researcher must be aware of potential pitfalls in these simple calculations. First, they apply to the initial reaction conditions. Each molecule of BHT traps two radicals. Therefore, although the calculated concentrations of BHT will trap 99% of the oxyl radicals initially produced, these initial BHT concentrations may be so low that the BHT is very quickly consumed by the metal/hydroperoxide radical generating system. In principle, the problem can be circumvented by using better antioxidants than BHT and using them at higher concentrations. However, if inhibition of the oxidation is *not* observed upon addition of an appropriate concentration of an antioxidant, it would be prudent to demonstrate that the antioxidant is still present at its initial concentration at the end of the reaction. Second, for simplicity we have ignored kinetic solvent effects (KSE's). Hydrogen atom abstractions from hydrocarbons by alkoxyl and alkylperoxyl radicals are not subject to KSE's.[6] However, hydrogen atom abstractions from phenols, particularly sterically non-hindered phenols, are subject to very significant KSE's with k_{AH}^{RO} and k_{AH}^{ROO} declining dramatically in hydrogen bond accepting (HBA) solvents.[6,20,21] Diphenylamine is a much poorer hydrogen bond donor than phenol and is therefore much less subject to KSE's in its radical-trapping reactions.[17]

(ii) Traps for carbon-centered radicals

For hydrocarbon oxidations carried out under conditions where dioxygen is absent or present only at a very low concentration throughout the reaction (as can be achieved by continuous purging with argon), mechanistic probes based on known traps for carbon-centered radicals become valuable. Two such traps, CCl_3Br and stable nitroxides of the TEMPO class, are particularly useful because they do not react with oxygen-centered radicals but yield identifiable and characteristic products with carbon-centered radicals. Thus, if the reaction in question actually does occur *via* **free** radicals, new products will be formed, i.e.

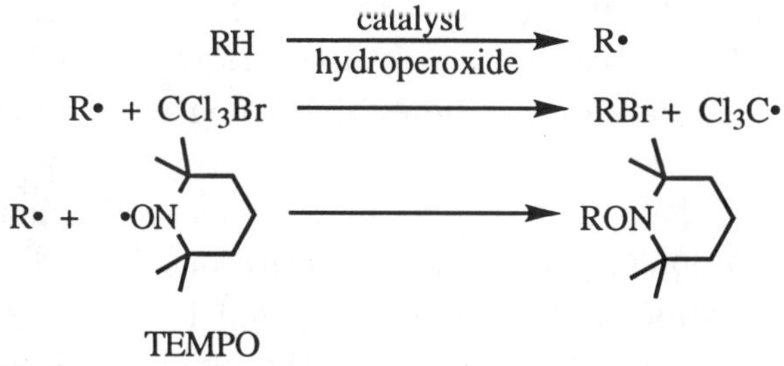

In analyzing for bromides by GC, care must be taken to avoid dehydrobromination (generally in the injection port) and, for resonance-stabilized R•, the homolytic cleavage of the R–O bond in the hydroxylamine (i.e. the reverse of the radical-trapping step).

Nitrones are another class of radical-trapping agents which may be employed in mechanistic investigations. These compounds are known as spin traps and convert transient radicals (e.g. R•, RO•, etc.) into persistent nitroxide radicals which can be identified by ESR spectroscopy, e.g.

$$R\bullet + PhCH{=}N(O)C(CH_3)_3 \longrightarrow PhCH(R)N(\dot{O})C(CH_3)_3 \qquad (2.14)$$

Despite their popularity (which can be attributed to the simplicity of such experiments), we do not recommend the use of nitrone spin traps in mechanistic studies of hydrocarbon oxidation by catalyst/hydroperoxide systems because nitroxide ESR signals can be produced by processes other than the capture of a **free** R• (RO•) radical.

4. Competitive Kinetics

In principle, competitive kinetic studies provide a simple method for distinguishing between **free** radical and radical-free hydrocarbon oxidations. In practice, this procedure is bedevilled by problems (and probably should be used only as a "last resort"). For example, the relative reactivities of two hydrocarbons are very dependent on the **free** radical which attacks them. Earlier in the chapter, we saw that the relative reactivities of cyclohexene : cyclohexane are $5.7 \times 10^6 : 1.2 \times 10^6 = 4.8$ towards alkoxyl radicals and $6.1 : 0.048 = 127$ towards alkylperoxyl radicals. In an experimental system which involved both alkoxyl and peroxyl radicals, the measured cyclohexene/cyclohexane relative reactivities could vary between 4.8 and 127 depending on just how the reaction was carried out! Would a non-radical biomimetic oxidation of these two hydrocarbons yield a reactivity ratio outside these limits? It seems somewhat improbable.

The same problems arise with competitions between normal and perdeuterated hydrocarbons, e.g. $c\text{-}C_6H_{12}$ and $c\text{-}C_6D_{12}$. The Deuterium Kinetic Isotope Effect (DKIE) could vary between ca. 1 (HO•), ca. 5 (RO•)

to ca. 10 ~ 20 (ROO•) depending on the reaction conditions. Again, one must ask what should be the DKIE for a non-radical oxidation by the particular catalyst employed? Undoubtedly competitive kinetic procedures using different substrates can provide useful mechanistic insights. However, we do not believe they are capable of distinguishing between **free** radical and radical-free hydrocarbon oxidations and recommend that, at most, they be used only as supporting evidence.

Some Selected Case Studies

The field of metal-catalyzed oxidations has been greatly expanded in recent years by several research groups who have developed a variety of systems that are capable of oxidizing hydrocarbons. As was mentioned earlier, most of these attempts to mimic monooxygenase chemistry involve an Fe^{III} (or Fe^{II}) salt or chelate and employ hydrogen peroxide or an organic hydroperoxide as the oxygen source. In this section, we will describe how some physical-organic chemists have used aspects of the methodologies described above to distinguish between high-valent iron-oxo oxidation mechanisms and simple **free** radical chemistry.

1. Gif Chemistry

Barton and co-workers have developed numerous "Gif" systems that are capable of oxidizing cycloalkanes to ketones.[22] These systems generally incorporate a simple iron salt (e.g. ferric nitrate) together with picolinic acid and either TBHP, hydrogen peroxide or superoxide (as the oxygen source) in a pyridine/acetic acid solvent system. The general mechanism proposed for these oxidations involved the formation of an Fe^{V}=O species which inserts into a (generally unactivated) carbon-hydrogen bond to form an intermediate containing an iron-carbon bond. Dioxygen is then assumed to insert into the iron-carbon bond and subsequent reactions lead to the products (usually ketones from methylene groups and alcohols from methine groups):

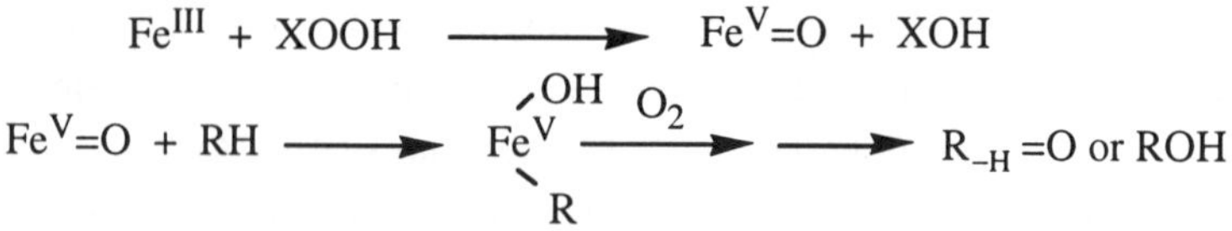

An alternative, but basically similar mechanism, has been recently proposed[23], in which an $Fe^{III}OOFe^{III}$ complex is the precursor to the $Fe^{V}=O$ species. The subsequent R-$Fe^{V}OOFe^{III}$ intermediate rearranges to give an alkylperoxoiron(III) species which then undergoes heterolytic cleavage to yield ketone:

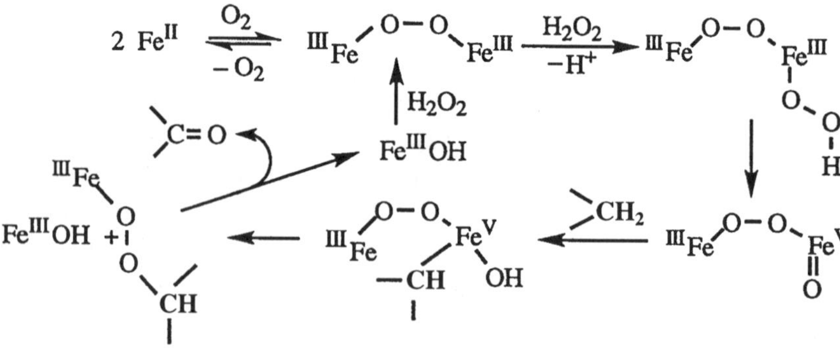

A much simpler mechanism can explain "Gif" chemistry. This is, of course, a **free** radical mechanism.[24] The radicals are produced by the reaction of the hydroperoxide with the iron. In this simple mechanism, the hydrogen atom abstracting species is an oxygen-centered radical (HO• or RO•) and **free** carbon-centered radicals are produced rather than an Fe^{V}-R intermediate. This **free** radical mechanism is strongly supported by radical clock methodologies.

The presence of **free** alkyl radicals during "Gif" oxidations was most clearly proven by Newcomb and co-workers[25] using the hypersensitive cyclopropane-based radical clocks shown below:

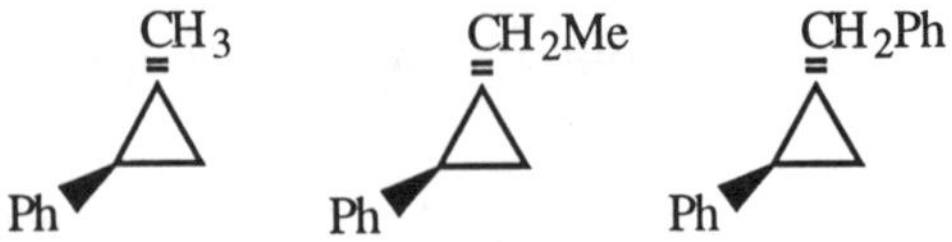

1 **2** **3**

The oxidation of **1** and **2** under GoAgg^{III} (H_2O_2) and GoAgg^{V} (TBHP) conditions resulted in products derived solely from the rearranged radicals and none from the initially formed (according to the radical mechanism) cyclopropyl carbinyl radicals. Oxidation of **1** by genuine enzymatic systems (cytochrome P450 and methane monooxygenase) gave mostly, and in some cases solely, un-rearranged alcohols as products. The small yields of rearranged products that are produced using cytochrome P450 are believed to arise from cationic species formed during the reaction. "Gif" oxidation of **3** gave both rearranged and un-rearranged products. The results with all three cyclopropanes are entirely consistent with the formation of **free** carbon-centered radicals once the rates of the ring opening of the different cyclopropyl carbinyl radicals and the rates of alkyl radical trapping by dioxygen are considered. The rates of ring opening for the cyclopropyl carbinyl radicals derived from **1**, **2** and **3** are 3×10^{11} s^{-1}, ~ 3×10^{11} s^{-1} and 1×10^{8} s^{-1}, respectively. The rate constant for the alkyl radical/dioxygen reaction is approximately 10^{10} $M^{-1}s^{-1}$. Therefore, at an oxygen concentration of (at most) 2 mM, the effective rate constant for oxygen trapping will be 2×10^{7} s^{-1}. Thus, the radicals derived from **1** and **2** will rearrange far too rapidly to react with O_2. However, trapping by oxygen competes with rearrangement of the cyclopropyl carbinyl radical derived from **3**. It is also worth noting that phenolic products were observed with each of the probes during oxidation under GoAgg^{III} conditions, i.e. with hydrogen peroxide as the oxidant, but not under GoAgg^{V} conditions where TBHP is used. This is further evidence that these two Gif systems operate via different active species, viz. HO• radicals and Me_3CO• radicals, respectively, and not via a common Fe^{V} species.

The formation of **free** carbon-centered radicals is also consistent with the formation of alkyl halides during "Gif" oxidations using $FeCl_3$ or $FeBr_3$ catalysts because these will be formed via the facile attack of alkyl radicals on iron halides as has been demonstrated by Minisci[26] and Kochi[27], e.g.

$$R\bullet + Fe^{III}Cl_3 \longrightarrow RCl + Fe^{II}Cl_2 \qquad (2.15)$$

and with the observation that alkyl radicals can be trapped by protonated aromatic bases:[28]

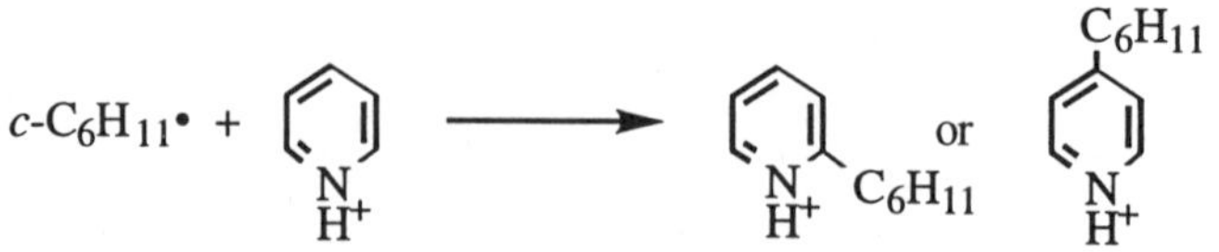

(2.16)

The nature of the hydrogen atom abstracting species in $GoAgg^{V}$ has also been investigated using the alkoxyl radical clock methodology.[29] The original recipe used TBHP and cycloalkanes which were oxidized to the corresponding ketones. Replacement of TBHP by MPPH gave **no** cycloalkane oxidation products but, instead, products which were clearly derived from benzyl radicals (see Reaction 2.9). The active species in $GoAgg^{V}$ must therefore be the alkoxyl radical derived from the hydroperoxide and not an Fe^{V}=O species.

The replacement of TBHP by MPPH in $GoAgg^{V}$ chemistry and in related work (*vide infra*) and the conclusion that **free** radicals were responsible for alkane oxidations has led to both implicit[30] and explicit[31] suggestions that MPPH was predisposed to free radical chemistry and therefore was not a competent replacement for TBHP. This is not the case. Recent work has demonstrated that MPPH is, indeed, a perfectly competent substitute for TBHP in hydrocarbon oxidations which are genuine two electron, radical-free processes.[5a,32]

The conclusions drawn from the MPPH results in the $GoAgg^{V}$ system were further confirmed by using other "probe" hydroperoxides. With *ortho*-methyl cumene hydroperoxide (*o*-MCH), cycloalkane oxidation products were formed in only minor amounts, with the main products being derived from the alkoxyl radical (see Table 2.2). With cumene hydroperoxide (CH), cycloalkane oxidation products and acetophenone (arising from the β-scission of the cumyloxyl radical) were formed in comparable amounts.

As mentioned earlier, these results arise because the different hydroperoxides yield alkoxyl radicals having intrinsic lifetimes (in acetonitrile at room temperature) for unimolecular decay processes which vary from less than 5 ns for $PhCH_2CMe_2O\bullet$, to 400 ns for *ortho*-methylcumyloxyl and 1.6 μs for cumyloxyl radicals, respectively. The intrinsic lifetime for *tert*-butoxyl is 12 μs. The ratio of cycloalkane derived products to alkoxyl radical decomposition products increases dramatically as the lifetime of the alkoxyl radical increases. In the case of CH, the rate of β-scission of cumyloxyl and of its abstraction of a hydrogen atom from the 0.14 M cycloalkane are comparable:

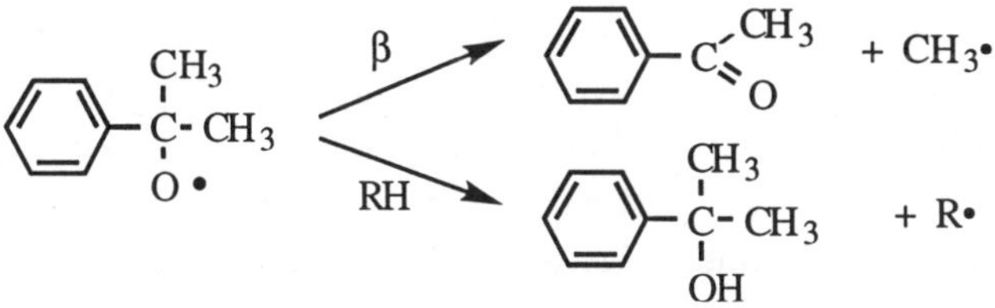

The foregoing proof that *tert*-butoxyl radicals are the active oxidants in $GoAgg^V$ chemistry is buttressed by the work of Minisci *et al.*[33] who showed that these radicals could be trapped with electron rich alkenes (e.g. styrene) and the resultant carbon-centered radicals could then be trapped by the iron chloride catalyst to yield chloroethers:

$$Me_3C\dot{O} + CH_2{=}CHPh \longrightarrow Me_3COCH_2\dot{C}HPh \xrightarrow{Fe^{III}Cl_3} Me_3COCH_2CH(Cl)Ph + Fe^{II}Cl_2 \quad (2.17)$$

The addition of cyclohexane to the reaction mixture yielded cyclohexyl chloride and the ratio of cyclohexyl chloride to the chloroether was found to be in good agreement with the relative rates for abstraction from cyclohexane to addition to styrene by *tert*-butoxyl radicals. It has subsequently been shown that once all the solvent effects on the various **free** radical reactions are fully appreciated,[34] that these "Gif" systems can be used in the synthesis of alkyl halides and mixed peroxides.[35,36] Finally, Barton has recently excommunicated $GoAgg^V$ chemistry from the "world of Gif" and accepts that TBHP chemistry (though not H_2O_2-based "Gif" chemistry) occurs via **free** radicals.[37]

All of the foregoing results are consistent with a **free** radical mechanism for the "Gif" systems considered in which an oxygen-centered radical derived from the hydroperoxide is the hydrogen atom abstracting species, not some high-valent iron species. It is clear that the products are formed from **free** carbon-centered radicals and are not derived from an intermediate containing an iron-carbon bond. These conclusions regarding "Gif" chemistry illustrate the need for careful mechanistic investigations because the products (mainly ketone from cycloalkanes) are not at first sight indicative of **free** radical chemistry. Under metal-free conditions, it would be expected that the products from **free** secondary alkyl radicals that are trapped by dioxygen would be alcohol and ketone in roughly equal amounts (together, possibly, with some mixed peroxides; see following section). However, the radical clock work outlined above, the formation of alkylpyridines upon addition of pyridine, the formation of alkyl chlorides in the presence of ferric chloride and so on provides overwhelming evidence favoring the operation of a **free** radical mechanism. Such a **free** radical mechanism is further supported by the identification of cycloalkyl hydroperoxides as intermediates in some "Gif" systems[38] and the proof that the oxygen in these hydroperoxides and their decomposition products are derived from dioxygen.[39–40] The fact that ketone is the major product is probably due to reaction of the hydroperoxide with the iron catalysts, followed by a concerted decomposition, i.e.

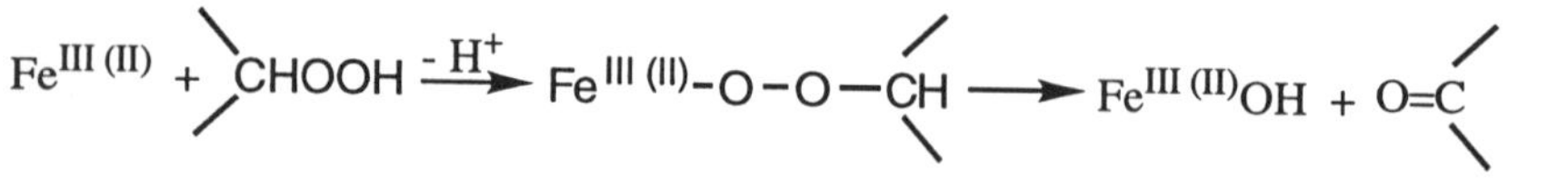

(2.18)

2. Iron-TPA Chemistry

Another class of non-heme, iron-based alkane oxidation catalysts containing the tetradenate, tripodal *tris*-(pyridylmethyl)amine (TPA) ligand has been developed by Que and coworkers.[42] Both mono-iron complexes, e.g. $[Fe^{III}Cl_2(TPA)]^+$ and μ-oxo containing di-iron complexes, e.g. $[Fe^{III}_2O(OAc)(TPA)_2]^{3+}$, have been reacted with an excess of TBHP (usually 140 equivalents based on catalysts at 0.7 mM) and an excess of cyclohexane (0.77 M) in acetonitrile under an atmosphere of argon at room temperature. The products were cyclohexanol, cyclohexanone and cyclohexyl *tert*-butyl peroxide which were formed in a roughly 1:1:1 ratio in more than stoichiometric quantities. With $[Fe^{III}Cl_2(TPA)]^+$ roughly one equivalent of cyclohexyl chloride was also formed.

It was proposed that the hydroperoxide reacted with the catalysts to form a high-valent iron-oxo species which abstracted a hydrogen atom from the cyclohexane. The alcohol (and the halide) were then supposed to be formed via a Groves-type of oxygen (or halogen) rebound mechanism:[43]

$$Me_3COOH + (ClFe^{III}) \longrightarrow (ClFe^{V}{=}O) + Me_3COH$$

$$(ClFe^{V}{=}O) + c\text{-}C_6H_{12} \longrightarrow (ClFe^{IV}\text{-}OH) + c\text{-}C_6H_{11}\bullet$$

$$(ClFe^{IV}\text{-}OH) + c\text{-}C_6H_{11}\bullet \longrightarrow (ClFe^{III}) + c\text{-}C_6H_{11}OH$$

$$(ClFe^{IV}\text{-}OH) + c\text{-}C_6H_{11}\bullet \longrightarrow (Fe^{III}\text{-}OH) + c\text{-}C_6H_{11}Cl$$

No mechanism for ketone formation was suggested.

The formation of roughly equal yields of ketone and alcohol are suggestive of a Russell-type[44] of bimolecular self-reaction of secondary alkylperoxyl radicals. If this is the case, the product forming reactions

$$2\ {>}CHOO^{\bullet} \rightleftharpoons {>}CHOOOOCH{<} \longrightarrow {>}CHOH + {>}C{=}O + O_2 \qquad (2.19)$$

must involve freely diffusing radicals rather than the caged radicals previously suggested. A mechanism is illustrated below that would readily account for all the products:

$$(ClFe^{III}) + Me_3COOH \longrightarrow (ClFe^{IV}{=}O) + H^+ + Me_3CO\bullet \qquad (2.20)$$

$$(ClFe^{IV}{=}O) + Me_3COOH \longrightarrow (ClFe^{III}) + {}^-OH + Me_3COO\bullet \qquad (2.21)$$

$$Me_3CO\bullet + c\text{-}C_6H_{12} \longrightarrow Me_3COH + c\text{-}C_6H_{11}\bullet \quad (2.22)$$

$$Me_3CO\bullet + Me_3COOH \longrightarrow Me_3COH + Me_3COO\bullet \quad (2.23)$$

$$c\text{-}C_6H_{11}\bullet + Me_3COO\bullet \longrightarrow c\text{-}C_6H_{11}OOCMe_3 \quad (2.24)$$

$$c\text{-}C_6H_{11}\bullet + Me_3COO\bullet \longrightarrow c\text{-}C_6H_{10} + Me_3COOH \quad (2.25)$$

$$c\text{-}C_6H_{11}\bullet + (ClFe^{III}) \longrightarrow c\text{-}C_6H_{11}Cl + (Fe^{II}) \quad (2.26)$$

$$(Fe^{II}) + Me_3COOH \longrightarrow Me_3CO\bullet + (Fe^{III}OH) \quad (2.27)$$

$$2\,Me_3COO\bullet \longrightarrow 2\,Me_3CO\bullet + O_2 \quad (2.28)$$

$$c\text{-}C_6H_{11}\bullet + O_2 \longrightarrow c\text{-}C_6H_{11}OO\bullet \quad (2.29)$$

$$c\text{-}C_6H_{11}OO\bullet + Me_3COO\bullet \longrightarrow c\text{-}C_6H_{11}OOCMe_3 + O_2 \quad (2.30)$$

$$2\,c\text{-}C_6H_{11}OO\bullet \longrightarrow c\text{-}C_6H_{11}OH + c\text{-}C_6H_{10}O + O_2 \quad (2.31)$$

The crucial feature of the mechanism is the formation of **free** *tert*-butoxyl radicals which are responsible for the formation of **free** cyclohexyl radicals (Reaction 2.22) and oxygen (Reaction 2.23 followed by Reaction 2.28). It should be noted that under the conditions chosen, i.e. $[c\text{-}C_6H_{12}] = 0.77$ M and [TBHP] = 0.1 M, Reactions 2.22 and 2.23 proceed at almost identical rates. That is, as the rate constants for hydrogen atom abstraction from cyclohexane and TBHP by an alkoxyl radical in acetonitrile at room temperature are $1.2 \times 10^6\ M^{-1}s^{-1}$ and $8.7 \times 10^6\ M^{-1}s^{-1}$,[6,16] respectively, the rates of formation of cyclohexyl radicals and *tert*-butylperoxyl radicals are $1.2 \times 10^6\ M^{-1}s^{-1} \times 0.77\ M = 9.2 \times 10^5\ s^{-1}$ and $8.7 \times 10^6\ M^{-1}s^{-1} \times 0.1\ M = 8.7 \times 10^5\ s^{-1}$. At the start of the reaction, the oxygen concentration is extremely low and therefore the cyclohexyl radicals are consumed via Reactions 2.24 and 2.25 and, for $[Fe^{III}Cl_2(TPA)]^+$, also by Reaction 2.26. The bimolecular self-reaction of cyclohexyl radicals:

$$2\,c\text{-}C_6H_{11}\bullet \longrightarrow (c\text{-}C_6H_{11})_2 + (c\text{-}C_6H_{12} + c\text{-}C_6H_{10}) \quad (2.32)$$

occurs at (or close to) the diffusion-controlled limit (~10^{10} $M^{-1}s^{-1}$) whereas the bimolecular self-reaction of *tert*-butylperoxyl radicals (Reaction 2.28) is slow (~10^3 $M^{-1}s^{1}$). It is now well established that when two different radicals are generated together at roughly equal rates and one radical is transient (i.e. undergoes a fast bimolecular self-reaction) and the other is persistent (i.e. undergoes a slow bimolecular self-reaction) the Ingold–Fischer "Persistent Radical Effect"[9,36,45] will dominate product formation and the major products will be formed via the cross-reaction between the transient and the persistent radicals. In the system under consideration, the *tert*-butylperoxyl radicals will rapidly reach a relatively high steady state concentration and will then trap most of the cyclohexyl radicals, i.e. Reactions 2.24 and 2.25 will dominate product formation. This prediction was demonstrated to be correct when product formation was monitored over time: the major *initial* product was the mixed peroxide.[46] As the oxygen concentration from the bimolecular self-reactions of the *tert*-butylperoxyl radicals increases, the trapping of the cyclohexyl radicals by oxygen, Reaction 2.29, begins to compete and eventually it dominates product formation.

An important finding from monitoring the time course of the reaction was that the $[Fe^{III}Cl_2(TPA)]^+$ -catalyzed reactions have a very distinct induction period after which the products are formed extremely rapidly.[9] This is because Reaction 2.26 yields Fe^{II} and this is a very much better catalyst for hydroperoxide decomposition (Reaction 2.27) than Fe^{III} as was proven by performing the reaction under air. In this situation, the reaction is strongly inhibited because oxygen traps the cycloalkyl radicals and prevents the reduction of the Fe^{III} catalyst. Reactions catalyzed with $[Fe^{III}_2O(OAc)(TPA)_2]^{3+}$, on the other hand, have a relatively short induction period and product formation occurs over a longer period of time than with the mono-iron catalyst. In addition, when these reactions are carried out under air, the reaction products and their profile (i.e. relative yields) are essentially the same as under argon. In this case, the actual Fe^{III} catalyst rather than the reduced Fe^{II} species is relatively active towards the TBHP. Furthermore, there is no univalent ligand which could be abstracted by the cycloalkyl radical and hence, there is no equivalent of Reaction 2.26 and little autocatalysis.

Proof that the product profiles obtained with both the mono- and di-Fe^{III} TPA catalysts could result from alkoxyl radical-driven chemistry was obtained by carrying out the reaction in the absence of an iron catalyst using di-*tert*-butyl hyponitrite (BONNOB) as a thermal source of *tert*-butoxyl radicals:

$$Me_3CON{=}NOCMe_3 \xrightarrow{\Delta} 2\,Me_3CO\bullet + N_2 \qquad (2.33)$$

These reactions led to the same three oxygen containing products and a similar product profile to those of the metal-catalyzed reactions. These results also demonstrated that mixed peroxides formed during metal/ROOH oxidation reactions do not have to arise from a ligand transfer reaction (as proposed by Kochi[47]):

$$R\bullet + (Fe^{III}OOCMe_3) \longrightarrow ROOCMe_3 + (Fe^{II}) \qquad (2.34)$$

They are much more likely to be formed via the persistent radical effect, i.e. Reactions 2.24 and 2.30. The observation that 76% of the mixed peroxide contained one atom of ^{18}O when the reaction of $[Fe^{III}Cl_2(TPA)]^+$ and TBHP with cyclooctane was performed under an atmosphere of $^{18}O_2$, provides further support for Reactions 2.29 and 2.30:[9]

$$c\text{-}C_6H_{11}\bullet + {}^{18}O_2 \longrightarrow c\text{-}C_6H_{11}{}^{18}O^{18}O\bullet \qquad (2.29')$$

$$c\text{-}C_6H_{11}{}^{18}O^{18}O\bullet + Me_3C^{16}O^{16}O\bullet \longrightarrow c\text{-}C_6H_{11}{}^{18}O^{16}OCMe_3 + {}^{18}O\text{-}{}^{16}O \qquad (2.30')$$

Initial publications claiming a non-radical mechanism and the involvement of an iron(V)-oxo species relied on the results obtained with two additives, BHT and dimethyl sulfide. Firstly, it was argued that **free** radicals could not be present because 15 mM BHT had no effect on the formation of one equivalent of cyclohexyl bromide from the reaction of 3.0 mM $[Fe^{III}Br_2(TPA)]^+$ with 1.15 M cyclohexane and 15.0 mM TBHP. As was pointed out earlier in this chapter, BHT is not an effective scavenger of alkyl radicals but it does react with oxygen-centered radicals. The rate constant for hydrogen atom abstraction by *tert*-butoxyl radicals from BHT in acetonitrile at 295 K is 2.0×10^7 $M^{-1}s^{-1}$, whereas that from cyclohexane is 1.2×10^6 $M^{-1}s^{-1}$. Therefore, under the conditions employed, the rates for reaction of the *tert*-butoxyl radicals with BHT and cyclohexane will be 2.0×10^7 $M^{-1}s^{-1}$ x 0.015 M = 3.0×10^5 s^{-1} and 1.2×10^6 $M^{-1}s^{-1} \times 1.15$ M = 1.38×10^6 s^{-1}, respectively. Obviously, 15 mM BHT can have very little effect on the reaction products. Secondly, dimethyl sulfoxide was formed during the oxidation of cyclohexane by TBHP and $[Fe^{III}Br_2(TPA)]^+$ in the presence of dimethyl sulfide. However, addition of dimethyl sulfide to a BONNOB initiated oxidation of cyclohexane by TBHP also led to the formation of dimethyl sulfoxide.[9] As mentioned earlier, Me_2S is not a diagnostic probe for Fe^V=O species in metal/ROOH-catalyzed oxidations.

The nature of the hydrogen atom abstracting agent in Fe^{III}TPA systems has been thoroughly explored using probe hydroperoxides.[8,9] There was no cycloalkane oxidation when TBHP was replaced by MPPH and *o*-MCH, but products arising from the unimolecular reactions of the corresponding alkoxyl radicals were obtained. With CH, both cyclooctane oxidation products and acetophenone were obtained. These observations prove that the active species is an alkoxyl radical not a high-valent iron-oxo species.

Interestingly, what appeared at first sight to be monooxygenase chemistry[31] was achieved in 1996 with a new and extremely active catalyst (10 min reaction time versus 2 h for the older FeTPA catalysts). When 10 equivalents of TBHP was added continuously by syringe pump to a deoxygenated acetonitrile solution of $[Fe^{III}_2O(H_2O)_2(TPA)_2]^{3+}$ at room temperature, cyclohexane was oxidized only to cyclohexanol (four equivalents).[31] This was an exciting result because cyclohexanol would be the sole product from a genuine monooxygenase mimic. It was proposed that the hydrogen atom abstracting species was an alkyl iron(III) peroxo species and that the alcohol was formed via a slow hydroxyl group rebound.[31] However, replacement of TBHP by MPPH under identical conditions gave no cyclohexanol which indicated to Kim *et al.* that the MPPH was not a competent replacement for TBHP.[31] However, to the authors of this chapter, it indicated that the chemistry must involve freely diffusing alkoxyl radicals. Further investigation[48] showed that in addition to the alcohol there were small amounts of mixed peroxide, a product which also implies **free** radical chemistry. In fact, the yield of mixed peroxide was dramatically dependent on the method by which the TBHP was added to the other reactants. If the needle of the syringe containing the TBHP (ten equivalents) was placed below the surface of a stirred solution, then alcohol was by far the major product. However, if the needle was held above the solution so that ten equivalents of TBHP was added dropwise, but at precisely the same rate, the mixed peroxide became the major product. Finally, if ten equivalents of TBHP was added all-at-once the mixed peroxide was the only product.

Additional experiments confirmed that $[Fe^{III}_2O(H_2O)_2(TPA)_2]^{3+}$ plus *tert*-alkyl hydroperoxides (ten equivalents) added by syringe pump (directly into the solution) gave **free** radical chemistry. Thus, only cycloalkyl bromides were formed upon the addition of a low concentration of CCl_3Br (30.5 mM). At this concentration, the CCl_3Br cannot capture the cycloalkyl radicals while they are still in the solvent cage in which they were formed. This result is proof that **free** cycloalkyl radicals are formed:

$$c\text{-}C_nH_{2n-1}\bullet + CCl_3Br \longrightarrow c\text{-}C_nH_{2n-1}Br + CCl_3\bullet \qquad (2.36)$$

In competitive experiments using cyclooctane and cyclohexane in the presence of 30.5 mM CCl_3Br, the $[c\text{-}C_8H_{15}Br]/[c\text{-}C_6H_{11}Br]$ ratio was about 2.1. This value corresponds to the independently measured value of 2.2 for hydrogen atom abstraction from the two cycloalkanes by BONNOB generated *tert*-butoxyl radicals in the presence of 30.5 mM CCl_3Br, and is consistent with the ratio of the absolute rate constants for hydrogen atom abstraction by *tert*-alkoxyl radicals from cyclooctane and cyclohexane, viz. $2.7 \times 10^6\ M^{-1}s^{-1}/1.2 \times 10^6\ M^{-1}s^{-1} = 2.3$. In the absence of CCl_3Br, the ratio of the yield of cyclooctane-derived oxidation products to cyclohexane-derived products was also approximately 2.1:1.0.

In the original publication[31], a value of 10 for the $c\text{-}C_6H_{12}/c\text{-}C_6D_{12}$ Deuterium Kinetic Isotope Effect (DKIE) was used as evidence that the active species was a metal-based oxidant because similar high values had been reported for metalloporphyrin-catalyzed oxidations (see following section). However, a DKIE of 10 could not be reproduced either in the presence or absence of CCl_3Br.[48] Instead, DKIE values of about 6 were obtained which are reasonably consistent with the expected DKIE of ca. 5 for genuine *tert*-butoxyl radicals.[48]

The foregoing experiments prove that syringe pump addition of TBHP to the $[Fe^{III}_2O(H_2O)_2(TPA)_2]^{3+}$ catalyst generates freely diffusing *tert*-butoxyl radicals which are responsible for the subsequent chemistry. The reason for the high selectivity for alcohol formation when ten equivalents of TBHP are added by syringe pump is that of the extremely low TBHP concentration due to the high activity of this catalyst. As a consequence, relatively few *tert*-butylperoxyl radicals will be formed via Reactions 2.21 and 2.23. With no oxygen present initially and little or none formed in the reaction via Reaction 2.28, the cycloalkyl radicals must be hydroxylated by the $HO\text{-}[Fe^{IV}]^+$ (or $HO\text{-}[Fe^{III}]$) formed in Reaction 2.20 (or 2.27) via a simple ligand transfer reaction, i.e. a slow oxygen rebound. (Independent experiments showed that cyclohexyl radicals were not hydroxylated by the initial catalyst.) In the case of dropwise addition, and even more in the case of the "all-at-once" addition of the TBHP, there are relatively high local concentrations of TBHP and hence more *tert*-butylperoxyl radicals are produced and these trap the cyclohexyl radicals to yield the mixed peroxide.

The chemistry of some iron-TPA complexes plus TBHP in aqueous solution has been investigated by Rabion *et al.*[49] by examining products formed from this hydroperoxide. With $[Fe^{III}_2O(\eta^1\text{-}H_2O)(\eta^1OAc)(TPA)_2]^{3+}$, the main TBHP decomposition product was acetone together with small amounts of methanol, formaldehyde, *tert*-butanol and *tert*-butyl methylperoxide. This product profile can *only* result from the formation of **free** *tert*-butoxyl radicals.[49] In addition, a **free** radical mechanism was supported by k_H/k_D values of 2.6 – 4.5 for the oxidation of some alcohols to ketones by $[Fe^{III}_2O(\eta^1\text{-}OAc)(TPA)_2]^{3+}$ and TBHP.[49] In aqueous solution, an initial reduction of these catalysts is favored because of their very positive reduction potentials, e.g. +366 mV versus Ag/AgCl for $[Fe^{III}_2O(\eta^1\text{–}H_2O)(\eta^1\text{-}OAc)(TPA)_2]^{3+}$. In contrast, the same $Fe^{3+}\text{-}O\text{-}Fe^{3+}/Fe^{3+}\text{-}O\text{-}Fe^{2+}$ reduction potential for $[Fe^{III}_2O(\eta^1\text{-}H_2O)(\mu\text{-}OAc)(TPA)_2]^{3+}$ in acetonitrile was found to be –0.05 mV versus Ag/AgCl. It was therefore proposed that in water, peroxyl radicals are formed initially from the TBHP following which the reduced form of the catalyst reacts with TBHP to produce alkoxyl radicals:

$$Fe^{3+}\text{-O-}Fe^{3+} + Me_3COOH \longrightarrow Fe^{3+}\text{-}\overset{H}{O}\text{-}Fe^{2+} + Me_3COO\bullet$$

$$Fe^{3+}\text{-}\overset{H}{O}\text{-}Fe^{2+} + Me_3COOH \longrightarrow Fe^{3+}\text{-O-}Fe^{3+} + Me_3CO\bullet + H_2O$$

These studies were later extended[50] to hydrocarbon oxidations by $[Fe^{III}_2O(\eta^1\text{-}H_2O)(\eta^1\text{-}OAc)(TPA)_2]^{3+}$ and TBHP in an aqueous micelle system. Cyclohexane was oxidized cyclohexanol and cyclohexanone. Cyclohexene gave various allylic oxidation products including 3-(*tert*-butylperoxy)cyclohexene. The TBHP gave both acetone and *tert*-butanol. The results were again properly interpreted in terms of a **free** radical mechanism initiated by the favorable reduction potentials of the catalysts.

The entire iron-TPA case study provides a useful object lesson since it demonstrates that even when the reaction product is that expected from a genuine monooxygenase mimic, e.g. alcohol as the sole product from an alkane, additional mechanism-diagnostic experiments are *essential*. The usefulness and validity of MPPH as a mechanistic probe has been further confirmed by demonstrating that this hydroperoxide can function as a two-electron oxidant in genuine two-electron oxidation chemistry.[5a,32]

3. Iron Porphyrin-Catalyzed Oxidations

A variety of heme enzymes including the cytochrome P450 monooxygenases, peroxidases and catalase contain iron protoporphyrin IX at the active site and for this reason the most widely studied biomimetic catalysts are iron-based and contain a porphyrin ligand. Initial attempts to use the protoporphyrin IX complex itself were unsuccessful because hydroxylation at the meso positions induced oxidative bleaching of the ligand.[51] This led to the development of porphyrin rings substituted at the four meso positions with phenyl groups which were more resistant to oxidative degradation. Iron(III) tetraphenyl porphyrin ($Fe^{III}TPP$)[52] has since been used extensively in attempts to mimic the monooxygenase reactions of cytochrome P450, in particular.

The incorporation of an oxygen atom into organic substrates by the cytochrome P450 enzymes occurs via the catalytic cycle shown below.[53] In its resting state, the iron(III) porphyrin is located in a hydrophobic pocket in the enzyme. Upon entry of the substrate into this pocket, the enzyme is activated and the iron(III) is reduced to iron(II), to which dioxygen binds to form a peroxy complex. Addition of a second electron and two protons results in the formation of the active oxidant which is believed to be an oxo-iron(IV) porphyrin π-radical cation ($^{\bullet+}PFe^{IV}=O$). It is this oxidant which "inserts" an oxygen atom into the substrate. The oxidized substrate is more hydrophilic than the starting compound and is released from the hydrophobic pocket. The active oxidant, ($^{\bullet+}PFe^{IV}=O$), is capable of a variety of reactions including alkane hydroxylation and alkene epoxidation, but **free** substrate-derived radicals are not involved in the process.

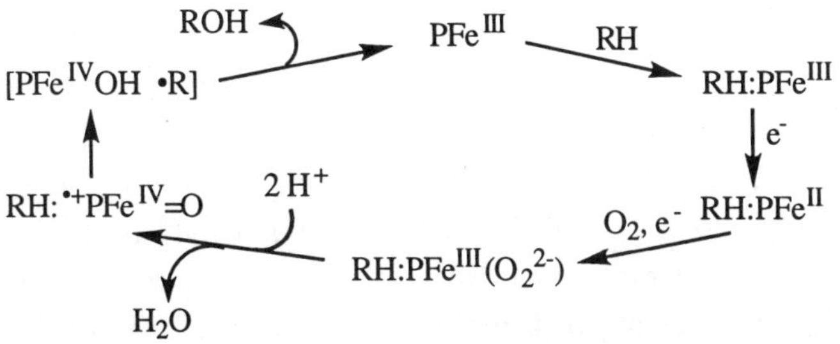

The oxidation of organic substrates by peroxidase enzymes differ from those of the cytochrome P450 monooxygenases in two main ways.[54] First, although the active species, called compound I, is a $^{\bullet+}PFe^{IV}=O$ complex, it is generated from hydrogen peroxide or an organic hydroperoxide rather than dioxygen. Second, oxidation occurs via single electron transfer rather than by oxygen atom "insertion". Compound I oxidizes substrates such as phenols or aromatic amines to give an oxo-iron(IV) porphyrin, called compound II. This is capable of a second one-electron oxidation and the resting state of the enzyme is regenerated. These differences in mechanism are not due to changes in the metal or the porphyrin ligand. The precise mechanism is determined by the proximal ligand, the distal amino acid residues and the nature of the protein surrounding the active site.

The development of metalloporphyrins as oxidation catalysts has generally followed the peroxidase route, the common practice being to use oxygen atom donors other than dioxygen. The first oxygen atom donor to be used was iodosylbenzene (PhIO) which with Fe^{III}(TPP)Cl catalyst was shown by Groves and Nemo[55] to epoxidize alkenes in a manner very similar to that of cytochrome P450. The epoxidation of *cis*-stilbene gave only *cis*-stilbene oxide implying a concerted oxygen atom insertion; no *trans*-stilbene oxide (indicative of a **free** radical reaction) was detected. However, cyclohexene gave both cyclohexene oxide (55% yield) and the allylic alcohol (15%).[55] This alcohol may be formed by a **free** radical process rather than by an oxygen rebound mechanism. (It should be noted that the oxidation of cyclohexene by cytochrome P450 also yields both the epoxide and allylic oxidation products.) The yield of allylic oxidation products formed during the epoxidation of cyclic alkenes with Fe^{III} tetra-(dichlorophenyl)-porphyrin and PhIO depends on the structure of the alkene.[56]

Alkane hydroxylations catalyzed by Fe^{III}(TPP)Cl/PhIO are also enzyme mimetic.[57] For example, adamantane gave 1-adamantanol and 2-adamantanol but no ketone.[57] Similarly, cyclohexane gave primarily cyclohexanol, and a high DKIE of 12.9 for c-C_6H_{12}/c-C_6D_{12} which may be more consistent with oxidation via a high-valent iron-oxo species than by **free** radicals.[57] The formation of cycloheptyl bromide [18% versus 27.6% for the combined alcohol (24%) and ketone (3.6%) products] during the oxidation of cycloheptane in the presence of 1.7 M CCl_3Br [57] does not *automatically* rule out an in-cage oxygen rebound mechanism because, at such high concentrations, one or more molecules of CCl_3Br are likely to be present in the cage "wall" and these could compete with the iron-oxo species for the cycloheptyl radicals. The rate constant for the alkyl radical/CCl_3Br reaction is approximately 1.4×10^8 $M^{-1}s^{-1}$.[1,58] If we assume that, on average, there is one CCl_3Br molecule present in the cage wall then bromination could, *in principle*, compete with alcohol formation if the effective rate constant for hydroxyl group rebound was ca. $(27.6/18) \times 1.4 \times 10^8\ s^{-1}$. Thus, the effective rate constant for in-cage trapping of the cycloheptyl radical is $(1.4 + 2.0) = 3.4 \times 10^8\ s^{-1}$ which is much too slow to compete effectively with diffusion of this radical out of the cage. We must, therefore, conclude that at least some free cycloheptyl radicals were formed in the system containing CCl_3Br. Further evidence that alkane oxidation with Fe^{III}(TPP)Cl/PhIO gives at least some free carbon-centered radicals which escape the solvent cage, comes from the observation that *cis*-decalin yields both *cis*-decalol (ca. 90%) and *trans*-decalol (ca. 10%).[57] Further mechanistic studies on this interesting system, particularly in the presence of nitrogenous bases, *vide infra*, are clearly warranted.

Although a variety of single oxygen atom donors other than PhIO are also capable of generating the high-valent iron-oxo species[59] (e.g. peracids and potassium monopersulfate), many workers have tried to make these oxidizing species by employing either hydrogen peroxide or an organic hydroperoxide, usually TBHP. The problem is that hydroperoxides are very susceptible to one-electron reduction by transition metals in their lower valence state. This leads to O–O bond homolysis and the formation of **free** radicals rather than the two-electron reduction required to generate the $^{\bullet+}PFe^{IV}{=}O$ species. Nevertheless, there is evidence that heterolytic cleavage of hydroperoxides can be achieved with synthetic metalloporphyrins provided a nitrogenous base is present that can ligate to the metal. For example, Mansuy and co-workers found that the epoxidation of *cis*-stilbene by Fe^{III}(TPP)Cl and cumene hydroperoxide in the presence of imidazole gave mainly *cis*-stilbene oxide with only a small amount of the *trans*-epoxide.[60]

The effect of added imidazole on the mechanism of O–O bond cleavage has also been investigated by Labeque and Marnett using 10-hydroperoxy-8,12-octadienoic (**4**) acid as a probe hydroperoxide.[14] The one-electron reduction of this hydroperoxide yields an alkoxyl radical which undergoes either a rapid β-scission to yield 10-oxodec-8-enoic acid (**5**) or is oxidized to 10-oxooctadec-8,12-dienoic acid (**6**). In contrast, two-electron reduction with heterolytic cleavage of the O–O bond would yield 10-hydroxyoctadec-8,12-dienoic acid (**7**):

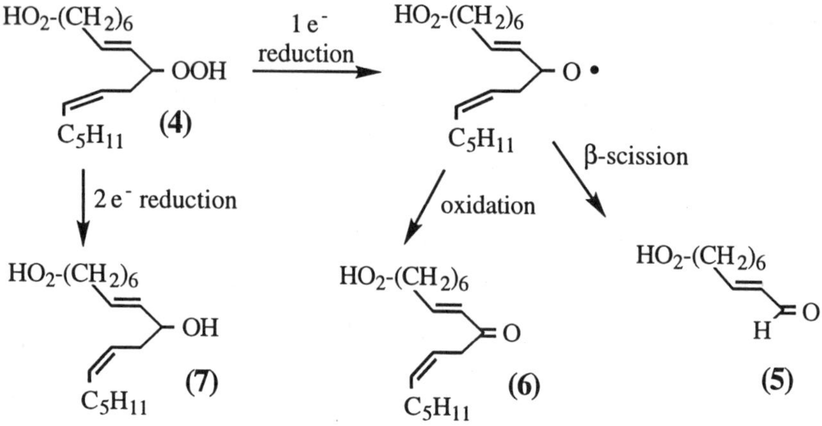

The major products formed from this hydroperoxide and Fe^{III}(TPP)Cl in dichloromethane were those derived from the alkoxyl radical with only traces of the product of heterolysis being detected. However, in the presence of added imidazole the major product became the heterolytically derived alcohol but there were also significant amounts of alkoxyl radical-derived products. Obviously hydroperoxide decomposition now occurs by both heterolytic and homolytic mechanisms. This conclusion received further support from a study of the epoxidation of *cis*-stilbene. In the absence of imidazole, the *trans*-epoxide was the major product. In the presence of imidazole, equal amounts of *cis*- and *trans*-stilbene oxide were formed, and the addition of p-methoxythioanisole selectively inhibited formation of the *cis*-epoxide.

In the absence of a nitrogenous base to stabilize the high-valent iron-oxo species[60] product studies by Bruice[61] and Lindsay Smith[62] and co-workers indicate hydroperoxide O–O bond homolysis predominates. The reaction of a variety of water-soluble iron(III) porphyrin complexes, e.g. 5,10,15,20-tetra(2,6-dimethyl-3-sulphonato)porphyrinato iron(III) or 5,10,15,20-tetra(*N*-methyl-4-pyridyl)porphyrinato iron(III), with TBHP in aqueous solution gave acetone, methanol, formaldehyde and *tert*-butylmethyl peroxide.[62b] These products require the initial formation of *tert*-butoxyl radicals which undergo β-scission to give acetone and methyl radicals. *Tert*-butyl alcohol (which would be formed if a heterolytic O–O bond cleavage had occurred) was only detected in minor amounts unless the reaction was carried out in the presence of 2,2'-azinobis (3-ethylbenzthiazoline-6-sulfonate) (ABTS), a good one-electron reducing agent, in which case the yield of *tert*-butanol was 98%. The replacement of TBHP by cumene hydroperoxide gave acetophenone as a product as required for a homolytic mechanism.

The simple and self-consistent **free** radical (homolytic) mechanism for O–O bond scission in the absence of nitrogenous bases has been challenged by Traylor and co-workers[30,63] who have claimed that the results are consistent with the heterolytic cleavage of the peroxide O–O bond. The resultant high-valent iron oxene species was hypothesized to abstract a hydrogen atom from the hydroperoxide to form peroxyl radicals and an iron-oxo species:

$$PFe^{V}=O + ROOH \longrightarrow PFe^{IV}OH + ROO\bullet \qquad (2.37)$$

in preference to inserting an oxygen atom into an organic substrate. Therefore, the radical-derived products were presumed to arise from side reactions rather than from the initial reaction of the hydroperoxide with the iron catalyst. High yields of cyclooctene epoxide when using TBHP with electron-deficient iron(III) porphyrin complexes, e.g. iron(III) 5,10,15,20-tetra(pentafluorophenyl) porphyrin [$Fe^{III}(F_{20}TPP)Cl$], have been cited in support of this mechanism (but see: *2. Ambiguous Alkene Substrates* for the absence of decisiveness in this particular diagnostic test). More tellingly, the epoxidation of *cis*-β-methylstyrene with TBHP gave predominantly the *trans*-epoxide, whereas PhIO gave mostly the *cis*-epoxide. Even more conclusive proof that these catalysts effect homolytic O–O bond scission rather than the claimed heterolytic scission was obtained when TBHP was replaced by MPPH (and related hydroperoxide probes) which caused a dramatic reduction in epoxide yields.

Additional clear evidence favoring iron/hydroperoxide **free** radical chemistry comes from studies using, e.g. *meso*-tetrakis (2,6-dichlorophenyl)-porphyrinato iron(III) chloride [$(Cl_8TPP)Fe^{III}Cl$] to catalyze the epoxidation of *cis*-stilbene and (Z)-1,2-bis(*trans*-2, *trans*-3-diphenylcyclopropyl) ethene.[5b,64] Stereospecific epoxidation of both olefins was observed when pentafluoroiodosyl benzene was used as the oxidant. However, with TBHP the predominant products were those derived from **free** radical reactions.[65]

Attempts to mimic the cytochrome P450 enzymes using dioxygen and a sacrificial reductant to activate the metalloporphyrin have achieved only limited success. For example, Mn^{III}TPP with sodium borohydride oxidized cyclohexene mainly to cyclohexanol[66] and, although Fe^{III}(TPP)Cl has been reported to catalyze the epoxidation of propylene in the presence of propionaldehyde[67], it seems highly probable that the active oxidant is formed by autoxidation of the aldehyde (see: 2. *Ambiguous Alkene Substrates*).

Claims by Ellis and Lyons that alkanes can be oxidized with metalloporphyrin catalysts and dioxygen without a reducing agent have not stood the test of time. Initially, isobutane was oxidized to *tert*-butyl alcohol using metalloporphyrin complexes such as $Fe^{III}(TPP)N_3$ at 80 °C.[68] With more stable halogenated porphyrins, e.g. 2,3,7,8,12,13,17,18-octabromo-5,10,15,20-tetra-(pentafluorophenyl)-porphyrinatoiron(III) chloride, $[Fe^{III}(TFPPBr_8)Cl]$, product yields increased and the reaction could be carried out at lower temperatures.[69] The mechanism postulated involved initial reduction of the iron(III) catalyst, followed by a reaction of the iron(II) complex with dioxygen to form a peroxo-bridged dimer which subsequently formed a high-valent iron-oxo species. This hypothetical complex was believed to abstract a hydrogen atom from the substrate with the alcohol being formed by oxygen-rebound:

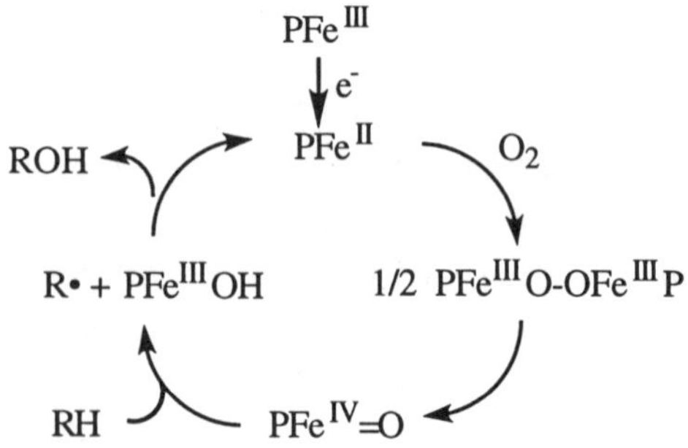

There is, by now, overwhelming evidence that the Ellis and Lyons reaction is a **free** radical autoxidation of isobutane catalyzed by the metalloporphyrin. Thus, the formation of acetone (~10%) is proof that free *tert*-butoxyl radicals are formed, and the formation of di-*tert*-butyl peroxide[70] is proof that **free** *tert*-butylperoxyl radicals are formed. The observed rate of reaction is consistent with a metal-catalyzed autoxidation mechanism.[71] It has also been demonstrated that although highly substituted, electron-deficient, iron(II) porphyrins are very efficient catalysts for hydroperoxide decomposition[72], they are unreactive towards dioxygen and therefore the mechanism outlined above is untenable.[73] Metal-catalyzed hydrocarbon oxidations always exhibit an induction period or strong autocatalysis during the time required for the hydroperoxide to reach its maximum steady-state concentration. It is therefore no surprise to learn that these reactions show an induction period, at the end of which significant changes in the nature of the catalyst were observed (by NMR) with the concomitant formation of oxidized products.[70] BHT inhibits these oxidations which also indicates a **free** radical autoxidation chain mechanism[73] (see: *3. Effects of Additives*, (i)).

This iron porphyrin-catalyzed oxidations case study indicates that these reactions *can* follow a biomimetic, radical-free mechanism (provided the iron complex has the right ligand environment) but that **free** radical mechanisms will normally dominate the chemistry.

4. Oxygenated Fenton Chemistry

The final iron catalyst plus hydroperoxide system to be considered in detail is that developed by Sawyer and co-workers[74], called "Oxygenated Fenton Chemistry." This involves the oxidation of a hydrocarbon by an iron catalyst (usually Fe^{II}) and H_2O_2 or TBHP under an atmosphere of oxygen at room temperature.[74] The reactions usually yield ketone and small quantities of other oxidized material, such as alcohol. The proposed mechanism (shown below for H_2O_2) involved formation of peroxy-iron(II) species which then reacts with dioxygen to form another intermediate which then oxidizes the substrate:

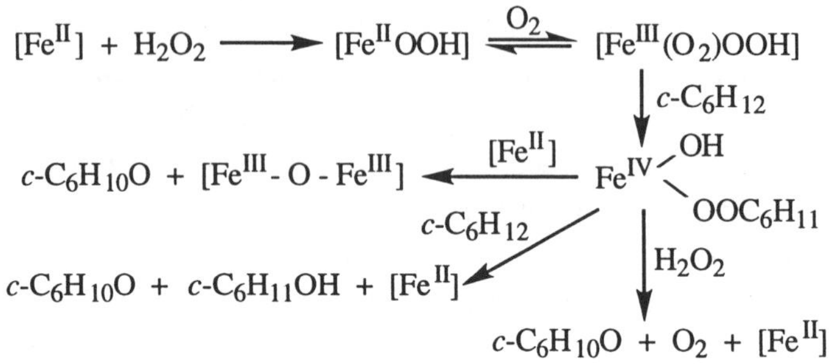

Sawyer's mechanism stands in complete contrast to the classical and overwhelmingly established Fenton mechanism[75] which involves the formation of **free** hydroxyl radicals from iron(II)/hydrogen peroxide systems, and hence **free** carbon-centered radicals from the substrate. The Sawyer mechanism was based mainly on differences in product distributions from those obtained when hydroxyl radicals have been generated by pulse radiolysis (PR — gives dimers as the main product from hydrocarbon substrates). Under PR conditions, the •OH radicals and the derived alkyl radicals are formed in the absence of metal ions and, as we have seen earlier in this section, it is unwise to draw mechanistic conclusions on the basis of small changes in product distributions.

The formation of carbon-centered radicals by Fenton ($Fe^{II} + H_2O_2$)[76] and Fenton-type (e.g. $Ti^{III} + H_2O_2$)[76b,77] reagents has been demonstrated quite conclusively using a flow system to obtain EPR spectra of the intermediates formed in these reactions. Spectra were reported of radicals derived from both aliphatic and aromatic substrates which were identical with those reported from PR experiments.[78] The differences in product distributions between PR and Fenton systems are due to the presence of metal ions in the latter because these metal ions may oxidize or reduce the carbon-centered radicals. For example, the oxidation of *tert*-butyl alcohol under Fenton conditions gives the dimer as the sole product because the primary alkyl radicals, $\bullet CH_2C(CH_3)_2OH$, are not oxidized by iron(III). However, if copper(II) (which is a much better oxidizing agent) is added, 2-methyl-1,2-propanediol becomes the major product.[79] Similarly, during the oxidation of 2-propanol in a flow EPR system with H_2O_2, the major EPR signal detected was that of the 2-hydroxyprop-1-yl radical using an iron catalyst but with a titanium(III) catalyst, the major signal was that of the 2-hydroxyprop-2-yl radical.[80] It was later shown that with the iron catalyst, just as with the titanium catalyst, the major radical formed was the 2-hydroxyprop-2-yl radical. However, in the presence of Fe^{III} this radical was immediately oxidized to acetone.[79]

The formation of hydroxyl radicals as the active oxidant in Fenton systems is more difficult to prove than the presence of **free** carbon-centered radicals. However, the relative reactivities for a range of substrates subjected to Fenton chemistry are in agreement with the relative rates of the reactions for **free** hydroxyl radicals generated in metal-free systems by PR.[75] The formation of alkoxyl radicals in Fenton-like systems using alkyl hydroperoxides has been demonstrated on several occasions, e.g. with 1-hydroxycyclohexyl hydroperoxide, a C_{12} opened-ring dibasic acid is formed, quite obviously from β-scission of the corresponding alkoxyl radicals, followed by coupling of the derived primary alkyl radicals:[81]

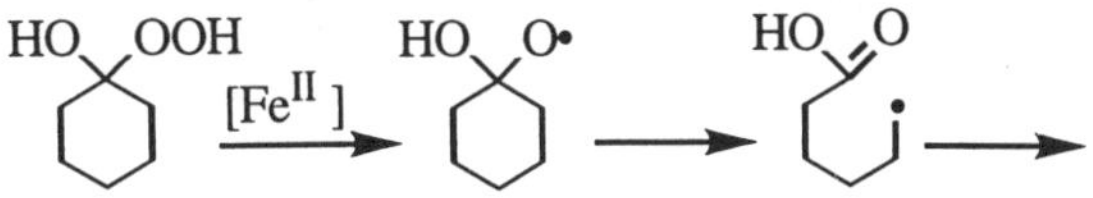

$HO_2C\text{-}(CH_2)_{10}\text{-}CO_2H$ (2.38)

The nature of the hydrogen atom abstracting species in "Oxygenated Fenton Chemistry" using TBHP has recently been investigated using both radical clock methodology and measuring the effect on product yields of BHT and diphenylamine (see: *3. Effects of Additives (i)*).[82] The oxidation of cyclohexane, ethylbenzene and cyclohexene (1 M) by ferric chloride, iron(II) tetrakis(triphenylphosphine oxide), iron(II) bis(2,2'-bipyridyl) or iron(II) bis(picolinate) (10 mM) with 20 mM TBHP led to average product yields of 7.5 mM, 25.6 mM and 86.0 mM, respectively. The reason for the higher yields of oxidized products with ethylbenzene and cyclohexene compared to cyclohexane is that these two substrates contain activated hydrogen atoms which render them subject to attack by peroxyl radicals as well as by alkoxyl radicals. The rate constants for peroxyl radical attack on the three substrates are 0.048, 1.1 and 6.1 $M^{-1}s^{-1}$ for cyclohexane,[11] ethylbenzene[83] and cyclohexene,[83] respectively. Indeed, if a rough correction is made for alkoxyl radical-induced hydrocarbon oxidation by subtracting the average cyclohexane product yield, the ratio of cyclohexene/ethylbenzene products arising from peroxyl radical attack is (86.1 – 7.5)/(25.6 – 7.5) = 4.3. This ratio agrees satisfactorily with the expected ratio based on the ratio of the rate constants for peroxyl radical attack on these two substrates, viz. 6.1/1.1 = 5.5. A simplified mechanism involving free radical autoxidation initiated by alkoxyl and peroxyl radicals is shown below:

Cyclohexane

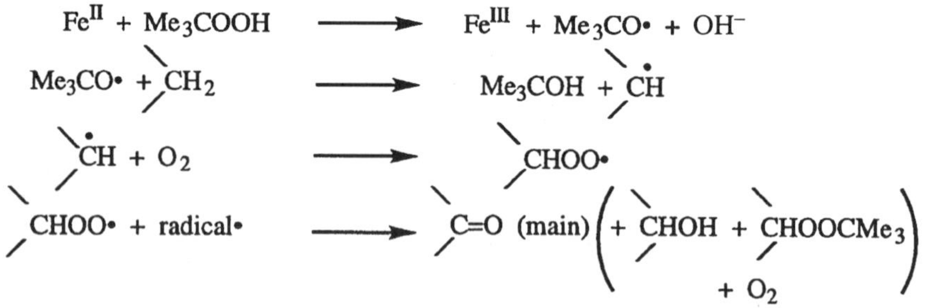

Cyclohexene and Ethylbenzene

The role of alkoxyl radicals in the cyclohexane oxidations was demonstrated by replacing TBHP with MPPH. This led to essentially no cyclohexane oxidation. The main product was benzaldehyde with minor amounts of benzyl alcohol. In contrast, the use of MPPH with cyclohexene or ethyl benzene gave significant quantities of the oxidized products because benzylperoxyl radicals were formed which initiated the autoxidation:

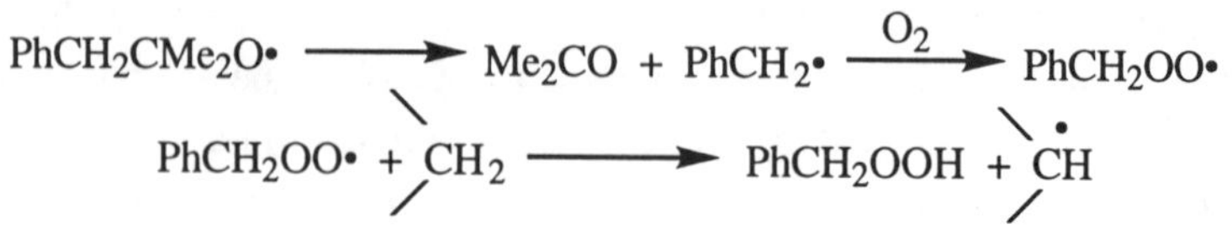

Proof that the oxidation of cyclohexene and ethylbenzene was initiated by *tert*-butoxyl radicals from TBHP and by benzylperoxyl radicals from

MPPH was obtained by the addition of BHT or diphenylamine. As mentioned earlier, these antioxidants are reactive towards *tert*-alkoxyl radicals, the rate constants in acetonitrile being 2.0×10^7 $M^{-1}s^{-1}$ for BHT[8] and 3.0×10^8 $M^{-1}s^{-1}$ for Ph_2NH,[17] and towards peroxyl radicals, with rate constants of 1.4×10^4 $M^{-1}s^{-1}$ (BHT)[18] and ~ 4×10^4 $M^{-1}s^{-1}$ (Ph_2NH).[19] Knowing these rate constants, predictions can be made about the effect of BHT and Ph_2NH on the yields of oxidized products. Thus, 100 mM Ph_2NH should completely inhibit the oxidation of ethylbenzene by MPPH because all the benzylperoxyl radicals will be trapped: $k_{ROO\bullet}[Ph_2NH](= 4 \times 10^4\ M^{-1}s^{-1} \times 0.1\ M = 4 \times 10^3\ s^{-1}) >> k_{ROO\bullet}[PhEt]\ (= 1.1\ M^{-1}s^{-1} \times 1\ M = 1.1\ s^{-1})$. Indeed, even 20 mM BHT should be sufficient to inhibit the reaction very strongly: $k_{ROO\bullet}[BHT](= 1.4 \times 10^4\ M^{-1}s^{-1} \times 0.02\ M = 2.8 \times 10^2\ s^{-1}) >> k_{ROO\bullet}[PhEt]$. Both predictions were confirmed experimentally using all four catalysts. Similarly, 100 mM Ph_2NH should almost completely inhibit the oxidation of ethylbenzene by TBHP: $k_{RO\bullet}[Ph_2NH]\ (= 3 \times 10^8\ M^{-1}s^{-1} \times 0.1\ M = 3 \times 10^7\ s^{-1}) > k_{RO\bullet}[PhEt]\ (= 1 \times 10^6\ M^{-1}s^{-1} \times 1\ M = 1 \times 10^6\ s^{-1})$. However, 50 mM BHT should only provide about 50% inhibition: $k_{RO\bullet}[BHT]\ (= 2 \times 10^7\ M^{-1}s^{-1} \times 0.05\ M) = 1 \times 10^6\ s^{-1}$, and 20 mM BHT should provide even less inhibition. Again, all of these predictions were confirmed experimentally.

The new experimental results described above plus all the evidence acquired by earlier workers regarding the formation of **free** radicals in Fenton systems is more than sufficient to prove that "Oxygenated Fenton Chemistry" operates by simple **free** radical chemistry, not by the suggested non-radical mechanism. This case study further illustrates the obvious: mechanistic conclusions should only be drawn when a system has been subjected to at least two of the tests described in this chapter. It should be noted that the results obtained with MPPH and ethylbenzene or cyclohexene could be misleading without a proper understanding of **free** radical chemistry. This case study also demonstrates that with a knowledge of the absolute kinetics of free radical reactions, quantitative predictions can be made regarding the effects of the addition of free radical traps and that these predictions can then be tested experimentally.

Addendum

In addition to the four case studies selected for detailed review and proof that claimed radical-free chemistries were frequently **free** radical chemistries there have been numerous other publications on iron-catalyzed hydrocarbon oxidations some of which have been claimed to operate via non-free radical mechanisms, while others are acknowledged to involve freely diffusing radicals. We have no arguments with any of the latter[84–89] but we believe that at least some of the former should be noted herein because the involvement of **free** radicals cannot be ruled out on the basis of the reported experimental evidence,[90–92] particularly in cases where TBHP has been employed.[86,93–96] We hope that these mechanistic proposals[86,90–96] do not stand unchallenged and that they will be put to the test using the procedures outlined in this chapter. On the other hand, there are claims for radical-free mechanisms which appear to be fully justified because valid mechanistic diagnostic experiments were performed.[97,98] In addition, there are some reports in which the authors clearly want it both ways, i.e. they have proposed that radical-free and **free** radical mechanisms operate simultaneously.[99–101] To us, this appears to be extremely improbable because the high reactivity of oxygen-centered radicals ensure that they will quickly dominate the overall chemistry.

Finally, a mechanistic probe which distinguishes oxidation by an oxo-iron (IV) porphyrin π-radical cation from oxidation via *tert*-butylperoxyl radicals has recently been reported.[102] The general utility of this probe in mechanistic studies remains to be determined.

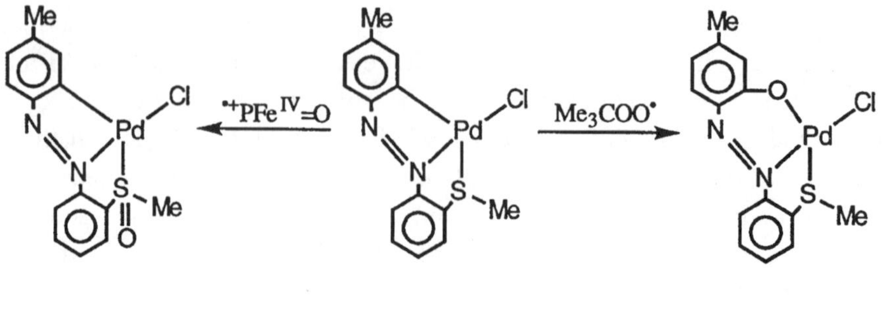

References

1. *Radical Reaction Rates in Liquids*, Landolt Börnstein Group II, Vols. 13 and 18, Subvol. A–D, (ed.) Fischer, H. Springer-Verlag, Berlin, 1994 and 1984.
2. Griller, D. and Ingold, K.U. *Acc. Chem. Res.* 1980, **13:** 317–323.
3. Newcomb, M. *Tetrahedron* 1993, **49:** 1151–1176.
4. Groves, J.T. and Subramanian, D.V. *J. Am. Chem. Soc.* 1984, **106:** 2177–2181.
5. (a) Oldroyd, R.D. *et al. Angew. Chem. Int. Ed. Engl.* 1996, **35:** 2787–2790; (b) Castellino, A.J. and Bruice, T.C. *J. Am. Chem. Soc.* 1988, **110:** 158–162.
6. Avila, D.V., Brown, C.E., Ingold, K.U. and Lusztyk, J. *J. Am. Chem. Soc.* 1993, **115:** 466–470.
7. Paul, H., Small, R.D. and Scaiano, J.C. *J. Am. Chem. Soc.* 1978, **100:** 4520–4527.
8. Arends, I.W.C.E., Ingold, K.U. and Wayner, D.D.M. *J. Am. Chem. Soc.* 1995, **117:** 4710–4711.
9. MacFaul, P.A., Arends, I.W.C.E., Ingold, K.U. and Wayner, D.D.M. *J. Chem. Soc., Perkin Trans. 2* 1997, 135–145.
10. Hartung, J. and Gallon, F. *J. Org. Chem.* 1995, **60:** 6706–6716.
11. Korcek, S., Chernier, J.H.B., Howard, J.A. and Ingold, K.U. *Can. J. Chem.* 1972, **50:** 2285–2297.
12. Mayo, F.R. *Acc. Chem. Res.* 1968, **1:** 193–201.
13. (a) Van Sickle, D.E., Mayo, F.R. and Arluck, R.M. *J. Am. Chem. Soc.* 1965, **87:** 4824–4832; (b) Van Sickle, D.E., Mayo, F.R., Gould, E.S. and Arluck, R.M. *J. Am. Chem. Soc.* 1967, **89:** 977–984.
14. Labeque, R. and Marnett, L.J. *J. Am. Chem. Soc.* 1989, **111:** 6621–6627.
15. Bateman, L. and Hargrave, K.R. *Proc. R. Soc. London,* 1954, **A224:** 399–411.
16. Avila, D.V., Ingold, K.U. and Lusztyk, J. *J. Am. Chem. Soc.* 1995, **117:** 2929–2930.
17. MacFaul, P.A., Ingold, K.U. and Lusztyk, J. *J. Org. Chem.* 1996, **61:** 1316–1321.
18. Burton, G.W. *et al. J. Am. Chem. Soc.* 1985, **107:** 7053–7065.
19. Brownlie, I.T. and Ingold, K.U. *Can. J. Chem.* 1967, **48:** 2419–2425.
20. Valgimigli, L., Banks, J.T., Ingold, K.U. and Lusztyk, J. *J. Am. Chem. Soc.* 1995, **117:** 9966–9971.
21. Unpublished results from this laboratory.

22. (a) Barton, D.H.R and Doller, D. *Acc. Chem. Res.* 1992, **25:** 504–512; (b) Barton, D.H.R. and Chavasiri, W., *Tetrahedron* 1994, **50:** 19–30.
23. Barton, D.H.R., Hu, B., Taylor, D.K. and Rojas Wahl, R.U. *J. Chem. Soc., Perkin Trans. 2* 1996, 1031–1041.
24. Perkins, M.J. *Chem. Soc. Rev.* 1996, 229–236.
25. Newcomb, M., Simankov, P.A. and Park, S-H., *Tet. Lett.* 1996, **37:** 819–822.
26. Minisci, F. *Angew. Chem.* 1958, **70:** 599.
27. Kochi, J.K. in *Free Radicals* (ed.) Kochi, J.K. John Wiley and Sons, Inc., New York, 1973, pp. 591–683.
28. Minisci, F. and Fontana F. *Tet. Lett.* 1994, **35:** 1427–1430.
29. Snelgrove, D.W., MacFaul, P.A., Ingold, K.U. and Wayner, D.D.M. *Tet. Lett.* 1996, **37:** 823–826.
30. Traylor, T.G. *et al. J. Am. Chem. Soc.* 1996, **117:** 3468–3474.
31. Kim, J., Harrison, R.G., Kim, C. and Que, L. Jr. *J. Am. Chem. Soc.* 1996, **118:** 4373–4379.
32. Ingold, K.U. *et al. Catal. Lett.* 1997, **48:** 21–24.
33. Minisci, F., Fontana, F., Araneo, S. and Recupero, F. *Tet. Lett.* 1994, **35:** 3759–3762.
34. Minisci, F. *et al. J. Am. Chem. Soc.* 1995, **117:** 226–232.
35. Minisci, F. *et al. Synlett.* 1996, 119–125.
36. Bravo, A. *et al. J. Org. Chem.* 1997, **62:** 3849–3857.
37. Barton, D.H.R. *Synlett.* 1997, 229–230.
38. Barton, D.H.R. Csuhai, E., Doller, D. and Balavoine, G. *J. Chem. Soc., Chem. Comm.* 1990, 1787–1789.
39. Knight, C. and Perkins, M.J. *J. Chem. Soc., Chem. Commun.* 1991, 925–927.
40. Barton, D.H.R. *et al. J. Am. Chem. Soc.* 1992, **114:** 2147–2156.
41. Barton, D.H.R. *et al. Tetrahedron Lett.* 1992, **33:** 5473–5476.
42. (a) Leising, R.A., Norman, R.E. and Que, L. Jr. *Inorg. Chem.* 1990, **29:** 2555–2557; (b) Leising, R.A., Zang, Y. and Que, L. Jr. *J. Am. Chem. Soc.* 1991, **113:** 8555–8557; (c) Leising, R.A., Kim, J., Pérez, M.A. and Que, L. Jr. *J. Am. Chem. Soc.* 1993, **115:** 9524–9530.
43. Groves, J.T. *J. Chem. Educ.* 1985, **62:** 928–931.
44. Russell, G.A. *J. Am. Chem. Soc.* 1957, **79:** 3871–3877.
45. (a) Daikh, B.E. and Finke, R.G. *J. Am. Chem. Soc.* 1992, **114:** 2938–2943; (b) Fischer, H. *J. Am. Chem. Soc.* 1986, **108:** 3925–3927.
46. Arends, I.W.C.E. *et al.* in *Free Radicals in Biology and Environment* (ed.) Minisci, F. Kluwer Academic Press, The Netherlands, 1997, pp. 79–90.

47. (a) Kochi, J.K. *Tetrahedron* 1962, **18:** 483–497; (b) Kochi, J.K. and Mains H.E. *J. Org. Chem.* 1965, **30:** 1862–1872.
48. MacFaul, P.A., Ingold, K.U., Wayner, D.D.M. and Que, L. Jr. *J. Am. Chem. Soc.,* in press.
49. Rabion, A. *et al. J. Am. Chem. Soc.* 1995, **117:** 12356–12357.
50. Rabion, A. Buchanan R.M., Seris, J.-L. and Fish, R.H. *J. Mol. Catal. A: Chemical* 1997, **116:** 43–47.
51. Chang, C.K. *J. Am. Chem. Soc.* 1979, **101:** 3413–3415.
52. Groves, J.T., Nemo, T.E. and Myers, R.S. *J. Am. Chem. Soc.* 1979, **101:** 1032–1033.
53. Groves, J.T. and Han, Y.-Z. in *Cytochrome P450: Structure, Mechanism and Biochemistry*, (ed.) Ortiz de Montellano, P.R. Plenum Press, New York, 1995, pp. 3–48.
54. *Peroxidases in Chemistry and Biology*, (eds.) Everse, J., Everse, K.E. and Grisham, M.B. CRC Press, Boca Raton, 1991.
55. Groves, J.T. and Nemo, T.E. *J. Am. Chem. Soc.* 1983, **105:** 5786–5791.
56. Appleton, A., Evans, S. and Lindsay Smith, J.R. *J. Chem. Soc., Perkin Trans. 2* 1996, 281–285.
57. Groves, J.T. and Nemo, T.E. *J. Am. Chem. Soc.* 1983, **105:** 6243–6248.
58. Mathew, L. and Warkentin, J. *Can. J. Chem.* 1988, **66:** 11–16.
59. Meunier, B. *Chem. Rev.* 1992, **92:** 1411–1456.
60. Mansuy, D., Battioni, P. and Renaud, J.-P. *J. Chem. Soc., Chem. Commun.* 1984, 1255–1257.
61. (a) Almarsson, Ö. and Bruice, T.C. *J. Am. Chem. Soc.* 1995, **117:** 4533–4544; (b) Bruice, T.C. *Acc. Chem. Res.* 1991, **24:** 243–249; (c) Balasubramanian, P.N. *et al. J. Am. Chem. Soc.* 1989, **111:** 1477–1483.
62. (a) Colclough, N. and Lindsay Smith, J.R. *J. Chem. Soc., Perkin Trans. 2* 1995, 235–241; (b) Lindsay Smith, J.R. and Lower, R.J. *J. Chem. Soc., Perkin Trans. 2* 1991, 31–39.
63. Traylor, T.G., Tsuchiya, S., Byun, Y.-S. and Kim, C. *J. Am. Chem. Soc.* 1993, **115:** 2775–2781.
64. Castellino, A.J. and Bruice, T.C. *J. Am. Chem. Soc.* 1988, **110:** 7512–7519.
65. He, G.-X. and Bruice, T.C. *J. Am. Chem. Soc.* 1991, **113:** 2747–2753.
66. Tabushi, I. and Koya, N. *J. Am. Chem. Soc.* 1979, **101:** 6456–6548.
67. Haber, J., Mlodnicka, T. and Poltowicz, J. *J. Mol. Catal.* 1989, **54:** 451–461.
68. Ellis, P.E. and Lyons, J.E. *J. Chem. Soc., Chem. Commun.* 1989, 1189–1190.
69. Ellis, P.E. and Lyons, J.E. *Coordn. Chem. Rev.* 1990, **105:** 181–193.

70. Moore, K.T., Horváth, I.T. and Therien, M.J. *J. Am. Chem. Soc.* 1997, **119:** 1791–1792.
71. Labinger, J.A., *Catal. Lett.* 1994, **26:** 95–99. Walling, C., private communication.
72. Bottcher, A. *et al. J. Mol. Catal. A: Chemical* 1997, **117:** 229–242.
73. Grinstaff, M.W., Hill M.G., Labinger, J.A. and Gray, H.B. *Science* 1994, **264:** 1311–1313.
74. Sawyer, D.T., Sobkowiak, A. and Matsushita, T. *Acc. Chem. Res.* 1996, **29:** 409–416.
75. Walling, C. *Acc. Chem. Res.* 1975, **8:** 125–131.
76. (a) Shiga, T., Boukhors, A. and Douzou, P. *J. Phys. Chem.* 1967, **71:** 3559–3565; (b) Croft, S., Gilbert, B.C., Lindsay Smith, J.R. and Whitwood, A.C. *Free Rad. Res. Comms.* 1992, **17:** 21–39.
77. Dixon, W.T. and Norman, R.O.C. *Nature (London)* 1962, **196:** 891–892.
78. Norman, R.O.C. *Adv. Phys. Org. Chem.* 1967, **5:** 53–119.
79. Walling, C. and Kato, S. *J. Am. Chem. Soc.* 1971, **93:** 4275–4281.
80. Shiga, T. *J. Phys. Chem.* 1965, **69:** 3805–3814.
81. Hawkins, E.G.E. *J. Chem. Soc.* 1955, 3463–3467.
82. MacFaul, P.A., Wayner, D.D.M. and Ingold K.U. *Acc. Chem. Res.*, 1998, **31:** 159–162.
83. Howard, J.A. and Ingold, K.U. *Can. J. Chem.* 1966, **44:** 1119–1130.
84. Tetard, D. and Verlhac, J.-B. *J. Mol. Catal. A: Chemical* 1996, **113:** 223–230.
85. Vol'pin, M.E. *et al. Russ. Chem. Bull.* 1996, **45:** 2000–2007.
86. Parton, R.F. *et al. J. Mol. Catal. A: Chemical* 1996, **113:** 445–454.
87. Nguyen, C.R., Guajardo, J. and Mascharak, P.K. *Inorg. Chem.* 1996, **35:** 6273–6281.
88. Padbury, G., Sligar, S.G., Labeque, R. and Marnett, L.J. *Biochemistry* 1988, **27:** 7846–7852.
89. Chakrabarti, S. *et al. Free Radical Biology and Medicine* 1996, **20:** 777–783.
90. Kodera, M., Shimakoshi, H. and Kano, K. *J. Chem. Soc., Chem. Commun.* 1996, 1737–1738.
91. Sobolev, A., Babushkin, D. and Talsi, E.P. *Mendeleev Commun.* 1996, 33–34.
92. Roelfes, G. *et al. J. Mol. Catal. A: Chemical* 1997, **117:** 223–227.
93. Parton, R.F. *et al. Nature* 1994, **370:** 541–544.
94. Parton, R.F. *et al. J. Catal.* 1996, **164:** 341–346.

95. Vankelecom, I.V.F. *et al. J. Catal.* 1996, **163:** 457–464.
96. Lee, K.A. and Nam, W. *J. Am. Chem. Soc.* 1997, **119:** 1916–1922.
97. Kim, C., Chen, K., Kim, J. and Que, L., Jr. *J. Am. Chem. Soc.* 1997, **119:** 5964–5965.
98. Wessel, J. and Crabtree, R.H. *J. Mol. Catal. A: Chemical* 1996, **113:** 13-22.
99. Ménage, S., Vincent, J.-M., Lambeau, C. and Fontecave, M. *J. Mol. Catal. A: Chemical* 1996, **113:** 61–75.
100. Kulikova, V., Gritsenko, O. and Shteinman, A.H. *Mendeleev Commun.* 1996, 119–120.
101. Gritsenko, O.N., Nesterenko, G.N. and Shteinman, A.A. *Russ. Bull. Chem.* 1995, **44:** 2415–2417.
102. Kamaraj, K. and Bandyopadhyay, D. *J. Am. Chem. Soc.* 1997, **119:** 8099–8100.

Chapter 3

Biomimetic Oxygenations Related to Cytochrome P450: Metal-Oxo and Metal-Peroxo Intermediates

Jennifer L. McLain, Jinbo Lee & John T. Groves

Department of Chemistry,
Princeton University Princeton,
New Jersey 08544

3.1 Oxometalloporphyrin Intermediates in Enzymatic Processes

Metalloenzymes catalyze a host of biological oxidation reactions that are crucial to life processes. The cytochromes P450, an important family of heme-iron metalloenzymes, are oxidoreductases that activate molecular oxygen (O_2) and incorporate one of the oxygen atoms into a wide variety of biological substrates, with concurrent two-electron reduction of the other oxygen atom to H_2O.[1] The structure, biochemistry, molecular biology and the chemistry of cytochrome P450 and related model systems was extensively reviewed recently.[1] Cytochrome P450 enzymes have been isolated from numerous mammalian tissues (e.g. liver, kidney, lung, intestine and adrenal cortex), as well as insects, plants, yeast and bacteria,[2] and are known to catalyze hydroxylations, epoxidations, *N*-, *S*- and *O*-dealkylations, *N*-oxidations, sulfoxidations, and dehalogenations.[1] These oxygenation reactions play a central role in carcinogen activation, drug and xenobiotic detoxification, and steroid and prostaglandin metabolism. Surprisingly, the active site of the enzyme, known in detail from several X-ray crystal structures, is extraordinarily simple, with an iron protoporphyrin IX (**3.1**; Fig. 3.1) buried deep in a hydrophobic, substrate-binding pocket of

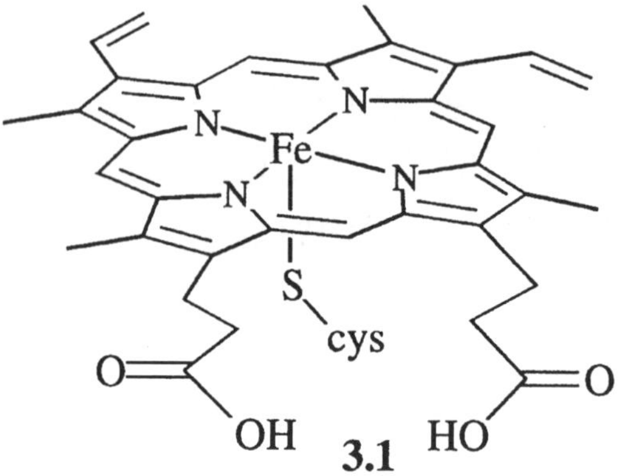

3.1

Fig. 3.1 Iron protoporphyrin IX with a cysteinate as the axial ligand, which is typical of cytochrome P450 enzymes.

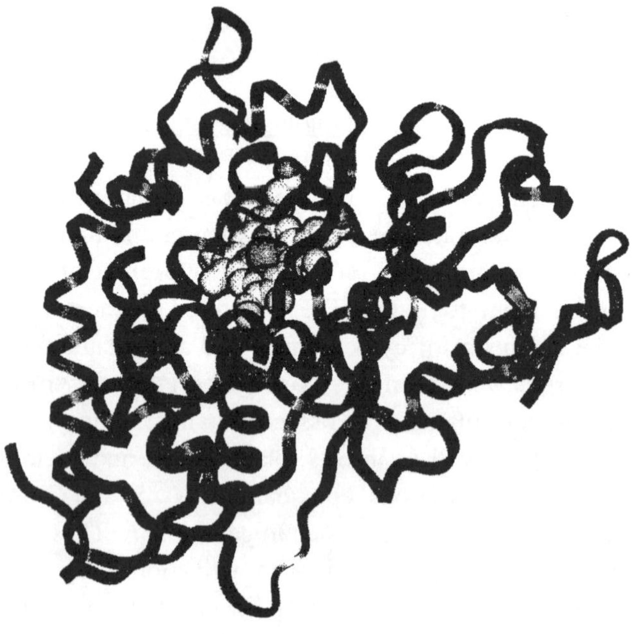

Fig. 3.2 Crystal structure of cytochrome $P450_{cam}$ (E.C.1.14.15.1) enzyme. Protein framework is shown as ribbons, and the heme is buried at a hydrophobic substrate-binding pocket. Adapted from Raag, R., Li, H., Jones, A.C. and Poulos, T.L. *Biochem.* **1993**, **32:** 4571.

the protein scaffold (Fig. 3.2). The protoporphyrin prosthetic group has a cysteinate as the 5th ligand and an open coordination site to bind and activate molecular oxygen (Fig. 3.1).

The principal features of the catalytic cycle of cytochrome P450 have been well-accepted for years; the cycle is known to involve the following steps for oxygen activation and transfer to substrates (Scheme 3.1): (1) Binding of substrate to the enzyme resting state which often converts the resting, low-spin ferric heme to the high-spin state; (2) Reduction of the ferric cytochrome P450 by one electron to the ferrous cytochrome P450; (3) Binding of molecular oxygen to the ferrous heme gives a ferrous cytochrome P450-dioxygen complex; (4) A second electron is transferred to this complex to afford a peroxoiron(III) complex; (5) Protonation and

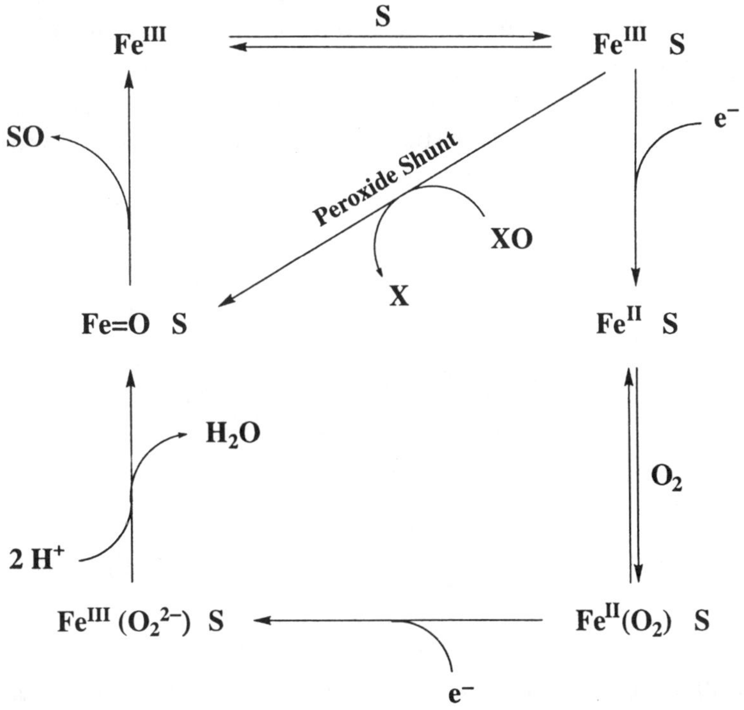

Scheme 3.1 Catalytic cycle for oxygen activation and transfer by cytochrome P450.

heterolytic cleavage of the O–O bond with concurrent production of a water molecule results in the formation of a reactive iron-oxo intermediate; (6) The oxygen atom is transferred from this oxo complex to the bound substrate; (7) Dissociation of the product completes the catalytic cycle. A shortcut of the catalytic pathway, known as the *Peroxide Shunt*, has also been demonstrated with the use of other oxygen donors such as peroxycarboxylic acids, iodosylbenzene, hypochlorite, periodate, amine *N*-oxides, and alkylhydroperoxides, among others (Scheme 3.1).[3] It is this oxygen atom transfer from the oxygen donor to the heme-iron and, subsequently, to the substrate that has been termed *Oxygen Rebound*.[3]

The iron-oxo intermediate is believed to be an oxoiron(IV) porphyrin cation radical (**3.2**; Fig. 3.3) in analogy to Compound I of horseradish peroxidase (HRP),[4–8] though no definitive characterization of this transient species is presently available for cytochrome P450 itself. More convincing evidence for the identity of the iron-oxo intermediate has come from model studies using synthetic porphyrin analogs (*vide infra*); the use of synthetic porphyrins to mimic metalloenzymes has long been shown to be an effective strategy in elucidating the sophisticated mechanisms of biological processes.[9]

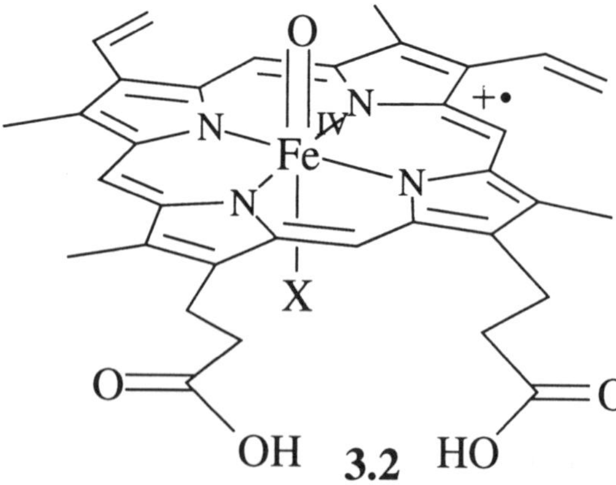

3.2

Fig. 3.3 Oxoiron(IV) cation radical of protoporphyrin IX. X = histidine imidazole for horseradish peroxidase and cytochrome oxidase; cysteine thiolate for cytochrome P450, chloroperoxidase and nitric oxide synthase.

In addition to their crucial roles in cytochrome P450 and HRP, high-valent oxometalloporphyrin complexes also play important roles in reactions catalyzed by chloroperoxidase[10] and nitric oxide synthase (NOS).[11–13] An oxomanganese(IV) protein radical species has been identified for the unnatural manganese HRP.[14] Heme iron-oxo intermediates have also been identified in the oxygen reduction cycle of cytochrome oxidase, which is the mitochondrial proton pump.[15,16]

3.2 Synthetic Oxometalloporphyrins as Models for Cytochrome P450 and Efficient Process Catalysts

3.2.1 Oxoiron porphyrins

Studies using synthetic metalloporphyrins as models for cytochrome P450 have afforded important insights into the nature of these enzymatic processes.[3,17] In fact, each of the intermediates shown in Scheme 3.1 has been independently identified by model studies using synthetic analogs, especially *meso*-tetraaryl porphyrins.[3,18–23] The first synthetic system which closely mimicked the chemistry of cytochrome P450 was chloro-α,β,γ,δ-tetraphenylporphinatoiron(III) (Fe(III)TPP(Cl), **3.3**) with iodosylbenzene as the oxidant to effect the stereospecific epoxidation of olefins and hydroxylation of hydrocarbons (Scheme 3.2).[24]

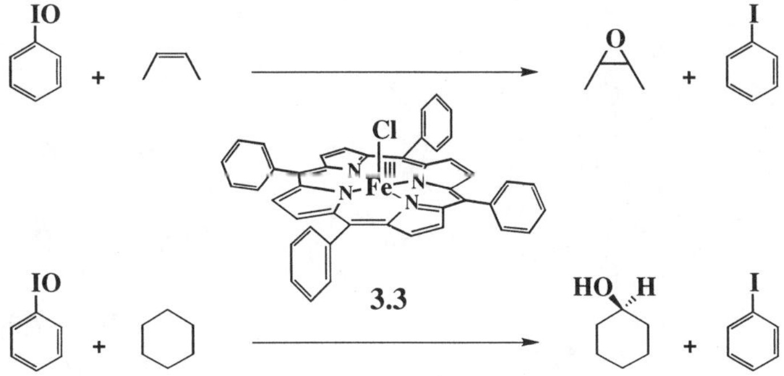

Scheme 3.2 Olefin epoxidation and alkane hydroxylation catalyzed by an iron porphyrin, Fe(III)TPP(Cl).

It was soon discovered that the reactive intermediates in the iron porphyrin model systems were high-valent oxoironporphyrin complexes.[23a] A green oxoiron(IV) porphyrin cation radical species (**3.4**) has been well-characterized by various spectroscopic techniques, including visible spectroscopy, NMR, EPR, Mössbauer and EXAFS (Fig. 3.4).[23a,25,26] A family of such complexes with different axial ligands has been reported, each of which displayed a characteristic ^{1}H-NMR for the pyrrole protons (toluene-d_8 at –80 °C). Addition of methanol replaced each of these ligands with a solvent molecule ($\delta = -22.8$).[23b] A recent report of the imidazole and *p*-nitrophenolate complexes of **3.4** has also appeared, affording closer model complexes for the Compounds I of peroxidase and catalase, respectively.[28c] The oxoiron(IV) porphyrin cation radical complexes (**3.4**) were shown to be as highly reactive as oxygen atom transfer agents toward olefins and hydrocarbons, i.e. reacting with olefins to afford epoxides and with alkanes to give alcohols.[18,23] Notably, **3.4**-4-Me-Im was also reactive toward norbornene, giving a 67% yield of the corresponding epoxide.[23b] Significant variations were observed in the reactivity of the complex **3.4** as a function of the axial ligand.[28d] By contrast, the reactivity and stereospecificity of the corresponding oxoiron(IV) porphyrin complexes were low.[3,29]

The mechanism leading to the formation of complex **3.4** has been elucidated. A reactive acylperoxoiron(III) porphyrin (**3.5**) was prepared and

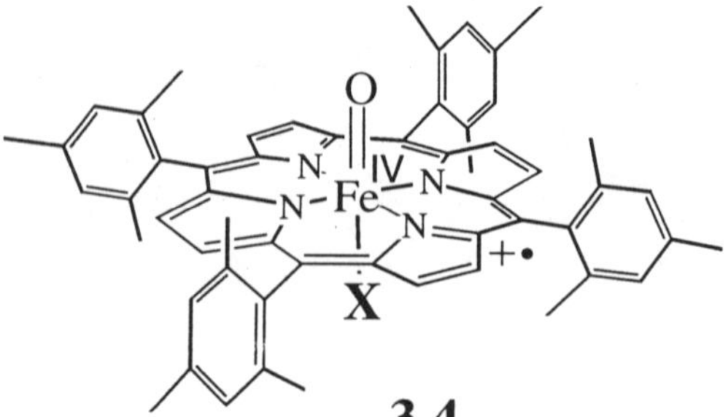

3.4

X =		
	Cl	$\delta = -9.01$
	O-Bz	$\delta = -15.6$
	OMe	$\delta = -16.2$
	4-Me-Im	$\delta = -13.7$

Fig. 3.4 The family of oxoiron(IV) porphyrin cation radical species, $Fe^{IV}(O)TMP^{+\cdot}(X)$.

characterized at low temperature, and the scission of the O–O bond was directly observed.[18] The oxoiron(IV) porphyrin cation radical **(3.4)** was produced by heterolytic cleavage of the O–O bond of complex **3.5**, which was assisted by an axial ligand and a proton (Scheme 3.3). The temperature dependence of this process revealed a remarkably low enthalpic barrier for O–O bond cleavage and a very large, negative entropy term characteristic of an associative process with the coordination of the axial ligand and placement of the proton. Significantly, the acylperoxyiron(III) adduct **(3.5)** was also shown to be a reactive electrophile.[18] Thus, Watanabe showed that epoxidation selectivities were sensitive to the structure of the peroxyacid in nonpolar solvents such as toluene. Accordingly, the iron-coordinated peroxyacid **3.5** was proposed to be the reactive species under these conditions. By contrast, the same oxoiron(IV)porphyrin cation radical **(3.4)** was formed with a variety of peroxyacids in methylene chloride.[27] Similar conclusions were reported subsequently by other groups.[28]

Other high-valent oxometalloporphyrin complexes have also been isolated and characterized in attempts to mimic cytochrome P450 enzymes and to develop practical process catalysts based on this paradigm.[3,30,31] The oxotitanium (IV) **3.6**,[32] oxovanadium(IV) **3.7**,[33] oxoniobium(IV) **3.8**,[34] *trans*-dioxoosmium(VI) **3.9**[35,36] and *cis*-dioxomolybdenum(VI) **3.10**[37] porphyrin complexes were quite stable and displayed little oxidative activity

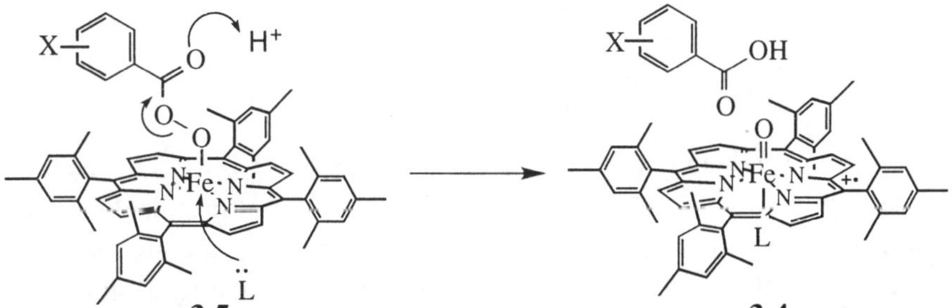

3.5 **3.4**

Scheme 3.3 Schematic illustration of the heterolytic cleavage of the O–O bond of an acylperoxoiron(III) porphyrin complex and subsequent generation of an oxoiron(IV) cation radical species.

(Fig. 3.5). By contrast, an oxochromium(V) porphyrin complex **(3.11),** capable of oxidizing a variety of organic substrates including olefins and alkanes,[38–41] has been characterized (Fig. 3.6). By analogy to the oxoiron(IV) porphyrin species,[3,29] the oxochromium(IV) porphyrin complex was unreactive towards olefins, though it could transfer its oxygen atom to triphenylphosphine.[42]

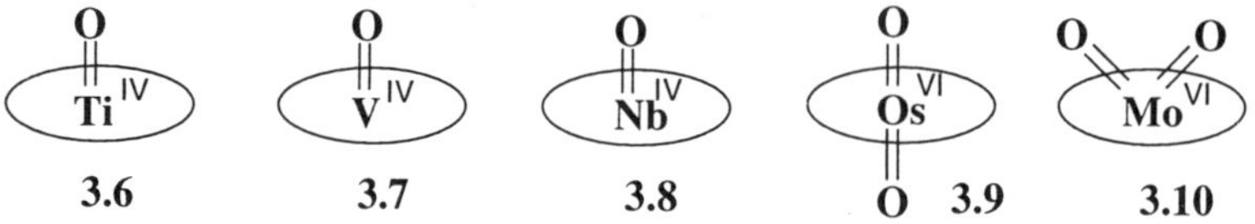

3.6 **3.7** **3.8** **3.9** **3.10**

Fig. 3.5 Some stable high-valent oxometalloporphyrin species.

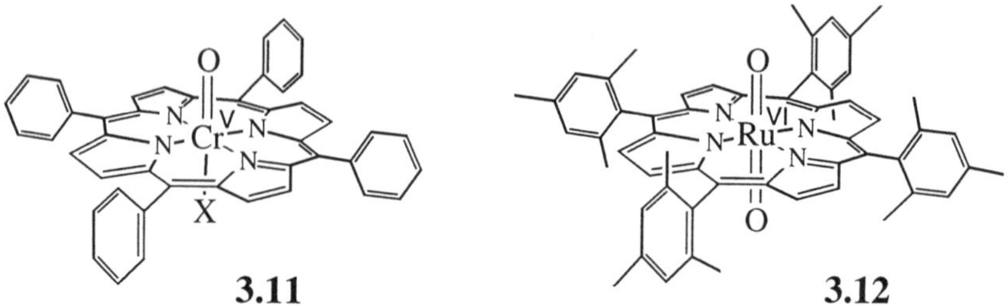

3.11 **3.12**

Fig. 3.6 Oxochromium(V) and *trans*-dioxoruthenium(VI) porphyrin complexes.

3.2.2 Oxoruthenium porphyrins

A *trans*-dioxo-ruthenium(VI) porphyrin complex (**3.12**; Fig. 3.6), coupled with molecular oxygen, has been demonstrated as the first genuinely aerobic metalloporphyrin epoxidation system without the need for a co-reductant.[43a] The reader is referred to important early work in this field by Dolphin and James.[44] More recently, ruthenium porphyrins showed promise for the activation of N_2O as the oxidant to epoxidize olefins, which would convert an industrial waste gas and an environmental hazard into useful products.[43b] A key observation in the turnover of this system was the *simultaneous*

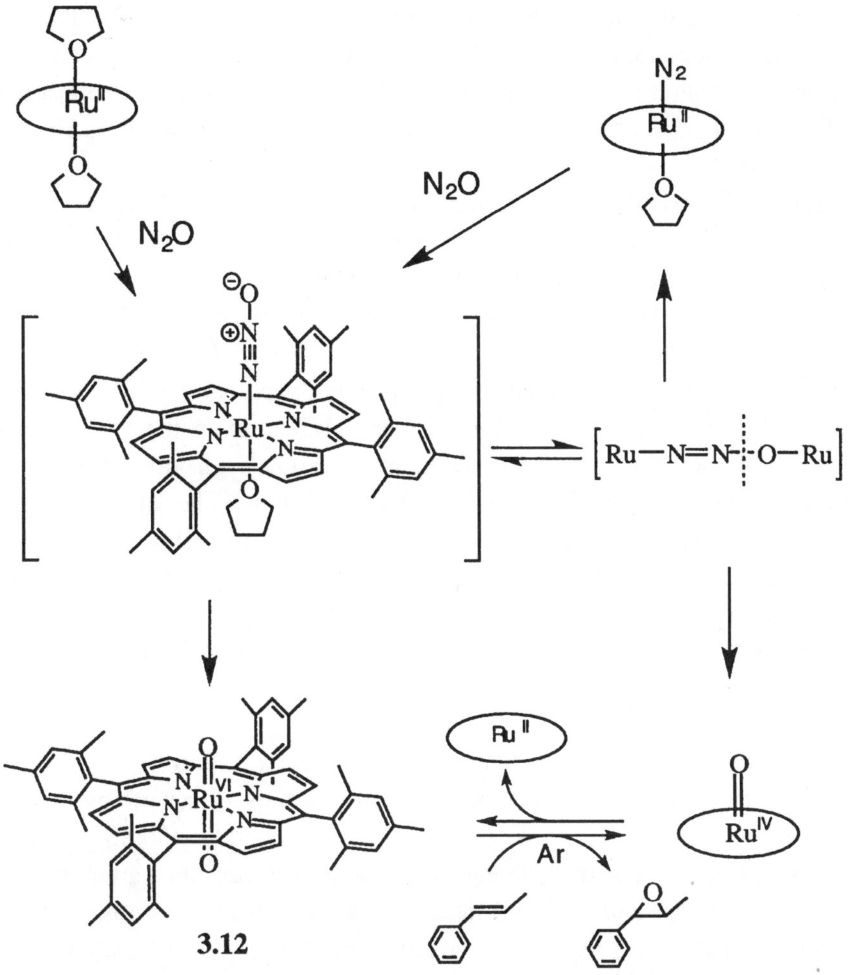

Scheme 3.4 Activation of N_2O by a ruthenium porphyrin.

appearance of oxoruthenium(IV) and dinitrogen ruthenium(II) intermediates, indicating that the cleavage of N_2O was bimolecular in ruthenium porphyrin as shown in Scheme 3.4. Thus, the mechanism proposed involves the initial binding of N_2O to ruthenium(II) through the nitrogen. Apparently, this adduct is a reactive species since it was not observed as a precursor of the cleavage products. Oxygen transfer from the rutheium(II)-dinitrogen (Ru(II)-N_2O) adduct to a second ruthenium(II) porphyrin would afford the observed products. The alternative binding of N_2O through its oxygen and the ejection of *free* dinitrogen is considered less likely since the Ru(II)-N_2O adduct is unstable to the loss of N_2 under the reaction conditions.

High activities have been found for **3.12** with 2,6-dichloropyridine-*N*-oxide (pyCl_2NO) as the oxygen donor [45] in the presence of HCl or HBr. An oxo,chlororuthenium(VI) intermediate was suggested for this system by Hirobe, *et al.*[45] Remarkably, carbonyl (5,10,15,20-tetrapentafluorophenylporphyrinato)-ruthenium(II) (Ru(II)TPFPP(CO), **3.13**) was shown to give extraordinary turnover rates approaching 1000 min^{-1}.[46] Hydroxylations of alkanes, olefin epoxidations and even the oxidation of benzene were achieved with high selectivity, high rates and high efficiency with **3.13** under mild, aprotic conditions. The rapid catalysis with this ruthenium porphyrin system was accomplished by a new and fast catalytic cycle involving the Ru(V)-Ru(III) couple (Scheme 3.5). The mechanism of action clearly deviated from the conventional and slower dioxoRu(VI)-oxoRu(IV) couple, because pyCl_2NO was incapable of oxidizing Ru(IV) to Ru(VI) rapidly under the conditions described.[46]

The kinetics of product evolution in a typical reaction of adamantane hydroxylation showed an initial induction period followed by a fast, apparently zero order phase with the maximum rate and the highest efficiencies. Deviation from linear behavior took place only after 90% of the oxygen donor and 80% of the substrate had been consumed. When Ru(VI)(TPFPP)$(O)_2$, prepared by reaction of Ru(II)(TPFPP)(CO) with 3-chloroperoxybenzoic acid, was used as the catalyst in air, no induction time was detected and zero order kinetics were observed as well. The induction period returned in the absence of oxygen. The well-defined and characteristic UV-vis spectra of metalloporphyrins provided an invaluable tool for mechanistic studies. Thus, monitoring the state of the metalloporphyrin catalysts during the course of catalytic oxygenation by UV-vis spectroscopy revealed that the initial form of the catalyst remained the predominant one throughout the oxidation, i.e. in the Ru(II)(TPFPP)(CO)-catalyzed reaction, at least 90% of the porphyrin catalyst existed as Ru(II)(TPFPP)(CO) and in Ru(VI)(TPFPP)$(O)_2$-catalyzed reaction, more than 90% of the catalyst was still in the form of Ru(VI)(TPFPP)$(O)_2$ despite the high turnover numbers reached (~ 400 TO). The fact that Ru(II)(TPFPP)(CO) demonstrated a similar and even higher maximum turnover rate of 4.9 TO/min in adamantane hydroxylation versus 4.0 TO/min for Ru(VI)(TPFPP)$(O)_2$

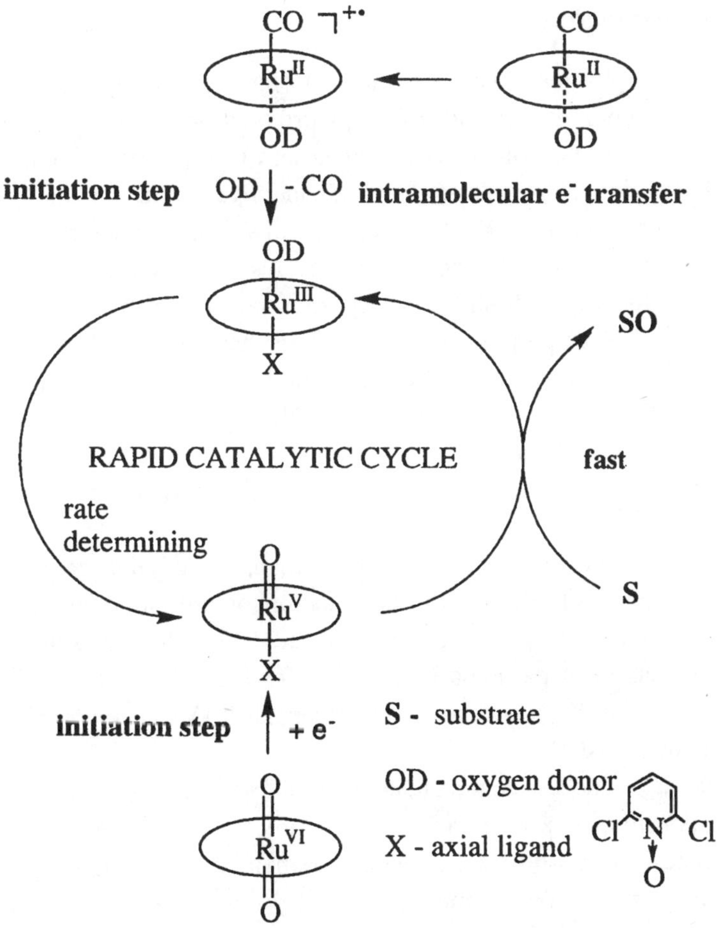

Scheme 3.5 Fast catalytic oxygenation with a ruthenium porphyrin, Ru(II)TPFPP(CO).

under the same conditions indicated that an active catalyst species other than Ru(VI)(TPFPP)$(O)_2$ or Ru(II)(TPFPP)(CO) was involved in this fast catalytic hydroxylation. The proposed mechanism for this interesting catalytic system is described below (Scheme 3.5). The induction periods observed represent the time it takes to generate a low, steady state concentration of ruthenium(III), the epr spectrum of which was detected.

3.2.3 Oxomanganese porphyrins

Manganese porphyrins have been shown to have unusually high reactivity toward olefin epoxidation and alkane hydroxylation.[47] Mounting evidence suggests the existence of an oxomanganese(V) porphyrin complex in the reactions of olefin epoxidation and alkane hydroxylation catalyzed by manganese porphyrins.[47–53] However, the physical characteristics of the putative oxomanganese(V) porphyrin species has remained particularly elusive,[54,55] due to its high reactivity and transient nature. By contrast, nitridomanganese(V) porphyrins (such as **3.14**) have been well-characterized (Fig. 3.7).[56–58] These complexes are stable and inert toward typical substrates, although *N*-acylation of **3.14** produced a reactive nitrene transfer agent.[59–62] Stable oxomanganese(V) complexes are few, the only examples involving the use of tetraanionic ligands to stabilize the high-valent manganese center (**3.15** and **3.16**).[63–65]

Structurally related to the porphyrins, manganese salen catalysts have shown wide applicability for the epoxidation of unfunctionalized olefins. These systems have been thoroughly reviewed.[66] First described by Kochi,[66a] this system has been particularly effective for the asymmetric epoxidation of prochiral olefins.[66b–c] Recently, an oxomanganese(V) salen complex (**3.13**) has been identified.[66]

Despite these successes in characterizing nitridomanganese(V) porphyrins and non-porphyrin oxomanganese(V) complexes, characterization of an oxomanganese(V) porphyrin species has remained a challenging task. The intermediacy of reactive oxomanganese(V) porphyrin complexes has long been implicated in olefin epoxidation and alkane hydroxylation because of the distinct reactivity patterns and $H_2{}^{18}O$-exchange behavior,[22,54,55,67–68] as

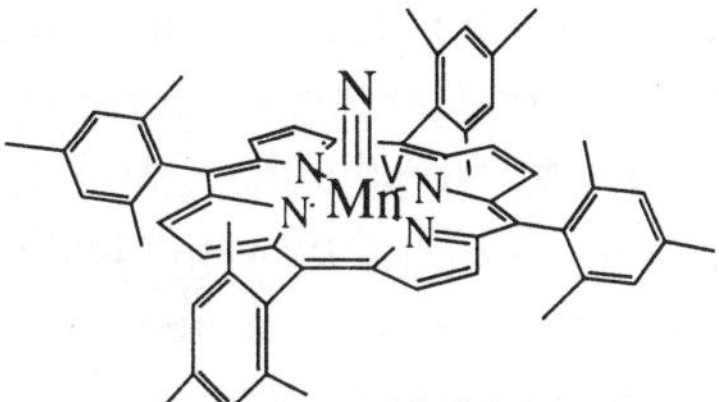

3.14

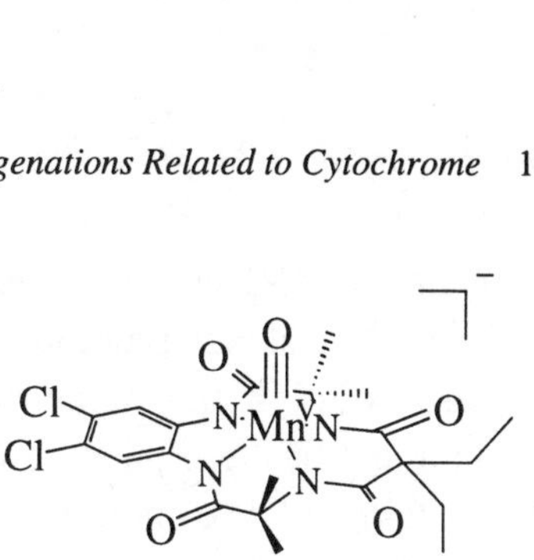

3.15

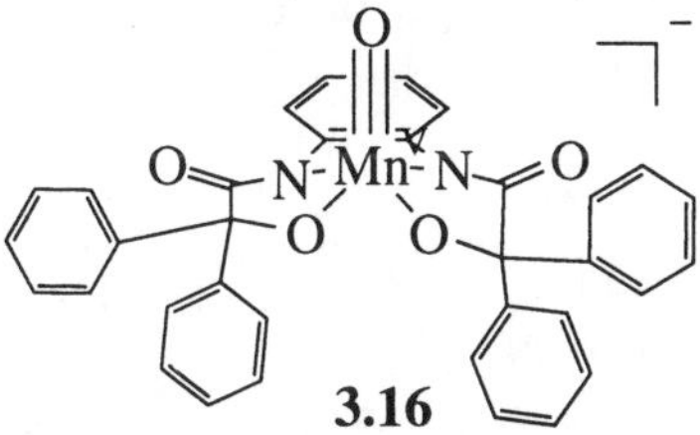

3.16

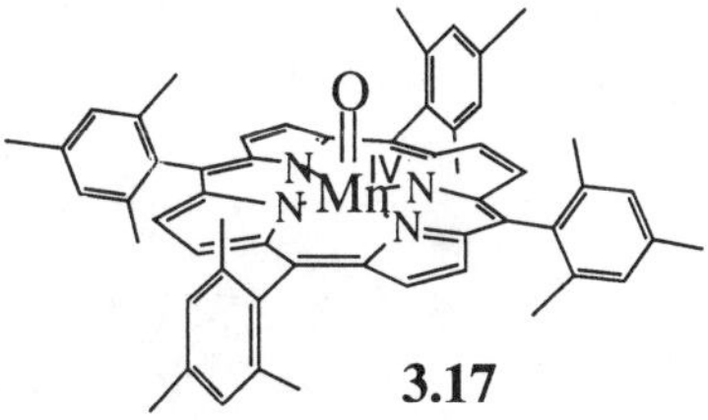

3.17

Fig. 3.7 Nitridomanganese(V) and oxomanganese(IV) porphyrin complexes, and two stable non-porphyrin oxomanganese(V) complexes.

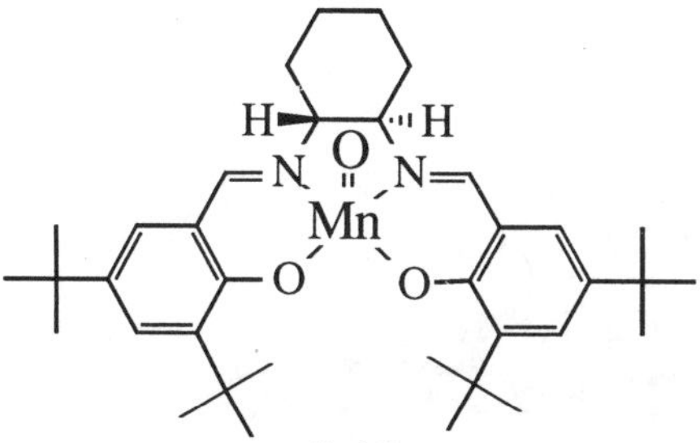

3.13

compared to that of the relatively stable oxomanganese(IV) porphyrin (**3.17**) intermediates which have been isolated and well-characterized.[55,70] Oxomanganese(IV) porphyrin complexes transferred oxygen to olefins with little stereoselectivity, while a transient oxomanganese(V) complex underwent oxygen transfer to olefins with predominant retention of configuration (Scheme 3.6).[55,71] Further, the oxomanganese(IV) species exchanged the oxo ligand with water slowly while the positively-charged oxomanganese(V) complex readily exchanged the oxo ligand with added ^{18}O water.[55] Characterization of an authentic oxomanganese(V) porphyrin species would help in the understanding of the mechanism and the development of these potentially useful catalysts. Indeed, the first detection of a reactive oxomanganese(V) intermediate using rapid-mixing stopped-flow techniques under ambient, catalytic conditions was described recently. A direct assessment of its reactivity in both one-electron and oxygen transfer processes was made possible by these observations.[72]

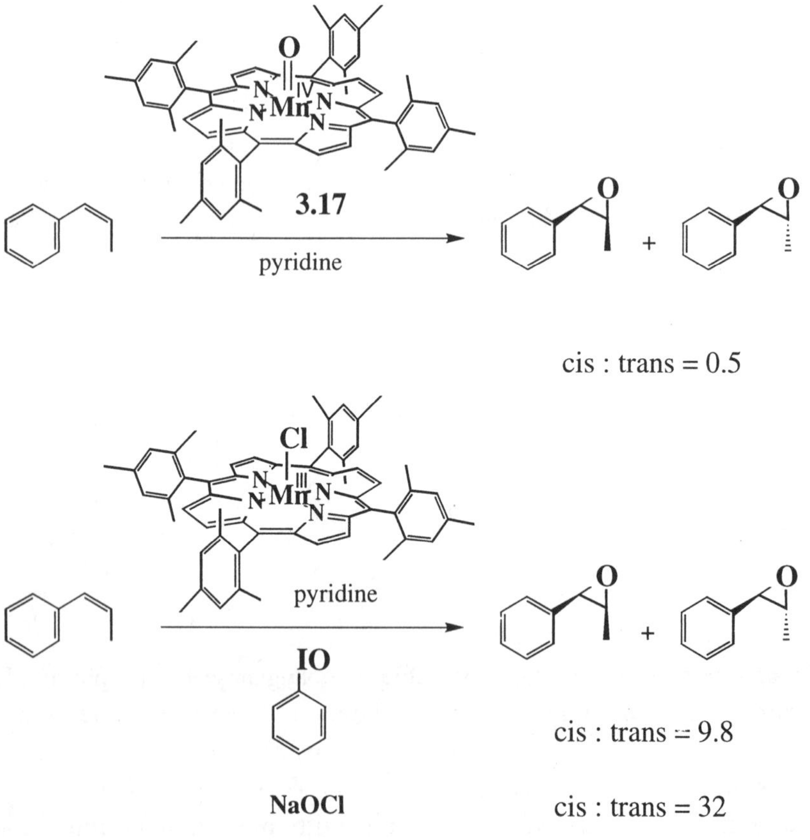

Scheme 3.6 Reactivity and stereoselectivity of oxomanganese(IV) and oxomanganese(V) porphyrins (generated by Mn(III) and oxidants *in situ*) in olefin epoxidation.

The reaction of tetra-N-methylpyridylporphyrinatomanganese(III) [Mn(III)TMPyP] with a variety of oxidants, *m*-CPBA, HSO_5^-, and ClO^-, was shown to produce the same, short-lived intermediate. An oxoMn(V) porphyrin structure was assigned to this intermediate. The Soret maximum of oxoMn(V) was found at 443 nm, intermediate between that of oxoMn(IV) (428 nm) and Mn(III)TMPyP (462 nm), thus facilitating its detection. The rate of formation of oxoMn(V) from Mn(III)TMPyP followed second order kinetics, first order in Mn(III) porphyrin and first order in oxidant. The rate constants have the following order: *m*-CPBA (2.7×10^7 $M^{-1}s^{-1}$) > HSO_5^- (6.9×10^5 $M^{-1}s^{-1}$) ~ ClO^- (6.3×10^5 $M^{-1}s^{-1}$). Once formed, the intermediate oxoMn(V) species was rapidly converted to oxoMn(IV) by one-electron reduction with a first order rate constant of 5.7 s^{-1}. The oxoMn(IV) species was relatively stable under the reaction conditions, decaying slowly to Mn(III)TMPyP. The identity of the intermediate as an oxomanganese(V) complex was indicated by its reactivity. The one-electron reduction of oxoMn(V) to oxoMn(IV) was greatly accelerated by nitrite ion ($k = 1.5 \times 10^7$ $M^{-1}s^{-1}$). However, the reaction between nitrite and oxoMn(IV) was much slower. The oxoMn(V) intermediate was shown to be highly reactive towards olefins, affording epoxide products. By contrast, oxoMn(IV) was not capable of effecting the same reaction under these conditions. In the presence of carbamazepine, a water-soluble olefin, efficient oxygen transfer from the highly reactive oxoMn(V) to the olefin resulted in the conversion of oxoMn(V) directly back to Mn(III)TMPyP. This oxygen transfer was extraordinarily rapid with a second order rate constant of 6.5×10^5 $M^{-1}s^{-1}$. With *m*-CPBA as the oxidant in the presence of $H_2{}^{18}O$, the product epoxide was shown to contain 35% ^{18}O, consistent with an O-exchange-labile oxoMn(V) intermediate. Nitrite ion inhibited the epoxidation reaction competitively by one electron reduction of the oxoMn(V) intermediate to the unreactive oxoMn(IV). Thus, it was possible to show that the oxo-aqua exchange rate was about 10^3 s^{-1} for the *coordinated water*, a process which occurred by prototropy in accord with the observations of Meunier.[68] Significantly, bulk water exchange was slower than oxo-transfer to the olefin under these conditions. These simple facts explain a body of often confusing data and claims in the literature

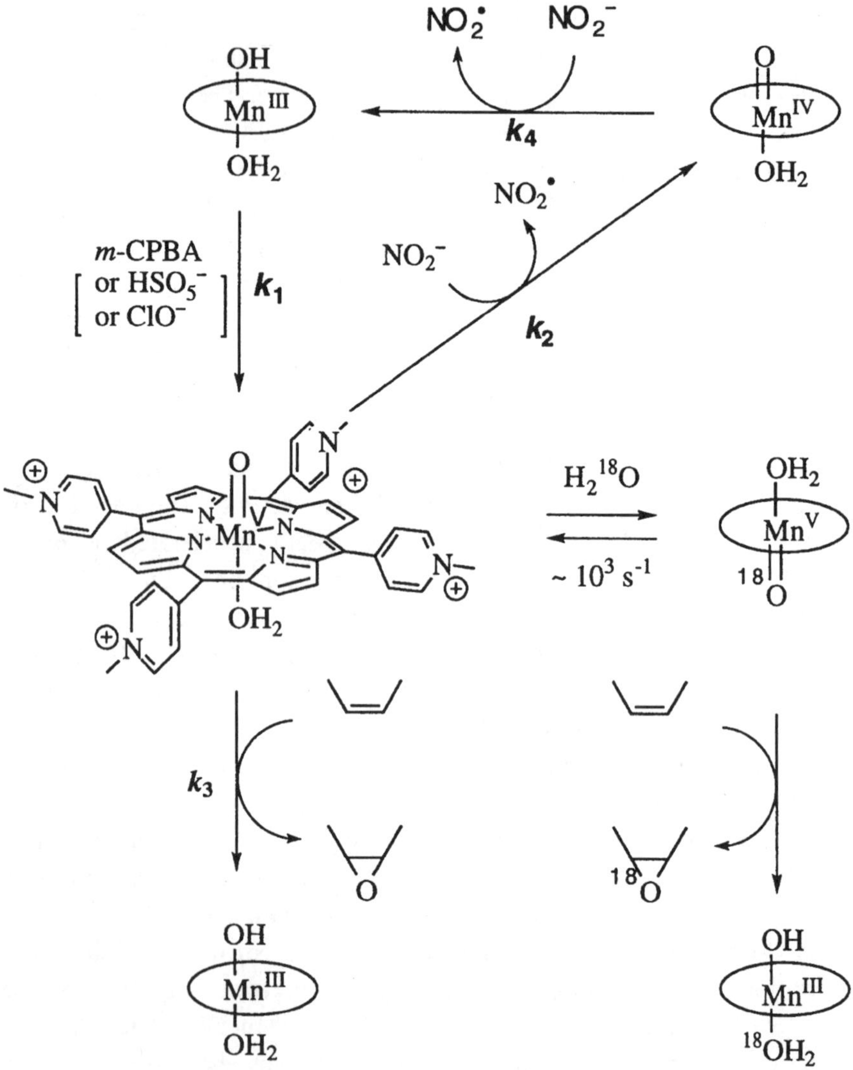

Scheme 3.7 (a) Rate constants: $k_1 = 2.7 \times 10^7$ $M^{-1}s^{-1}$; $k_2 = 1.5 \times 10^7$ $M^{-1}s^{-1}$; $k_3 = 6.5 \times 10^5$ $M^{-1}s^{-1}$; $k_4 = 1.4 \times 10^2$ $M^{-1}s^{-1}$; ^{18}O exchange ~ 10^3 s^{-1}. (b) ^{18}O incorporation: 35% ^{18}O labeling from the solvent (95% $H_2{}^{18}O$) into the epoxide.

of this field. Peroxyacids will not exchange the peroxo oxygen with water at all and neither does the product epoxide. Thus, in those cases for which ^{18}O-incorporation is observed into the products, an oxo-metal species is strongly implicated as the oxygen donor. However, failure to see ^{18}O-exchange is not a definitive result since the exchange rate may simply be too slow to compete with oxygen atom transfer. These transformations and the ^{18}O-exchange process are summarized in the Scheme 3.7 below.

3.2.4 *On the mechanism of hydroxylation by cytochrome P450 and related model systems.*

The cytochrome P450-catalyzed hydroxylation of hydrocarbons and those of model iron and manganese porphyrins are considered to occur by a mechanism involving hydrogen atom abstraction from the substrate (R-H) followed by rapid transfer of the metal-bound hydroxy radical to an intermediate caged alkyl radical (R·) (Scheme 3.8).[3] The intermediate radical can be trapped in the model reactions by bromotrichloromethane. This so-called *Oxygen rebound* mechanism[18a,b] is consistent with the stereochemical, regiochemical and allylic scrambling results observed in the oxidation of

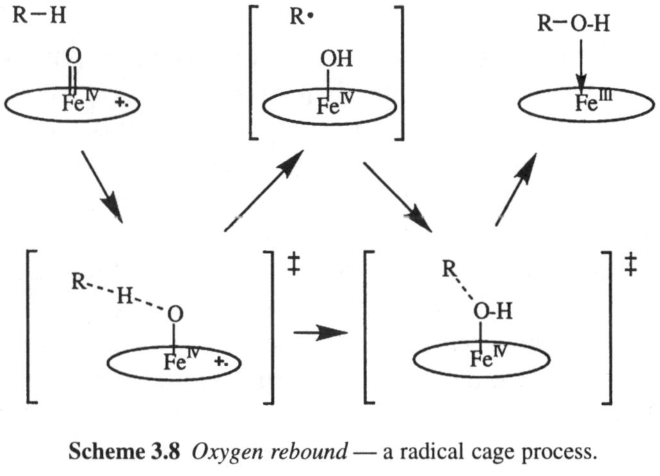

Scheme 3.8 *Oxygen rebound* — a radical cage process.

norbornane, camphor and cyclohexene by cytochrome P450, which are characterized by *partial* positional or stereochemical scrambling. Significantly, simple metalloporphyrins gave similar results.[3,17c] The results rule out *freely* diffusing radicals, for example, and have been reviewed in detail elsewhere.[3] The reader is also referred to Chap. 2 in the monograph by Ingold and MacFaul which considers carefully and critically the methods to discriminate between authentic, biomimetic oxygenations and reactions of freely diffusing radicals. Thermodynamic effects in such reactions are examined in Chap. 1 by Mayer.

The "radical clock" method[69a] has been applied effectively to alkane hydroxylations using P450 and related model systems to measure the rate of the rebound process.[69b] For example, bicyclo[2.1.0]pentane was oxidized by rat liver microsomes to a 7:1 mixture of *endo*-2-hydroxybicyclo[2.1.0]pentane and 3-cyclopenten-1-ol. The rate constant for the rearrangement of bicyclo[2.1.0]pent-2-yl radical to 3-cyclopenten-1-yl radical was determined to be 2.4×10^9 s^{-1} at room temperature by using laser flash photolysis techniques.[69c] Thus, a rate constant of $k_{OH} = 1.7 \times 10^{10}$ $M^{-1}s^{-1}$ was estimated for the rebound process, more rapid than many typical conformational and configurational rearrangements. The numerous reported cases of retention of configuration and stereochemistry in cytochrome P450-catalyzed hydroxylations, and in particular, the remarkable regioselectivity observed in the hydroxylation of bicyclo[2.1.0]pentane (which undergoes abstraction and rebound only at the *endo* face) can be explained in these terms.

Some "fast" radical clocks have shown *less* rearrangement than slower clocks in these P450 and model metalloporphyrin-catalyzed oxygenations.[69d] A radical cage process can accommodate results of this type. As shown in Scheme 3.8, the transition state for hydrogen abstraction will position the active oxygen only a few tenths of an Ångstrom *farther* from the hydroxylated carbon atom than the transition state for the ultimate C–O bond formation. Thus, the extent of radical rearrangement would be expected to depend critically on the *tightness* of the radical cage and the ensemble of steric forces experienced by the incipient radical within the cage. The extent of rearrangement detected by a particular probe may simply reflect a facile molecular trajectory from the hydrogen abstraction transition state to the hydroxylation transition state. Indeed, in cases for which the radical is produced within the attractive potential well of the new C–O bond, the timing of radical rearrangements, and thus the timing of the radical clocks, are likely to be affected. Newcomb has termed this scenario a *non-synchronous concerted* pathway.[69d] Interestingly, the effect is most pronounced for hydroxylation at the C–H bond of methyl groups, with a very strong C–H bond and a small steric size. Both effects would push the reaction coordinate toward a tighter radical cage.

A clear case of differential reactivity of superficially similar radical intermediates was observed in the hydroxylation of ethyl benzene by a chiral iron porphyrin (Fig. 3.8).[70b] The intrinsic enantioselectivity for the

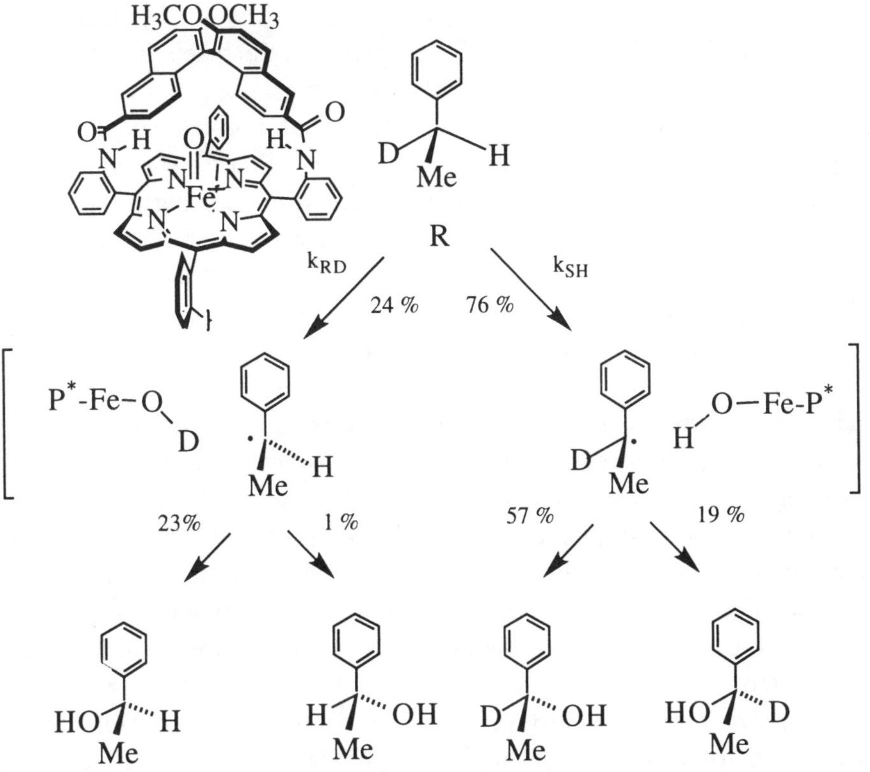

Fig. 3.8 *Oxygen rebound* and differential stereochemical outcomes with a chiral iron porphyrin.

enantiotopic protons of ethyl benzene was shown to be modest, about 2:1. However, the hydrogen-deuterium inventory of the products showed that while the benzylic radical produced by abstraction of the pro-*R* deuterium rebounded to produce the alcohol (with 95% *retention* of configuration at carbon), significant *racemization* was observed following abstraction of the pro-*S* hydrogen. Clearly, the same carbon radical was produced in both cases, but the effective *lifetime* of the pro-*R* radical, as revealed by the very small amount of rearrangement, must be shorter than that of the pro-*S* radical due to the asymmetry of the steric environment.

There may well be other, subtle and interesting features of these radical cage hydroxylation reactions that are in need of more experimental data and better fundamental understanding. One of them could be interactions of the incipient carbon radical with the π-cloud of the porphyrin ring. The effect is similar to the solvent effects observed on the reactivity of chlorine atoms.[69e] Another effect is molecular dynamics. Recent evidence was presented by Carpenter for the interception of non-statistical reactive trajectories in a singlet biradical ring closure reaction.[69f] The existence of two populations for the biradical intermediate with differing lifetimes was invoked to explain the unusual stereochemical outcome for the closure of cyclopentane 1,3-diyl to bicyclopentane. The effect was attributed to dynamic energy redistribution effects which are incompletely randomized in the sub-picosecond time regime. This is the time scale of the fastest radical clocks and biradical reactions of this type have obvious analogies to the radical cage process invoked for these hydroxylation reactions.

Another effect recently suggested by Schwarz focuses on spin-orbit coupling effects and the spin states of reacting intermediates.[69g] Thus, the efficiency of crossing between high- and low-spin states along the hydroxylation reaction trajectory could affect the rate of *oxygen rebound* and, accordingly, the lifetime of any radical intermediate. While such effects are common for first row elements, "spin forbiddenness' is usually discounted for organometallic reactions. However, while intersystem crossing rates in such systems are fast, $> 10^5\ s^{-1}$, and CO trapping rates by a triplet cobalt(II) complex to afford a singlet Co(II) carbonyl has a rate constant of $10^9\ M^{-1}s^{-1}$,[69h,i] these rates are still *slow* compared to the time scale of the fastest radical clocks. Thus, spin state effects such as this may have to be considered to have a full view of *oxygen rebound.*

Unusual effects on the rate of cyclohexane oxidation by the system of chloropentafluorophenylporphyrinato)-iron(III) [Cl-Fe(III)TPFPP] and iodosylbenzene were reported upon changing the head gas.[69j] Thus, the reaction rate slowed perceptibly when methane or hydrogen was used instead of argon at modest partial pressures (0.3–1.0 atm). Overall relative rate changes were less than two-fold over this range of pressure, and hydrogen was more effective than methane. The authors suggested a retarding effect due to reversible interception of the reactive iron-oxo intermediate by methane or hydrogen, even though neither reacted signifcantly under these conditions. Other explanations exist, however. The head gas might have a weak interaction with the resting Fe(III) porphyrin which is the most prevalent form of the catalyst inventory during turnover. Further, the head gas might interfere with the rate-limiting reaction of iodosylbenzene with the metalloporphyrin in some subtle way. More data will be needed to explain this curious result.

3.3 Metalloporphyrins as Detectors and Decomposition Catalysts for Peroxynitrite, an Important Biological Oxidant

The fast reactions between metalloporphyrins and oxidants provide the foundation for their use in catalyzing oxygenations of a variety of organic substrates.[3] The rapid rates of these reactions and the intermediacy of observable oxometalloporphyrin species, especially in the reactions of manganese porphyrins with various oxidants (*m*-CPBA, NaOCl, $KHSO_5$),[72] have also inspired the use of these porphyrins as detectors for a transient biological oxidant, peroxynitrite ($ONOO^-$). $ONOO^-$, a strong one- and two-electron oxidant, has been implicated in a number of pathological conditions.[73] Structurally, $ONOO^-$ is the conjugate base of peroxynitrous acid and, like other similar peroxides, it rapidly oxidizes manganese porphyrins.[67,74,75] Thus, its reactions with manganese porphyrins to generate oxomanganese intermediates allowed the sensitive detection of this transient endotoxic molecule under cell-like conditions.[67,74] Using manganese porphyrins as $ONOO^-$ detectors, it was demonstrated that $ONOO^-$ can quickly diffuse across biological membranes and react with its biological targets.[74]

The fast oxidation of manganese(III) porphyrins to oxomanganese(IV) species by $ONOO^-$ satisfies the prerequisite for these porphyrins to be efficient $ONOO^-$ decomposition catalysts.[74] Indeed, manganese porphyrins, redox-coupled with biological antioxidants (such as ascorbate, glutathione, Trolox®) or superoxide ion, rapidly reduce $ONOO^-$ and thus, prevent the oxidation and nitration of biological substrates by this toxic oxidant[75,76] (Scheme 3.9). Amphiphilic analogs of the water-soluble metalloporphyrins[75] that are suitable for liposomal delivery have also been developed. Further, these amphiphilic metalloporphyrins in sterically stabilized liposomes are highly active in decomposing $ONOO^-$,[77] and thus are

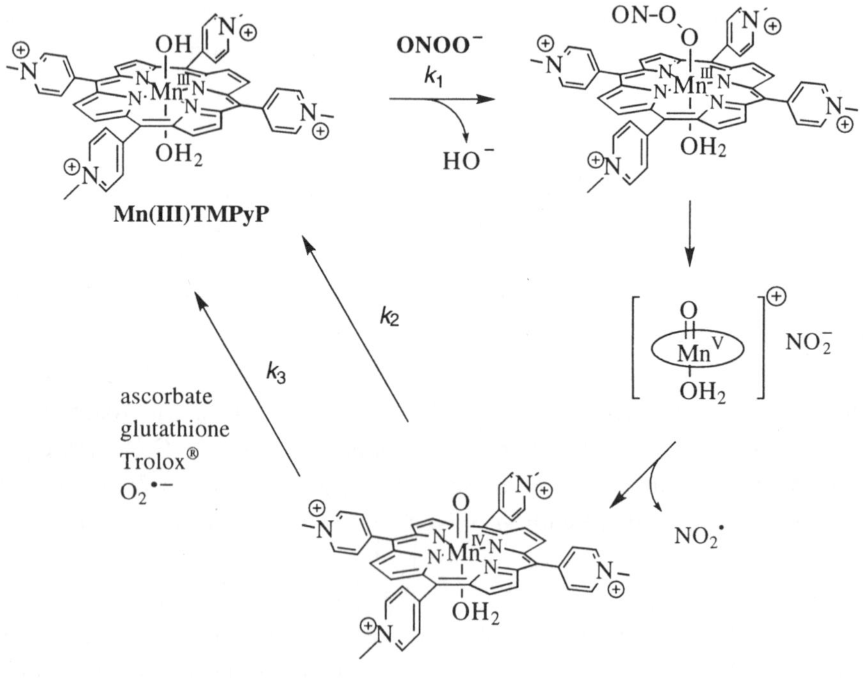

Scheme 3.9 Reaction of peroxynitrite with Mn(III)TMPyP.

potential therapeutic agents for the treatment of $ONOO^-$-related diseases. Iron(III) porphyrins are also rapidly oxidized by $ONOO^-$, and Stern *et al.* reported the significant discovery that water-soluble iron porphyrins have the ability to decompose $ONOO^-$ by isomerizing it to nitrate.[78a] These compounds, moreover, have shown significant biological responses such as reducing edema in animals.[78b] Presently, a consensus has yet to be reached about how these iron porphyrins catalyze the isomerization of $ONOO^-$. The intermediacy of iron(IV) is implicated[78a] but the precise details appear to depend on the structure of the porphyrin.

3.4 Synthetic Metalloporphyrins as Selective Catalysts and Building Blocks of Supramolecular Assemblies

Understanding the mechanism of cytochrome P450 through the study of synthetic model systems offers an opportunity to develop practical regioselective and stereoselective catalysts. Such catalytic systems have been extensively surveyed in recent reviews,[3,79] which have shown that the stereoselectivity of metalloporphyrin-catalyzed oxidations is sensitive to the metal, the porphyrin ligand, its superstructure, the sixth axial ligand, and the host matrix. Space does not permit a discussion of all the elegant catalysts developed, but a few systems are shown here as examples.

By analogy to the natural enzymes which utilize a protein scaffold to effect substrate recognition and stereoselectivity, special steric features have been introduced into the synthetic porphyrin catalysts to achieve regio- and enantio-selective oxidation. The success of such systems relies on the steric interactions between the substrates and porphyrin catalysts, which position the substrates specifically toward the reactive metal-oxo center. Breslow *et al.*[80] recently reported a remarkably selective, catalytic steroid hydroxylation using an artificial cytochrome P450 enzyme. The synthetic strategy to induce selectivity in the model system was the attachment of four cyclodextrin appendages to a synthetic manganese porphyrin used in place of the heme center of the natural enzyme. These donut-shaped heptamylose sugars have a hydrophobic central cavity which is known to bind aromatic molecules. The substrate steroid is modified with such an aromatic group at either end of the molecule. The host-guest complex obtained from these designed partners (**3.18**) displays a limited region of the substrate steroid in the vicinity of the catalytic manganese center. What is most significant about this artificial enzyme is the fact that hydroxylation occurs *only* at carbon 6 to give the 6α-hydroxysteroid, even though there are many sites in this molecule with similar intrinsic reactivity. This high selectivity is found to depend critically on the precise arrangement of the aromatic groups of the substrate. Moreover, the manganese porphyrin-cyclodextrin construct is able to release the product sterol and hydroxylate a number of steroid molecules. Catalytic turnover with such high positional selectivity is highly unusual for this kind of model system.[81]

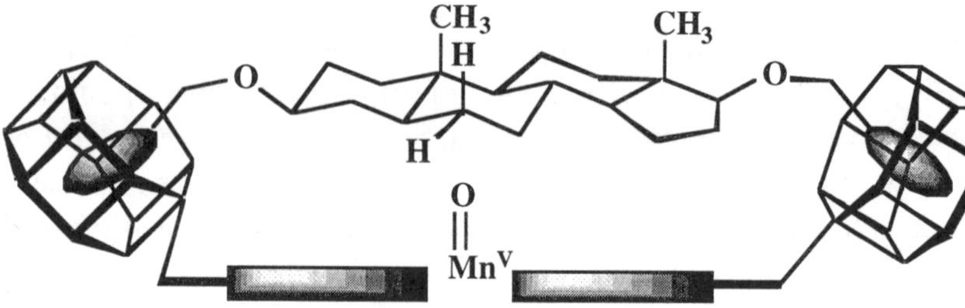

3.18

The first use of a chiral porphyrin (**3.19**; Fig. 3.9) to carry out asymmetric epoxidation was reported in 1983, giving 50% *ee* with *p*-chlorostyrene.[84] The *ee* was improved to ~70% for epoxidation of *cis*-β-methylstyrene with the use of a very robust chiral vaulted binaphthyl porphyrin (**3.20**; Fig. 3.10).[85,86] More significantly, this catalyst afforded the first case of catalytic asymmetric *hydroxylation* by a model system, giving a ~70% *ee* for hydroxylation of ethylbenzene and related hydrocarbons (see discussion in Sec. 3.2.4 above).[85,86]

Recent developments of this system have led to studies of the binaphthyl-peptide-strapped porphyrin (**3.21**).[86b] For styrene epoxidation, this chiral catalyst afforded an *ee* greater than 90% in the initial stages of the reaction, with the (*R*)-styrene oxide predominating. NMR T_1 relaxation studies with the copper(II) derivative of the same ligand showed that it was the protons

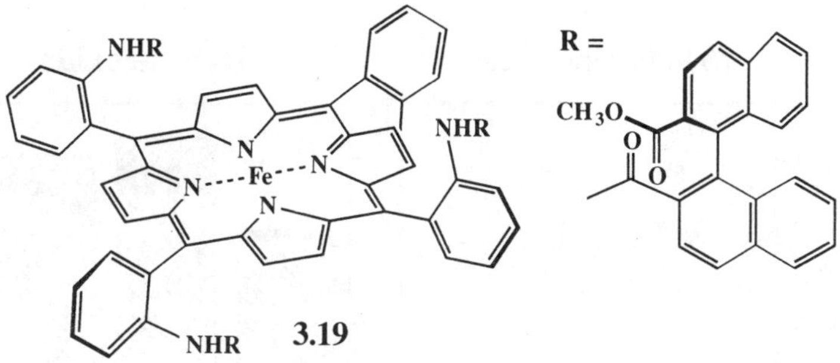

3.19

Fig. 3.9 The first chiral porphyrin designed for asymmetric epoxidation of olefins.

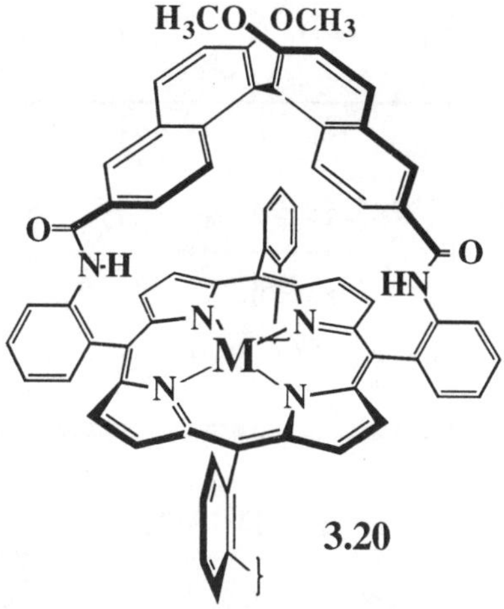

3.20

Fig. 3.10 A chiral, vaulted binaphthyl metalloporphyrin.

of major enantiomer of the product epoxide that were more efficiently relaxed in an inversion-recovery experiment with **Cu-3.21**. Comparative T_1 data are shown in Fig. 3.11. Accordingly, the major product must fit better into the chiral cavity of this catalyst. Thus, one can conclude that the enantioselectivity observed can be attributed to a simple lock and key process in which the transition state for epoxidation must look rather like the coordinated epoxide.

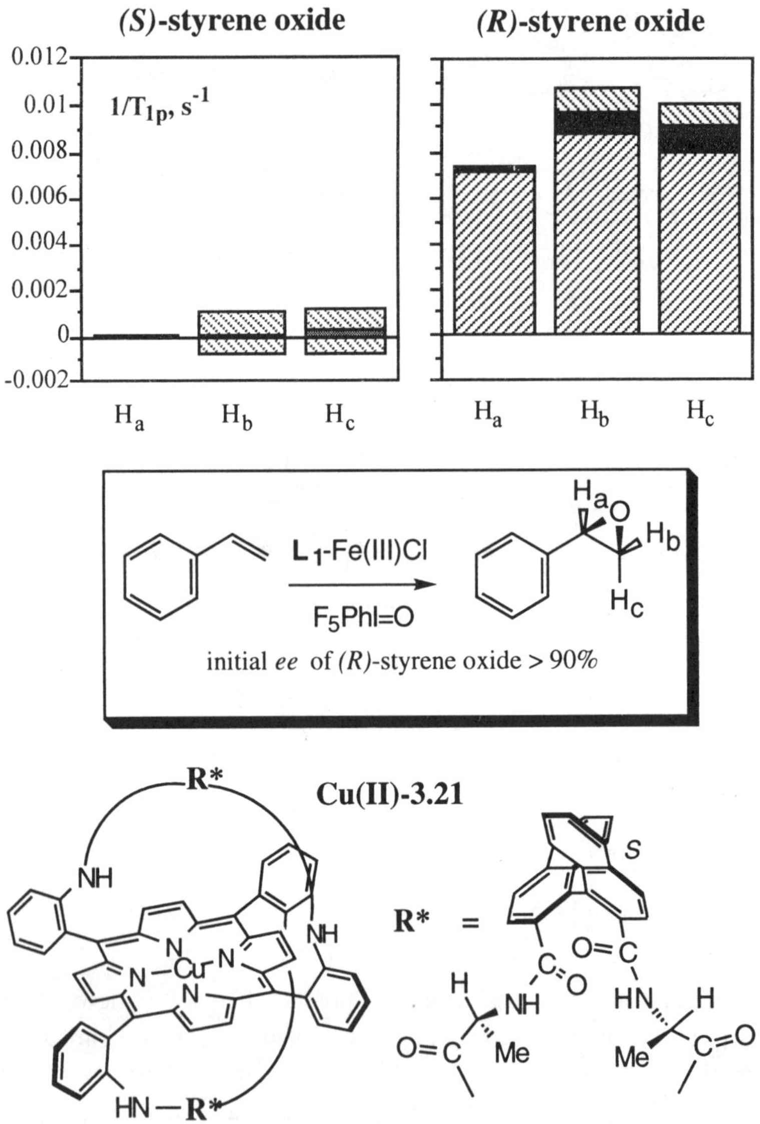

Fig. 3.11 Comparison of ^{1}H NMR relaxivity and epoxidation *ee*.

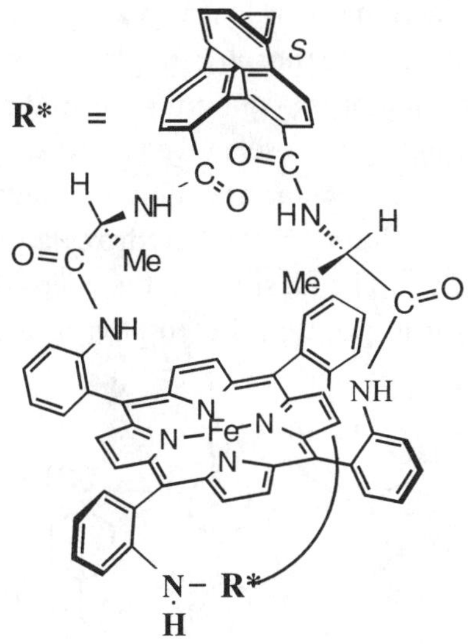

Fe-3.21

Collman *et al.* reported a series of "picket basket" porphyrins (**3.22**), which show extremely high shape selectivity (> 1000 for *cis*-2-octene versus *cis*-cyclooctene) and relatively high enantioselectivity in olefin expoxidation (*ee* around 70–85%).[82,83]

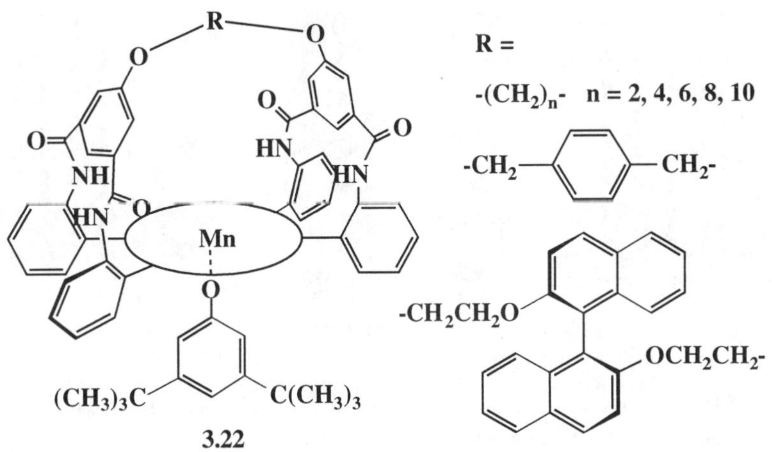

3.22

The synthesis of a steroidal porphyrin (**3.23**) with two steroid moieties appended to each side of the porphyrin plane was described.[87] Upon intercalation of this membrane-spanning porphyrin into a synthetic phospholipid bilayer, a highly regioselective catalytic system was assembled (Fig. 3.12). This catalytic vesicular assembly epoxidized diolefinic sterols exclusively at the side chain[87,88] and hydroxylated cholesterol at the C-25 tertiary carbon.[87,89] The design of the steroidal porphyrin (**3.23**) allowed the lipid bilayer to position the porphyrin in a spatially controlled

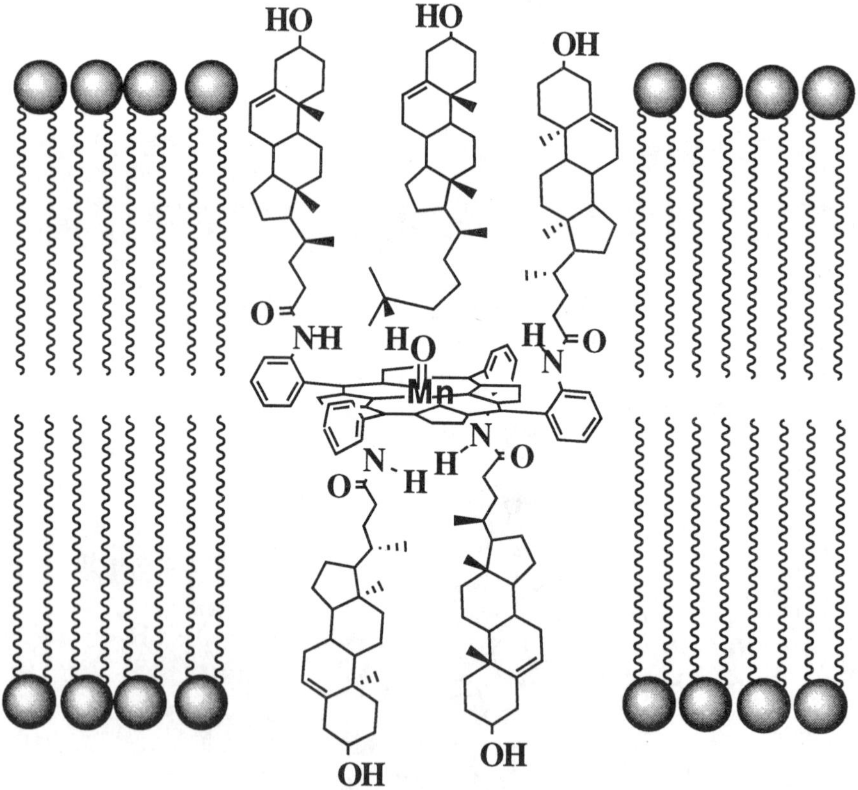

3.23

Fig. 3.12 Idealized molecular bilayer assembly of oxomanganese steroidal porphyrin, cholesterol, and dimyristoylphosphocholine.

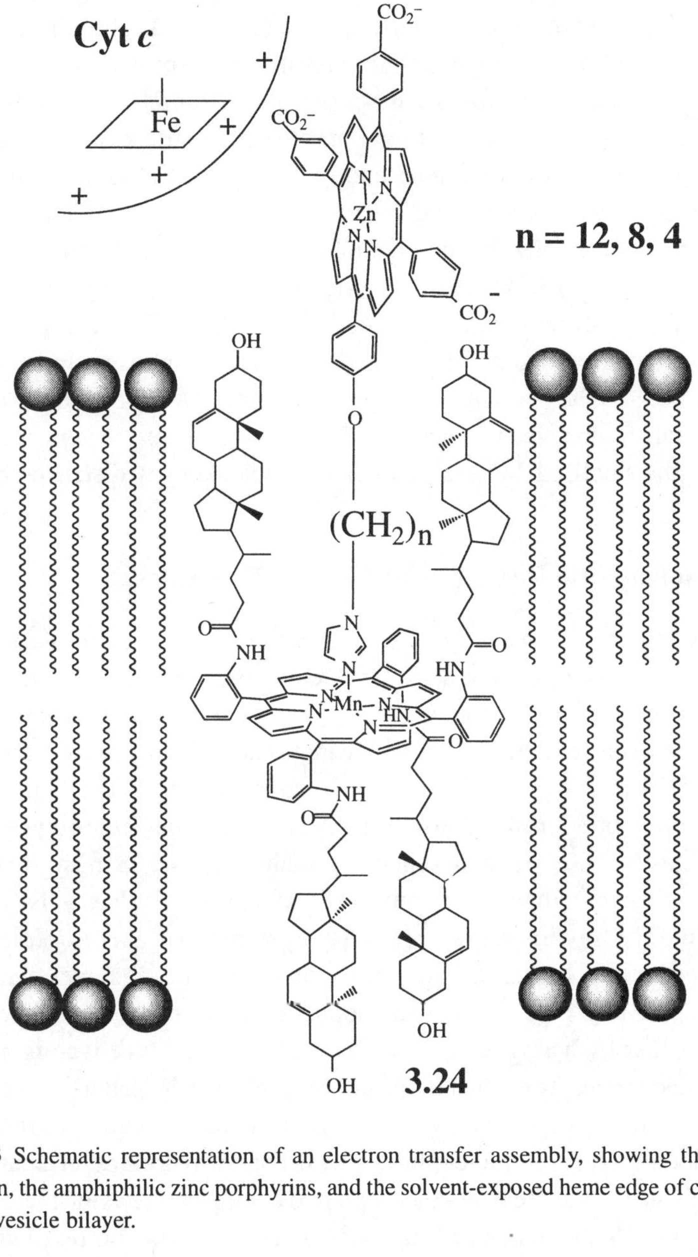

3.24

Fig. 3.13 Schematic representation of an electron transfer assembly, showing the steroidal porphyrin, the amphiphilic zinc porphyrins, and the solvent-exposed heme edge of cytochrome *c* in the vesicle bilayer.

manner, with the porphyrin macrocycle at the center of the membrane bilayer and parallel to the membrane surface (Fig. 3.12). The self-assembled and highly ordered nature of phospholipid membranes offered the opportunity to construct novel supramolecular assemblies, using specially designed porphyrins (e.g. the steroidal porphyrin) as structural and functional components. A multicomponent catalytic vesicular assembly has been described in which electrons derived from the enzymatic decarboxylation of pyruvic acid, mediated by an amphiphilic flavin, reduced a manganese(III) steroidal porphyrin (**3.24**) and thus effected subsequent oxygen activation and transfer.[90] More recently, a bis-heme assembly was reported that recruited cytochrome *c* to the outside surface of phospholipid vesicles (Fig. 3.13).[91,92] The structure of the construct was probed by observing electron transfer between cytochrome *c* and the redox centers embedded within a phospholipid bilayer at approximately known distances.

3.5 Nucleophilic Reactions of Metallo-Peroxides

Metal-peroxo species recently gained attention as nucleophiles capable of acyl transfer and related reactions. Mechanisms based on nucleophilic iron-bound peroxides were postulated for cytochrome P450 aromatase and nitric oxide synthase enzymes.[93,94] Meares reported site-specific hydrolytic cleavage of a peptide bond by an iron peroxo moiety.[95] The high stability of the amide bond toward hydrolysis is of crucial importance to the sustainability of life since it renders proteins robust. In neutral aqueous solution at room temperature, amide bond hydrolysis has a half-life of approximately seven years.[96] Living systems are able to selectively hydrolyze peptide bonds with a series of hydrolytic enzymes which are capable of rapidly cleaving the linkage. Metallopeptidases catalyze amide cleavage either by acting as a Lewis acid catalyst or by activating a metal-bound nucleophile. The most familiar example of the latter is the metal-hydroxide mechanism by which a metal-bound water molecule is deprotenated to form a nucleophilic metal-bound hydroxide at neutral pH. This mechanism has been modeled extensively and reviews of these efforts are available.[97] This section will describe active metal-bound peroxides with an emphasis on iron. The discussion begins with an analysis of peroxides as nucleophiles. Next, the structure and aqueous chemistry of iron and peroxide will be considered. From both heme and non-heme systems, the most salient examples of enzymatic mechanisms in which mononuclear iron peroxide species are active participants will then be presented. The biological activity of the cancer-fighting antibiotic bleomycin and the biopolymer cleavage abilities of some iron chelates in the presence of hydrogen peroxide will also be discussed. Model studies provide valuable information for understanding the mechanisms of metalloenzymes and for designing effective artificial metalloenzymes. The last two sections will overview both heme and non-heme mononuclear metal peroxides with an emphasis on those hat have shown nucleophilic activity.

3.6 Peroxide as a Nucleophile

Numerous studies have confirmed that the hydroperoxide anion is highly nucleophilic. In a comparison of the reactivity of hydroxyl ion and hydroperoxide ion toward benzonitrile, HOO^- was found to be 10^4 times more reactive than HO^-.[98] The rates of hydrolysis of *p*-nitrophenyl acetate with the peroxide anions HOO^- and CH_3OO^- were 10^2 times faster than with hydroxide.[99] The same rate difference was found with organophosphorus ester substrates.[100] Hydrogen peroxide is also capable of cleaving some activated amide bonds. Cleavage of acetylimidazole with hydroperoxide anion at neutral pH has a half-life of ~75 secs.[101] The mechanism for peroxide hydrolysis may proceed as is depicted in Scheme 3.10. Nucleophilic attack of a peroxide anion on an amide carbonyl forms a tetrahedral intermediate which breaks down into a peroxy acid and an amine.

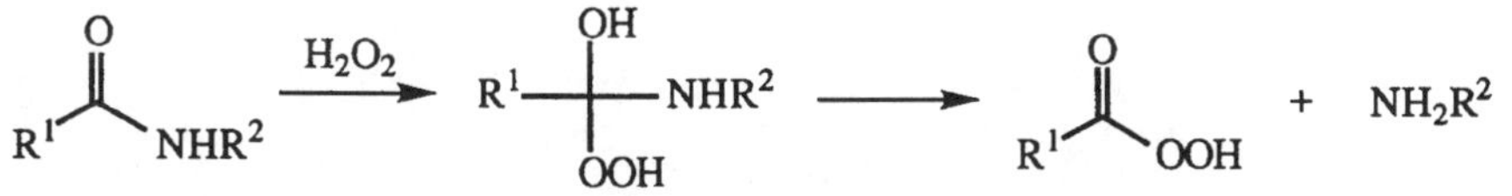

Scheme 3.10

The nucleophilicity of the hydrogen peroxide anion compared to hydroxide is attributed to the alpha effect.[10] The alpha effect arises from the presence of one or more pairs of unshared electrons adjacent to the nucleophilic atom. Due to the electron pair, the reactivity of these nucleophilic reagents, such as HOO^-, is therefore greater than would be expected from basicity criteria alone.

Analogously, superoxide has been shown to act as an effective nucleophile. In a reaction with superoxide, alkyl halides and tosylates were rapidly converted to alcohols with inversion of configuration.[103] Peroxo intermediates were implicated in the nucleophilic cleavage of both carboxylic and phosphate esters by superoxide.[104]

The alpha effect renders the hydroperoxy anion much more nucleophilic than hydroxide. Accordingly, a metal-coordinated peroxide is expected to be more nucleophilic than a metal-coordinated hydroxide.

The pK_a of H_2O_2 is 11.75.[105] Upon coordination of H_2O_2 to a metal, the pK_a of the proximal oxygen will be lowered by an amount comparable to the pK_a decrease found for metal coordination of water, about 6 to 8 pH units.[106] Deprotonation of the distal oxygen may or may not be necessary prior to nucleophilic attack. Based on the side-on complex formed by (EDTA)Fe^{3+} with hydrogen peroxide in alkaline pH,[107] the distal oxygen has an estimated pK_a of ~10. Therefore at neutral pH, an iron peroxide complex is expected to be formulated as Fe-OOH and should be an efficient nucleophile due to the alpha effect.

3.7 Iron Dioxygen Bonding Structure and Fenton Chemistry

Mononuclear iron peroxide compounds may be characterized by the dioxygen binding mode which is either end-on (η^1) or side-on (η^2):

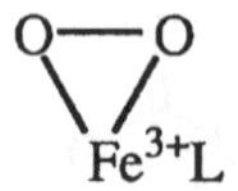

η^2 side-on

η^1 end-on

Both modes of coordination are known for dioxygen complexes of various transitions metals.[108] Ferrous hemoglobin is known to bind dioxygen in an (η^1) conformation.[109] A side-on peroxide structure was found for the Fe^{3+}(OEP)-peroxo complex.[110] EPR[111] and Raman[112] spectroscopic evidence for side-on binding of peroxide to Fe^{3+}-EDTA at high pH exists. EPR also indicates that at pH 7.6, hydrogen peroxide binds end-on to the ferric complex.[111]

Theoretical calculations of iron dioxygen and peroxide binding structure have been done. A generalized valence-bond calculation favored end-on bonding in a model of dioxygen bound to a ferrous porphyrin.[113] Calculations of iron peroxides are conflicting. High-level *ab initio* calculations on a ferric hydroperoxide afforded a cyclic side-on structure.[114] The interaction of molecular oxygen with iron atoms in an argon matrix has been explored by a number of investigators. Both side-on and end-on dioxygen complexes were detected by infrared bands at 946–956 cm^{-1} and 1204 cm^{-1}, respectively.[115]

The combination of iron and hydrogen peroxide in aqueous solution rapidly yields radicals via Fenton's reagent.[116] The postulated steps of the redox radical chain are as follows:[117]

$$Fe^{3+} + H_2O_2 \rightarrow Fe^{2+} + HO_2\bullet + H^+ \qquad (3.1)$$

$$Fe^{2+} + H_2O_2 \rightarrow Fe^{3+} + HO^- + HO\bullet \qquad (3.2)$$

$$HO\bullet + H_2O_2 \rightarrow H_2O + HO_2\bullet \qquad (3.3)$$

$$Fe^{3+} + HO_2\bullet \rightarrow Fe^{2+} + O_2 + H^+ \qquad (3.4)$$

$$Fe^{2+} + HO_2\bullet \rightarrow Fe^{3+} + HO_2^- \qquad (3.5)$$

In acidic solution, the rate of reaction between ferrous ion and hydrogen peroxide to produce hydroxyl radicals (Eq. 3.2) is 76 $M^{-1}s^{-1}$.[118] At neutral pH, H_2O_2 and aminocarboxylate chelates of Fe^{II} produce hydroxyl radicals with rates greater than 10^4 $M^{-1}s^{-1}$.[119] It is evident that an iron-peroxo intermediate must act as a nucleophile faster than this rate of decomposition.

Alternatively, a mechanism may be envisioned in which peroxide nucleophilically attacks the substrate prior to the reaction of the resulting tetrahedral intermediate with iron. A characterizable tetrahedral intermediate (**A**) was formed when 2-chloro- or 2-bromo-cyclohexanone was treated with aqueous hydrogen peroxide.[28] Redox decomposition of the peroxide catalyzed by ferrous ion yielded 1,12-dodecanedioic acid via dimerization of the intermediate primary carbon radical (Scheme 3.11). The addition of H_2O_2 to ethyl pyruvate similarly resulted in the oxyhydroperoxide which was then degraded to the alkoxycarbonyl radical.[29] In the presence of a hydrolyzable acyl group however, the iron may assist in cleavage rather than in decomposing the peroxide.

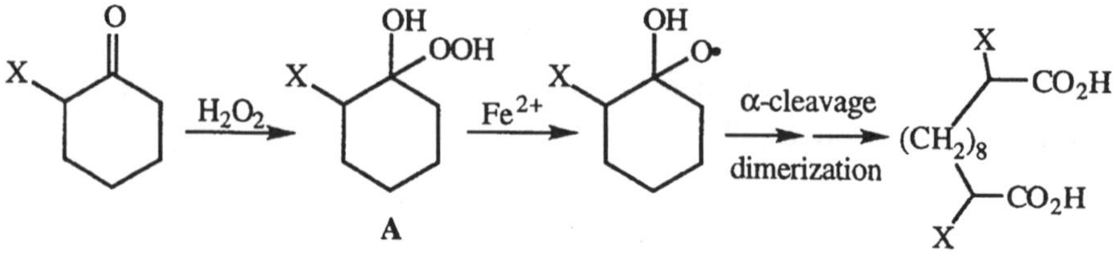

Scheme 3.11

3.8 Active Iron-Peroxy Intermediates in Heme Proteins

3.8.1 Cytochrome P450, oxidase, and peroxidase

An iron-peroxy compound is postulated to be an intermediate in the mechanisms of a group of related heme proteins, cytochromes P450, cytochrome oxidase, and the peroxidases. As described above, the cytochromes P450 are a large group of monooxygenases which utilize oxygen to oxidize a variety of metabolites and xenobiotics. A typical example of P450 activity is the hydroxylation of organic substrates according to Eq. (3.6).

$$NADPH + H^+ + O_2 + RH \rightarrow ROH + H_2O + NADP^+ \qquad (3.6)$$

Cytochrome oxidase is the terminal enzyme in the respiratory chain of aerobic organisms in which dioxygen reduction occurs at an active site consisting of a heme iron and a copper ion. The peroxidases oxidize a wide variety of inorganic and organic substrates at a heme center utilizing hydrogen peroxide. Although the reaction endpoints differ, the steps by which oxygen is activated by these enzymes are remarkably similar as is illustrated in (Scheme 3.12).

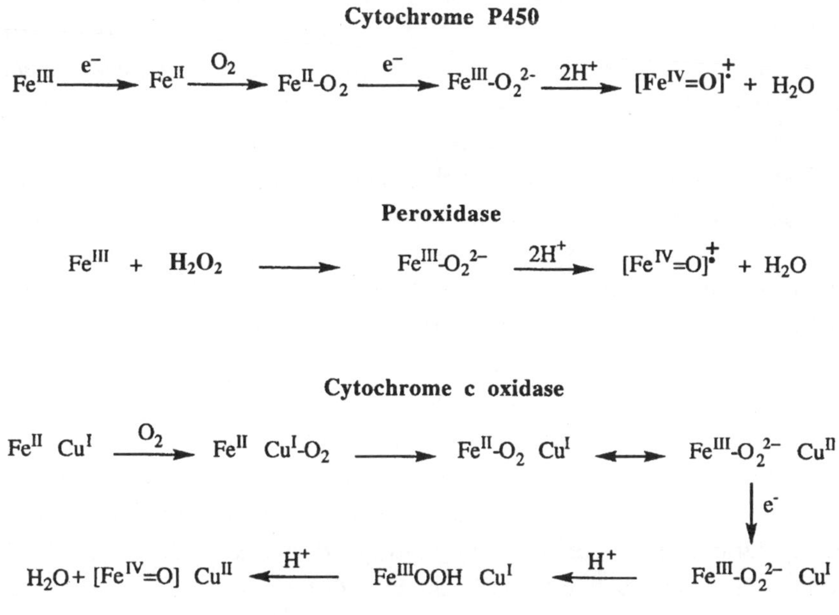

Scheme 3.12

In the P450 reaction cycle, the slow step is the second electron-injection step. Thus, the concentrations of the catalytically active species are never significant enough for isolation or characterization.[122] Hence, evidence for these intermediates comes both from analogy with peroxidase chemistry and model chemistry. The hydroxylation reaction has been shown to involve an initial hydrogen abstraction followed by an oxygen rebound and incorporation into product.[123] The ferryl oxygen complex postulated from comparison to Compound I in peroxidase chemistry was thus corroborated. The preceding peroxy intermediate in P450 is not considered to be the active oxygen species in the reaction and evidence for its formation remains scant. P450 does generate reversible complexes with organic peroxides that have been characterized by UV-vis spectroscopy.[124]

The rate-limiting step for peroxidases is peroxide binding. The presumed iron-peroxy intermediate therefore undergoes rapid conversion to the ferryl with no detectable build-up of the peroxy complex even at –30 °C.[125] Conversion of the peroxy complex is estimated to be greater than 1×10^3 s^{-1}.[126] A recent low temperature stopped flow experiment detected an intermediate preceding the ferryl species which may be the peroxy intermediate.[127] Its structure remains to be established.

The intermediates in the cytochrome oxidase reaction cycle build to detectable levels and have been characterized. The rates of formation of the iron-peroxy and iron-hydroperoxy intermediates are estimated to be 3×10^4 s^{-1} and 1×10^4 s^{-1} respectively.[122] Time-resolved resonance Roman methods have identified among others, the iron-hydroperoxy intermediate.[128] The peroxy species has also been characterized optically.[129]

Although the iron-peroxy intermediate of P450 has for the most part been ruled out as the active oxidizing species for most oxygenations mediated by this enzyme, it may still partake in some interesting chemistry. Hamilton suggested that one mechanism of dioxygen activation could be the acylation of the ferric iron-coordinated peroxide complex.[130] This hypothesis entails a nucleophilic attack of the iron-peroxy moiety on a carboxylate or an amide.[131] Although Hamilton believed that the peroxyacid itself could be the oxidizing species, postulating a bound peracid or peramide does not exclude the ferryl complex. The acylated peroxide can generate the active ferryl intermediate by either homolytic or heterolytic cleavage of the peroxide bond.[131,132] Sligar reported the incorporation of ^{18}O from molecular oxygen into lipoic acid following a single turnover of bacterial cytochrome P450.[133] These results suggest that the carboxylate is subjected to a nucleophilic attack by bound dioxygen. However, acylation has never been demonstrated in these protein systems and remains debatable.

3.8.2 Aromatase related enzymes

As was discussed in the preceding section, the cytochrome P450 family is usually associated with hydroxylation reactions involving a postulated ferryl species as the active oxidant. However, a subset of P450 enzymes exists that catalyzes not only the conventional hydroxylation reaction but also an acyl carbon bond cleavage reaction. It has been proposed that a nucleophilic iron-peroxo species is the active intermediate in this group of enzymes.

Aromatase is the P450 enzyme that catalyzes the NADPH- and O_2-dependent conversion of androgens to estrogens in three oxidative steps. The first two steps are relatively well-understood by classical P450 mechanisms. In the third step, the bond between C-10 and the aldehyde C-19 is cleaved to afford the aromatic ring and formic acid (Fig. 3.14). It was demonstrated by Akhtar in isotopic labeling studies that an oxygen atom from the equivalent of molecular oxygen consumed in the third step,

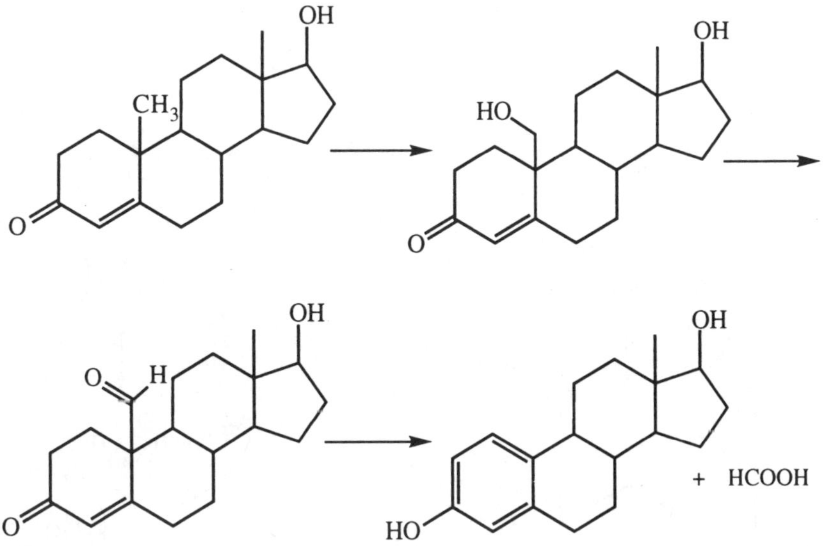

Fig. 3.14 Pathway of the conversion of an androgen to an estrogen by the enzyme P450 aromatase. The final step is believed to proceed via nucleophilic attack of a peroxo iron complex on the aldehyde carbon.

as well as one of the C-19 hydrogens, are incorporated into the formic acid.[95] Based on this data, Akhtar proposed nucleophilic attack of an enzyme-bound ferric peroxide on the 19-aldehyde group. Similar studies on the related enzymes, lanosterol 14α-demethylase, progesterone 17,20-lyase, and pregnenolone 17,20-lyase-Δ^{16}-desaturase supported the theory that the iron-peroxy P450 intermediate reacted with the aldehyde before it could cleave to form the familiar ferryl intermediate.[134]

The iron-peroxy mechanism has since been favored by most investigators. Attack of the iron-peroxo nucleophile on the carbonyl is proposed to form a tetrahedral intermediate which then breaks down by either a homolytic or heterolytic pathway (Scheme 3.13). Many studies to date have been done

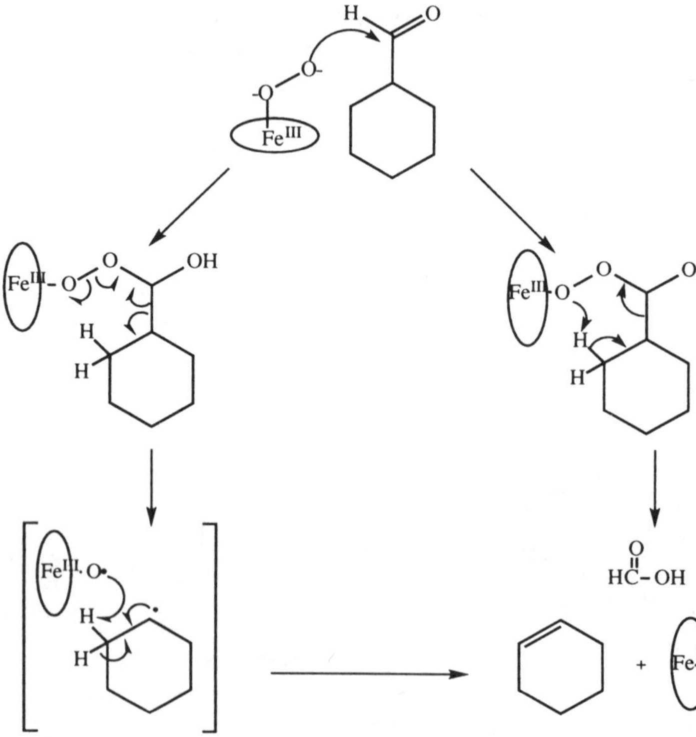

Scheme 3.13

using steroid analogs which are aromatized in an analogous manner to the native substrate by the enzyme.[135] Recently, a steroid analog was shown to be oxidized with the same stereochemical specificity as the aromatase reaction.[136] P450 enzymes are also competent for the oxidative deformylation of a variety of xenobiotic aldehydes to olefins and formate.[137,138] Evidence against the participation of a ferryl moiety in the carbon-carbon bond cleavage step was obtained when it was demonstrated that several oxidants such as, iodosobenzene and *m*-chloroperbenzoic acid, known to support P450 hydroxylations did not produce olefins. Olefins were only formed in the presence of O_2 and NADPH or H_2O_2.[137]

Support for the iron-peroxy mechanism came from mechanistic studies on steroid analogs using H_2O_2 in place of aromatase and dioxygen.[139] The reaction between an androgen 19-aldehyde analog and hydrogen peroxide was found to be facilitated by enolization of the 3-ketone group. The intermediacy of a 19,19-hydroxyhydroperoxide was postulated in the aromatization which gave the corresponding estrogen derivative along with formic acid.[140,141]

It remains an enigma as to how the same P450 enzyme that hydroxylates substrates with a ferryl oxidant, switches pathways in the presence of certain substrates to perform carbon-carbon bond cleavages mediated by a nucleophilic iron-peroxide (Fig. 3.15). Molecular orbital calculations supported the iron-peroxy mechanism over the ferryl mechanism when acyl groups are present.[142] Some researchers suggested that the functional group of the substrate chooses a compatible iron-oxygen species ($Fe^{IV}O$ or $Fe^{III}OOH$) for further reaction.[143] Evidence for this comes from an investigation using cytochrome $P450_{17\alpha}$. This enzyme catalyzes a hydroxylation reaction and two different acyl carbon bond cleavages involving ketones. When a steroid analog with an aldehyde in place of the ketone was used as a substrate, the hydroxylation activity was suppressed and olefin was the predominant product.[144] This study suggests that the more electrophilic aldehyde guided the enzyme to favor the peroxide pathway.

It was also suggested that protons regulate oxene versus peroxide chemistry. A mutation in cytochrome $P450_{cam}$ that replaced threonine-252

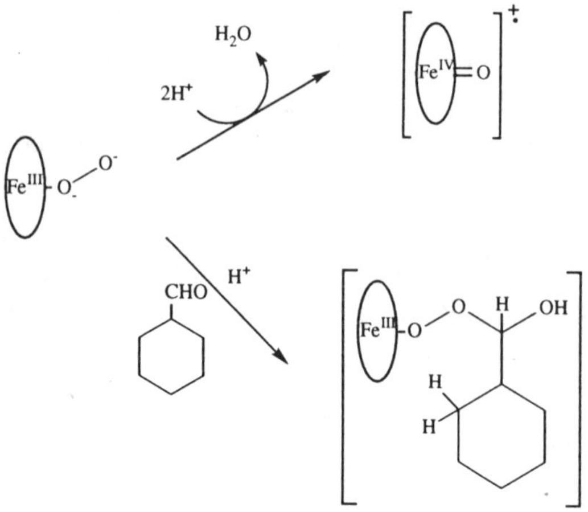

Fig. 3.15 Pathways by which the ferric peroxide intermediate may react. The peroxide may cleave to yield an active oxene intermediate. In the presence of a carbonyl, the peroxide nucleophilically attacks the aldehyde to form a peroxy tetrahedral intermediate. Adapted from Ref. 146.

with alanine or valine caused monooxygenase function to be suppressed.[145] Most of the O_2 consumed was recovered as H_2O_2. This discovery led to the hypothesis that the hydroxy amino acid was involved in oxygen activation. When an analogous mutation was made in the cytochrome P450 2B4 enzyme, it was discovered that in the presence of substrates like cyclohexane peroxide, formation was increased and hydroxylated product was decreased.[146] However, deformylation of cyclohexane carboxaldehyde was increased. The enzyme cytochrome P450 CYP17 catalyzes oxidations via both oxene and peroxide chemistry. Solvent isotope effects and pH studies done with this enzyme suggested that protons were important in determining the oxidant.[147]

3.9 Nitric Oxide Synthase

Nitric oxide (•NO) is a ubiquitous metabolic intermediate and an important signal molecule. Nitric oxide synthases (NOSs) are highly analogous to the cytochrome P450 family of enzymes, sharing the axial thiolate ligation[96] and similar redox cofactor arrangements (although NOS has a tetrahydrobiopterin and a calmodulin-binding site). NOSs catalyze the oxidation of L-arginine to •NO and citrulline in the presence of NADPH and O_2 (Fig. 3.16).[147]

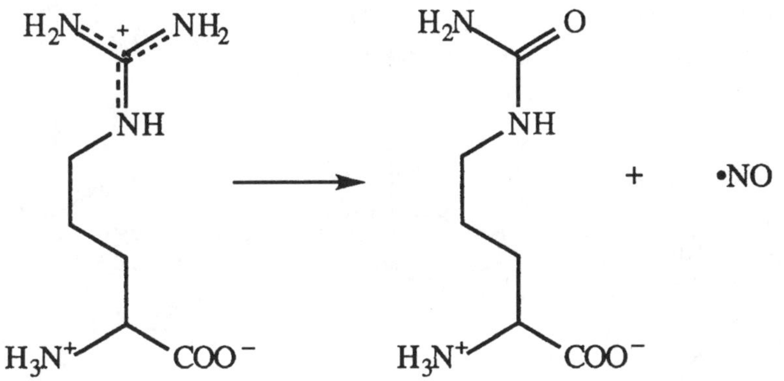

Fig. 3.16 The conversion of L-arginine to •NO and citrulline catalyzed by nitric oxide synthase.

The first step in the reaction is a typical P450 hydroxylation of one of the amidine nitrogen atoms of arginine to produce N^{ω}-hydroxy-L-arginine (NHA).[96] The conversion of NHA to citrulline remains mechanistically ambiguous. Isotope labeling studies determined that the oxygen atoms in citrulline and •NO were derived from molecular oxygen.[148] In addition, the nitrogen in •NO was derived exclusively from the hydroxylated amidine nitrogen in NHA.[149] The currently accepted phenomenological mechanism was proposed by Marletta[96] and its general characteristics were refined by Ingold.[150] The mechanism is analogous to that described above for aromatase. Following oxidation of and proton loss from NHA, the guanidino carbon of NHA is attacked by a iron peroxide nucleophile. This leads to a tetrahedral intermediate that decomposes to yield citrulline and •NO (Scheme 3.14 adapted from Ref. 94).

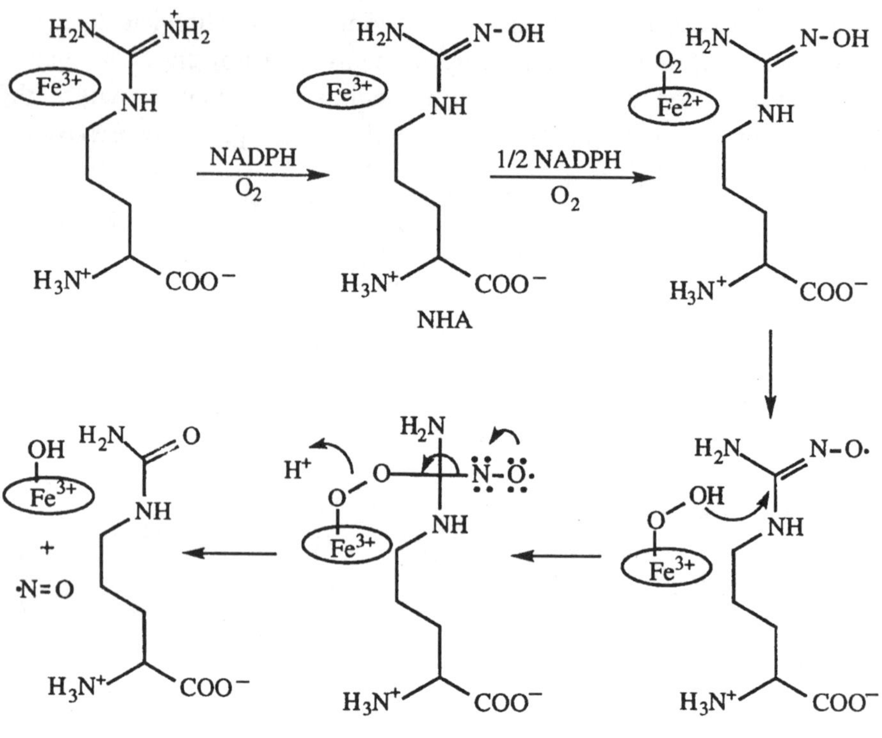

Scheme 3.14

In support of the iron-peroxy mechanism, it has been shown that H_2O_2 can substitute for O_2 and NADPH in the conversion of NHA to citrulline.[151] Oxidants typically competent for P450 hydroxylations, *tert*-butylhydroperoxide and cumene hydroperoxide did not support product formation. Oxygen from labeled H_2O_2 was shown to reside in the citrulline carbonyl.

In summary, as the aromatase and nitric oxide synthase examples illustrate, P450 enzymes capable of monooxygenation via a ferryl may also be able to exhibit alternate modes of reactivity including nucleophilic iron-peroxide chemistry. Intriguingly, however, model reactions of these types have yet to be illucidated and alternative mechanisms should also be considered. Particularly awkward in Scheme 3.14 above is that Fe(III) plays only the role of a Lewis acid whereas homolysis of the O–O bond in the final step to afford a ferryl species appears much more likely.

3.10 Iron Peroxides in Non-Heme Proteins

3.10.1 Intradiol catechol dioxygenases

The catechol dioxygenases catalyze the degradation of aromatic compounds. These enzymes cleave catechols to yield either muconic acids (intradiol dioxygenases) or muconic semialdehydes (extradiol dioxygenases) (Fig. 3.17). The extradiol enzymes are less understood than the intradiol enzymes. The extradiol enzymes contain a ferrous active site which binds dioxygen without oxidizing to a ferric superoxide intermediate. Subsequent attack of the dioxygen on the iron-bound substrate leads to catabolism.[152] Since the extradiol enzymes maintain a ferrous center throughout the reaction, they are not of primary concern in this review. The most extensively studied of the intradiol dioxygenases are catechol 1,2-dioxygenase (CTD) and protochatechuate 3,4-dioxygenase (PCD). The active sites have been characterized by a variety of spectroscopic techniques. Evidence of a high-spin ferric center coordinated to two tyrosines, two histidines and a water, comes from EPR, electronic, and Mössbauer spectroscopies.[153]

Steady-state kinetic studies on CTD and PCD indicate that the substrate binds to the enzyme prior to oxygen[153a], i.e.

$$E + S \rightarrow ES + O_2 \rightarrow ESO_2 \rightarrow E + P \tag{3.7}$$

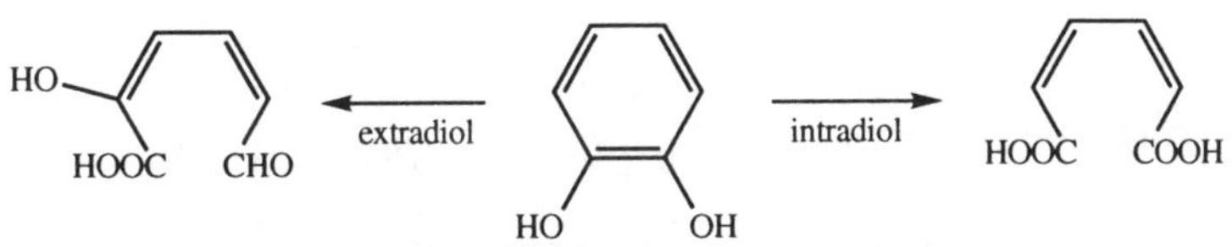

Fig. 3.17 The reactions catalyzed by extradiol and intradiol catechol dioxygenases.

The proposed mechanism is shown in Scheme 3.15.[153a,154] The substrate is activated for reaction with dioxygen by binding to the ferric center. The catecholate is rendered susceptible to oxygen attack because of a delocalization of unpaired spin density from the iron.[155] A ferric-peroxide complex results and the product is subsequently decomposed via an anhydride intermediate. The plausibility of the peroxide intermediate is supported by a crystal structure of an iridium peroxy-catecholate compound which contains the same peroxo bridge.[156]

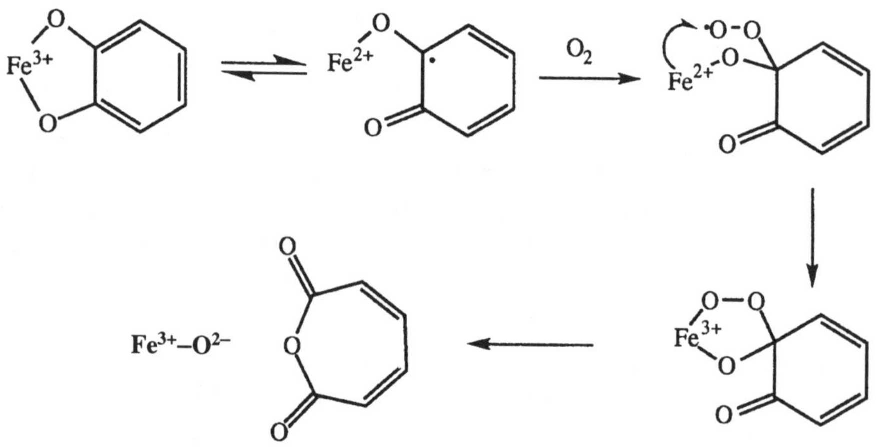

Scheme 3.15

The proposed mechanism of intradiol dioxygenases does not involve a nucleophilic attack of an iron-bound peroxide. However, the suspected involvement of an iron-peroxy intermediate is intriguing.

3.10.2 4-Methoxybenzoate o-demethylase (putidamonooxin)

Putidamonooxin is the terminal oxygenase of the 4-methoxybenzoate monooxygenase isolated from *Pseudomonas putida*. It catalyzes the conversion of the methoxy-derivatized acid into 4-hydroxybenzoic acid and formaldehyde (Scheme 3.16).[157] The enzyme contains both an [2Fe-2S] cluster and a non-heme iron site.[152] The coordination sphere of the iron has not been fully characterized but it is known that exogenous ligand-binding sites exist as has been demonstrated by NO binding.[158]

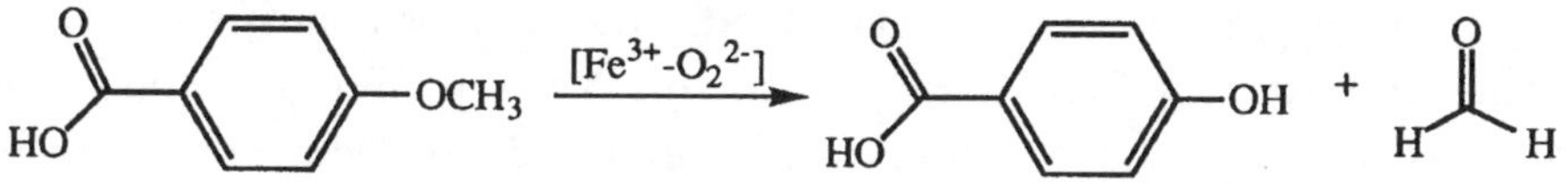

Scheme 3.16

Substrates which are not oxidized during enzyme turnover are classified as being uncoupled. In the presence of uncoupled substrates, more than 80% of the dioxygen consumed by putidamonooxin is converted to H_2O_2.[159] In addition, Mössbauer[160], EPR,[161] and kinetic studies indicate that dioxygen is activated by the uptake of two electrons, one from the iron-sulfur cluster and the other from the iron active site. The enzyme is active when the iron active site is in its ferrous state and substrate and dioxygen are present. The catalytic process is postulated as follows:[158]

$$Fe^{III} \xrightarrow{e^-} Fe^{II} \xrightarrow{O_2} Fe^{II}\text{-}O_2 \longrightarrow Fe^{III}\text{-}O_2^- \xrightarrow{e^-} Fe^{II}\text{-}O_2^- \longrightarrow$$

Dioxygen reacts with the ferrous center and following uptake of the second electron, an iron-peroxide species is formed. Based on the results with uncoupled substrates and the requirement for two electrons, the active species is presumed to be the ferric peroxide. Additional evidence for the ferric peroxide intermediate was obtained when 4-vinylbenzoate was used as a substrate for putidamonooxin in the presence of $^{18}O_2$.[162] It was discovered that both oxygens in the product, 4-(1,2-dihydroxyethyl)benzoate were derived from the labeled dioxygen. The specific mechanism of iron-peroxy substrate transformation has yet to be determined.

3.10.3 α-Ketoacid dependent dioxygenases

Hutton first determined that prolyl 4-hydroxylase, the enzyme which catalyzes the hydroxylation of proline, required α-ketoglutarate for its activity.[163] Since then, several other enzymes have also been shown to be dependent on α-ketoglutarate. These enzymes also require a ferrous ion, dioxygen, and ascorbate to maintain the ferrous state of iron.[152] α-ketoacid dependent dioxygenases catalyze a wide variety of oxidations. For hydroxylases, the dioxygen atoms are incorporated into the hydroxyl group of the product and the carboxylate group of the product succinate.[164] A typical reaction is depicted in Scheme 3.17.

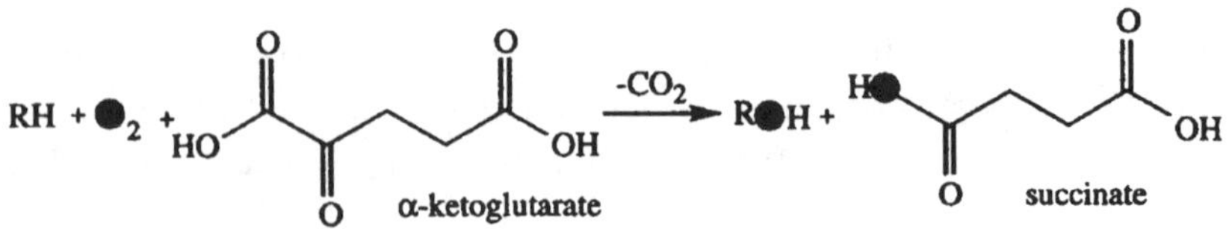

Scheme 3.17

The long-standing accepted mechanism for α-ketoacid dependent dioxygenases is based on that proposed for prolylhydroxylase.[165] The mechanism consists of (1) dioxygen binding to the ferrous center to form a ferric superoxide species; (2) nucleophilic attack of this complex on the bound α-ketoacid; (3) decarboxylation of the α-ketoacid with concomitant generation of the oxidizing species (speculated to be a ferryl); (4) oxidation of the substrate; and (5) product release. The mechanism has since been corroborated with model studies.[166] For the purposes of this review, the pertinent step is step number 2 (steps 1–3 are illustrated in Scheme 3.18). Although the nucleophile is a bound superoxo instead of a peroxo, the general characteristics of a nucleophilic iron-bound dioxygen molecule are exemplified by this group of enzymes.

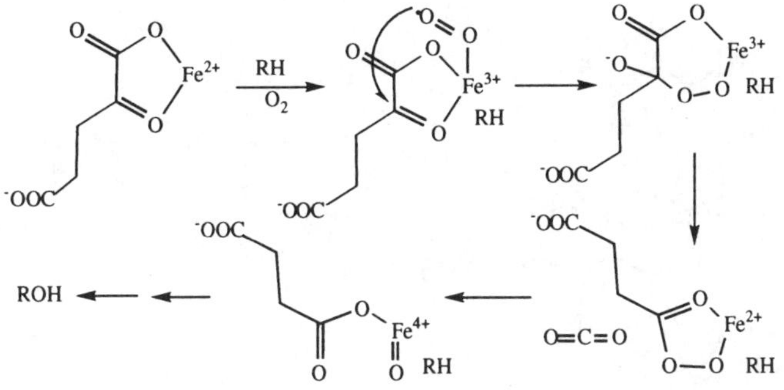

Scheme 3.18

3.11 Pseudo-Metalloenzymes

3.11.1 Bleomycin

Bleomycin (BLM) is a glycopeptide antibiotic used in the treatment of cancers. Its therapeutic activity involves DNA binding and selective cleavage in the presence of ferrous ions and dioxygen. Its structure is shown in Fig. 3.18. BLM is divided into metal- and DNA-binding domains linked by a peptide fragment.[165] Although iron is the active metal, many other metals have been complexed by BLM for investigatory purposes. Iron is ligated to five nitrogens from histidine, pyrimidine, the peptide backbone, and two amines, and its coordination environment is square-pyramidal.[168] The metal-binding domain is hypothesized to interact specifically within the minor groove and abstract a 4'-hydrogen, ultimately producing base propenals.[169,170]

The ferrous chelate of BLM is oxygenated and reduced by one-electron to form "activated bleomycin." BLM can also be activated with ferric ions and hydrogen peroxide.[171,172] EPR[171] and Mössbauer[173] studies have shown that the iron of activated BLM is low-spin ferric complex. Electrospray mass spectrometry of iron BLM demonstrated that activated bleomycin is a ferric peroxide complex.[174] This conclusion agrees with the Fe^{3+}-OOH EPR assignment made in model studies on a dioxygen iron bleomycin

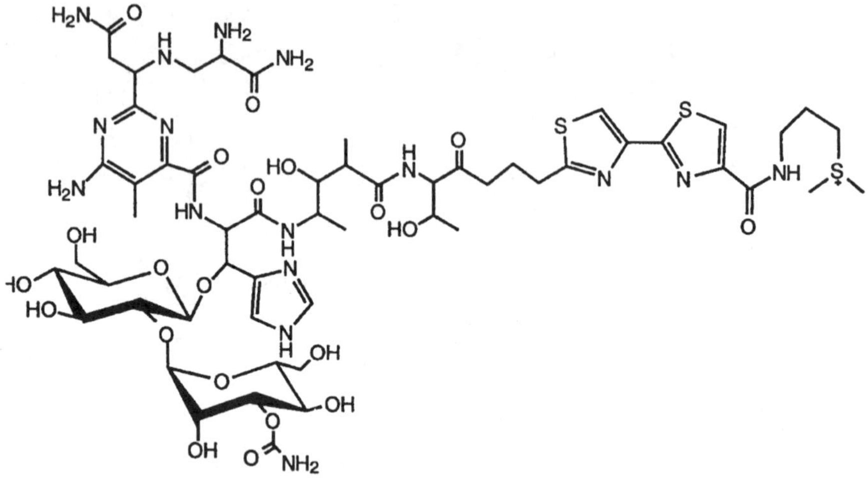

Fig. 3.18 The structure of bleomycin-A_2.

analog.[175] Scheme 3.19 illustrates the pathway to activated BLM, binding of dioxygen results in a ferric superoxide complex that has been identified by Mössbauer.[173] Reduction by a reducing agent or another equivalent of ferrous BLM results in peroxyferric BLM.

$$\mathrm{Fe^{2+}\text{-}BLM} \xrightarrow{O_2} \mathrm{BLM\text{-}Fe^{2+}\text{-}O_2^-} \xrightarrow{e^-,\,H^+} \underset{\text{"activated BLM"}}{\mathrm{BLM\text{-}Fe^{3+}\text{-}O_2H}} \xleftarrow{H_2O_2} \mathrm{Fe^{3+}\text{-}BLM}$$

Scheme 3.19

The structure of activated BLM was recently investigated by X-ray absorption spectroscopy (XAS) and NMR of the stable cobalt analog. XAS found no evidence for a short iron-oxo bond which would indicate a ferryl oxidant. Thus, the ferric peroxide formalism was upheld. The structure of the hydroperoxide form of Co•BLM and its complex with a DNA oligonucleotide was determined by NMR.[176] The proton associated with the hydroperoxide of Co•BLM was observed to have intermolecular NOEs to the minor groove protons of the cleavage site base.[176c] Molecular modeling of this complex predicted the distal oxygen of the hydroperoxide to be 2.5 Å from the 4'-hydrogen of the base. Intramolecular hydrogen bonds between the hydroperoxide ligand and the peptide linker region of BLM were predicted to be instrumental in positioning the peroxide near the 4'-hydrogen.

Although a ferric peroxide complex is the reactive species of BLM, it functions oxidatively by abstracting a hydrogen. No nucleophilic characteristics have been ascribed to this ferric peroxy intermediate. The BLM peroxide is not believed to oxidize DNA via hydroxyl radicals, the cleavage is specific and radical scavengers are ineffective.[177] It is known that the DNA products resulting from BLM cleavage incorporate oxygen atoms from solvent water, not the oxidant.[178] However, this does not rule out the involvement of a ferryl intermediate, since the oxygen rebound step may not be kinetically competitive.

3.11.2 Metal-peroxo hydrolysis of proteins and DNA

Quite a bit of structural information can be gained from the site-specific cleavage of DNA and proteins by metal complexes. These characterizations are typically performed with a bound metal complex which in the presence of oxygen or hydrogen peroxide and a reducing agent, releases diffusible oxidants such as hydroxyl radicals which in turn cleave the biopolymer in the vicinity of their site of generation.[179]

Surprisingly, using conditions commonly employed by others, Meares *et al.* observed site-specific *hydrolysis* of a protein backbone instead of the expected oxidative cleavage. For site specific cleavage of bovine serum albumin, a ferrous EDTA derivative (BABE: 1-(*p*-bromoacetamidobenzyl)-EDTA) was introduced at a cysteine in the polypeptide chain. After treatment with H_2O_2 and ascorbate, three peptide fragments which accounted for the entire protein were recovered.[180] The amino acids at the sites of cleavage were found to be unaltered, the damage typically found with oxidative cleavage was not detected. In order to gain more insight into the proximity of the cleavage site to the bound iron, a similar experiment was performed on human carbonic anhydrase I (HCAI) whose structure is known.[181] Subsequent to conjugation of the iron chelate to HCAI, the protein was cleaved at one site in the presence of ascorbate and H_2O_2.[182] One hundred percent cleavage was accomplished in 10 secs by one equivalent of each reagent. Hydroxyl radical scavengers had no effect on the reaction. Again Edman sequencing of the fragments indicated that hydrolysis, not oxidative cleavage had occurred. Comparison of the cleavage site to the crystal structure of HCAI showed that the peptide bond cleaved was in close proximity to the iron chelate site. Significant evidence for a hydrolytic mechanism was obtained when reactions were run in the presence of $[^{18}O]H_2O_2$. One atom of ^{18}O was incorporated into the carboxyl group of the C-terminal fragment with ~93% efficiency.[5]

These results indicated unprecedented nucleophilic cleavage of a polypeptide backbone by an iron-bound oxygen species. Meares *et al.* postulated that the reactive intermediate was an iron-peroxo group positioned in precisely the proper distance from and orientation to the scissile amide bond. This positioning would enable the carbonyl to trap the peroxy nucleophile prior to its decomposition. In the suggested reaction mechanism (Scheme 3.20, adapted from Ref. 5), the iron-bound peroxide attacks the amide carbonyl forming a tetrahedral intermediate. This intermediate could break down heterolytically as shown, or homolytically to yield the new carboxyl terminus. The oxidation state of iron throughout the mechanism is unknown as is the purpose of ascorbate.

A similar mechanism was suggested for cleavage of the tetracycline repressor protein.[183] Protein cleavage was initiated by incubating the repressor with an iron-tetracyline complex and 10 equivalents of ascorbate and H_2O_2 for 15 secs. Amino acids at the newly formed N-terminal sites were not oxidatively damaged. As Meares observed, cleavage in the presence of radical scavengers had no effect.

Usually, this peptide cleavage method of incubating a protein with a bound iron complex in the presence of hydrogen peroxide and ascorbate results in hydroxyl radical oxidative cleavage.[184]

Hydrogen peroxide was found to be competent for the cleavage of a specific phosphodiester bond in DNA with assistance from Flp

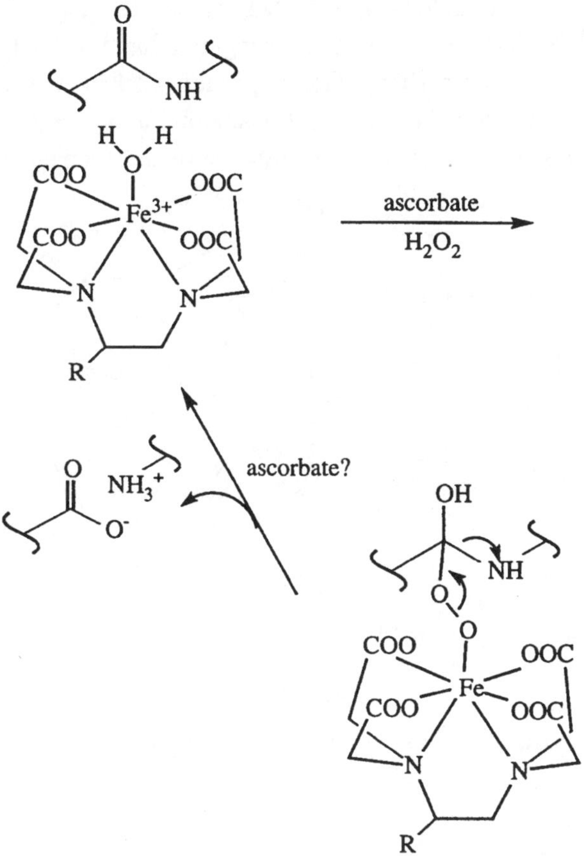

Scheme 3.20

recombinase.[185] Flp recombinase activates the phosphodiester bond for attack but does not itself cleave DNA at the observed cleavage site. The nucleophilic attack results in the production of one phosphate and one hydroxyl end without the loss of nucleotides. In this cleavage reaction by a nucleophilic peroxide, no metals are necessary for activation.

Non-oxidative DNA cleavage was also found to be effected by Fe_2-(HPTB)(OH)$(NO_3)_4$ (HPTB=N,N,N',N'-tetrakis(2-benzimidazolylmethyl)-2-hydroxy-1,3-diaminopropane) in the presence of excess H_2O_2 or O_2 and a reductant.[186] The linearized plasmid pBR322 contains 3'-OH and 5'-OPO_3

ends and could be efficiently religated by T4 DNA ligase. The cleavage reaction is not inhibited by hydroxyl radical scavengers. The Fe_2-HPTB complex forms a (μ-1,2-peroxo)diiron(III) species (**3.25**) which could electrostatically interact with the phosphate backbone upon binding, **3.25** attacks the phosphodiester bond via a nucleophilic peroxide mechanism.

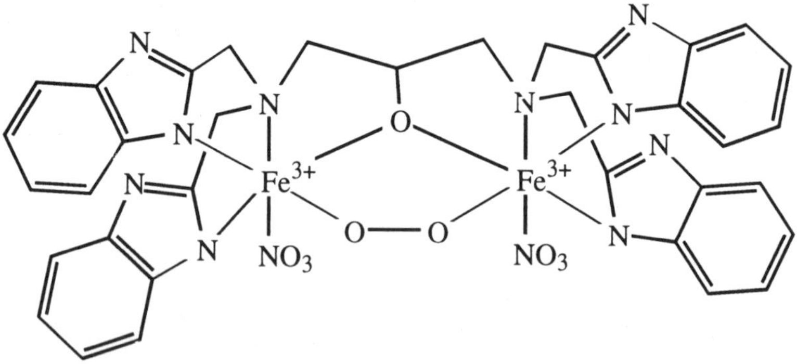

3.25

3.12 Biomimetic Monomeric Metal Peroxide Complexes

3.12.1 Transition metal peroxides

The general topic of transition metal peroxides has been extensively reviewed.[187] This section will discuss structures and general reactivity of metal-peroxo complexes as a background to the more in-depth look at mononuclear iron dioxygen complexes that follows. The most widely known metal peroxides are the η^2-peroxo complexes. Transition metals which form side-on dioxygen complexes include:[187e-g] Ti, V, Nb, Cr, Mo, W, Fe,[188] Ru, Os, Co, Rh, Ir, Ni, Pd, Pt, and U. Mononuclear η^1-dioxygen metal complexes are believed to be formed with Co,[189] Cr,[190] Fe,[96] Rh,[191] Ir,[192] and Sm.[193]

Metal-peroxo complexes can react with both electrophilic and nucleophilic substrates although not all peroxo complexes are reactive.

The reactivity of peroxo-metal complexes is influenced by the identity of the metal atom, the number of peroxo ligands, and the nature of the other ligands in the compound.[187c] Typically peroxo-metal complexes act as oxygen transfer agents with a large variety of substrates. A general reaction mechanism is depicted in Fig. 3.19.[187a] Mononuclear cobalt hydroperoxy complexes undergo nucleophilic attack of substrate on the proximal oxygen. An example is shown in Scheme 3.21.[189b] Group VIII peroxo complexes react with carbonyl compounds to form metalloozonides (Scheme 3.22).[187e,g] Nucleophilic peroxo complexes react with electrophiles such as cyano olefins to yield cyclic peroxidic adducts (Scheme 3.23).[194] Peroxo-metal complexes act as nucleophiles in reactions with alkyl halides. S_N2 nucleophilic attack on the carbon atom occurs with inversion of configuration to form alkyl peroxidic complexes (Scheme 3.24).[187e]

Thus, metal-peroxo complexes mostly act as oxidants. Nucleophilic reactivity is seen with alkyl halides and electron-poor olefins but in

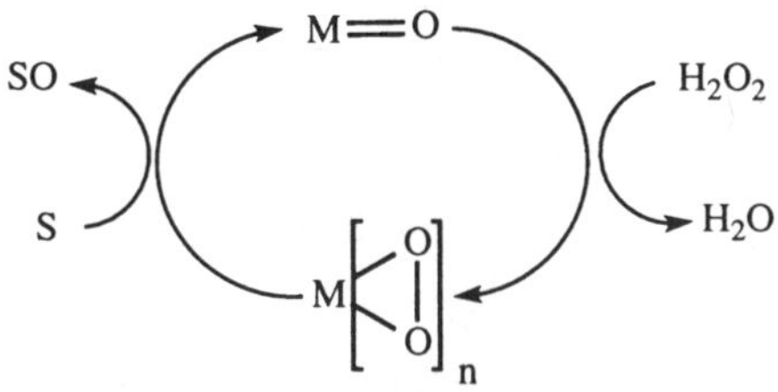

Fig. 3.19 General mechanism of substrate (S) oxidation catalysis by transition metal-peroxo complexes (M = V, Mo, W). Typically, n = 1 or 2 (adapted from Ref. 187a).

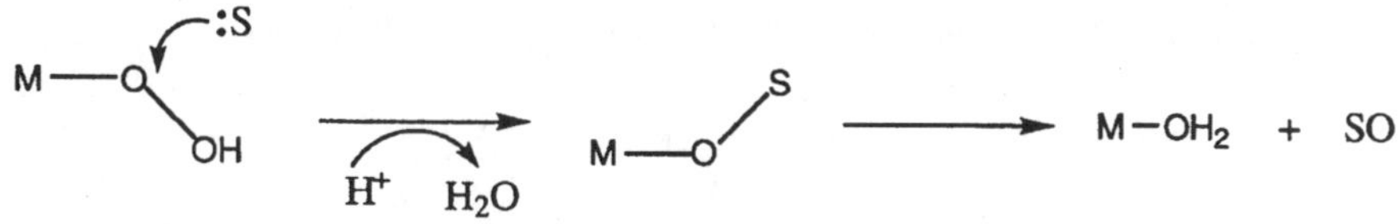

Scheme 3.21

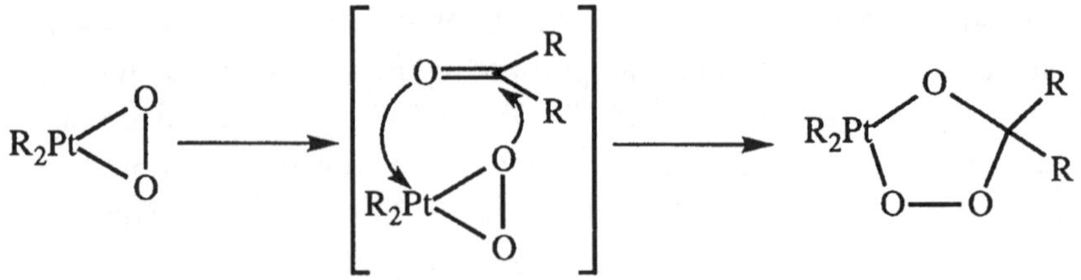

Scheme 3.22

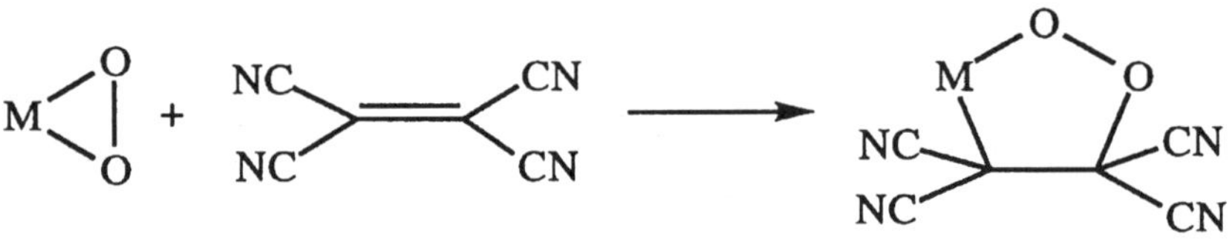

Scheme 3.23

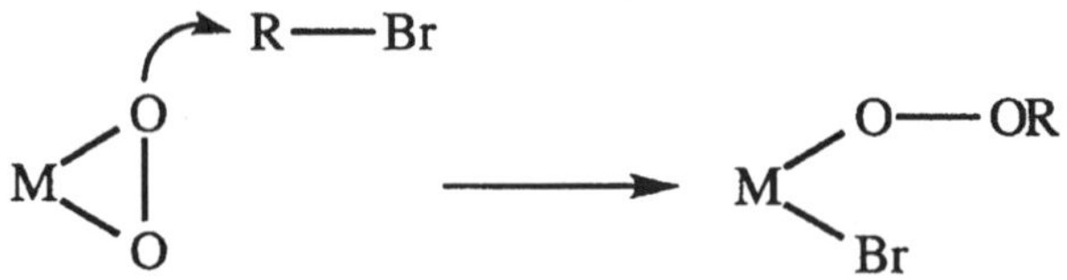

Scheme 3.24

the presence of substrates with carbonyl functional groups, only Pd- and Pt-peroxo complexes are reactive, and hydrolysis is not the mechanism observed.

3.12.2 Heme and pseudo-heme iron dioxygen complexes

Under the proper conditions, ferrous porphyrins bind molecular oxygen reversibly:

$$Fe^{2+}(Por) + O_2 \rightleftharpoons Fe(Por)(O_2) \tag{3.9}$$

In simple porphyrin systems, this 1:1 adduct undergoes rapid autoxidation through a μ-peroxo-diiron intermediate (Por)Fe^{3+}-O_2-Fe^{3+}(Por) to the μ-oxo dimer.[195] The peroxo intermediate was detected by visible, ^{1}H NMR,[196] and resonance Raman[197] spectroscopies.

The formation of 1:1 iron dioxygen complexes in model systems is accomplished by either synthesizing porphyrins that inhibited dimerization or working at low temperatures or both. One of the first synthetic iron oxygen carriers is a pyrroporphyrin complex with an imidazoyl propyl side chain that binds oxygen reversibly at -45 °C.[198] The first porphyrin with the ability to complex dioxygen at room temperature is Collman's picket fence porphyrin, TP(piv)P (tetrakis(pivaloylphenyl)porphyrin) (Fig. 3.20).[199] An X-ray crystal structure showed that the dioxygen is bound η^1 end-on.[200] Numerous other examples of sterically hindered iron porphyrin dioxygen carriers have since been synthesized, including Baldwin's capped porphyrin complex[201] and Collman's recent picnic basket porphyrin.[202]

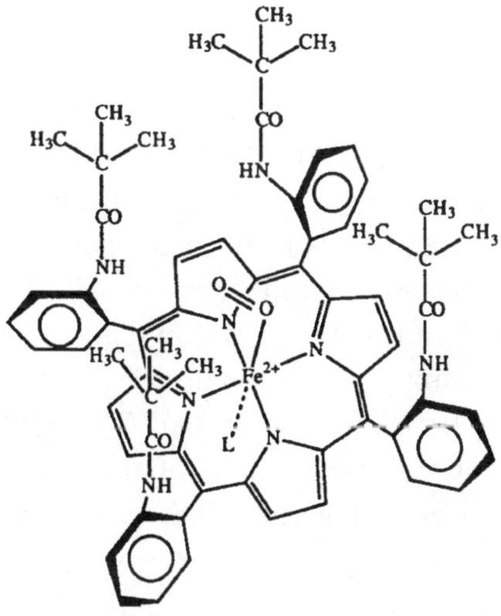

Fig. 3.20 Structure of the oxygen adduct of iron picket fence porphyrin (tetrakis(pivaloylphenyl)porphyrin) Fe(TpivPP)(L)(O_2).

Valentine *et al.* reported the first mononuclear ferric porphyrin peroxo complex $[Fe^{3+}(Por)(O_2)]^-$.[203] The peroxo species was formed by adding superoxide to ferrous TPP (meso-tetraphenylporphyrin) or OEP (octaethylporphyrin). The visible absorption spectra of the peroxo complexes exhibited Soret and α,β bands at unusually low energies. EPR spectra taken at 77 K showed a sharp resonance at $g = 4.2$ and weak resonances at $g = 2$ and 8 typical of a rhombic high-spin ferric complex. The peroxide was characterized by an infrared band at 806 cm^{-1} that shifted to 759 cm^{-1} when $^{18}O_2^-$ was used. A similar shift was observed with Raman spectroscopy. Identical ferric porphyrin peroxo complexes were also synthesized by the electrochemical reduction of $Fe^{2+}(OEP)(O_2)$[204] and the addition of dioxygen to $Fe^{1+}(OEP)$[205]. Ferric porphyrin peroxide complexes were also characterized by ^2H NMR[206] and Mössbauer spectroscopies.[207] Although no crystal structure of a ferric porphyrin peroxo complex exists, spectroscopic characteristics indicate that the structure is analogous to that seen by X-ray crystallography for $(Por)Ti^{IV}O_2$[208] and $[PMn^{III}O_2]^-$[209] in which peroxide is bound in an η^2 side-on manner.

Recently, porphyrin and pseudo-porphyrin peroxo iron complexes in which the peroxide is believed to be bound end-on were reported. In the autoxidation of ferrous cyclidenes, a ferric peroxo intermediate was detected.[210] This stable intermediate was even detectable at ambient temperatures. Its EPR spectrum indicated that the complex was in a low-spin ferric state with a very low anisotropy ($g = 2$). The first porphyrin hydroperoxide adduct was formed upon addition of H_2O_2 to Fe^{3+}-TPP in the presence of alkaline reagents.[119] It also exhibited an EPR spectrum (77 K) indicative of a low-spin ferric complex with anomalously small g anisotropy. The peroxide complexes Fe^{3+}-OEP(OOH)[212] and Fe^{3+}–TMP(OOH)[213] were also reported. The EPR spectra of these peroxide complexes contrasted with those of the high-spin ferric peroxo porphyrin complexes discussed earlier and were consistent with those of Fe^{III}(hemoglobin)(OOH) and Fe^{III}(horseradish peroxidase)(OOH). In addition, similar EPR spectra were obtained from mononuclear ferric heme coordinated alkyl peroxides which are necessarily bound end-on.[214] Alkyl peroxide complexes of iron porphyrins are exceedingly unstable and have been detected only at low temperatures, primarily by NMR spectroscopy.[215]

Early studies of η^2 peroxo iron porphyrin complexes demonstrated that they did not display typical electrophilic reactivities such as hydroxylation or epoxidation.[216] The reactivity of these peroxo complexes was more similar to the group VIII "nucleophilic" peroxo complexes. A reaction typically characteristic of group VIII peroxo complexes is that with sulfur dioxide to yield metal-sulfato complexes. Analogously, addition of sulfur dioxide to $[Fe^{3+}(P)(O_2)]^-$ (P = TPP, OEP) yielded sulfate.[217]

The nucleophilic character of η^2-peroxo porphyrin complexes was demonstrated further by the acylation of manganese and iron-peroxo porphyrins complexes. The acylation of a TMPMn^{III}-O_2^- complex with *m*-chlorobenzoyl chloride was described.[218] Treatment of TMPMn^{III}Cl with *m*-chloroperoxybenzoate also resulted in the acylperoxymanganese species (Scheme 3.25). In the reaction of

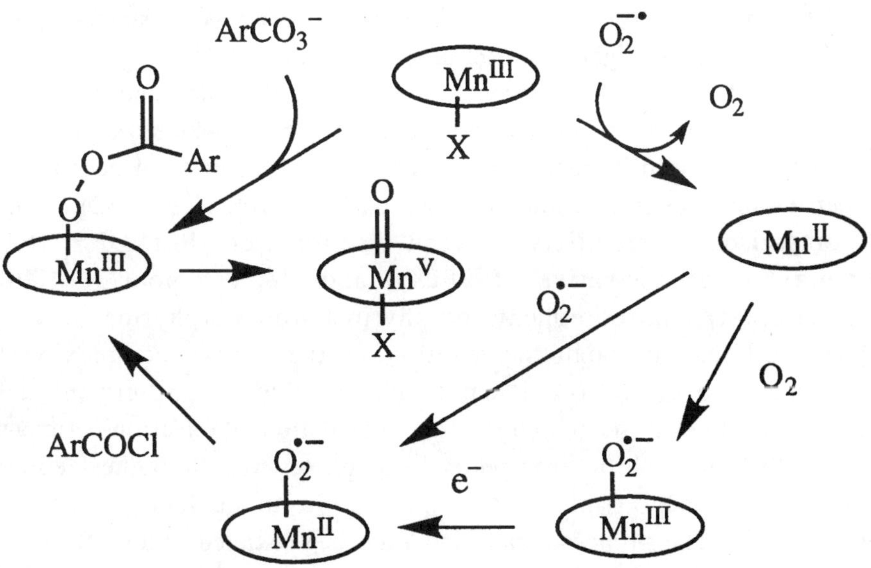

Scheme 3.25

$[Fe(P)O_2]^-$ (P = TPP, TMP) with acetic anhydride, a peroxyacetate intermediate formulated as $PFe^{III}OOAc$ was detected.[219] An EPR signal for this complex at $g = 5.9$ was obtained. The peroxyacetate species was also synthesized by the addition of peroxyacetic acid to the ferric porphyrin. Similarly, $[TP(piv)PFe^{III}O_2]^-$ was acylated by exposure to an excess of carbon dioxide.[220] The oxidation of the ferric porphyrin with peroxycarbonate also produced the acylperoxo intermediate. Manganese superoxo porphyrins react with carbon dioxide in the same manner.[221]

Iron-peroxo porphyrin complexes react with aldehydes and 3-methyl-indole in oxidation reactions. The mechanism of aromatization of steroid analogs has not yet been determined but is postulated to proceed through a nucleophilic attack of an iron peroxide on an aldehyde, thereby modeling the proposed mechanism for aromatase.[222] In the conversion of 3-methyl indole to *o*-formamidoacetophenone by Fe^{III}TPP in the presence of oxygen, an intermediate species was detected by EPR and assigned as the Fe^{III}TPP-OO-substrate complex.[223]

Valentine recently reported the nucleophilic oxygen transfer from a bound iron-peroxo ligand to electron-deficient olefins. Peroxo complexes of ferric TPP and TMP porphyrins reacted with enones and quinones to produce the corresponding epoxides. Up to 76% epoxide was formed in these nucleophilic reactions which are proposed to proceed via a peroxy intermediate (Scheme 3.26 adapted from Ref. 132).[224] In a comparison of reactivity between electron rich [PPIXDME (the dimethyl ester derivative of PPIX)] and electron poor [F_{20}TPP (5,10,15,20-(pentafluorophenyl)porphyrin)] iron porphyrins, it was discovered that although the stability of the peroxo complex was increased with electron-deficient ligands, the nucleophilicity of the peroxo moiety was decreased.[225] The epoxidation of electron-deficient olefins by nucleophilic iron-peroxo complexes was hypothesized to involve ring opening of the η^2-peroxo prior to nucleophilic attack of the olefin. The proposed transition state is illustrated in Fig. 3.21.

In summary, it was demonstrated by a number of investigators that iron-peroxo complexes are capable of behaving as nucleophiles.

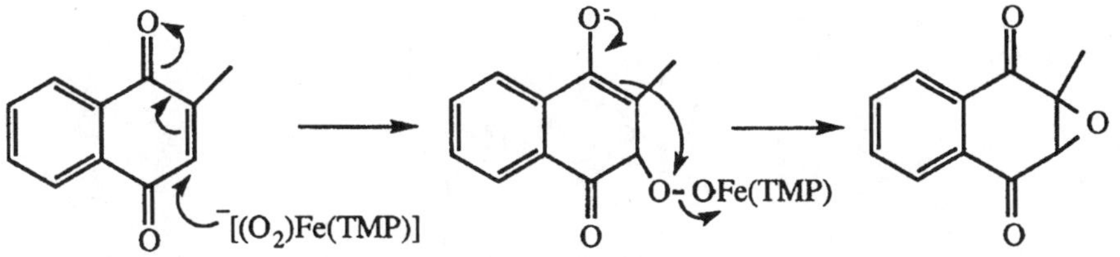

Scheme 3.26

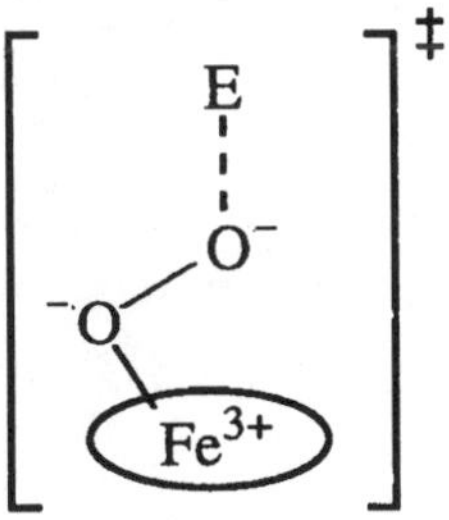

Fig. 3.21 Open peroxo transition state proposed in the nucleophilic attack of iron-peroxo complexes on electron poor olefins **E** (adapted from Ref. 133).

3.12.3 Non-heme mononuclear iron peroxides

Perhaps the most well-known non-heme mononuclear iron-peroxo complex formed upon reaction of ferric ethylenediaminetetraacetic acid (EDTA) with hydrogen peroxide at high pH.[226] This purple complex (λ_{max} 520 nm, $\varepsilon = 528\ M^{-1}cm^{-1}$) is best formulated as $[Fe^{3+}\text{-EDTA-}O_2]^{3-}$ (**3.26**) and the stoichiometry of its formation is 1:1 in Fe^{3+}-EDTA and peroxide.[227] The purple complex can also be formed by the addition of superoxide to Fe^{2+}-EDTA.[107,228] Both EPR[111] and Raman[112] spectroscopic evidence for side-on binding of peroxide to Fe^{3+}-EDTA at high pH exists. In addition, the EPR spectrum of Fe^{3+}-EDTA at pH 7.6 indicates end-on binding of a hydroperoxide.[112]

The intermediacy of a hydroperoxide iron adduct in the "Gif" systems for the functionalization of saturated hydrocarbons has been proposed.[229]

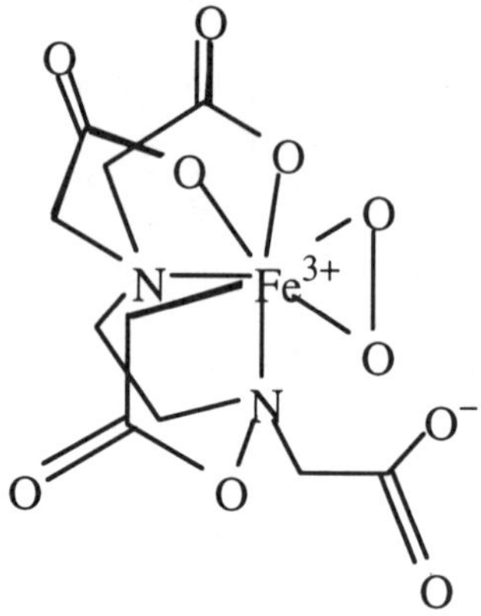

3.26

Recently, EPR evidence of low-spin ferric complexes assigned as $Fe(bpy)_2(OOH)Py$ [bpy, bipyridine; Py, pyridine] and $Fe(phen)_2(OOH)Py$ [phen, phenanthroline] was reported.[230] Based on the observation that the products obtained in Fenton systems are often different from those obtained from reaction with hydroxyl radicals, Sawyer proposed that a nucleophilic ferric peroxo complex, not free HO• is the dominant reactant.[231] It should be noted that this hypothesis is widely debated.[232]

Mononuclear non-heme iron dioxygen complexes are often too unstable to be characterized or isolated. In addition to the ferric-EDTA purple complex, there are a few notable exceptions. An iron (II) complex of a pentadentate aminopyridyl ligand based on ethylene diamine (**3.3**) was proposed to form an unstable end-on bound peroxo complex upon reaction with *t*-butyl hydroperoxide or H_2O_2.[233] Compound **3.27** formed a purple complex with hydrogen peroxide. This complex gave a rhombic EPR signal ($g = 2.19, 2.12, 1.95$) indicative of a low-spin ferric peroxo derivative. The observed EPR spectrum is similar to those reported for low-spin heme ferric peroxo complexes in which the peroxide is bound end-on (*vide infra*).

Que *et al.* observed an unstable pink species upon reaction of *t*-butyl hydroperoxide with $[Fe(tmpma)(O_2CPh)]ClO_4$ (tmpma, tris[(6-methyl-2-pyridyl)methyl]amine at -60 °C.[234] This pink species exhibits a band at

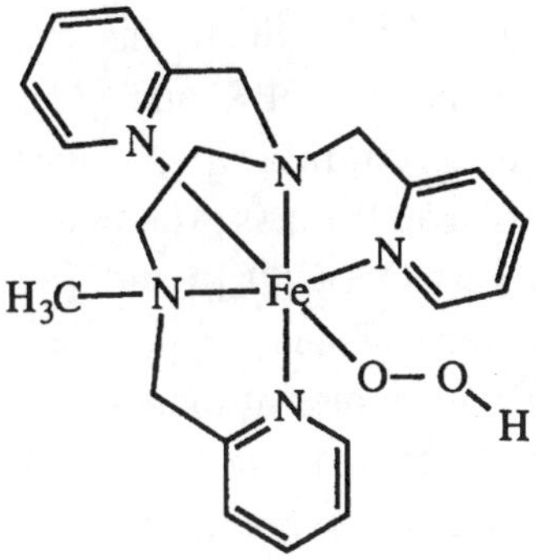

3.27 (OOH)

510 nm in the electronic spectrum which may indicate a side-on peroxide coordination analogous to the ferric-EDTA peroxo complex.

Iron complexes of cyclam (1,4,8,11-tetraazacyclotetradecane) were found to be effective catalysts for cyclohexene epoxidation by hydrogen peroxide.[235] Alkyl hydroperoxides did not epoxidize cyclohexene under the same conditions. A reaction mechanism was postulated in which an HOO^- complex of iron cyclam was the reactive intermediate.

Recently, a mononuclear ferric hydroperoxide intermediate was characterized. Reaction of an iron(II) complex of the pentadentate ligand **3.28** with H_2O_2 at −8 °C generated a transient purple species

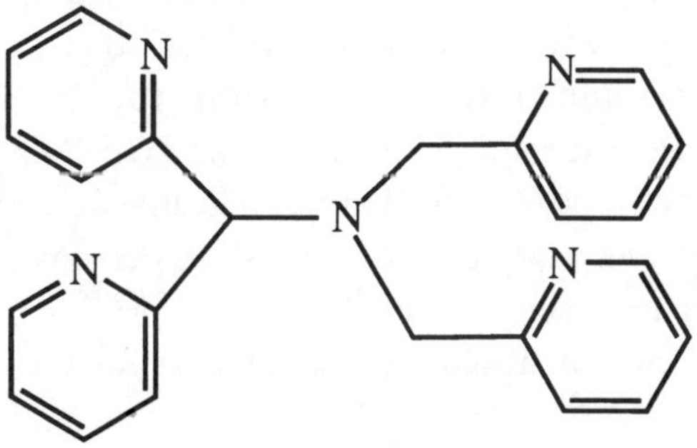

3.28

(λ_{max} 530 nm $\varepsilon \sim 1100$ $M^{-1}cm^{-1}$) which was characterized as a low-spin hydroperoxoiron(III) complex by EPR and electrospray ionization mass spectroscopy.[236] The EPR spectrum gave *g* values at 2.17, 2.12, and 1.98. Mass spectroscopy afforded mass spectra with prominent peaks corresponding to [**3.28**-Fe(OOH)ClO_4]$^+$ and [**3.28**-FeOOH)ClO_4]$^-$. This complex was able to oxidize alkanes.

In summary, only a few examples of mononuclear iron-peroxo complexes have been characterized. Most of these unstable complexes are competent for oxidation chemistry but have not as yet displayed any nucleophilic capabilities.

3.12.4 Hydrolytic metalloperoxides

Although no simple non-heme iron-peroxo model complexes are able to hydrolyze phosphate esters or amides, a few peroxo complexes of other metals have been reported to do so. Upon exposure to oxygen, a dinuclear copper complex (**3.29**) formed a purple intermediate formulated as **3.29**O_2 which effected hydrolysis of solvent DMF to yield a dinuclear copper bound formate.[237] Electrospray ionization and FAB mass spectrometric analysis of the reaction product with $^{18}O_2$ indicated that one ^{18}O atom was inserted into the formate. The reaction was proposed to proceed via a nucleophilic Cu-peroxo intermediate (Scheme 3.27). Alternatively, the reaction may go through a Cu-hydroxo intermediate.

The lanthanides: La(III), Pr(III), Nd(III), and Eu(III) have been shown to combine with hydrogen peroxide to provide almost nine orders of magnitude rate acceleration for hydrolyzing BNPP (bis(*p*-nitrophenyl)phosphate) at pH 7 and 25 °C.[238] The rate of reaction was second order in [La(III)] and [H_2O_2], indicative of a dimetallo-diperoxo active center. Labeling experiments with $H_2{}^{18}O_2$ demonstrated incorporation of ^{18}O into the phosphate product. Chin *et al.*[258,259] concluded that a La bound peroxide acted as a nucleophilic catalyst for the diester hydrolysis (Scheme 3.28).

A dinuclear Ce(III) complex was proposed to cleave the dinucleotide dApdA (2'-deoxyadenyl(3'-5')-2'-deoxy-adenosine) in the presence of

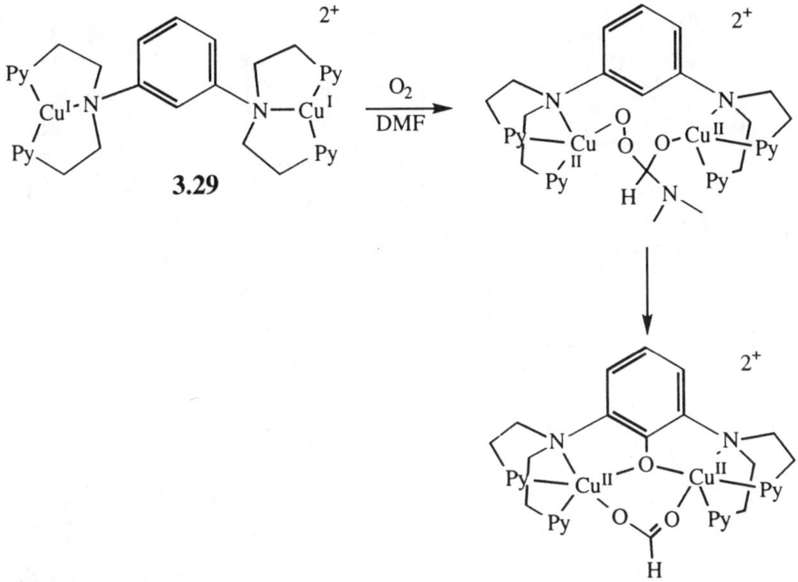

Scheme 3.27

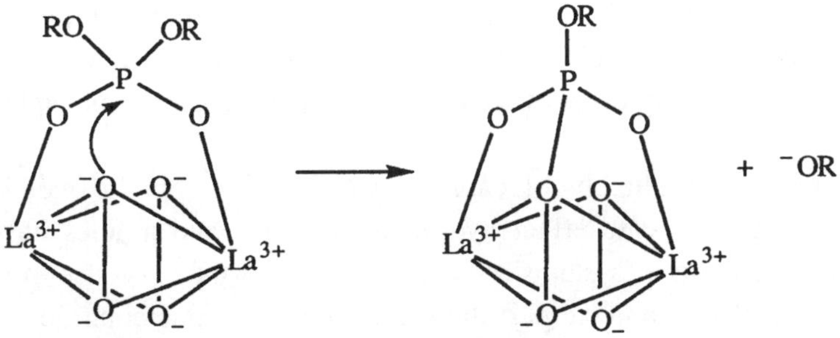

Scheme 3.28

dioxygen to give dA and inorganic phosphate.[239] No other lanthanide ions tested were competent for this reaction. In the proposed mechanism, two Ce(III) ions were oxidized to Ce(IV) by molecular oxygen thus yielding a dicerium peroxo complex. The peroxide then acted as a nucleophile by

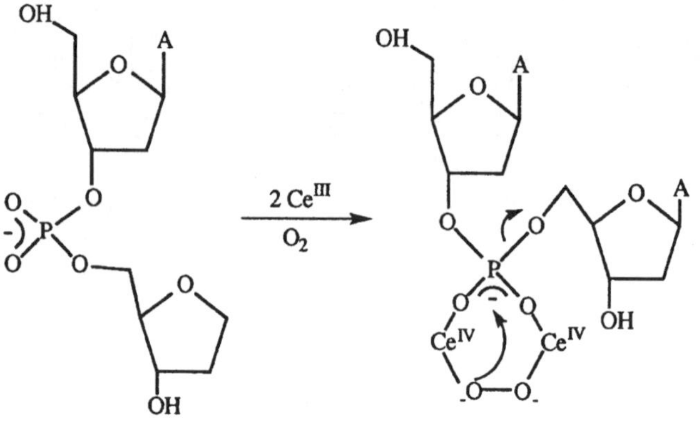

Scheme 3.29

attacking the coordinated phosphate diester (Scheme 3.29). A cryptate Ce(IV) peroxo complex was also implicated as a possible nucleophile in the catalytic hydrolysis of BNPP.[240]

Catalysis of the BNPP hydrolysis was observed by the La(III) complex of a cyclodextrin bipyridine dimer in the presence of H_2O_2.[241] Unlike the lanthanide reactions studied by Chin *et al.*, the stoichiometry in the transition state for this phosphate diester hydrolysis was determined to be 1:1 La(III): H_2O_2. Again, nucleophilic attack of a bound peroxo on the phosphate was postulated.

In summary, a number of cases in which a nucleophilic metal-bound peroxide is proposed to effect hydrolysis of esters and amides have been reported. These model systems serve as evidence for the existence of reactive peroxo iron species and support the *possibility* of the nucleophilic reactions proposed for these species.

3.13 Conclusions

The goal of biomimetic chemistry[242] is to provide *bona fide* examples of biological reactions in simple, isolable, synthetic compounds. The varied and often enigmatic transformations mediated by cytochrome P450 have inspired a vigorous two-decade long effort to understand and mimic these interesting reactions. Reactive metal-oxo and metal-peroxo intermediates have now been thoroughly characterized in model porphyrin environments and their involvement in a wide range of heme proteins is well-accepted. The successes with metalloporphyrin model systems have stimulated efforts to find examples of this chemistry in non-heme iron proteins and related model compounds. Further, knowledge of the enzymatic mechanisms offers the promise of new biomimetic processes for industrial scale catalysis.

Acknowledgments

The authors thank the many coworkers and colleagues at Princeton and around the world whose discoveries and insights have led to the development of this rich and fruitful field. Efforts at Princeton were generously supported over the years by grants from the National Institutes of Health, the National Science Foundation, the Monsanto Chemical Company, the ARCO Chemical Company, the Mobil Foundation and Solvay-Interox.

References

1. Ortiz de Montellano, P.R. (ed.) *Cytochrome P450. Structure, Mechanism and Biochemistry*. Plenum Press, New York, 1995.
2. Dawson, J.H. and Eble, K.S. *Adv. Inorg. Bioinorg. Mech.* 1986, **4:** 1–64.
3. Groves, J.T. and Han, Y.-Z. in *Cytochrome P450. Structure, Mechanism and Biochemistry* (ed.) Montellano, P.R.O. Plenum Press, New York, 1995, pp. 3–48.
4. Mann, T. in *Oxidases and Related Redox Systems* (eds.) King, T.E., Mason, H.S. and Morrison, M. Liss, New York, 1988, pp. 29 49.
5. Dolphin, D., Forman, A., Borg, D.C., Fayer, J. and Felton, R.H. *Proc. Natl. Acad. Sci. USA* 1971, **68:** 614–618.
6. Dolphin, D. and Felton, R.H. *Acc. Chem. Res.* 1974, **7:** 26–32.
7. Peisach, J., Blumberg, W.E., Wittenberg, B.A. and Wittenberg, J.B. *J. Biol. Chem.* 1968, **243:** 1871–1880.
8. Penner-Hahn, J.E., McMurry, T.J., Renner, M.L.-G., Eble, L. and Davis, K.S., 1983, **258:** 12761–12764.

9. Hewson, W.D., Hager, L.P. in *The Porphyrins* (eds.) Dolphin, D. Academic Press, New York, 1979 Vol. 8, pp. 295–332.
10. Fann, Y.-C., Gerber, N.C., Osmulski, P.A., Hager, L.P., Sligar, S.G. and Hoffmann, B.M. *J. Am. Chem. Soc.* 1994, **116:** 5989–5990.
11. Marletta, M.A. *J. Biol. Chem.* 1993, **268:** 12231.
12. White, K.A. and Marletta, M.A. *Biochemistry,* 1992, **31:** 6627.
13. McMillan, K., Bredt, D.S., Hirsch, D.J., Snyder, S.H., Clark, J.E. and Masters, B.S.S. *Proc. Natl. Acad. Sci. USA* 1992, **89:** 1141–1145.
14. Nick, R.J., Ray, G.B., Fish, K.M., Spiro, T.G. and Groves, J.T. *J. Am. Chem. Soc.* 1991, **113:** 1838–1840.
15. Varotsis, C., Zhang, Y., Appleman, E.H. and Babcock, G.T. *Proc. Natl. Acad. Sci. USA* 1993, **90:** 237–241.
16. Ogura, T., Hirota, S., Proshlyakov, D.A., Shinzawa-Itoh, K., Yoshikawa, S. and Kitagawa, T. *J. Am. Chem. Soc.* 1996, **118:** 5443–5449.
17. (a) Davies, J.A. *Selective Hydrocarbon Activation: Principle and Progress* VCH, New York, 1994; (b) Traylor, T.G. *Pure Appl. Chem.* 1991. **63:** 265–274; (c) Traylor, T.G., Hill, K.W., Fann, W.P., Tsuchiya, S. and Dunlap, B.E. *J. Am. Chem. Soc.* 1992, **114:** 1308; (d) Ostovic, D. and Bruice, T.C. *Acc. Chem. Res.* 1992, **25:** 314–320; (e) Meunier, B. *Chem. Rev.* 1992, **92:** 1411–1456; (f) Mansuy, D. *Coordin. Chem. Rev.* 1993, **125:** 129–141.
18. Groves, J.T. and Watanabe, Y. *J. Am. Chem. Soc.* 1988, **110:** 8443–8452.
19. Groves, J.T. and Watanabe, Y. *Inorg. Chem.* 1987, **26:** 785–786.
20. Groves, J.T. and Watanabe, Y. *J. Am. Chem. Soc.* 1986, **108:** 7834–7836.
21. Groves, J.T. and Watanabe, Y. *J. Am. Chem. Soc.* 1986, **108:** 507.
22. Groves, J.T. and Watanabe, Y. *Inorg. Chem.* 1986, **25:** 4808.
23. (a) Groves, J.T., Haushalter, R.C., Nakamura, M., Nemo, T.E. and Evans, B.J. *J. Am. Chem. Soc.* 1981, **102:** 2884–2886; (b) Groves, J.T. and McMurry, T.J. *Rev. Port. Chim.* 1985, **27:** 102–103.
24. Groves, J.T., Nemo, T.E. and Myers, R.S. *J. Am. Chem. Soc.* 1979, **101:** 1032–1033.
25. Penner-Hahn, J.E., McMurry, T.J., Renner, M., Latos-Grazynsky, L., Eble, K.S., Davis, I.M., Balch, A.L., Groves, J.T., Dawson, J.R. and Hodgson, K. *J. Biol. Chem.* 1983, **258:** 12761–12764.
26. (a) Penner-Hahn, J.E., Eble, K.S., McMurry, T.J., Renner, M., Balch, A.L., Groves, J.T., Dawson, J.H. and Hodgson, K.O. *J. Am. Chem. Soc.* 1986, **108:** 7819–7825; (b) Groves, J.T., Quinn, R., McMurry, T.J., Lang, G. and

Boso, B. *J. Chem. Soc., Chem. Commun.* 1984, 1455–1456; (c) Boso, B., Lang, G., McMurry, T.J. and Groves, J.T *J. Chem. Phys.* 1983, **79:** 1122–1126.
27. (a) Machii, K., Watanabe, Y. and Morishima, I. *J. Am. Chem. Soc.* 1995, **117:** 6691–6697; (b) Tsurumaki, H., Watanabe, Y. and Morishima, I. *J. Am. Chem. Soc.* 1993, **115 :** 11784–11788.
28. (a) Lee, K.H. and Nam, W. *J. Am. Chem. Soc.* 1997, **119:** 1916–1922; (b) Kamaraj, K. and Bandyopadhyay, D. *J. Am. Chem. Soc.* 1997, **119:** 8099–8100; (c) Fujii, H., Yoshimura, T. and Kamada, H. *Inorg. Chem.* 1997, **36:** 6142–6143; (d) Gross, Z. and Nimri, S. *Inorg. Chem.* 1994, **33:** 1731–1732; (e) Gross, Z. and Ini, S. *J. Org. Chem.* 1997, **62:** 5514–5521.
29. Groves, J.T., Gross, Z. and Stern, M.K. *Inorg. Chem.* 1994, **33:** 5065–5072.
30. Sheldon, R.A. *Metalloporphyrins in Catalytic Oxidations*, M. Dekker, New York, 1994.
31. Ellis, P.E. and Lyons, J.E. *Coord. Chem. Rev.* 1990, **105:** 181–193.
32. Eaton, S.S. and Eaton, G.R. *J. Am. Chem. Soc.* 1975, **97:** 3660–3666.
33. Petterson, T.C. and Alexander, L.E. *J. Am. Chem. Soc.* 1968, **90:** 3873–3875.
34. Guilard, R., Richard, P., Borai, M. and Laviron, E. *J. Chem. Soc., Chem. Commun.* 1980, 516–518.
35. Antipas, A., Buchler, J.W., Gouterman, M. and Smith, P.D. *J. Am. Chem. Soc.* 1980, **102:** 198–207.
36. Buchler, J.W. and Smith, P.D. *Angew. Chem., Int. Ed. Engl.* 1974, **13:** 341.
37. Ledon, H. and Bonnett, M.J. *J. Chem. Soc., Chem. Commun.* 1979, 702–704.
38. Groves, J.T. and Kruper, W.J. *J. Am. Chem. Soc.* 1979, **101:** 7613–7614.
39. Groves, J.T. and Haushalter, R.C. *J. Chem. Soc., Chem. Commun.* 1981, 1165–1166.
40. Creager, S.E. and Murray, R.W. *Inorg. Chem.* 1985, **24:** 3824–3828.
41. Garrison, J.M. and Bruice, T.C. *J. Am. Chem. Soc.* 1989, **111:** 191–198.
42. Takahashi, T. Ph.D. Dissertation, University of Michigan, 1985.
43. (a) Groves, J.T. and Quinn, R. *Inorg. Chem.* 1984, **23:** 3844–3846; (b) Groves, J.T. and Quinn, R. *J. Am. Chem. Soc.* 1985, **107:** 5790–5792; (c) Groves, J.T. and Roman, J.S. *J. Am. Chem. Soc.* 1995, **117:** 5594–5595.
44. (a) Dolphin, D., James, B.R. and Leung, T. *Inorg. Chim. Acta* 1983, **79:** 25–27; (b) Leung, T., James, B.R. and Dolphin, D. *Inorg. Chim. Acta* 1983, **79:** 180–181; (c) Barley, M., Becker, J.Y., Domazetis, G., Dolphin, D. and James, B.R. *Can. J. Chem.* 1983, **61:** 2389–2396; (d) James, B.R., Dolphin, D., Leung, T.W., Einstein, F.W. and Willis, A.C *Can. J. Chem.* 1984, **62:** 1238–1245; (e) James, B.R., Mikkelsen, S.R., Leung, T.W., Williams, G.M. and

Wong, R. *Inorg. Chim. Acta (B)* 1984, **85:** 209–213; (f) Cheng, S.Y.S. and James, B.R. *J. Mol. Catal. A.* 1997, **117:** 91–102; (g) For reviews, see (f) James, B.R. in *Fundamentals of Research in Homogeneous Catalysis* (ed.) Shilov, A.E. Gordon Breach, NY, 1986, pp. 309–324; (g) Mlodnika, T. and James, B.R. in *Metalloporphyrin Catalyzed Oxidations* (eds.) Montanari. F. and Casella, L. Kluwer; Dordrecht, The Netherlands, 1994, pp. 121.

45. (a) Higuchi, T., Satake, C. and Hirobe, M. *J. Am. Chem. Soc.* 1995, **117:** 8879–8880; (b) Ohtake, H., Higuchi, T. and Hirobe, M. *Heterocycles,* 1995 and references therein; (c) Higuchi, T. *J. Synth. Org. Chem. Jpn.* 1995, **53:** 644–644.
46. (a) Groves, J.T., Bonchio, M., Carofiglio, T. and Shalyaev, K. *J. Am. Chem. Soc.* 1996, **118:** 8961–8962; (b) Groves, J.T., Shalyaev, K.V., Bonchio, M. and Carofiglio, T in *Studies in Surface Science and Catalysis, Vol. 110; Proceedings of the Third World Congress on Oxidation Catalysis* (eds.) Grasselli, R.K., Oyama, S.T., Gafney, A.M. and Lyons, J.E. Elsevier, 1997, pp. 865–872.
47. (a) Groves, J.T., Kruper, W.J. and Haushalter, R.C. *J. Am. Chem. Soc.* 1980, **102:** 6377–6380; (b) McMurry, T.J. Ph.D. Thesis, University of Michigan, 1983.
48. (a) Hill, C.L. and Schardt, B.C. *J. Am. Chem. Soc.* 1980, **102:** 6374–6375; (b) Carnieri, N., Harriman, A. and Porter, G. *J. Chem. Soc. Dalton Trans.* 1982, 931–938; (c) Carnieri, N., Harriman, A. and Porter, G. *J. Chem. Soc. Dalton Trans.* 1982, 123–1238; (d) Kamp, N.W.J. and Lindsay Smith, J.R. *J. Mol. Catal. A: Chem.* 1996, **113:** 131–145.
49. For a discussion of the oxygenation of toluene by permanganate and chromyl reagents, see (a) Gardner, K.A. and Mayer, J.M. *Science*, 1995, **269:** 1849–1851; (b) Gardner K.A., Kuehnert, L.L. and Mayer, J.M. *Inorg. Chem.* 1997, **36:** 2069–2078; (c) Wang, K. and Mayer, J.M. *J. Org. Chem.* 1997, **62:** 4248–4252.
50. (a) De Porter, B. and Meunier, B. *J. Chem. Soc., Perkin Trans. II* 1985, 1735–1740; (b) Meunier, B., Guilmet, E., De Carvalho, M.E. and Poilblanc, R. *J. Am. Chem. Soc.* 1984, **106:** 6668–6676.
51. Collman, J.P., Brauman, J.I. and Meunier, B. *Proc. Natl. Acad. Sci. USA* 1984, **81:** 3245–3248.
52. Collman, J.P., Brauman, J.I., Meunier, B., Hayashi, T., Kodadek, T. and Raybuck, S.A. *J. Am. Chem. Soc.* 1985, **107:** 2000–2005.
53. Collman, J.P., Kodadek, T. and Brauman, J.I. *J. Am. Chem. Soc.* 1986, **108:** 2588–2594.

54. Groves, J.T., Watanabe, Y. and McMurry, T.J. *J. Am. Chem. Soc.* 1983, **105:** 4489–4490.
55. Groves, J.T. and Stern, M.K. *J. Am. Chem. Soc.* 1988, **110:** 8628–8638.
56. Buchler, J.W., Dreher, C., Lay, K.L., Lee, Y.J.A. and Scheidt, W.R. *Inorg. Chem.* 1983, **22:** 888–891.
57. Hill, C.L. and Hollander, F.J. *J. Am. Chem. Soc.* 1982, **104:** 7318–7319.
58. Bottomley, L.A. and Neely, F.L. *Inorg. Chem.* 1997, **36:** 5435–5439.
59. Groves, J.T. and Takahashi, T. *J. Am. Chem. Soc.* 1983, **105:** 2073–2074.
60. Bottomley, L.A. and Neely, F.L. *J. Am. Chem. Soc.* 1988, **110:** 6748–6752.
61. Bottomley, L.A. and Neely, F.L. *J. Am. Chem. Soc.* 1989, **111:** 5955–5957.
62. Woo, L.K., Goll, G.G., Czapla, D.J. and Hays, J.A. *J. Am. Chem. Soc.* 1991, **113:** 8478–8484.
63. Collins, T.J. and Gordon-Wylie, S.W. *J. Am. Chem. Soc.* 1989, **111:** 4511–4513.
64. Collins, T.J., Powell, R.D., Slebodnick, C. and Uffelman, E.S. *J. Am. Chem. Soc.* 1990, **112:** 899–901.
65. MacDonnell, F.M., Fackler, N.L.P., Stern, C. and O'Halloran, T.V. *J. Am. Chem. Soc.* 1994, **116:** 7431–7432.
66. (a) Srinivasan, K., Michaud, P. and Kochi, J.K. *J. Am. Chem. Soc.* 1986, **108:** 2309; (b) Jacobsen, E.N. in *Comprehensive Organometallic Chemistry II* (eds.) Wilkinson, G., Stone, F.G.A., Abel, E.W. and Hegedus, L.S. Pergamon, New York, 1995, Vol. 12, chap. 11; (c) Katsuki, T. *Coord. Chem. Rev.* 1995, **140:** 189; (d) Palucki, M., Finney, N.S., Pospisil, P.J., Güler, M.L., Ishida, T. and Jacobsen, E.N. *J. Am. Chem. Soc.* 1998, **120:** 948–954 and references therein; (e) Feichtinger, D. and Plattner, D.A. *Angew. Chem. Int. Edit.* 1997, **36:** 1718–1719.
67. Groves, J.T. and Marla, S.S. *J. Am. Chem. Soc.* 1995, **117:** 9578–9579.
68. (a) Bernadou, J., Fabiano, A.-S., Robert, A. and Meunier, B. *J. Am. Chem. Soc.* 1994, **116:** 9375–9376; (b) Pitie, M., Bernadou, J. and Meunier, B. *J. Am. Chem. Soc.* 1995, **117:** 2935–2936; (c) Balahura, R.J., Sorokin, A., Bernadou, J. and Meunier, B. *Inorg. Chem.* 1997, **36:** 3488–3492.
69. (a) Griller, D. and Ingold, K.U. *Acc. Chem. Res.* 1980, **13:** 317–323; (b) Ortiz de Montellano, P.R. and Stearns, R.A. *J. Am. Chem. Soc.* 1987, **109:** 3415–3420; (c) Bowry, V.W., Lusztyk, J. and Ingold, K.U. *J. Am. Chem. Soc.* 1989, **111:** 1927–1928; (d) Newcomb, M., Le Tadic-Biadatti, M.H., Chestney, D.L., Roberts, E.S. and Hollenberg, P.F. *J. Am. Chem. Soc.* 1995, **111:** 1927–1928 and references therein; (e) Huyser, E. *Adv. Free-Radical Chem.* 1965, **1:** 77–135; (f) Reyes, M.B. and Carpenter, B.K. *J. Am.*

Chem. Soc. 1998, **120:** 1641–1642; (g) Shaik, S., Filatov, M., Schröder, D. and Schwatz H. *Chem. Eur. J.* 1998, **4:** 193–199; (h) Detrich, J.L., Reinaud, O.M., Rheingold, A.L. and Theopold, K.H. *J. Am. Chem. Soc.* 1995, **117:** 11745–11748; (i) Siegbahn, P.E.M. *J. Am. Chem. Soc.*, 1996, **118:** 1487–1496; (j) Collman, J.P., Chien, A.S., Eberspacher, T.A. and Brauman, J.I. *J. Am. Chem. Soc.* 1998, **120:** 425–426.

70. (a) Czernuszewicz, R.S., Su, Y.O., Stern, M.K., Macor, K.A., Kim, D., Groves, J.T. and Spiro, T.G. *J. Am. Chem. Soc.* 1988, **110:** 4158–4165; (b) Cf. Ref. 85.
71. Arasasingham, R.D., He, G.X. and Bruice, T.C. *J. Am. Chem. Soc.* 1993, **115:** 7985–7991.
72. Groves, J.T., Lee, J. and Marla, S.S. *J. Am. Chem. Soc.* 1997, **119:** 6269–6273.
73. Beckman, J.S. and Koppenol, W.H. *Am. J. Physiol.* 1996, **271:** C1424–C1437.
74. Marla, S.S., Lee, J. and Groves, J.T. *Proc. Natl. Acad. Sci. USA* 1997, **94:** 14243.
75. Lee, J., Hunt, J.A. and Groves, J.T. *Bioorg. Med. Chem. Lett.* 1997, **7:** 2913–2918.
76. Lee, J., Hunt, J.A. and Groves, J.T. *J. Am. Chem. Soc.* 1998, in press.
77. Hunt, J.A., Lee, J. and Groves, J.T. *Chem. Biol.* 1997, **4:** 845–858.
78. (a) Stern, M.K., Jensen, M.P. and Kramer, K. *J. Am. Chem. Soc.* 1996, **118:** 8735–8736; (b) Salvemini, D., Wang, Z.-G., Stern, M.K., Currie, M.G. and Misko, T.P. *Proc. Natl. Acad. Sci. USA* 1998, **94:** 2659–2663.
79. Collman, J.P., Zhang, X., Lee, V. J., Uffelman, E. S. and Brauman, J.I. *Science* 1993, **261:** 1404–1411.
80. (a) Breslow, R., Zhang, X. and Huang, Y. *J. Am. Chem. Soc.* 1997, **119:** 4535–4536; (b) Breslow, R., Huang, Y. and Zhang, X.J. *Proc. Natl. Acad. Sci.* 1997, **94:** 11156–11158.
81. Groves, J.T. *Nature* 1997, **389:** 329.
82. Collman, J.P., Zhang, X., Hembre, R.T. and Brauman, J.I. *J. Am. Chem. Soc.* 1990, **112:** 5356–5357.
83. Collman, J.P., Lee, V.J., Zhang, X., Ibers, J.A. and Brauman, J.I. *J. Am. Chem. Soc.* 1993, **115:** 3834–3835.
84. Groves, J.T. and Myers, R.S. *J. Am. Chem. Soc.* 1983, **105:** 5791–5796.
85. Groves, J.T. and Viski, P. *J. Am. Chem. Soc.* 1989, **111:** 8537–8538.
86. (a) Groves, J.T. and Viski, P. *J. Org. Chem.* 1990, **55:** 3628–3634; (b) Groves, J.T., Crowley, S.J. and Shalyaev, K.V. *Chirality*, 1998, **10:** 106–119.
87. Groves, J.T. and Neumann, R. *J. Am. Chem. Soc.* 1989, **111:** 2900–2909.
88. Groves, J.T. and Neumann, R. *J. Am. Chem. Soc.* 1987, **109:** 5045.

89. Groves, J.T. and Neumann, R. *J. Org. Chem.* 1988, **53:** 3891–3893.
90. Groves, J.T. and Ungashe, S.B. *J. Am. Chem. Soc.* 1990, **112:** 7796–7797.
91. Groves, J.T., Fate, G.D. and Lahiri, J. *J. Am. Chem. Soc.* 1994, **116:** 5477–5478.
92. Lahiri, J., Fate, G.D., Ungashe, S.B. and Groves, J.T. *J. Am. Chem. Soc.* 1996, **118:** 2347–2358.
93. Akhtar, M., Calder, M.R., Corina, D.L. and Wright, J.N. *Biochem. J.* 1982, **201:** 569.
94. Marletta, M.A. *J. Biol. Chem.* 1993, **268:** 12231.
95. Rana, T.M. and Meares, C.F. *Proc. Natl. Acad. Sci. USA* 1991, **88:** 10578.
96. Kahne, D. and Still, W.C. *J. Am. Chem. Soc.* 1988, **110:** 7529.
97. (a) Sutton, P.A. and Buckingham, D.A. *Acc. Chem. Res.* 1987, **20:** 357; (b) Chin, J. *Acc. Chem. Res.* 1991, **24:** 145; (c) Chin, J., Banaszcyk, M., Jubian, and Kim, J.H. *Bioorg. Chem.* 1991, **2:** 175; (d) Suh, J. *Acc. Chem. Res.* 1992, **25:** 273.
98. Wiberg, K.B. *J. Am. Chem. Soc.* 1955, **77:** 2519.
99. Jencks, W.P. and Carriuolo, J. *J. Am. Chem. Soc.* 1960, **82:** 1778.
100. Behrman, E.J., Biallas, M.J., Brass, H.J., Edwards, J.O. and Isaks, M. *J. Org. Chem.* 1970, **35:** 3069.
101. Jenks, W.P. and Oakenfull, D.G. *J. Am. Chem. Soc.* 1971, **93:** 178.
102. Edwards, J.O. and Pearson, R.G. *J. Am. Chem. Soc.* 1962, **84:** 16.
103. San Fillippo, J. Jr., Chern, C.-I. and Valentine, J.S. *J. Org. Chem.* 1975, **40:** 1678.
104. San Fillippo, J. Jr., Romano, L.J., Chern, C.-I. and Valentine, J.S. *J. Org. Chem.* 1976, **41:** 586.
105. Evans, M.G. and Uri, N. *Trans. Faraday Soc.* 1949, **45:** 224.
106. Buckingham, D.A. and Engelhardt, L.M. *J. Am. Chem. Soc.* 1975, **97:** 5915.
107. McClune, G.J., Fee, J.A., McClusky, G.A. and Groves, J.T. *J. Am. Chem. Soc.* 1977, **99:** 5220.
108. (a) Valentine, J.S. *Chem. Rev.* 1973, **73:** 235; (b) Jones, R.D., Summerville, D.A. and Basolo, F. *Chem. Rev.* 1979, **79:** 139.
109. Rifkind, J.M. *Adv. Inorg. Biochem.* 1988, **7:** 155.
110. McCandlish, E., Miksztal, A.R., Nappa, M., Sprenger, A.Q., Valentine, J.S., Strong, J.D. and Spiro, T.G. *J. Am. Chem. Soc.* 1980, **102:** 4268.
111. Fujii, S., Ohya-Nishiguchi, H. and Hirota, N. *Inorg. Chim. Acta* 1990, **175:** 27.
112. Ahmad, A., McCallum, J.D., Shiemke, A.K., Appelman, E.H., Loehr, T.M. and Sanders-Loehr, J. *Inorg. Chem.* 1988, **27:** 2230.

113. Goddard, W.A., III and Olafson, B.D. *Proc. Natl. Acad. Sci. USA.* 1975, **72:** 2335.
114. Bach, R.D., Su, M-.D., Andres, J.L. and Schlegel, H.B. *J. Am. Chem. Soc.* 1993, **115:** 8763.
115. (a) Andrews, L., Chertinin, G.V., Ricca, A. and Bauschlicher, C.W. Jr. *J. Am. Chem. Soc.* 1996, **118:** 467; (b) Chang, S., Blyholder, G. and Fernandez, J. *Inorg. Chem.* 1981, **20:** 2813; (c) Abramowitz, S., Acquista, N. and Levin, I. *Chem. Phys. Lett.* 1977, **50:** 423.
116. Haber, F. and Weiss, J.J. *Proc. Roy. Soc. London, Ser. A.* 1934 **147:** 332.
117. Barb, W.G., Baxendale, J.H., George, P. and Hargrave, K.R. *Trans. Faraday Soc.* 1951, **47:** 591.
118. Walling, C. *Acc. Chem. Res.* 1975, **8:** 125.
119. Rush, J.D. and Koppenol, W.H. *J. Am. Chem. Soc.* 1988, **110:** 4957.
120. (a) Kharasch, M.S. and Sosnovsky, G. *J. Org. Chem.* 1958, **23:** 1322; (b) Schulz, M., Kirschke, K. and Hohne, E. *Chem. Ber.* 1967, **100:** 2242.
121. (a) Bernardi, R., Caronna, T., Galli, F., Minisci, F. and Perchinunno, M. *Tet. Lett.* 1973, **9:** 645; (b) Minisci, F. *Topics in Current Chemistry*, 1976, **62:** 1.
122. Babcock, G.T., Floris, R., Nilsson, T., Pressler, M., Varotsis, C. and Vollenbroek, E. *Inorg. Chim. Acta.* 1996, **243:** 345.
123. Groves, J.T. and McClusky, J. *J. Am. Chem. Soc.* 1976, **98:** 859.
124. Blake, R.C. and Coon, M.J. *J. Biol. Chem.* 1980, **255:** 4100.
125. Balny, C., Travers, F., Barman, T. and Douzou, P. *Eur. Biophys. J.* 1987, **14:** 375.
126. Erman, J.E., Hager, L.P. and Sligar, S.G. *Adv. Inorg. Biochem.* 1994, **10:** 71.
127. Back, H.K and Van Wart, H.E. *Biochem.* 1989, **28:** 5714.
128. (a) Ogura, T., Takahashi, S., Shinzawa-Itoh, K., Yoshikawa, S. and Kitagawa, T. *Bull. Chem. Soc. Jpn.* 1991, **64:** 2901; (b) Ogura, T., Takahashi, S., Hirota, S., Shinzawa-Itoh, K., Yoshikawa, S., Appelman, E.H. and Kitagawa, T. *J. Am. Chem. Soc.* 1993, **115:** 8527; (c) Einarsdottir, O. *Biochim. Biophys. Acta* 1995, **1229:** 129.
129. Wikstrom, M. and Morgan, J.E. *J. Biol. Chem.* 1992, **267:** 10266.
130. Hamilton, G.A. in *Molecular Mechanism of Oxygen Activation* (ed.) Hayaishi, O. Academic Press, New York, 1974, p. 405.
131. White, R.E. and Coon, M.J. *Ann. Rev. Biochem.* 1980, **49:** 315.
132. Sligar, S.G., Kennedy, K.A. and Pearson, D.C. in *Oxidases and Related Redox Systems* (eds.) King, T.E., Mason, H.A., Morrison, M. Pergamon Press, New York, 1982, p. 837.

133. Lipscomb, J., Sligar, S., Namtvedt, M. and Gunsalus, I.C. *J. Biol. Chem.*, 1976, **251:** 1116.
134. Stevenson, D.E., Wright, J.N. and Akhtar, M. *J. Chem. Soc., Perkin Trans. I* 1988: 2043.
135. (a) Watanabe, Y. and Ishimura, Y. *J. Am. Chem. Soc.* 1989, **111:** 8047; (b) Watanabe, Y. and Ishimura, Y. *J. Am. Chem. Soc.* 1989, **111:** 410; (c) Cole, P.A., Bean, J.M. and Robinson, C.H. *Proc. Natl. Acad. Sci. USA* 1990, **87:** 2999.
136. Vaz, A.D.N., Kessell, K.J. and Coon, M.J. *Biochem.* 1994, **33:** 13651.
137. Vaz, A.D.N., Roberts, E.S. and Coon, M.J. *J. Am. Chem. Soc.* 1991, **113:** 5886.
138. Roberts, E.S., Vaz, A.D.N. and Coon, M.J. *Proc. Natl. Acad. Sci. USA* 1991, **88:** 8963.
139. Cole, P.A. and Robinson, C.H. *J. Med. Chem.* 1990, **33:** 2933.
140. Cole, P.A. and Robinson, C.H. *J. Am. Chem. Soc.* 1988, **110:** 1284.
141. Cole, P.A. and Robinson, C.H. *J. Am. Chem. Soc.* 1991, **113:** 8130.
142. Korzekwa, K.R., Trager, W.F., Mancewicz, J. and Osawa, Y. *J. Steroid Biochem. Molec. Biol.* 1993, **44:** 367.
143. Akhtar, M., Wright, J.N., Shyadehi, A.Z. and Robichaud, P. *Pure & Appl. Chem.* 1994, **66:** 2387.
144. Robichqud, P., Wright, J.N. and Akhtar, M. *J. Chem. Soc., Chem. Commun.* 1994, 1501.
145. Imai, M., Shimada, H., Watanabe, Y., Matsushima-Hibiya, Y., Makino, R., Koga, H., Horiuchi, T. and Ishimura, Y. *Proc. Natl. Acad. Sci. USA* 1989, 7823.
146. Vaz, A.D.N., Pernecky, S.J., Raner, G.M. and Coon, M.J. *Proc. Natl. Acad. Sci. USA* 1996, **93:** 4644.
147. Swinney, D.C. and Mak, A.Y. *Biochem.* 1994, **33:** 2185.
148. Leone, A.M., Palmer, R.M., Knowles, R.G., Francis, P.L., Ashton, D.S. and Moncada, S. *J. Biol. Chem.* 1991, **266:** 23790.
149. (a) Stuer, D.J., Kwon, N.S., Nathan, C.F., Griffith, O.W., Feldman, P.L. and Wiseman, J. *J. Biol. Chem.* 1991, **266:** 6259; (b) Pufahl, R.A., Nanjappan, P.G., Woodard, R.W. and Marletta, M.A. *Biochem.* 1992, **31:** 6822.
150. (a) Feldman, P.L., Griffith, O.W. and Stuehr, D.J. *Chem. Eng. News* 1993, **71:** 26; (b) Korth, H.-G., Sustmann, R., Thater, C., Butler, A.R. and Ingold, K.U. *J. Biol. Chem.* 1994, **269:** 17776.
151. Pufahl, R.A., Wishnok, J.S. and Marletta, M.A. *Biochem.* 1995, **34:** 1930.
152. Feig, A.L. and Lippard, S.J. *Chem. Rev.* 1994, **94:** 759.

153. (a) Que, L., Jr. in *Iron Carriers and Iron Proteins*, (ed.) Loehr, T.M., VCH, New York, 1989 Vol. 5, pp. 467–524; (b) Que, L., Jr. *Adv. Inorg. Biochem.* 1983, **5:** 167.
154. (a) Cox, D.D. and Que, L., Jr. *J. Am. Chem. Soc.* 1988, **110:** 8085; (b) Jang, H.G., Cox, D.D. and Que, L., Jr. *J. Am. Chem. Soc.* 1991, **113:** 9200.
155. (a) Que, L., Jr., Lipscomb, J.D., Münck, E. and Wood, J.M. *Biochim. Biophys. Acta* 1977, **485:** 60; (b) Que, L., Jr. Lauffer, R.B., Lynch, J.B., Murch, B.P. and Pyrz, J.W. *J. Am. Chem. Soc.* 1987, **109:** 5381.
156. Barbaro, P., Bianchini, C., Mealli, C. and Meli, A. *J. Am. Chem. Soc.* 1991, **113:** 3181.
157. Bernhardt F.-H., Staudinger, H. and Ullrich, V. *Hoppe-Seyler's Z. Physiol. Chem.* 1970, **351:** 467.
158. Twilfer, H., Bernhardt, F.-H. and Gersonde, K. *Eur. J. Biochem.* 1985, **147:** 171.
159. Bernhardt, F.-H. and Kuthan, H. *Eur. J. Biochem.* 1981, **120:** 547.
160. (a) Bill, E., Bernhardt, F.-H., Marathe, V.R. and Trautwein, A. *Cienc. Biol: Biol. Mol. Cel.* 1980, **5:** 141; (b) Bill, E., Bernhardt, F.-H., Marathe, V.R. and Trautwein, A. *J. Phys.* 1980, **41:** C1–485; (c) Bill, E., Bernhardt, F.-H. and Trautwein, A. *Eur. J. Biochem.* 1981, **121:** 39.
161. Twilfer, H., Bernhardt, F.-H. and Gersonde, K. *Eur. J. Biochem.* 1981, **119:** 595.
162. Wende, P., Pflegler, K. and Bernhardt. F.-H. *Biochem. Biophys. Res. Commun.* 1982, **104:** 527.
163. Hutton, J.J., Tappel, A.L. and Udenfriend, S. *Arch. Biochem. Biophys.* 1967, **118:** 231.
164. Kivirikko, K.I., Myllyla, R. and Pihlajaniemi, T. *FASEB* 1989, **3:** 1609.
165. Hanauske-Abel, H.M. and Gunzler, V. *J. Theor. Biol.* 1982, **94:** 421.
166. (a) Ha, H.H., Ho, R.Y.N., Kisiel, J.F. and Valentine, J.S. *Inorg. Chem.* 1995, **34:** 2265; (b) Chiou, Y.-M. and Que, L., Jr. *J. Am. Chem. Soc.* 1995, **117:** 3999.
167. (a) Stubbe, J. and Kozarich, J.W. *Chem. Rev.* 1987, **87:** 1107; (b) Vanderwall, D.E., Lui, S.M., Wu. W., Turner, C.J., Kozarich. J.W. and Stubbe, J. *Chem. Biol.* 1997, **4:** 373–387.
168. Sugiura, Y. *Biochem. Biophys. Res. Commun.* 1979, **87:** 643.
169. Wu, J.C., Stubbe, J. and Kozarich, J.W. *Biochem.* 1985, **24:** 7569.
170. Giloni, L., Takeshita, M., Johnson, F., Iden, C. and Grollman, A.P.J. *Biol. Chem.* 1981, 8608.
171. Burger, R.M., Peisach, J. and Horwitz, S.B. *J. Biol. Chem.* 1981, **256:** 11636.
172. Natrajan, A., Hecht, S.M., van der Marel, G.A. and van Boom, J.H. *J. Am. Chem. Soc.* 1990, 3997.

173. Burger, R.M., Kent, T.A., Horwitz, S.B., Münck, E. and Peisach, J. *J. Biol. Chem.* 1983, **258:** 1559.
174. Sam, J.W., Tang, X.-J. and Peisach, J. *J. Am. Chem. Soc.* 1994, **116:** 5250.
175. Guajardo, R.J., Hudson, S.E., Brown, S.J. and Mascharak, P.K. *J. Am. Chem. Soc.* 1993, **115:** 7971.
176. (a) Xu, R.X., Nettesheim, D., Otvos, J.D. and Petering, D.H. *Biochem.*, 1994, **33:** 907; (b) Wu, W., Vanderwall, D.E., Lui, S.M., Tang, X-.J., Turner, C.J., Kozarich, J.W. and Stubbe, J. *J. Am. Chem. Soc.* 1996, **118:** 1268; (c) Wu, W., Vanderwall, D.E., Turner, C.J., Kozarich, J.W. and Stubbe, J. *J. Am. Chem. Soc.* 1996, **118:** 1281.
177. Rodriguez, L.O. and Hecht, S.M. *Biochem. Biophys. Res. Commun.* 1982, **104:** 1470.
178. Rabow, L.W., McGall, G.H., Stubbe, J. and Kozarich, J.W. *J. Am. Chem. Soc.* 1990, **112:** 3203.
179. (a) Strobel, S.A. and Dervan, P.B. *Science* 1990, **249:** 73; (b) Sitlani, A., Long, E.C., Pyle, A.M. and Barton, J.K. *J. Am. Chem. Soc.* 1992, **114:** 2303; (c) Chen, C.B. and Sigman, D.S. *Science* 1987, **237:** 1197; (d) Mack, D.P. and Dervan, P.B. *J. Am. Chem. Soc.* 1990, **112:** 4604; (e) Chow, C.S. and Barton, J.K. *J. Am. Chem. Soc.* 1990, **112:** 2839; (f) Schepartz, A. and Cuenoud, B. *J. Am. Chem. Soc.* 1990, **112:** 3247. Hoyer, D., Cho, H. and Schultz, P.G. *J. Am. Chem. Soc.* 1990, **112:** 3249.
180. Rana, T.M. and Meares, C.F. *J. Am. Chem. Soc.* 1990, **112:** 2457.
181. Kannan, K.K., Notstrand, B., Fridborg, K., Ohlsson, A. and Petef, M. *Proc. Natl. Acad. Sci. USA* 1975, **72:** 51.
182. Rana, T.M. and Meares, C.F. *J. Am. Chem. Soc.* 1991, **113:** 1859.
183. Ettner, N., Hillen, W. and Ellestad, G.A. *J. Am. Chem. Soc.* 1993, **115:** 2546.
184. (a) Platis, I.E., Ermacora, M.R. and Fox, R.O. *Biochem.* 1993, **32:** 12761; (b) Ermacora, M.R., Ledman, D.W., Hellinga, H.W., Hsu, G.W. and Fox, R.O. *Biochem.* 1994, **33:** 3625; (c) Cuenoud, B., Tarasow, T.M. and Schepartz, A. *Tett. Lett.* 1992, **33:** 895; (d) Ermacora, M.R., Delfino, J.M., Cuenoud, B., Shepartz, A. and Fox, R.O. *Proc. Natl. Acad. Sci., USA* 1992, **89:** 6383; (e) Shepartz, A. and Cuenoud, B. *J. Am. Chem. Soc.* 1990, **112:** 3247; (f) Hoyer, D., Cho, H. and Schultz, P.G. *J. Am. Chem. Soc.* 1990, **112:** 3249.
185. Kimball, A.S., Lee, J., Jayaram, M. and Tullius, T.D. *Biochem.* 1993, **32:** 4698.
186. Schnaith, L.M.T., Hanson, R.S. and Que, L. Jr. *Proc. Natl. Acad. Sci. USA*, 1994, **91:** 569.

187. (a) Conte, V., Di Furia, F. and Moro, S. *J. Phys. Org. Chem.* 1996, **9:** 329; (b) Reynolds, M.S. and Butler, A. *Inorg. Chem.* 1996, **35:** 2378; (c) Butler, A., Claugue M.J. and Meister, G.E. *Chem. Rev.* 1994, **94:** 625; (d) Taube, H. *Prog. Inorg. Chem.* 1986, **34:** 607; (e) Mimoun, H. in *The Chemistry of Functional Groups, Peroxides* (ed.) Patai, S. John Wiley & Sons Ltd, New York, 1983 p. 463; (f) Vaska, L. *Acc. Chem. Res.* 1976, **9:** 175; (g) Valentine, J.S. *Chem. Rev.* 1973, **73:** 235.
188. Fujii, S., Ohya-Nishiguchi, H. and Hirota, N. *Inorg. Chem. Acta* 1990, **175:** 27.
189. (a) Bakac, A., Assink, B., Espenson, J.H. and Wang, W.-D. *Inorg. Chem.* 1996, **35:** 788; (b) Mirza, S.A., Bocquet, B., Robyr, C., Thomi, S. and Williams, A.F. *Inorg. Chem.* 1996, **35:** 1332; (c) Bakac, A. *Prog. Inorg. Chem.* 1995, **43:** 267.
190. Wang, W.-D., Bakac, A. and Espenson, J.H. *Inorg. Chem.* 1993, **32:** 5034.
191. Suzuki, H., Matsuura, S., Moro-Oka, Y. and Ikawa, T. *J. Organomet. Chem.* 1985, **286:** 247.
192. Carmona, D., Lamata, M.P., Ferrer, J., Modrego, J., Perales, M., Lahoz, F.J., Atencio, R. and Oro, L. A. *J. Chem. Soc., Chem. Commun.* 1994, 575.
193. Zhang, X., Loppnow, G.R., McDonald, R. and Takats, J. *J. Am. Chem. Soc.* 1995, **117:** 7828.
194. Sheldon, R.A. and van Doorn, J. *J. Organomet. Chem.* 1975, **94:** 115.
195. Jones, R.D., Summerville, D.A. and Basolo, F. *Chem. Rev.* 1979, **79:** 139.
196. Chin, D.-H., La Mar, G.N. and Balch, A.L. *J. Am. Chem. Soc.* 1980, **102:** 4344.
197. Paeng, I.R., Shiwaku, H. and Nakamoto, K. *J. Am. Chem. Soc.* 1988, **110:** 1995.
198. Chang, C.K. and Traylor, T.G. *J. Am. Chem. Soc.* 1973, **95:** 5810.
199. (a) Collman, J.P., Gagne, R.R., Halbert, T.R., Marchon, J.-C. and Reed, C.A. *J. Am. Chem. Soc.* 1973, **95:** 7868; (b) Collman, J.P., Gagne, R.R., Reed, C.A., Halbert, T.R., Lang, G. and Robinson, W.T. *J. Am. Chem. Soc.* 1975, **97:** 1427; (c) Collman, J.P. *Acc. Chem. Res.* 1977, **10:** 265.
200. (a) Collman, J.P., Gagne, R.R., Reed, C.A., Robinson, W.T. and Rodley, G.A. *Proc. Natl. Acad. Sci. USA*, 1974, **71:** 1326; (b) Jameson, G.B., Rodley, G.A., Robinson, W.T., Gagne, R.R., Reed, C.A. and Collman, J.P. *Inorg. Chem.* 1978, **17:** 850.
201. (a) Almog, J., Baldwin, J., Dyer, R.L. and Peters, M. *J. Am. Chem. Soc.* 1975, **97:** 226; (b) Almog, J., Baldwin, J. and Huff, J. *J. Am. Chem. Soc.* 1975, **97:** 227.
202. Collman, J.P., Zhang, X.M., Wong, K. and Brauman, J.I. *J. Am. Chem. Soc.* 1994, **116:** 6245–6251.

203. (a) Almog, J.A., Miksztal, A.R., Shaevitz, B.A., Lang, G. and Valentine, J.S. *J. Am. Chem. Soc.*; Strong, J.D. and Spiro, T.G. *J. Am. Chem. Soc.* 1980, **102:** 4268.
204. Welborn, C.H., Dolphin, D. and James, B.R. *J. Am. Chem. Soc.* 1981, **103:** 2869.
205. Reed, C.A. in *Electrochemical and Spectrochemical Studies of Biological Redox Compounds* (eds.) Kadish, K.M. Advances in Chemistry 201, American Chemical Society, Washington, DC, 1982, p. 333.
206. Shirazi, A. and Goff, H.M. *J. Am. Chem. Soc.* 1982, **104:** 6318.
207. Burstyn, J.N., Roe, J.A., Miksztal, A.R., Shaevitz, B.A., Lang, G. and Valentine, J.S. *J. Am. Chem. Soc.* 1988, **110:** 1382.
208. Guilard, R., Latour, J.M., Lecomte, C., Marchon, J.C., Protas, J. and Ripoll, D. *Inorg. Chem.* 1978, **17:** 1228.
209. van Atta, R.B., Strouse, C.E., Hanson, L.K. and Valentine, J.S. *J. Am. Chem. Soc.* 1987, **109:** 1425.
210. (a) Sauer-Masarwa, A., Herron, N., Fendrick, C.M. and Busch, D.H. *Inorg. Chem.* 1993, **32:** 1086; (b) Sauer-Masarwa, A., Dickerson, L.D., Herron, N. and Busch, D.H. *Coord. Chem. Rev.* 1993, **128:** 117.
211. Tajima, K. *Inorg. Chim. Acta.* 1989, **163:** 115.
212. Tajima, K., Shigematsu, M., Jinno, J., Ishizu, K. and Ohya-Nishiguchi, H. *J. Chem. Soc., Chem. Commun.* 1990, 144.
213. Tajima, K., Oka, S., Edo, T., Miyake, S., Mano, H., Mukai, K., Sakurai, H. and Ishiuz, K. *J. Chem. Soc., Chem. Commun.* 1995, 1507.
214. Tajima, K., Jinno, J., Ishizu, K., Sakurai, H. and Ohya-Nishiguchi, H. *Inorg. Chem.* 1989, **28:** 709.
215. Balch, A.L. *Inorg. Chim. Acta.* 1992, **297:** 198-200.
216. Valentine, J.S., Burstyn, J.A. and Margerum, L.D. *Mechanisms of Dioxygen Activation in Metal-Containing Monooxygenases: Enzymes and Model Systems.* Plenum, New York, 1988.
217. Miksztal, A.R., Valentine, J.S. *Inorg. Chem.* 1984, **23:** 3548.
218. (a) Groves, J.T., Watanabe, Y. and McMurry, T.J. *J. Am. Chem. Soc.* 1983, **105:** 4489; (b) Groves, J.T. and Watanabe, Y. *Inorg. Chem.* 1986, **27:** 4808; (c) Watanabe, Y. and Groves, J.T. in *The Role of Oxygen in Chemistry and Biochemistry* (eds.) Ando, W. and Mooro-Oka, Y. Elsevier Science, Amsterdam, 1988, p. 471.
219. Khenkin, A.M. and Shteinman, A.A. *J. Chem. Soc., Chem. Commun.* 1984, 1219.
220. Schappacher, M., Weiss, R., Montiel-Montoya, R. and Trautwein, A. *J. Am. Chem. Soc.* 1985, **107:** 3736.

221. Schappacher, M. and Weiss, R. *Inorg. Chem.* 1987, **26:** 1190.
222. (a) Watanabe, Y. and Ishimura, Y. *J. Am. Chem. Soc.* 1989, **111:** 410; (b) Watanabe, Y. and Ishimura, Y. *J. Am. Chem. Soc.* 1989, **111:** 8047.
223. Tajima, K., Yoshino, M., Mikami, K., Edo, T., Ishizu, K. and Ohya-Nishiguchi, H. *Inorg. Chim. Acta* 1990, **172:** 83.
224. (a) Sisemore, M.F., Burstyn, J.N. and Valentine, J.S. *Angew. Chem. Int. Ed. Engl.* 1996, **35:** 206–208; (b) Selke, M. and Valentine, J.S. *J. Am. Chem. Soc.* 1998, **120:** 2652.
225. Selke, M., Sisemore, M.F. and Valentine, J.S. *J. Am. Chem. Soc.* 1996, **118:** 2008.
226. Cheng, K.L. and Lott, P.F. *Anal. Chem.* 1956, **28:** 462.
227. Walling, C., Kurz, M. and Schugar, H.J. *Inorg. Chem.* 1970, **9:** 931.
228. Bull, C., McClune, G.J. and Fee, J.A. *J. Am. Chem. Soc.* 1983, **105:** 5290.
229. Barton, D.H.R. and Taylor, C.K. *Pure & Appl. Chem.* 1996, **68:** 497.
230. Sobolev, A.P., Babushkin, D.E. and Talsi, E.P. *Mendeleev Commun.* 1996, **1:** 33.
231. (a) Sawyer, D.T., Sobkowiak, A. and Matsushita, T. *Acc. Chem. Res.* 1996, **29:** 409; (b) Hage, J.P., Llobet, A. and Sawyer, D.T. *Bioorg. & Med. Chem.* 1995, **3:** 1383; (c) Sawyer, D.T., Kang, C., Llobet, A. and Redman, C. *J. Am. Chem. Soc.* 1993, **115:** 5817.
232. (a) Minisci, F., Fontana, F., Araneo, S., Recupero, F., Banfi, S. and Quici, S. *J. Am. Chem. Soc.* 1995, **117:** 226; (b) Newcomb, M., Simakov, P.A. and Park, S.-U. *Tett. Lett.* 1996, **37:** 819; (c) Snelgrove, D.W., MacFaul, P.A., Ingold, K.U. and Wayner, D.D.M. *Tett. Lett.* 1996, **37:** 823.
233. Bernal, I., Jensen, I.M., Jensen, K.B., McKenzie, C.J., Toftlund, H. and Tuchagues, J.-P. *J. Chem. Soc. Dalton Trans.* 1995, 3667.
234. Zang, Y., Elgren, T.E., Dong, Y. and Que, L. Jr. *J. Am. Chem. Soc.* 1993, **115:** 811.
235. Nam, W., Ho, R. and Valentine, J.S. *J. Am. Chem. Soc.* 1991, **113:** 7052.
236. Lubben, M., Meetsma, A., Wilkinson, E.C., Feringa, B. and Que, L. Jr. *Angew. Chem. Int. Ed. Engl.* 1995, **34:** 1512.
237. (a) Murthy, N.N., Mahroof-Tahir, M. and Karlin, K. *J. Am. Chem. Soc.* 1993, **115:** 10405; (b) Murthy, N.N. and Karlin, K.D. in *Mechanistic Bioinorganic Chemistry* (eds.) Thorp, H.H. and Pecoraro, V.L. American Chemical Society, Washington DC, 1995, p. 165.
238. (a) Takasaki, B.K. and Chin, J. *J. Am. Chem. Soc.* 1993, **115:** 9337; (b) Takasaki, B.K. and Chin, J. *J. Am. Chem. Soc.* 1995, **117:** 8582.

239. Takasaki, B.K. and Chin, J. *J. Am. Chem. Soc.* 1994, **116:** 1121.
240. Oh, S.J., Song, K.H., Whang, D., Kim, K., Yoon, T.H., Moon, H. and Park, J.W. *Inorg. Chem.* 1996, **35:** 3780.
241. Breslow, R. and Zhang, B. *J. Am. Chem. Soc.* 1994, **116:** 7893.
242. Breslow, R. *Chem. Biol.* 1998, **5:** R27–R28.

Chapter 4

Models of Heme Peroxidases and Catalases

Bernard Meunier

Laboratoire de Chimie de Coordination du CNRS,
205 route de Narbonne, 31077 Toulouse cedex 4, France

1. Introduction

Peroxidases were among the first enzymes to be discovered. In fact, the oxidation of guaiacol by hydrogen peroxide was observed by Schonbein in 1855. These enzymes are found in all plants and animals, and are essential for living systems.[1] Their key role is to catalyze the abstraction of one or two electrons (usually two electrons) via a single electron transfer from an organic substrate, hydrogen peroxide being used as an electron acceptor according to Eq. (1).

$$AH_2 + H_2O_2 \text{ ------ peroxidase -------> } A + 2H_2O \qquad (1)$$

The prosthetic group of peroxidases is often composed of a heme with a histidine residue as proximal ligand. Horseradish peroxidase (the paradigm of plant peroxidases), ligninase and manganese peroxidase are a few examples of heme-enzymes belonging to the superfamily of plant peroxidases.[2] Myeloperoxidase, lactoperoxidase, salivary peroxidase and thyroid peroxidase are examples of human heme-peroxidases.[1] However, it must be noted that glutathione peroxidase catalyzes the elimination of peroxides in human blood with a selenium ion at the active site.[3] After a short description of the structure and catalytic activity of heme-peroxidases and heme-catalases, we will provide a survey on models of these heme-containing enzymes. Models of non-heme peroxidases and catalases will be presented in the next chapter.

2. Structure and Catalytic Activity of Selected Heme-Peroxidases

Within the scope of the present book, we will limit the presentation of peroxidase structures and activities to a few classical examples which have been extensively studied during recent decades: horseradish peroxidase and three fungal peroxidases: chloroperoxidase, ligninase and manganese peroxidase.

2.1 Horseradish peroxidase

The largest number of publications on peroxidases have been devoted to horseradish peroxidase (HRP).[4–6] The different oxidation states and the properties of this enzyme have been well-studied.

Horseradish peroxidase contains at least seven major isoenzymes, but the majority of the physico-chemical data have been obtained with the mixture of isoenzymes B and C (C being the major component in highly purified samples). The prosthetic group of HRP is a ferric-protoporphyrin IX with a histidine as proximal ligand (His-170). No high-resolution X-ray structure of HRP is presently available, but the essential amino acids of the distal site have been identified (the structure of cytochrome *c* peroxidase CCP was the first heme-peroxidase structure to be determined by X-ray diffraction).[7] The high-valent iron-oxo entity of HRP-Compound I is generated by hydrogen peroxide via its heterolytic cleavage involving the guanidinium function of arginine-38 (acting as proton source for the formation of the water molecule) and histidine-42 acting as base in the deprotonation step of H_2O_2 (Scheme 1).[8]

The oxidation of HRP by hydrogen peroxide generates a reactive intermediate (Compound I) containing two redox equivalents (above the resting state of the native enzyme) and has been characterized by several physico-chemical methods. Compounds I and II have a Soret band at 400 nm (broad) and 420 nm, respectively. Compound I consists of a iron(IV)-oxo species with a radical cation on the porphyrin ring (Scheme 2). The formation of Compound I is a chemically controlled reaction with a

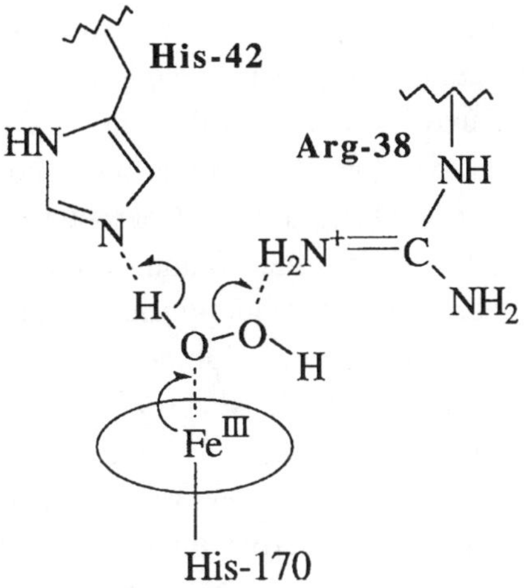

Scheme 1 Distal amino acids essential in the heterolytic cleavage of the O–O bond of hydrogen peroxide by HRP.

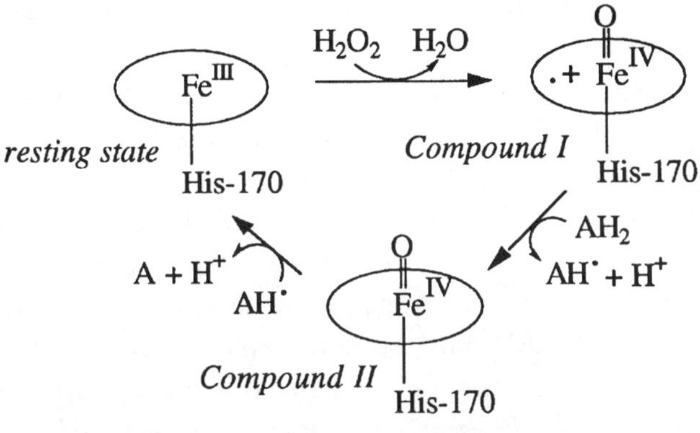

Scheme 2 Catalytic cycle of a heme-peroxidase (AH_2 is a two-electron donor).

pre-equilibrium of reactants to form a precursor complex (named Compound 0) which has been observed in stopped-flow UV-vis experiments. This Compound 0 is a ferric complex with a hyperporphyrin spectrum having a molecule of H_2O_2 within the active site[9] (for calculations on a putative ferric-hydroperoxide intermediate, see Ref. 10).

Compound I has been characterized by X-ray absorption and Raman spectroscopies. Both methods confirmed the presence of an iron(IV)-oxo entity in both Compounds I and II with a short Fe=O bond of 1.6 Å consistent with a ferryl structure.[11] More recently, Powers and colleagues suggested that the Fe–O bond distance of Compound II was longer: 1.9 Å at pH 7 and 1.7 at pH 10.[12] Resonance Raman studies confirmed the Fe=O structure of Compounds I and II with vibrations at 737 and 776 cm^{-1}, respectively.[13] The π-radical cation of Compound I has a predominant $^2A_{2u}$ character, indicative of an electron abstraction from the a_{2u} orbital of the porphyrin ligand.[14] This porphyrin radical is ferromagnetically coupled with the spin $S = 1$ of the ferryl state, as evidenced by EPR and Mössbauer studies.[15] Compound III is an inactive intermediate corresponding to an iron(III)-peroxo state resulting from the addition of dioxygen to the ferrous state of HRP or from an excess of hydrogen peroxide with the ferric state of the enzyme.[16] The redox potentials (E_0' values) of Compounds I and II have been determined as being 0.88 V and 0.90 V, respectively for the couples Compound I/Compound II and Compound II/ferric state (two very similar values).[17]

The high-valent iron-oxo species of reactive intermediates of HRP are not directly accessible to the different substrates. Alkylation of the δ-*meso* position of the heme group by alkylhydrazines indicates that substrates approach the active site of HRP from one particular edge of the prosthetic group.[18] Additional data on the substrate-binding site of HRP have been obtained by proton NMR spectroscopy and molecular dynamics.[19–21] The weak binding site of HRP must be considered as an open pocket which can accommodate a large range of substrates, explaining why this enzyme is so useful as a catalyst in the H_2O_2-oxidation of many different electron-donating molecules as soon as their redox potentials are compatible with that of Compounds I and II. For example, this enzyme has also been used for modeling the extra-hepatic oxidation (not oxygenation) of exogens (including many different drugs).[22] Studies on the mechanism of O-demethylation of aromatic substrates catalyzed by HRP have shown that the methoxyl substituent is removed as methanol and not as formaldehyde (as in cytochrome P450 mediated oxidation).[23,24] Despite several claims,[25]

HRP is unable to catalyze the transfer of an oxygen atom from hydrogen peroxide to a substrate molecule. HRP is unable to oxidize styrene whereas chloroperoxydase is able to catalyze the oxidation of this olefin to a mixture of the corresponding epoxide and phenylacetaldehyde.[26] A study on the co-oxidation of styrene and 4-methylphenol catalyzed by HRP indicates that the epoxide results from the reaction with hydroperoxide derived from the addition of dioxygen to a phenolic radical intermediate.[27]

2.2 *Chloroperoxidase*

Chloroperoxidase (CPO) is among the few peroxidases which are able to catalyze the oxidative chlorination of substrates using H_2O_2 and Cl^- (myeloperoxidase MPO is another example) according to Eq. (2).[28]

$$AH + H_2O_2 + Cl^- + H^+ \text{ ------ CPO -------> } ACl + 2H_2O \quad (2)$$

This halogen atom substitution is catalyzed by CPO on substrates containing an activated carbon-hydrogen bond such as β-diketones (chlorodimedone is a classical substrate for CPO activity assays). This heme-enzyme can be easily obtained and purified from the fungus *Caldariomyces fumago*. A noticeable difference between CPO and HRP is the presence of a cysteine residue as proximal ligand (Cys-29) in CPO.[29,30] The X-ray structure has been determined recently and confirms the presence of a manganese(II) ion bound to a heme propionate and also surrounded by His-105, Glu-104 and Ser-108 (this manganese ion may be the binding site for a chloride ion).[31] The structure of myeloperoxidase has also been determined and a calcium binding site was located relatively distant from the heme. This peroxidase, secreted by activated phagocytes, catalyzes the formation of hypochlorous acid which is one of the key factors in the antimicrobial activity of these neutrophiles.[32]

The addition of hydrogen peroxide to the ferric state of the enzyme generates CPO-Compound I, the only detectable intermediate having a Fe(IV)=O bond with a Raman stretching band at 790 cm^{-1}.[33,34] The radical-cation of Compound I may be delocalized on the macrocycle or on the axial ligand (see Ref. 35 for a proposal on the role of the S-ligand in high-valent cysteinato-heme enzyme). The addition of chloride to Compound I might generate a Fe(III)-OCl entity, a candidate to explain the substrate chlorination.[36] The alternative mechanism is the formation of free HOCl.[28] The same mixture of chloroanisole isomers was observed in the chlorination of anisole by free HOCl or by the CPO chlorinating system. Another alternative mechanism is to consider that the manganese site of CPO is the binding site of Cl^-. The short distance between this Mn site and the heme should facilitate the electron removal from the chloride ion and then Cl^+ could be transferred to the substrate present at the active site. Such a hypothesis would explain the absence of formation of free HOCl during the CPO catalytic cycle without invoking the formation of an iron(III)hypochlorito entity.

Chloroperoxidase is inactivated by formation of an N-phenyl-heme in the oxidation of phenylhydrazine or an N-alkyl-heme in the oxidation of terminal olefins.[37,38] These data suggest that different substrates have relatively easy access to the active site of CPO. In fact, this enzyme is able to catalyze the oxidation of a large range of substrates (tertiary amines, olefins, heterocycles, ...). The mechanism of N-demethylation of tertiary amines catalyzed by CPO probably involves an electron abstraction from the nitrogen atom followed by a proton elimination of the methyl group (a primary isotope effect has been observed, $k_H/k_D = 2.5$). An alternative mechanism is the abstraction of a hydrogen atom from the methyl group as observed for the hydroxylation reaction catalyzed by cytochrome P450, but no hydroxylase activity has been reported so far for CPO.[24,28] Contrarily to HRP or MPO, CPO is able to catalyze the epoxidation of different olefins (styrene, propylene, allyl chloride).[39] The oxygen atom of the epoxide arises from the primary oxidant as evidenced by using ^{18}O-labeled hydrogen peroxide.[26] Recent studies have demonstrated that the CPO-catalyzed epoxidation of *cis*-disubstituted olefins is highly enantioselective: enantiomeric excesses (*ee* values) range from 33 to 85%.[40] To avoid the irreversible degradation of CPO in the presence of H_2O_2, the concentration of this oxidant must be maintained as low as possible during the course of epoxidation. In these conditions, it has been possible to obtain 4200 catalytic cycles with an *ee* value of 94% in the epoxidation of methylallyl propionate.[41] The other useful reaction catalyzed by CPO is the enantioselective oxidation of sulfides to provide chiral sulfoxides with *ee*'s up to 90–95%.[42–44] Chiral sulfoxides are important synthons in the stereoselective synthesis of natural products. From data obtained with a series of *para*-substituted thioanisoles, the detailed mechanism of the CPO-catalyzed S-oxygenation was proposed to involve a catalyzed oxygen atom transfer from Compound I to the substrate rather than a one-electron oxidation of the sulfide followed by the addition of H_2O_2 on the intermediate S-radical-cation, as proposed for sulfide oxidations catalyzed by HRP.[45]

2.3 Ligninase

The possible development of enzymatic degradation of lignin from wood pulp in the cellulose industry has stimulated many studies during the last two decades on ligninolytic fungi. In particular, one fungus has dominated the attention of several research groups in universities and industries: *Phanerochaete chrysosporium*, which is able to degrade lignin by producing two extracellular peroxidases: ligninase (LiP) and manganese peroxidase

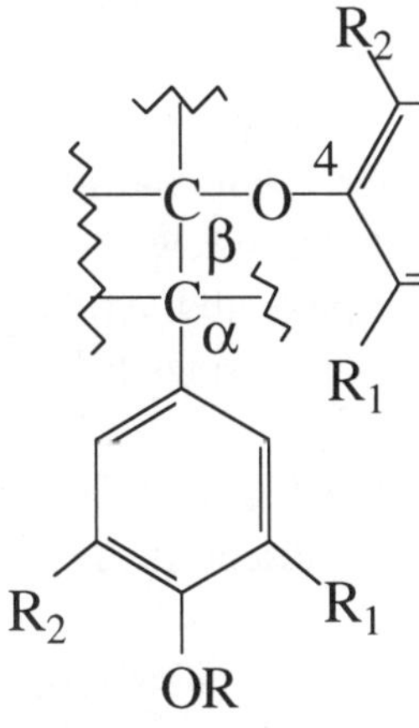

$R = R_1 = R_2 = H$ p-hydroxyphenyl type

$R = R_2 = H, R_1 = OMe$ guaiacyl type

$R = H, R_1 = R_2 = OMe$ syringyl type

Scheme 3 Structure of a phenylpropanoid subunit of lignin with an arylglycerol-β-arylether linkage (β-O-4).

(MnP). Both enzymes are able to oxidize lignin (the ultimate degradation product of lignin being CO_2) via the formation of radical-cation intermediates on the aromatic rings of this natural polymer resulting from the polymerization of *p*-coumarylic, coniferyl or sinapylic alcohols (these cinnamic alcohols are all derived from phenylalanine).[46–48] The main target in lignin oxidation is the oxidation of phenylpropanoid units which constitute nearly 50% of the different linkages between these subunits of lignin (Scheme 3).

In particular, the oxidative cleavage of the C_α–C_β bond of these arylglycerol-β-arylether motifs is the key step in both enzymatic and chemical degradation of lignin. This C–C bond cleavage is triggered by the initial formation of radical-cation intermediates on one of the methoxylated aromatic rings as mentioned above.

The purification of ligninase in 1983 by the groups of both Kirk and Gold has largely contributed to the understanding of the mechanism of action of this heme-enzyme.[49,50] Ligninase contained in the extracellular fluid of ligninolytic cultures of *Phanerochaete chrysosporium* has been purified by gel permeation chromatography and six isozymes have been characterized. The main one (H8) has been used for physico-chemical studies. LiP has a molecular weight of 42 000 Da and this glycoprotein contains one iron-protoporphyrin IX entity as prosthetic group. Early studies on LiP based on ^{18}O incorporation into oxidation products (when performed under ^{18}O-labeled molecular oxygen) led to the classification of ligninase as an oxygenase.[51] However, additional studies indicated that ^{18}O from labeled water was incorporated within the degradation products as expected for a peroxidase-mediated oxidation (the initial radical-cation generated on methoxylated aromatic rings is highly electrophilic and reacts with water).[52]

The resting state of the enzyme is a high-spin iron(III)-porphyrin with a histidine as proximal residue.[53] No water molecule is present as the sixth ligand at room temperature. The X-ray structure of LiP confirms the presence of a proximal histidine (His-175), the classical distal amino acids (His-47 and Arg-43) and a large pocket on the heme edge.[54] The overall fold of ligninase resembles that of cytochrome *c* peroxidase. As for other peroxidases, a high-valent iron-oxo entity associated with a porphyrin π-radical-cation, Compound I, is generated by the addition of hydrogen peroxide (see Scheme 4 for the catalytic cycle of LiP). This green compound (Soret band at 408 nm) has UV-vis and EPR characteristics similar to that of HRP-Compound I.[55,56]

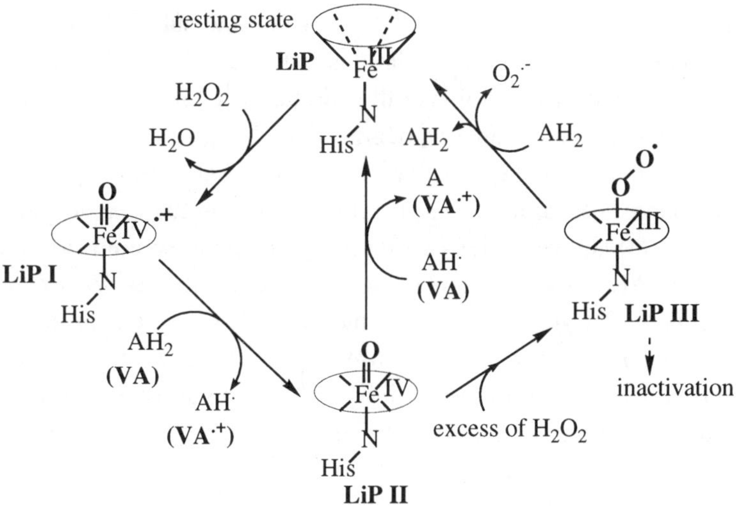

Scheme 4 Catalytic cycle of ligninase (AH_2 = a two-electron reductant, VA = veratryl alcohol).

The addition of 0.5 equivalent of a two-electron reducing agent like veratryl alcohol generates LiP-Compound II (Soret band at 420 nm) which corresponds to the iron(IV)-oxo without oxidation of the porphyrin ligand.

Unexpectedly, LiP is a fragile enzyme when exposed to more than 20 equivalents of hydrogen peroxide: besides the two catalytically active forms (Compounds I and II), two inactive forms LiP III and LiP III* have been obtained in the presence of an excess of hydrogen peroxide or by addition of molecular oxygen to the reduced form of LiP (ferrous state).[57] The difference between these two inactive forms is due to the presence of an extra molecule of H_2O_2, probably in the vicinity of the prosthetic group in the case of LiP III*. Any adventitious electron transfer from the ferrous state of the enzyme to this extra hydrogen peroxide molecule would generate a hydroxyl radical, thus creating an irreversible inactivation of LiP.[58] The fragility of LiP in the presence of an excess of hydrogen peroxide is the limiting factor for the industrial use of ligninase in delignification or pollutant oxidation.

Veratryl alcohol is a secondary metabolite in delignification mediated by *Phanerochaete chrysosporium* and is thus a good substrate for ligninase. It also enhances the rates of oxidations catalyzed by LiP.[59] It has been proposed that veratryl alcohol, via the formation of its radical-cation, is a chemical mediator in the oxidative degradation of lignin. The veratryl alcohol radical-cation mediates the transfer of electrons between the enzyme and lignin, two large macromolecules (for discussions on the kinetics and redox data, see Ref. 60 and 61). Many oxidation products generated from veratryl alcohol by LiP/H_2O_2 have been identified and quantified. Veratraldehyde, two quinones (2-hydroxymethyl-5-methoxy-1,4-benzoquinone and 2-methoxy-1,4-benzoquinone) and three lactones have been isolated.[62] Oxidation products of lignin model dimers having an arylglycerol-3-arylether moiety have also been identified.[63,64] The C_α–C_β bond cleavage is mediated by the initial formation of a radical-cation on one of the two aromatic rings. In the case of 1-(3,4-dimethoxyphenyl)-2-(2-methoxyphenoxy) propane-1,3-diol, the main oxidation products generated by LiP/H_2O_2 are veratraldehyde, guaiacol and glycolaldehyde.[65] This latter compound is oxidized by the fungal glyoxal oxidase to produce hydrogen peroxide which can be used by LiP.[66] Experiments performed under $^{18}O_2$ and/or with $H_2{}^{18}O$ provided details on the mechanisms of formation of these different oxidation products.[67]

One of the most promising properties of ligninase is its capacity to oxidize various recalcitrant organic molecules. Microorganisms in soils and water convert many man-made chemicals into inorganic products.[68] However, some molecules are recalcitrant to biotransformation and are subject to biomagnification and accumulation in plants or animal tissues. Among the compounds resisting microbial degradation, DDT and lindane are two classical examples. Both are slowly transformed by microbial reduction[69] and few extracellular enzymes are able to catalyze their oxidation. In order to be effective, these enzymes must have higher redox capabilities than horseradish peroxidase. *Phanerochaete chrysosporium* is able to oxidize trichlorophenols and benzodioxines.[70–73] This oxidation is catalyzed by ligninase and can be extended to other polycyclic aromatics like benzo[a]pyrene and 9,10-dimethylanthracene.[70,74]

2.4 *Manganese peroxidase*

The culture medium of *Phanerochaete chrysosporium* contains another extracellular peroxidase. This enzyme is manganese-dependent and four isozymes have been identified, but none of them are able to catalyze the oxidation of veratryl alcohol in the presence of H_2O_2.[75] Manganese peroxidase is a heme-glycoprotein of 46 000 Da which requires manganese(II) for its activity (this peroxidase also has a calcium ion to prevent the thermal inactivation of the enzyme).[76–78] The primary amino acid sequence of MnP has been deduced from its cDNA.[79] This peroxidase exhibits a proximal histidine as well as distal histidine and arginine that catalyze the heterolytic cleavage of the peroxidic O–O bond. The X-ray structure has confirmed these features and it was proposed that the binding site of the manganese(II) ion consisted of a heme propionate and three acidic ligands, Glu-35, Glu-39 and Asp-179.[80]

The native manganese peroxidase is oxidized by hydrogen peroxide to a Compound I species similar to HRP-Compound I. Mn^{II} ion binds to the enzyme and reduces MnP-Compound I to MnP-Compound II. This latter oxidized form of MnP is only reduced to native MnP by manganese(II) but not by phenols. The oxidation of Mn^{II} to Mn^{III} by MnP is dependent on the presence of chelating agents in the reaction mixture. These chelating agents, which are α-hydroxyacids (lactate, succinate, malonate) probably have two different roles: they stimulate the catalytic activity by facilitating the dissociation of the enzyme-manganese complex by chelating Mn^{III} and they stabilize manganese at the oxidation state III, since Mn^{III} salts have a tendency to disproportionate to Mn^{II} and Mn^{IV}.[81,82] However, pyrophosphate, a strong chelating agent of manganese salts inhibits the MnP activity, suggesting that the enzymatic oxidation of manganese occurs with free Mn^{II} ion. The binding site of manganese is located on the heme edge close to the δ-*meso* carbon of the prosthetic group.[77,80,83] Manganese ions cannot be replaced by iron(III), copper(II) or cobalt(II) salts. Finally, the addition of a large excess of H_2O_2 to the ferric state of the enzyme results in the formation of a Compound III species (see Scheme 5 for the catalytic cycle of MnP).

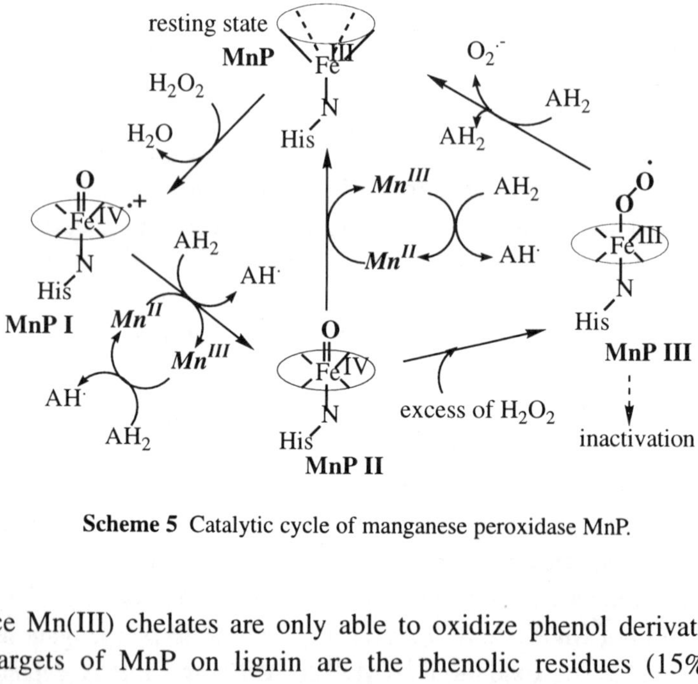

Scheme 5 Catalytic cycle of manganese peroxidase MnP.

Since Mn(III) chelates are only able to oxidize phenol derivatives, the main targets of MnP on lignin are the phenolic residues (15% of the phenylpropane units of lignin have a free hydroxyl group in the *para* position).

A detailed study on a lignin model dimer with a phenol function, such as 1-(3,5-dimethoxy-4-hydroxyphenol)-2-(4-methoxy)propane-1,3-diol, indicated that such a dimer was cleaved by $MnP/H_2O_2/Mn^{II}$ to the following monomers: 2,6-dimethoxybenzoquinone, 2-hydroxy-4'-methoxyacetophenone and 1-(4-methoxyphenyl)ethane-1,2-diol. Labeling studies with $^{18}O_2$ and

$H_2{}^{18}O$ confirmed that the key reaction was the initial formation of a phenoxy radical.[84] Partial depolymerization of lignin by MnP has been observed *in vitro*.[85]

The ligninolytic activity of *Phanerochaete chrysosporium* is dependent on both enzymes, LiP and MnP, and of the presence of molecular oxygen in the culture medium to facilitate the propagation of free radical chain reactions. Otherwise, anaerobic conditions favor the recombination of lignin-centered radicals to generate new C–C or C–O bonds within lignin, leading to further reticulation of lignin, not to its degradation.

3. Models of Heme-Peroxidases

3.1 Models of horseradish peroxidase

Early peroxidase models were based on simple metal complexes or modified hemin derivatives.[86] Hemin liberated from hemoglobin is not soluble in neutral aqueous solution or usual organic solvents, so hemin has been attached to a polyethylene glycol polymer in order to obtain a catalyst having peroxidase activity in organic solvents as well as in aqueous mixtures.[87] The peroxidase activity was higher in trichloroethane compared to buffered water at pH 7: rate constants being 3×10^1 $M^{-1}s^{-1}$ and 2.3×10^3 $M^{-1}s^{-1}$, respectively. Anionic metalloporphyrins adsorbed on ion-exchange resins have an efficient peroxidase activity which has been monitored by titration of the quinoid dye ($\lambda_{max} = 505$ nm) generated in the co-oxidation of 4-aminoantipyrine and phenol in the presence of hydrogen peroxide.[88] Oxidation of phenols and tertiary amines has been performed with different metalloporphyrins and various oxidants (not only hydrogen peroxide, but also alkylhydroperoxides, organic peracids or aromatic amine *N*-oxides).[89–93] All these different metalloporphyrin complexes exhibit catalytic activities similar to those of peroxidases: oxidation of phenols, *N*-dealkylation of tertiary amines. In this latter case, the metalloporphyrin-mediated dealkylation proceeds by an initial one-electron abstraction when using iodosylbenzene as primary oxidant or by a hydrogen atom abstraction with *t*-butylhydroperoxide[89] (however, the active species with this oxidant might be an alkoxyl radical, not a metal-oxo entity). It must be noted that reaction conditions (ionic strength, proton transfer, ligand concentration,... etc.) are best controlled and kinetic and electrochemical data more easily interpreted in aqueous media.[93] In particular, the aqua ligand *trans* to the metal-oxo species is more acidic when the oxidation state of the metal center increases. Iron(III) or manganese(III) complexes of *meso*-tetrakis(2,6-dimethyl-3-sulfonatophenyl)porphyrin or *meso*-tetrakis(2,6-dichloro-3-sulfonatophenyl)porphyrin are suitable water-soluble metalloporphyrin derivatives for these studies, since these metal complexes do not form μ-oxo dimers nor aggregate due to steric hindrance created by the *ortho* substituents.

Microperoxidases-8 or -11 (MP-8 or MP-11) have been used as peroxidase models. These compounds consist of a peptide fragment (8 or 11 residues) containing a covalently bound heme which are obtained by enzymatic hydrolysis of cytochrome *c*.[94–96] MP-8 and -11 are oxidized by hydrogen peroxide[94] and are able to catalyze the *N*-demethylation of amines and oxidation of sulfides in the same manner as peroxidases.[95,96]

Artificial peroxidases have also been designed using the catalytic antibody strategy. The groups of Schultz and Harada initiated the field of antibody-metalloporphyrin complexes.[97–99] Immunization with *meso*-tetrakis(4-carboxyphenyl)porphyrin (TCPP) provides an antibody which binds very strongly to Mn(III)-TCPP or to Fe(III)TCPP ($K_d = 4.1$ and 1.5×10^{-7} M, respectively). The oxidation of pyrogallol catalyzed by these antibody-metalloporphyrins has been studied in detail and the K_m and k_{cat} values for the iron antibody are 4.0 mM and 50 min^{-1}, respectively.[99] Higher affinities for metalloporphyrins have been obtained using the iron(III) complex α,α,α,β atropoisomer of the *meso*-tetrakis(2-carboxyphenyl)porphyrin.[100] An IgG_1 antibody of this iron complex has a k_{cat} value of 100 min^{-1}.

Peroxidase models based on metalloporphyrins have been used for the *in vitro* oxidation of drugs or drug candidates, since these biomimetic methods can be considered as a promising complementary approach to traditional techniques used to carry out oxidative activation of drugs (perfused organs, isolated cells, tissue homogenates or purified enzymes). These biomimetic catalysts should facilitate the preparation and the isolation of reactive metabolites, due to the absence of proteins in the reaction mixture (electrophilic metabolites are usually trapped by nucleophilic sites of proteins). Several groups have developed the use of these metalloporphyrins in organic or aqueous solutions to catalyze the oxidation of drugs such as acetaminophen,[101] piperidine derivatives,[102] ellipticine derivatives,[103] lidocaine[104] and an antagonist of a vasopressin receptor.[105] When using sterically hindered metalloporphyrins and monopersulfate as primary oxidant, the catalytic activities of these peroxidase-type oxidations are very close to that observed with horseradish peroxidase.[103] These metalloporphyrin-mediated oxidations are highly dependent upon the drug structure. When a drug presents easily oxidizable sites, peroxidase-type oxidations are favored with respect to oxygenations (P450-type hydroxylations) of particular C–H bonds of the drug.[105–107]

The horseradish peroxidase-catalyzed oxidation of luminol is widely used in the field of non-radioactive immunologic assays. Iron and manganese porphyrins are also highly efficient catalysts when using perborate as oxidant. The chemiluminescence produced by nanomolar concentrations of Fe-TPPS can be over three times as strong as that produced by HRP with a signal-to-noise ratio increasing up to 200.[108]

3.2 Models of chloroperoxidase

Only a few reports have been devoted to the modeling of chloroperoxidase with metalloporphyrin catalysts. In the absence of protein around the active site to control the electron transfer, the oxidation of chloride ions competes with the direct oxidation of the organic substrate by the high-valent metal-oxo species. The first CPO model was based on the association of $NaClO_2$ with an iron porphyrin generating HOCl *in situ* which reproduced the main features of CPO-catalyzed chlorinations.[109] A second CPO model, using hydrogen peroxide and chloride ions, has been developed with an anionic manganese porphyrin supported on a cationic ion-exchange resin (the structures of these supported catalysts will be discussed in the next paragraph on ligninase models).[110] MnTMPS supported on a polyvinylpyridine polymer (PVP) was able to catalyze the oxidation of dimedone, a classical CPO substrate, to chlorodimedone and hydroxydimedone, a side-product resulting from the direct oxidation of dimedone by the intermediate manganese-oxo species (see Scheme 6).

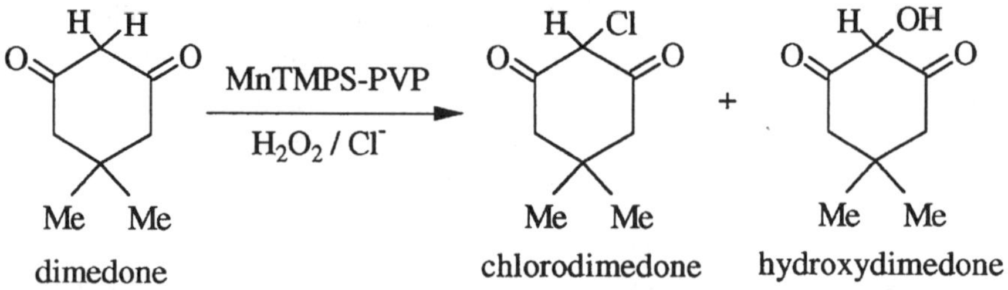

Scheme 6 MnTMPS-PVP as CPO model.

A synthetic iron porphyrin with a thiolato ligand in axial position has been oxidized by hydrogen peroxide in the presence of chloride. The shift of the Soret band from 408 nm to 404 nm might be due to the formation of a putative Fe(III)OCl entity which can generate free HOCl after protonation.[36,111] This thiolato-iron porphyrin complex was able to catalyze the chlorination of chlorodimedone to dichlorodimedone with a turnover number up to 154 and with a very low loading of catalyst (0.7%).

3.3 *Models of ligninase*

This survey is limited to model systems based on synthetic water-soluble metalloporphyrins. Early models of ligninase have been developed with iron protoporphyrin IX in combination with *t*-butylhydroperoxide[112,113] or Fe(TPP)Cl in combination with O_2 and a dihydropyridine as electron source.[114] However, these metalloporphyrins are not water-soluble and are easily destroyed under oxidizing conditions.

3.3.1 Models based on water-soluble metalloporphyrins

Water-soluble sulfonated metalloporphyrins have been found to be suitable catalysts in ligninase model studies.[115–118] In one case, sodium hypochlorite was used as primary oxidant, but this oxidant cannot be developed in aqueous reaction mixtures, since hypochlorite only exists at elevated pH values and is only an efficient oxygen atom donor in hydrophobic media.[119,120] Aromatic chlorinations are observed with hypochlorite below pH 8–9. In this context, potassium monopersulfate seems to be a more suitable oxidant in ligninase modeling (for a review on potassium monopersulfate as oxygen atom donor, see Ref. 121). The best oxidant, i.e. the "clean oxidant," is of course hydrogen peroxide. However, the activation of H_2O_2 by synthetic metalloporphyrins in the absence of cofactors (imidazole or acid derivatives) usually leads to the homolytic cleavage of

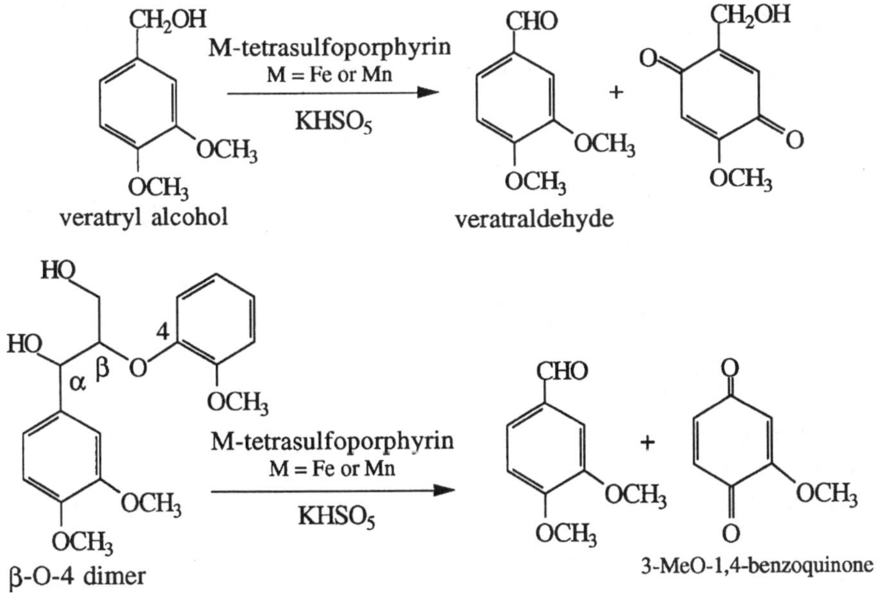

Scheme 7 Oxidation of small molecules used as lignin models by $KHSO_5$/Fe- or Mn-tetrasulfonatoporphyrin systems.

the peroxide O–O bond and the resulting hydroxyl radicals are strong bleaching agents for metalloporphyrins. The activity of "$KHSO_5$/sulfonated metalloporphyrin" models has been checked with the usual lignin model molecules: veratryl alcohol and 1-(3,4-dimethoxyphenyl)-2-(2-methoxyphenoxy)propane-1,3-diol. We found that iron(III) *meso*-tetrakis(4-sulfonatophenyl)porphyrin (FeTPPS) in combination with potassium monopersulfate in an acetonitrile/pH 3 buffer mixture was an efficient catalyst for veratryl alcohol oxidation to veratraldehyde and 2-hydroxymethyl-5-methoxy-1,4-benzoquinone, the same two oxidation products generated by ligninase (see Scheme 7).[118]

The β-O-4 dimer with one arylglycerol-β-arylether linkage was also cleaved at the C_α–C_β bond to generate veratraldehyde and 3-methoxy-1,4-benzoquinone. Catalytic activities ranged from 1 to 5 cycles sec^{-1}.[117,118] The corresponding manganese complex, MnTPPS was less active, with an optimum catalytic activity around pH 6.

Robust sulfonated iron and manganese porphyrins have also been used in ligninase modeling e.g. M-TMPS,[122a] M-TDCPPS[122b] and M-Br_8TMPS[122a] (see Scheme 8 for the structures of the first two catalysts).

These sterically hindered water-soluble complexes are among the most efficient metalloporphyrin catalysts. The degradation products of lignin models are similar to those obtained in ligninase-catalyzed oxidations. These iron catalysts[122–124] are able to decolorize Poly B-411, a blue dye

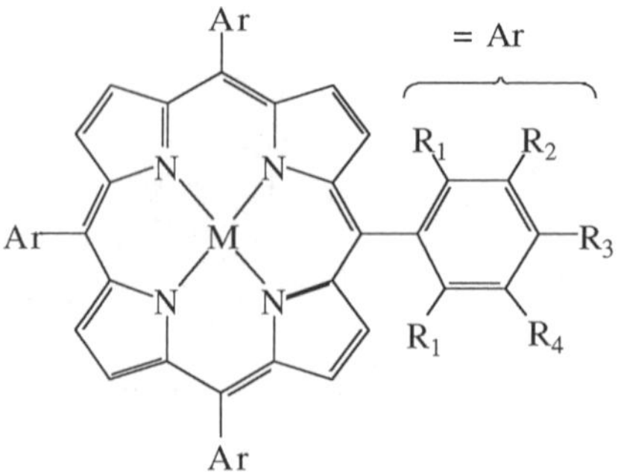

TMPS: R_1 = Me, R_2 = R_4 = SO_3^-, R_3 = Me

TDCPPS: R_1 = Cl, R_2 = H, R_4 = SO_3^-, R_3 = H

M = Fe(III) or Mn(III)$(H_2O)_2$

Scheme 8 Structures of M-TMPS and M-TDCPPS.

used to monitor the activity of ligninase.[125] Iron(III) *meso*-tetrakis (pentafluorophenyl)-β-tetrasulfonatoporphyrin is also an efficient catalyst for the oxidation of lignin models with hydrogen peroxide or magnesium monoperphthalate.[127] Water-soluble metallophthalocyanines were compared to the metalloporphyrins mentioned above as ligninase models. The most active catalyst was FePcS with hydrogen peroxide at low pH values (2.6–4.7) and at 85 °C.[128,129] In these rather drastic conditions, 70% of veratryl alcohol was converted to a mixture of products containing veratraldehyde and 2-hydroxymethyl-5-methoxy-1,4-benzoquinone (both products have also been obtained with metalloporphyrin catalysts (see Scheme 7).

3.3.2 Models based on supported metalloporphyrins

Sulfonated metalloporphyrins are also efficient catalysts in ligninase modeling when they are supported on ion-exchange resins. For examples, FeTPPS or MnTPPS strongly interact with the nitrate form of cationic Amberlite IRA 900 (an ion-exchange resin derived from polyvinylbenzene bearing ammonium residues) to give solid supported catalysts that are more stable than the corresponding soluble catalysts.[117,118] When the MnTPPS-Amberlite catalyst was reused in a second run, its catalytic activity was still 95% of the first run.[117]

The major problem with supported manganese sulfonatoporphyrin catalysts is the necessary presence of pyridine in the reaction mixture in order to increase the catalyst activity (for the role of pyridine and imidazole derivatives to boost the catalytic activity of metalloporphyrins by a proximal effect, see Refs. 119 and 120). With a polyvinylpyridine (PVP) support, in contrast to polyvinylbenzene, it has been possible to take advantage of the proximal effect of a pyridine ligand without free pyridine present in the reaction mixture. Such supported manganese catalysts are easily prepared from the commercially available poly(4-vinylpyridinium) polymer crosslinked with divinylbenzene (see Scheme 9 for a representation of supported manganese porphyrin on PVP).[130] Deprotonation of poly(4-vinylpyridinium) provide the necessary free pyridine units for axial coordination to manganese sulfonatoporphyrins.

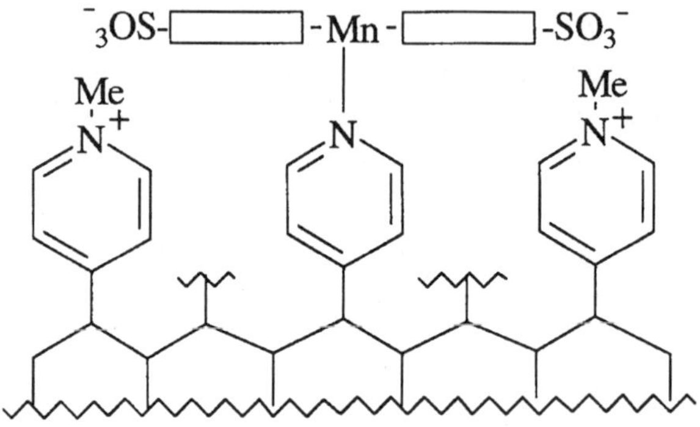

Scheme 9 Representation of the supported MnTMPS-PVP catalyst with a pyridine from the PVP polymer acting as proximal ligand.

The coordination of pyridine arising from PVP to the metalloporphyrin was monitored by UV-vis spectroscopy and was also studied by molecular mechanics.[131,132] However, since the only proximal pyridine interaction is insufficient to avoid leaching of MnTMPS from PVP during catalytic oxidations, an additional treatment of MnTMPS-PVP by acid or by a methylating agent was necessary to recreate pyridinium units and to reinforce the catalyst-polymer interactions by additional strong electrostatic interactions between the peripheral sulfonato groups of the metalloporphyrin and the pyridinium units of the cationized PVP. Ninety per cent of the initial run activity was still obtained when MnTMPS-protonated PVP was recycled for the third time in the oxidation of 1-(3,4-dimethoxyphenyl)-2-(2-methoxyphenoxyl)-propane-1,3-diol.[130]

3.4 *Oxidation of pollutants by ligninase models*

Since ligninase is the only heme peroxidase capable of oxidizing poorly oxidizable molecules recalcitrant to biodegradation (see above, Sec. 2.3), the ligninase models described above have been evaluated as catalysts for pollutant oxidation. Sulfonated iron or manganese porphyrin catalysts with $KHSO_5$ or H_2O_2 as primary oxidants were able to effectively promote the oxidative dechlorination of 2,4,6-trichlorophenol (TCP) to 2,6-dichlorobenzoquinone. Catalytic activities of up to 20 cycles sec^{-1} were observed with monopersulfate as oxidant and a very low loading of FeTPPS catalyst (0.1–0.3%).[133] However, changing this oxidant for hydrogen peroxide provoked a decrease in the catalytic activity. By using various methoxylated aromatics, it has been possible to establish a comparative scale for the catalytic activities of HRP, LiP and FeTPPS. While HRP was unable to catalyze the oxidation of organic substrates having a $E_{1/2}$ potential value above 1.2 V (Ag/AgCl), LiP was active up to 1.4 V and FeTPPS/$KHSO_5$ up to 1.7 V.[133]

However, the limits of this catalytic reaction are a lower catalytic activity with H_2O_2, the cost of metalloporphyrins and the absence of extensive oxidation of the generated quinones. Consequently, a biomimetic system able to degrade trichlorophenols should be based on a cheap and readily available catalyst with H_2O_2 as oxidant (the "clean" oxidant, water being the only by-product after oxidation) and also, the oxidation of TCP should give ring cleavage products and CO_2, like in biomineralization processes.

We recently found that iron(III) tetrasulfonatophthalocyanine (FePcS, see Scheme 10 for structure) was able to catalyze the H_2O_2 oxidation of TCP according to the reaction conditions listed above.[134–138]

Since phthalocyanines are industrial dyes, metallophthalocyanines are an attractive alternative as readily available biomimetic (in fact "bioinspired") oxidation catalysts. Iron and manganese phthalocyanine catalysts have been found to be less active in oxygenation reactions than metalloporphyrins.[139] FePcS was prepared in high yield (70%) in a one-step reaction by improving the Weber-Busch method.[137,140] The analysis was consistent with a μ-oxo Fe^{III}-O-Fe^{III} dimer structure with an intense visible band at 632 nm.[136,137]

With 3.7% of FePcS and five equivalents of H_2O_2 with respect to 2,4,6-trichlorophenol in a mixture of acetonitrile/buffered water (1/3, v/v) at pH 7, TCP was quantitatively converted at room temperature within several minutes (TCP concentration = 2000 ppm).[134] The transient formation of 2,6-dichloroquinone (DCQ) and the release of two Cl^- ions per consumed

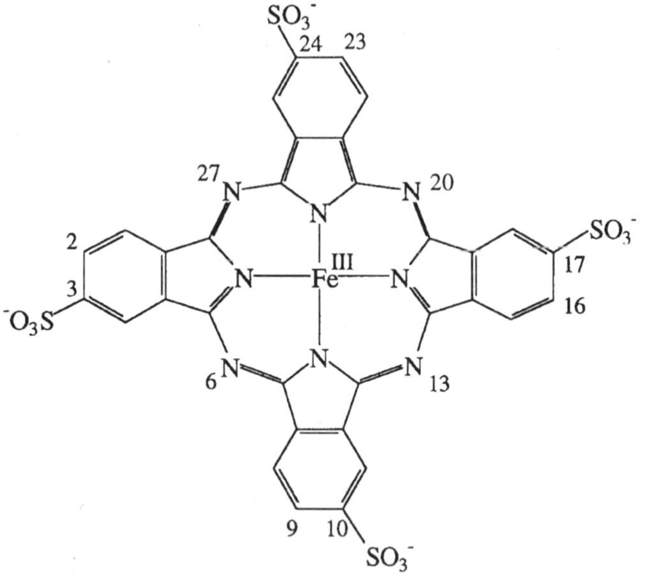

Scheme 10 Structure of iron(III) tetrasulfonatophthalocyanine FePcS. Only the 3,10,17,24 isomer of the four possible regioisomers is depicted. See the text for a discussion on the formation of dimers via μ-oxo bridges.

TCP molecule indicated that TCP was converted to ring cleavage products. Two categories of products were identified: five products corresponding to aromatic ring cleavage and four ones resulting from oxidative coupling of TCP (Scheme 11).[134,136] The total yield of identified products was 60–70% after 60 mins of reaction at room temperature.

The major TCP cleavage product was chloromaleic acid (yield = 24%). Chlorofumaric, maleic and fumaric acids were only minor products. Oxalic acid was among the products of TCP oxidation.[136] This C_2 diacid was expected since chloro-C_4 diacids are a signature of the oxidative ring cleavage of TCP.

Analyses of TCP oxidation products generated by $FePcS/H_2O_2$ in $H_2{}^{18}O$ suggested that the main source of oxygen atom in chloromaleic acid was not water but H_2O_2, since only one labeled oxygen atom was detected on one of its carboxylic functions. Under $^{18}O_2$, only 11% of chloromaleic acid

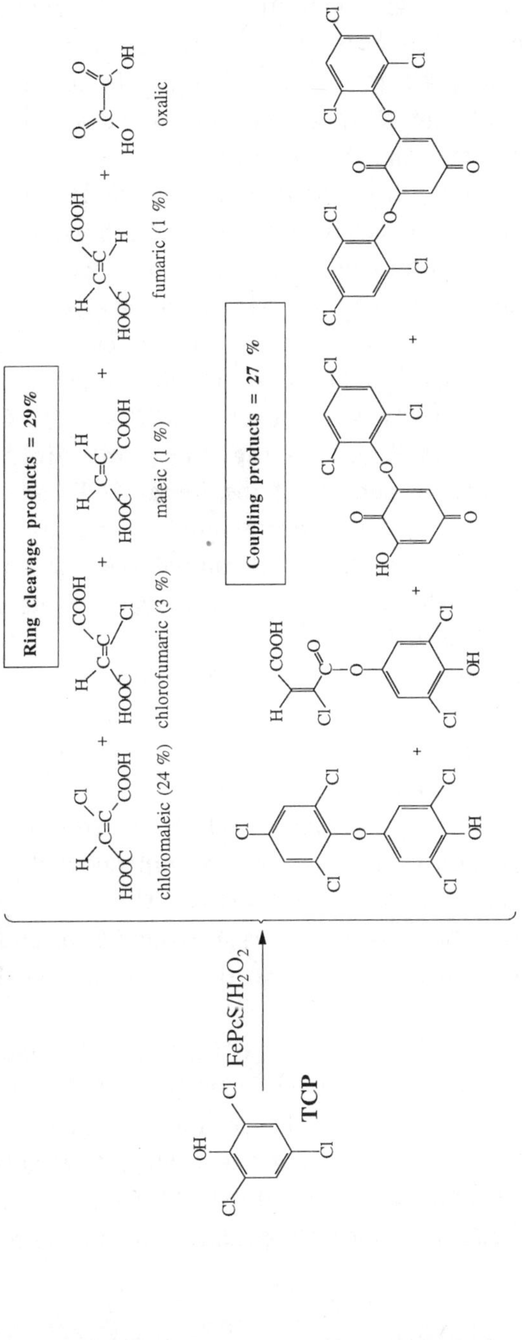

Scheme 11 Identified products of the H_2O_2 oxidation of 2,4,6-trichlorophenol (TCP) catalyzed by FePcS.

was labeled with one ^{18}O atom, indicating that oxygen atom incorporation from dioxygen was low compared to that arising from H_2O_2 or H_2O.[136] These oxygen labeling experiments suggested that the oxygen atom source in the oxidative TCP ring cleavage products were, in decreasing order, hydrogen peroxide, water and then dioxygen.

The C_4-diacids resulting from the ring cleavage of TCP by FePcS/H_2O_2 were also degraded by the catalytic system. The possible formation of carbon dioxide as the ultimate degradation product of these diacids was investigated by monitoring the release of labeled CO_2 during the oxidation of a (U-^{14}C)-TCP by FePcS/H_2O_2 (in order to have a complete material balance of oxidation products).[135] After 90 min at room temperature, 11%of the initial radioactivity was recovered as CO_2. The ^{14}CO yield was 3% of the starting TCP, increasing the total yield of carbon oxides to 14%. Sixty-nine percent of the initial TCP radioactivity was detected in the aqueous phase (corresponding mainly to C_4-diacid intermediates). Only 13% of the TCP radioactivity was recovered as hydrophobic organic products resulting from oxidative coupling reactions. This study with radiolabeled TCP gave a complete overview of the oxidative degradation of TCP since 96% of the initial radioactivity was recovered. The FePcS/H_2O_2 catalytic system was not only able to cleave the TCP aromatic ring but also had the capacity to mineralize this poorly biodegradable molecule to CO_2 and inorganics.

Control experiments suggested that the TCP oxidation by FePcS/H_2O_2 was not mediated by hydroxyl radical resulting from a possible metal-assisted homolysis of the weak peroxidic bond of H_2O_2. The initial step before the formation of active metal-oxygen species in oxidation reactions catalyzed by FePcS is the cleavage of the μ-oxo dimer to an iron(III) monomer (Scheme 12).[137]

The formation of FePcS monomers was favored (i) by acidic pH values (ii) by the presence of a strong anionic ligand (like TCP phenolate) or (iii) by a dissociating solvent (acetonitrile, acetone, alcohols,... etc.). In the absence of such a co-solvent, dimers of FePcS are catalytically inactive.[137] This fact is not a drawback of the FePcS/H_2O_2 system since it may be possible to use this catalytic system to oxidize pollutants in process waters

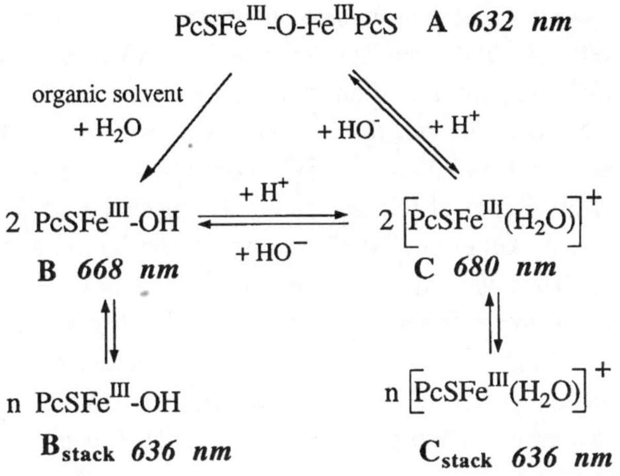

Scheme 12 Different possible forms of FePcS in solution. (A: dimer, B and C: monomers; B_{stack} and C_{stack} are stacked monomers)

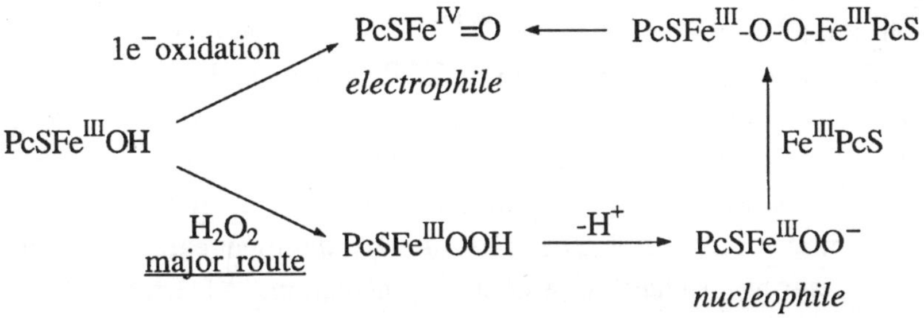

Scheme 13 Active species generated by oxidation of FePcS by hydrogen peroxide.

still containing 10–20% of organic solvents (in conditions where microorganisms are unable to operate).

From the different studies on the oxidation of chlorophenols[138] or polyaromatics[141] (anthracene and phenanthrene) by FePcS/H_2O_2, the oxidation of FePcS might generate two different iron-centered oxidizing species: a nucleophilic iron-peroxo entity Fe(III)-OOH and an electrophilic iron-oxo Fe(IV)=O (Scheme 13).

This latter species is able to oxidize anthracene, but is a poor epoxidizing agent compared to high-valent iron or manganese(V)-oxo complexes.

When $KHSO_5$ was used as oxidant in the presence of FePcS, TCP was quickly oxidized to DCQ, but this quinone was then oxidized at a slower rate compared to the initial rate of TCP conversion. The fast degradation of quinones with FePcS/H_2O_2 suggested the presence of a nucleophilic oxidizing species like $PcSFe^{III}$-OOH which is probably deprotonated at pH 7. Such Fe(III)-OOH entities are able to oxidize quinone derivatives by nucleophilic addition.[136,142] Such a complex cannot be produced by $KHSO_5$, thus explaining why the FePcS/$KHSO_5$ system was unable to quickly oxidize DCQ after the two-electron oxidation of TCP. DCQ resulted from the one-electron oxidation of the phenolate form of TCP at pH 7 by $PcSFe^{III}$-OOH or $PcSFe^{IV}$=O.

The current hypotheses on the oxidative degradation of TCP by FePcS/H_2O_2 can be summarized as follows (Scheme 14).
The TCP anion is oxidized to a phenoxy radical and then to an aromatic carbocation which reacts with water to give rise to the corresponding DCQ with the release of one Cl^- ion.[136,138] The oxidation of DCQ, the key step in the biomimetic mineralization of TCP, is mediated by PcSFeOOH acting as a nucleophilic oxidant to produce intermediate violet quinones (only one of these quinones, "identified quinone," is depicted in Scheme 14). The C–C bond cleavage of quinone intermediates may be due to the nucleophilic attack of PcSFeOOH on one quinone carbonyl group leading to open-ring product via a Grob fragmentation.[143] Decarbonylation and dioxygen incorporation would lead to chloromaleic acid, the major C4-diacid identified in the TCP oxidation mixture.

Pentachlorophenol (PCP) was also oxidized by FePcS/H_2O_2 (96% converted within 5 min with 1% of catalyst/PCP).[136] Dechlorination of PCP ranged from 0.34 to 1.43 Cl^- released per converted PCP molecule. Dichloromaleic anhydride was the only identified product resulting from the oxidative cleavage of PCP. Iron-, manganese- and cobalt sulfonatophthalocyanines were also able to efficiently catalyze the oxidation of various catechols with hydrogen peroxide.[144] Tetrachlorocatechol was oxidized to dichloromaleic acid, a ring cleavage product.

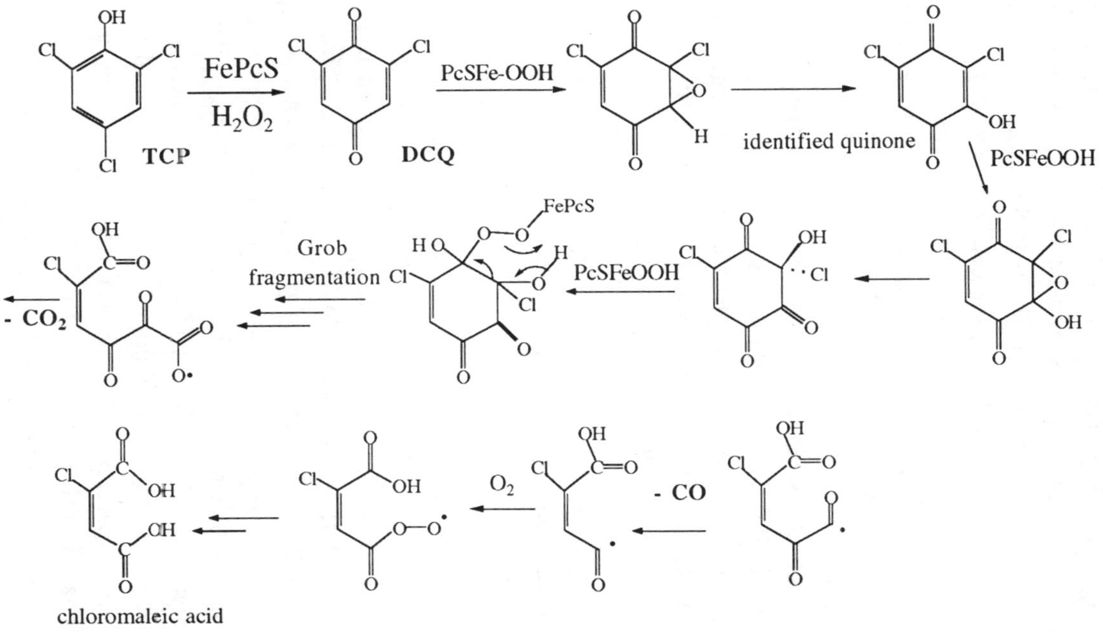

Scheme 14 Possible degradation pathways from TCP to chloromaleic acid by the FePcS/H_2O_2 system.

FePcS can also be adsorbed on cationic ion-exchange resins, like Amberlites, as previously observed for sulfonated metalloporphyrins (see above, paragraph 3.3.2). These FePcS-Amberlite supported catalysts (based on a cationic resin) should be pre-saturated with substrates which are able to adsorb onto the resin itself (as was the case for TCP), before adding the oxidant. In these conditions, the catalytic oxidation of TCP was observed even when the reaction mixture did not contain the organic co-solvent usually required to obtain a catalytic activity with FePcS in solution.[145] The dispersion of FePcS onto the support avoided the formation of inactive dimers.

In conclusion of this section, it can be noted that the easy-to-handle catalytic system FePcS/H_2O_2, might have a future in the oxidative degradation of pollutants, especially in the treatment of large amounts of concentrated pollutants, even in the presence of residual organic solvents where microorganisms are unable to operate.

3.5 Models of manganese peroxidase

The modeling of a non-specific peroxidase like ligninase is relatively simple compared to the design of metalloporphyrin catalysts capable of selectively oxidizing manganese(II) salts (in the presence of chelating agents) and of easily oxidized phenolic substrates. MnP has a manganese-binding site, on the edge of the prosthetic group, thus able to control the selective oxidation of manganese(II) salts in the presence of phenols. How can we reproduce the same selective oxidation with a synthetic metalloporphyrin without building an appropriate cavity i.e able to selectively accomodate Mn(II) around the catalytic site? Modeling of MnP is obviously more challenging than that of LiP. Cui and Dolphin reported in 1990 the role of manganese sulfate in the decolorization of Poly B-411 dye catalyzed by a sterically hindered sulfonated iron porphyrin, FeTDCPPS, in the presence of *m*-chloroperbenzoic acid.[146] In the absence of $MnSO_4$, the dye decolorization stopped after 1 min because of the fast catalyst bleaching. Addition of the manganese salt improved the catalytic reaction over a period of 7 min. The catalytic oxidation of Mn(II) chelates i.e. the key step performed by MnP, has been mimicked using synthetic metalloporphyrins.[148] Oxidation of Mn(II)pyrophosphate to the corresponding Mn(III) chelate was performed with catalytic amounts of iron or manganese sulfonated porphyrins. As expected, sterically hindered iron complexes were the most active catalysts (8 and 3 cycles/min for FeTDCPPS and FeTMPS, respectively). The catalytic activities of these biomimetic catalysts are well below the activity of MnP itself (8200 cycles/min[149]), but this difference of activity between the enzyme and the models is related to the absence of Mn(II)-binding site on these simple metalloporphyrins. In addition, since these catalysts have several negatively-charged sulfonato groups at the periphery of the ligand, the substrate $[Mn(II)(H_2P_2O_7)_2]^{2-}$ (which is also negatively-charged) has a tendency to be repelled from the catalytic site, in contrast to the enzyme. A better situation would be obtained with cationic metalloporphyrins.

A striking feature of the metalloporphyrin-catalyzed oxidation of manganese pyrophosphate was the enhancement of the catalytic activity by the addition of limited amounts of methoxylated benzene derivatives acting as co-substrates: 1,2-dimethoxybenzene, 3,4-dimethoxybenzyl alcohol (veratryl alcohol) or 1,2,4-trimethoxybenzene (see Ref. 149 for data on their enzyme-mediated oxidations). The catalytic activity of FeTMPS was strongly increased by 2–5% of one of these methoxylated benzenes. Two explanations have been proposed for these improved oxidations of Mn-pyrophosphate: (i) an inhibition of the metalloporphyrin bleaching due to the presence of an easily oxidized substrate in the reaction mixture (80–90% of the catalyst was still present after 10 min with 3% of one of these di- or trimethoxylated benzene derivatives instead of 20% of remaining catalyst in their absence), or (ii) a role of redox mediator for these methoxylated benzene derivatives, as reported in the oxidation of Mn salts by lignin peroxidase (50% of the starting organic mediator was still present in the reaction medium as evidenced by HPLC; in the absence of manganese pyrophosphate, the organic mediator was completely oxidized within a few minutes). The radical-cation of these methoxylated aromatics generated by the high-valent oxo-metalloporphyrin complex might behave as a strong oxidant with respect to Mn(II)-pyrophosphate and thus enhance its oxidation. These results with a biomimetic model system further illustrate the key role of low molecular weight organic mediators in the oxidative degradation of lignin by fungal peroxidases.

In order to mimic the manganese-binding site of MnP, a metalloporphyrin bearing a manganese chelating ligand has been recently prepared.[150] This complex is able to oxidize Mn(II) to Mn(III) when the iron center is oxidized by pentafluoroiodosylbenzene.

4. Structure and Activity of Heme-Catalases

Catalases, along with superoxide dismutases, are key enzymes for controlling the level of reduced oxygen species in living systems[151,152] (they catalyze the dismutation of hydrogen peroxide and superoxide anion, respectively). Only heme-catalases will be discussed here, see Chap. 5 for the presentation and the modeling of non-heme catalases. Heme-catalases usually have an iron(III) protoporphyrin IX as prothestic group or modified porphyrins as in the two catalases of *E. coli* designated HPI and HPII.[153] Within peroxisomes of the same plant (sunflower), two catalases have been identified with different primary structures.[154] A particular case is the catalase-peroxidase KatG of *Mycobacterium tuberculosis* which is able to catalyze the dismutation of hydrogen peroxide and the oxidation of substrates like isoniazid, one of the most widely used anti-tuberculosis drug.[155,156]

All these tetrameric heme-catalases are highly efficient catalysts for hydrogen peroxide dismutation to dioxygen and water at a very high rate of 10^5 cycles sec^{-1} (Eq. 3).

$$2H_2O_2 \text{ ------ catalase -------> } O_2 + 2H_2O \tag{3}$$

Catalase is also able to oxidize a few small molecules, in particular ethanol (the pro-*R*-hydrogen atom is stereospecifically removed as in the alcohol dehydrogenase-mediated oxidation).[157] Several heme-catalase three-dimensional structures are available: from the fungus *Penicillium vitale*,[158] from beef liver[159] or from the bacterium *Micrococcus lysodeikticus*.[160] The proximal ligand site of these heme-catalases is occupied by a tyrosine residue (Tyr-357 and Tyr-339 in beef liver and bacterial catalases, respectively) with a Fe-O distance of 1.9 Å. The sixth coordination position of iron is occupied by a water molecule. The distal side of these catalases exhibit the usual histidine and arginine residues required for the heterolytic activation of hydrogen peroxide. In addition, the binding site of NADPH has been located.[160] All heme catalases are tightly bounded to NADPH, which is necessary to reduce the Compound II of catalase (which does not take part in the catalytic cycle of the H_2O_2 dismutation), although its accumulation leads to enzyme inactivation. Calculations suggested that the electron tunneling pathway between the heme and NADPH also involved a tyrosine (Tyr-214) which is near the nicotinamide-binding site.[161]

The oxidation of the iron(III) resting state of catalase by hydrogen peroxide generates a high-valent metal-oxo species considered to be an iron(IV)-oxo with a radical-cation delocalized on the porphyrin ring according to UV-vis, Raman and EPR studies.[162–164]

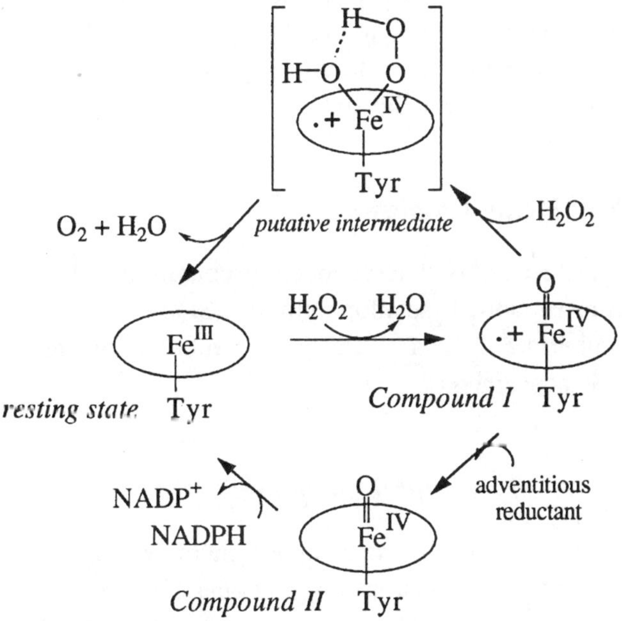

Scheme 15 Intermediates in the catalytic cycle of catalase.

Differences between Cat-Compound I and other peroxidase-Compound I species have been noticed: the exchange coupling between $S = 1$ of the Fe(IV)-oxo and $S' = 1/2$ of the porphyrin radical-cation is ferromagnetic (not antiferromagnetic) and exhibits a predominant $^{2}A_{2u}$ character. The sign and magnitude of the exchange parameter J is certainly influenced by the nature of the proximal ligand in catalase (a tyrosine instead of a histidine). The motion of the iron with respect to the heme plane has been monitored by time-resolved crystallography: the iron is 0.1 Å below, 0.3 Å above and within the plane in Cat-resting state, Cat-Compound I and Cat-Compound II, respectively.[165]

The addition of hydrogen peroxide to Cat-Compound I might generate a putative intermediate with a hydroperoxo and a hydroxo ligand on the distal side (see Scheme 15). Labeling studies with doubly ^{18}O-labeled H_2O_2 proved that molecular oxygen is generated from the intact O–O bond in hydrogen peroxide.[166] Kinetic data indicated that the rate constant of Compound I formation is $10^7\ M^{-1}s^{-1}$, below the value for a simple diffusion-controlled bimolecular reaction ($10^{10}\ M^{-1}s^{-1}$).[167] Cat-Compound II is generated by an adventitious reductant originating from the solvent or the protein matrix probably via the oxidation of a tyrosine residue.[168]

5. Models of Heme-Catalases

Models of heme-catalases have been prepared using soluble metalloporphyrins or supported metalloporphyrins. Studies on catalase modeling with non-porphyrinic ligands will also be mentioned where mononuclear complexes are concerned.

5.1 Models based on soluble complexes

Early studies on the catalytic decomposition of hydrogen peroxide have been performed with various iron(III) proto-and deuteroporphyrins and also with metalloporphyrins soluble in organic solvents.[86,169] The dismutation of hydrogen peroxide using this oxidant in the presence of metalloporphyrins is an easy reaction, and involves the formation of a high-valent iron-oxo species. The formation of the analog of Cat-Compound I has been well-studied with non-aggregating iron(III) or manganese(III) sulfonated porphyrins.[93,170–172] ABTS, 2,2'-azinobis(3-ethylbenzthiazolinesulfonic acid, is a suitable one-electron donor to monitor the formation of the metal(IV)-oxo radical-cation. With a manganese complex, the rate constant of the formation of Compound I analog is modulated by the nature of the axial ligand: $k = 5.6 \times 10^5$ $M^{-1}s^{-1}$ without axial ligand and a ten-fold increase with imidazole.[171] A large solvent kinetic isotope effect of 3 has been observed with iron-porphyrins in the formation of the Cat-Compound I analog, in agreement with H–O bond-breaking accompanying O–O bond cleavage.[172]

The influence of the proximal ligand on the catalase activity of synthetic metalloporphyrins has been studied with complexes having an imidazole, an alkoxo or a phenoxo ligand in the fifth coordination position (see Scheme 16 for the structure of this latter catalase model with two "basket-handles").[173,174]

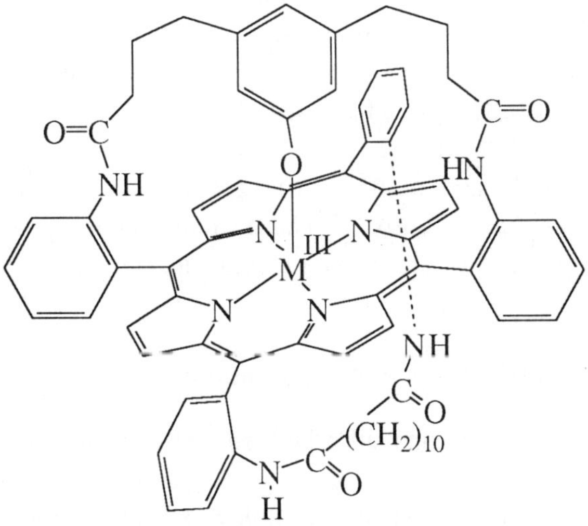

Scheme 16 Catalase model with a phenoxo proximal ligand (M = Fe or Mn).

Two reactions have been studied with these catalase models: (i) the catalytic dismutation of hydrogen peroxide and (ii) the catalytic epoxidation of olefins. Manganese complexes were always more efficient for both oxygenase and dismutase activities than the corresponding iron complexes. However, whereas complexes having a proximal nitrogen-ligand catalyzed both reactions (dismutation and oxygenation), molecules with a proximal oxygen-ligand catalyzed preferentially the dismutation of H_2O_2 in the case of manganese derivatives and *only* this reaction in the case of iron complexes (the "true" models of catalase).[174]

Non-porphyrinic metal complexes have been used as catalase models. A water-soluble Ru(II) complex of a macrocycle with tertiary amines was found to be very efficient in the catalytic dismutation of hydrogen peroxide.[175] The initial rate of hydrogen peroxide decomposition by this ruthenium complex was 6×10^{-4} $M^{-1}s^{-1}$.

Oxidant-mediated injuries to endothelial cells contribute to the pathogenesis of a variety of diseases (atherosclerosis, reperfusion injury after myocardial ischemia and stroke, adult respiratory distress syndrome,... etc.) originating from hydrogen peroxide produced by inflammatory cells such as neutrophils. Consequently, in order to treat such disorders, catalase mimicks based on manganese Schiff base complexes have been developed as potential therapeutic agents. These manganese complexes have been successfully applied against the β-amyloid toxicity related to Alzheimer disease,[176] the protection of rat kidneys from ischemia-reperfusion induced damage,[177] and also for the prevention and suppression of autoimmune encephalomyelitis.[178] These small catalase models, easy to prepare, might have a future as therapeutic agents.

5.2 *Models based on supported complexes*

Sulfonated metalloporphyrins adsorbed on ion-exchange resins are good catalase models and they can be recycled with little loss of catalytic activity.[179] Fe(III) deuteroporphyrin linked to polylysine via the axial ligation of an amino side chain of the lysine residues can also catalyze the dismutation of hydrogen peroxide.[180] More recently, iron-phthalocyanine complexes supported on carbon black or zeolite are also able to dismute H_2O_2, zeolite being a more appropriate support because of its higher specific surface.[181]

6. Concluding Remarks

This chapter must not be considered as an exhaustive review on heme-peroxidase and heme-catalase models, but as a short presentation of the main features in a particular field of biomimetic catalysis. The cited articles should allow the reader to investigate this field further and to develop new efficient peroxidase and catalase models. Hydrogen peroxide (as a source of hydroxyl radicals) plays a key role in tissue damage. Therefore, stable and efficient catalase models may be important as a new class of drugs used for the treatment of diseases related to oxidative stress. Most of the presently available models exhibit insufficient catalase activity (several orders of magnitude below the activity of catalase itself), thus cannot be considered as drug candidates.

Acknowledgments

The author is indebted to all talented co-authors and collaborators who have contributed to the work on peroxidases and peroxidase models performed in Toulouse. Financial support and post-doctoral fellowships were provided by CNRS, ELF-Atochem, SANOFI and EERO (European Environmental Research Organisation, Netherlands). Dr Steven Ross is gratefully acknowledged for editing this contribution.

Abbreviations

Br_8TMPS: dianion of *meso*-tetrakis(3,5-disulfonatomesityl)-β-octabromoporphyrin; CCP: cytochrome *c* peroxidase; DCQ: 2,6-dichloro-1,4-benzoquinone; DDT: 1,1,1-trichloro-2,2-bis(*p*-chlorophenyl)ethane; EPR: electron paramagnetic resonance; HRP: horseradish peroxidase; lindane:

1α,2a,3β,4α,5α,6β-hexachlorocyclohexane; LiP: ligninase; MnP: manganese peroxidase; NADPH: reduced form of nicotinamide adenine dinucleotide phosphate; Pc: phthalocyanine; PcS: tetrasulfonatophthalocyanine; TCP: 2,4,6-trichlorophenol; TDCPPS: dianion of *meso*-tetrakis(3-sulfonato-2, 6-dichlorophenyl)porphyrin; TMPS: dianion of *meso*-tetrakis(3,5-disulfonatomesityl)porphyrin, TPPS: dianion of *meso*-tetrakis(4-sulfonatophenyl)-porphyrin; VA: veratryl alcohol.

References

1. *Peroxidases in Chemistry and Biology* (eds.) Everse, J., Everse, K.E. and Grisham, M.B. CRC Press, Boca Raton, 1991, 2 Vols.
2. Welinder, K.G., *Curr. Opin. Struct. Biol. 1992,* **2:** 388–393.
3. Spallholz, J.E. and Boylan, in Ref. 1, Vol. 1, 259–291.
4. Dunford, H.B. in Ref. 1, Vol. 2, 1–24.
5. English, A.M. and Tsaprailis, G. *Adv. Inorg. Chem.* 1995, **43:** 79–125.
6. Dawson, J.H. *Science* 1988, **240:** 433–439.
7. Poulos, T.L., Freer, S.T., Alden, R.A., Edwards, S.L., Skogland, U., Takio, K., Eriksson, B., Miller, M.A., Yonetani, T. and Kraut, J. *Biochemistry* 1980, **255:** 575–580.
8. Nagano, S., Tanaka, M., Ishimori, K., Watanabe, Y. and Morishima, I. *Biochemistry* 1996, **35:** 14251–14258.
9. Baek, H.K. and Van Wart, H.E., *Biochemistry* 1989, **28:** 5714–5719.
10. Harris, D.L. and Loew, G.H. *J. Am. Chem. Soc.* 1996, **118:** 10588–10594.
11. Penner-Hahn, J.E., Eble, K.S., McMurry, T.J., Renner, M., Balch, A.L., Groves, J.T., Dawson, J.H. and Hodgson, K.O. *J. Am. Chem. Soc.* 1986, **108:** 7819–7825.
12. Chang, C.S., Yamazaki, I., Sinclair, R., Khalid, S. and Powers, L. *Biochemistry* 1993, **32:** 923–928.
13. Paeng, K.-J. and Kincaid, J.R. *J. Am. Chem. Soc.* 1988, **110:** 7913–7915.
14. Chuang, W.-J. and Van Wart, H.E. *J. Biol. Chem.* 1992, **267:** 13293–13301.
15. Rutter, R., Hager, L.P., Dohonau, H., Hindruch, M., Valentine, M. and Debrunner, P. *Biochemistry* 1984, **23:** 6809–6816.
16. Nakajima, R. and Yamazaki, I. *J. Biol. Chem.* 1987, **262:** 2576–2581.
17. He, B., Sinclair, R., Copeland, B.R., Makino, R., Powers, L.S. and Yamazaki, I. *Biochemistry* 1996, **35:** 2413–2420.

18. Ator, M.A., David, S.K. and Ortiz, de Montellano, P.R. *J. Biol. Chem.* 1987, **262:** 14954–14960.
19. Veitch, N.C. and Williams, R.J.P. *Eur. J. Biochem.* 1990, **189:** 351–362.
20 Banci, L., Bertini, I., Bini, T., Tien, M. and Turano, P. *Biochemistry* 1993, **32:** 5825–5831.
21. Banci, L., Carloni, P. and Savellini, G.G. *Biochemistry* 1994, **33:** 12356–12366.
22. Meunier, B. *Biochimie* 1987, **69:** 3–9.
23. Meunier, G. and Meunier, B. *J. Am. Chem. Soc.* 1985, **107:** 2558–2560.
24. Meunier, B. in Ref. 1, Vol. 2, 201–217.
25. Klibanov, A.M., Berman, Z. and Alberti, B.N. *J. Am. Chem. Soc.* 1981, **103:** 6263–6264.
26. Ortiz de Montellano, P.R., Choe, Y.S., DePillis, G. and Catalano, C.E. *J. Biol. Chem.* 1987, **262:** 11641–11646.
27. Ortiz de Montellano, P.R. and Grab, L.A., *Biochemistry* 1987, **26:** 5310–5314.
28. Griffin, B.W. in Ref. 1, Vol. 2, 85–137.
29. Blanke, S.R. and Hager, L.P. *J. Biol. Chem.* 1998, **263:** 18739–18743.
30. Fann, Y.-C., Gerber, N.C., Osmulski, P.A., Hager, L.P., Sligar, S.G. and Hoffman, B.M. *J. Am. Chem. Soc.* 1994, **116:** 5989–5990.
31. Sundaramoorthy, M., Terner, J. and Poulos, T.L. *Structure* 1995, **3:** 1367–1377.
32. Zeng, J. and Fenna, R.E. *J. Mol. Biol.* 1992, **226:** 185–207.
33. Dunford, H.B., Lambeir, A.-M., Kashem, M.A. and Pickard, M. *Arch. Biochem. Biophys.* 1987, **252:** 292–302.
34. Egawa, T., Miki, H., Ogura, T., Makino, R., Ishimura, Y. and Kitagawa, T. *FEBS Let.* 1992, **305:** 206–208.
35. Bernadou, J., Fabiano, A.S., Robert, A. and Meunier, B. *J. Am. Chem. Soc.* 1994, **116:** 9375–9376.
36. Wagenknecht, H.-A. and Woggon, W.-D. *Chem. & Biol.* 1997, **4:** 367–372.
37. Samokyszyn, V.M. and Ortiz de Montellano, P.R. *Biochemistry* 1991, **30:** 11646–11653.
38. Debrunner, P.G., Dexter, A.F., Schulz, C.E., Xia, Y.-M. and Hager, L.P. *Proc. Natl. Acad. Sci. USA* 1996, **93:** 12791–12798.
39. Geigert, J., Lee, T.D., Dalietos, D.J., Hirano, D.S. and Neidleman, S.L. *Biochem. Biophys. Res. Commun.* 1986, **136:** 778–782.
40. Allain, E.J., Hager, L.P., Deng, L. and Jacobsen, E.N. *J. Am. Chem. Soc.* 1993, **115:** 4415–4416.
41. Dexter, A.F., Lakner, F.J., Campbell, R.A. and Hager, L.P. *J. Am. Chem. Soc.* 1995, **117:** 6412–6413.

42. Colonna, S., Gaggero, N., Manfredi, A., Casella, L., Gullotti, M., Carrea, G. and Pasta, P. *Biochemistry* 1990, **29:** 10465–10468.
43. Casella, L., Gullotti, M., Ghezzi, R., Poli, S., Beringhelli, T., Colonna, S. and Carrea, G. *Biochemistry* 1992, **31:** 9451–9459.
44. Colonna, S. and Gaggero, N. *Phosph. Sulf. Silicon* 1994, **95-96:** 103–111.
45. Kobayashi, S., Nakano, M., Kimura, T. and Schaap, A.P. *Biochemistry* 1987, **26:** 5019–5022.
46. *Lignin Biodegradation: Microbiology, Chemistry and Potential Applications* (eds.) Kirk, T.K., Higuchi, T. and Cheng, H.M. CRC Press, Boca Raton, 1980, Vols. 1 and 2.
47. *Biomass, Part B, Lignin, Pectin and Chitin* (eds.) Wood, W.A. and Kellog, S.T. in "Methods in Enzymology." Academic Press, New York, 1988, Vol. 161.
48. Labat, G. and Meunier, B. *Bull. Soc. Chim. Fr.* 1990, **127:** 553–564.
49. Tien, M. and Kirk, T.K. *Science* 1983, **221:** 661–663.
50. Glenn, J.K., Morgan, M.A., Mayfield, M.B., Kuwahara, M. and Gold, M.H. *Biochem., Biophys. Res. Commun.* 1983, **114:** 1077–1083.
51. Gold, M.H., Kuwahara, M., Chui, A.A. and Glenn, J.K. *Arch. Biochem. Biophys.* 1984, **234:** 353–362.
52. Kersten, P.J., Tien, M., Kalyanaraman, B. and Kirk, T.K. *J. Biol. Chem.* 1985, **260:** 2609–2612.
53. Andersson, A.L., Renganathan, V., Chui, A.A., Loehr, T.M. and Gold, M.H. *J. Biol. Chem.* 1985, **260:** 6080–6087.
54. Poulos, T.L., Edwards, S.L., Wariishi, H. and Gold, M.H. *J. Biol. Chem.* 1993, **268:** 4429–4440.
55. Renganathan, V. and Gold, M.H. *Biochemistry* 1986, **25:** 1626–1631.
56. Khindaria, A. and Aust, S.D. *Biochemistry* 1996, **35:** 13107–13111.
57. Wariishi, H. and Gold, M.H. *J. Biol. Chem.* 1990, **265:** 2070–2077.
58. Meunier, B. *Chemtracts Inorg. Chem.* 1991, **3:** 199–202.
59. Harvey, P.J., Schoemaker, H.E. and Palmer, J.M. *FEBS Lett.* 1986, **195:** 242–246.
60. Khindaria, A., Yamazaki, I. and Aust, S.D. *Biochemistry* 1995, **34:** 16860–16869.
61. Schoemaker, H.E., Lundell, T.K., Hatakka, A.I. and Piontek, K. *FEMS Microbiol. Rev.* 1994, **13:** 321–332.
62. Schmidt, H.W.H., Haemmerli, S.D., Schoemaker, H.E. and Leisola, M.S.A. *Biochemistry* 1989, **28:** 1776–1783.

63. Miki, K., Renganathan, V. and Gold, M.H. *Biochemistry* 1986, **25:** 4790–4796.
64. Hammel, K.E., Tien, M., Kalyanaranam, B. and Kirk, T.K. *J. Biol. Chem.* 1985, **260:** 8348–8353.
65. Kirk, T.K., Tien, M., Fee, J.A. and Ondias, M.R. *Biochem. J.* 1986, **236:** 279–287.
66. Hammel, K.E., Mozuch, M.D., Jensen, K.A. and Kersten, P.J. *Biochemistry* 1994, **33:** 13349–13354.
67. Miki, K., Kondo, R., Renganathan, V., Mayfield, M.B. and Gold, M.H. *Biochemistry* 1988, **27:** 4787–4794.
68. Alexander, M. *Biodegradation and Bioremediation* Academic Press, New York, 1994.
70. Bumpus, J.A., Tien, M., Wright, D. and Aust, S.D., *Science* 1985, **228:** 1434–1436.
71. Hammel, K.E. and Tardone, P.J. *Biochemistry* 1988, **27:** 6563–6568.
72. Joshi, D.K. and Gold, M.H. *Appl. Environm. Microbiol.* 1993, **59:** 1779–1785.
73. Joshi, D.K. and Gold, M.H. *Biochemistry* 1994, **33:** 110969–10976.
74. Haemmerli, S.D, Leisola, M.S.A., Sanglard, D. and Fiechter, A. *J. Biol. Chem.* 1986, **261:** 6900–6903.
75. Kuwahara, M., Glenn, J.K., Morgan, M.A. and Gold, M.H., *FEBS Lett.* 1984, **169:** 247–250.
76. Mino, Y., Wariishi, H., Blackburn, N.J., Loehr, T.M. and Gold, M.H. *J. Biol. Chem.* 1988, **263:** 7029–7036.
77. Harris, R.Z., Wariishi, H., Gold, M.H. and Ortiz de Montellano, P.R. *J. Biol. Chem.* 1991, **266:** 8751–8758.
78. Sutherland, G.R.J. and Aust, S.D. *Biochemistry* 1997, **36:** 8567–8573.
79. Pribnow, D., Mayfield, M.B., Nipper, V.J., Brown, J.A. and Gold, M.H. *J. Biol. Chem.* 1989, **264:** 5036–5040.
80. Sundaramoorthy, M., Kishi, K., Gold, M.H. and Poulos, T.L. *J. Biol. Chem.* 1997, **272:** 17574–17580.
81. Wariishi, H., Dunford, H.B., MacDonald, I.D. and Gold, M.H. *J. Biol. Chem.* 1989, **264:** 3335–3340.
82. Kishi, K., Wariishi, H., Marquez, L., Dunford, H.B. and Gold, M.H. *Biochemistry* 1994, **33:** 8694–8701.
83. Kishi, K., Kusters-van Someren, M., Mayfield, M.B., Sun, J., Loehr, T.M. and Gold, M.H. *Biochemistry* 1996, **35:** 8986–8994.
84. Wariishi, H., Valli, K. and Gold, M.H. *Biochemistry* 1989, **28:** 6017–6023.

85. Wariishi, H., Valli, K. and Gold, M.H. *Biochem. Biophys. Res. Commun.* 1991, **176:** 269–275.
86. Jones, P., Robson, T. and Brown, S.B. *Biochem. J.* 1973, **135:** 353–359.
87. Takahashi, K., Matsushima, A., Saito, Y. and Inada, Y. *Biochem., Biophys. Res. Commun.* 1986, **138:** 283–288.
88. Saito, Y., Mifune, M., Nakashima, S., Tanaka, Y., Chikuma, M. and Tanaka, H. *Chem. Pharm. Bull.* 1986, **34:** 5016–5019.
89. Lindsay Smith, J.R. and Mortimer, D.N. *J. Chem. Soc., Chem. Commun.* 1985, 64–65.
90. Traylor, T.G., Lee, W.A. and Stynes, D.V., *Tetrahedron* 1984, **40:** 553–568.
91. Wong, W.H., Ostovic, D. and Bruice, T.C. *J. Am. Chem. Soc.* 1987, **109:** 3428–3436.
92. Colclough, N. and Lindsay Smith, J.R. *J. Chem. Soc. Perkin Trans. 2,* 1994, 1–11.
93. Bruice, T.C. *Acc. Chem. Res.* 1991, **24:** 243–249.
94. Wang, J.-S., Baek, H.K. and Van Wart, H.E. *Biochem. Biophys. Res. Commun.* 1991, **179:** 1320–1324.
95. Nakamura, S., Mashino, T. and Hirobe, M. *Tetrahedron Lett.* 1992, **33:** 5409–5412.
96. Colonna, S., Gaggero, N., Carrea, G. and Pasta, P. *Tetrahedron Lett.* 1994, **35:** 9103–9104.
97. Cochran, A.G. and Schultz, P.G. *J. Am. Chem. Soc.* 1990, **112:** 9414–9415.
98. Harada, A., Okamoto, K. and Kamachi, M. *Chem. Lett.* 1991, 953–956.
99. Harada, A., Fukushima, H., Shiotsuki, K, Yamaguchi, H., Oka, F. and Kamachi, M. *Inorg. Chem.* 1997, **36:** 6099–6102.
100. Quilez, R., De Lauzon, S., Desfosses, B., Mansuy, D. and Mahy, J.P. *FEBS Lett.* 1996, **395:** 73–76.
101. Bernadou, J., Bonnafous, M., Labat, G., Loiseau, P. and Meunier, B. *Drug Met. Dispos.* 1991, **19:** 360–365.
102. Masumoto, H., Ohta, S. and Hirobe, M. *Drug Met. Dispos.* 1991, **19:** 768–780.
103. Vidal, M., Bonnafous, M., Defrance, S., Loiseau, P., Bernadou, J. and Meunier, B. *Drug Met. Dispos.* 1993, **21:** 811–817.
104. Carrier, M.N., Battioni, P. and Mansuy, D. *Bull. Soc. Chim. Fr.* 1993, **130:** 405–416.
105. Gaggero, N., Robert, A., Bernadou, J. and Meunier, B. *Bull. Soc. Chim. Fr.* 1994, **131:** 706–712.

106. Chorghade, M.S., Hill, D.R., Lee, E.C., Pariza, R.J., Dolphin, D.H., Hino, F. and Zhang, L.Y. *Pure Appl. Chem.* 1996, **68:** 753–756.
107. Higuchi, T. and Hirobe, M. *J. Mol. Cat. A: Chem.* 1996, **113:** 403–422.
108. Adam, Y., Bernadou, J. and Meunier, B. *New J. Chem.* 1992, **16:** 525–528.
109. Wilson, I., Bretscher, K.R., Chea, C.K. and Kelly, H.C. *J. Inorg. Biochem.* 1983, **19:** 345–357.
110. Labat, G. and Meunier, B. *J. Chem. Soc., Chem. Commun.* 1990, 1414–1416.
111. Wagenknecht, H.A. and Woggon, W.D. *Angew. Chem. Int. Ed. Engl.* 1997, **36:** 390–392.
112. Habe, T., Shimida, M., Okamoto, T., Panijpan, B. and Higuchi, T. *J. Chem. Soc., Chem. Commun.* 1985, 1323–1324.
113. Paszczynski, A., Crawford, R.L. and Blanchette, R.A. *Appl. Environ. Microbiol.* 1988, **54:** 62–68.
114. Okamoto, T., Sasaki, K. and Oka, S. *J. Am. Chem. Soc.* 1988, **110:** 1187–1196.
115. Dolphin, D., Nakano, T., Maione, T.E., Kirk, T.K. and Farrell, R. in *Lignin Enzymic and Microbiol. Degradation* (INRA Ed.) Paris, 1987, **40:** 157–162; *Chem. Abst.* 1988, **109:** 69402u.
116. Meunier, B., Labat, G. and Séris, J.L. French Patent 88.09169 (1988); *Chem. Abst.* 1990, **113:** 152001r.
117. Labat, G. and Meunier, B. *New J. Chem.* 1989, **13:** 801–804.
118. Labat, G. and Meunier, B. *J. Org. Chem.* 1989, **54:** 5008–5011.
119. Meunier, B., Guilmet, E., De Carvalho, M.E. and Poilblanc, R. *J. Am. Chem. Soc.* 1984, **106:** 6668–6676.
120. Meunier, B. *Chem. Rev.* 1992, **92:** 1411–1456.
121. Meunier, B. *New J. Chem.* 1992, **16:** 203–211.
122. (a) Hoffman, P., Labat, G., Robert, A. and Meunier, B. *Tetrahedron Lett.* 1990, **31:** 1991–1994; Cui, F. and Dolphin, D. *ACS Symp. Ser.* 1989, **399:** 519–528.
123. Cui, F. and Dolphin, D. *Bioorg., Med. Chem.* 1994, **2:** 735–742.
124. Dolphin, D., Traylor, T.G. and Xie, L.Y. *Acc. Chem. Res.* 1997, **30:** 251–259.
125. Glenn, J.K. and Gold, M.H. *Appl. Environ. Microbiol.* 1983, **45:** 1741–1745.
126. Artaud, I., Ben-Aziza, K. and Mansuy, D. *J. Org. Chem.* 1993, **58:** 3373–3380.
127. Zhu, W. and Ford, W.T. *J. Mol. Cat.* 1993, **78:** 367–378.
129. Hampton, K.W. and Ford, W.T. *J. Mol. Cat. A: Chem.* 1996, **113:** 167–174.
130. Labat, G. and Meunier, B. *C. R. Acad. Sci. Paris,* Part II 1990, **311:** 625–630.
131. Campestrini, S. and Meunier, B. *Inorg. Chem.* 1992, **31:** 1999–2006.

132. Pattou, D., Labat, G., Defrance, S., Séris, J.L. and Meunier, B. *Bull. Soc. Chim. Fr.* 1994, **131:** 78–88.
133. Labat, G., Séris, J.L. and Meunier, B. *Angew. Chem. Int. Ed. Engl.* 1990, **29:** 1471–1473.
134. Sorokin, A., Séris, J.L. and Meunier, B. *Science* 1995, **268:** 1163–1166.
135. Sorokin, A., De Suzzoni-Dezard, S., Poullain, D., Noël, J.P. and Meunier, B. *J. Am. Chem. Soc.* 1996, **118:** 7410–7411.
136. Sorokin, A. and Meunier, B. *Chem. Eur. J.* 1996, **2:** 1308–1317.
137. Hadasch, A, Sorokin, A., Rabion, A. and Meunier, B. *New J. Chem.* 1998, **22:** 45–51.
138. Meunier, B. and Sorokin, A. *Acc. Chem. Res.* 1997, **30:** 470–476.
139. Belal, R. and Meunier, B. *J. Mol. Cat.* 1988, **44:** 187–190.
140. Weber, J.H. and Busch, D.H. *Inorg. Chem.* 1965, **4:** 469–471.
141. Sorokin, A. and Meunier, B. *Eur. J. Inorg. Chem.* 1998, 1269–1281.
142. Sisemore, M.F., Burstyn, J.N. and Valentine, J.S. *Angew. Chem. Int. Ed. Engl.* 1996, **35:** 206–208.
143. Grob, C.A. *Angew. Chem. Int. Ed. Engl.* 1969, **8:** 535–546.
144. Sorokin, A., Fraisse, L., Rabion, A. and Meunier, B. *J. Mol. Cat. A: Chem.* 1997, **117:** 103–114.
145. Hadasch, A., Sorokin, A., Rabion, A., Fraisse, L. and Meunier, B. *Bull. Soc. Chim. Fr.* 1998, **134:** 1025–1032.
146. Cui, F. and Dolphin, D. *Holzforschung* 1991, **45:** 31–35.
147. Defrance, S., Séris, J.L. and Meunier, B. *New J. Chem.*, 1992, **16:** 1015–1016.
148. Glenn, J.K., Akileswaran, L. and Gold, M.H. *Archiv. Biochem. Biophys.* 1986, **251:** 688–696.
149. Kersten, P.J., Kalyanaraman, B., Hammel, K.E., Reihammar, B. and Kirk, T.K. *Biochem. J.* 1990, **268:** 475–480.
150. Policar, C., Artaud, I and Mansuy, D. *Inorg. Chem.* 1996, **35:** 210–216.
151. Schonbaum, G.R. and Chance, B. in *The Enzymes* (ed.) Boyer P.D. Academic Press, New York, 1976, Vol. 13, Chap. 7, pp. 363–408.
152. Kimbrough, D.R., Magoun, M.A. and Langfur, M. *J. Chem. Educ.* 1997, **74:** 210–212.
153. Chiu, J.T., Loewen, P.C., Switala, J., Gennis, R.B., Timkovich, R. *J. Am. Chem. Soc.* 1989, **111:** 7046–7050.
154. Kleff, S., Sander, S., Mielke, G. and Eising, R. *Eur. J. Biochem.* 1997, **245:** 402–410.

155. Johnsson, K., Froland, W.A. and Schultz, P.G. *J. Biol. Chem.* 1997, **272:** 2834–2840.
156. Magliozzo, R.S. and Marcinkeviciene, J.A. *J. Am. Chem. Soc.* 1996, **118:** 1303–11304.
157. Corrall, R.J.M., Rodman, H.M., Margolis, J. and Landau, B.R. *J. Biol. Chem.* 1974, **249:** 3181–3182.
158. Vainshtein, B.K., Melik-Adamyan, W.R., Barynin, V.V., Vagin, A.A., Grebenko, A.I., Borisov, V.V., Bartels, K.S., Fita, I. and Rossmann, M.G. *J. Mol. Biol.* 1986, **188:** 49–61.
159. Fita, I. and Rossmann, M.G. *J. Mol. Biol.* 1985, **185:** 21–37.
160. Murshudov, G.N., Melik-Adamyan, W.R., Grebenko, A.I., Barynin, V.V., Vagin, A.A., Vainshtein, B.K., Dauter, Z. and Wilson, K.S. *FEBS Lett.* 1992, **312:** 127–131.
161. Olson, L.P. and Bruice, T.C. *Biochemistry* 1995, **34:** 7335–7347.
162. Dolphin, D., Froman, A., Borg, D.C., Fajer, J. and Felton, R.H. *Proc. Natl. Acad. Sci. USA* 1971, **68:** 614–618.
163. Chuang, W.J. and Van Wart, H.E. *J. Biol. Chem.* 1992, **267:** 13293–13301.
164. Benecky, M.J., Frew, J.E., Scowen, N., Jones, P. and Hoffman, B. *Biochemistry* 1993, **32:** 1929–11933.
165. Gouet, P., Jouve, H.M., Williams, P.A., Andersson, I., Andreoletti, P., Nussaume, L. and Hadju, J. *Nature Struct. Biol.* 1996, **3:** 951–956.
166. Jarnagin, R.C. and Wang, J.H. *J. Am. Chem. Soc.* 1958, **80:** 786–787.
167. Jones, P. and Dunford, H.B. *J. Theor. Biol.* 1977, **69:** 457–470.
168. Ivancich, A., Jouve, H.M. and Gaillard, J. *J. Am. Chem. Soc.* 1996, **118:** 12852–12853.
169. Traylor, T.G. and Xu, F. *J. Am. Chem. Soc.* 1987, **109:** 6201–6202.
170. Zipplies, M.F., Lee, W.A. and Bruice, T.C. *J. Am. Chem. Soc.* 1986, **108:** 4433–4445.
171. Balasubramanian, P.N., Schmidt, E.S. and Bruice, T.C. *J. Am. Chem. Soc.* 1987, **109:** 7865–7873.
172. Panicucci, R. and Bruice, T.C. *J. Am. Chem. Soc.* 1990, **112:** 6063–6071.
173. Belal, R., Momenteau, M. and Meunier, B. *J. Chem. Soc., Chem. Commun.* 1989, 412–414.
174. Robert, A., Loock, B., Momenteau, M. and Meunier, B. *Inorg. Chem.* 1991, **30:** 706–711.
175. Choua, S., Pacheco, P., Coquelet, C. and Bienvenüe, E. *J. Inorg. Biochem.* 1997, **65:** 79–85.

176. Bruce, A. J., Malfroy, B. and Baudry, M. *Proc. Natl. Acad. Sci. USA* 1996, **93:** 2312–2316.
177. Gianello, P., Saliez, A., Bufkens, X., Pettinger, R., Misseleyn, D., Hori, S. and Malfroy, B. *Transplantation* 1996, **62:** 1664–1666.
178. Malfroy, B., Doctrow, S.R., Orr, P.L., Tocco, G., Fedeseyeva, E.V. and Benichou, G. *Cell. Immunol.* 1997, **177:** 62–68.
179. Saito, Y., Mifune, M., Nakashima, S., Kawaguchi, T., Odo, J., Tanaka, Y., Chikuma, M. and Tanaka, H. *Chem. Pharm. Bull.* 1986, **34:** 2885–2889.
180. Barteri, M., Jones, P. and Mantovani, O. *J. Chem. Soc. Dalton Trans.* 1986, 333–336.
181. Sosa, R.C., Parton, R.F., Neys, P.E., Lardinois, O., Jacobs, P.A. and Rouxhet, P.G. *J. Mol. Cat. A: Chem.* 1996, **110:** 141–151.

Chapter 5

Non-Heme Peroxidases and Catalases: Mechanistic Implications from the Studies of Manganese and Vanadium Model Compounds

Carla Slebodnick, Neil A. Law & Vincent L. Pecoraro

Contribution from the Department of Chemistry,
The Willard H. Dow Laboratories,
University of Michigan,
Ann Arbor, MI 48109-1055

1. Introduction

It is common for nature to develop multiple mechanisms for achieving the same, or similar goals. Peroxidase and catalase enzymes, which are most commonly thought of as heme enzymes, are no exception. This paper, which was completed in January, 1998, aims to compare and contrast the structure and reactivity of non-heme vanadium and manganese enzymes that exhibit the same reactivity as certain better known heme peroxidases and catalases. As we will show, these non-heme containing enzymes react through very different mechanisms. Our discussion concentrates primarily on vanadium haloperoxidases (VHPO) and manganese catalases and their model compounds, although other biologically relevant vanadium and manganese systems are also discussed. Special emphasis is placed on the properties of the metals and how they define the mechanisms of activity. As chapters covering heme peroxidases and catalases are included in this book, only minimal discussion of these systems, necessary for understanding the discussion on non-heme systems, is included in this chapter.

2. Heme Catalases and Peroxidases

Peroxidases[1–4] catalyze the net two-electron oxidation of a substrate (A) or two sequential one-electron oxidations by peroxide to form water and the oxidized substrate ($2A^{\cdot}$ or A^{2+}), according to Eq. (5.1) and (5.2). Possible substrates include hydrocarbons, amines, thiols, sulfonates, and halides. Heme peroxidases include horseradish and spinach peroxidases, lignin peroxidase, haloperoxidase, cytochrome *c* peroxidase, and P450 (under some circumstances). They are involved in a variety of roles including synthesis of biomolecules (thyroid peroxidases) and producing oxidants, such as singlet oxygen and oxidized halides as an immune system response designed to fend off attacks by microorganisms. They are also involved in drug metabolism and have potential applications in the paper industry (bleaching and lignin breakdown) and diagnostic industry.[1,2,5,6] The Fe(porphyrin)(L) (L = his, cys, tyr) environment provides access to multiple oxidation states enabling the oxidation reactions to proceed by redox mechanisms involving both one electron and two electrons. The general mechanism for all the heme peroxidases is outlined in Fig. 5.1. The first step involves reaction of Fe^{III}(porphyrin) with H_2O_2 to form a high-valent Fe^{IV}(O)(porpyrin$^{+\cdot}$). This highly electrophilic oxo then reacts through two consecutive one-electron steps to form two one-electron oxidized substrates or one two-electron step forming a two-electron oxidized substrate and the Fe^{III}(porphyrin). The reactivity of the specific enzyme is a function of the proximal ligand, the hydrophobicity of the active site environment, and substrate accessibility.[1–4]

$$H_2O_2 + 2A \longrightarrow 2A^{\cdot} + [O] + H_2O \qquad (5.1)$$

$$H_2O_2 + A \longrightarrow A^{2+} + [O] + H_2O \qquad (5.2)$$

Catalases are similar to peroxidases, except that peroxide is the sole substrate, resulting in the net disproportionation of hydrogen peroxide to form O_2 and H_2O according to Eq. (5.3). Catalases are required to protect organisms against oxidative damage from hydrogen peroxide, for example by the reaction of H_2O_2 with Fe^{II} to form hydroxyl radicals via Fenton

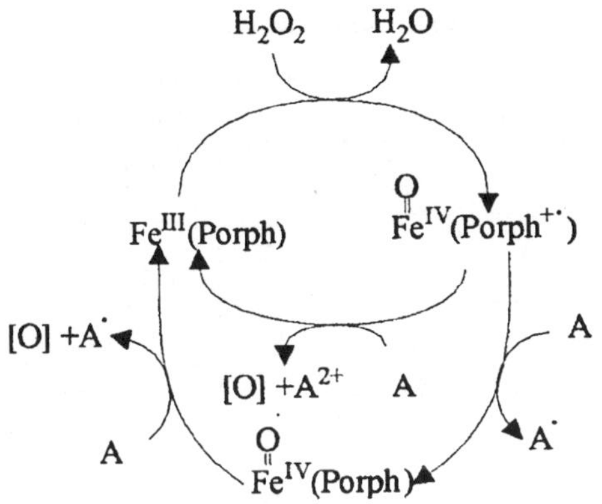

Fig. 5.1 General mechanism of heme peroxidases.

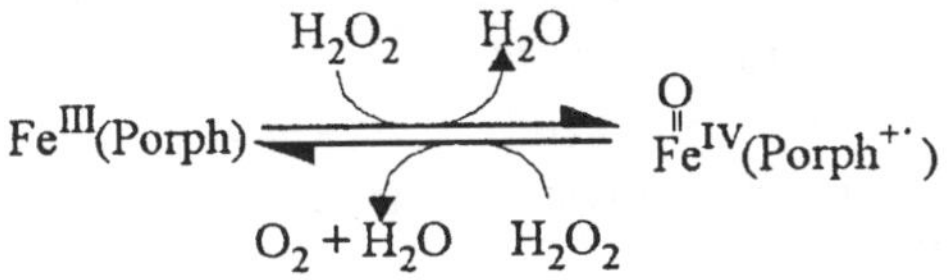

Fig. 5.2 General scheme for the Fe(porphyrin) peroxidases.

chemistry. All known Fe(porphyrin) peroxidases are capable of carrying out catalase activity via the mechanism outlined in Fig. 5.2.[2–4,7] Table 5.1 summarizes the relative catalytic rates for a number of different catalases, peroxidases, and model compounds discussed in this review.

$$2H_2O_2 \longrightarrow O_2 + 2H_2O \tag{5.3}$$

3. Vanadium Haloperoxidases

3.1 Vanadium haloperoxidase

Vanadium haloperoxidases are isolated primarily from marine algae, although they have also been found in fungi and one species of lichen. They may be partially responsible for production of a variety of halogenated

Table 5.1 Catalase activity.[a]

Enzyme or Complex	k_{cat} (s^{-1})	k_{cat}/K_m (M^{-1}•s^{-1})	Reference
Bovine liver **catalase**	3 600 000		127
Thermus thermophilus **catalase**	260 000	3.2×10^6	129–131, 169
Lactobacillus plantarum **catalase**	200 000	5.7×10^5	117
Thermoleophilum album **catalase**	26 000	1.7×10^6	128
A. nodosum **Vanadium Haloperoxidase** (pH 4)	300^b		36
$[Mn^{IV}(salpn)(\mu\text{-}O)]_2$	250	1×10^3	158
$[Mn^{II}(2\text{-}OHpicpn)]_4(ClO_4)_4$	150	70	161
Arginase (rat liver)	30	10.9	166
$[Mn^{II}(2\text{-}OHsalpn)]_2^{2-}$	22	350	141
$[Mn^{III}Mn^{IV}(\mu\text{-}O)_2(\mu\text{-}OAc)(tacn)(bipy)(MeOH)](ClO_4)_2$	13.2	---	152
$[Mn^{III}Mn^{IV}(\mu\text{-}O)_2(\mu\text{-}OAc)(tacn)(OAc)_2](ClO_4)_2$	5.5	---	152
$[Mn_2^{III}(anthracenediporphyrin)]Cl_2$	5.4	---	154
$[Mn_2^{II}(2\text{-}OHtbpn)(\mu\text{-}OAc)(n\text{-}BuOH)](ClO_4)_2$	1.2	---	146
$(Ba^{2+},Ca^{2+})[Mn_3^{III}Mn^{II}(\mu\text{-}O)(\mu\text{-}OH)(\mu\text{-}OAc)_2(2\text{-}OHtapn)_3]$	1.04	---	160
$[Mn_2^{II}LI(\mu\text{-}C_6H_5CO_2)_2(NCS)]$	0.79	---	153, 155
$[Mn^{III}(tetraphenylporphyrin)]^-$	0.013	---	154
$[Mn^{II}(H_2O)_6](ClO_4)_2$	0.0063	---	159

[a)]The rates observed for the systems above are for conditions which vary from study to study, whereas the k_{cat} values are presumably maximal rates with saturating hydrogen peroxide. [b)]This rate is for substrate bromination, but researchers later showed that the rates of bromination and O_2 formation are the same.[40]

antifungal and antibacterial agents as well as many of the naturally generated halocarbons thought to be responsible for ozone depletion in the outer atmosphere. A maximum of one vanadium binds per 60–70 kDa protein subunit and vanadium remains in the +5 oxidation state throughout the entire catalytic cycle. VHPOs are described as chloroperoxidases (ClPO), bromoperoxidases (BrPO), or iodoperoxidases (IPO), depending on the most electronegative halide they can oxidize. Thus, VClPOs are the most reactive, oxidizing Cl^-, Br^- and I^-.[8–11] There are no fluoroperoxidases because hydrogen peroxide does not have the potential to oxidize fluoride.

3.1.1 Structure

X-ray techniques have been the primary method of characterizing the active site structure of the V^V form of VHPOs, including EXAFS, XANES,[12–14] and three single crystal X-ray structures of V^V forms of VClPO from *Curvularia inaequalis* at pH 8. There are two vanadate structures (native- and azide-inhibited) and a peroxovanadate structure.[15,16] In all three crystal structures, histidine is the only protein residue bound directly to the vanadium, although several strong hydrogen bonding interactions with amino acids and water solvates also assist in holding the vanadium tightly in place. In the two vanadate structures, vanadium adopts a distorted trigonal bipyramidal geometry with three equatorial oxygen donors, an axially bound histidine, and either an azide (azide-inhibited form), or a hydroxide as the fifth ligand. In the native structure of the VClPO, His-404 is hydrogen-bonded to a vanadate oxygen and has been proposed to function as a crucial acid-base catalyst. The peroxovanadate structure adopts a distorted square pyramidal geometry with the peroxide hydrogen-bonded to Lys-353, which may also function as an acid-base catalyst.

The tertiary structure of VClPO is predominantly α-helical, with a bundle of eight-helices serving as the primary motif. The vanadate resides at the top of four of these helices and a channel to this site exists allowing access of H_2O_2, X^-,[15,16] and possibly select organic substrates.[17] One side of this channel is composed primarily of hydrophobic residues consisting of Pro, Phe, and Trp residues and resembles halogen and halide binding sites on other halide utilizing proteins. The other side is primarily hydrophilic.[15,16]

The strong sequence homology[9,16,18] of the active site amino acid residues between VClPO (*C. inaequalis*) and VBrPO (*A. nodosum*) suggests that VBrPO adopts a structure analogous to the crystallographically characterized VClPO. The vanadium bound histidine (His-496 of VClPO) and all amino acid residues involved in hydrogen bonding at the active site are conserved between VClPO (*C. inaequalis*) and VBrPO (*A. nodosum*), except for Lys-353, which is replaced with an arginine in VBrPO. Lys-353 may be involved in acid-base catalysis, so this substitution may play a crucial role in the differing reactivities of VClPO and VBrPO. Another important difference is the replacement of a His for Phe-397 in the hydrophobic region of the protein where halide binding is believed to occur.[11] Structural similarities between VClPO and VBrPO are also supported by the metrical parameters derived from the EXAFS data for VBrPO (*A. nodosum*), which are consistent with a V^{V}-oxo with four or five oxygen or nitrogen donors.[12]

Based on the X-ray data, the active site structure of VClPO is very similar to a number of vanadate-inhibited soluble acid phosphatases.[19–22] Although there is no sequence homology between these acid phosphatases and VHPOs, the vanadate bound structures all adopt trigonal bipyramidal vanadium coordination with a single amino acid (cysteine or histidine) and four non-protein oxygen donors coordinated to vanadium. In addition, three families of membrane bound and bacterially secreted acid phosphatases do have sequence homology with the active site of VHPOs.[23–25] This raises the questions of (1) do VHPOs possess phosphatase activity and is this their original role and (2) can vanadium pose a risk to humans, not just by inhibiting phosphatases, but by generating a new and potentially detrimental VHPO action? It has been shown that apo-VClPO can dephosphorylate *p*-nitrophenylphosphate, thus exhibiting phosphatase activity.[23] However, whether or not any vanadate-inhibited acid phosphatases are capable of haloperoxidase activity has yet to be determined. This is especially important to consider with the explosion of interest in using vanadium to treat diabetes.[26–28] The insulin mimetic action of vanadium is presumed to arise from vanadium inhibition of protein tyrosine phosphatases that are responsible for deactivating the insulin receptor kinase and other kinases that are vital to regulating insulin's effects.[26,27,29–31]

A variety of spectroscopic studies have also been performed on the catalytically inactive V^{IV} form of VBrPO. The EXAFS spectrum of the V^{IV} VHPO suggests octahedral geometry with one short oxo bond and five longer oxygen or nitrogen donors, consistent with the proposed structure.[12] EPR spectroscopy of VBrPO (*A. nodosum*)[32] and VClPO[33] at 90 K give axially symmetric spectra (two sets of eight lines), consistent with monomeric vanadyl. The ESEEM spectrum gives four peaks resulting from nitrogen hyperfine coupling.[34] The higher frequency peaks suggest that at least one imidazole is coordinated in the equatorial plane. Based on hyperfine coupling values for a series of small molecule V^{IV} complexes with axial and/or equatorial nitrogen donors, the lower frequency peaks in the spectrum of the enzyme are consistent with the presence of a nitrogen donor in the axial position. Binding the acid-base catalyst (His-404) to the vanadyl to form a six-coordinate octahedral complex is one explanation for the ESEEM spectrum observed for the enzyme.[35] Reactivation of the enzyme to the V^{V} form with H_2O_2 is known to be slow. If the resultant conformational changes of the protein backbone upon coordination of His-404 closes the substrate channel or changes the redox potential of the vanadium center, this could explain the inactivity of the reduced enzyme, even in the presence of oxidants such as H_2O_2.[32]

3.1.2 Reactivity

VHPOs oxidize halides by a very different mechanism than heme HPOs. A redox mechanism is unlikely because vanadium is already in its highest and most stable oxidation state (+5, d^0) when it binds peroxide. It is conceivable that vanadium shuttles between V^{III} and V^{V} during this cycle but the fact that oxo ligands are not common structural features in V^{III} but are to V^{V} systems means that major, non-reversible coordination changes would be required to shuttle between these two oxidation states. Instead, V^{V}'s oxophilicity and Lewis acid strength make it ideal for carrying out halide oxidations by a Lewis acid mechanism, binding and activating H_2O_2 towards nucleophilic attack by the electron rich halide. The two-electron oxidized halogen, formally referred to as "X^+," is most likely an equilibrium mixture of an enzyme bound species and freely diffusable HOX, X_2, and X_3^-, with concentrations of each species depending on the halide, halide concentration, and pH, Eq. (5.4). Once formed, the oxidized halide can react with electron rich organic substrates to form halogenated organic compounds, Eq. (5.5), with the net reaction shown in Eq. (5.6). Alternatively, the halide can oxidize a second equivalent of H_2O_2, Eq. (5.7). In this case, the net reaction is that of heme catalase, Eq. (5.3).[9–11]

$$H_2O_2 + X^- + H^+ \xrightarrow{\text{VHPO}} \text{“}X^{+}\text{”} + OH^- + H_2O \tag{5.4}$$

$$(X^+ = Enz - X^+, HOX, X_2, \text{ or } X_3^-)$$

$$X^+ + R - H \longrightarrow R - X + H^+ \tag{5.5}$$

$$H_2O_2 + X^- + H^+ + R - H \longrightarrow R - X + 2H_2O \tag{5.6}$$

$$H_2O_2 + X^+ \longrightarrow O_2 + X^- + 2H^+ \tag{5.7}$$

Kinetic studies have been performed on three species of VBrPO[36, 37] as well as VClPO from *C. inaequalis*,[38] all giving results consistent with a sequentially ordered mechanism involving peroxide binding to the vanadium followed by halide binding and oxidation (Fig. 5.3). The pH profile for the peroxide binding step indicates the presence of an acid-base catalyst with pK_a of 5.5–6.5 (depending on VHPO species) that must be deprotonated for H_2O_2 to bind. Presumably, this acid is His-404 of VClPO (or the equivalent His residue in the VBrPOs). Bromide inhibits H_2O_2 binding to the VBrPOs by a non-competitive mechanism above ca. pH 4, reaching a maximum K_i^{Br} at ca. pH 5.25 and then decreasing again.[36,37] Chloride inhibits H_2O_2 binding to VClPO by a competitive mechanism at pH 3.1 and a non-competitive mechanism above pH 4.1, similar to Br^- inhibition of VBrPOs.[38]

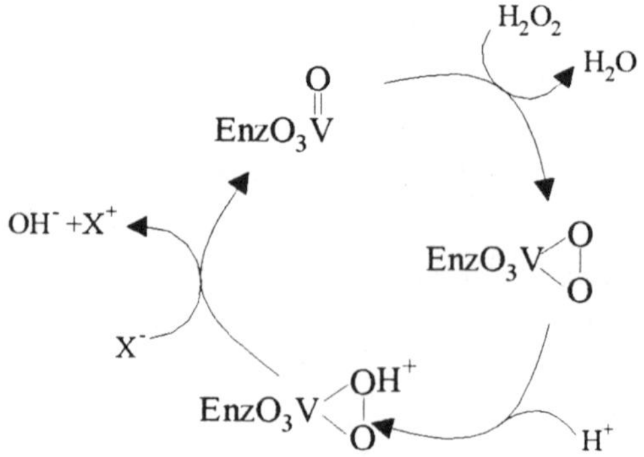

Fig. 5.3 Proposed mechanism of vanadium haloperoxidase (reproduced from 167 with permission from Springer-Verlag).

Table 5.2 Second order rate constants for hydrogen peroxide binding and bromide oxidation in VBrPO enzymes and models.

Peroxovanadate Species	Solvent	$k_{H_2O_2}$ ($M^{-1}s^{-1}$)	k^{Br-} ($M^{-1}s^{-1}$)	Reference
VBrPO (*A. nodosum*)[a]	H_2O, pH 4	9.5×10^4	1.75×10^5	36
	H_2O, pH 7.9	2.5×10^6	2800	
VBrPO (*M. Pyrifera*)	H_2O, pH >6.5	99		171
VBrPO (*F. distichus*)	H_2O, pH >6.5	112		171
$V_2O_2(O_2)_3$	H_2O or 25% EtOH/H_2O		4.1 ± 0.1	168
$VO(O_2)(Hnta)^-$	CH_3CN		170 ± 30	68
$VO(O_2)(Hheida)^-$	CH_3CN		280 ± 40	68
$VO(O_2)(ada)^-$	CH_3CN		220 ± 30	68
$VO(O_2)(Hbpg)^+$	CH_3CN	11 ± 2[b]	21 ± 3	66, 68
$VO(O_2)(tpa)^+$	CH_3CN		100 ± 30	68
H_2O_2	CH_3CN		3.7 ± 0.9	68

[a]rate constants represent estimated lower limit based on k_{cat}/K_m.
[b]$k_{H_2O_2}$ measured in the presence of ≤ 1 equivalent of acid.

The estimated lower limit for the second order rate constant of peroxide binding ($K_m^{H_2O_2}/V$), along with other kinetic parameters for the enzymes and VHPO model complexes are summarized in Table 5.2.

In the second step of the proposed mechanism (Fig. 5.3), the halide attacks the enzyme bound peroxide. This reaction shows Michaelis-Menton type saturation, indicating that the halide binds to the enzyme. Unlike the peroxide binding step, halide binding is facilitated by more acidic conditions. Although the enzymology studies provide no direct evidence for actual acid-base chemistry at the vanadium active site, studies with model compounds strongly suggest that a site at the vanadium center (presumably the peroxide) must be protonated prior to nucleophilic attack by the halide (*vide infra*). Based on the X-ray crystal structure of peroxide bound VClPO, Lys-353 (or the corresponding Arg of VBrPO, *A. nodosum*) is the most likely source of this proton.[15] Protonation of the peroxide is consistent with the proposed electrophilic mechanism. It has been suggested that the halide binds directly to the vanadium just prior to oxidation,[15] but considering the EXAFS data, which shows no indication of halide binding under turnover conditions, halide binding directly to vanadium seems unlikely.[12] In addition, the proposed electrophilic mechanism also makes halide binding at the vanadium center unlikely since this would reduce halide nucleophilicity and substantially decrease reactivity. A halide binding pocket near the vanadium binding site is, however, reasonable.

Once formed, the oxidized halide reacts further with an organic substrate (Eq. 5.4) or an additional equivalent of H_2O_2 (Eq. 5.6) to form a halogenated substrate or singlet oxygen, respectively. Although the net reaction for O_2 formation is the same as for heme catalases (Eq. 5.2), VHPOs are not "true catalases" in that halide must be present for the reaction to proceed. Heme HPOs can catalyze H_2O_2 disproportionation through either direct oxidation of H_2O_2 by $Fe^{IV}(O)(porphyrin^{+\bullet})$ or the halide assisted mechanism characteristic of VHPOs. In the presence of an organic substrate, substrate halogenation and oxygen formation are competitive; formation of singlet oxygen is favored at higher H_2O_2 concentrations and higher pH. The rates of substrate halogenation and dioxygen formation are the same, indicating the rate-determining step is halide oxidation and both substrate halogenation and O_2 formation are extremely facile.[12,37,39–41]

VHPOs are capable of oxidizing substrates other than halides, including the pseudohalides (thiocyanate[42] and cyanide[43]) and histidine.[44] Studies with peroxovanadate model compounds suggest the range of substrates available to VHPOs may be much larger than currently known. In theory, any sufficiently nucleophilic reagent that can access the VHPO active site with an oxidation potential less than H_2O_2's reduction potential will be oxidized. Possible substrates of biological significance include sulfides and unsaturated hydrocarbons (alkenes and aromatics).

3.2 Amavadin

Amavadin is a vanadium containing small molecule natural product isolated from *Amanita muscaria* mushrooms.[45,46] Amavadin is a V^{IV} complex that adopts the structure shown in Fig. 5.4.[47,48,172] The lack of a vanadium oxo bond in both the V^{IV} and V^{V} forms of this molecule make it quite unique. The function of amavadin is currently unknown. Most researchers have been pursuing studies of amavadin's ability to act as a one-electron oxidant,

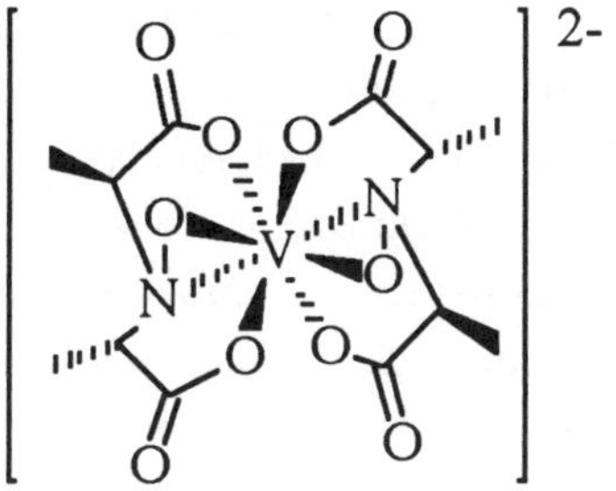

Fig. 5.4 Drawing of amavadin's coordination environment.

since it can be reversibly oxidized ($E° = 0.53$ V vs. SCE in H_2O)[49] and since structural changes between the V^{IV} and V^{V} forms are minimal,[47,48] making it an ideal candidate for redox reactions. Researchers recently demonstrated that certain thiols can be oxidized to disulfides using an amavadin model complex. The model compound, abbreviated $V(hida)_2^{2-}$ is the same as amavadin, except it lacks the chiral methyl group of the ligand. Subsequent electrochemical reduction of $V^{V}(hida)^-$ regenerated an active oxidant, $V^{IV}(hida)_2^{2-}$, allowing for multiple turnovers.[50]

More recent studies suggest that amavadin may act as a VIPO. The impetus for these experiments was the realization that the bidentate hydroxylamine coordination to vanadium is analogous to peroxide coordination. Studies of the oxidized model compound $V(hida)_2^-$ (oxidized with H_2O_2 or Ce^{4+}) show significant rates for iodide oxidation. By monitoring the spectroscopic formation of I_3^-, the rate of I^- oxidation by $V(hida)_2^-$ was determined to be approximately 100 times faster than the rate of I^- oxidation by H_2O_2 under the same conditions.[51]

3.3 *VHPO model compounds*

Although a number of VHPO model studies have been performed, few have concentrated on the mechanistic details of halide oxidation by peroxovanadate complexes. Most model studies have focused on structural and spectroscopic models,[35,51–56] or product analysis.[57–62] In these studies, there was little or no effort to characterize the active vanadate and peroxovanadate species or measure reaction rates. One reason for this is the complex equilibria that exist for vanadates and peroxovanadates in solution. In the absence of organic ligands to stabilize monomeric vanadium complexes, vanadate establishes a number of oligomeric and protonic equilibria, depending on acid and vanadium concentrations. Similarly, peroxovanadates form monomeric and dimeric species and can bind as many as four peroxides.[57–62] Even in the presence of coordinated ligand, dimerization reactions and ligand dissociation often create difficulties.

The first detailed mechanistic study of a vanadium-assisted halide oxidation by H_2O_2 was with VO_2^+ in aqueous solution.[63,64] With this system, a mechanism involving H_2O_2 binding to V, followed by bromide oxidation was proposed, similar to VHPOs. Two important differences between this system and VHPO are that a dimeric peroxovanadate species, $V_2O_2(O_2)_3$, was implicated as the active catalyst and strongly acidic conditions are required for activity ($pH \leq 2$). The identification of a dimeric active species was based on the facts that (1) the rate of halide oxidations showed a clear second order dependence on [V], and (2) a maximum rate was observed under pH and H_2O concentrations where $VO(O_2)^+$ and $VO(O_2)_2^-$ were equal, and thus, the concentration of the dimeric $V_2O_2(O_2)_3$ is maximized, Eq. (5.8)–(5.10). The proposed mechanism for this reaction is outlined in Fig. 5.5.

$$VO_2^+ + H_2O_2 \longrightarrow VO(O_2)^+ + H_2O \quad K_1 \tag{5.8}$$

$$VO(O_2)^+ + H_2O_2 \longrightarrow VO(O_2)_2^- + 2H^+ \quad K_2 \tag{5.9}$$

$$VO(O_2)^+ + VO(O_2)_2^- \longrightarrow V_2O_2(O_2)_3 \quad K_3 \tag{5.10}$$

The rate of halide oxidation is first order in [Br^-], implicating the halide-binding step as the rate-determining step. The rates of halide oxidation by $V_2O_2(O_2)_3$ were also measured in mixed ethanol/water (0–25% ethanol v/v) and a clear solvent dependence was observed. It was shown that this change in rate was a result of the solvent dependence of K_2, and K_3, Eqs. (5.9) and (5.10) and not a change in the rate constant for halide oxidation, Eq. (5.11).[63,64]

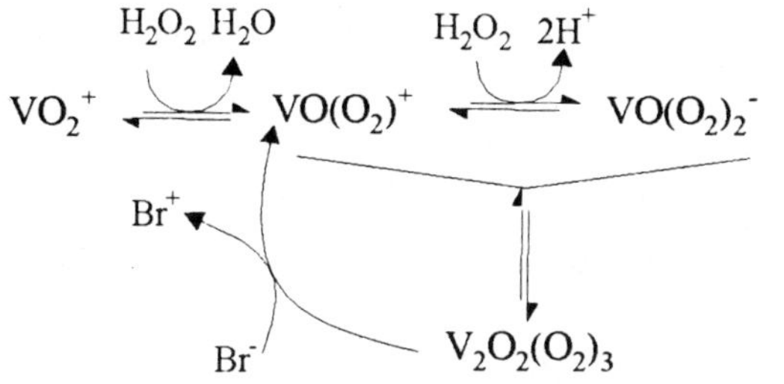

Fig. 5.5 Proposed mechanism of halide oxidation by $V_2O_2(O_2)_3$ (adapted from Ref. 168).

$$\text{rate} = k[V_2O_2(O_2)_3][Br^-] = k[VO(O_2)^+][VO(O_2)_2^-][Br^-] \tag{5.11}$$

In a separate study, a $V_2O_2(O_2)_3$ core in aqueous solution was also implicated, but in this case, glycine was coordinated to the metal to form $V_2O_2(O_2)_3(Hgly)_2(H_2O)_2$.[59] Unfortunately, detailed analysis of this system was not performed and it is not clear if the dimer remains intact in solution or if glycine remains bound during the reaction cycle.

Some of the most successful reactivity studies have been performed in non-aqueous solution. The use of a less polar solvent enables researchers to study the behavior of complexes in more hydrophobic environments. The energetics of protein folding dictate that the more hydrophobic residues reside on the interior of the protein where the active site is often located. As hydrophobic pockets cannot be created with ease using model compounds, the use of a less polar solvent is a useful (although somewhat extreme) method of mimicking this non-polar environment.

The first study reported in non-aqueous solution was with Schiff base complexes of the form LVO(OEt)(EtOH) (L = N-(2-hydroxyphenyl)salicylideneamine or N-(2-carboxyphenyl)salicylideneamine.[65] ^{51}V NMR and UV-vis spectroscopic studies, in combination with limited kinetic studies, were used to identify the active catalyst as LVO(OH). The LVO(OEt)(EtOH) complex forms as many as five species in solution which are in rapid equilibrium, including the active LVO(OH) catalyst. Addition of H_2O_2 results in quantitative formation of $LVO(O_2)$, the active peroxovanadate species. The proposed mechanism is outlined in Fig. 5.6.

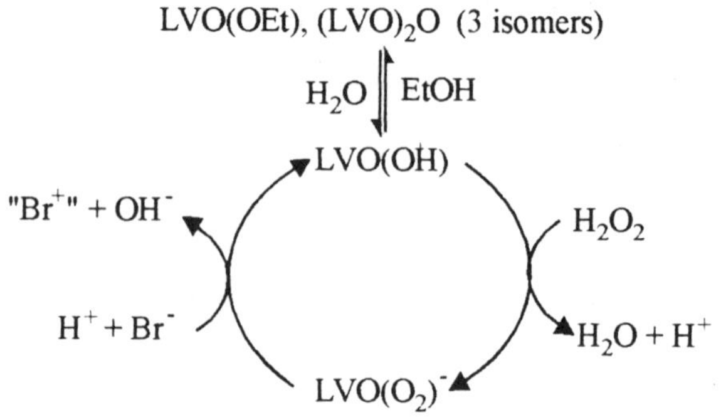

Fig. 5.6 Proposed mechanism of halide oxidation by LVO(OH) (adapted from Ref. 65).

The most complete study on model compounds has been with tetradentate tertiary amine complexes containing different combinations of carboxylate, amine, alcohol, amide, or pyridyl donors (Fig. 5.7 A–D). The starting dioxovanadate complexes (VO_2L', L' = tetradentate ligand) and oxoperoxovanadate complex ($VO(O_2)L'$), analogous to the peroxide bound intermediate in the enzyme bound system have been isolated and structurally characterized. In addition, the kinetics of peroxide binding to the complex[66] and the kinetics of subsequent halide oxidation[67,68] have been measured and a detailed mechanism proposed. These systems also catalyze O_2 formation in the absence of an organic substrate, similar to enzymes. One key point with the kinetic studies of the VO_2L' complexes is that the reactions were run in acetonitrile. This less polar solvent was vital, since chloride, bromide and iodide oxidation were not observed under any aqueous conditions. Both bromide and iodide can be oxidized in acetonitrile, but chloride oxidation was not achieved in water or acetonitrile.

The kinetics of peroxide binding to the VO_2L' complexes to form the oxoperoxo complex, $VO(O_2)L'$, are first order in VO_2L'. In the absence of acid, Michaelis-Menton behavior was observed upon H_2O_2 addition, indicating the formation of a reversible $VO_2L' \cdot H_2O_2$ intermediate. Upon addition of acid, the Michaelis-Menton behavior is greatly diminished or disappears completely (depending on ligand and acid concentration), and a first order dependence on H_2O_2 concentration is observed. The proposed

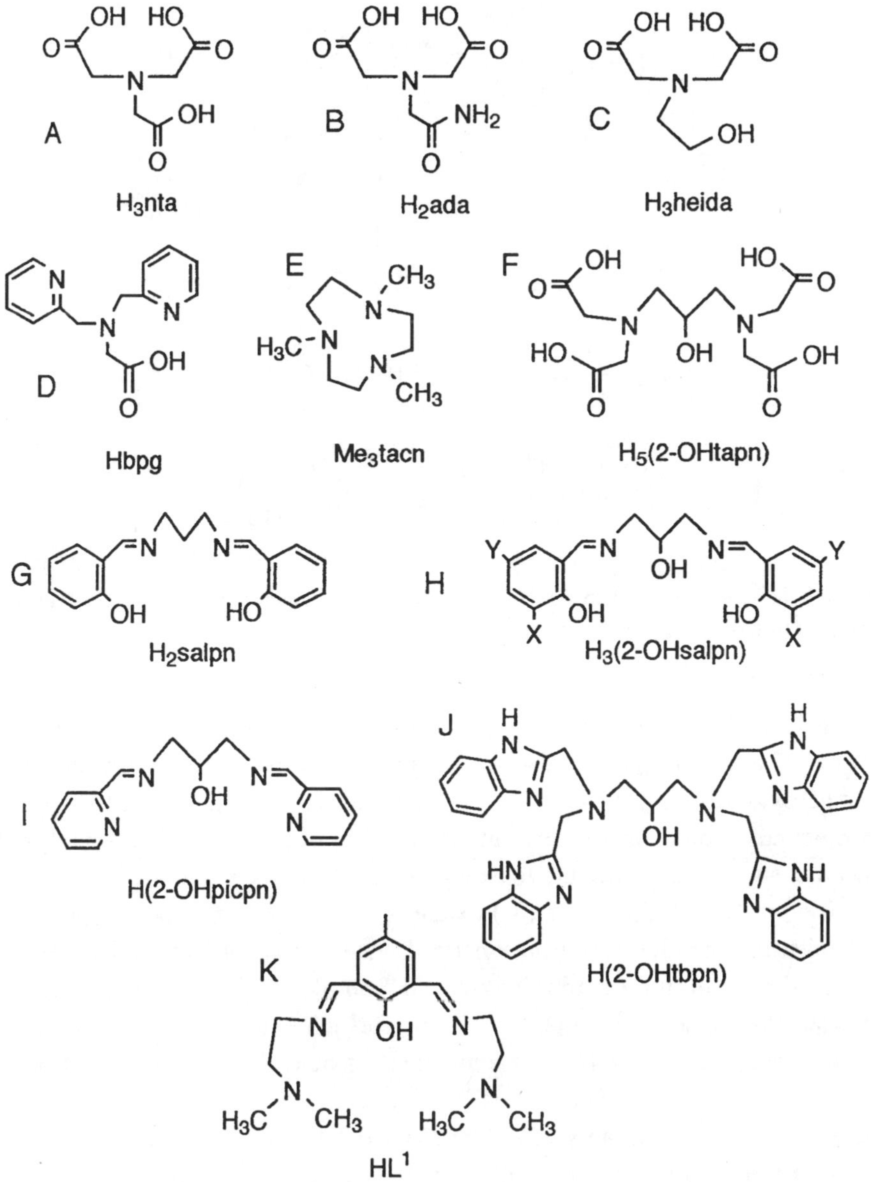

Fig. 5.7 A selection of ligands from the synthetic model systems described in the text.

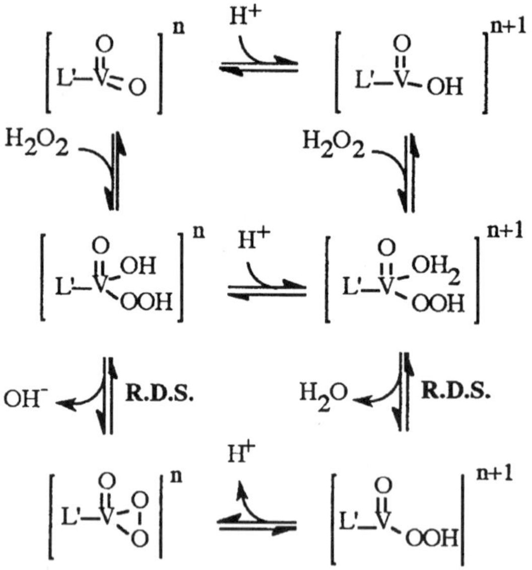

Fig. 5.8 Proposed mechanism for H_2O_2 binding to tetradentate tertiary amine complexes (adapted from Ref. 66).

mechanism for peroxide binding is summarized in Fig. 5.8.[66] Since it is known that an acid-base catalyst with a pK_a of 5.5–6.0 must be deprotonated for peroxide to bind in the VHPOs, base-catalyzed studies with the model compounds would prove very interesting. Unfortunately, a base-catalyzed mechanism is very difficult to study in the CH_3CN solution.

The kinetics of halide (Br^-, I^-) oxidation are first order in peroxovanadate complex and halide. Based on kinetic, UV-vis and NMR analysis, there is no evidence of halide binding to the complexes at any point during the reaction. One very important factor for achieving halide oxidation in acetonitrile is the need for one equivalent of acid. In the absence of acid, no reaction occurs. These results suggest that the complexes must be protonated (presumably at the peroxide) to become active. By studying the rate dependence on acid, estimated pK_a's for the protonated peroxovanadate complexes in acetonitrile have been obtained and range from 5.5–6.0, depending on the ligand. For comparison, the pK_a's of HI and HBr in acetonitrile are also in this range, while the pK_a of HCl is ca. 9. Because of the pK_a dependence of complex and acid on solvent conditions, three equilibrium states are possible. Thus, one likely explanation for the lack of chloride oxidation in acetonitrile is that the chloride buffers the reaction, preventing the formation of the active protonated peroxovanadate complex. By comparing the relative pK_a's of various acids in acetonitrile and water, the pK_a of the protonated peroxovanadates in water is estimated to be near 0. The high acidity of the complex explains the absence of Cl^-, Br^-, or I^- oxidation in aqueous solution, since the active protonated peroxovanadate species is not formed under conditions where the complex remains intact.[67,68]

Based on these studies, the mechanism in Fig. 5.9 has been proposed for the VO_2L' model complexes. This mechanism is also consistent with enzymatic studies, where it was shown that the halide oxidation step proceeds faster at lower pH where protonation of the peroxide is favored.[32] In addition, His-404 and Lys-353 are both located near the vanadium binding site and may participate in activating the complex through hydrogen bonding interactions. With regard to the other model studies, no evidence for complex

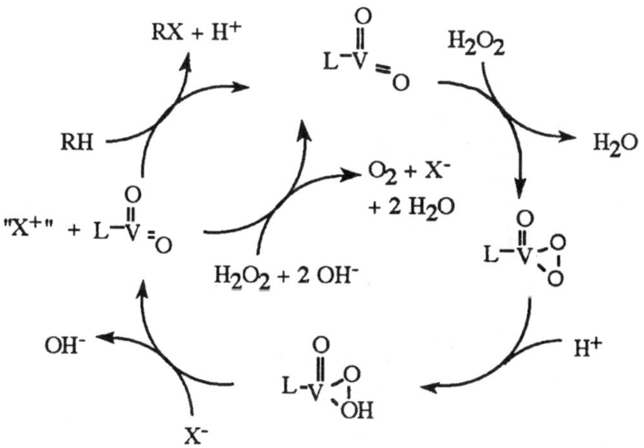

Fig. 5.9 Proposed mechanism for halide oxidation by tetradentate tertiary amines (adapted from Ref. 68).

protonation was provided, but the reactions were all run under acidic conditions because of the need for protons to drive the net reaction, Eq. (5.4). In the cases where the active peroxovanadate species contains a $V_2O_2(O_2)_3$ core, NMR studies in acetonitrile have shown that this species is an asymmetric dimer.[22] In the crystal structure of a tetraperoxo dimer, the asymmetry is caused by a bridging peroxide that is bidentate to one vanadium and monodentate to the other (Fig. 5.10a).[69,70] Based on this data, the $V_2O_2(O_2)_3$ dimer is proposed to have an asymmetrically bridging peroxide of the form shown in Fig. 5.10b. In this reaction, the second vanadium could very well function as a "giant proton," activating the diperoxovanadate complex toward nucleophilic attack by halide in a similar manner to the proton in the monomeric complexes.

Although much progress has been made in recent years toward achieving bromide oxidation with model complexes, Cl^- oxidation has remained relatively elusive. It has been observed with $V_2O_2(O_2)_3$ in aqueous solution but no rates were reported, presumably because the rates were too slow to measure.[63] As with the tetradentate tertiary amine complexes, this can be explained if equilibrium state A of Fig. 5.11 is favored. This is true in aqueous solution, where both HCl and $HVO(O_2)$ are strong acids. In

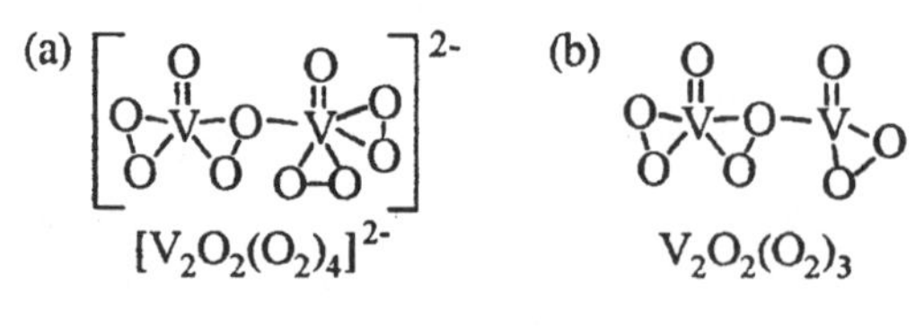

Fig. 5.10 Drawing of (a) $V_2O_2(O_2)_4^{2-}$ and (b) proposed coordination of $V_2O_2(O_2)_3$.

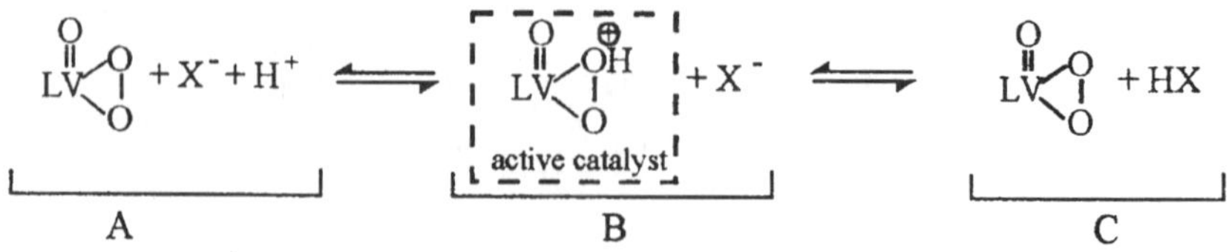

Fig. 5.11 Possible protonic equilibria between peroxovanadates and halides, depending on solvent and halide (reproduced from Ref. 22 with permission from Elsevier Science).

acetonitrile, Cl^- is significantly more basic than the peroxovanadate complexes and buffers the solution (Fig. 5.11, equilibrium C). In order to achieve chloride oxidation, conditions must be obtained where the peroxovanadate is both a strong enough base to be protonated under reaction conditions and a stronger base than the halide.

Chloride oxidation has been achieved by using mixed water/acetonitrile solvent.[71] By adjusting water concentration in acetonitrile, vanadate has been shown to catalyze chloride oxidation, with a maximum rate occurring in 6% H_2O/94% CH_3CN. Presumably, this water concentration represents conditions where the concentration of the active protonated peroxovanadate species is at a maximum. Unfortunately, ^{51}V NMR shows that the equilibrium under these conditions is very complex, and the active peroxovanadate species has not yet been identified. Approximate first order dependence of the rate of halide oxidation on [V] strongly supports a monomeric complex as the active species. The rate dependence on Cl^- concentration has not been determined because Cl^- acts as both a substrate and an inhibitor. However, the rate of bromide oxidation is first order in both [Br^-] and [V]. ^{51}V NMR studies, in combination with further kinetic studies, are currently being used to understand the details of this reaction better.

3.4 Peroxidase models

Although there is no direct evidence that VHPOs can oxidize substrates other than halides and pseudo halides (CN^-[43] and SCN^-[42]), in theory, electron rich substrates that can access the substrate channel of VHPOs should be able to nucleophilically attack the bound halide and be oxidized. Oxidation of variety of substrates, including alkenes, alkanes, aromatics, phosphines, and sulfides, in the presence of vanadium complexes and ROOH (R = H, alkyl, or acyl) is already well-documented. These reactions proceed by an electrophilic or radical mechanism depending on the vanadium complex, solvent, and substrate.[72–75] Two substrates associated with electrophilic oxidation by peroxovanadate or alkylperoxovanadate complexes are alkenes and sulfides.

The oxidation of alkenes by peroxo and alkylperoxovanadate complexes can proceed by both electrophilic and radical mechanisms, depending on the vanadium complex and the electron richness of the substrate. For those reactions that are electrophilic, epoxide is the only product, while the radical-based reactions results in formation of a mixture of epoxide, alcohol, and ketone (or aldehyde). Based on product analysis studies, it has been demonstrated that epoxide formation is favored with electron rich substrates while radical products are favored with electron poor substrates. In addition, kinetic studies show that the epoxidation reactions proceed at faster rates as alkene nucleophilicity increases, supporting an electrophilic mechanism.[72–75] With regard to sulfide oxidation, a number of reactivity studies with a variety of peroxovanadate complexes have been performed and all support an electrophilic oxidation mechanism.[72,76–78] For example, the substrate thianthrene-5-oxide, which contains both an electron rich sulfide and an electron poor sulfoxide, is oxidized only at the sulfide under acidic conditions.[76] The sulfide oxidation reactions are acid-catalyzed and show a first order dependence on [V] and [sulfide], as is expected for an electrophilic mechanism.[72,76–78]

The possibility that VHPOs may be able to catalyze a variety of reactions beyond halide oxidation is interesting in itself, but the possibility that some vanadium-inhibited phosphatases with known sequences or structural similarities to VHPOs could possess this reactivity implies a much broader impact. It has already been proposed that peroxovanadates have insulin mimetic action because they bind at and permanently deactivate the active site of protein tyrosine phosphatases (PTPs) involved in halting the insulin cycle. The proposed mechanism of deactivation is oxidation of an active site cysteine to a sulfoxide. Introduction of small molecule peroxovanadates in diabetes treatment could also result in unwanted reactivity through oxidation of other cysteines, thus targeting and damaging systems in addition to the PTPs involved in the insulin cycle. If this does not occur there is still the possibility that the peroxovanadates will bind and inhibit other PTPs in the body. This could lead to new and potentially detrimental activities such as VHPOs or peroxidases.

4. Manganese Systems

4.1 Manganese peroxidase, MnP

Although there are no known naturally occurring enzymes which employ a Mn(porphyrin), a manganese-dependent peroxidase (MnP) does exist. MnP is a 46 kDa extracellular enzyme that involves an iron protoporphyrin IX, manganese and oxalate as cofactors for the oxidation of lignin.[79–85] This enzyme is one of many components excreted by white-rot fungi as part of the lignin degrading systems. Other components include lignin peroxidase (LiP), a hydrogen peroxide generating system, and based on newer reports, in some cases, a laccase enzyme.[93,94]

Lignin is found in many types of vascular plants and woody plants in particular, where it comprises 15–36% of a plant cell wall mass.[86–90] It provides support and protection for plant tissue by forming a "matrix" around the cellulose.[86,88–90] Lignin, the second most common biomaterial after cellulose, is a random three-dimensional biopolymer of phenolic monomers connected to one another by a variety of linkage motifs. The phenolic monomers utilized to build up lignin include 4-hydroxy-3-methylcinnamyl alcohol (coniferyl), p-hydroxycinnamyl alcohol (coumaryl), and 3,5-dimethoxy-4-hydroxycinnamyl (sinapyl) alcohol with the overall percent content of each monomer in lignin and the structure varying from one plant or cell type to another.[86,88–90] The random structural motif gives lignin the ability to support and protect plant tissue, but it also makes lignin rather difficult to degrade. Furthermore, the decomposition of the inner biopolymer, cellulose, into smaller subunits is hindered by the presence of lignin.[86,88,89] It has been suggested that lignin's interference with fungal access to the cellulose caused the evolution of the system of enzymes designed to biodegrade lignin. White-rot fungi comprise one of three classes of fungi known to competently biodegrade lignin (brown-rot and soft-rot, being the others).[91] This is an extremely important task in light of the retarded degradation of cellulose by the presence of lignin and by the fact that cellulose is the most prevalent biopolymer on the planet. This set of observations has led to the suggestion that the ability of these fungi to degrade lignin may be quite important to the cycling of carbon.[86–89,92]

MnP is one of only two known enzymes that degrades lignin (lignin peroxidase (LiP) being the other).[81,86] Now, however, there are reports that some fungi may also employ a laccase for lignin degradation.[93,94] Both MnP and LiP contain a heme that is oxidized by H_2O_2. A H_2O_2 generation system is also believed to be excreted with these peroxidases.[92,95,96] Both enzymes function by one-electron oxidations of substrates that diffuse from the enzyme into the lignin substrate.[86] These small diffusable oxidants initiate a cascade of radical reactions in lignin, that eventually leads to the oxidation and cleavage of the linkages that connect the lignin components to one another.[86,92] In LiP, veratryl alcohol serves as the substrate that is oxidized by heme, while in MnP, this role is filled by the Mn^{II} ions (*vide infra*).

4.1.1 Structure

In 1994, a crystal structure at 2.06 Å resolution for the manganese peroxidase from the white-rot basidiomycete *Phanerocaete chrysosporium* was reported (Fig. 5.12).[85] This structure showed that Mn binds to one of the two propionate arms that extends from the heme group. Overall, the MnP enzyme structure is comprised predominantly of α-helical motifs with two domains that wrap around the heme in a manner analogous to other plant and fungal peroxidases.[85] MnP is similar to LiP in structure, amino acid sequence homology, and the presence of two Ca^{2+} ions in the structure. The heme active site shows a proximal bound histidine, with another unbound histidine and a water molecule on the distal side. Both histidines are suggested to be involved in hydrogen bonding networks. Additional polar residues are also observed near the heme. Of great interest is the fact that a Mn^{II} ion is bound to one propionate arm of the heme. In addition, four protein residues participate directly in binding Mn^{II} to the enzyme, via the carboxylates of two glutamates, an aspartate, and an arginine that engages in hydrogen bonding to one of the manganese ligating glutamates.[85,97,98] This arginine is believed to mitigate the overall charge at the site. No small organic chelators, such as oxalate, are bound to Mn^{II} in the structure.

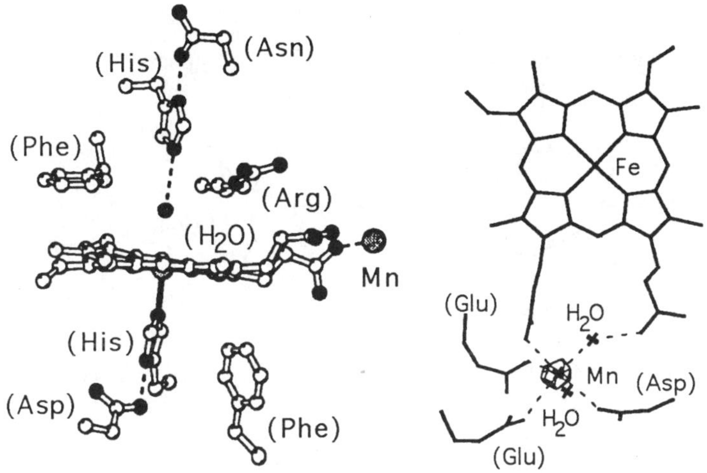

Fig. 5.12 Active site of the manganese peroxidase enzyme from *P. chrysosporium* (reproduced from Ref. 85 with permission from The American Chemical Society).

4.1.2 Catalytic cycle

The catalytic cycle of MnP, shown in Fig. 5.13, has been elucidated over the past 13 years.[80,81,83,86,99–101] The reaction to catalyze lignin decomposition begins with the oxidation of the resting state Fe^{III}(porphyrin) to Compound I, Fe^{IV}(O)(porphyrin$^{+\bullet}$), by hydrogen peroxide. An equivalent of Mn^{II} is then oxidized to Mn^{III}. Based on the crystallographic data,[85] it is believed that this first equivalent of manganese is bound to at least one of the propionate arms on the heme. This Mn^{III} then diffuses away from the enzyme where it oxidizes lignin by one electron. (It is not yet known whether oxalate or another small chelating molecule is necessary to remove the Mn^{III} as a Mn^{III}-chelate from the enzyme.) Compound I, upon oxidation of the manganese is converted to Compound II, an Fe^{IV}(O)(porphyrin). A second equivalent of Mn^{II} is then oxidized to Mn^{III} to return the enzyme to its resting state. Although it has been shown that MnP will oxidize substrates other that Mn^{II} for the first oxidation step, the second has an absolute requirement for manganese.[81,102] In particular, the second oxidation occurs preferentially with Mn^{II} chelated by oxalate (*vide infra*).[83,84] In LiP,

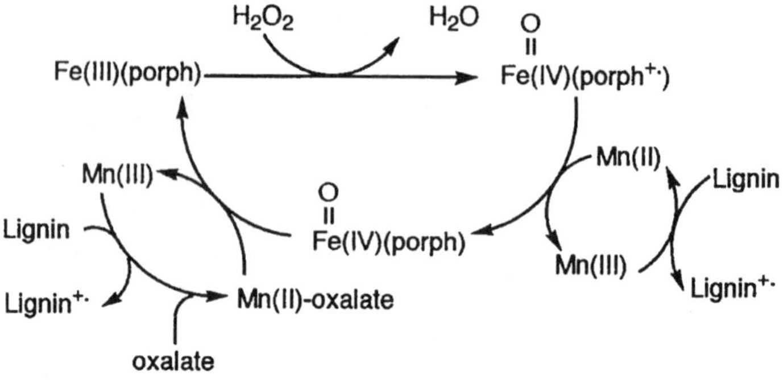

Fig. 5.13 A proposed mechanism of manganese peroxidase activity (based on Refs. 83, 85, 86, 100, 102).

veratryl alcohol is co-produced with the enzyme, and LiP has been shown to readily oxidize this material.[103,104] Presumably, veratryl alcohol acts in the same capacity as Mn^{III} with respect to lignin oxidation. Another interesting result relating LiP to MnP is the recent discovery that LiP is capable of oxidizing Mn^{II} to Mn^{III}, but it is not yet known whether this occurs in a natural setting.[105]

As mentioned above, oxalate plays a key role in the function of MnP.[81,83,84,102,103,106] In 1989, it was observed that small molecule α-hydroxy acids, such as citric and lactic acid, or organic acids, such as malonic or oxalic acids, promoted MnP activity in the step of the reaction cycle going from Compound I to Compound II.[102] These ligands are proposed to stabilize Mn^{III} ions and thus promote diffusion of Mn^{III} from the enzyme.[81,101,102] In the absence of these ligands, disproportionation with another Mn^{III} to form Mn^{II} and Mn^{IV} could occur.[81] Stronger chelators were later observed to inhibit MnP activity, and it was proposed that the better chelators competed with the enzyme for Mn^{II}.[100] At this time, it was also demonstrated that both malonate and oxalate are produced in active cultures.[100] A second critical role for oxalate was proposed based on new kinetic data showing that the reduction of Compound II to the resting state requires a Mn^{II}-oxalate complex, presumably formed from exogenous oxalate and Mn^{II}.[83] This complex strongly facilitates the reduction of Compound II to the Fe^{III}(porphyrin), reaching a maximum at oxalate concentrations favoring a 1:1 manganese:oxalate ratio.[83] It is possible that malonate and lactate can also promote MnP activity in a similar manner.[81,83,100] Free Mn^{II} may be able to reduce the heme, but not with the efficiency of the oxalate complex.[83]

4.1.3 Models

A few model systems have been prepared for MnP, addressing the structure and reactivity of the manganese center as well as electron transfer reaction from the heme to Mn^{II} via the propionate arms. Initial structure and reactivity models for the Mn center were based on the observation that α-hydroxy acids stimulate the activity of MnP. Complexes of Mn^{III} and α-hydroxy acids were prepared *in situ*, and their reactivity with lignin analogs established that they could initiate the degradation of lignin by a radical process.[81] In 1991, Saadeh *et al.* reported the preparation, isolation and structural characterization of complexes with α-hydroxy acids in the Mn^{III} and Mn^{IV} oxidation states.[107] This work included complexes with lactic acid, 2-hydroxyisobutyric acid (HIB) and 2-hydroxy-2-ethylbutyric acid (HEB) in monomeric or chain-like structures.[107] Such well-characterized complexes were then tested for peroxidase activity with the lignin analog and test substrate, vanillylacetone.[108] The pyruvaldehyde oxidation product from vanillylacetone is consistent with reactions initiated by one-electron chemistry.[109] In support of this, Mn^{II} was identified by EPR spectroscopy in these reaction mixtures, indicating that manganese reduction had indeed occurred. These studies supported the hypothesis that the presence of α-hydroxy acids might stimulate the activity of MnP by stabilizing the Mn^{III} with respect to disproportionation to Mn^{II} and Mn^{IV}, presumably $Mn^{IV}O_2$, and yet would not produce a complex that would be too stable with respect to oxidation of lignin.[109]

Electron transfer from the heme to the manganese ion has also been modeled.[110] A porphyrin with pentafluorophenyl substituents on three meso positions was prepared. The fourth meso position was functionalized with an alkyl arm that terminated in a 6-(methylamino)-2,2'-bipyridine functional group to create a manganese binding site. EPR spectroscopy confirmed that both Fe^{III} and Mn^{II} were bound in the final product, and that they were not magnetically coupled to each other. Upon addition of the oxidant pentafluoroiodosylbenzene, the Mn^{II} EPR signal disappeared. In a test of reactivity, 2,6-dimethylphenol, a standard test substrate for MnP, was added to the system and oxidized to 2,2',6,6'-tetramethyldiphenoquinone. The reaction coincided with reappearance of the Mn^{II} EPR signal. This process was shown to be catalytic. Control experiments showed that electron transfer occurs in the presence of peroxide, heme, and Mn^{II}, but not if one of the metal species was removed, supporting a mechanism where substrate oxidation is mediated by the oxidized manganese and not directly by an oxidized heme species.

4.1.4 Mn chlorination reactions

Unlike iron and vanadium, there are no known manganese haloperoxidase enzymes. Figure 5.14 presents a hypothetical cycle for manganese haloperoxidase reaction that is based on chlorination reactions that have been observed with small molecule Mn complexes. A detailed analysis of the different steps in the cycle, however, makes it clear that a catalytic reaction is not possible. To generate the dimeric complex $[Mn^{IV}(salpn)(\mu\text{-}O)]_2$, a $[Mn^{III}salpn]^+$ in the presence of a base and an oxidant such as hydrogen peroxide is required.[111–113] This in turn reacts with HCl to form a high-valent $Mn^{IV}(salpn)Cl_2$ complex, which is capable of halogenating alkenes. The resulting $[Mn^{III}(salpn)Cl]$ is then recycled back to the beginning of the reaction. All of the individual steps above have precedence in our laboratories, but when one analyzes the component reactions, it becomes clear that a catalytic cycle, as outlined above, is not feasible. The first reaction to form $[Mn^{IV}(salpn)(\mu\text{-}O)]_2$ involves the oxidation of manganese in a base-assisted manner in the presence of the oxidant. The formation of the $Mn^{IV}(salpn)Cl_2$ complex is acid-dependent and requires the addition of four equivalents of acid.[114,115] This proceeds via a redox reaction that converts the $[Mn^{IV}(salpn)(\mu\text{-}O)]_2$ to two $[Mn^{III}(salpn)]^+$, which then disproportionate in the presence of acid and

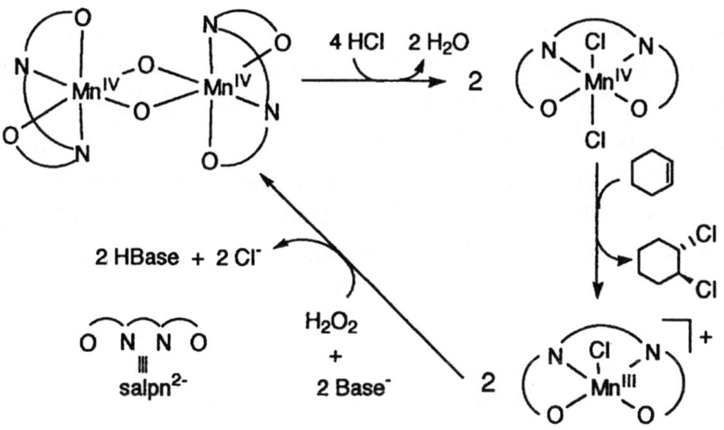

Fig. 5.14 Scheme for a hypothetical "manganese haloperoxidase".

chloride to yield Mn^{IV}(salpn)Cl_2 and a Mn^{II} species.[116] It is known that Mn^{II}(salpn) will rapidly comproportionate with Mn^{IV}(salpn)Cl_2, therefore Mn^{II}(salpn) is not the Mn^{II} product, and it seems likely that a ligand degradation process is involved.[116] The halogenation of the alkene then requires two equivalents of Mn^{IV}(salpn)Cl_2.[114]

Overall, the Mn chlorination reactions appear to involve some form of "caged-radical" mechanism.[114] This is inconsistent with both the vanadium and heme HPO systems in that no "X^+" species is formed as a reactive intermediate. Furthermore, by the end of the reaction in Fig. 5.14, half of the manganese complexes from the first step have been "lost" from the cycle as well as some of the initial ligand, because ligand oxidation is currently believed to be involved in the reduction of the dioxo dimer when it reacts with acid.[116] Thus, the "MnHPO" has been deconstructed.

A mechanism similar to that of VHPO is not possible with Mn because it would require the presence of the strongly oxidizing permanganate ion (to be strictly analogous mechanistically) which would likely destroy essential protein residues. Vanadate, on the other hand, is relatively redox stable and bioavailable. Lower valent manganese systems (Mn^{II}, Mn^{III}, Mn^{IV}) tend to react directly with peroxide in a redox process because of the number of stable oxidation states available to manganese. For example, Mn^{II} is most often associated with Lewis acid reactions, but this ion is oxidized in the presence of H_2O_2. Since the vanadium is already in its highest oxidation state, further oxidation by H_2O_2 is not a problem.

A heme-type mechanism is not likely with monomeric manganese species because, in all known biological manganese systems, one Mn is employed for each electron required: there are no known examples of mononuclear high-valent biological manganese systems capable of carrying out multiple electron reactions similar to the heme enzymes. Since HPOs catalyze the net two-electron oxidation of a halide, a system capable of accumulating two electron equivalents is necessary. Dimeric manganese systems are also not ideal for halide oxidation, presumably because oxidation equivalents are distributed between two Mn ions and a charge potential large enough to oxidize the halide is not accessible. This point is supported by the fact that Mn catalases are strongly inhibited by chloride.[117] In addition, Cl^- is involved in the highly oxidizing OEC cycle, but formation of oxidized halide has never been reported.[118,119]

4.2 Manganese catalase

Many of the known manganese-containing redox enzymes interact with dioxygen or its reduced forms, including the Oxygen Evolving Complex,[120–123] manganese superoxide dismutase[124] and catalases.[117,120,121] Mn catalases react with H_2O_2, catalyzing the disproportionation of hydrogen peroxide to dioxygen and two molecules of water, according to Eq. (5.3).[117] Like the heme catalases, a redox mechanism is proposed for the Mn catalases.[117] The heme and Mn catalase metal sites differ in several important ways. First, the heme enzymes contain a monomeric center capable of two-electron chemistry whereas the Mn catalases contain a dinuclear site which distributes redox equivalents across both centers. Second, porphyrin oxidation is essential to heme enzyme reactivity while organic groups are not oxidized in Mn catalase. Third, heme enzymes achieve relatively high oxidation levels for intermediates ($Fe^{IV}P$, $Fe^{IV}P^{+\bullet}$), where Mn catalases employ standard oxidation levels for Mn (e.g. Mn^{II} and Mn^{III}).[117,125]

4.2.1 Structure

To date, three manganese-containing catalases have been reported (*Lactobacillus plantarum*,[126,127] *Thermoleophilum album*,[128] and *Thermus thermophilus*[129,130]). The *T. thermophilus* enzyme has been crystallographically characterized, and is composed of a homohexamer of 36 kDa subunits, with 2 Mn per subunit.[129–131]

Mn catalases have been known for several years before the presence of manganese was confirmed in 1983.[127] A low resolution crystal structure of enzyme from *T. thermophilus* in 1985 shows a primarily α-helix motif with a dinuclear manganese site located within a four helix bundle. The Mn-Mn separation is about 3.6 Å.[129,130] Optical spectroscopy is also consistent with an active site that contains two manganese ions bridged by a minimum of one oxo (μ-O^{2-}) and one carboxylate ligand. The basis for this assignment is the similarity of the UV-vis spectra of the enzyme to known dinuclear model compounds with bridging oxo and carboxylate ligands.[117,132,175] In addition, structurally characterized members of this group of compounds typically exhibited Mn-Mn vectors on the order of 3.1–3.6 Å.[111,132] Resonance Raman and MCD data also support the dinuclear active site.[176,177,178]

Until 1997, no new reports of a Mn catalase structure appeared, and EPR[117] and EXAFS[117,133] are used not only to confirm the presence of a dimeric Mn species at the active site, but also to estimate Mn-Mn distances. The multiline signal in the EPR confirms the presence of Mn-Mn coupling[134,135] and coupling values have been used to estimate Mn-Mn distances.[136] XAS studies suggest that the first coordination sphere about the manganese of the catalase consists primarily of O/N donors.[117,133,173,174] Data on the superoxidized species show a Mn-Mn vector of 2.7 Å, which is consistent with a $Mn_2(\mu_2\text{-}O)_2$ core. A second shell of oxygen donors at about 1.8 Å is also resolved, which is also consistent with such a core. These results confirm the presence of dinuclear manganese center in Mn catalase.

In 1997, the remaining questions of coordination environment around the enzyme active site were resolved with the report of the crystal structures of the reduced, Mn^{II}_2, and the oxidized, Mn^{III}_2, forms of the enzyme from

T. thermophilus at 1.6 Å and 1.4 Å, respectively.[137,138] The active site assembly of the reduced Mn^{II} form of the enzyme (Fig. 5.15) is comprised of two five-coordinate manganese ions bridged by one carboxylate from a protein residue, one hydroxide ion, and a water molecule. One histidine and one carboxylate from protein residues complete the coordination environment forming square-pyramidal manganese centers. An additional water is located above one of the manganese, near a lysine. In the oxidized Mn^{III}_2 form of the enzyme, the general arrangement of the Mn site is the

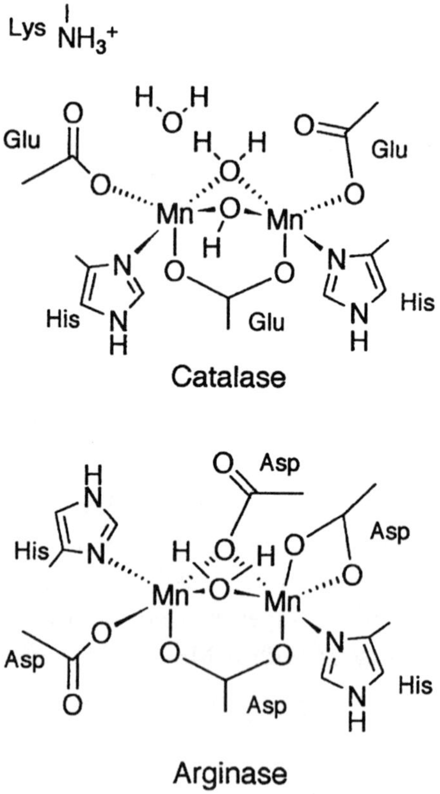

Fig. 5.15 Representations of the active site structure of *T. thermophilus* Mn catalase (top) and rat liver arginase (bottom).[137,138,164]

same as the Mn^{II}_2 with the exception of a bridging oxo instead of the hydroxo. The Mn-Mn vector in the reduced Mn^{II}_2 enzyme is 3.18 Å while it is 3.14 Å for the oxidized Mn^{III}_2 enzyme.[137,138]

4.2.2 Catalytic reaction

Mn catalases shuttle between the Mn^{II}_2 oxidation states and the Mn^{III}_2 oxidation states in two two-electron reactions.[125,139,176] A proposal for the catalytic cycle of catalase based on the Penner-Hahn proposal and the recent crystallographic data is shown in Fig. 5.16.[117] The rates of disproportionation are presented in Table 5.1. As isolated, *T. thermophilus* exists as a mixture of enzyme oxidation states including Mn^{II}_2, $Mn^{II}Mn^{III}$, Mn^{III}_2, and $Mn^{III}Mn^{IV}$.[117,136] The $Mn^{II}Mn^{III}$ and $Mn^{III}Mn^{IV}$ oxidation states

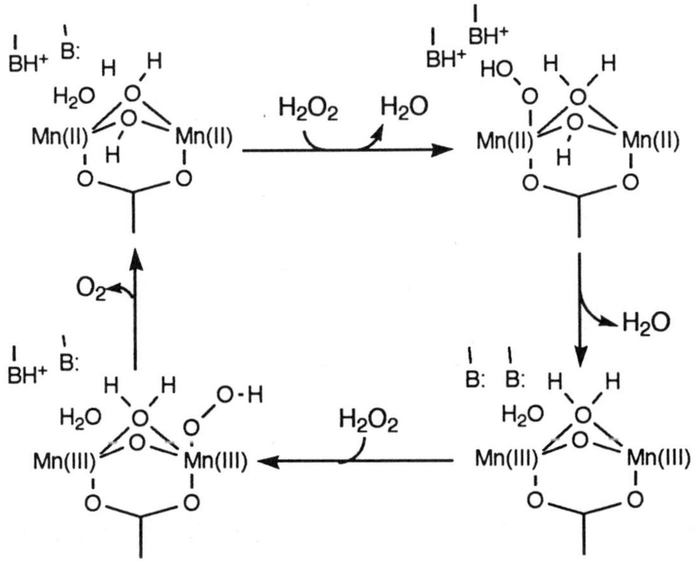

Fig. 5.16 A proposed catalytic mechanism for Mn catalase. This scheme is based on the scheme proposed by Penner-Hahn[117] for Mn catalase, and the crystal structure of the *T. thermophilus* Mn catalase.[137,138]

are catalytically inactive states, presumably resulting from misfiring of the enzyme. The enzyme can be reactivated through treatment with hydroxylamine, a one-electron reductant, which converts the overall oxidation state of the manganese in this enzyme to homogeneous Mn^{II}_2.[117] If the excess hydroxylamine is not removed following reduction, subsequent addition of hydrogen peroxide results in reformation of the "off" oxidation states (Mn^{II}/Mn^{III} and Mn^{III}/Mn^{IV}). Formation of the II/III form of catalase is explained by the one-electron reduction of the Mn^{III}_2 form under turnover conditions. The two-electron oxidation of this $Mn^{II}Mn^{III}$ species by H_2O_2 then converts the enzyme to the superoxidized $Mn^{III}Mn^{IV}$ form of the enzyme.[117] The precedent for this chemistry is also backed by model compounds (*vide infra*).[140,141] Mn catalases exhibit saturation kinetics with hydrogen peroxide, and are inhibited by azide, chloride, fluoride and thiocyanate.[117,173] There is significant controversy with respect to the number of fluorides and whether their mode of inhibition is through a terminal or a bridging species. A final note on reactivity is that catalase from *T. album* was shown to react in a limited capacity as a peroxidase with *p*-phenylenediamine.[128]

Other types of manganese-containing enzymes can also exhibit catalase activity. For example, in some of the lower oxidation states of the Oxygen Evolving Complex, a catalase-like reaction with hydrogen peroxide has been reported[142] (although this observation has been contested recently[170]). It likely involves the Mn^{III} and Mn^{IV} oxidation states. This reaction will not be further discussed here, and the reader is directed to one of several reviews for more information.[120,121,123,143]

4.2.3 Models

Many chemists have produced catalase models over the past several years and several reviews have appeared.[120,121,132,144,145] Some of the rates of reaction of these compounds are presented in Table 5.1. The systems tend to be divided into two classes: low-valent manganese complexes and high-valent manganese complexes. Representative examples from both classes are summarized below.

In 1994, Pessiki *et al.* reported studies of catalase mimics based on the ligand 2-OHtbpn (N,N,N',N'-tetrakis-(benzimidazol-2-ylmethyl)-1,3-diaminopropane-2-ol; Fig. 5.7).[146,147] This study followed up on earlier catalysts reported by Mathur *et al.* with the same ligand.[148] This dinucleating ligand was used to prepare two forms of Mn^{II} dimers, both with an acetate bridge. The 1994 complex also had an *n*-butanol bound to one of the Mn^{II} ions. During the proposed catalytic cycle, hydrogen peroxide displaces the acetate and butanol, bridging the Mn^{II}_2 center and is then cleaved to produce a $(Mn^{III}\text{-}OH)_2$ moiety. A second hydrogen peroxide is then deprotonated by the hydroxides and the dimer is re-reduced with concomitant generation of dioxygen. Displacement of the two coordinated waters by H_2O_2 restarts the cycle (Fig. 5.17).[146,147]

A similar mechanism was proposed by Pecoraro and co-workers with one of the most efficient catalase model systems. These manganese dimers were formed with a series of 2-OHsalpn ligands containing a variety of

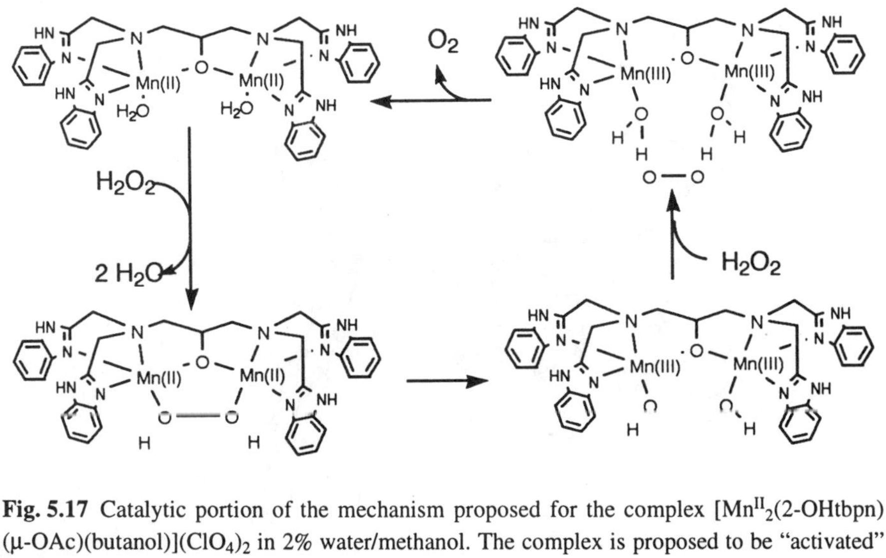

Fig. 5.17 Catalytic portion of the mechanism proposed for the complex $[Mn^{II}_2(2\text{-OHtbpn})(\mu\text{-OAc})(butanol)](ClO_4)_2$ in 2% water/methanol. The complex is proposed to be "activated" by displacement of the acetate bridge and butanol by two water molecules to form the complex shown in the upper-left-hand corner of the scheme, $[Mn^{II}_2(2\text{-OHtbpn})(H_2O)]^{3+}$ (adapted with permission from Ref. 146)

electron-donating and electron-withdrawing substituents at the 3 and 5 positions of the ligand phenolate groups (Fig. 5.7; for the remainder of this section, the abbreviation 2-OHsalpn will represent all substituent variations of N,N'-(salicylidenimine)-1,3-diaminopropan-2-ol).[149] This family of complexes has proven to be a unique and versatile model system, providing both detailed structural and mechanistic information.

One of the most useful properties of these complexes is their stability over a wide range of oxidation states. As a result, they have been prepared, isolated, and crystallographically characterized in all oxidation states relevant to the Mn catalase enzymes (Mn^{II}_2, $Mn^{II}Mn^{III}$, Mn^{III}_2, and $Mn^{III}Mn^{IV}$), allowing for direct analysis of how oxidation state affects the crystal structure.[140] The crystal structure for a typical member of this class of compound is shown in Fig. 5.18. In all structures, no change in coordination environment is observed. The only structural changes observed are changes in bond lengths that are consistent with the oxidation state changes.

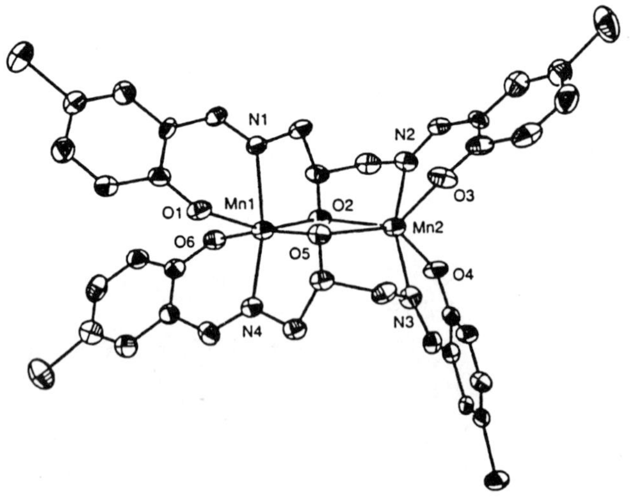

Fig. 5.18 $[(C_2H_5)_4N][Mn^{II}Mn^{III}(2\text{-}OH(5\text{-}Clsal)pn)_2]$. A representative member of the family of complexes synthesized with the H_2(2-OHsalpn) ligands (reproduced from Ref. 140 with permission from The American Chemical Society).

Both the stoichiometric and catalytic reactivity of these model compounds have been studied and related to the enzyme reactivity. Stoichiometric studies have been used to address the oxidation state cycle of Mn catalases as well as enzyme deactivation. Spectroscopic studies show that, under conditions that inhibit the oxidation of H_2O_2, addition of H_2O_2 to $[Mn^{II}(2\text{-OHsalpn})]_2^{2-}$ results in clean formation of $Mn^{III}_2(2\text{-OHsalpn})_2$. Similarly, under suitable conditions, the addition of H_2O_2 to $[Mn^{III}(2\text{-OHsalpn})]_2$, results in formation of $[Mn^{II}(2\text{-OHsalpn})]_2^{2-}$. This observation shows that this system will disproportionate hydrogen peroxide as a discrete unit that cycles between the Mn^{II}_2 oxidation state and the Mn^{III}_2 oxidation state, which is consistent with the mechanism proposed for Mn catalases. The formation of $[Mn^{III/IV}(2\text{-OHsalpn})]_2^{1+}$ is achieved by introducing one equivalent of H_2O_2 to $[Mn^{II/III}(2\text{-OHsalpn})]_2^{1-}$ in DMF solution, supporting the theory that this is probably the mechanism of enzyme deactivation.[140,141]

The catalytic behavior of these systems was also studied, yielding the kinetic data summarized in Table 5.1. The H_2O_2 disproportionation reactions were run in acetonitrile, and their progress followed using an oxygen electrode. This family of complexes exhibited saturation kinetics with hydrogen peroxide. The substituent on the phenolate ring affected both peroxide binding strength (K_m) and the rate constant for the reaction (k_{cat}). The more electron donating groups slow the rate-determining step of H_2O_2 oxidation by stabilizing the III/III oxidation state. The rates observed for the complexes ranged from 4.2 to 21.9 s^{-1}. This correlates with the mechanism of action proposed for these compounds (Fig. 5.19).[141] The proposed mechanistic cycle begins with the binding of a bridging hydrogen peroxide followed by reduction and cleavage of the O–O bond to form two OH^- and concomitant oxidation of the manganese dimer to Mn^{III}_2. In the second step, the base deprotonates a second equivalent of H_2O_2 and the hydroperoxide anion binds terminally to $[Mn^{III}(2\text{-OHsalpn})]_2$. The terminal coordination is proposed based on the binding mode observed for other small molecules to these dimers, such as THF or methanol (Fig. 5.20).[150,151] Upon binding of HOO^- to the manganese, one of the alkoxide bridges shifts to become a terminal ligand. The first deprotonation also leads to the formation of the first water molecule of the disproportionation. A second equivalent of base then deprotonates the hydroperoxide monoanion unit leading to the formation and release of dioxygen, the concomitant reduction of the dimer to Mn^{II}_2, and formation of the second water. Isotope labeling showed that both of the oxygens of the dioxygen formed in this process are from the same molecule of hydrogen peroxide. In addition, the lack of ligand exchange when two different 2-OHsalpn ligands were introduced under turnover conditions show that these dimeric species do not dissociate in solution.

Several models that invoke higher valent manganese oxidation states in their catalytic cycles have also been prepared. These include the asymmetric complexes with N,N',N''-trimethyl-1,4,7-triazacyclononane (Me_3tacn) of Wieghardt,[152] the dinucleating ligand systems of Okawa[153], the $[Mn^{IV}(salpn)(\mu\text{-}O)]_2$ complex, and the linked porphyrin dimers of Naruta.[154]

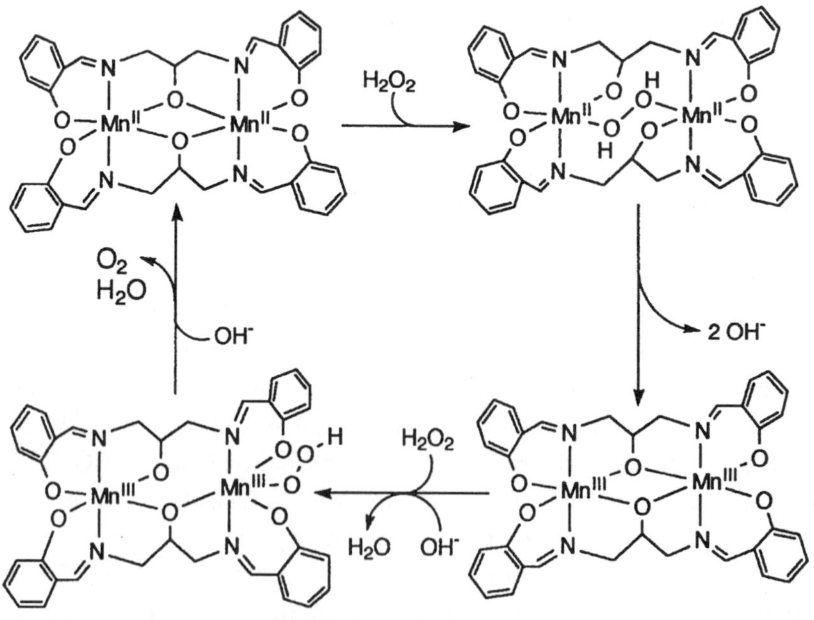

Fig. 5.19 A proposed catalytic scheme for catalytic hydrogen peroxide disproportionation by the family of complexes synthesized with the H_2(2-OHsalpn) ligands (adapted from Ref. 141).

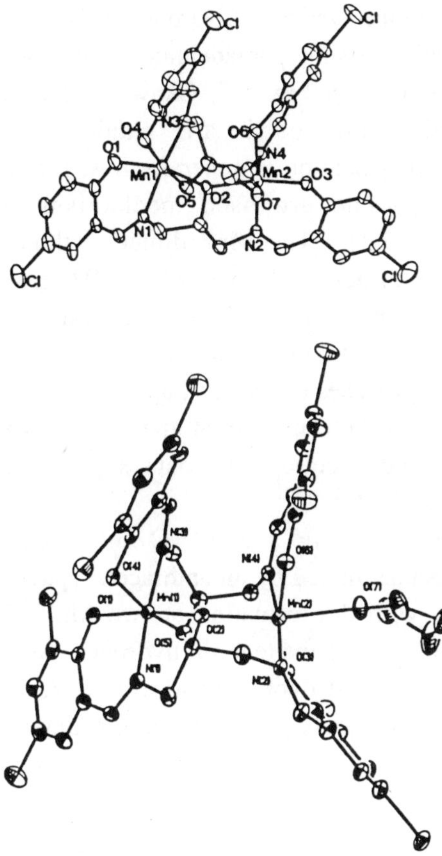

Fig. 5.20 [$Mn^{III}Mn^{IV}$(2-OHsalpn)$_2$(THF)]. A crystal structure representative of the "asymmetric" (2-OHsalpn)-based dimers wherein a bridging alkoxide has shifted to become a terminal alkoxide ligand, leaving an open site for an additional ligand on one of the manganese ions (reproduced from Ref. 150 with permission from The American Chemical Society)

Wieghardt's $Mn^{III}Mn^{IV}$ tacn ligands form the asymmetric complexes, [$Mn^{III}Mn^{IV}(\mu\text{-O})_2(\mu\text{-OAc})$(tacn)(bipy)(MeOH)]$(ClO_4)_2$ and [$Mn^{III}Mn^{IV}$ $(\mu\text{-O})_2(\mu\text{-OAc})$(tacn)$(OAc)_2$]$(ClO_4)_2$, with the unique ligands binding to the Mn^{III} ion.[152] These systems presumably cycle between the $Mn^{III}Mn^{IV}$ and $Mn^{II}Mn^{III}$ oxidation states giving the catalase rates shown in Table 5.1.

The Mn catalase mimic systems of Naruta and Okawa invoke high-valent manganese-oxo species. Using pentadentate or heptadentate dinucleating ligands, for example HL^1 in Fig. 5.7, Okawa and co-workers[153,155] have prepared functional mimics of the Mn catalases. The rate at which these catalyze disproportionation is generally low, for example 0.79 s^{-1}, as shown in Table 5.1. The proposed mechanism of this process is quite intriguing (Fig. 5.21). The Mn^{II} dimer is thought to be oxidized by H_2O_2 forming a dimer with two Mn^{III}-OH groups, which is the active catalyst. Reaction of an additional equivalent of H_2O_2 with this complex yields two Mn^{IV}=O moieties, and two molecules of water. A second equivalent of H_2O_2 approaches the catalyst and transfers its protons, resulting in dioxygen release and reformation of the initial $[Mn^{III}\text{-}OH]_2$ complex. Mass spectrometric and spectroscopic data support the presence of the Mn^{IV}=O intermediate.[153, 155]

Another system invoking a Mn^{IV}=O intermediate employs two Mn^{III}porphyrin molecules linked by an anthracene spacer.[154,156,157] Catalysis in this system starts with the formation of two Mn^{IV}-O (hydroxo or oxo) porphyrins from the first equivalent of hydrogen peroxide. Imidazole is required in this reaction, and likely acts as a base. A second equivalent of hydrogen peroxide then reacts with this intermediate to yield dioxygen,

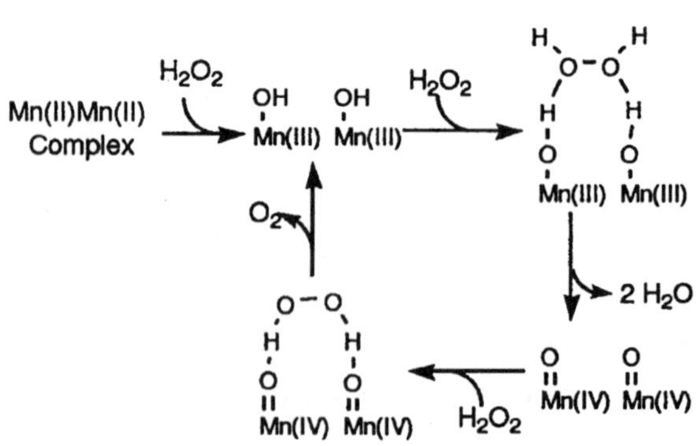

Fig. 5.21 A proposed catalytic scheme for the catalase activity of the complexes with ligand HL^1 (adapted from Refs. 153 and 155).

two molecules of water, and the starting catalyst. Isotope labeling studies with 1:1 mixtures of $H_2{}^{16}O_2$ and $H_2{}^{18}O_2$ showed no mixed isotope dioxygen, suggesting that dioxygen is formed from a single molecule of hydrogen peroxide. In 1997, Naruta *et al.* reported the synthesis of a Mn^{IV}porphyrin(OR) (R = OMe or OH) dimer, and concluded that an analogous species is formed during H_2O_2 disproportionation.[157] The proposed catalytic cycle is presented in Fig. 5.22.

The high-valent catalase mimic, $[Mn^{IV}(salpn)(\mu_2\text{-}O)]_2$, is the most efficient catalase mimic known to date, Table 5.1.[112,158] Its relevance to catalase activity lies mostly in its ability to mimic the proposed catalase-like reaction of the Oxygen Evolving Complex. This system has also been extensively reviewed.[120–123] The catalytic cycle for this complex involves shuttling between the $Mn^{IV}{}_2$ dimer and discrete $[Mn^{III}L]^+$ monomers (Fig. 5.23). Isotope labeling studies show that the dioxygen evolved during this reaction comes from the same molecule of hydrogen peroxide.

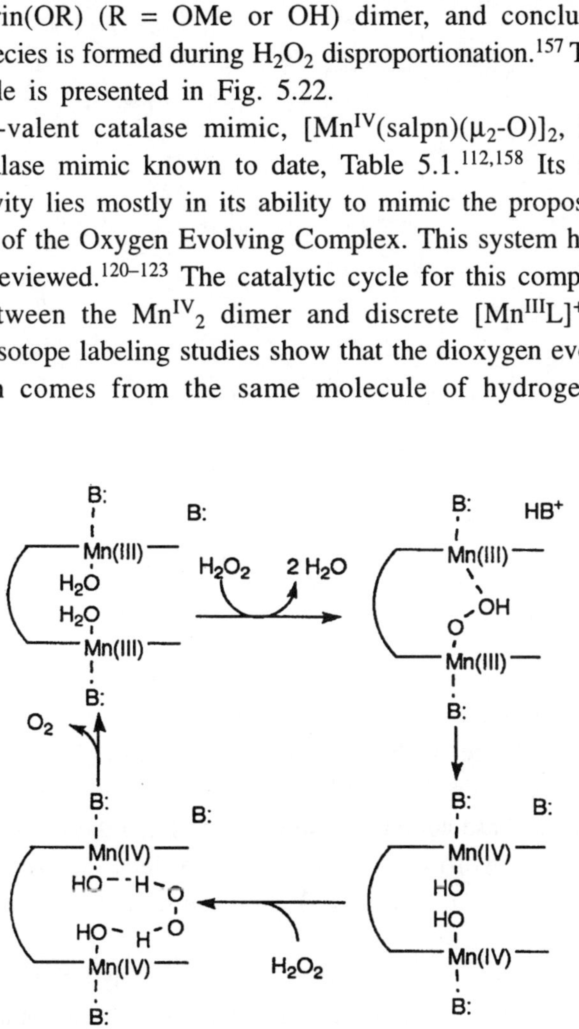

Fig. 5.22 A proposed catalytic scheme for the linked Mn(porphyrin) dimers. (adapted from Refs. 154 and 157).

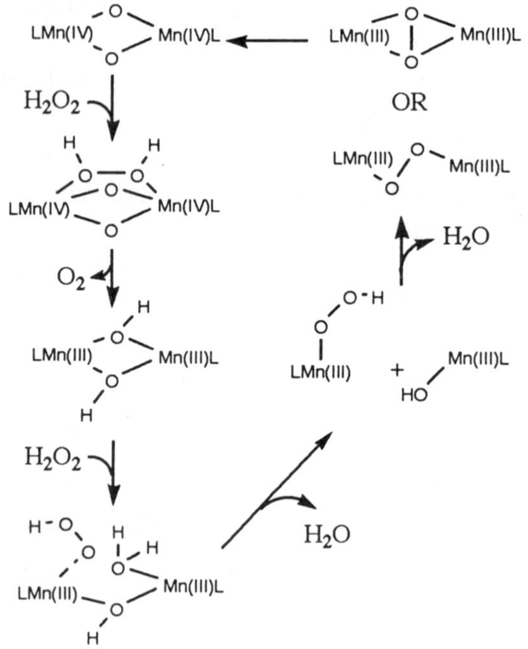

Fig. 5.23 A proposed catalytic scheme for catalase activity by $[Mn^{IV}(salpn)(\mu\text{-}O)]_2$. L = $salpn^{2-}$ (adapted from Ref. 112).

Re-oxidation of the two $[Mn^{III}L]^+$ consumes a second equivalent of H_2O_2. The bridging oxo-'s react with the protons from the first and second equivalents of H_2O_2 to form water. The synthesis of $[Mn^{IV}(salpn)(\mu_2\text{-}O)]_2$ from hydrogen peroxide and $[Mn^{III}L]^+$ is known to require base equivalents. Presumably, these base equivalents are formed from the initial oxo bridges. This complex disproportionates hydrogen peroxide at the fastest rate known for any model complex (250 s^{-1}) and saturation kinetics have been observed.[112,158,159]

Tetranuclear complexes that disproportionate hydrogen peroxide are also known. The first was reported in 1990 by Stibrany and Gorun who prepared a complex from the ligand 2-hydroxy-1,3-diaminopropanetetraacetic acid (Fig. 5.24) to form a complex with an overall oxidation state of $Mn^{II}Mn^{III}_3$.[160] It disproportionates hydrogen peroxide with a rate of

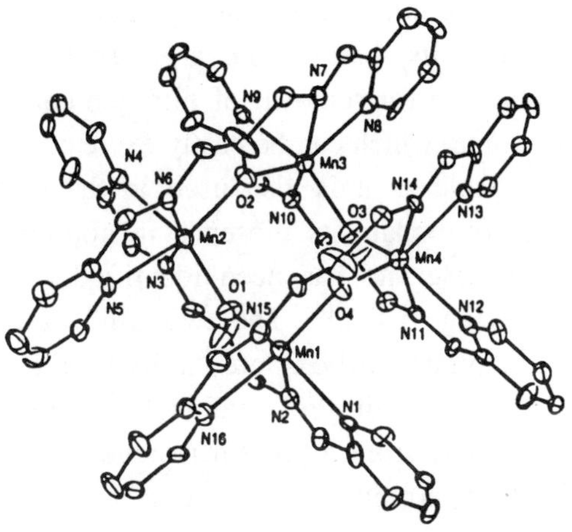

Fig. 5.24 $[Mn^{II}(2\text{-OHpicpn})]_4$. Crystal structure of a tetranuclear Mn catalase mimic. (reproduced from Ref. 161 with permission from The American Chemical Society).

1.04 s^{-1}. A second tetramer, reported by Gelasco *et al.* in 1995, disproportionates hydrogen peroxide at a rate of 150 s^{-1}. It is the second fastest catalase mimic known.[161] This complex is comprised of four Mn ions that are bridged by the ligand N,N'-(picolylideneimino)-1,3-diamino propan-2-ol to form a box-shaped molecule. The complex is proposed to react via a mechanism in which the hydrogen peroxide coordinates to the cluster in the middle of the box.[161]

4.3 Arginase

4.3.1 Structure and reactivity

A new "Mn catalase" has recently been discovered in the guise of the enzyme arginase, specifically rat liver arginase. This enzyme is a metallohydrolase that contains a binuclear manganese active site.[145,162,163] The crystal structure of rat liver arginase at a resolution of 2.1 Å was recently reported (Fig. 5.15).[164] Arginase catalyzes the hydrolytic conversion of arginine to urea and ornithine and requires manganese as a cofactor to function. Widespread in its distribution, arginase is important to many organisms as a mechanism for removal of nitrogen-containing compounds through formation of urea, which can be easily excreted. Arginase, however, is not limited to this role as it is also used in the syntheses of several polyamines, which may explain its presence in non-ureolytic species.[145]

A dinuclear manganese site has been identified in the recent crystal structure of arginase (Fig. 5.15).[164] The two Mn^{II} of the active site are bridged by a monodentate carboxylate, a hydroxide and a bidentate carboxylate. The remainder of the ligands for each manganese are a histidine and a carboxylate from a protein residue. One of the manganese ions is five-coordinate, approximately square pyramidal, while the other is six-coordinate due to a terminal bidentate carboxylate. Unlike Mn catalase, the Mn of arginase is not located in a four-helix bundle, but near the surface of the enzyme at the bottom of a 15 Å-deep cleft, surrounded by loops from beta-sheets and α-helices of the enzyme. Only two non-metal bound polar residues that may engage directly in reactivity are observed near the active site, a histidine and an aspartate.

Arginase catalyzes the hydrolysis of arginine to ornithine and urea.[164] More recently it has been shown to disproportionate hydrogen peroxide, although this is severely attenuated compared to naturally occurring Mn catalases and even some model compounds.[165,166] A peroxidase reaction has also been observed with *p*-phenylenediamine.[166]

4.3.2 *Comparison of structures — arginase and Mn catalase*

Both the EXAFS[165] data and the crystal structures of Mn catalase[137,138] and rat liver arginase[164] indicate major structural similarities between the active sites of these enzymes, although they are designed to catalyze quite different reactions. The active sites of both enzymes contain a dinuclear arrangement of Mn ions that are bridged in very similar manners as shown in Fig. 5.15. The primary difference in these two structures is a monodentate bridging carboxylate and an increased Mn-Mn distance (3.3 Å) in arginase[164] compared to a bridging water and a Mn-Mn distance of 3.18 Å in Mn catalase.[137,138] This structural similarity leads one to ask whether these enzymes can catalyze similar reactions and, if not, what attributes of the system leads to their differentiation. It has recently been reported that rat liver arginase catalyzes the disproportionation of H_2O_2 at pH 9.5,[165,166] although this rate is considerably less than authentic Mn catalase and even some model systems (Table 5.1). In addition, attempts to form a superoxidized arginase enzyme with hydroxylamine and hydrogen peroxide were not successful.[166] The structures of these two enzymes provide clues as to why their chemistries differ, and some discussion on this topic has appeared in the literature.[166]

Both active sites are located in hydrophobic pockets, and the protein structures that provide the manganese-binding sites are quite different, α-helices versus a loop and turn region. The numbers and types of charged residues at the active site probably also play a role, but it seems likely that most of the differentiation results from the small differences in the active site structures themselves. First, the coordination number and geometry about the manganese active site differ in the two enzymes. Mn catalase has two five-coordinate Mn ions which leaves two open sites for peroxide binding while arginase has one five-coordinate and one six-coordinate manganese. Second, the nature of the ligands with respect to the role of the enzyme provides clues to this reactivity difference. The Mn ions of arginase function as Lewis acids and not redox centers, activating water for the hydrolysis of arginine to ornithine and urea. Catalase chemistry on the other hand is a redox process and the ligation environment at the manganese centers need to be able to support oxidation state changes. The bridging oxo- (or perhaps two hydroxides) in Mn catalase is better able to support manganese oxidation than the monodentate carboxylate found in arginase. In general, the hydroxide and oxo donors are better ligands for the higher valent manganese. Therefore, the "lesion" in the catalase activity of arginase compared to true Mn catalase is presumably the result of a lack of stabilization for the oxidized Mn^{III}_2 arginase. Once oxidized, reduction of arginase to the stable Mn^{II}_2 form by H_2O_2 should be extremely facile.

5. Summary

While heme proteins have been recognized for most of the 20th century, it is only recently that scientists have discovered that early first row transition elements can participate in important biochemical transformations. In this chapter, we have shown that manganese and vanadium are now established players as metallocofactors in a variety of biological systems. Manganese often substitutes for similar activities that are known for iron enzymes. In some cases, such as the ribonucleotide reductases, catechol dioxygenases and superoxide dismutases, there is a direct structural resemblance between Mn and Fe active sites. This is not the case in the Mn catalases which substitute a dinuclear center for a heme moiety. Nonetheless, there is still similarity between Mn and Fe as there are many known cases where dinuclear Fe centers are competent to carry out reactions that are also associated with heme enzymes. In contrast, there are no examples where vanadium is known to mimic the function or structure of non-heme iron enzymes (with the possible exception of the FeMo cofactor in nitrogenase which can be replaced by FeVCo). This is not too surprising, as Mn is adjacent to iron in the periodic table and has oxidation levels that are similar to Fe, whereas vanadium tends to form structurally and functionally different oxoanions that are far more redox-stable in higher oxidation states. Thus, vanadium reacts with peroxide to activate the molecule for nucleophilic attack by halides in a non-metal centered redox process, while manganese invariably undergoes redox reactions with peroxide. Metabolism of dioxygen or its reduced forms is a prominent theme when one thinks of manganese redox enzymes. In contrast, halide metabolism and oxidation is the basis for much of the biological chemistry of vanadium.

The enzymatic disproportionation of hydrogen peroxide is an interesting case study for these systems. There are now known heme catalases and peroxidases, Mn catalases and arginases, and vanadium haloperoxidases capable of hydrogen peroxide disproportionation chemistry. Heme enzymes destroy peroxide by forming Compound I intermediates that require extreme valence states for iron and the porphyrin. Manganese enzymes distribute the oxidation and reduction chemistry across two metal centers that cycle between common aqueous oxidation levels for manganese.

Vanadium recruits halides to assist in peroxide decomposition since V(V) does not participate in redox reactions with peroxide. Thus, nature has found three mechanistically distinct approaches to solve the problem of peroxide toxicity.

Despite these differences, there are interesting similarities between the biological chemistries of manganese and vanadium. Both elements are thought to be used by fungi as diffusable, small molecule oxidants. For manganese, the goal is to degrade lignin presumably through a direct redox reaction, or possibly as a precursor to a reactive organic radical. Presumably, the lignin degradation aids in the growth process for the fungus. The role of amavadin, the small molecule vanadium complex, is a bit more enigmatic. Maybe it serves as a general oxidant; however, another possibility is that it may function as the most rudimentary haloperoxidase to ward off other organisms. It is clear that both vanadium and manganese are not the first choice of living systems for mediating redox conversions on substrates as iron enzymes clearly prevail in absolute numbers. However, when iron is limited, as in the sea, vanadium and manganese enzymes emerge to satisfy the necessities of life. As scientists explore more unique life forms found in more demanding environments, it is likely that manganese, vanadium and other less common elements will be at the forefront of interesting new biological cofactors.

6. Acknowledgments

We gratefully acknowledge Prof. V.V. Barynin and Drs. M. Tyler Caudle, Michael J. Baldwin and Brent J. Hamstra for useful discussion and the NIH (Grants GM39406 and GM 42703) for funding.

References

1. Yamazaki, I. in *Metalloproteins: Chemical Properties and Biological Effects* (eds.) Otsuka, S. and Yamanaka, T. Elsevier, New York, 1988, pp. 218–224.
2. Everse, J., Everse, K.E. and Grisham, M.B. (eds.) *Peroxidases in Chemistry and Biology*, Vol. CRC Press, Inc., Boca Raton, FL, 1991.

3. Xie, L. and Dolphin, D. in *Metalloporphyrin Catalyzed Oxidations* (eds.) Montanari, F. and Casella, L. Kluwer Academic, Dordrecht, 1994, pp. 269–306.
4. Chapman, S.K., Daff, S. and Munro, A.W. *Struct. Bonding (Berlin)* **88:** 1997, 39–70.
5. Foye, W.O., Lemke, T.L. and Williams, D.A. (eds.) *Principles of Medicinal Chemistry*. Williams & Wilkins, Baltimore, 1995.
6. Rob, A., Ball, A.S., Tuncer, M. and Wison, M.T. *J. Chem. Educ.* 1997, **74:** 212.
7. Yamazaki, I. in *Metalloproteins: Chemical Properties and Biological Effects* (eds.) Otsuka, S. and Yamanaka, T. Elsevier, New York, 1988, 224–229.
8. Vilter, H. *Phytochem.* 1984, **23:** 1387–1390.
9. Vilter, H. *Metal Ions in Biological Systems* 1995, **31:** 325–362.
10. Butler, A. and Walker, J.V. *Chem. Rev.* 1993, **93:** 1937–1944.
11. Butler, A. and Baldwin, A.H. *Struct. Bonding (Berlin)* 1997, **89:** 109–132.
12. Arber, J.M. *et al. Biochemistry* 1989, **28:** 7968–7973.
13. Kusthardt, U. *et al. FEBS Lett.* 1993, **329:** 5–8.
14. Hormes, J. *et al. Biochim. Biophys. Acta* 1988, **956:** 293–299.
15. Messerschmidt, A. and Wever, R. *J. Biol. Chem.* 1997, **378:** 309–315.
16. Messerschmidt, A. and Wever, R. *Proc. Nat. Acad. Sci. USA* 1996, **93:** 392–396.
17. Tschirret-Guth, R.A. and Butler, A. *J. Am. Chem. Soc.* 1994, **116:** 411–412.
18. Simons, B.H. *et al. Eur. J. Biochem.* 1995, **229:** 566–574.
19. Lindqvist, Y., Schneider, G. and Vihko, P. *Eur. J. Biochem.* 1994, **221:** 139–142.
20. Denu, J.M. *et al. Proc. Nat. Acad. Sci. USA* 1996, **93:** 2493–2498.
21. Zhang, M., Zhou, M., Van Etten, R.L. and Stauffacher, C.V. *Biochemistry* 1997, **36:** 15–23.
22. Slebodnick, C. and Pecoraro, V.L. *Inorg. Chim. Acta* 1998, **283**, 37–43.
23. Hemrika, W. *et al. Proc. Nat. Acad. Sci. USA* 1997, **94:** 2145–2149.
24. Stukey, J. and Carman, G.M. *Protein Sci.* 1997, **6:** 469–472.
25. Neuwald, A.F. *Protein Sci.* 1997, **6:** 1764–1767.
26. Shechter, Y. *Diabetes* 1990, **39:** 1–5.
27. Shechter, Y. *et al.* in *Vanadium in Biological Systems* (ed.) Chasteen N.D. Kluwer Academic Publishers, Boston, 1990, pp. 129–142.
28. Orvig, C., Thompson, K.H., Battell, M. and McNeill, J.H. *Metal Ions in Biological Systems* 1995, **31:** 575–594.

29. Shechter, Y. *et al. Mol. Cell. Biochem.* 1995, **153:** 39–47.
30. Shaver, A., Ng, J.B., Hall, D.A. and Posner, B.I. *Mol. Cell. Biochem.* 1995, **153:** 5–15.
31. Li, J., Elberg, G., Crans, D.C. and Schechter, Y. *Biochemistry* 1996, **35:** 8314–8318.
32. de Boer, E., Boon, K. and Wever, R. *Biochemistry* 1988, **27:** 1629–1635.
33. van Schijndel, J., Vollenbroek, E. and Wever, R. *Biochim. Biophys. Acta* 1993, **1161:** 249–256.
34. de Boer, E. *et al. FEBS Lett.* 1988, **235:** 93–97.
35. (a) LoBrutto, R. *et al. J. Am. Chem. Soc.* 1998, **120:** 4410–4416; (b) Fukui, K., Ohya-Nishiguchi, H., Kamada, H. *Inorg. Chem.* 1998, **37:** 2326–2327; (c) Smith II, T.S., Root, C.A., Kampf, J.W., Rasmussen, P.G., Pecoraro, V.L., to be submitted to *J. Am. Chem. Soc.*
36. de Boer, E. and Wever, R. *J. Biol. Chem.* 1988, **263:** 12326–12332.
37. Soedjak, H.S. and Butler, A. *Biochim. Biophys. Acta* 1991, **1079:** 1–7.
38. Van Schijndel, J.W.P.M. *et al. Eur. J. Biochem.* 1994, **225:** 151–157.
39. Everett, R.R. and Butler, A. *Inorg. Chem.* 1989, **28:** 393–395.
40. Everett, R.R., Soedjak, H.S. and Butler, A. *J. Biol. Chem.* 1990, **265:** 15671–15679.
41. Everett, R.R., Kanofsky, J.R. and Butler, A. *J. Biol. Chem.* 1990, **265:** 4908–4914.
42. Walker, J.V. and Butler, A. *Inorg. Chim. Acta* 1996, **243:** 201–206.
43. Wever, R. and Kustin, K. *Adv. Inorg. Chem.* 1990, **35:** 81–115.
44. Winter, G.E.M. and Butler, A. *Biochemistry* 1996, **35:** 11805–11811.
45. ter Meulen, H. *Recl. Trav. Chim. Pays-Bas* 1931, **50:** 491.
46. Kneifel, H. and Bayer, E. *Angew. Chem. Int. Ed. Engl.* 1973, **12:** 508.
47. Carrondo, M.A.A.F.d.C.T., Duarte, M.T.L.S., da Silva, J.J.R.F. and da Silva, J.A.L. *Struct. Chem.* 1992, **3:** 113–119.
48. Armstrong, E.M. *et al. J. Am. Chem. Soc.* 1993, **115:** 807–808.
49. Nawi, M.A. and Riechel T.L. *Inorg. Chim. Acta* 1987, **136:** 33–39.
50. Guedes da Silva, M.F.C. *et al. J. Am. Chem. Soc.* 1996, **118:** 7568–7573.
51. Hamstra, B.J. Ph. D. Dissertation, University of Michigan, 1998.
52. Carrano, C.J. *et al. Inorg. Chem.* 1990, **29:** 944–951.
53. Holmes, S. and Carrano, C.J. *Inorg. Chem.* 1991, **30:** 1231–1235.
54. Cornman, C.R., Kampf, J. and Pecoraro, V.L. *Inorg. Chem.* 1992, **31:** 1981–1983.
55. Asgedom, G. *et al. J. Chem. Soc., Dalton Trans.* 1995, 2459–2466.

56. Hamstra, B. J. *et al. Inorg. Chem.* 1997, **36:** 4866–4874.
57. Conte, V., Di Furia, F. and Moro, S. *Tetrahedron Lett.* 1994, **35:** 7429–7432.
58. Dinesh, C.U., Kumar, R., Pandey, B. and Kumar, P. *J. Chem. Soc., Chem. Commun.* 1995, 611–612.
59. Bhattacharjee, M., Ganguly, S. and Mukherjee, J. *J. Chem. Res. (S)* 1995, 80–81.
60. Andersson, M., Conte, V., Di Furia, F. and Moro, S. *Tetrahedron Lett.* 1995, **36:** 2675–2678.
61. Conte, V., Di Furia, F., Moro, S. and Rabbolini, S. *J. Mol. Catal. A* 1996, **113:** 175–184.
62. Rao, A.V.S., Ravishankar, H.N. and Ramasarma, T. *Arch. Biochem. Biophys.* 1996, **1:** 121–134.
63. de la Rosa, R.I., Clague, M.J. and Butler, A. *J. Am. Chem. Soc.* 1992, **114:** 760–761.
64. Clague, M.J. and Butler, A. in *Adv. Inorg. Biochem.* (ed.) Marzilli, L.G.E., Eichorn, G.L. Elsevier, New York, 1994, pp. 219–243.
65. Clague, M.J., Keder, N.L. and Butler, A. *Inorg. Chem.* 1993, **32:** 4754–4761.
66. Hamstra, B.J., Colpas, G.J. and Pecoraro, V.L. *Inorg. Chem.* 1998, **37:** 949–955.
67. Colpas, G.J., Hamstra, B.H., Kampf, J.W. and Pecoraro, V.L. *J. Am. Chem. Soc.* 1994, **116:** 3627–3628.
68. Colpas, G.J., Hamstra, B.J., Kampf, J.W. and Pecoraro, V.L. *J. Am. Chem. Soc.* 1996, **118:** 3469–3478.
69. Lapshin, A.E., Smolin, Y.I. and Sheplev, Y.F. *Acta Cryst.* 1989, **C45:** 1477–1479.
70. Lapshin, A.E., Smolin, Y.I. and Sheplev, Y.F. *Acta Cryst.* 1990, **C46:** 738–741.
71. Slebodnick, C., Colpas, G.J. and Pecoraro, V.L. unpublished results.
72. Butler, A., Clague, M.J. and Meister, G.E. *Chem. Rev.* 1994, **94:** 625–638.
73. Mimoun, H., Mignard, M., Brechot, P. and Saussine, L. *J. Am. Chem. Soc.* 1986, **108:** 3711–3718.
74. Mimoun, H. *et al. J. Am. Chem. Soc.* 1983, **105:** 3101–3110.
75. Sharpless, K.B. and Verhoeven, T.R. *Aldrichimica Acta* 1979, **12:** 63–74.
76. Ballisteri, F.P. *et al. J. Am. Chem. Soc.* 1991, **113:** 6209–6212.
77. Vollarova, O., Benko, J. and Sivak, M. *Collect. Czech. Chem. Commun.* 1996, **61:** 574–588.
78. Ghiron, A.F. and Thompson, R.C. *Inorg. Chem.* 1990, **29:** 4457–4461.

79. Pasczcynski, A., Huyhn, V.-B. and Crawford, R.L. *Arch. Biochem. Biophys.* 1986, **244:** 750–760.
80. Glenn, J.K. and Gold, M.H. *Arch. Biochem. Biophys.* 1985, **242:** 329–341.
81. Glenn, J.K., Akileswaran, L. and Gold, M.H. *Arch. Biochem. Biophys.* 1986, **251:** 688–696.
82. Kadota, S. *et al. Biochem. Biophys. Res. Commun.* 1987, **147:** 259–266.
83. Kuan, I.-C., Johnson, K.A. and Tien, M. *J. Biol. Chem.* 1993, **268:** 20064–20070.
84. Kuan, I.-C. and Tien, M. *Proc. Nat. Acad. Sci. USA* 1993, **90:** 1242–1246.
85. Sundaramoorthy, M., Kishi, K., Gold, M.H. and Poulos, T.L. *J. Biol. Chem.* 1994, **269:** 32759–32767.
86. Gold, M.H., Wariishi, H. and Valli, K. in *Biocatalysis in Agricultural Biotechnology* (eds.) Whitaker, J.R. and Sonnet, P.E. American Chemical Society, Washington, DC, 1989, pp. 127–140.
87. Hammel, K.E. in *Degradation of Environmental Pollutants by Microorganisms and Their Metalloenzymes* (eds.) Sigel, H. and Sigel, A. Marcel Dekker, Inc., New York, 1992, pp. 41–60.
88. Crawford, R.L. *Lignin Biodegradation and Transformation.* John Wiley & Sons, New York, 1981.
89. Sarkanen, K.V. and Ludwig, C.H. *Lignins: Occurrence, Formation, Structure and Reactions.* Wiley-Interscience, New York, 1971.
90. Douglas, J.C. *Trends Plant Sci.* 1996, **1:** 171–178.
91. Shimada, M. *et al. J. Biotechnol.* 1997, **53:** 103–113.
92. Kirk, T.K. and Farrell, R.L. *Annu. Rev. Microbiol.* 1987, **41:** 507–541.
93. Eggert, C., Temp, U. and Eriksson, K.-E.L. *FEBS Lett.* 1997, **407:** 89–92.
94. Youn, H.-D., Yung, C.H. and Sa-Ouk, K. *FEMS Microbiol. Lett.* 1995, **132:** 183–188.
95. Kersten, P.J. *Proc. Nat. Acad. Sci. USA* 1990, **87:** 2936–2940.
96. Kersten, P.J. and Kirk, T.K. *J. Bacteriol.* 1987, **169:** 2195–2202.
97. Sundaramoorthy, M., Kishi, K., Gold, M.H. and Poulos, T.L. *J. Biol. Chem* 1997, **272:** 17574–17580.
98. Whitwam, R.E. *et al. Biochemistry* 1997, **36:** 9766–9773.
99. Wariishi, H., Valli, K. and Gold, M.H. *Biochemistry* 1989, **28:** 6017–6023.
100. Wariishi, H., Valli, K. and Gold, M.H. *J. Biol. Chem.* 1992, **267:** 23688–23695.
101. Wariishi, H., Akileswaran, L. and Gold, M.H. *Biochemistry* 1988, **27:** 5365–5370.

102. Wariishi, H., Dunford, H.B., MacDonald, I.D. and Gold, M.H. *J. Biol. Chem.* 1989, **264:** 3335–3340.
103. Zapanta, L.S. and Tien, M. *J. Biotechnol.* 1997, **53:** 93–102.
104. Khindaria, A., Yamazaki, I. and Aust, S.D. *Biochemistry* 1995, **34:** 16860–16869.
105. Khindaria, A., Barr, D.P. and Aust, S.D. *Biochemistry* 1995, **34:** 7773–7779.
106. Timofeevsky, S.L. and Aust, S.D. *Biochem. Biophys. Res. Commun.* 1997, **239:** 645–649.
107. Saadeh, S.M., Lah, M.S. and Pecoraro, V.L. *Inorg. Chem.* 1991, **30:** 8–15.
108. Huynh, V.-B. and Crawford, R.L. *FEMS Microbiol. Lett.* 1985, **28:** 119–123.
109. Saadeh, S.M. Ph. D. Dissertation, University of Michigan, 1992.
110. Policar, C., Artaud, I. and Mansuy, D. *Inorg. Chem.* 1996, **35:** 210–216.
111. Larson, E. *et al. Inorg. Chem.* 1992, **31:** 373–378.
112. Larson, E.J. and Pecoraro, V.L. *J. Am. Chem. Soc.* 1991, **113:** 3810–3818.
113. Boucher, L.J. and Coe, C.G. *Inorg. Chem.* 1975, **14:** 1289–1295.
114. Law, N.A. *et al. J. Chem. Soc., Chem. Commun.* 1995, 2015–2016.
115. Matsushita, T., Fujiwara, M. and Shono, T. *Chem. Lett.* 1981, 631–634.
116. Law, N.A., Ph. D. Dissertation, University of Michigan, 1999.
117. Penner-Hahn, J.E. in *Manganese Redox Enzymes* (ed.) Pecoraro, V.L. VCH Publishers, Inc., New York, 1992, pp. 29–45.
118. Yocum, C.F. in *Manganese Redox Enzymes* (ed.) Pecoraro, V.L. VCH, New York, 1992, pp. 71–83.
119. Wincencjusz, H., Van Gorkom, H. and Yocum, C.F. *Biochemistry* 1997, **36:** 3663–3670.
120. Pecoraro, V.L. (eds.) *Manganese Redox Enzymes*. VCH Publishers, Inc., New York, 1992.
121. (a) Pecoraro, V.L., Baldwin, M.J. and Gelasco, A. *Chem. Rev.* 1994, **94:** 807–826; (b) Law, N.A., Caudle, M.T. and Pecoraro, V.L. in *Advances in Organic Chemistry* (ed.) Sykes, A.G. Academic Press, Inc., New York, 1999, pp. 305–440; (c) Vocum, C.F. and Pecoraro, V.L. *Curr. Opin. Chem. Biol.* 1999, **3:** 182–187; (d) Pecoraro, V.L. and Hsieh, W.-Y. in *Manganese and its Role in Biological Processes* (eds.) Sigel, A. and Sigel, H. *Met. Ions Biol. Syst.* Marcel Dekker, New York, Vol. 37, in press.
122. Ort, D.R. and Yocum, C.F. (eds.) *Oxygenic Photosynthesis: The Light Reactions.* Kluwer Academic Publishers, Boston, 1996.
123. Debus, R.J. *Biochim. Biophys. Acta* 1992, **1102:** 269–352.
124. Ludwig, M.L., Pattridge, K.A. and Stallings, W.C. in *Manganese in Metabolism*

and Enzyme Function (eds.) Schramm, V.L. and Wedler, F.C. Academic Press, Inc., New York, 1986, pp. 405–430.
125. Waldo, G.S. and Penner-Hahn, J.E. *Biochemistry* 1995, **34:** 1507–1512.
126. Beyer, W.F., Jr. and Fridovich, I. *Biochemistry* 1985, **24:** 6460–6467.
127. Kono, Y. and Fridovich, I. *J. Biol. Chem.* 1983, **258:** 6015–6019.
128. Algood, G.S. and Perry, J.J. *J. Bacteriol.* 1986, **168:** 563–567.
129. Barynin, V.V. *et al. Dolk. Akad. Nauk. S.S.S.R* 1986, **288:** 877–880.
130. Barynin, V.V. and Grebenko, A.I. *Dokl. Akad. Nauk. S.S.S.R* 1986, **286:** 461–464.
131. Vainshtein, B.K., Melik-Adamyan, W.R., Barynin, V.V. and Vagin, A.A. in *Progress in Bioorganic Chemistry and Molecular Biology* (ed.) Ovchinnikov, Y.A. Elsevier Science Publishers, New York, 1984, pp. 117–25.
132. Wieghardt, K. *Angew. Chem. Int. Ed. Engl.* 1989, **28:** 1153–1172.
133. Riggs-Gelasco, P.J., Stemmler, T. and Penner-Hahn, J.E. *Coord. Chem. Rev.* 1995, **144:** 245–286.
134. (a) Fronko, R.M. and Penner-Hahn, J.E. *J. Am. Chem. Soc.* 1988, **110:** 7554–7555; (b) Haddy, A. *et al. Inorg. Chem.* 1994, **33:** 2677–2682.
135. Khangulov, S.V. *et al. J. Chem. Soc., Chem. Commun* 1989, 1755–1756.
136. Khangulov, S.V. *et al. Biochemistry* 1995, **34:** 2015–2025.
137. Barynin, V.V., personal communication.
138. Barynin, V.V. *et al. J. Inorg. Biochem.* 1997, **67:** 196.
139. Waldo, G.S., Fronko, R.M. and Penner-Hahn, J.E. *Biochemistry* 1991, **30:** 10486–10490.
140. Gelasco, A., Kirk, M.L., Kampf, J.W. and Pecoraro, V.L. *Inorg. Chem.* 1997, **36:** 1829–1837.
141. Gelasco, A., Bensiek, S. and Pecoraro, V.L. *Inorg. Chem.* 1998, **37:** 3301–3309.
142. Frasch, W.D. and Mei, R. *Biochim. Biophys. Acta* 1987, **891:** 8–14.
143. Yachandra, V.K., Sauer, K. and Klein, M.P. *Chem. Rev.* 1996, **96:** 2927–2950.
144. Pecoraro, V.L., Gelasco, A. and Baldwin, M.J. in *Mechanistic Bioinorganic Chemistry* (eds.) Thorp, H.H. and Pecoraro, V.L. ACS Books, Washington D.C., 1995, pp. 265–301.
145. Dismukes, G.C. *Chem. Rev.* 1996, **96:** 2909–2926.
146. Pessiki, P.J. and Dismukes, G.C. *J. Am. Chem. Soc.* 1994, **116:** 898–903.
147. Pessiki, P.J., Khangulov, S.V., Ho, D.M. and Dismukes, G.C. *J. Am. Chem. Soc.* 1994, **116:** 891–897.

148. Mathur, P., Crowder, M. and Dismukes, G.C. *J. Am. Chem. Soc.* 1987, **109:** 5227–5233.
149. Gelasco, A. and Pecoraro, V.L. *J. Am. Chem. Soc.* 1993, **115:** 7928–7929.
150. Larson, E. *et al. J. Am. Chem. Soc.* 1992, **114:** 6263–6265.
151. Bonadies, J. A. *et al. Inorg. Chem.* 1989, **28:** 2037–2044.
152. Bossek, U., Saher, M., Weyhermuller, T. and Wieghardt, K. *J. Chem. Soc., Chem. Commun.* 1992, pp. 1780–1782.
153. Sakiyama, H., Okawa, H. and Isobe, R. *J. Chem. Soc., Chem. Commun.* 1993, 882–884.
154. Naruta, Y. and Maruyama, K. *J. Am. Chem. Soc.* 1991, **113:** 3595–3596.
155. Sakiyama, H. *et al. J. Chem. Soc., Dalton Trans.* 1993, 591–596.
156. Naruta, Y., Sasayama, M. and Sasaki, T. *Angew. Chem. Int. Ed. Engl.* 1994, **33:** 1839–1841.
157. Naruta, Y., Sasayama, M.-A. and Ichihara, K. *J. Mol. Catal. A* 1997, **117:** 115–121.
158. Larson, E.J. and Pecoraro, V.L. *J. Am. Chem. Soc.* 1991, **113:** 7809–7810.
159. Gelasco, A.K. Ph. D. Dissertation, University of Michigan, 1995.
160. Stibrany, R.T. and Gorun, S.M. *Angew. Chem. Int. Ed. Engl.* 1990, **29:** 1156–1158.
161. Gelasco, A., Askenas, A. and Pecoraro, V.L. *Inorg. Chem.* 1996, **35:** 1419–1420.
162. Wilcox, D. *Chem. Rev.* 1996, **96:** 2435–2458.
163. Christianson, D.W. *Prog. Biophys. Mol. Biol.* 1997, **67:** 217–252.
164. Kanyo, Z.F., Scolnick, L.R., Ash, D.E. and Christianson, D.W. *Nature* 1996, **383:** 554–557.
165. Stemmler, T.L. *et al. Biochemistry* 1997, **36:** 9847–9858.
166. Sossong Jr., T.M. *et al. JBIC* 1997, **2:** 433–443.
167. Slebodnick, C., Hamstra, B.J. and Pecoraro, V.L. *Struct. Bonding (Berlin)* 1997, **89:** 51–108.
168. Clague, M.J. and Butler, A. *J. Am. Chem. Soc.* 1995, **117:** 3475–3484.
169. Shank, M., Barynin, V.V. and Dismukes, G.C. *Biochemistry* 1994, **33:** 15433–15436.
170. Sheptovitsky, Y.G. and Brudrig, G.W. *Biochemistry* 1998, **37:** 5052–5059.
171. Soedjak, H.S. and Butler, A. *Biochemistry* 1990, **29:** 7974–7981.
172. Berry, R.E. *et al. Angew Chem. Int. Ed. Engl.* 1999, **38:** 795–797.
173. Stemmler, T.L. *et al. J. Am. Chem. Soc.* 1997, **119:** 9215–9225.
174. Grush, M.M. *et al. J. Am. Chem. Soc.* 1996, **118:** 65–69.

175. Sheats, J.E. *et al. J. Am. Chem. Soc.* 1987, **109:** 1435.
176. Brunold, T.C. *et al. J. Am. Chem. Soc.* 1998, **120:** 8724–8738.
177. Gamelin, D.R. *et al. J. Am. Chem. Soc.* 1994, **116:** 2392–2399.
178. Michaud-Soret, I. *et al. Inorg. Chem.* 1998, **37:** 3874–3876.

Chapter 6

Methane Monooxygenase Models

Zhengbo Hu & Sergiu M. Gorun*

Department of Chemistry, Brown University
Providence, RI 02912, USA

I Introduction

The main commercial process for methanol production steam-reforms natural gas to produce synthesis gas intermediate:

$$CH_4(g) + H_2O(g) \rightarrow CO(g) + 3H_2(g) \quad \Delta H^\circ = 49.3\,\text{Kcal} \qquad \text{(I-1)}$$

Catalysts: alumina supported Ni, Re, etc.

$$CO(g) + 2H_2(g) \rightarrow CH_3OH(g) \quad \Delta H^\circ = -21.7\,\text{Kcal} \qquad \text{(I-2)}$$

Catalysts: zeolites, Cu/ZnO

The excess hydrogen is used to produce the required steam, thus improving the overall thermal efficiency of this process. Interestingly, in addition to the conventional steam-reforming Reaction (I-1), a direct oxidation process of CH_4 with O_2 or air and yielding syngas has been reported recently:[1]

$$CH_4 + 1/2O_2 \rightarrow CO + 2H_2 \qquad \text{(I-1')}$$

This reaction, which takes place over alumina-supported noble metals (Rh, Pt) seems to occur via C–H bond dissociation (to yield H_2) and preferential reaction of O_2 with adsorbed C species to give CO. Other commercial sources of methanol utilize coal or biological waste.

A direct, one-step process that avoids synthesis gas production is the partial oxidation of methane:

$$CH_4(g)+1/2O_2(g) \rightarrow CH_3OH(liq.) \quad \Delta H^\circ = -30.7 \text{ Kcal} \qquad \text{(I-3)}$$

The early recognition of the possibility of direct oxidation of methane led to a significant effort that resulted in numerous gas-phase reaction schemes solid catalysts combinations. In general, for gas-phase reactions, the yield of methanol seems to decrease with high O_2 concentration, increase with pressure and temperature, and vary only slightly with the residence time.

Due to low CH_3OH yields and relatively high temperature and pressure operating conditions, gas-phase systems are currently deemed commercially non-viable. Another type of CH_4 oxidation is known in biology:

$$CH_4 + O_2 + 2H^+ + 2e^- \rightarrow CH_3OH + H_2O \qquad \text{(I-4)}$$

This process is used by methanotrophic bacteria as a source of energy and carbon. Importantly, only one methanol molecule is produced for each dioxygen molecule. It is interesting to note the paradoxical nature of this reaction in which two reducing equivalents are needed for the oxidation. There is no *overall* thermodynamic need for reducing equivalents since Reaction (I-3) is thermodynamically downhill. However, the active site of methanotrophic bacteria, a collection of proteins termed methane monooxygenases (MMOs), seem to be able to insert only one of the two oxygen atoms of dioxygen into a C–H bond. In this respect, the MMOs resemble cytochrome P450 with the important difference that the latter is not known to hydroxylate methane. The attractive features of this oxidation pathway are the room temperature and 1 atm reaction conditions. The problem of supplying reducing equivalents in a commercially viable process is an issue which, however, does not preclude the search for patent protection for the hydroxylase (see below) components of MMOs.[2] The key for low temperature oxygenation of C–H bonds (Reaction I-4) is the catalyst. Thus, despite the fact that MMOs do not perform Reaction (I-3), their structures and mechanisms of oxygen insertion into C–H bonds have been researched intensely. This chapter will briefly review MMOs and discuss relevant model chemistry, focusing primarily on the relationships between structure and reactivity.

II Methane Monooxygenases: Structure and Function

Currently, two types of MMOs are known to be present in methanotrophs:[3]

(i) The particulate, pMMO, whose hydroxylase component contains a trinuclear copper center.
(ii) The soluble, sMMO, whose hydroxylase contains a dinuclear iron center.

While it has been suggested that iron may also be present in a homo- or heterometallic center in pMMO,[4] only the sMMO, for which a dinuclear iron center has been unequivocally established, will be discussed here.

The sMMO, abbreviated heretoforth MMO, is an apparatus comprised of a 40 kDa reductase, a 245 kDa hydroxylase and a 15 kDa component called Component B. The reductase, which contains iron-sulfur and flavin adenine dinucleotide cofactors,[5] acts as a battery by storing two NADH-provided electrons and delivering them to the hydroxylase. In addition, it may have a regulatory ("gating") role.[3]

A detailed analysis of the structures of two crystal forms of the hydroxylase of MMO from *M. capsulatus* (Bath) reveals[6] that a change in the crystal packing results in a Leu 110 side-chain change, which could allow (and restrict) substrate (and product) access to the active site. Such a change could also be induced by Component B binding resulting in a "gating" effect.

In contrast, Component B does not seem to play a role in transferring electrons, yet it forms a complex with the hydroxylase. Its binding modifies the structure of the dinuclear iron center, as evidenced *inter alia* by the six-fold reduction in the antiferromagnetically coupling of the Fe(II)-Fe(III) mixed-valence state (see below) from $-30\ cm^{-1}$ to $-5\ cm^{-1}$.[7] In addition, hydroxylation rates vary non-linearly with the relative Component B/hydroxylase (and reductase/hydroxylase) ratios, a fact ascribed to the formation of several binary and ternary complexes of the three components.[8] The regioselectivity of MMO seems to be influenced by the Component B too. Thus, gradual addition of Component B to B-depleted MMO systems enhanced the hydroxylation of primary carbons when isopentane was used as the substrate.[9]

The above suggests the catalytic importance of secondary interactions, in addition to the direct modification/tuning of the dinuclear metal center of the hydroxylase by the coordinated ligands. Such interactions also tune the selectivity within the cytochrome P450 family. Thus, while the expected order of reactivity tertiary > secondary > primary C–H bonds is prevalent, the ω-hydroxylases preferentially attack primary C–H bonds.

The hydroxylases of *M. capsulatus* (Bath) and *M. trichosporium OB3b* (abbreviated as MMOH) have been isolated and structurally characterized recently[10] in at least two oxidation states: $MMOH_{red}$, Fe(II)Fe(II) and $MMOH_{ox}$, Fe(III)Fe(III). The resting, diferric state, can be reduced by one electron to yield a mixed-valence state, $MMOH_{mv}$.

The structures of the reduced and oxidized form of MMOH confirm the presence of a dinuclear iron center, bridged by carboxylate and oxo groups. The oxygen rich metal environment is typical for oxygen activating diiron enzymes such as MMO and ribonucleotide reductase R2 (See Fig. II-1), although hemerythrin, some membrane-bound monooxygenases

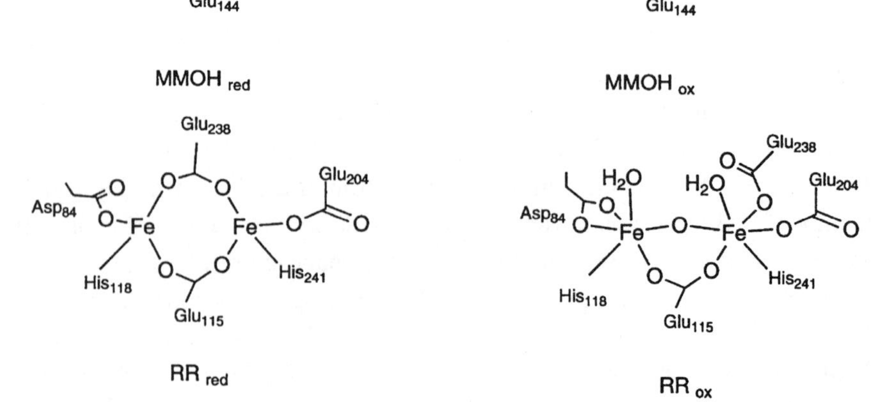

Fig. II-1 Structure of the diiron active centers in reduced and oxidized MMOH and RR R2.

and desaturases contain nitrogen rich ligation.[11] The coordinatively unsaturated but electron rich diferrous enzymes bind O_2, yielding an active “iron-oxo” catalyst after several steps. The intermediates include Component P,[12] a complex whose iron-peroxo structure may be remarkably similar to one of the iron-peroxo models.[13] The Mössbauer spectrum of this model closely matches that of Component P. Intermediate P rearranges to Q, a higher valent $Fe^{IV}Fe^{IV}$ bis μ-oxo structure termed the “diamond core”[14] which abstracts one hydrogen from CH_4. Ribonucleotide reductase may follow a related oxygen activation pathway in abstracting a hydrogen from a tyrosine to form a tyrosyl radical.[15] Importantly, a manganese-based RR performing the same chemistry is known.[16] Thus synthetic chemists have employed both iron and manganese in trying to mimic the chemistry of biological C–H oxygenation.

III Synthetic Models

Many diiron complexes with μ-carboxylato and/or μ-oxo or μ-OH bridges which model the structural and spectroscopic properties of the metal sites in diiron-oxo proteins have been synthesized.[17,18] Few of these species, however, display catalytic activity which mimic the alkane functionalization chemistry of MMO.[19] We will focus on recent (last 5 years) advances on functional iron models for MMO and especially, on their catalytic oxidation of alkanes and alkenes. The literature prior to 1993, which has been covered in several reviews, will be considered sparingly. Manganese analogues will be included considering that manganese model compounds show similar catalytic activity, or even better efficiency compared with coordinatively similar iron complexes. Structurally characterized dinuclear iron and manganese complexes will be discussed in detail, the catalytic properties of some related mono- and polynuclear related species will be considered insofar they shed light on the dinuclear systems. Consequently, some well-known non-heme iron catalytic systems, including Gif and Fenton chemistry will not be discussed in this chapter, as they have been reviewed recently.[20,21] (including Chap. 2 of this monograph).

The efficiency and selectivity of catalytic oxidation systems, i.e. their yields, turnover numbers and distribution of the products, depend on many factors, including oxygen source (molecular oxygen in the presence of protons and electron donors, hydrogen peroxide, alkyl peroxide or other reagents such as PhIO), solvent, substrate, oxidation state of metal center, ligands, coordination geometry, temperature and pressure, etc. The catalytic effects of these factors will be discussed systematically in this chapter.

A possible classification of model dinuclear complexes takes into account the number of supporting bridges linking two metals. Thus, singly-bridged (Type I), doubly-bridged (Type II) and triply-bridged complexes (Type III) are represented in Fig. III-1. The oxo and carboxylato groups are representative examples of bridging anions. In addition, Type IV complexes include complexes with dinucleating ligands. A further classification refinement considers the differences in the terminal ligands, both chelating and monodentate. For example, both Type I-a and I-b exhibit N_4XO donor set but the N_4 subset is a tetradentate ligand in the former and a bis bidentate one in the latter.

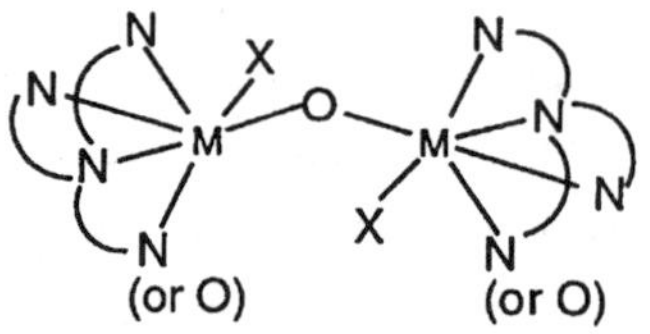

Type I-a

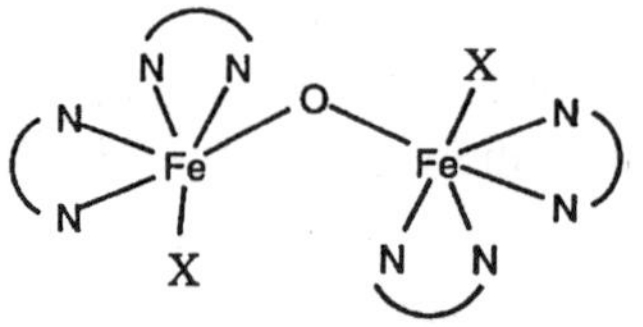

Type I-b

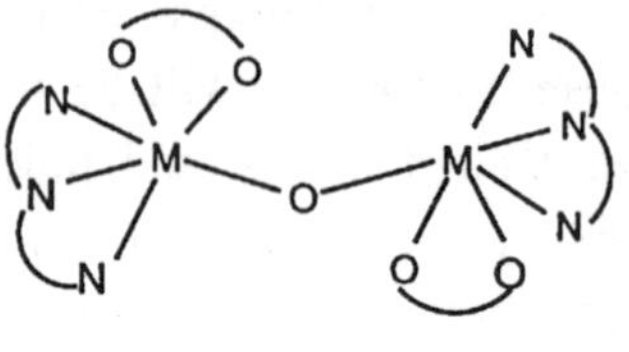

Type I-c

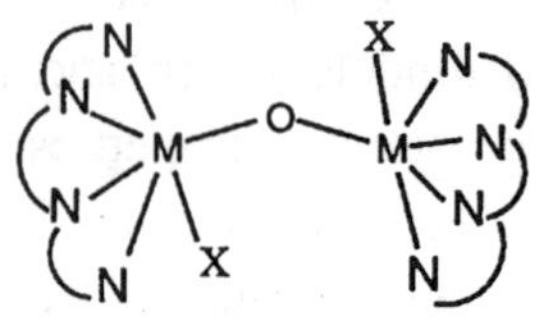

Type I-d

Type I: Single bridge and mononucleating ligand.

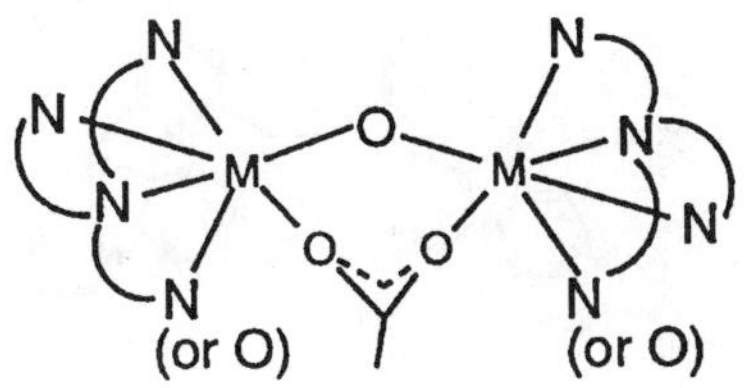

Type II-a

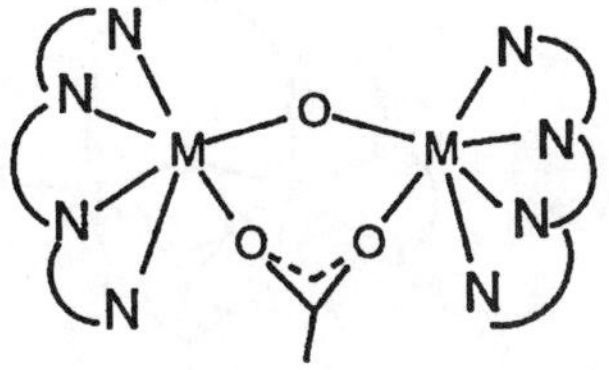

Type II-b

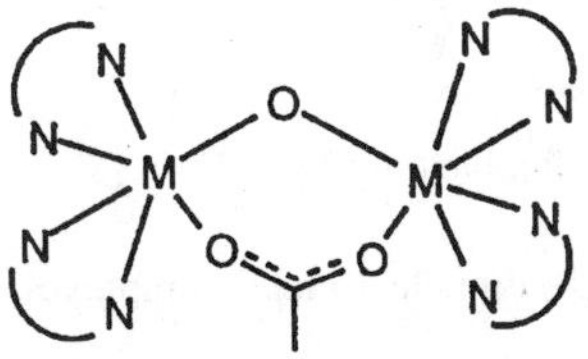

Type II-c

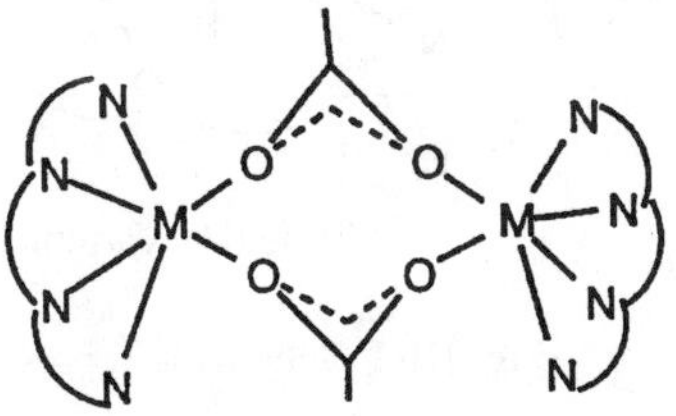

Type II-d

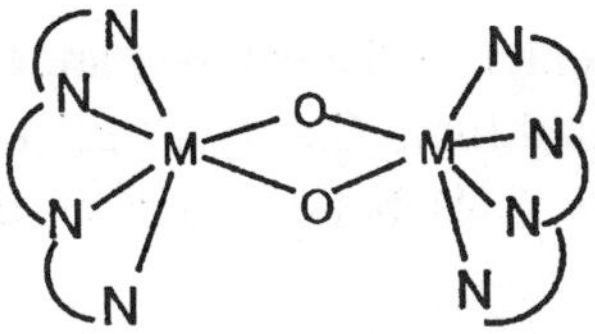

Type II-e

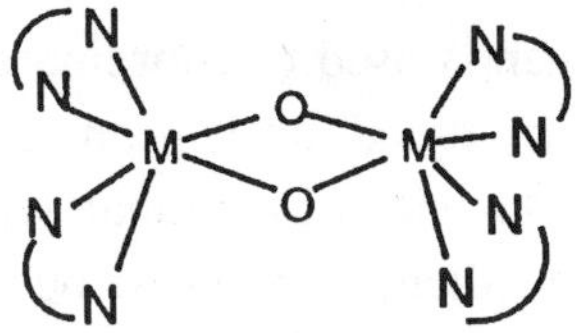

Type II-f

Type II: Double bridge and mononucleating ligand.

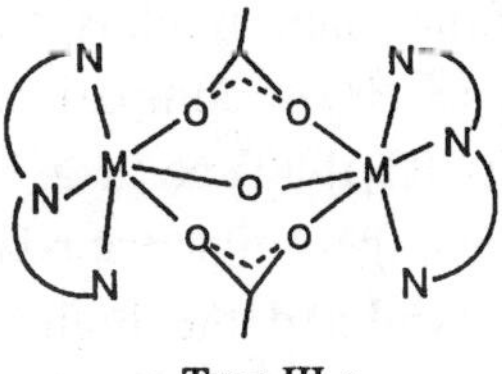

Type III-a

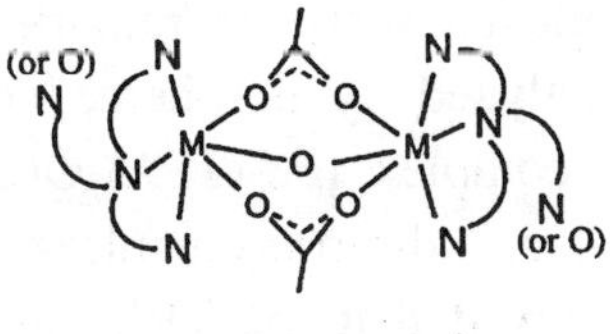

Type III-b

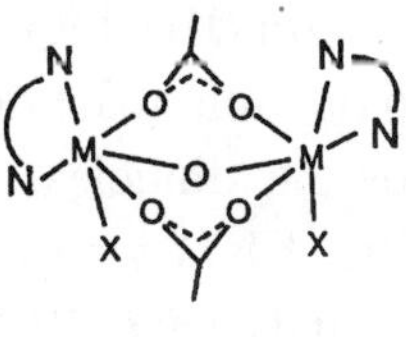

Type III-c

Type III: Triple bridge and mononucleating ligand.

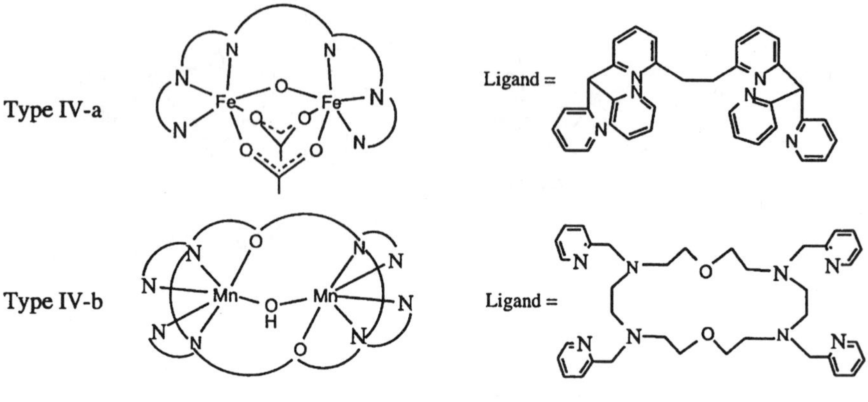

Type IV: Variable bridges and dinucleating ligand.

Fig. III-1 Schematic representations of the dinuclear metal complexes.

Terminal ligand effects

The ligands used to coordinate the metal center of the catalyst are essential for both metal activation and catalyst stability. Catalyst resistance to oxidation becomes critical when refractory C–H bonds, such as those of primary alkanes, need to be activated.

A systematic study of cyclohexane oxidation with *t*-butyl hydroperoxide (TBHP) in the presence of dinuclear iron complexes employed the ligands shown in Fig. III-2.[22]

The results, including the distribution of the major products, cyclohexanone (CyONE), cyclohexanol (CyOH) and cyclohexyl *t*-butyl peroxide (CyOOtBu) are summarized in Table III-1.

Iron(III)-TPA complexes serve as remarkable functional models for oxidation reactions catalyzed by non-heme enzymes.[22–25] A comparison of the dinuclear TPA complex $[Fe_2(TPA)_2O(OAc)](ClO_4)]$ (Type II-a in Fig. III-1) with other μ-oxo diferric complexes containing similar tripodal ligands (Fig. III-2), showed that the TPA complex exhibited the highest catalytic efficiency.[22] The TPA complex was stable (confirmed by ^{1}H NMR and UV-Vis spectroscopy) and yielded the same products upon the addition of a second, similar amount of TBHP after the completion of the first run.

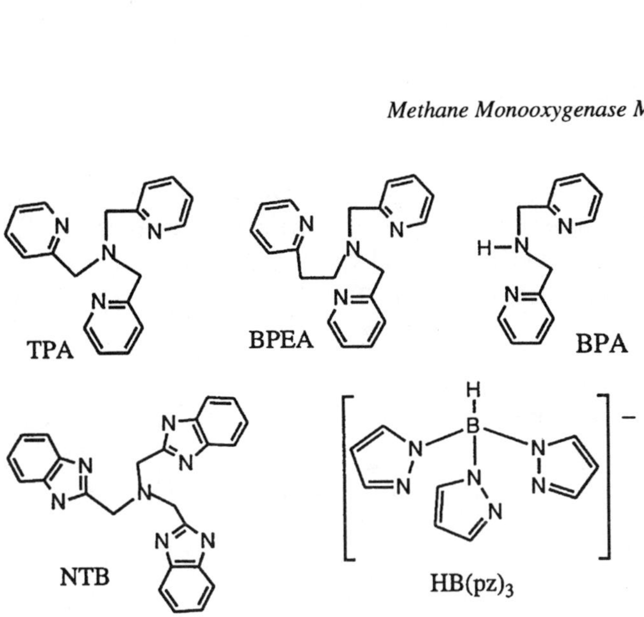

Fig. III-2 Schematic representations of the ligands of Ref. 22.

Table III-1 Product distributions for oxidation of cyclohexane (0.77M) catalyzed by diferric complexes (0.7 mM)[a]

Fe catalyst	Reaction Time	Product[b]		
		CyOH	CyONE	CyOOtBu
$[Fe_2(TPA)_2O(OAc)](ClO_4)_3$	0.25	9	11	16
$[Fe_2(BPEA)_2O(OAc)_2](ClO_4)_3$	1	10	14	8
$[Fe_2(BPA)_2O(OBz)_2](ClO_4)_2$	16	15	13	10
$[Fe_2(NTB)_2(OAc)](ClO_4)_3$	16	8	7	0.7
$[Fe_2(HB(Pz)_3)_2O(OAc)_2)]$	16	0	0	0

[a] Reactions were carried out in CH_3CN under 1 atm of Ar at 25°C. TBHP was used as the oxidant.
[b] Moles of product per mole of catalyst.

When one of the pyridyl methyl arms in TPA was replaced by a pyridyl ethyl arm (BPEA ligand), a diferric complex of the same type was obtained. The effect of this one-carbon homologation in the ligand resulted in a 400% increase in the reaction time. Replacement of the pyridines in TPA by benzimidazoles (NTB ligand) slowed down the catalytic activity even more. A Type III-a complex, $[Fe_2(HB(pz)_3)_2O(OAc)_2]$ showed no catalytic activity with TBHP as the oxidant but, interestingly, displayed limited catalytic activity using dioxygen as the oxygen source in the presence of Zn powder (electron donor) and acetic acid (proton donor).[26]

Further modifications of the TPA ligand[27–28] are shown in Fig. III-3.

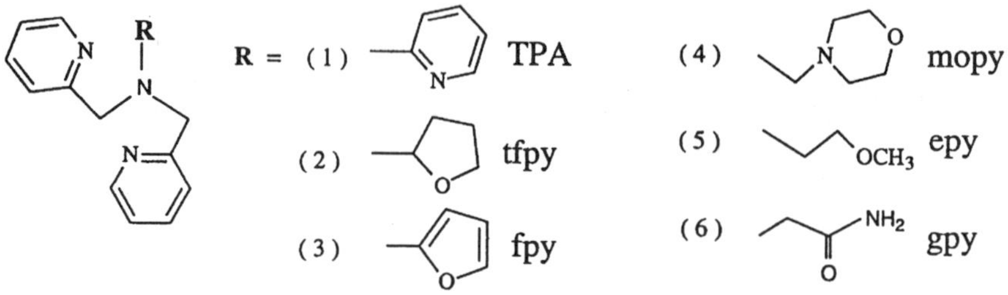

Fig. III-3 Schematic representations of modified TPA ligands.[27,28]

The Type I-a dinuclear iron(III) complex (linear oxo bridge), $[Fe_2OCl_2(TPA)_2](ClO_4)_4$, and the similar dinuclear iron complexes with mopy and fpy ligands, exhibited high catalase-like activity but negligible oxygenation of cyclohexane with H_2O_2. In contrast, the similar complexes with tfpy and epy ligands showed little H_2O_2 disproportionation activity, but high efficiency for oxygenation of cyclohexane. The structurally characterized complexes seem to maintain their integrity in solution, as judged by the lack of spectral changes upon the addition of hydrogen peroxide.

In order to explain the reactivity differences, several Fe-OO(H) intermediates are considered in Scheme III-1.

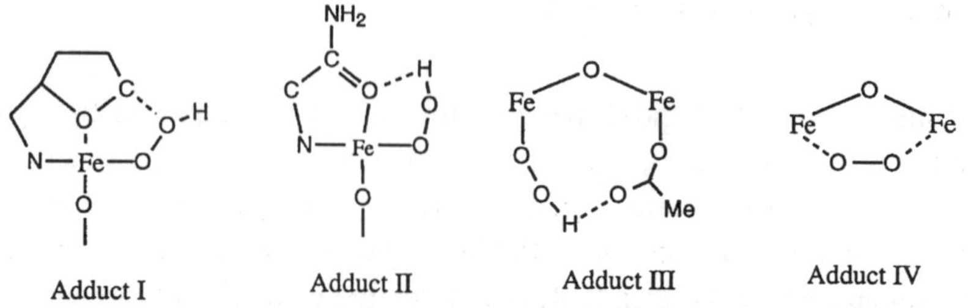

Scheme III-1

The key assumption in formulating adducts I-III is that the peroxide anion replaces one of the coordinated chloride anions, binding in a η^1-coordination mode.[29,30] Further stabilization/activation occurs by weak electrophilic interaction with a carbon of the tetrahydrofuran ring (see Adduct I). This activation is the key feature for the facile hydroxylation of cyclohexane. The negligible catalytic activity of the mopy complex is explained by the steric effect of mopy ligand, viz. the steric repulsion between the pyridine and morpholine rings renders the interaction between the peroxide adduct and a C–H bond of the mopy ligand impossible.

Replacing one of three pyridyl by an amide group, $CH_2C(O)NH_2$, also results in a Type I-a complex (characterized crystallographically), which exhibits high activity for both hydroxylation of cyclohexane and the decomposition of H_2O_2.[28] Another η^1 peroxide iron(III) adduct, Adduct II, is proposed as the active species for cyclohexane oxygenation, but a (μ-η^1: η^1-peroxide)(μ-oxo) diiron(III) intermediate (Adduct IV in Scheme III-1) is responsible for catalase-like activity. The role of the amide group was further probed via extended Hückel Molecular Orbital (EHMO) calculations of the peroxide and hydroperoxide model compounds $[Fe^{III}(NH_3)_4(HCONH_2)(HO_2)]^{2+}$ and $[Fe^{III}(NH_3)_4(py)(HO_2)]^{2+}$. Weak interactions between the peroxide oxygen atoms and the amide group were noted, but no interaction occurs between peroxide and py in the second compound. The amide-hydroperoxide interaction is thus deemed essential for oxidation of cyclohexane by a coordinated hydroperoxide, and illustrates the direct role terminal ligands may play in stabilizing catalytic intermediates.

Two related Type II dinuclear iron(III) compounds with a bent oxo and single acetate bridges were prepared using TPA (Type II-a complex) and mep (Type II-b complex) ligands[31] (see Fig. III-4 for the mep ligand).

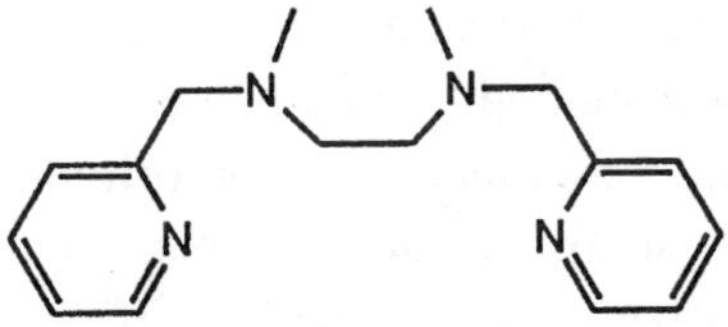

Fig. III-4 mep ligand.

The diiron cores were similar (shown by X-ray diffraction), but the two complexes exhibited extremely different reactivity with hydrogen peroxide in solutions. The TPA complex, which was stable in solution, showed high catalase-like activity, while the catalytically inactive mep complex oxygenated cyclohexane to cyclohexanol and cyclohexanone. The UV-Vis spectrum of the TPA complex (after the addition of H_2O_2), was consistent with the proposed (μ-η^1:η^1-peroxide)(μ-oxo) diiron(III) intermediate (Adduct IV), deemed essential for catalase-like activity. On the other hand, UV-Vis data on the mep complex in the presence of H_2O_2 suggested a linear oxo-bridged diiron(III) species[17] with an end-on (as opposed to bridging) peroxide anion.

The same ligand family can also produce triply-bridged diiron cores. Type III-b, bis(μ-acetato)(μ-oxo) diiron(III) complexes with tridentate fpy (X-ray structure), tfpy (X-ray structure), epy, mopy or TPA, also functionalize cyclohexane in the presence of H_2O_2.[32] The TPA and epy complexes still have catalase-like activity (perhaps via Adduct IV, Scheme III-1) but are poor cyclohexane oxygenation catalysts. In contrast to Type II complexes, where an end-on coordinated HOO^- group forms hydrogen bonds with the terminal amido group, Type III complexes are too sterically crowded to favor such interactions. Instead, a carboxylate shift[33] may allow the formation of another type of hydrogen bonding with a carboxylate group, resulting in the formation of Adduct III. Like Adduct II, this species was proposed to be the active intermediate for the oxygenation reaction, but not for decomposition of H_2O_2. Both peroxide Adduct III and IV could exist in the solution in the presence of the H_2O_2, but their ratio depends upon the steric effect of the R group of the ligand (Fig. III-2). Thus, a sterically regulated dual catalase/oxygenation activity is observed.

The competition between the oxygenation of organic substrates and decomposition of hydrogen peroxide or alkyl peroxides is often observed in catalytic systems. Although additional evidence is needed to support the above proposed peroxide intermediates for different catalytic pathways, it is clear that relatively small changes in metal coordination environments can induce dramatic differences in catalytic efficiency and even reaction pathways. This is consistent with the noted steric and electronic effect of ligands in forming and stabilizing metal dioxygen adducts[13,34,35] viz. steric bulkiness around the metal center promotes dioxygen-binding reversibility, while reduced electron density at the metal center stabilizes a dioxygen adduct. Weak ligand-peroxide interactions could also play an important catalytic role, as shown above. Subtle effects such as the size and shape of "catalytic pockets" in MMOH perhaps influence the mode of O_2 binding.

Preparing MMO models mimicking subtle steric effects, models which should ultimately include chiral dinuclear complexes, represents a challenge for the coordination chemists. To this end, a chiral bipy-based ligand, (-)4,5-pinene bipyridine (pbipy), was shown to produce the diferric complex $[Fe_2O(pbipy)_4(H_2O)_2](ClO_4)_4$ (Type I-c).[36] While the parent complex, $[Fe_2O(bipy)_4(H_2O)_2](ClO_4)_4$, is converted rapidly to an inactive mononuclear ferrous complex by H_2O_2, this robust complex (at least 50 catalytic turnovers) also catalyzes the oxidation of alkanes, alkenes and sulfides. The oxidation of dimethylsulfide afforded the sulfoxide, while, importantly, oxidation of *p*-bromo(phenyl)methylsulfide yielded the R enantiomer in 38% enantiomeric excess. Not surprisingly, $[Fe_2O(bipy)_4(H_2O)_2](ClO_4)_4$ generated only a racemic mixture of the sulfoxides from the above substrate. This selectivity is interpreted to suggest a non-radical autoxidation pathway, yet a radical-free mechanism seems unlikely, as discussed in Chap. II.

Ancillary ligand effects

The term "ancillary ligand" refers to the terminal or bridging groups which coordinate the metal *in addition* to the terminal chelating ligands. These terminal or bridging ligands are often more labile than their chelating, terminal counterparts and have been shown to directly influence the reactivity of metal centers. For example, it is well-known that one or more open or accessible metal sites are essential for catalytic oxidations, since both the oxidant and, to a certain extent, the substrate need to bind to the metal center or in its immediate vicinity.

This point is illustrated by a series of hydrocarbon oxidations with TBHP catalyzed by diferric complexes in which the ancillary ligands vary.[37,38] The results of this study are summarized in Table III-2.

Table III-2 Effect of ancillary ligands: oxidation of hydrocarbons catalyzed by diferric complexes[a]

Fe Catalyst	Substrate	Turnover No. of Oxygenated Products[b]	Yield[c] (%)
$[Fe_2O(H_2O)_2(bipy)_4](ClO_4)_4$	cyclohexane	CyOH: 17, CyONE: 19 CyOOtBu: 4	44
	adamantane	1-adamantanol: 24 2-adamantanol: 3 2-adamantanone: 4	24
	toluene	benzaldehyde: 6 benzyl alcohol: 1	10
	benzyl alcohol	benzaldehyde: 100	70
$[Fe_2O(H_2O)_2(dmbipy)_4](ClO_4)_4$	cyclohexane	CyOH: 17, CyONE: 19 CYOOtBu: 7	44
$[Fe_2O(H_2O)_2(dmbipy)_4](ClO_4)_4$	cyclohexane	CyOH: 16, CyONE: 19 CyOOtBu: 8	44
$[Fe_2O(OAc)(bipy)_2](ClO_4)_2$	cyclohexane	CyOH: 18, CyONE: 15 CyOOtBu: 6	38
$[Fe_2O(OAc)(phen)_2](ClO_4)_2$	cyclohexane	CyOH: 18, CyONE: 13 CyOOtBu: 5	35
$[Fe_2O(OAc)(dmbipy)_2](ClO_4)_2$	cyclohexane	CyOH: 27, CyONE: 24 CyOOtBu: 7	57
$[Fe_2OCl_2(L)_2](ClO_4)_2$ (L = bipy, phen or dmbipy)	cyclohexane	CyOH: 1, CyONE: 1 CyOOtBu: 1	< 2
$[Fe_2O(CF_3COO)_2(L)_2](ClO_4)_2$ (L = bipy or phen)	cyclohexane	CyOH: 1, CyONE: 1 CyOOtBu: 1	< 2
$[Fe_2O(CF_3COO)_2(dmbipy)_2](ClO_4)_2$	cyclohexane	CyOH: 7, CyONE: 5 CyOOtBu: 3	15

[a] The oxidation was performed in CH_3CN under an Ar atmosphere. TBHP was used as the oxygen source. Ratios of catalyst: TBHP:substrate are 1 : 140 : 1100 for oxidation of cyclohexane, and 1 : 140 : 115 for oxidation of adamantane.

[b] Turnover number = mole of product per mole of catalyst.

[c] Total yield is based on oxidant. The ketone yields are molar yields multiplied by 2.

In an earlier work[37] it was found that a Type I complex, $[Fe_2OCl_6](NEt_4)_2$, is inactive toward oxidation of cyclohexane. However, $[Fe_2O(H_2O)_2(phen)_4](ClO_4)_4$, Type I-b, X = H_2O (Fig. III-1), is an efficient catalyst, the addition of imidazole increasing the ketone/alcohol ratio. This difference may be attributed to an electronic effect, viz. a nitrogen rich metal coordination provided by phen, and enhanced by the addition of imidazole, being required or preferred. The coordination number of the metal centers is also of particular interest. One might expect the four-coordinated iron centers in $[Fe_2OCl_6]^{2-}$ to favor additional binding relative to the six-coordinated iron centers of the second catalyst. In reality, the presence of labile H_2O ligands in the latter may allow the formation of open coordination sites, thus allowing beneficial electronic effects to manifest themselves.

The complexes of Table III-2 extend the above work. In addition to a μ-oxo bridge, ancillary H_2O, Cl^-, CH_3COO^- or CF_3COO^- ligands bind to the metal centers. Crystallographic studies of $[Fe^{III}{}_2O(H_2O)_2(phen)_4](ClO_4)_4$, $[Fe^{III}{}_2O(Cl)_2(phen)_4](ClO_4)_4$ and $[Fe^{III}{}_2O(CF_3COO)_2(dmbipy)_4](ClO_4)_2$ reveal that CF_3COO^-, H_2O and Cl^- are attached to each iron(III) as monodentate ligands (Type I-c), but the acetate group bridges the diiron(III) core (Type II-c).[38]

The catalytic activities of all these complexes were investigated using TBHP in CH_3CN under anaerobic conditions.[38] Dioxygen is not involved in the reaction since cyclohexyl hydroperoxide could not be detected in aerated solutions. The complexes with the same X but different bidentate nitrogen donor ligands show little variation in activity, except that $[Fe^{III}{}_2O(CF_3COO)_2(dmbipy)_2](ClO_4)_2$ is more active than $[Fe^{III}{}_2O(CF_3COO)_2(L)_2](ClO_4)_2$ (L = bipy or phen). The higher activity of $[Fe^{III}{}_2O(CF_3COO)_2(dmbipy)_2](ClO_4)_2$ was explained by the higher basicity of the methyl-substituted bipyridine ligand. Importantly, the lability of the exchangeable ligand correlates directly with the efficiency of the catalysts. H_2O and CH_3COO^- are found to be substitutionally labile, but the CF_3COO^- group is less exchangeable and Cl^- is harder or even impossible to replace. The order of catalytic activity of these complexes thus correlates inversely with the bond strengths between the ancillary ligand and the metal center, viz. $H_2O > CH_3COO^- > CF_3COO^- > Cl^-$. The stability of the complexes, however, tends to decrease with increased catalytic activity. Thus, metal complexes with terminal or bridging H_2O or CH_3COO^- ligands are reduced slowly in solution to a ferrous $[FeL_3]^{2+}$ species, a process which is accelerated by TBHP.

In addition to the effects induced by the modifications of the *terminal* TPA ligand (mentioned previously, Table III-1), the effect of various *bridging* ligands on the catalytic efficiency of $[Fe_2(TPA)_2O(X)](ClO_4)_n$ (n = 2 or 3) complexes is summarized in Table III-3.[22]

Table III-3 Oxidation of cyclohexane catalyzed by TPA diferric complexes with various ancillary ligands.[a]

Fe Catalyst (Structural Types)		Reaction Time (h)	Product[b]		
			CyOH	CyONE	CyOOtBu
$[Fe_2(TPA)_2O(OAc)](ClO_4)_3$	(Type II-a)	0.25	9	11	16
$[Fe_2(TPA)_2O(OBz)](ClO_4)_3$	(Type II-a)	0.25	11	12	14
$[Fe_2(TPA)_2O(O_2CC_6H_4\text{-}4\text{-}OCH_3)](ClO_4)_3$	(Type II-a)	1	11	13	18
$[Fe_2(TPA)_2O(O_2P(OPh)_2)](ClO_4)_3$	(Type II)	2	12	12	18
$[Fe_2(TPA)_2O(CO_3)](ClO_4)_2$	(Type II)	4	8	8	6
$[Fe_2(TPA)_2O(\text{phthalate})](ClO_4)_2$	(Type II)	20	15	14	5

[a] Reactions were carried out in CH_3CN under 1 atm of Ar atmosphere at 25°C. TBHP was used as the oxidant.
[b] Moles of product per mole of catalyst.

While the acetate- and benzoate-bridged complexes exhibit similar reactivity, the more electron donating (4-OCH_3) benzoate bridge decreases the catalytic activity. This effect is probably due to a Fe^{III}/Fe^{II} redox potential reduction which disfavors the iron-catalyzed oxidation of the hydroperoxide to its radical (Haber-Weiss mechanism). Diphenyl phosphate, carbonate or phthalate bridges have similar effects. Generally, bridging monoanions gave rise to shorter reaction times than bridging dianions, which enforce shorter Fe–O bonds.

Oxygen sources for catalysis

Producing a biomimetic, catalytically active iron-oxo intermediate can, in principle, be achieved in several ways, depending upon the oxidation state of the oxygen source. Thus, functional models for methane monooxygenase are often obtained via two basic routes:

(1) Oxygenase route: O_2^0 starting point, from Fe(II) + O_2 + a hydrogen donor.
(2) Peroxidase route: O_2^{2-} starting point, from Fe(III) + H_2O_2 (or another single-oxygen donor such as PhIO, ROOH and peroxy acids).

Of the two routes, the first one is closely related to its biological counterpart, considering that MMO uses O_2 exclusively as its oxygen source.

The catalytic activities of $[Fe_2O(HBpz_3)_2(OAc)_2]$-Zn-HOAc and related system for oxidation of hydrocarbons using molecular oxygen are summarized in Table III-4.[26,39]

Complex $[Fe_2O(HBpz_3)_2(OAc)_2]$ exhibits two N_3O_3 coordinated octahedral iron (III) centers bridged by one μ-oxo and two μ-acetato groups (Type III-a)[40]. The metal centers thus exhibit a nitrogen rich environment, unlike the oxygen rich environment of the metal centers in MMOH. Both the proton (HOAc) and electron (Zn powder) donors are required for catalysis, but a large excess of acetic acid results in the formation of inactive $[Fe(HBpz_3)_2]^+$. Adamantane gives 1-adamantanol (2.48 yield based on the complex), 2-adamantanol (0.50 yield) and 2-adamantanone (1.08 yield), suggesting a low 2.2 3°/2° carbon reactivity ratio.[26]

The alcohol yield increases significantly when a related complex $[Fe_2O(HBpz_3)_2(hfacac)_2]$ is used.[39] Unlike the acetate group in $[Fe_2O(HBpz_3)_2(OAc)_2]$, the bidentate hfacac (hexafluoro acetylacetonate) coordinates to each Fe^{III} in a terminal fashion leaving a single μ-oxo bridge to support a dinuclear center (Type I-c) with a more open structure, as reflected by its Fe-O-Fe angle of 169°, versus 125° in $[Fe_2O(HBpz_3)_2(OAc)_2]$. The $[Fe_2O(HBpz_3)_2(hfacac)_2]$-Zn-Hhfacac system in CH_2Cl_2 (excess protonated ligand is the source of protons) oxidizes alkanes and even aromatic compounds (see Table III-4) with a high turnover number of 48.3 versus 2.3 in the case of adamantane. Importantly, 1-adamantanol

Table III-4 Hydrocarbon oxidation catalyzed by $[Fe_2O(HBpz_3)_2(OAc)_2]$ and $[Fe_2O(HBpz_3)_2(hfacac)_2]$ using dioxygen.[a]

Catalyst (mmol)	Substrate (mmol)	Time (h)	Zn (g)	Proton Source[b]	Product Yield[c]
$[Fe_2O(HBpz_3)_2(OAc)_2]$ (0.030)	cyclohexane (3.7)	30	0.5	0.05 ml	cyclohexanol (0.99) cyclohexanone (0.84)
	adamantane (3.9)	30	0.5	0.05 ml	1-adamantanol (0.50) 2-adamantanol (1.08) 2-adamantanone (2.48)
	cyclohexene (3.7)	30	0.5	0.05 ml	cyclohexenol (0.54) cyclohexanone (0.73) cyclohexene epoxide (0.20)
$[Fe_2O(HBpz_3)_2(hfacac)_2]$ (0.0015)	cyclohexane (1.66)	30	0.5	0.43 mmol	cyclohexanol (4.5) cyclohexanone (0.2)
	adamantane (1.66)	30	0.5	0.43 mmol	1-adamantanol (46.6) 2-adamantanol (1.7) 2-adamantanone (trace)
	pentane (1.66)	30	0.5	0.43 mmol	2-pentanol (21.0)
	benzene (1.66)	30	0.5	0.43 mmol	phenol (12.1)
	toluene (1.66)	30	0.5	0.43 mmol	*p*-anisole (6.3) *o*-anisole (2.7)
	chlorobenzene (1.66)	30	0.5	0.43 mmol	*p*-chlorophenol (2.2) *o*-chlorophenol (0.7)

[a] Reactions are performed under 1 atm of O_2 at room temperature in 20 ml of CH_2Cl_2.
[b] The proton source is HOAc when $[Fe_2O(HBpz_3)_2(OAc)_2]$ is the catalyst, or Hhfacac when $[Fe_2O(HBpz_3)_2(hfacac)_2]$ is the catalyst.
[c] Based on the catalyst.

dominates the product distribution while CyOH is the major oxidation product of cyclohexane. No benzylic oxidation is observed in the hydroxylation of toluene. Although the oxidation state and structure of the active species and the detailed mechanism of the catalytic reaction were not reported, this work demonstrates that the hydroxylation of arenes (such as benzene, toluene and chlorobenzene) with dioxygen could be catalyzed by metal complexes in the presence of H^+ and e^- donors.

The oxidation of hydrocarbons is also catalyzed by *diferrous* complexes with oxygen rich environments,[41–43] a coordination type which is closer to that of MMOH. The ligands and their complexes are shown in Fig. III-5.

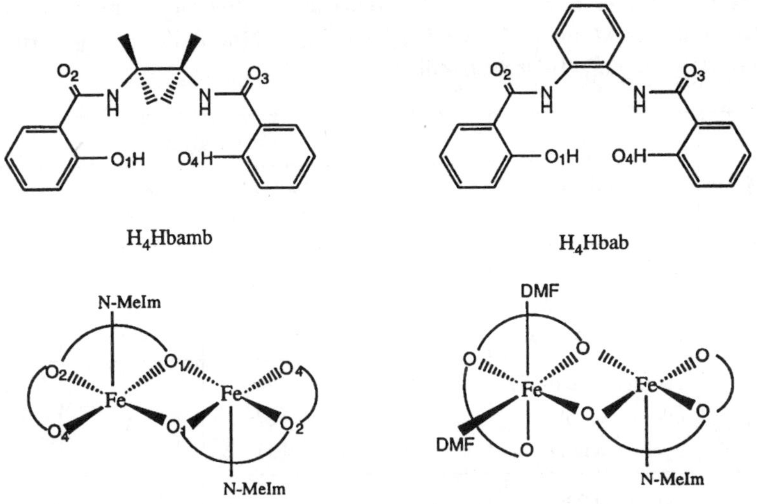

Fig. III-5 Schematic representations of the ligands and their dinuclear iron-complexes. L stands for either of the two ligands, H_4Hbamb or H_4Hbab.

The diferrous complex, obtained anaerobically from the reaction of *trans*-Fe(*N*-MeIm)$_2$(Cl)$_2$(MeOH)$_2$ with the dilithium salt of H_4Hbab (H_2L) in MeOH, is formulated as Fe^{II}_2(L)$_2$(*N*-MeIm)$_2$, based upon its X-ray structure (Fig. III-5).[41,42] Both metals exhibit a symmetric NO_4 environment, which is retained in the analogous complex of the ligand H_4Hbab (see Fig. III-5). Recrystallization of the latter complex from DMF yields the asymmetric complex [$Fe^{II}_2(H_2Hbab)_2(N\text{-MeIm})(DMF)_2$], in which one iron center retains its NO_4 environment, but the other one exhibits a O_6 octahedral geometry. Both the nuclearity and asymmetry of the DMF complex is maintained in solution, as demonstrated by solution molecular weight determinations and Mössbauer spectroscopy. Oxidations of cyclohexene catalyzed by $Fe^{II}_2(H_2Hbab)_2(N\text{-MeIm})_2$ and related complexes (see below), were carried out anaerobically in 1% DMF/CH_2Cl_2, using iodosylbenzene (PhIO). Due to the exclusion of air, PhIO was the only oxygen source. The results are presented in Table III-5.

Table III-5 Turnover numbers for the oxidation of hydrocarbons catalyzed by $Fe^{II}_2(H_2Hbamb)_2(N\text{-MeIm})_2$, $[Fe^{II}Fe^{III}(H_2Hbamb)_2(N\text{-MeIm})_2]I^-$, $[Fe^{III}_2(H_2Hbamb)_2(N\text{-MeIm})_2](I^-)_2$ and $Fe^{III}_2(H_2Hbab)_2(N\text{-MeIm})_2$.[a]

Substrate	Products	Catalyst			
		[$Fe^{2+}Fe^{2+}$] (H_2 Hbamb)	[$Fe^{2+}Fe^{3+}$] (H_2 Hbamb)	[$Fe^{3+}Fe^{3+}$] (H_2 Hbamb)	[$Fe^{2+}Fe^{2+}$] (H_2 Hbab)
cyclohexane	cyclohexanol	57	1	–	Not reported
	cyclohexanone	1	1	–	
	chlorocyclohexane	49	38	13	
cyclohexene	cyclohexene oxide	83	78	–	Total T.N. 30[b]
	cyclohexenol	100	127	5	
	cyclohexenone	100	151	–	
toluene	benzyl alcohol	23	1	1	Not reported
	benzaldehyde	5	–	1	
	o-cresol	1	1	–	
	m-cresol	1	19	–	
	p-cresol	1	1	–	

[a] Reaction conditions: for H_2Hbamb complexes, catalyst: PhIO: substrate = 1 : 500 : 2500, [catalyst] = 1 mM; all reactions were performed in 10% DMF/CH_2Cl_2 for 12 h. For the diferrous H2Hbab complex, catalyst: PhIO: substrate = 1 : 2000 : 10000, [catalyst] = 0.05 mM; the reaction was performed in 1% DMF/CH_2Cl_2 for 6 h.

[b] Products distribution: cyclohexene epoxide (12.5%), cyclohexanol (25%), cyclohexenone (50%) and cyclohexanone (12.5%).

The cyclohexene product distribution ratio was cyclohexene epoxide : cyclohexenol:cyclohexenone:cyclohexanone = 1:2:4:1 for a total turnover number of 30. No chlorinated products were observed in CH_2Cl_2, but bromine incorporation into the organic products was detected when CH_2Br_2 was used as the solvent.[19]

A similar ligand (H_4Hbamb) (see Fig. III-5) was used to prepare dinuclear iron-complexes with different oxidation states, $Fe^{II}_2(H_2Hbamb)_2(N\text{-}MeIm)_2$, $Fe^{II}Fe^{III}(H_2Hbamb)_2(N\text{-}MeIm)_2(I^-)$ and Fe^{III}_2 $(H_2Hbamb)_2(N\text{-}MeIm)_2$ $(I^-)_2$[43]. Cyclic voltammetry experiments show for $[Fe^{II}_2(H_2Hbamb)_2$ $(N\text{-}MeIm)_2]$ a quasi-reversible wave at −690 mV (NHE) and an electrochemically reversible couple at −310 mV (NHE), corresponding to $[Fe^{II}Fe^{II}] \rightarrow [Fe^{II}Fe^{III}]$ and $[Fe^{II}Fe^{III}] \rightarrow [Fe^{III}Fe^{III}]$ transitions, respectively. The mixed-valence and diferric complexes were subsequently obtained by chemical oxidation with I_2. These complexes allowed a systematic study of the catalytic oxidation chemistry of cyclohexane, cyclohexene and toluene (see Table III-5). When the substrate is cyclohexane, complex $[Fe^{II}Fe^{II}]$ produces cyclohexanol and only a trace of cyclohexanone. In this case, halogen incorporation is observed when CH_2Cl_2 is used as the solvent. In the case of toluene, benzylic oxidation predominates with complex $[Fe^{II}Fe^{II}]$, while aromatic ring oxidation occurs almost exclusively with the mixed-valence $[Fe^{II}Fe^{III}]$ catalyst. The $[Fe^{III}Fe^{III}]$ complex was almost inactive, which suggests that at least one ferrous center was necessary for efficient oxygen transfer using PhIO as the oxygen source. Despite the exclusion of molecular oxygen and the use of the oxo transfer agent PhIO, it is likely that a free radical mechanism operates in this case, as in the case for cyclohexane oxidation with TBHP catalyzed by $[Fe_2O(TPA)_2(H_2O)_2]^{4+}$.[25,44]

The results of catalytic oxidation studies of hydrocarbons by complex $Fe^{III}_2O(OAc)_2Cl_2(bipy)_2$, Type III-c, and related complexes using different oxygen sources are shown in Table III-6.[45–48]

When dioxygen molecule was used to oxidize cyclohexane with $Fe_2O(OAc)_2Cl_2(bipy)_2$ in the presence of Zn powder and glacial acetic acid[45], the only product was cyclohexanone (turnover number 2.5), while the alcohol is the only product of MMO. This biomimetic result also differs from those obtained with the $Fe_2O(OAc)_2(HBpz_3)_2/O_2/Zn/HOAc$ system (Table III-4), in which, "normal", i.e. comparable quantities of CyOH and CyONE were detected. Using TBHP instead of $O_2/Zn/HOAc$ with $Fe_2O(OAc)_2Cl_2(bipy)_2$ gives the "normal" result (see Table III-6) and higher turnover numbers.

Systematic mechanistic studies were carried out with H_2O_2 under an O_2 atmosphere, another type of oxygen source.[46] Complexes $Fe_2O(OAc)_2Cl_2(bipy)_2$, $[Fe_2O(OAc)(tmima)_2](ClO_4)_3$ and $[Fe_4O_2(OAc)_7(bipy)_2](ClO_4)$ are used as catalysts (see Table III-6). Comparable catalytic activities are observed for $Fe_2O(OAc)_2Cl_2(bipy)_2$ and $[Fe_2O(OAc)(tmima)_2](ClO_4)_3$, less

Table III-6 Oxidation of hydrocarbons catalyzed by $Fe_2O(OAc)_2Cl_2(bipy)_2$ (Tpe III-c), $[Fe_2O(OAc)(tmima)_2](ClO_4)_3$ (Type II-d), $[Fe_2O(H_2O)_2(tmima)_2](ClO_4)_4$ (Type I-d) and $[Fe_4O_2(OAc)_7(bipy)_2](ClO_4)$.

Catalyst	Oxygen Source	Substrate	Products (see the notes for details)	Notes
$Fe_2O(OAc)_2Cl_2(bipy)_2$	TBHP	ethane propane cyclohexane	ethanol (< 1) isopropanol (8.8) n-propanol (1.5) CyOH (32) CyONE (22)	*a*
$Fe_2O(OAc)_2Cl_2(bipy)_2$	O_2	cyclohexane	CyONE (T.N. = 2.5), no CyOH	*b*
$[Fe_4O_2(OAc)_7Cl_2(bipy)_2]^+$	TBHP	cyclohexane	CyOH (64) CyONE (33)	*c*
$Fe_2O(OAc)_2Cl_2(bipy)_2$ $[Fe_2O(OAc)(tmima)_2]^{3+}$ $[Fe_4O_2(OAc)_7(bipy)_2]$	H_2O_2	cyclohexane	CyOH (5.7) CyONE (6.1) CyOH (6.8) CyONE (7.3) CyOH (2.3) CyONE (2.9)	*d*
$Fe_2O(OAc)_2Cl_2(bipy)_2$ $[Fe_2O(OAc)(tmima)_2]^{3+}$ $[Fe_4O_2(OAc)_7(bipy)_2]^+$	H_2O_2	adamantane	1-ol (19) 2-ol (9.1) 2-one (7.2) 1-ol (16.8) 2-ol (8.5) 2-one (6.4) 1-ol (8.7) 2-ol (4.8) 2-one (3.1)	*e*
$Fe_2O(OAc)_2Cl_2(bipy)_2$ $[Fe_2O(OAc)(tmima)_2]^{3+}$ $[Fe_4O_2(OAc)_7(bipy)_2]^+$	H_2O_2	toluene	PhCHO (3.5) $PhCH_2$ OH (1.4) cresols (1.5) PhCHO (3.0) $PhCH_2$ OH (0.9) cresols (4.2) PhCHO (1.9) $PhCH_2$ OH (0.6) cresols (0.6)	*f*

Table III-6 (*Continued*).

Catalyst	Oxygen Source	Substrate	Products (see the notes for details)	Notes
$[Fe_2O(OAc)(tmima)_2]^{3+}$	TBHP	ethane	ethanol (1) acetaldehyde (2) propionaldehyde (1) acetone (6)	*g*
		propane	2-propanol (3) 1-propanol (1)	
		cyclohexane	CyOH (18) CyONE (21)	
		adamantane	1-ol (4) 2-ol (1) 2-one (1)	
		toluene	benzaldehyde (16) benzoic acid (5)	
$[Fe_2O(H_2O)_2(tmima)_2]^{4+}$	TBHP	ethane	ethanol (0.2) acetaldehyde (0.12) acetic acid (0.5)	*h*
		propane	propionaldehyde (0.4) acetone (4) 2-propanol (3) 1-propanol (0.19)	
		cyclohexane	CyOH (3.7) CyONE (3.8)	
		adamantane	1-ol (1.6) 2-ol (0.14) 2-one (0.33)	
		toluene	benzaldehyde (3.5) benzoic acid (0.4)	

[a] Yields are based on TBHP consumed (iodometric titration). Reaction conditions: at room temperature in CH_3CN, with catalyst: TBHP = 1 : 150, [catalyst] = 7×10^{-4} M. Reaction times: 3 days for oxidation of ethane and cyclohexane, 2 days for propane.

[b] Reaction conditions: room temperature, under 1 atm O_2, catalyst: Zn : HOAc : substrate ratio of 1 : 191 : 199 : 397 ([catalyst] = 0.0007 M in CH_3CN).

[c] Yields are based on TBHP consumed. Similar reaction conditions as in note (a) except lasting for 5 h.

[d] Mmol of product per mmol of complex. Reaction conditions: room temperature in CH_3CN in air.

[e] Percent of starting materials (%). 1-ol is 1-adamantanol. 2-ol is 2-adamantanol. 2-one is 2-adamantanone. Reaction conditions: room temperature in CH_3CN in air with catalyst: H_2O_2 = 1 : 150, [catalyst] = 1mM.

[f] Percent of starting materials (%). Reaction conditions: same as in note (c).

[g] Mmol product per mmol of complex. Reaction conditions: catalyst (0.005 mmol) and substrate (4.5 mmol) were dissolved in 5 ml of CH_3CN and stirred vigorously while TBHP (0.75 mmol) was added at room temperature in the presence of oxygen gas.

[h] Mmol of product per mmol of complex per hour. Reaction conditions: room temperature in CH_3CN under 1 atm of O_2.

for $[Fe_4O_2(OAc)_7(bipy)_2](ClO_4)$. Interestingly, molecular oxygen is essential for the reaction with the dinuclear iron species since bubbling argon through the solutions resulted in a dramatic, 69% to 100%, decrease in the turnover numbers. The free radical trap 2,4,6-tri-*tert*-butylphenol (TBP) quenches the oxidation, consistent with a proposed autoxidation mechanism in which molecular oxygen traps a cyclohexyl radical to give CyOOH. The latter was identified in the reaction mixture and characterized by ^{13}C NMR spectroscopy. CyOOH subsequently decomposes to CyOH and CyONE, but CyOH is not the precursor of CyONE. Further support for this mechanism comes from the observation that in the absence of both substrate and molecular O_2, $Fe_2O(OAc)_2Cl_2(bipy)_2$ and H_2O_2 decompose CyOOH to give CyOH/CyONE in 0.7 ratio, comparable with the observed 0.9 catalytic ratio.

A similar but not identical mechanism was proposed for the $[Fe_2O(OAc)(tmima)_2](ClO_4)_3$ -catalyzed oxidation with TBHP under O_2 atmosphere (see Table III-6).[47] The proposed mechanism assumes the formation of an $Fe^{IV}OFe^{IV}=O$ intermediate thought to produce the initial alkyl radicals. Classical trapping of this radical by dioxygen gives a peroxyl radical, the precursor for the simultaneous production of alcohol and ketone. Unlike the H_2O_2/O_2 oxygen source for the $[Fe_2O(OAc)(tmima)_2](ClO_4)_3/H_2O_2/O_2$ system, TBHP can further oxidize the alcohol to ketone even without O_2 involvement, perhaps via Reaction III-1 (ligands omitted from the catalyst's formula):

$$R-OH + FeOFe=O \longrightarrow R=O + FeOFe + H_2O \qquad \text{(III-1)}$$

Despite this important mechanistic difference, the CyOH/CyONE ratio seems to be independent of the source of oxygen.

The effect of ancillary ligand upon the oxidation efficiency of the Fe/tmima complexes was probed by replacing the μ-OAc bridge in $[Fe_2O(OAc)(tmima)_2](ClO_4)_3$ (Type II-b) with two terminal H_2O ligands to form the complex $[Fe_2O(H_2O)_2(tmima)_2](ClO_4)_4$ (Type I-a).[48] The aqua complex is approximately twice as active (Table III-6), suggesting that the substitution of the ancillary ligand by the oxidant (TBHP) might be rate-limiting. This observation is consistent with the lability order established for complexes of Table III-2 (see above).

Solvent effects

The solvent affects not only the solubility and stability of catalysts, but sometimes can even influence the formation of active species in the catalytic reactions.

For example, the TBHP oxidation of cyclohexane catalyzed by diferric TPA complexes is solvent-dependent.[22] When CH_3CN, benzonitrile or pyridine are used, the oxidation products are invariably CyOH, CyONE and CyOOtBu. However, benzonitrile favors the formation of CyOH and CyONE, while CyOOtBu is the major product when pyridine is the solvent.

For the catalysts $Fe^{II}{}_2(H_2Hbamb)_2(N\text{-MeIm})_2$ and $[Fe^{II}Fe^{II}(H_2Hbamb)_2(N\text{-MeIm})_2](I)$, polar solvents such as CH_3CN quench the catalytic oxidation,[43] an effect ascribed to the competition between solvent and the oxygen donor for the metal center. The $Fe_2O(OAc)_2(HBpz_3)_2/O_2/Zn/HOAc$ system shows the same behavior.[26] CH_2Cl_2 is the preferred solvent, because CH_2Cl_2, unlike CH_3CN or pyridine, has little or no ligating capability.

An intriguing problem of MMO biomimetic chemistry is the oxidation of alkanes in aqueous media. In MMOH, the immediate active site environment is somewhat hydrophobic (see above), thus suggesting biomimetic oxidations in organic solvents. Indeed, the majority of biomimetic studies have been performed in such media.

Yet the *aqueous* chemistry component, including the interactions of substrates, catalysts and oxidation agents need to be understood. To this end, a catalytic decomposition study of TBHP by $[Fe_2O(\mu\text{-OAc})(L)_2]^{3+}$ (L = TPA, BPIA or BIPA) in aqueous solution at pH = 4.2 suggests that the homolytic cleavage of the O–OH or OO–H bonds of TBHP predominates over other pathways.[49] The formation of $(CH_3)_2CO$ (75–85%), CH_3OH (10–25%), CH_2O (> 10%) and *t*-BuOH (15–24%) suggests the intermediacy of *t*-BuO• radicals. The aqueous catalytic oxidation of cyclohexanol with the above systems gives cyclohexanone as the predominant product. The reactions proceed with similar turnover numbers, 15–32, and similar K_H/K_D values (≈ 3.1) for all three catalysts, consistent with the notion that the *t*-BuO• radicals initiates homolytic C–H bond abstraction from the alcohol substrate to generate the products. Free radical pathways are thus likely to dominate the aqueous hydrocarbon oxidation, consistent with the extensively studied decomposition of H_2O_2 and other peroxo–type oxidants.

Dinucleating ligands

The introduction of covalent linkages between ligands enhances their chelating effect and thus entropically stabilizes the complexes. Both the ancillary (usually carboxylates) and terminal ligands have been modified as shown in the following two examples.

An interesting oxygen-rich dinucleating ligand (H_2XDK) and its derivative (H_2PXDK) have been designed to produce diiron complexes (see Fig. III-6).[50,51]

The diferrous complexes [Fe_2(μ-XDK)(μ-OBz)$(ImH)_2$(OBz)(MeOH)] and [Fe_2(μ-$O_2CC(CH_3)_3$)(μ-PXDK)$(N\text{-MeIm})_2$($O_2C\acute{C}(CH_3)_3$)] exhibit iron(II) coordination environments close to those of MMOH. Mössbauer spectroscopy and X-ray crystallography revealed that the iron centers in both XDK and PXDK complexes are inequivalent, one being six coordinate and the other four-coordinate. The reaction of the diferrous complexes with O_2 in THF yielded intermediates whose electronic spectroscopy resembled that of the iron-peroxide intermediate P of MMOH. The catalytic activity of these complexes has not been reported yet, but this dinucleating ligand may be suitable for the stabilization and characterization of catalytic intermediates.

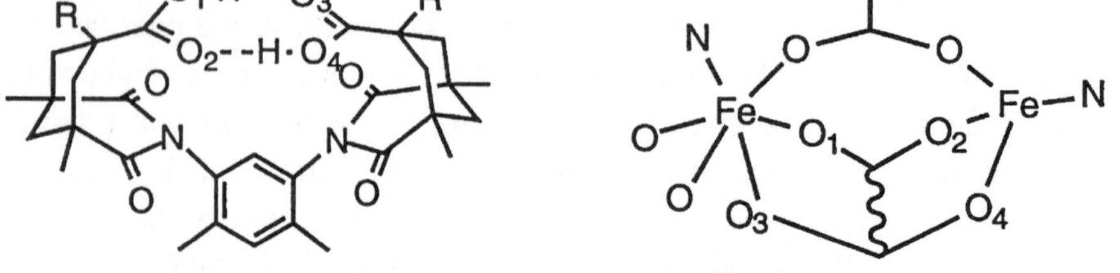

Fig. III-6 Schematic representation of the ligands H_2XDK, R = CH_3 and H_2PXDK, R = $(CH_2)_2CH_3$ (left) and the first coordination sphere of their metal complexes (right).

The first claimed example of a discrete diferric complex with $[Fe_2(\mu\text{-}O)(\mu\text{-}OAc)_2]^{2+}$ core supported by a dinucleating ligand, 1,2-bis [2-di(2-pyridyl)methyl)-6-pyridyl]ethane[52] is shown in Fig. III-1 (Type IV-a). The diferric complex is stable in polar solvents such as CH_3CN or DMF. Oxidation of alkanes at room temperature by *m*-chloroperbenzoic acid (*m*-CPBA) in CH_3CN/CH_2Cl_2 solution under Ar produces the results shown in Table III-7.[53]

Since the presence of O_2 does not affect the yield and turnover number, the authors suggest that the oxidation does not proceed via a classical radical-chain mechanism, yet the product distribution is consistent with such a mechanism. Based upon the finding that the tertiary : secondary : primary C–H bond oxygenation ratio for methylcyclohexane is 150 : 15 : 1 and the tertiary : secondary C-H one for adamantane is 12 : 1, the authors suggest a radical-rebound mechanism, proposed to be similar to that of MMOH. The presence of chlorocyclohexane among the products also supports a radical mechanism, but the distinction between a radical-chain and a radical-rebound mechanism is still unclear in this case.

Table III-7 Product distribution for oxidation alkane catalyzed by $[Fe_2(\mu\text{-}O)(\mu\text{-}OAc)_2(L)](ClO_4)_2$, L = 1,2-bis[2-di(2-pyridyl)methyl-6-pyridyl]ethane [a]

Substrate	Reaction Time (min)	Product (% Yield[b], Turnover Number)
cyclohexane	5	CyOH (40, 164); CyONE (17, 68); ε-caprolactone (12, 48); CyCl (3, 12)
adamantane	20	1-ol (41, 163); 2-ol (10, 39); adamantanone (6, 24)
methylcyclohexane	15	1-methyl-CyOH (26, 104); 2-, 3-and 4-methyl-CyOH (25, 100) cyclohexylmethanol (0.5, 2) methyl-CyONE (12, 48)

[a] Reaction conditions: [catalyst] = 2.0 mM, [substrate] = 3.0 M, [oxidant] = 0.8 M in a mixture of CH_2Cl_2 (3 ml) and CH_3CN (0.3 ml).
[b] Yields are based on *m*-CPBA used.

Considering the above, it seems that the introduction of dinucleating ligands induces additional stabilizing effects, like the steric and electronic effects which are qualitatively similar to those of other chelating ligands.

Effect of substrate upon catalytic efficiency

The C–H bond strength relates directly to the hydroxylation efficiency of hydrocarbons. As expected, oxidation of ethane, propane and cyclohexane catalyzed by non-heme metal aggregates results in lower yield and slower rates for the oxidation, as the number of carbons in the alkanes decreases.[45,54]

Although cyclohexane is the most common substrate used to test the catalytic properties of dinuclear metal complexes, the oxidation of smaller hydrocarbons (such as methane and ethane) is more important and certainly more challenging. The oxidation of methane, ethane and propane was accomplished with the $Fe_2O(OAc)_2Cl_2(bipy)_2/H_2O_2/O_2$ system at room temperature in CH_3CN.[46] The formation of methanol, ethanol and propanol was verified qualitatively by GC/MS analysis of their benzoate esters but was not quantified due to the very low alcohol yields. The deuterated benzoate ester forms exclusively from CD_4, thus eliminating the ligands or other impurities as methanol precursors.

Manganese models of MMO

Although a manganese MMO is not currently known, similarities have been drawn between iron and manganese coordination chemistry and several native or modified metalloenzymes (superoxide dismutase, ribonucleotide reductases, etc.). Similar oxygen functionality with either manganese or iron, prompted synthetic chemists to explore the possibility of building manganese-based MMO models. Several functional models, mostly dinuclear, will be discussed in this section to illustrate the parallel iron and manganese chemistry.

The dinuclear manganese complex $[Mn^{III}{}_2O(OAc)_2(bipy)_2](ClO_4)_2$ which is isostructural with the previously mentioned $[Fe^{III}{}_2O(OAc)_2(bipy)_2](ClO_4)_2$, catalyzes the oxidation of cyclohexane by TBHP. The manganese complex is more efficient (turnover number 2 versus 0.15 for the Fe complex)[55] while also being more stable.

The $[Mn_2O(OAc)(tmima)_2](ClO_4)_3$, was compared with its iron analogue[47] and showed to be similar with respect to solvent effects, oxidant type, catalyst stability, substrate effects and rate of oxygenation.[56] The latter two effects led to the proposal that both manganese[56] and iron[47] catalysts, operate via similar mechanisms (see above).

A dinucleating macrocyclic ligand L1 [1,4,10,13-(2-pyridylmethyl)-1,4,10,13-tetraaza-7,16-dioxacyclooctadecane] produces a Mn(II) complex with a μ-OH bridge type IV-b.[57] This complex, $[Mn_2L1(OH)](ClO_4)_3$, has two symmetrically N_4O_2 -coordinated metal centers, with one of the coordinated oxygen from the hydroxo bridge and the other from the macrocyclic ligand. The latter oxygen is proposed to be labile and replaceable by a peroxide. This complex is not stable in the presence of H_2O_2 in aqueous solution (MnO_2 is formed) but it catalyzes the oxidation of cyclohexane by TBHP in CH_3CN at ambient conditions. The reaction (catalyst: TBHP : cyclohexane = 1 : 175 : 1000) is completed in 24 h affording CyOH and CyONE in 34% combined yield, and turnover numbers of 18 for CyOH and 20 for CyONE.

Two new imidazole- or pyridine-containing ligands (Fig. III-7) form di-(μ-oxo) Mn(III)Mn(IV) and (μ-acetato)(μ-oxo) Fe(III)Fe(III) complexes.[58]

A comparison of their cyclohexane oxidation efficiencies in CH_3CN by using TBHP at ambient conditions is shown in Table III-8.

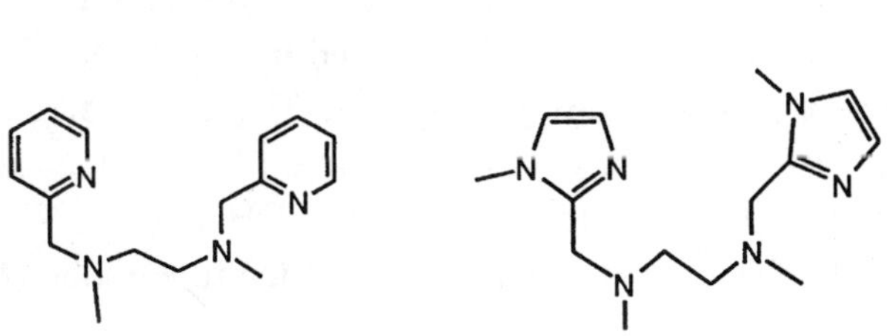

L1-1 = Mebispicen

L1-2 = Mebism

Fig. III-7 Molecular structures of the ligands Mebispcen(L1-1) and Mebism(L1-2).

Table III-8 Product distribution for cyclohexane oxidation catalyzed by dinuclear manganese and iron complexes of L1-1 and L1-2 ligands (substrate/TBHP/catalyst ratio is 1000 : 175 : 1).

Catalyst	Reaction Time	Product[a]			Yield[b]
		CyOH	CyONE	CyOOtBu	
$[Mn_2(L1\text{-}1)_2O_2](ClO_4)_3$	12	16	16	2	28
$[Mn_2(L1\text{-}2)_2O_2](ClO_4)_3$	12	14	13	1	29
$[Mn_2(TPA)_2O_2](ClO_4)_3$	12	2	2	–	3
$[Fe_2(L1\text{-}1)_2O(OAc)](ClO_4)_3$	1	17	19	–	33
$[Fe_2(L1\text{-}2)_2O(OAc)](ClO_4)_3$	1	4.5	5.5	–	11

[a] Moles of product per mole of catalyst.
[b] Yield based on TBHP consumed.

Unlike the (L1-2) complex, the (L1-1) complex is stable under reaction conditions that result in the formation of CyOH and CyONE in 1 : 1 ratio, in addition to small amounts of CyOOtBu. For comparison, the Mn TPA complex is much less active. Interestingly, adding HOAc or NaOAc to the (L1-1) complex / TBHP catalytic system enhances the reaction rate while keeping the CyOH/CyONE ratio 1 : 1. In contrast, Me_2S inhibits the formation of both CyOH and CyONE, while an excess of CH_2Br_2 produces CyBr. The above observations are consistent with an inner-sphere TBHP activation mechanism (acetate coordination affords labile sites) resulting in the formation of radical species, which are totally quenched (Me_2S) or partly captured (CH_2Br_2 or O_2) to give CyBr or CyOOtBu. For these reasons, the TBHP oxidation of adamantane catalyzed by the diferric L1-2 complex does not proceed under an Ar atmosphere. The effect of oxygen source variation was tested for cyclohexane oxidation by the manganese (L1-1) complex. Cumyl hydroperoxide (CHP) gave CyOH and CyONE but in lower yield than TBHP. 2-methyl-1-phenyl-2-propyl hydroperoxide (MPPH) gave neither CyOH nor CyONE (only MPPH-generated benzyl alcohol and benzaldehyde), while MCPBA gave only CyOH in relatively low yield. When TBHP was used as the oxidant, an initial induction period disappeared upon the addition of a second equivalent of TBHP, suggesting that the TBHP induces formation of the true catalyst. Indeed, low temperature UV-Vis and EPR spectroscopy indicated the reduction of the Mn(III/IV) cluster, while a $g = 2$ signal was ascribed to an organic radical, probably *t*-BuOO•.

Free radical chemistry was also invoked to explain the reactivity of a bis (μ-OAc) (Type II-d) complex of the chelating tertiary mep amine ligand (see also Fig. III-4).[59] Catalytic oxidation of cyclohexane using either PhIO or TBHP as the oxidant in CH_3CN, in the presence of CCl_4, yields mainly cyclohexyl chloride, proving that cyclohexyl radicals are formed during the reaction. Adamantane gives exclusively 1-adamantanol with PhIO and almost exclusively the same product with TBHP, but the catalyst decomposes in several hours to an inactive form in the presence of PhIO.

The stoichiometric initiation of free radical chemistry can be achieved also with manganese complexes of dinucleating ligands, such as Mn^{III}_2 (2-OHsalpn)$_2$ (Complex **1** in Scheme III-2.[60,61])

Complex **1** forms the solvated form which catalyzes the oxidation of cyclohexane with TBHP in CH_3CN.[60] High catalytic turnovers and yields are observed for the formation of CyOH and CyONE in about 1 : 1 ratio. H_2O_2 is a less effective oxidant than TBHP, while dioxygen is inactive. Mechanistically, Complex **1** reacts with 1 equivalent of TBHP to give $Mn^{III}Mn^{IV}$(2-OHsalpn)$_2$OH (Complex **4** in Scheme III-2) and a *tert*-butoxy radical via a pre-equilibrium intermediate, **1**-ROOH (Complex 3).[61] Complex **2** / TBHP/ O_2 system exhibits a similar reactivity producing (i) 2-cyclohexen-1-one, 2-cyclohexen-1-ol, and traces of cyclohexene oxide from cyclohexene, (ii) 2-phenyl-2-propanol and acetophenone from cumene, and (iii) CyOH and CyONE from cyclohexane, even after Complex **1** was converted into the inactive Complex **4**. Complex **1** thus acts as a radical initiator, not a catalyst.

In the absence of dinucleating ligands, mono- and dinuclear complexes of the same ligand can simultaneously be catalytically active. Such situations occur, perhaps, more often than not and require careful control experiments, including catalyst stability studies under reaction conditions. For example, a mixture of $Mn(ClO_4)_2 \cdot 6H_2O$ and bipy efficiently catalyzes the oxidation of cyclohexane by TBHP in CH_3CN under Ar at room temperature.[62] The major products (see Table III-9) are CyOH and CyONE (1:1 ratio) and small amounts of *tert*-butyl cyclohexyl hydroperoxide.

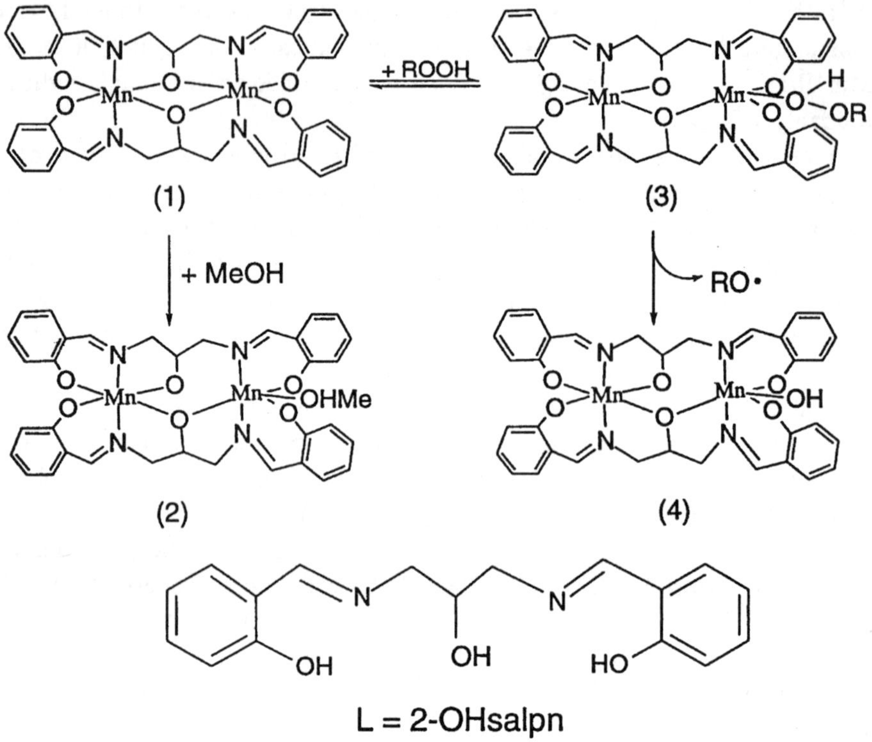

Scheme III-2

The yields reach a maximum when the bipy/Mn ratio is at least 2. $[Mn(bipy)_3]^{2+}$ was detected in the Mn^{2+}/bipy mixture in the absence of oxidant (TBHP), and an authentic sample was found to have the same catalytic efficiency as the mixture (see Table III-9). In the presence of 1.5 equivalents of TBHP however, the Mn/bipy mixture generates the bis (μ-oxo) mixed-valence complex, $[Mn_2O_2(bipy)_4]^{3+}$, identified by EPR, UV-Vis spectroscopy and cyclic volammetry. An authentic sample of $[Mn_2O_2(bipy)_4]^{3+}$ is also an efficient catalyst which maintains its nuclearity during the reaction. Similar results are obtained with phen complexes (see Table III-9), but $[Mn(phen)_3]^{2+}$ forms the dinuclear $[Mn_2O_2(phen)_4]^{3+}$; somewhat slower. TBHP and CHP are equally good oxidants but PhIO is

Table III-9 Oxidation of cyclohexane catalyzed by mononuclear and dinuclear manganese complexes.

Catalyst	[cat]/mM	Oxidant	CyOH[a]	CyONE[a]	Yield[b] (%)
Mn^{2+} / 2 or 3 bipy	0.7	TBHP	15–16	15	35–36
$[Mn(bipy)_3](BF_4)_2$	0.7	TBHP	14	16	37
$[Mn_2O_2(bipy)_4](ClO_4)_3$	0.35	TBHP	33	40	43
$[Mn(phen)_3](BF_4)_2$	0.7	TBHP	1	1	2[c]
$[Mn_2O_2(phen)_4](ClO_4)_3$	0.35	TBHP	33	40	40
$[Mn(bipy)_3](BF_4)_2$	0.7	PhIO	1.8	1.5	25
$[Mn_2O_2(phen)_4](ClO_4)_3$	0.35	CHP	18	20	21

[a] Turnover number = mmol of product per mmol of catalyst. Typical reaction conditions are catalyst:oxidant:substrate = 1:140:1100. Time: 90 mins.
[b] Total yields of CyOH + CyONE + dialkylperoxide.
[c] Complex $[Mn(phen)_3]^{2+}$ gives oxidation yields similar to $[Mn_2O_2(phen)_4]^{3+}$, but only after a few hours.

poor. The higher valence di-μ-oxo Mn(IV)Mn(IV) complex, $[Mn_2O_2(bipy)_4]^{4+}$; is inactive in the absence of the oxidant, suggesting that a dimanganese(IV) complex is not an intermediate. It is interesting to note that manganese is hexacoordinated in the dinuclear complexes of Table III-9, which suggests either an outer-sphere activation mechanism or a higher coordination number.

A number of higher nuclearity manganese complexes (see for example Refs. 63 and 64) catalyze the oxygenation of C–H bonds, but since these complexes have a nuclearity higher than 2, they are not discussed here. It is also worth mentioning that biomimetic, deep oxidation of C–H bonds is also of interest for handling hydrocarbon waste or bleaching.[65] This chemistry, however is not discussed here either since MMOH does not over-oxidize its natural substrate.

Conclusions

The catalytic oxidation of C–H bonds of hydrocarbon, especially those of small molecules such as CH_4, is of general interest.[66,67] Yet, CH_4 has seldom been used as a substrate in biomimetic oxidations of hydrocarbons with dinuclear iron and manganese MMO models. In general, the biomimetic field is significantly more advanced in reproducing structural and spectroscopic properties of the active site of MMO. As shown in this chapter, important lessons from structural and reactivity studies of both MMO and its models are being learned; these lessons will undoubtedly provide an impetus to the still fascinating and challenging modeling work.

Abbreviations

BIPhMe = bis(1-methylimidazol-2-yl)phenylmethoxymethane
bipy = 2,2'-bipyridyl
BIPA = bis[2-(1-methylimidazolyl)methyl][(2-pyridyl)methyl]amine
BPA = bis(2-pyridylmethyl)amine
BPEA = N-(2-pyridylethyl)-N,N-bis(2-pyridylmethyl)amine
BPIA = bis[2-(bis(pyridyl)methyl][2-(bis(methylimidazolyl)methyl]amine
CHP = cumyl hydroperoxide
CyOH = cyclohexanol
CyONE = cyclohexanone
CyOOH = cyclohexyl hydroperoxide
CyOOtBu = *tert*-butyl cyclohexyl peroxide
dmbipy = 4,4'-dimethyl-2,2'-bipyridine
EDTA = ethylenediaminetetraacetic acid
HB(pz)3 = hydrotris(pyrazolyl)borate
H4Hbab = 1,2-bis(2-hydroxylbenzamido)-benzene
H4Hbamb = 2,3-bis(2-hydroxylbenzamido)-2,3-dimethylbutane
Hhfacac = hexafluoroacetylacetone
HQ = 8-hydroxyquinoline
ImH = imidazole
MCPBA (or *m*-CPBA) = *m*-chloroperbenzoic acid
Mebispicen = N,N'-dimethyl-N,N'-bis(2-pyridylmethyl)-1,2-ethanediamine
Mebism = N,N'-dimethyl-N,N'-bis[bis(1-methyl imidazol-2-yl)methyl)-1,2-ethanediamine
MeIm = 1-methyl imidazole

mep = N,N'-dimethyl-N,N'-bis(2-pyridylmethyl)ethane-1,2-diamine
MMOH = hydroxylase of methane monooxygenase
mopy = N-{2-[bis(2-pyridylmethyl)amino]ethyl}morpholine
MPDP = *m*-phenylenedipropionate dianion
MPPH = 2-methyl-1-phenyl-2-propyl hydroperoxide
N-MeIm = *N*-methyl imidazole
NTB = *N,N,N*-tris(2-benzimidazolylmethyl)amine
OAc = acetate
OBz = benzoate
2-OHsalpn = 1,3-bis(salicylideneamino)-2-propanol
PA = picolinic acid
pbipy = (-)4,5-pinene-bipyridine
PDA = perdodecanoic acid
PPAA = phenylperacetic acid
py = pyridine
PyIm = 2-(2'-pyridyl)imidazole
RR = ribonucleotide reductase
salen = N,N'-ethylenebis(salicylideneamine)
TBHP = *tert*-butyl hydroperoxide
tmima = tris[(1-methylimidazol-2-yl)methyl]amine
TPA = tris(2-pyridylmethyl)amine

References

1. Hichman, D.A. and Schmidt, L.D. *Science* 1993, **259:** 343–346.
2. Lipscomb, J.D. and Wayzata, M. *Patent* 1993, US 5,192, 672.
3. Wallar, B.J. and Lipscomb, J.D. *Chem. Rev.* 1996, **96:** 2625–2657.
4. Zahn, J.A. and DiSpirito, A.A. *J. Bacteriol.* 1996, **178:** 1018–1029.
5. (a) Fox, B.G., Froland, W.A., Dege, J.E. and Lipscomb, J.D. *J. Biol. Chem.* 1989, **264:** 10023–10033; (b) Lund, J. and Dalton, H. *Eur. J. Biochem.* 1985, **147:** 291–296; (c) Prince, R.C. and Patel, R.N. *FEBS Lett.* 1986, **203:** 127–130.
6. Rosenzweig, A.C., Brandstetter, H., Whittington, D.A., Nordlund, P., Lippard, S.J. and Frederich, C.A. *Proteins: Structure, Function and Genetics* 1997, **29:** 141–152.

7. Liu, Y. Ph.D. Thesis, University of Minnesota, 1995.
8. Fox, B.G., Liu, Y., Dege, J.E. and Lipscomb, J.D. *J. Biol. Chem.* 1991, **266:** 540–550.
9. Froland, W.A., Andersson, K.K., Lee, S.K., Liu, Y. and Lipscomb, J.D. *J. Biol. Chem.* 1992, **267:** 17588–17597.
10. (a) Rosenzweig, A.C, Frederick, C.A., Lippard, S.J. and Nordlund, P. *Nature* 1993, **366:** 537–543; (b) Rosenzweig, A.C., Nordlund, P., Takahara, P.M., Frederick, C.A. and Lippard, S.J. *Chem. Biol.* 1995, **2:** 409–418; (c) Elango, N., Radhakrishna, R., Froland, W.A., Wallar, B.J., Earhart, C.A., Lipscomb, J.D. and Ohlendorf, D.H. *Protein Sci.* 1997, **6:** 556–568.
11. Shanklin, J., Achim, C., Schmidt, H., Fox, B.G. and Münck, E. *Proc. Natl. Acad. Sci. USA* 1997, **94:** 2981–2986.
12. (a) Liu, K.E., Wang, D., Huynh, B.H., Edmondson, D.E., Salifoglou, A. and Lippard, S.J. *J. Am. Chem. Soc.* 1994, **116:** 7465–7466; (b) Liu, K.E., Valentine, A.M., Qiu, D., Edmondson, D.E., Appelman, E.H., Spiro, T.G. and Lippard, S.J. *J. Am. Chem. Soc.* 1995, **117:** 4997–4998 and corrections. *J. Am. Chem. Soc.* 1997, **119:** 11134.
13. Kim, K. and Lippard, S.J. *J. Am. Chem. Soc.* 1996, **118:** 4914–4915.
14. Shu, L., Nesheim, J.C., Kauffmann, K., Münck, E., Lipscomb, J.D. and Que, L. Jr. *Science* 1997, **275:** 515–518.
15. Bolinger, J.M. Jr., Stubbe, J., Huynh, B.H. and Edmondson, D.E. *J. Am. Chem. Soc.* 1991, **113:** 6289–6291
16. Auling, G. and Follmann, H. in *Metal Ions in Biological Systems* (eds.) Sigel, H. and Sigel, A., Marcel Dekker, Inc., New York, 1994, Vol. 30, p. 131.
17. Que, L. Jr. and True, A.E. in *Progress in Inorganic Chemistry* (ed.) Lippard, S.J. Wiley, New York, 1990, Vol. 38, pp. 97–200
18. Feig, A.L. and Lippard, S.J. *Chem. Rev.* 1994, **94:** 759–805.
19. Caradonna, J.P. and Stassinopoulos, A. *Adv. Inorg. Biochem.* 1994, **9:** 245–315 and references therein.
20. Gif Chemistry: (a) Barton, D.H.R. and Doller, D. *Acc. Chem. Res.* 1992, **25:** 504–512 and references therein; (b) Singh B., Long, J.R., de Biani, F.F., Gatteschi, D. and Stavropoulos, P. *J. Am. Chem. Soc.* 1997, **119:** 7030–7047 and references therein.
21. Oxygenated Fenton Chemistry: (a) Sawyer, D.T., Sobkowiak, A. and Matsushita, T. *Acc. Chem. Res.* 1996, **29:** 409–416 and references therein; (b) Walling, C. *Acc. Chem. Res.* 1998, **31:** 155–157; (c) MacFaul, P.A., Wayner, D.D.M. and Ingold, K.U. *Acc. Chem. Res.* 1998, **31:** 159–162.

22. Leising, R.A., Kim, J., Perez, M.A. and Que, L. Jr. *J. Am. Chem. Soc.* 1993, **115:** 9524–9530.
23. Kojima, T., Leising, R.A., Yan, S. and Que, L. Jr. *J. Am. Chem. Soc.* 1993, **115:** 11328–11335.
24. Leising, R.A., Norman, R.E. and Que, L. Jr. *Inorg. Chem.* 1990, **29:** 2553–2555.
25. Kim, J., Harrison, R.G., Kim, C. and Que, L. Jr. *J. Am. Chem. Soc.* 1996, **118:** 4373–4379.
26. Kitajima, N., Fukui, H. and Moro-oka, Y. *J. Chem. Soc. Chem. Commun.* 1988, 485–486.
27. Nishida, Y., Okuno, T., Ito, S., Harada, A., Ohba, S., Matsushima, H. and Tokii, T. *Chem. Lett.* 1995, 885–886.
28. Ito, S., Okuno, T., Matsushima, H., Tokii, T. and Nishida, Y. *J. Chem. Soc., Dalton Trans.* 1996, 4479–4484.
29. Nishida, Y., Tanaka, N. and Okazaki, M. *Polyhedron* 1994, **13:** 2245–2249.
30. Nishida, Y., Itoh, H. and Yamazaki, A., *Polyhedron* 1994, **13:** 2473–2476.
31. Okuno, T., Ito, S., Ohba, S. and Nishida, Y. *J. Chem. Soc., Dalton Trans.* 1997, **19:** 3547–3551.
32. Ito, S., Okuno, T., Matsushima, H., Tokii, T. and Nishida, Y. *J. Chem. Soc., Dalton Trans.* 1996, 4037–4039.
33. Rardin, R.L., Tolman, W.B. and Lippard, S.J. *New J. Chem.* 1991, **15:** 417–430.
34. Que, L. Jr. *J. Chem. Soc., Dalton. Trans.* 1997, 3933–3940.
35. (a) Dong, Y., Yan, S., Yong, V.G. Jr. and Que, L. Jr. *Angew. Chem., Int. Ed. Engl.* 1996, **35:** 618–620; (b) Ookubo, T., Sugimoto, H., Nagayama, T., Masuda, H., Sato, T., Tanaka, K., Maeda, Y., Okawa, H., Hayashi, Y., Uehara, A. and Suzuki, M. *J. Am. Chem. Soc.* 1996, **118:** 701–702.
36. Duboc-Toia, C., Menage, S., Lambeaux, C. and Fontecave, M. *Tetrahedron Lett.* 1997, **38:** 3727–3730.
37. Fontecave, M., Roy, B. and Lambeaux, C. *J. Chem. Soc., Chem. Commun.* 1991, 939–940.
38. Menage, S., Vincent, J.M., Lambeaux, C., Chottard, G., Grand, A. and Fontecave, M., *Inorg. Chem.* 1993, **32:** 4766–4773.
39. Kitajima, N., Ito, M., Fukui, H. and Moro-oka, Y. *J. Chem. Soc., Chem. Commun.* 1991, 102–104.
40. Armstrong, W.H., Spool, A., Papaefthymiou, G.C., Frankel, R.B. and Lippard, S.J. *J. Am. Chem. Soc.* 1984, **106:** 3653–3667.

41. Stassinopoulos, A. and Caradonna, J.P. *J. Am. Chem. Soc.* 1990, **112:** 7071–7073.
42. Stassinopoulos, A., Schulte, G., Papaefthymiou, G.C. and Caradonna, J.P. *J. Am. Chem. Soc.* 1991, **113:** 8686–8697.
43. Mukerjee, S., Stassinopoulos, A. and Caradonna, J.P. *J. Am. Chem. Soc.* 1997, **119:** 8097–8098.
44. MacFaul, P.A., Ingold, K.U., Wayner, D.D.M. and Que, L. Jr. *J. Am. Chem. Soc.* 1997, **119:** 10594–10598.
45. Vincent, J.B., Huffman, J.C., Christou, G., Li, Q., Nanny, M.A., Hendrickson, D.N., Fong, R.H. and Fish, R.H. *J. Am. Chem. Soc.* 1988, **110:** 6898–6900.
46. Fish, R.H., Konings, M.S., Oberhausen, K.J., Fong, R.H., Yu, W.M., Christou, G., Vincent, J.B., Coggin, D.K. and Buchanan, R.M. *Inorg. Chem.* 1991, **30:** 3002–3006.
47. Fish, R.H., Oberhausen, K.J., Chen, S., Richardson, J.F., Pierce, W. and Buchanan, R.M. *Catal. Lett.* 1993, **18:** 357–365.
48. Buchanan, R.M., Chen, S., Richardson, J.F., Bressan, M., Forti, L., Morvillo, A. and Fish, R.H. *Inorg. Chem.* 1994, **33:** 3208–3209.
49. Rabion, A., Chen, S., Wang, J., Buchanan, R.M., Seris, J.-L. and Fish, R.H. *J. Am. Chem. Soc.* 1995, **117:** 12356–12357.
50. Herold, S., Pence, L.E. and Lippard, S.J. *J. Am. Chem. Soc.* 1995, **117:** 6134–6135.
51. Herold, S. and Lippard, S.J. *J. Am. Chem. Soc.* 1997, **119:** 145–156.
52. Kodera, M., Shimakoshi, H., Nishmura, M., Okawa, H., Iijima, S. and Kano, K. *Inorg. Chem.* 1996, **35:** 4967–4973.
53. Kodera, M., Shimakoshi, H. and Kano, K. *Chem. Commun. (Cambridge)* 1996, 1737–1738.
54. Fish, R.H., Fong, R.H., Vicent, J.B. and Christou, G. *J. Chem. Soc., Chem. Commun.* 1988, 1504–1506.
55. Vincent, J.M., Stephane, M., Lambeaux, C. and Fontecave, M. *Tetrahedron Lett.* 1994, **35:** 6287–6290.
56. Fish, R.H., Fong, R.H., Oberhausen, K.J., Konings, M.S., Vega, M.C., Christou, G., Vincent, J.B. and Buchanan, R.M. *New J. Chem.* 1992, **16:** 727–733.
57. Tetard, D., Rabion, A., Verlhac, J.-B. and Guilhem, J. *J. Chem. Soc., Chem. Commun.* 1995, 531–532.
58. Tetard, D. and Verlhac, J.-B. *J. Mol. Catalysis A: Chemical* 1996, **113:** 223–230.
59. Che, C.-M., Tang, W.-T., Wong, K.-Y., Wong W.-T. and Lai, T.-F. *J. Chem. Research (S)* 1991, 30.

60. Ganeshpure, P.A., Tembe, G.L. and Satish, S. *Tetrahedron Lett.* 1995, **36:** 8861–8864.
61. Caudle, M.T., Riggs-Gelasco, P., Gelasco, A.K., Penner-Hahn, J.E. and Pecoraro, V.L. *Inorg. Chem.* 1996, **35:** 3577–3584.
62. Menage, S., Collomb-Dunand-Sauthier, M.N., Lambeaux, C. and Fontecave, M. *J. Chem. Soc., Chem. Commun.* 1994, 1885–1886.
63. (a) Gorun, S. and Stibrany, R.T. Patent 1993, US 5,183,945; (b) Gorun, S. and Stibrany, R.T. *Patent* 1991, US 5,025,101; (c) Bond, J.E., Gorun, S.M., Stibrany, R.T., Schriver, G.W. and Vanderspurt, T.H. in *The Activation of Dioxygen and Homogeneous Catalytic Oxidation* (eds.) Barton, D.H.R. *et al.*, Plenum Press, New York, 1993, pp. 199–208.
64. (a) Wessel, J. and Crabtree, R.H. *J. Mol. Catal. A: Chemical* 1996, **113:** 13–22; (b) Sarneski, J.E., Michos, D., Thorp, H.H., Didiuk, M., Poon, T., Blewitt, J., Brudvig, G.W. and Crabtree, R.H. *Tetrahedron Lett.* 1991, **32:** 1153–1156; (c) Taft K.L., Kulawiec, R.J., Sarneski, J.E. and Crabtree, R.H. *Tetrahedron Lett.* 1989, **30:** 5689–5692.
65. Hage, R., Iburg, J.E., Kerschner, J., Koek, J.H., Lempers, E.L.M., Martens, R.J., Racherla, U.S., Russell, S.W., Swarthoff, T., van Vliet, M.R.P., Warnaar, J.B., van der Wolf, L. and Krijnen, B. *Nature* 1994, **369:** 637–639.
66. Labinger, J.A. *Fuel Proc. Tech.* 1995, **4:** 325–338.
67. (a) Lin, M., Hogan, T. and Sen, A. *J. Am. Chem. Soc.* 1997, **119:** 6048–6053; (b) Periana, R.A., Taube, D.J., Gamble, S., Taube, H., Satoh, T. and Fujii, H. *Science* 1998, **280:** 560–564.

Chapter 7

Models of Copper Enzymes and Heme-Copper Oxidases

Marie-Aude Kopf & Kenneth D. Karlin

Department of Chemistry,
The Johns Hopkins University, Baltimore,
MD 21218, USA

1. Introduction: Copper Proteins and Copper-Dioxygen Complexes

It is of continuing interest to develop reagents or catalysts for oxidative transformations, which are based on copper coordination complexes and biomimetic processes. Thus, it is the purpose of this chapter to highlight recent advances in (i) our understanding of the nature and function of copper proteins which effect oxidation or oxygenation reactions using O_2, (ii) the coordination chemistry of copper-dioxygen adducts, particularly structures and physical properties, (iii) biomimetic chemistry, with examples involving O_2 reactions with copper complexes leading to substrate hydroxylation (monooxygenase models), or alcohol oxidations (galactose oxidase models), and (iv) model compounds and chemistry relevant to the porphyrin iron/copper binuclear center in heme-copper oxidases.

1.1 Copper in biology

Copper ion is an essential trace element found in living systems, and its importance resides in its role as a protein or enzyme active site constituent.[1–11] Recently described chaperone proteins[12,13] and metallothioneins[2,11] may aid the cell trafficking of copper ion homeostasis, but the major role of copper proteins involve oxidation-reduction (i.e. "redox") activity. Ligand donors for copper which are typically available in protein matrices include the side chain imidazole group of histidine, the phenol oxygen donor of tyrosine, or the sulfur atom of the thiol group from cysteine. With these ligands, the Cu(I) and Cu(II) oxidation states are readily accessible and interconvertible under physiological conditions, using available oxidants (e.g. molecular oxygen [dioxygen; O_2]) or reducing agents such as glutathione (γ-Glu-Cys-Gly) or ascorbic acid (vitamin C).

Protein active site copper ions perform the functions of electron transfer (e.g. as electron carriers in photosynthetic organisms or in respiratory pathways in certain bacteria), reversible O_2-binding and transport, mono- or dioxygenation of organic substrates (*oxygenases*, incorporating one or both atoms of O_2), or oxidation/dehydrogenation of substrates accompanied by O_2 reduction to either hydrogen peroxide or water (*oxidases*). Comparable functions carried out by heme and non-heme iron enzymes are described in other chapters in this volume. Copper-mediated oxidative processes "gone wrong" have been discussed,[14–17] including in possible connection with aging, "Lou Gehrig's disease" (Cu/Zn superoxide dismutase mutation),[18] Alzheimer's disease,[19] and others.[20] Separate copper proteins participate in the biological nitrogen cycle, catalyzing the reduction of nitrite (e.g. $NO_2^- \rightarrow NO$)[21] and nitrous oxide (e.g. $N_2O \rightarrow N_2$).[22]

1.2 Copper proteins in dioxygen (O_2) processing

The major classes of copper proteins involved in O_2 processing are given in Table 1.

Hemocyanin and copper monooxygenases Dioxygen-transporting hemocyanins occur in the hemolymph of mollusks and arthropods. Dioxygen-binding occurs with a $Cu^I/O_2 = 2:1$ reaction stoichiometry, formally an oxidative addition reaction, with ligation in the "oxy" form occurring in a side-on μ-η^2:η^2-bridging peroxo (O_2^{2-}) fashion, as indicated in Scheme 1 and Fig. 7.1.[23,24]

Table 1 Copper proteins involved in dioxygen (O_2) processing.

Protein	Source	Biological Function	Refs.
Oxygen Carrier			
Hemocyanin*	molluscs and arthropods	O_2 transport	1, 23, 24
Copper Monooxygenases			1.25
Tyrosinase	fungal, mammal	tyrosine oxidation	
Dopamine β–monooxygenase (DβM)	adrenal, brain	dopamine → norepinephrine	25
Peptidylglycine α–amidating monooxygenase (PAM)*	pituitary, heart	oxidative N-dealkylation	25.26
Methane monooxygenase (MMO)	methanogenic bacteria	methane → methanol	1
Ammonia monooxygenase (AMO)	nitrifying bacteria	ammonia → hydroxylamine	
Copper Dioxygenases			
Quercetinase	fungal	quercetin oxidative cleavage	27
Copper Oxidases			1
"Blue" multi-copper oxidases (converting O_2 to $2H_2O$)			1
Laccase*	tree, fungal	phenol and diamine oxidation	28
Ascorbate oxidase*	plants	oxidation of L-ascorbate	29
Ceruloplasmin*	human, animal serum	Fe(II) → Fe(III) and/or radical scavenging	

Table 1 *(Continued)*

Protein	Source	Biological Function	Refs.
"Non-blue" oxidases (converting O2 to 2H2O2)			25
Amine oxidase*	most animals	elastin, collagen formation and alcohol oxidative deamination	
Galactose oxidase*	fungi	galactose oxidation alcohol → aldehyde	30
Glyoxal oxidase	fungi	aldehyde → carboxylic acid	
Phenoxazinone synthase	*Streptomyces*	phenoxazinone formation and oxidative coupling	1
Heme-copper oxidases	ubiquitous, aerobic organisms	terminal oxidases	31
Cytochrome c oxidase (CcO)*	mitochondria, eukaryotes	proton translocation (pump)	
Quinol oxidase	prokaryotes		
Others			
Superoxide dismutase (SOD)*	red blood cells, animals	O_2^- dismutation (detoxification)	32, 33

* X-ray structure(s) available.

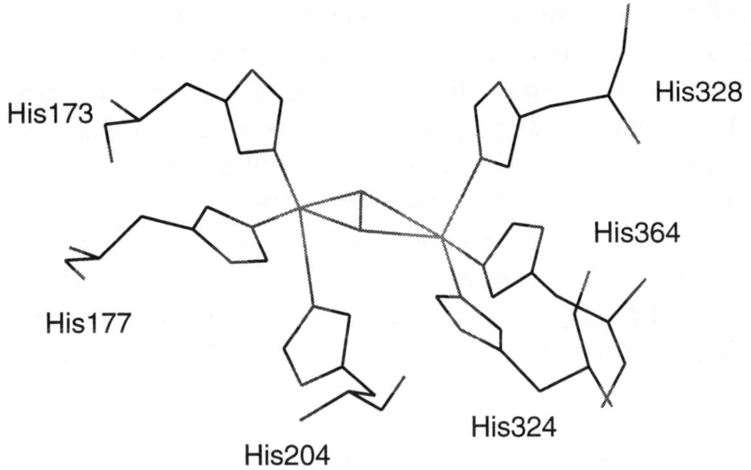

Fig. 7.1 Active site structure of oxy-hemocyanin, generated with RASMOL using X-ray coordinates, Protein Data Bank (PDB) code 1OXY.

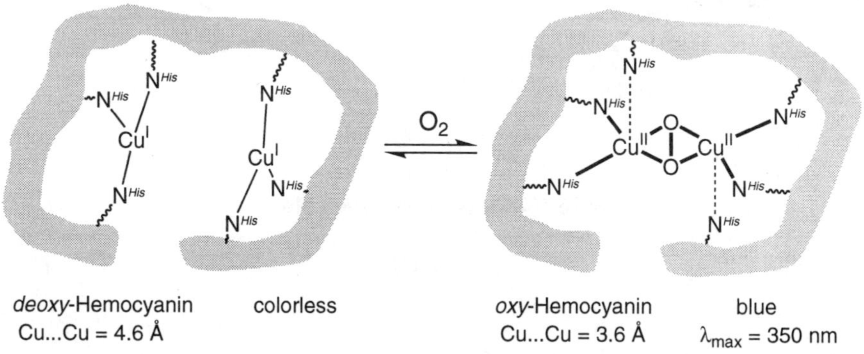

Scheme 1

An X-ray structure is not available for the monooxygenase tyrosinase, but spectroscopic and biochemical insights reveal its active site to be very similar to the binuclear site in hemocyanins. The dioxygen-adduct intermediate responsible for *ortho*-oxygenation of phenols is thus nearly identical to oxy-hemocyanin. A major difference in proteins is that the protected buried active site in oxy-hemocyanin is instead accessible to phenolic substrates in tyrosinase. Detailed insight into how the peroxo moiety transfers to phenol substrate still needs to be resolved; recent experiments on model systems[34] and the protein[35,36] suggest that tyrosinase effects a *direct* phenol to quinone conversion, rather than phenol to *o*-catechol transformation. This is further discussed below.

New insights as well as new questions have recently arisen in connection with the active site structures and chemistry in dopamine β-monooxygenase (DβM) (benzylic hydroxylation; dopamine → norepinephrine) and the related peptidylglycine α-amidating monooxygenase (PAM), Scheme 2.

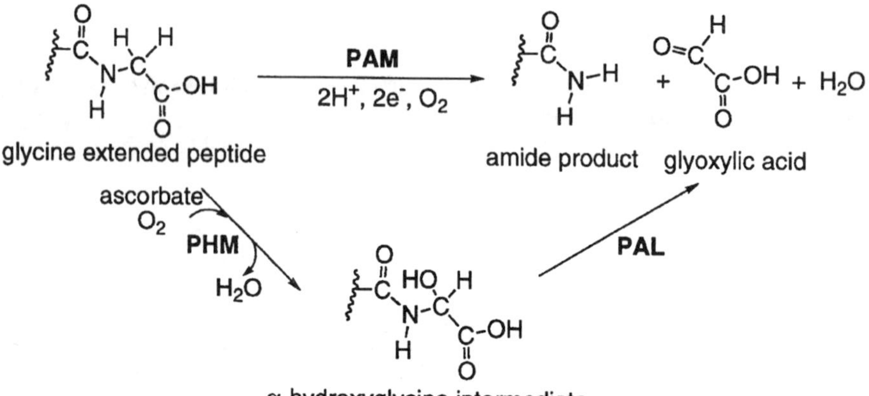

Scheme 2

A new X-ray structure (Fig. 7.2)[26] of the hydroxylating PHM domain of PAM confirms basic information derived from other spectroscopic probing, including the presence of inequivalent copper ions;[37] one (Cu_B) having a sulfur methionine ligand and a substantial Cu...Cu distance (~ 11 Å), thus confirming previous evidence which precludes electronic coupling between Cu(II) ions (when oxidized). A proposed mechanism,[26] which draws on previous mechanistic studies carried out on DβM,[25] involves formation of a Cu(II)-OOH (hydroperoxo) species at Cu_B, with the electrons

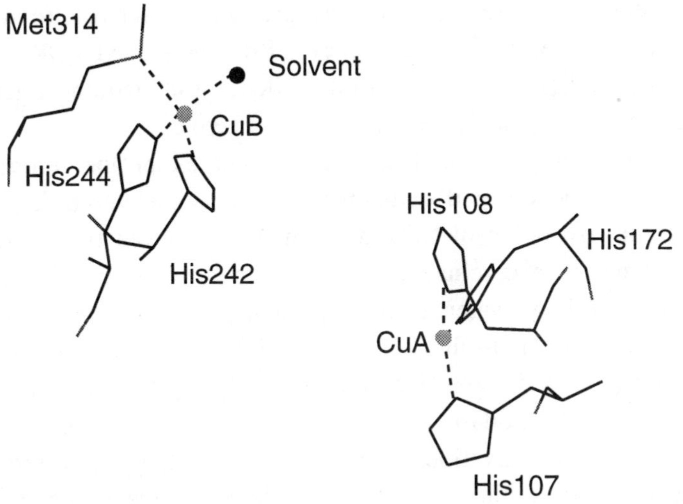

Fig. 7.2 Active site structure showing the ligation in PHM, generated with RASMOL using X-ray coordinates, PDB 1PHM.

drawn from the two Cu(I) ions in the active reduced state. A heterolytic cleavage of the O–O bond, coupled with abstraction of the pro S substrate hydrogen, could lead to a substrate radical, Cu-O• radical, plus water. Recombination of the radicals would lead to product. Other mechanisms should be considered, and among questions to consider for this mechanism are the nature of the electron transfer between copper ions, the identity of the proton source for Cu(II)-OOH (possibly from the substrate),[26] and whether a nearby tyrosine residue may assist O–O cleavage, as previously suggested for DβM.[25]

Less is known about the particulate, membrane-bound copper dependent methane monooxygenase. This exhibits a broad specificity, yet both regio- and stereoselectivity of reaction are observed.[38] Chan and co-workers[33] suggested that the enzyme may contain several types of copper active site clusters, including trinuclear centers, and recent emphasis has been given to methane C–H activation by high-valent μ-oxo dicopper species.[38] On the other hand, DiSpirito and co-workers[39] described a spectroscopically detectable required non-heme iron, and propose an active site with both iron and copper. Recent low frequency EPR[40] or pulsed EPR/ENDOR[41] spectroscopic studies provide the first evidence for histidine ligation, for at least some of the Cu(II) centers in this enzyme.

Copper oxidases As a class, these effect either the multiple one- (e.g. ascorbate, phenols and cytochrome *c*) or two-electron (e.g. amines and galactose) oxidation of substrates, mediated by reaction of dioxygen at single copper ions or clusters. In "blue" multi-copper oxidases, O_2 is reduced by $4e^-/4H^+$ to water, while hydrogen peroxide is the two-electron reduced product in "non-blue" oxidases (Table 1). While oxygen atom incorporation from dioxygen does not occur, O–O cleavage reactions do, thus mechanisms of oxidase processes are of interest within the general context of dioxygen activation by copper ion. Readers are referred to the references given in Table 1 for recent reviews pertinent to multi-copper oxidases, with their highly interesting trinuclear copper cluster, the site of O_2-binding and reduction. An adjacent type 1 "blue" copper ion shuttles electrons from the substrate (e.g. ascorbate, phenols, Fe(II)) to the tricopper center. X-ray structures are available on ascorbate oxidase,[28] ceruloplasmin,[42] and more recently on a fungal laccase[43] which is missing a type 2 Cu ion (part of the trinuclear cluster). Biochemical/spectroscopic studies from Solomon and co-workers[1] suggested that the four-electron reduction of O_2 in laccase may proceed via two two-electron steps, the first O_2-derived reaction intermediate being a hydroperoxo (i.e. ^-OOH)-bridged mixed-valent dicopper species.

Amine oxidases catalyze reactions involving functions such as the crosslinking of collagen and elastin, and the regulation of blood plasma biogenic amines.[25] A carbonyl-containing cofactor has been long known to be present, identified by Klinman and co-workers as topaquinone (TPQ) (**B**). Several protein X-ray crystal structures[44–47] confirm this conclusion, while also verifying the spectroscopically deduced coordination environment (**A**), with three histidine ligands. The TPQ lies close to the copper ion, but is not coordinated, and it is deduced that this group is conformationally flexible. The enzyme reaction involves initial Schiff-base formation of amine substrate with quinone.[24] The role of copper is to facilitate reoxidation by

O_2 of the reduced cofactor (after the substrate reactions), a Cu(I)-semiquinone intermediate (in equilibrium with Cu(II)-quinone) has been detected and is presumed to interact directly with O_2.[25] Another aspect of considerable interest is the self-processing of TPQ, known to be derived from a protein tyrosine precursor. The copper ion is implicated in this cofactor biogenesis and mechanistic schemes involving copper-dioxygen chemistry have been presented.[25,46,48]

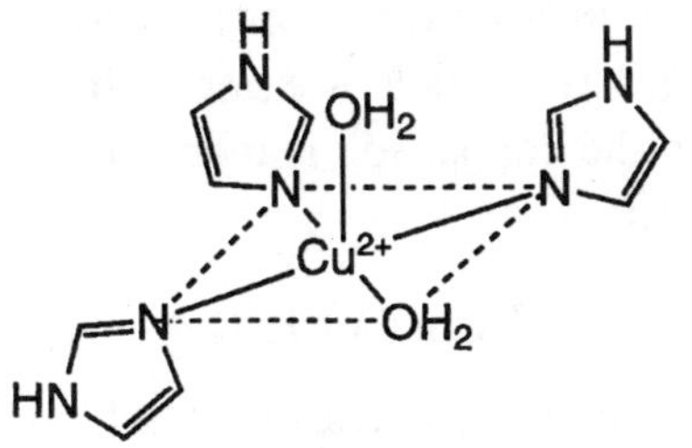

A

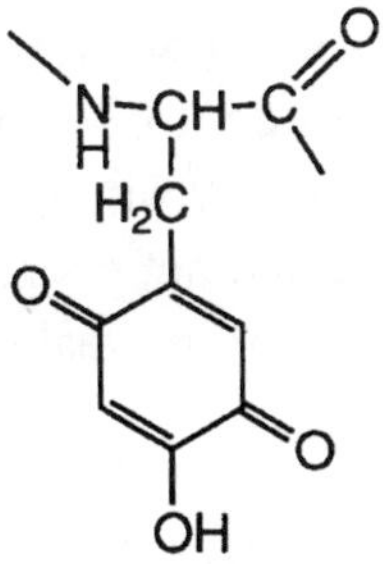

B

Galactose oxidase is an extracellular fungal enzyme, catalyzing the stereospecific two-electron oxidation of D-galactose and other primary alcohols.[49]

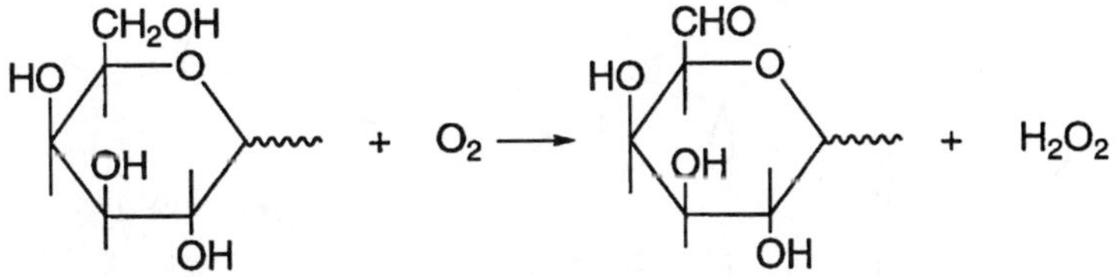

Prior to any knowledge from a protein X-ray structure, Whittaker and co-workers[49–51] deduced that a stabilized ligand-protein radical-cation was involved, which explained a long-standing puzzle of how a single active site copper ion could effect a two-electron process. The X-ray structure[30]

on the protein revealed a novel protein cofactor consisting of a thioether bond linking a cysteine to the *ortho*-position of a copper-coordinated tyrosine ligand, with the latter also undergoing a π-stacking interaction with a tryptophan residue, Fig. 7.3.

The reaction to oxidize alcohols can be thought of as proceeding by oxidation of a reduced protein and copper(I) center by dioxygen, in a two-electron process, to give oxidized copper(II) and protein-ligand radical cation $P^{+\bullet}$. The "stored" oxidizing equivalents can then oxidize a coordinated alcohol substrate, producing aldehyde and restoring the reduced enzyme center. Proton transfers from substrate to dioxygen (as product peroxide) are suggested to be mediated via protonation-deprotonation of the coordinated tyrosinate/tyrosine ligands, which are more likely to be protonated and not coordinated when the copper ion is reduced, or possibly because one is a weaker axial ligand.

$$(P)Cu(I) + O_2 + 2H^+ \rightarrow (P^{+\bullet})Cu(II) + H_2O_2$$
$$RCH_2OH + (P^{+\bullet})Cu(II) \rightarrow RC(O)H + (P)Cu(I) + 2H^+$$

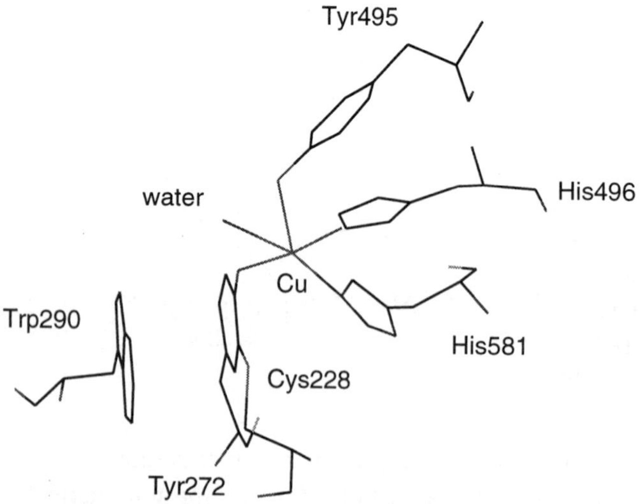

Fig. 7.3 Active site structure showing the ligation in galactose oxidase at pH 7, generated with RASMOL, using X-ray coordinates, PDB 1GOG.

Heme-copper oxidases will be discussed below, in connection with model studies.

2. Copper-Dioxygen Complexes and Monooxygenase/Oxidase Models

2.1 Discrete copper-dioxygen complexes

There are four principal binding modes previously established for the interaction of transition metal complexes with dioxygen (Chart 1), and all are suggested or proven for copper. These include end-on binding, as found in hemoglobin, and side-on binding, as in hemocyanin. Both may be combined with Cu^I/O_2 = 1:1 and Cu^I/O_2 = 2:1 stoichiometry, leading to a superoxo-copper(II) and peroxo-dicopper(II) products, respectively. In addition, in copper-dioxygen chemistry, a Cu^I/O_2 = 2:1 or Cu^I/O_2 = 3:1 stoichiometry can lead to complexes with a bis-μ-oxo dicopper or tricopper core containing copper(III), respectively, as described below.

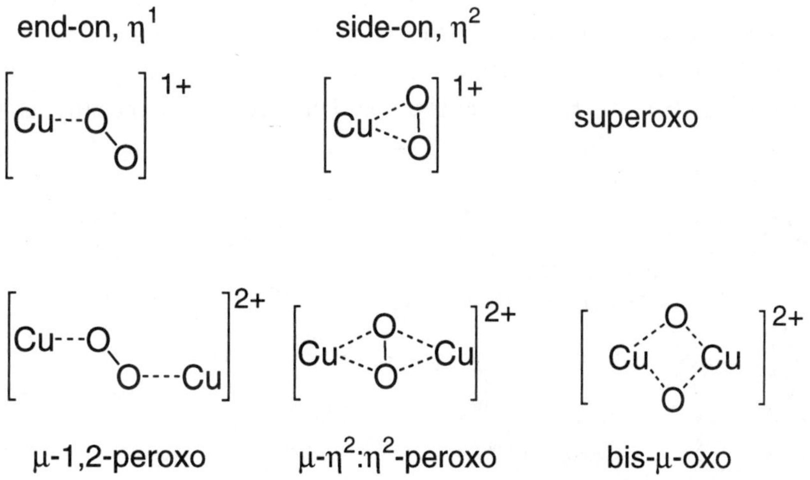

Chart 1

Summaries of recent advances and characterizations of copper-dioxygen complexes have been presented.[5,7,8,52] Here, we emphasize those characterized by X-ray diffraction, Fig. 7.4 and recent advances. The first X-ray crystallographic description for a copper-dioxygen adduct, synthetic or protein, came with our characterization of the 2:1 adduct $[\{(TMPA)Cu^{II}\}_2(O_2^{2-})]^{2+}$ (**1**), with PF_6^- or ClO_4^- counter-anions, formed by reversible oxygenation of copper(I) precursor complex $[(TMPA)Cu^I(RCN)]^+$ (**1a**, R = Me, Et, TMPA = tris[2-pyridylmethyl] amine).[53,54] The dication **1** is best described as a peroxo-dicopper(II) species. It is essentially diamagnetic, possessing a trans-μ-1,2-$Cu_2O_2^{2+}$ moiety with Cu...Cu = 4.36 Å and O–O = 1.432 Å. Other relevant physical properties are υ_{O-O} = 831 cm^{-1} (resonance Raman),[55] and UV-vis (i.e. peroxo-to-Cu(II) LMCT) bands at 440 nm (ε = 2000 $M^{-1}cm^{-1}$), λ_{max} = 525 nm (11 500), ~ 600 nm (sh, ~ 7600) and a *d-d* band at 1035 nm (180). Kinetics studies[56] reveal that $[\{(TMPA)Cu^I(RCN)]^+$ reacts with O_2 to give a transient (at –80 °C, λ_{max} = 410 nm) 1:1 superoxo-copper(II) adduct, $[\{(TMPA)Cu(O_2)]^+$ (**1b**), with suggested end-on η^1 coordination (Chart 1). The full kinetics and thermodynamics of these reactions, and those with complexes with ligands TMPAE, BPQA, BQPA, plus binucleating analogues D^1 and DO (Chart 2), have been elucidated and described.[56–59] We have shown that ligand DO confers room temperature stability to the resulting dioxygen adduct $[\{(DO)Cu_2(O_2)]^{2+}$.[57] Reedijk and co-workers were also recently able to

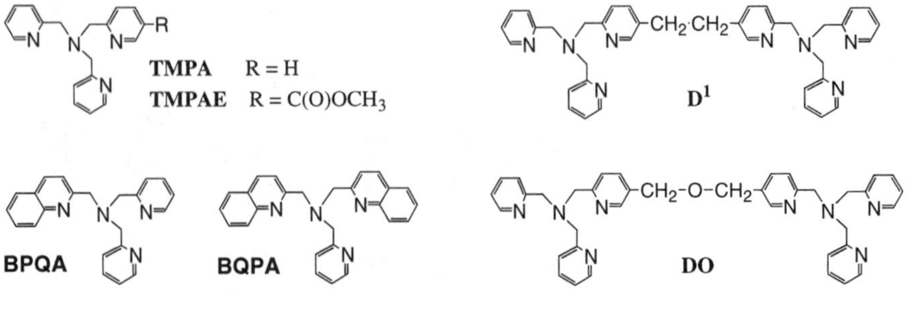

Chart 2

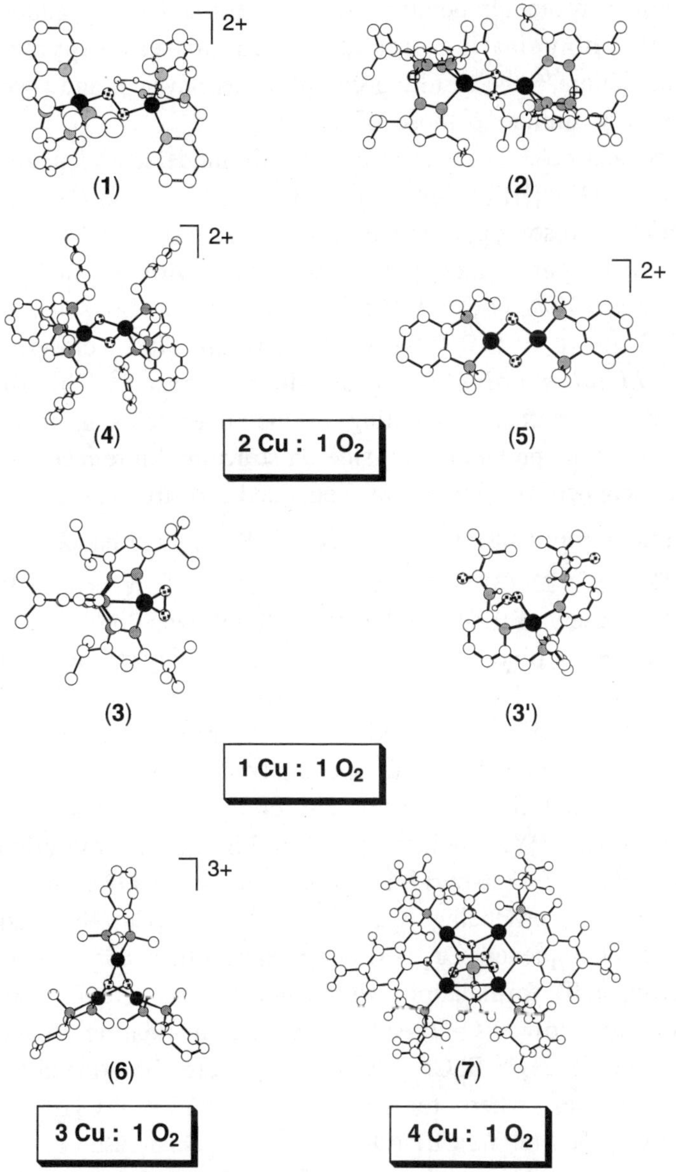

Fig. 7.4 Chem 3-D X-ray structure drawings of complexes of copper bound to O_2 and its derivatives.

obtain similar room temperature stability for a Cu_2-O_2 adduct with a macrocyclic binucleating ligand, which possesses pyrazolyl and alkylamine donors;[60] the resulting complex has physical and spectroscopic properties very similar to those of **1**.

Kitajima and co-workers used a sterically hindered tridentate ligand in characterizing $\{Cu^{II}[HB(3,5\text{-}iPr_2pz)_3]\}_2(O_2^{2-})$ (**2**, Fig. 7.4) ($HB(3,5\text{-}iPr_2pz)_3$ = hydrotris(3,5-diisopropyl-pyrazolyl)borate anion), which has a side-on ligated μ-η^2:η^2 peroxodicopper(II) structure, with physical properties (i.e. Cu···Cu = 3.56 Å, O-O = 1.412 Å, $\upsilon_{O\text{-}O}$ = 741 cm^{-1} (resonance Raman), λ_{max} = 349 nm (ε = 21,000 $M^{-1}cm^{-1}$), 551 nm (790)) closely matching those of *Limulus polyphemus* oxy-hemocyanin.[52] This important contribution to synthetic modeling of metalloproteins in fact preceded confirmation of the protein Cu_2O_2 side-on structure. More recently,[61] a side-on superoxocopper(II) complex was generated and structurally characterized by the same group; $Cu^{II}(O_2^-)(HB(3\text{-}tBu\text{-}5\text{-}iPrpz)_3)$ (**3**) has λ_{max} = 352 nm (2,330), O-O = 1.22 (3) Å and $\upsilon_{O\text{-}O}$ = 1111 cm^{-1} (resonance Raman).

In fact, a variety of tridentate ligands with nitrogen donors give copper(I) complexes which bind O_2 to give a side-on μ-η^2:η^2-bridging peroxo dicopper(II) product having the characteristic intense UV (345–365 nm) charge-transfer band.[52,62–69] Based on low temperature solution Extended X-ray Absorption Fine Structure spectroscopic (EXAFS) investigation,[67] we first proposed this for binuclear complexes containing $-(CH_2)_n$ (n = 3–5)-linked PY2 chelates (PY2 = bis[2-(2-pyridyl)ethyl]amine), i.e. ligand Nn (PY = 2-pyridyl). Dioxygen (and CO) reacts reversibly with $[Cu^I_2(Nn)]^{2+}$, producing $[Cu_2(Nn)(O_2)]^{2+}$ (or $[Cu_2(Nn)(CO)_2]^{2+}$).[68,69] The dioxygen complexes $[Cu_2(Nn)(O_2)]^{2+}$ have a bent side-on structure and peroxodicopper(II) formulation, with characteristic intense 350–365 nm absorption, and low ν(O-O) stretch.[70] Mononucleating ligands RPY2 (R = Me, Ph, $PhCH_2$;[62] $PhCH_2CH_2$ [71–73]) can also be utilized to generate similar complexes, where for example $[(MePY2)Cu^I(CH_3CN)]^+$ reacts with O_2 giving $[\{(MePY2)Cu\}_2(O_2)]^{2+}$.[62,73] A special case of these ligands and their chemistry occurs when the linker is a *m*-xylyl moiety, leading to a monooxygenase model system discussed below.

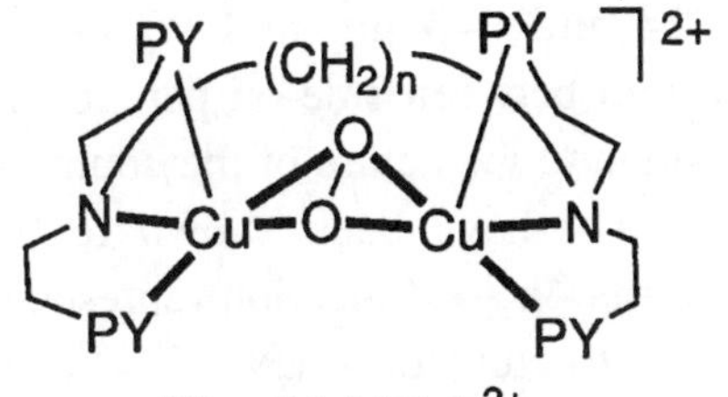

$[Cu_2(Nn)(O_2)]^{2+}$
λ_{max} = 350–365 nm
ν(O-O) = 740–765 cm^{-1}
Cu...Cu = 3.3-3.4 Å (EXAFS)

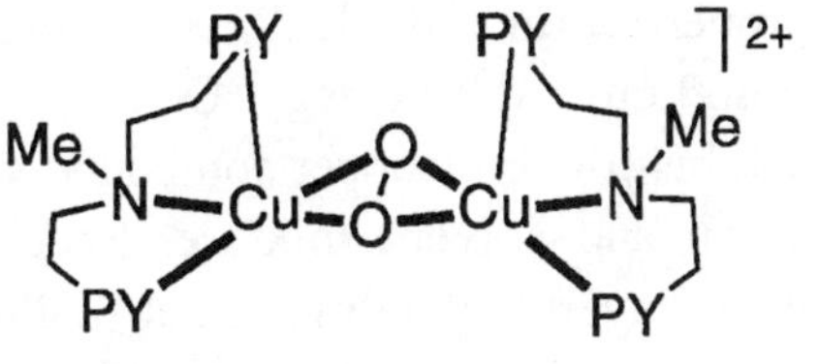

$[\{(MePY2)Cu\}_2(O_2)]^{2+}$
λ_{max} = 360 nm
ν(O-O) = 730 cm^{-1}

Others have shown that μ-η^2:η^2-peroxo-dicopper(II) species also form with tris(1-R-4-R′-imidazolyl)phosphines[64,65,74] or trisubstituted triazacyclononane (TACN) ligands. For the latter case, Tolman and co-workers[66,75] discovered very novel and important new chemistry. Oxygenation of **4a** can give both Cu/O_2 = 2:1 adducts **4b** or **4** (Scheme 3 and Fig. 7.4). For R = *i*Pr in CH_2Cl_2 solvent at low temperatures, **4b** (λ_{max} = 366 nm, ν(O-O) = 722 cm^{-1}) is formed reversibly. When the oxygenation is carried out in tetrahydrofuran solvent, **4** is obtained. Its

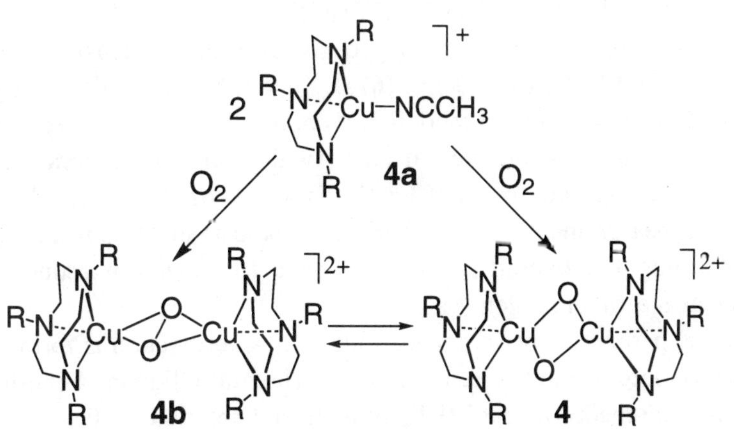

Scheme 3

properties include UV-vis features at 324 and 448 nm and $\nu(Cu_2O_2)$ = 600 cm^{-1} (580 using $^{18}O_2$). The equilibrium between side-on peroxo **4b** and bis-μ-oxo dicopper compound **4** is sensitive to the nature of the counter-anion and solvent mixture. Species **4a** and **4** have been shown to be interconvertible, an important finding, since the **4b** → **4** reaction represents a unique O–O bond forming reaction. The structure of **4** (Fig. 7.4, R = benzyl) reveals O...O = 2.287 Å, with a short value of Cu...Cu = 2.794 Å. Theoretical arguments suggest a $[Cu^{III}_2(\mu\text{-}O^{2-})_2]^{2+}$ assignment for **4**.

Stack and co-workers[76] also demonstrated O–O bond breaking reactions in copper(I)-dioxygen reactivity studies utilizing ligands having the simple ethylenediamine moiety. Three-coordinate copper(I) complexes, $[LCu(CH_3CN)]^+$ (**5a**, L = (1*R*,2*R*)-cyclohexanediamine), react with dioxygen at −80 °C in CH_2Cl_2 to afford diamagnetic species $[(LCu)_2(\mu\text{-}O)_2]^{2+}$ (**5**). An X-ray structure of one derivative (Fig. 7.4; R_1 = Me, R_2 = Et) reveals a square-planar N_2O_2 core surrounding copper, with very short Cu–O bonds (Average = 1.81 Å), Cu...Cu = 2.743 (1) Å, and O...O = 2.344(1) Å. The complexes are best described as bis(μ-oxo)dicopper(III) complexes, with Cu absorption K-edge data corroborating the formal +3 metal oxidation state.

A new Cu^I/O_2 reaction stoichiometry was also discovered by this same research group.[77] $[(TMCHD)Cu^I(CH_3CN)]^+$ (TMCHD = Me_4(1*R*,2*R*)-cyclohexanediamine) reacts with O_2 to give a trinuclear cluster compound $[(TMCHD)_3Cu^{II}_2Cu^{III}(\mu_3\text{-}O^{2-})_2)]^{3+}$ (**6**) (λ_{max} = 355 nm), with two μ_3-oxide ligands (O...O = 2.37 Å) and three copper ions. An X-ray structure (Fig. 7.4) reveals one unique Cu atom, with short Cu–O bond distances (~ 1.83 Å, compared to the Cu^{II}-O distances of ~ 2.00_{ave} Å). X-ray absorption measurements and comparisons[78] indicate that this is a copper(III) ion, affirming the overall reaction to be a four-electron reduction and O–O cleavage of dioxygen.

A novel peroxo-copper complex with μ_4-η^1-coordination is formed when copper(II) perchlorate is reacted with tridentate ligand 4-methyl-2,6-bis(pyrrolidinomethyl)phenol (HL) in methanol solution, in the presence of air, triethylamine base and 3,5-di-*tert*-butylcatechol.[79] The four copper(II) ions in $[Cu_4L_2(O_2)(OMe)_2(ClO_4)]^+$ (**7**, Fig. 7.4) form a near-planar rectangle, with alternate methoxo (Cu...Cu = 3.03 Å) and phenolato (Cu...Cu = 2.99 Å) bridging. Weaker axial ligation is completed by bridging perchlorato oxygen atoms. The peroxide ligand has a typical O–O bond length of 1.453(4) Å. Neither Cu^{I}_{n}/O_2 chemistry nor the way this complex is formed have been discussed yet.

A mononuclear Cu^{II}-OOH complex was recently characterized, structure **3'** (Fig. 7.4), possessing a ligand similar to TMPA, bis(6-pivalamide-2-pyridylmethyl)-(2-pyridylmethyl)amine. The compound forms from reaction of H_2O_2 with a copper(II) complex with this ligand; N-H hydrogen bonding from an amide ligand group to the hydroperoxo moiety provides stabilization.[80]

2.2 A monooxygenase model involving arene hydroxylation

Closely related to binuclear copper complexes $[Cu^{I}_{2}(Nn)]^{2+}$ and dioxygen adducts $[Cu_2(Nn)(O_2)]^{2+}$ mentioned above, are $[Cu^{I}_{2}(R\text{-}XYL\text{-}H))]^{2+}$

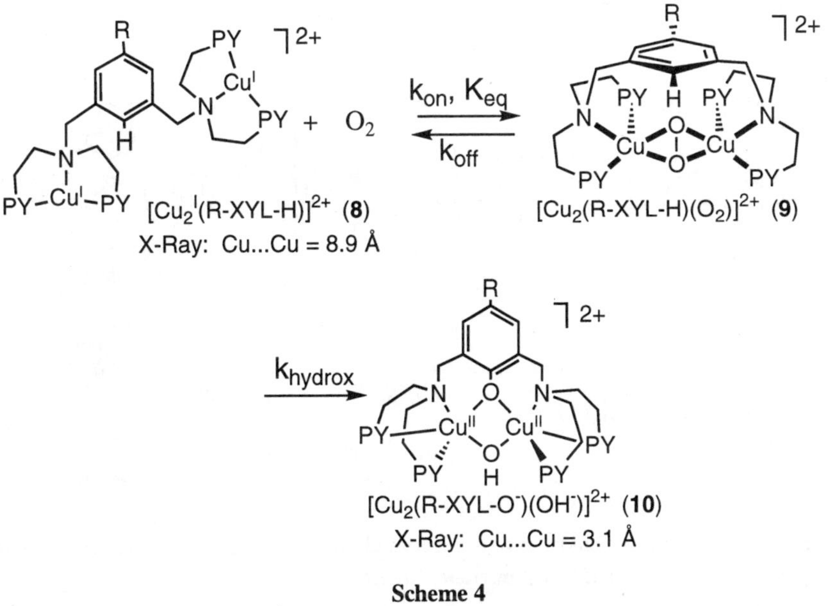

Scheme 4

(e.g. **8**, R = H), which undergo reactions with O_2 causing hydroxylation in the 2-position of the arene (*m*-xylyl) spacer, giving products $[Cu^{II}_2(R\text{-}XYL\text{-}O^-)(OH^-)]^{2+}$ (**10**), Scheme 4.[4,7,81–83] This system, proposed as a model for tyrosinase (having a dicopper active site), represents one of very few examples in the literature wherein dioxygen can be employed to effect a hydrocarbon oxygenation, under extremely mild conditions, i.e. < 1 atm O_2 pressure and at room temperature or below.[6] Other kinds of dicopper(I)/O_2 xylyl hydroxylations have subsequently been observed.[5,7,8,52,84–86]

The detailed proposed mechanism, Fig. 7.5, involves formation of a Cu_2-O_2 adduct $[Cu_2(R\text{-}XYL\text{-}O^-)(O_2)]^{2+}$ (**9**), followed by electrophilic attack of this side-on bound peroxo moiety, present in favorable proximity to the π-system of the arene substrate.[4,10,82] A summary of the data and observations pointing to this mechanism are:

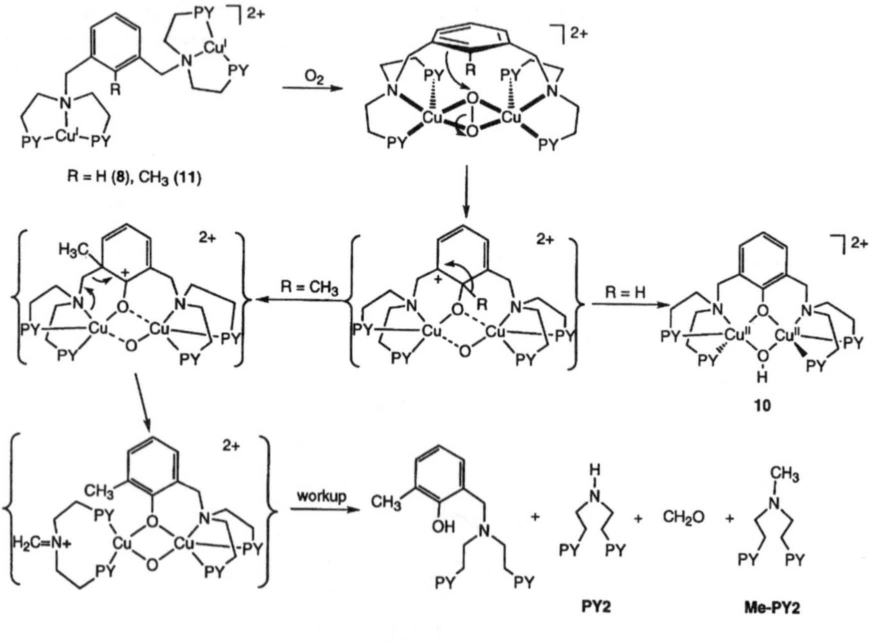

Fig. 7.5 A monooxygenase model system. Proposed mechanism of copper-mediated arene hydroxylation and "N.I.H. shift" (1,2-migration) reactions.

(1) The reaction of O_2 with $[Cu^I_2(XYL\text{-}H))]^{2+}$ (**8**) giving $[Cu^{II}_2(XYL\text{-}O^-)(OH^-)]^{2+}$ (**10**) (Scheme 4) proceeds quantitatively in solvents such as dichloromethane and dimethylformamide. Isotope labeling experiments using $^{18}O_2$, plus the stoichiometry of reaction ($Cu/O_2 = 2{:}1$; manometry), confirm incorporation of oxygen atoms, and the monooxygenase-like reaction, with one atom of O_2 inserting into the xylyl C–H bond, and the other ending up as coordinated hydroxide. The free product phenol is released by extraction of **10** using aqueous ammonia.

(2) Low temperature kinetic studies carried out on oxygenation reactions of $[Cu^I_2(R\text{-}XYL\text{-}H))]^{2+}$ (**8**); R = NO_2, H, $C(CH_3)_3$, F) reveal the kinetics-thermodynamics of formation of $Cu_2\text{-}O_2$ intermediates $[Cu_2(R\text{-}XYL\text{-}H)(O_2)]^{2+}$ (**9**); the subsequent first order irreversible step (k_{hydrox}) describes the oxygen atom insertion, producing $[Cu^{II}_2(R\text{-}XYL\text{-}O^-)(OH^-)]^{2+}$ (**10**).[58,83] The close similarity of the UV-vis spectra of **9** compared to $[Cu_2(Nn)(O_2)]^{2+}$ (*vide supra*), and recent resonance Raman spectroscopic data[87] obtained for $[Cu_2(NO_2\text{-}XYL\text{-}H)(O_2)]^{2+}$ point to these intermediates as having the $\mu\text{-}\eta^2\text{-}\eta^2$-peroxo-dicopper(II) formulation.

(3) A variety of information points to electrophilic attack on the π-system of the xylyl ligand by the side-on peroxo intermediates $[Cu_2(R\text{-}XYL\text{-}H)(O_2)]^{2+}$ (**9**). When a deuterium atom is placed in the 2-position (site of hydroxylation), k_{hydrox} (Scheme 4) for the hydroxylation of $[Cu_2(XYL\text{-}D)]^{2+}$ is within experimental error of that seen for the -H parent compound $Cu^I_2(XYL\text{-}H))]^{2+}$ (**8**), precluding C–H bond cleavage in the rate-determining step. We also find an increase in $\Delta H^{\neq}$ associated with k_{hydrox} with electron-withdrawing ability of R, along the series $[Cu_2(R\text{-}XYL\text{-}H)(O_2)]^{2+}$ (**9**, R = *t*Bu, F, H and NO_2).[83]

(4) Reactivity comparisons of three classes of peroxo-dicopper(II) complexes, $[Cu_2(N4)(O_2)]^{2+}$ (see above), $[\{(TMPA)Cu\}_2(O_2)]^{2+}$ (**1**), and $[Cu_2(XYL\text{-}O\text{-})(O_2)]^{+}$ (see diagram, below),[88] also point to a non-basic electrophilic relative behavior in reactions of the side-on peroxo group in the N4 complex, and thus by analogy $[Cu_2(R\text{-}XYL\text{-}H)(O_2)]^{2+}$ (**9**).[89] In reactions with H^+, CO_2 and PPh_3, $[Cu_2(N4)(O_2)]^{2+}$ does not readily protonate, it is unreactive toward CO_2 and it oxygenates PPh_3, while the other peroxo compounds with end-on coordination readily give H_2O_2 or carbonates or liberate O_2 (respectively) in reactions with the same reagents.

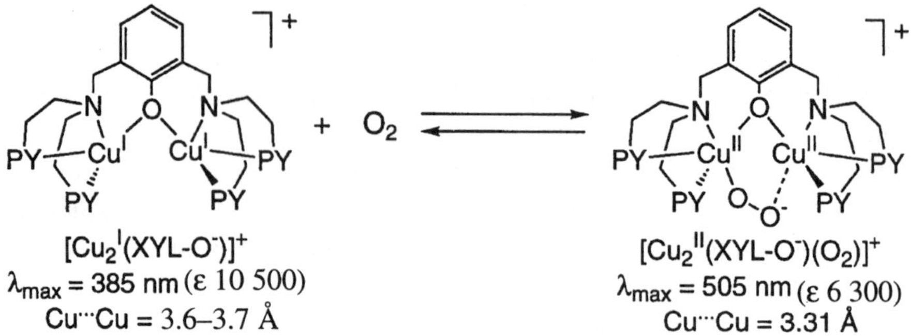

(5) An important insight into the *m*-xylyl hydroxylation mechanism comes from experiments in which a methyl group is placed into the 2-position of the ligand, complex $[Cu^{I}_{2}(H\text{-}XYL\text{-}CH_3)]^{2+}$ (**11**), Fig. 7.5.[82] Instead of causing benzylic hydroxylation or blocking ring attack, 2-hydroxylation occurs and the methyl group undergoes a 1,2-migration. Thus, when **11** is reacted with dioxygen in CH_2Cl_2 and the resulting solution is worked up for its organic products, a phenol, **MePY2**, **PY2** and formaldehyde are isolated or detected in excellent yield and with good material balance (Fig. 7.5). An isotope labeling experiment also established that the source of oxygen in the phenol product was O_2. The process observed is reminiscent of the "N.I.H shift", observed previously in iron hydroxylases, where electrophilic attack of a presumed metal-oxo species results in hydroxylation-induced migrations, due to the formation of carbonium ion intermediates.[7,8,10,82]

To summarize the complete proposed mechanism of action (Fig. 7.5), dicopper(I) complexes **8** or **11** first react with O_2 to give dioxygen-complex intermediates with μ-η^2-η^2-peroxo structures capable of attacking substrate as electrophiles. The xylyl ligand substrate is located in a favorable proximity to the peroxide group in $[Cu_2(R\text{-}XYL\text{-}H)(O_2)]^{2+}$ (**9**), and set up for *p*-π arene orbital attack.[87] This peroxo attack by the Cu_2O_2 unit generates a cationic intermediate, the methyl group (in **11**) undergoes a 1,2-migration, and re-aromatization with "assistance" of the amine nitrogen lone pair causes loss of an iminium ion in a retro-Mannich reaction. Under the experimental conditions employed, hydrolysis produces **PY2** and the formaldehyde observed. This mechanism can also be applied to the parent complex $Cu_2(XYL\text{-}H)]^{2+}$ (**8**) where the 2-H atom migrates and during re-aromatization, H^+ can be easily lost in preference to the ligand arm.

The discovery of the bis-μ-oxo dicopper(III) form of product from reaction of $Cu^I{}_n/O_2$ reactions (*vide supra*), leads to the question of whether it is this species or the μ-η^2-η^2-peroxo complex (in equilibrium, perhaps), which is responsible for the attack on the aromatic xylyl ligand.[1,86] A recent study[86] with a binucleating xylyl ligand with triazacyclononane donors indicates that both types of copper-oxygen species can form, depending on whether inter- or intramolecular Cu-O_2-Cu units form. Both in that system and in recent further investigations of the complexes discussed here,[87] the data are consistent with the view that it is the μ-η^2-η^2-peroxo-dicopper(II) complex which is responsible for aromatic hydroxylation. Further inquiries are required.

2.3 *Tyrosinase models: external phenol hydroxylation*

Phenol oxidative coupling and/or oxygenation chemistry is of commercial interest. Brackman and Havinga's work[90] involving early tyrosinase models is notable, and a number of other interesting systems have since been described.[8,91] Here, we highlight results, and examine a tyrosinase model system described by Casella and co-workers.[92,93] They utilized **L-66** (Fig. 7.6) and/or related binucleating ligands which vary in the number of methylene groups between alkylamine and benzimidazole donors. The dicopper(I) complex $[Cu^I{}_2($**L-66**$)]^{2+}$ (**12**, Fig. 7.6) mediates the ortho-hydroxylation of exogenous phenols by reaction with O_2. Labeling experiments indicated the expected incorporation of oxygen atoms into products, and a mononucleating analogue of **L-66** showed no tyrosinase-activity, consistent with the importance of a binuclear copper complex-mediated reaction. With methyl-4-hydroxybenzoate as a substrate phenol, Casella's experimental protocols indicated that appreciable quantities of catecholate product were obtained, and quenching of reactions at a low temperature suggested it to be the only product at early stages. Warming or longer reaction times led to the ultimate

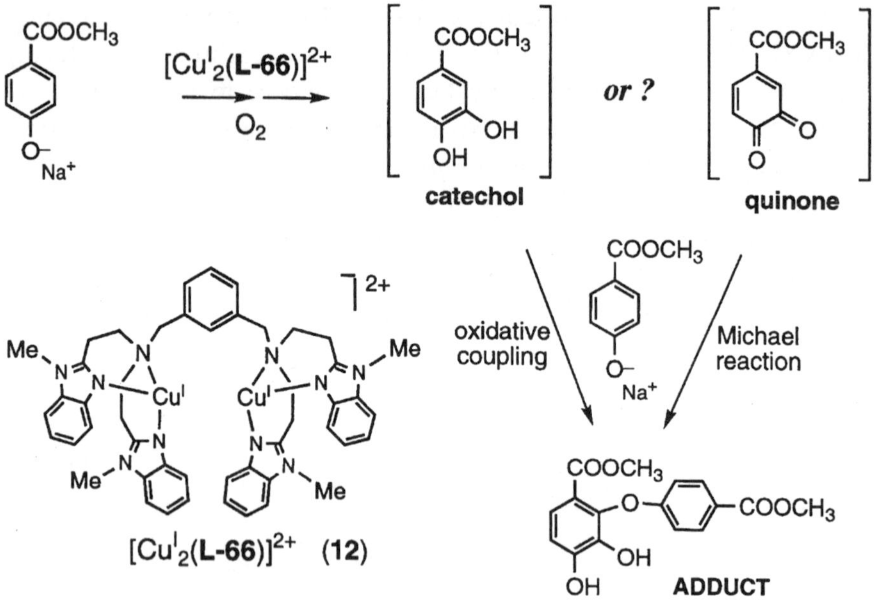

Fig. 7.6 A tyrosinase model system involving external phenol oxygenation. Questions concerning the nature of the primary product (catechol versus quinone) have arisen. See text for further explanation.

major product (**ADDUCT**), claimed to arise from a slower oxidative coupling of phenol and catechol. Maumy and Capdevielle[94] also published results supporting Casella's view that catechols are directly produced by copper ion-mediated hydroxylation reactions.

However, a controversy exists pertaining to the question of the identity of the primary reaction product and course of reaction. Sayre and Nadkarni[34] also carried out experiments using $[Cu^{I}_2($**L-66**$)]^{2+}$ (**12**), but they determined that the **ADDUCT** forms as a Michael addition product of phenol and 4-carbomethoxy-1,2-benzoquinone, noting that 4-carbomethoxy-1,2-catechol is inert to oxidation under the conditions employed. Thus, Sayre concluded that the benzoquinone is the *direct* product of oxygenation in the tyrosinase-like reaction, $[Cu^{I}_2]$ + phenol + $O_2 \rightarrow [Cu^{I}_2]$ + quinone + H_2O. Such a reaction thus proceeds like a dioxygenase, regenerating catalyst and not requiring external reductant. It is difficult to make conclusions on the mechanism owing to the fact that experimental finding of a catechol product does not prove its intermediacy, since it could form readily from reduction of quinone or semiquinone (e.g. by Cu(I)), and then be later converted to the apparent Michael product through a radical coupling reaction.[95]

It is notable that Kitajima[52] previously suggested a similar pathway and dioxygenase-like mechanism for tyrosinase reaction, based on observed reaction tendancies of phenol substrates with his own μ-η^2-η^2-peroxo-dicopper(II) complex **2** (Fig. 7.4). In fact, very recent investigations[35,36] suggested that, in the enzyme itself, catechols are produced in an indirect route not involving enzymatic reactions. These workers upheld the view that tyrosinase does *not* act as a tyrosine hydroxylase, but only produces *o*-quinones either from mono- or diphenol substrates. If correct, this may require rethinking of the detailed mode of dioxygen activation in binuclear complexes, at least with respect to phenolic substrates.

2.4 A model for dopamine β-monooxygenase

An important monooxygenase model system was described by Itoh and co-workers,[71,72] using a pyridyl-amine tridentate ligand similar to that used in our xylyl and Nn copper complexes (*vide supra*). Here, a phenylethyl pendant group undergoes specific and high yield hydroxylation at the benzylic position, following O_2 reaction with the copper(I) complex of this mononuclear compound $[(Py2phe)Cu^I]^+$ (**13**, PY = 2-pyridyl; Scheme 5). Oxygen atom incorporation into Py2Phe-OH using labeled O_2 has been confirmed, and a Cu^I/O_2 = 2:1 stoichiometry observed. The efficiency of the reaction is high (50% isolated yield), which represents a quantitative yield when considering the reaction as a monooxygenase process, with one atom of two from dioxygen being incorporated into product.

Spectroscopic evidence indicates formation of a μ-η^2:η^2 side-on peroxo intermediate $[\{Cu(Py2Phe)\}_2(O_2)]^{2+}$ (**14**), analogous to the RPY2 complexes we have previously studied,[62] *vide supra*. Further mechanistic studies[71] implicate C–H abstraction as occurring, and the authors suggest (without direct evidence) that a bis-μ-oxodicopper(III) intermediate species may carry

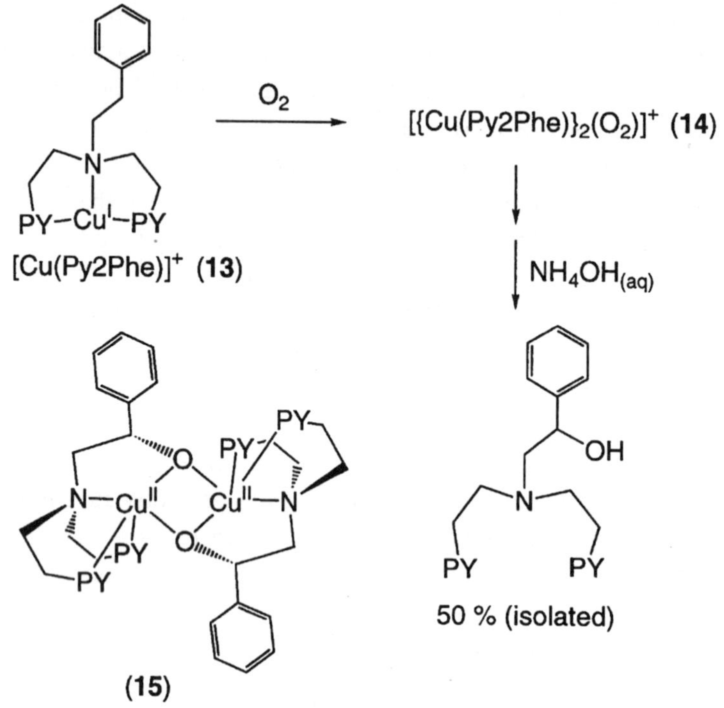

Scheme 5

out this chemistry. Another interesting feature is that the reaction can be made to go to 100% yield by using a reductant such as benzoin or hydroquinone in reaction with a copper(II) complex of the ligand, $[Cu^{II}(Py2Phe)(^{-}OClO_3)_2]$, followed by stirring in the presence of O_2. The stable bis-μ-alkoxo-dicopper(II) complex, $[\{Cu^{II}(Py2Phe\text{-}O^{-})\}_2]^{2+}$ (**15**), forms (Scheme 5), and its structure has been determined.[72]

Réglier and co-workers[96] recently described a system which is a close analogue of Itoh's, wherein an indane group was appended to the same pyridyl-amine tridentate ligand. Both regio- and stereoselective oxygen atom transfer to the benzylic C–H bond was observed, with retention of configuration. Large deuterium kinetic isotope effects (7.6, for Cu(II)/benzoin/Et_3N/O_2 reaction; 11, for Cu(I)/O_2 chemistry) were observed, implicating the intermediacy of different reactive copper-oxygen species. A stable bis-μ-alkoxo dicopper(II) complex oxygenation product was likewise observed.

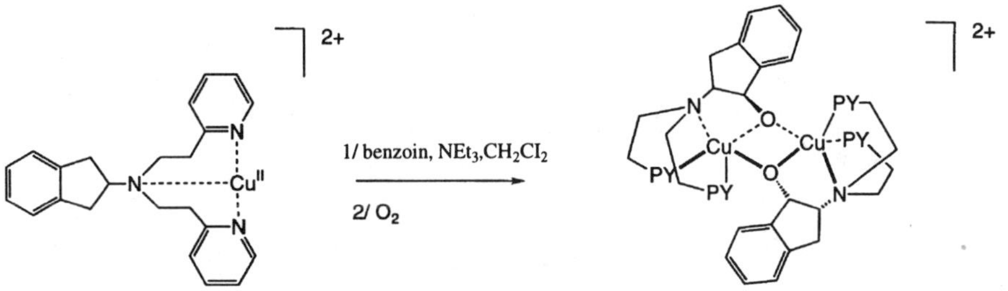

2.5 *Models for peptidylglycine α-amidating monooxygenase (PHM)*

There have been a number of model systems described for PHM, involving oxidative N-dealkylation chemistry mediated in copper complex oxidative processes.[62,97–101]

Sayre and co-workers[97] studied peptide-Cu(II) complexes containing glycine, alanine (Ala) or aminoisobutyric acid (Aib) at the C-terminus. These were shown to coordinate to the peptide nitrogen atom via N-deprotonation. Electrochemical and chemical studies indicated that oxidation to the level of Cu(III), electrochemically or using oxidants such as persulfate or $IrCl_6^{2-}$, resulted in oxidative decarboxylation and hydrolysis of the resulting N-acylimines to carboxamide or aldehyde/ketone. The optimal complex studied was a Cu(II) complex of picolinyl-Aib-Ala. Oxidation-mediated cleavage to picolinyl-Aib-NH_2 and pyruvic acid occurred; this reaction was shown to proceed through a Cu(III) intermediate, and a suggested initial one-electron N-oxidized intermediate. The authors discussed alternative pathways and mechanistic considerations for PHM in detail.

Capdevielle and Maumy[100] demonstrated the conversion of N-salicyloyl-glycine **C** into N-salicyloyl-α-hydroxyglycine **D**. The salicylamide **E** was also obtained. Three oxidant systems gave good product yields:

(i) Cu(0)/O_2/pyridine (py), (ii) Cu(II)/H_2O_2/py, and (iii) Cu(II) complex of C/trimethylamine oxide/CH_3CN. Although no direct evidence was obtained, the authors suggested trivalent copper-oxo species as important oxidizing intermediates.

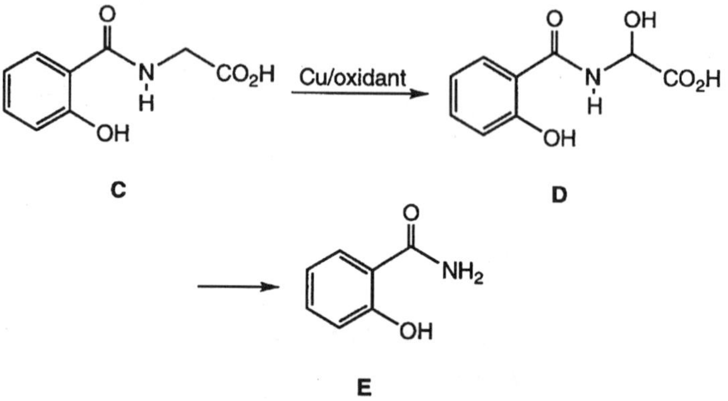

Again using RPY2 (R = benzyl) copper(I)/dioxygen chemistry, we[62] also showed that the dioxygen adduct of $[(PhCH_2PY2)Cu^I]^+$, $[\{(PhCH_2PY2)Cu\}_2(O_2)]^{2+}$ (**16**), reacted to give a moderate yield of oxidatively N-dealkylated ligand, suggested to occur by attack of the peroxo group oxygen atom on the benzylic hydrogen. Oxygen atom incorporation (18-O from $^{18}O_2$) occurred, as did deuterium transfer to the benzaldehyde product.

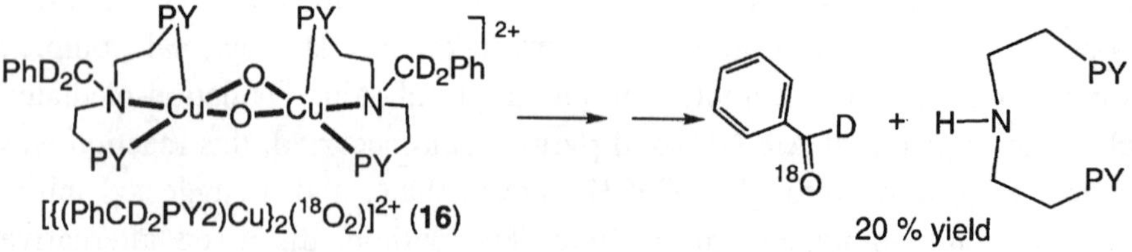

More recent detailed investigations were carried out by Tolman and co-workers, on well-characterized bis-μ-oxo dicopper(III) complexes $[\{LCu\}_2(O)_2]^{2+}$ (**4**), formed using substituted R_2R'-TACN ligands,[101] and their binucleating analogs wherein TACN was linked by an ethylene[102] or xylyl[86] group. For most of the complexes, oxygenation leads exclusively to cleavage of one of the substituents, producing either acetone or benzaldehyde (Fig. 7.7). All these reactions exhibit a remarkable kinetic isotope effect (KIE) of up to 40, clearly establishing C–H bond cleavage as the rate-determining step. Observed preferential isopropyl over benzyl N-dealkylation in $[\{(L^{iPr2Bn})Cu\}_2(O)_2]^{2+}$ may be due to equatorial disposition, and thus close contact of the reactive oxo group and the *i*Pr substrate.[86]

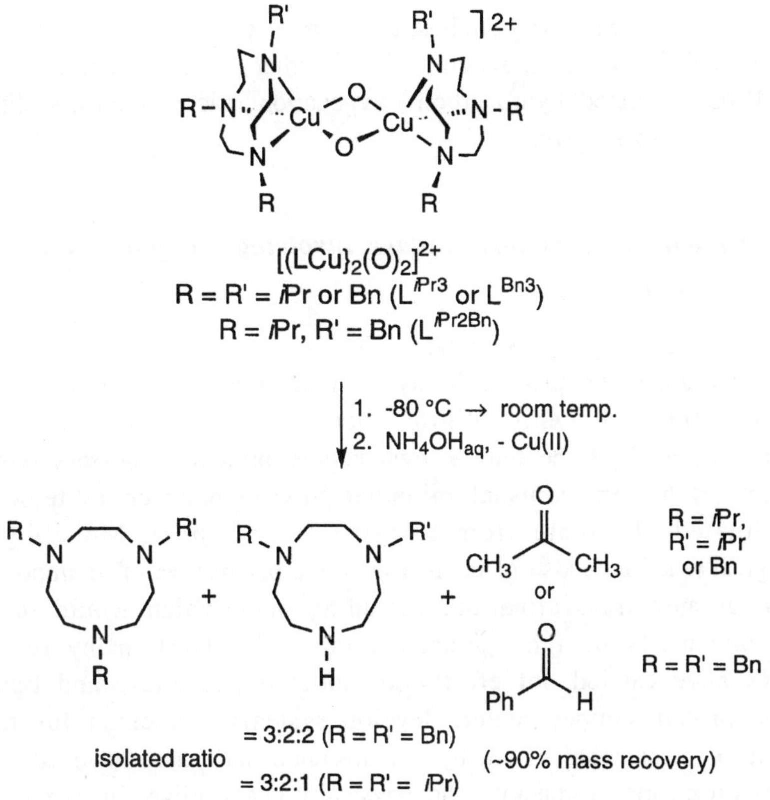

Fig. 7.7 Oxidative N-dealkylation reactions occurring in bis-μ-oxo-dicopper(III) complexes with trisubstituted TACN ligands. See text for further discussion.

Neither attack of the macrocyclic TACN ring structure nor oxidation of the ethylene tether is seen in that linked system.[102] With the *m*-xylyl-linked *i*Pr_2-TACN system, concentration-dependent spectra and decay rates are found for the oxygenated species, indicating preferential formation of bis-μ-oxo species at high concentrations, but a μ-η^2:η^2-peroxo complex in dilute solution. Detailed kinetic analysis showed that the latter leads to ring hydroxylation in 2-position, analogous to the original $[Cu^I_2(R\text{-XYL-H})]^{2+}$ (**8**, Scheme 4), with no KIE, whereas the former again leads to *i*Pr cleavage with a large KIE.[86] The results are explained by competition of inter- and intramolecular adduct formation and point to overall importance of proximity effects determining the relative reactivities.

In $LCu(I)_n/O_2$ reactivity studies, there are also recent reports describing oxidative O-dealkylation reactions.[85,103] Other recent examples[74,104] of $LCu(I)_n/O_2$-mediated hydrocarbon oxygenation reactions on coordinated ligands are also noteworthy.

2.6 *A model for galactose oxidase, involving catalytic alcohol oxidation*

The modeling of galactose oxidase has attracted considerable interest, especially since the protein X-ray confirmation of the unusual active site structural elements. While operating via a phenoxyl-radical active species, it is the only known copper protein to possess tyrosine ligation, it has the unusual thioether linkage between cysteine and coordinated phenolate from tyrosine (*vide supra*). Modeling the biological reaction is also of interest from the perspective of its importance as an organic transformation, including ones which would include environmentally benign synthetic catalysts.[105] Thus, many research groups have carried out efforts to model the structure and ligation of the protein copper center, develop systems mimicking functional aspects of the alcohol to aldehyde conversion, and there have also been successful efforts to generate and physically characterize Cu(II)-phenoxyl radical (i.e. Cu(II)-•OAr; diamagnetic) complexes or their paramagnetic zinc(II) analogues.[105–117]

Here, we focus on one very recently described system,[117] in which *catalytic* oxidization of benzylic and allylic alcohols to aldehydes (or ketones) occurs using O_2 under mild conditions; moreover, the mechanism observed appears to faithfully replicate that observed for galactose oxidase. Stack and co-workers synthesized and characterized a series of copper Schiff-base ligand complexes, possessing two phenolate donors, each with bulky *ortho*- and *para*-substituents. A drawing of the BSP (with *o*-thiophenyl and *p-tert*-butyl groups) copper complex is shown in Fig. 7.8.

A key feature of the [Cu(II)L] (**17**, L = dianionic tetradentate ligand) complexes is their distorted non-square planar structure, deduced from an X-ray structure and solution EPR spectra. One-electron oxidation (i.e. using $NO^+BF_4^-$) leads to EPR silent Cu(II)-phenoxyl radical

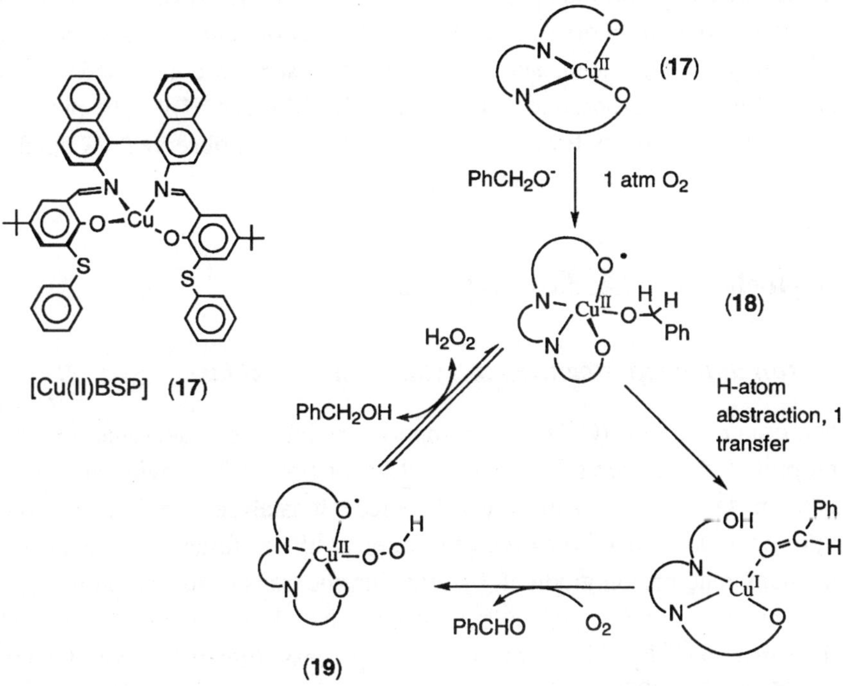

Fig. 7.8 Mechanism proposed for catalytic alcohol oxidation in a model system for galactose oxidase. See text for further explanation.

complexes, as inferred from (i) Cu K-edge X-ray absorption spectra and (ii) EPR spectra on analogous one-electron-oxidized zinc(II) complexes. The radical cation complexes $[Cu(II)L]^{+\bullet}$ bind alcoholates in a 1:1 stoichiometry, affording pentacoordinate $[Cu(II)L(^{-}OR)]^{\bullet}$ (**18**) species, and for R = $PhCH_2$, anaerobic conversion to benzaldehyde and a copper(I) complex occurs. Under a dioxygen atmosphere, neat alcohol substrates (with a catalytic amount of base) are converted to aldehydes or ketones, catalytically, with more than 1000 turnovers possible. The mechanistic scheme, Fig. 7.8, has been deduced from studies with [Cu(BSP)] and benzyl alcohol, using kinetic measurements and kinetic-isotope effects (KIE = 5.3), plus Hammet relationships ($\rho = -0.14$) with *para*-substituted benzylic alcohols. These data are consistent with a C^{α}–H bond cleavage in the rate-determining step. Hydrogen atom transfer is suggested to occur from substrate to coordinated phenoxyl-radical leading to de-coordination in the reduced Cu(I)···product complex. The two-electron oxidation of $[Cu(I)(BSP)]^{-}$ is suggested, and while no direct spectroscopic evidence is available for the hydroperoxo complex $[Cu(II)(BSP)(OOH)]^{\bullet}$ (**19**) product of dioxygen reaction, its formation is inferred from an observed O_2 uptake (aldehyde product:O_2 = 1:1) and EPR silence.

3. Cytochrome *c* Oxidase and Models

3.1 Iron and copper centers in cytochrome c oxidase

Cytochrome *c* oxidase (C*c*O) is a complex terminal respiratory membrane-bound protein that is composed of up to 13 subunits (in the bovine enzyme), but contains fewer subunits in bacterial sources. It catalyzes the four-electron, four-proton reduction of dioxygen to water, with its function to generate a transmembrane proton gradient, by effecting the translocation ("pumping") of four protons per O_2 molecule reduced. The electrochemical potential gradient generated by this proton pumping process ultimately provides the driving force for ATP synthesis.[31]

The primary activity of the enzyme appears to reside with subunits I and II (Fig. 7.9). Electrons from cytochrome *c* are first transferred to the

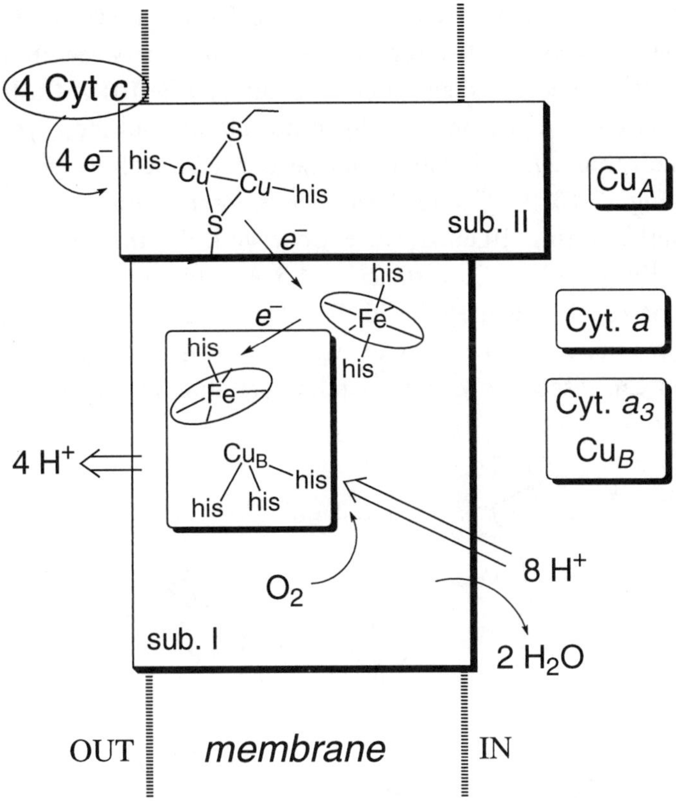

Fig. 7.9 Schematic representation of the metal centers in cytochrome *c* oxidase.

Cu_A center in the subunit II. This center consists of a binuclear mixed-valence (Cu(1.5)-Cu(1.5)) site, capable of accepting one electron, which is subsequently passed in subunit I to a low-spin $heme_a$ and to the key binuclear heme a_3– Cu_B center, where the molecular oxygen reduction occurs.

The Cu_B center is ligated by three imidazole groups and the high-spin heme a_3 is bound to an imidazole from histidine which is distal to copper. Much previous physical/spectroscopic and biochemical characterization have been carried out on fully oxidized state, which may be equivalent to the resting state and/or "as-isolated" enzyme form which is generated after the fourth electron transfer and protonation. Observations of EPR silence and suggested strong antiferromagnetic coupling between iron(III) and copper(II), led to earlier suggestions of a bridged Fe(III)-X-Cu(II) moiety, with X supposed to be an oxo, hydroxo, chloro, sulfido, cysteinato or imidazolato group. In the initial enzyme X-ray structure reports for both bovine (Fig. 7.10)[118,119] and *Paracoccus denitrificans* C*c*O,[120,121] no detectable bridging ligands were observed. In the bovine enzyme structure,[118,119] the Fe...Cu distance is 4.5 Å, while the original report on the bacterial C*c*O structure[120] (with Fe...Cu = 5.2 Å) was reinterpreted[121] to also have Fe...Cu = 4.5 Å, with suggested water and hydroxide ligands on Fe and Cu, respectively. Further reinterpretation of the electron density

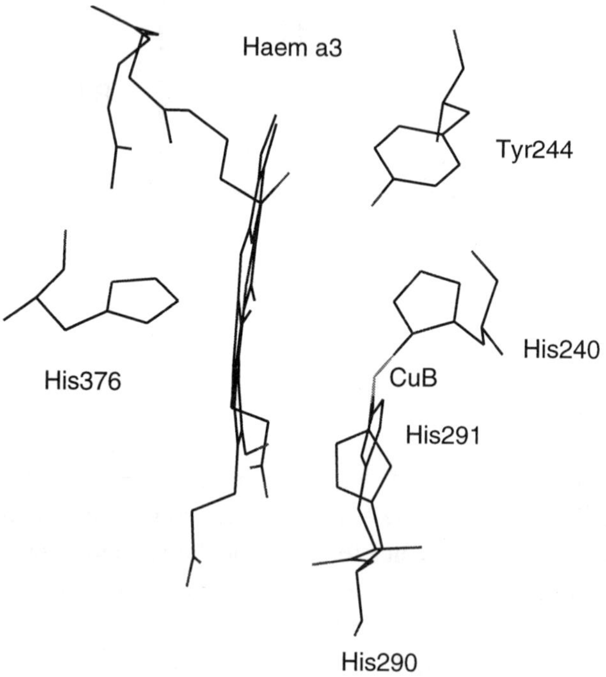

Fig. 7.10 Active site structure of the binuclear Fe-Cu (heme$_{a3}$-Cu$_B$) center in cytochrome *c* oxidase, generated with RASMOL, using X-ray coordinates, PDB 1OCC. The subsequently determined His-Tyr covalent link is not shown hare, see text.

around Cu_B led to suggestions of a covalent bond between His-240 and Tyr-244 (see Fig. 7.10) (His-276 and Tyr-280 in the bacterial structure),[121] now also seen in updates from the bovine enzyme structure.[122] The functional implications of this for the O_2 reduction and/or proton pumping mechanism have yet to be deduced, but one possibility suggested is that this His-Tyr "dimer" could donate an electron to dioxygen, thus becoming a phenoxyl radical.[121,122] Yoshikawa and co-workers[122] also obtained X-ray structural data on fully reduced, reduced CO-bound, oxidized with putative peroxo ligand, and azide-ligated oxidized forms.

Kinetics and spectroscopic investigations[31,123] revealed that Cu_B is the initial binding site for O_2,[124,125] thus highlighting the importance of Cu(I)/O_2 reactivity in the dioxygen reduction process. The O_2 ligand is then transferred to the heme a_3 iron forming an initial oxy Fe-O_2 adduct analogous to oxy-hemoglobin. From here, the exact identity of the subsequently formed species is controversial, with suggestions that electron injection leads to hydro(peroxo) metal and/or ferryl ($Fe^{IV}=O$) (or perhaps $Fe^{V}=O$) intermediates. The resting oxidized state is regenerated after the fourth electron-transfer and protonation to give water. Proton pumping accompanies formation of the late intermediates, peroxy and/or ferryl.

3.2 Models formed in a "self-assembly" strategy

Most previous attempts at modeling the heme a_3-Cu_B site have typically been directed at structural models for the oxidized "resting" state and the generation of mixed-metal iron(III)-Cu(II) complexes having various bridging ligands.[126,127] Even though the dioxygen reactivity of both iron(II) and copper(I) complexes are separately known, very few studies of the dioxygen reactivity with both of these reduced metal ions together have been carried out.[128,129]

In our initial attempts to delve into this type of chemistry, we choose to use TMPA-copper complexes, since the O_2 reactivity of [(TMPA)Cu^I(RCN)]$^+$ (**1a**), is well-understood.[8,58] When an equimolar mixture of (F_8-TPP)Fe^{II} (**20**) (or an analogue with axial piperidine ligands) and **1a** are allowed to react at –80 °C in CH_2Cl_2 in the presence of dioxygen and warmed up to 0 °C, μ-oxo complex $[(F_8\text{-TPP})Fe^{III}\text{-}(O^{2-})\text{-}Cu^{II}\text{-}(TMPA)]^+$ (**21**) is isolated. Interestingly, the acid-base reaction of equimolar amounts of $[(TMPA)Cu^{II}(CH_3CN)]^{2+}$ (**22**), $(F_8\text{-TPP})Fe^{III}\text{-}OH$ (**23**), and Et_3N in CH_3CN solvent under Ar gives the same complex based on its uv-vis and NMR spectroscopic properties (Fig. 7.11).

The X-ray structure of $[(F_8\text{-TPP})Fe^{III}\text{-}(O^{2-})\text{-}Cu^{II}(TMPA)]^+$ (**21**) revealed a linear Fe(III)-oxo-Cu(II) coordination (∠ Fe-O-Cu = 178.2°), with very short Fe-O (1.740 Å, similar to μ-oxo-bridged iron-porphyrin dimers) and Cu-O (1.856 Å) bond lengths.[130] Mössbauer spectroscopy and magnetic susceptibility measurements indicated **21** is an $S = 2$ spin system, with high-spin iron(III) strongly antiferromagnetically coupled

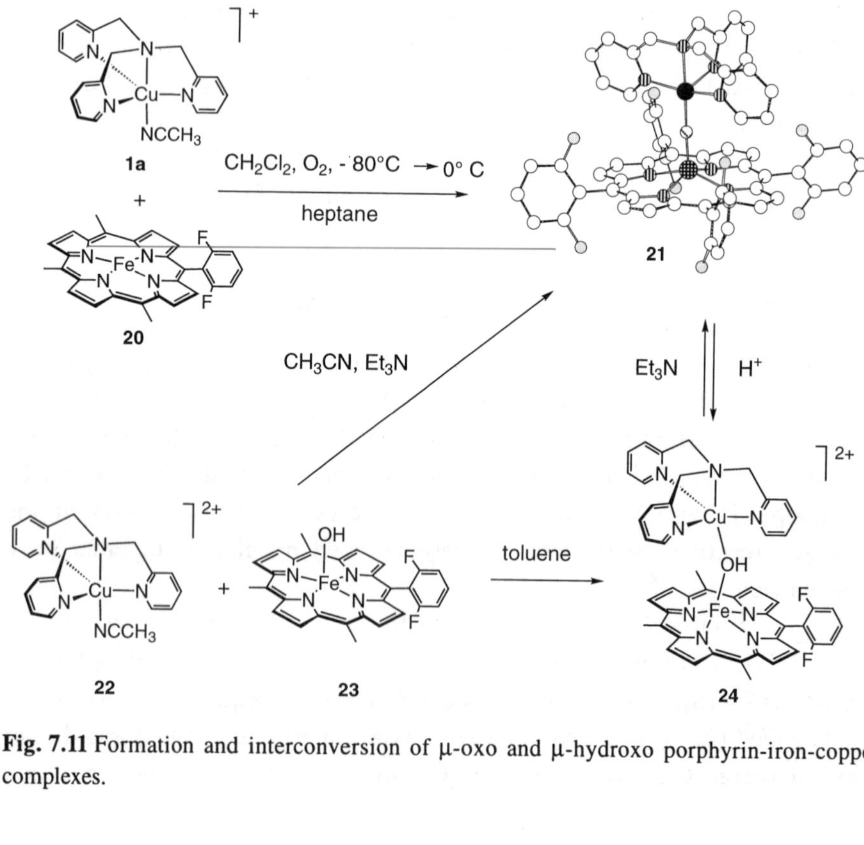

Fig. 7.11 Formation and interconversion of μ-oxo and μ-hydroxo porphyrin-iron-copper complexes.

($J = -\ 87\ cm^{-1}$ (solid), where $H = -\ 2JS_1{\bullet}S_2$) to an $S = 1/2$ Cu(II) ion center.[130] The ^{1}H-NMR spectroscopic showed an unusual spectrum at room temperature with resonances ranging from +65 ppm downfield to −104 ppm upfield. From ^{1}H and/or ^{2}H NMR studies, the spectrum has been fully assigned (Fig. 7.12).[131] The unusual behavior wherein the Cu^{II}(TMPA) ligands proton resonances shift upfield, whereas the pyrrole hydrogens on the iron porphyrin shifts downfield, is ascribed to the $S = 2$ ground state with strong antiferromagnetic coupling.[131] This behavior has also been observed in protein systems where two metals with different spins are antiferromagnetically coupled, i.e. Fe^{III}-X-Fe^{II} in ferredoxins or Cu^{II}-imidazolate-Co^{II} in cobalt-substituted Cu-Zn superoxide dismutase.[131]

Protonation of $[(F_8\text{-TPP})Fe^{III}\text{-}(O^{2-})\text{-}Cu^{II}(TMPA)]^+$ (**21**) by addition of one equivalent of triflic acid, produces the hydroxo-bridged complex, $[(F_8\text{-TPP})Fe\text{-}(OH)\text{-}Cu(TMPA)]^{2+}$ (**24**), Fig. 7.11.[132] This can also be isolated

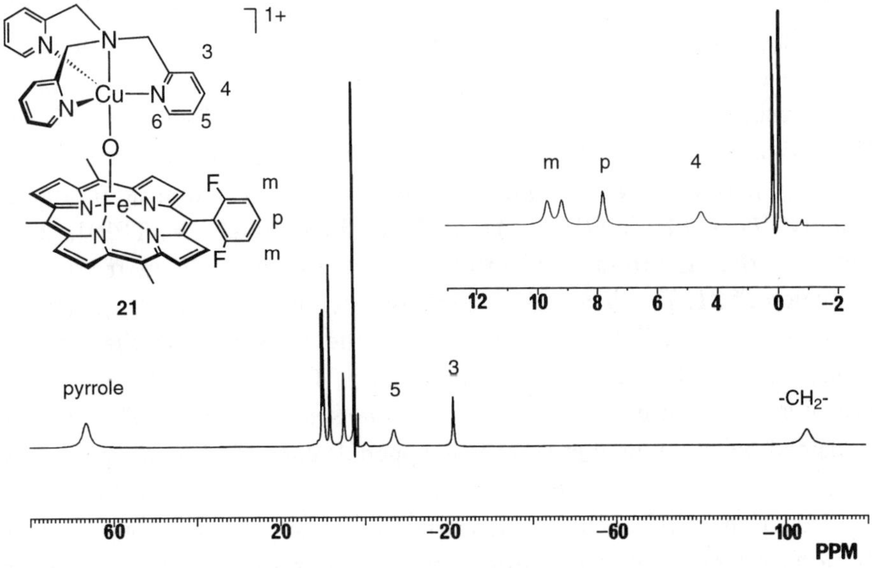

Fig. 7.12 ^{1}H-NMR spectrum of $[(F_8\text{-TPP})Fe^{III}\text{-}(O^{2-})\text{-}Cu^{II}(TMPA)]^+$.

by reacting an equimolar mixture of $[(TMPA)Cu^{II}(CH_3CN)]^{2+}$ (**22**) and $[(F_8\text{-}TPP)Fe^{III}\text{-}OH]$ (**23**). EXAFS measurements indicated that **24** possesses elongated Fe–OH and Cu–OH bonds and a bending such that ∠Fe-O-Cu at 157°.[132] This results in a loss of the outer shell strong multiple scattering observed in **21**, with its linear Fe-O-Cu moiety. We have suggested that **24** is a plausible model for the resting ("as-isolated") oxidized enzyme, with its supposed Fe-X-Cu antiferromagnetic coupling. Since protonation of enzyme turnover intermediates leads to product water and/or translocated protons, it is of interest that the oxo moiety in **21** is basic; we estimate that $14 < pK_a < 17$ in CH_3CN ($pK_a \sim 8 \pm 2.5$ in water).[132]

Isotopic labeling of the dioxygen reacted with $(F_8\text{-}TPP)Fe^{II}$ (**20**) and $[(TMPA)Cu^{I}(RCN)]^+$ (**1a**) to give $[(F_8\text{-}TPP)Fe^{III}\text{-}(O^{2-})\text{-}Cu^{II}(TMPA)]^+$ (**21**), thus indicating that the oxo group in the product is derived from O_2, as judged by a new isotope-sensitive IR band at 856 cm^{-1}. Manometry of O_2 uptake revealed an $Fe/Cu/O_2$ = 2:2:1 stoichiometry. We have carried out experiments trying to uncover aspects of the mechanism of the reaction which gives this kinetically stable μ-oxo product **21**, this being unstable with respect to thermodynamic sinks such as (P)Fe(III)-O-Fe(III)P (P = porphyrinate). We have shown[133] that the formation of **21** should proceed via μ-1,2-peroxo-dicopper adduct $[\{(TMPA)Cu^{II}\}_2(O_2^{2-})]^{2+}$ (**1**), and/or a decomposition product (e.g. Cu-O•), otherwise the products of the reaction with O_2 appear to be exclusively the iron products, namely $[(F_8\text{-}TPP)Fe\text{-}OH]$ (**23**) or $[(F_8\text{-}TPP)Fe\text{-}O\text{-}Fe(F_8\text{-}TPP)]$. This suggests the importance of prior copper-dioxygen chemistry in the formation of **21**. This heterodinuclear μ-oxo complex can also be generated by reacting $[(F_8\text{-}TPP)Fe^{III}\text{-}Cl]$ (**25**) in the presence of the peroxo adduct **1** in a Cu:Fe = 2:1 stoichiometry. But when a Cu:Fe = 2:2 stoichiometry is used, only the iron products are observed. This can be explained by assuming that peroxo-dicopper(II) complex **1** is in equilibrium with two equivalents of **1a**, and that the latter is capable of effecting the reduction of $[(F_8\text{-}TPP)Fe\text{-}Cl]$ (**25**). This possibility was confirmed by demonstrating that **1a** stoichiometrically reduces **25**, producing $[(TMPA)Cu^{II}Cl]^+$ (**26**) plus **20**, Fig. 7.13.[133]

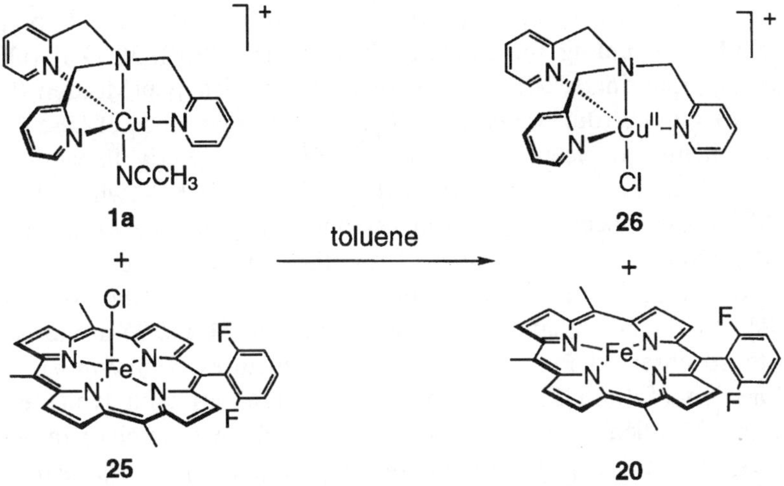

Fig. 7.13 Redox and atom-transfer reaction of $[(TMPA)Cu^{I}(CH_3CN)]^+$ with $[(F_8\text{-}TPP)Fe^{III}\text{-}Cl]$.

Also using a self-assembly strategy, Holm and co-workers prepared and characterized a series of heme-based molecular assemblies containing the Fe(III)-X-Cu(II) unit, possessing "unsupported" bridging groups X = O^{2-}, OH^-, HCO_2^- and CN^-.[134–139] The latter ligands are of interest for several reasons. Complexes with X = CN^- are of potential interest to deduce a molecular basis for cyanide toxicity. Also, cyanide has been used as a biochemically spectroscopic (infra red) enzyme probe.[140,141] In addition, two kinetically distinct forms of the oxidized bovine heart C*c*O, designated as "fast" and "slow", have been identified.[31,142] The properties of the slow and formate-bound forms have been shown to be very similar, suggesting that an endogenous carboxylate might be a bridging ligand in the binuclear site of the slow form. Finally, a re-investigation of the *Bacillus subtilis* enzyme by both Cu EXAFS and ENDOR spectroscopy detected the presence of an O/N Cu(II) ligand with an exchangeable proton.[143] This observation suggests an equilibrium with pH between the μ-oxo and the μ-hydroxo bridging species.

The oxo-bridged complex $[(OEP)Fe\text{-}O\text{-}Cu(Me_6tren)]^+$ (**27**) is prepared by acid-base coupling the iron(III) complex $[Fe(OEP)(OClO_3)]$ (OEP = octaethylporphyrinate(2-)) and the hydroxo species $[Cu(Me_6tren)(OH)]^+$ (Me_6tren) = tris(2-(dimethylamino)ethyl)amine) in presence of base.[137] The X-ray structure, depicted in Fig. 7.14, indicates a nearly linear bridging group, with ∠ Fe-O-Cu = 176.7° and Cu...Fe = 3.571 Å, while structural and Mössbauer spectroscopic data establish the presence of a high-spin iron(III) atom in this compound. Thus, the structural and spectroscopic properties closely match those observed for $(F_8\text{-TPP})Fe^{III}\text{-}(O^{2-})\text{-}Cu^{II}$ $(TMPA)]^+$ (**21**). Note that the Fe...Cu distances in **21** and **27** are far shorter than those observed in the enzyme X-ray structures (*vide supra*).

Reaction of $[Fe(OEP)(py)(CN)]$ (py = pyridine) with $[Cu(Me_5dien)(OH_2)]^{2+}$ (Me_5dien) = 1,1,4,7,7-pentamethyldiethylenetriamine) in acetone affords the corresponding bridging complex $[(py)(OEP)Fe\text{-}CN\text{-}Cu(Me_5dien)(OCMe_2)]^{2+}$ (**28**) relevant to cyanide-inhibited C*c*O. The X-ray structure (Fig. 7.14) shows a nearly linear Fe-CN fragment (bond angle 175.9°).[134] Cyanide stretching frequencies (ν_{CN}) in the solid state and in solution in dichloromethane (2175 cm^{-1} and 2176 cm^{-1} respectively) are nearly close to those for the cyanide-inhibited oxidized enzymes (2152–2146 cm^{-1}). Magnetic and Mössbauer spectroscopic results

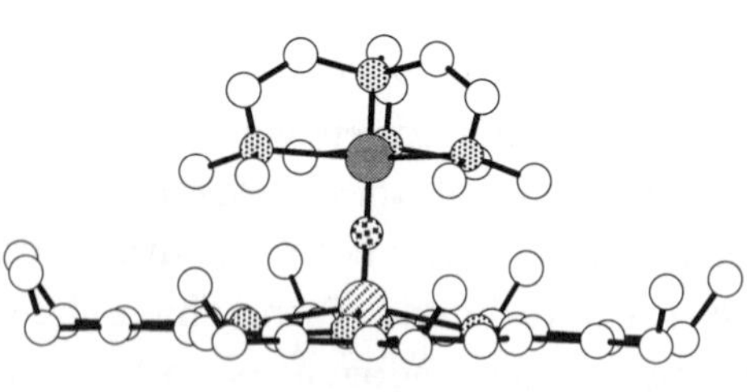

27

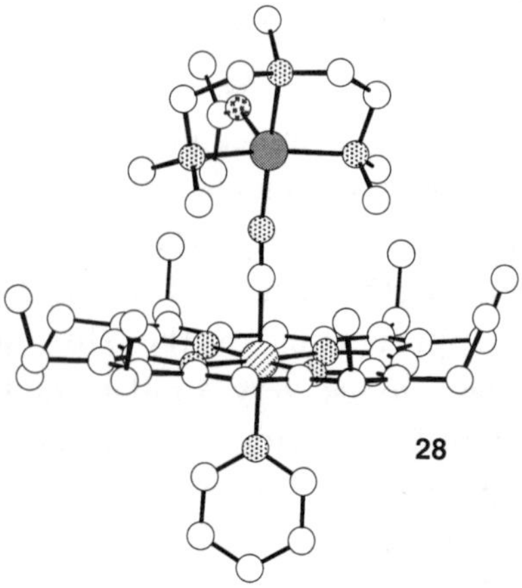

28

Fig. 7.14 Chem-3-D representations of X-ray structures of μ-oxo (**27**) and μ-CN^- (**28**)-bridged (porphyrinate)Fe(III)-Cu(II) complexes.

show ferromagnetically coupled low-spin Fe(III) and Cu(II) spins (S = 1) in **28**.[136] Using a variety of chelating tridentate or tetradentate ligands, Holm and co-workers[134,135] generated a whole series of (OEP)(py)Fe^{III}-CN-Cu^{II}(L) complexes, with low-spin iron(III) and S = 1 ground state. They found that increasing ν_{CN} correlates with decreasing Cu-N-C angles and increasing Cu–N bond distances, with ν_{CN} of the model compounds encompassing those observed for the enzymes (2152–2146 cm^{-1}). One compound isolated has copper(I) bound as (OEP)Fe^{III}-NC-Cu^{I}(dien), where ν_{CN} = 2100 cm^{-1}, iron is high-spin, and where the metals are coordinated as a linkage isomer (compared to the Fe^{III}-CN-Cu^{II}(L) complexes), i.e. N-to-Fe and C-to-Cu. The combined data lead to the conclusion that the enzyme possesses a tight [Fe^{III}-CN-Cu^{II}] bridge, rather than a binuclear site in which cyanide is bridged exclusively to one subsite.

3.3 *Models formed utilizing binucleating ligands*

A number of research groups have previously[126,127] or recently[144–151] utilized binucleating ligands to develop structural, spectroscopic and/or reactivity models for C*c*O. Here, we focus upon recent efforts and advances from the J.P. Collman's group[152] and our own.

In their attempt to mimic the binuclear site of the C*c*O, Collman and co-workers designed a new series of "capped" porphyrins, with the synthesis of complexes based on the multiple Michael addition of a 1,4,7-triazacyclonane (TACN) to a pre-metallated Fe-porphyrin. Addition of CuBr and an excess of bulky base L (e.g. 1,5-dicyclohexylimidazole) affords the reduced Fe^{II}/Cu^{I} complex **29** shown below.[153]

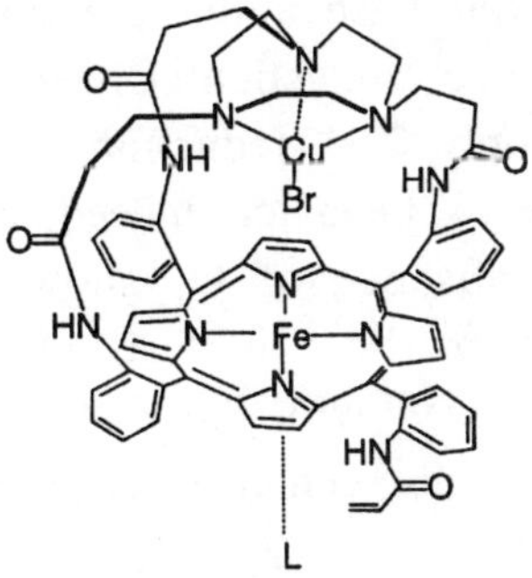

(29)

Dioxygen binding gives an O_2-adduct. As monitored by UV-vis spectroscopy, its formation appears to be irreversible, based on the lack of displacement of the bound O_2 unit under both CO and Ar purges. Notably, under the same conditions, the Fe(II)/copper free complex forms a reversible adduct,[154] suggesting an important role of copper in dioxygen binding. Several lines of evidence suggest that this oxygenated adduct could be a bridged peroxo species. A redox titration indicates that the complex requires exactly four equivalents of cobaltocene (4 e^-) for a complete reduction, a gas titration experiment shows that the fully reduced form takes up exactly one equivalent of O_2, and an electrospray mass spectrum is consistent with the formulated O_2-adduct product. Nevertheless, the resonance Raman studies of this O_2-adduct gives equivocal results, since the isotopic shift upon $^{18}O_2$ substitution (758 $\rightarrow$ 740 cm^{-1}) was found to be only about one-half the predicted value.[153]

In a cytochrome *c* oxidase model complex utilizing cobalt instead of iron, a related "peroxide" adduct did show a normal isotopic shift. In this complex, the researchers also covalently attached an imidazole axial base to the proximal face utilizing the remaining acrylamide group, Fig. 7.15. The Co(II)/Cu(I) complex **30** binds O_2 in a 1:1 stoichiometry by forming a very stable adduct **31**. Infrared studies indicate a peroxide-like $^{16}O_2$ band at 804 cm^{-1} which shifts to 756 cm^{-1} with $^{18}O_2$. While structural insights are lacking, once more, the redox titration with cobaltocene is consistent with a peroxide-bridging within the Co/Cu species **31**.[155]

Interestingly, this Co(II)/Cu(I) complex catalyzes the 4 e^- reduction of O_2 in aqueous solution at physiological pH. The researchers have shown that both the axial base and the redox active Cu center are required to achieve the 4 e^- reduction. Under the same conditions, the Fe/Cu analogue shows only 2 e^- reduction of O_2. This difference can be explained by their redox behavior. Indeed, cyclic voltammograms of this complexes in the absence of O_2 show that the Co(III)/Co(II) potential is higher than the Cu(II)/Cu(I) potential, whereas a reversed situation is observed with the Fe/Cu catalyst, meaning that a prior reduction of the copper center is required, and that Cu and its O_2-binding are critical.[155]

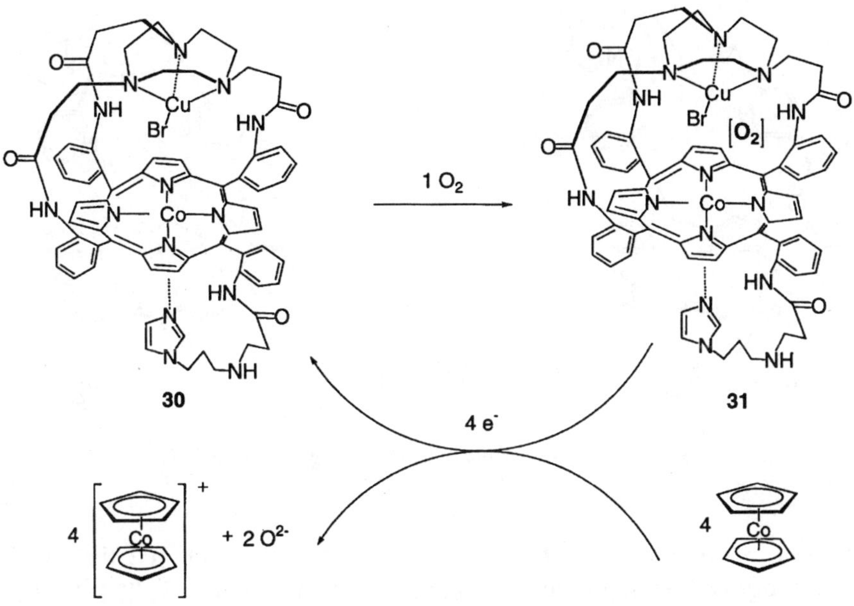

Fig. 7.15 Oxygenation of (porphyrinate)Co(II)-Cu(I) complex **30** affording a dioxygen-adduct **31** and cobaltacene reduction.

Returning to our own studies, the reaction of $[(TMPA)Cu^{I}(RCN)]^+$ (**1a**) and F_8-TPP)Fe^{II} (**20**) with O_2 (Fig. 7.11) giving $[(F_8\text{-TPP})Fe^{III}\text{-}(O^{2-})\text{-}Cu^{II}\text{-}(TMPA)]^+$ (**21**), represents a crude functional model for cytochrome *c* oxidase, since iron and copper chemistry together lead to O–O bond cleavage. Details of the mechanism of this reaction are of interest, however more recently, we have chosen focus our attention on structural, reactivity and mechanistic studies using analogous systems which employ binucleating ligands, such as ^{5}L and ^{6}L (Fig. 7.16), where a tripodal tetradentate TMPA moiety is tethered to the periphery of a tetraarylporphyrin (Ar = 2,6-difluorophenyl).[156,157]

With ^{5}L, metallation with iron $FeCl_2$, exposure to air, and workup, leads to an oxo-bridged heme/non-heme diiron complex $[^5LFe^{III}\text{-}O\text{-}Fe^{III}\text{-}Cl]^+$ (**32**), with high-spin iron(III), ∠ Fe-O-Fe = 157.3°, and Fe...Fe = 3.484 Å. Iron from the tetradenate TMPA portion of ^{5}L could be selectively removed,

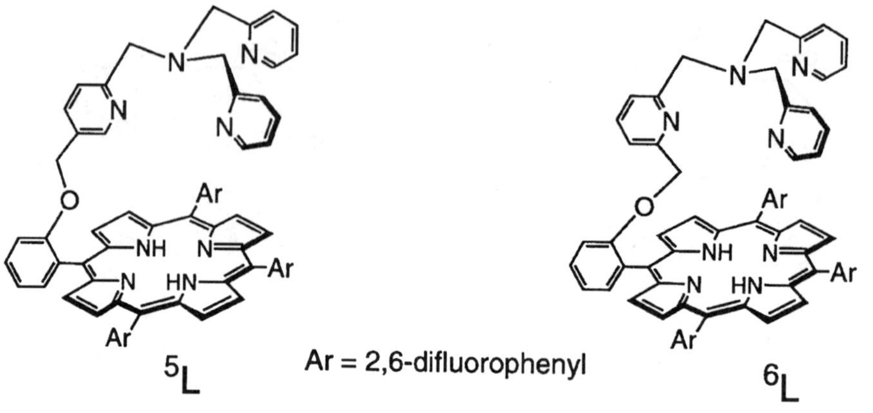

Fig. 7.16 Structure of the binucleating ligands **⁶L** and **⁵L**.

affording a high-spin iron(III) product with empty tether, $[(^5L)Fe^{III}\text{-}OH]$. An oxo-bridged analogue to the complex without tether, $[(F_8\text{-}TPP)Fe^{III}\text{-}O\text{-}Fe^{III}(TMPA)]^+$ could also be synthesized. The di-iron complexes may be models for the active site of nitric oxide reductase.[157] Recent work with the ^6L binucleating ligand indicated that analogous chemistry affords a di-iron species with ^6L (**33**) and $[(^6L)Fe^{III}\text{-}OH]$ (**34**), with empty tether, Scheme 6. Reaction of **34** with a Cu(II) salt in the presence of a base results in the formation of the new oxo-bridged complex $[(^6L)Fe^{III}\text{-}O\text{-}Cu^{II}]^+$ (**35**), which has been characterized by UV-vis and ^1H-NMR spectroscopies, mass spectrometry, and X-ray diffraction.[158,159]

Similar chemistry can be achieved with the ^5L binucleating ligand, leading to $[(^5L)Fe^{III}\text{-}O\text{-}Cu^{II}]^+$ (**36**). With both EXAFS spectroscopy and X-ray diffraction, our studies revealed significant structural differences in the $Fe^{III}\text{-}O\text{-}Cu^{II}$ complexes, as shown in Fig. 7.17.[158,159] Whereas ∠ Fe-O-Cu = 178.2° (near-linear) in the untethered complex **21**, the connection through the 5- or 6-pyridyl position of TMPA moiety to the *ortho* position of one of the porphyrin *meso*-aryl groups considerably alters the geometry of the oxo-bridged system, from a slightly bent geometry for the ^6L complex **35** (∠ Fe-O-Cu = 170.8°) to a very bent configuration in the case of the ^5L complex **36** (∠ Fe-O-Cu ~ 139°; EXAFS spectroscopic

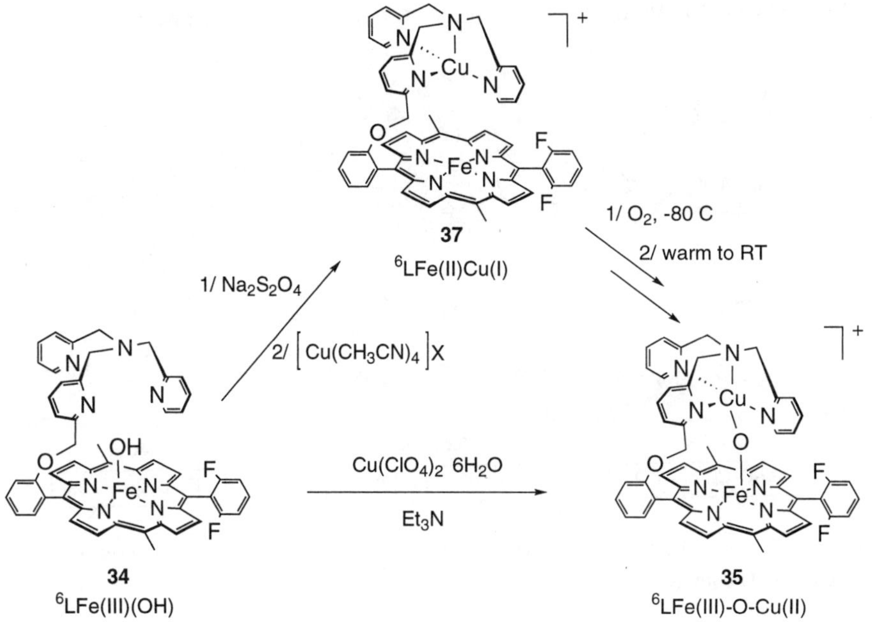

Scheme 6

analysis). These results showed how the variation in ligand design and resulting ligand constraints can cause differences in structural properties of similar μ-oxo complexes. It will be of interest to explore whether these structural changes result in differences in spectroscopic or chemical properties, such as ^{1}H-NMR spectroscopy and magnetic coupling between Fe(III) and Cu(II); the protonation chemistry of the oxo bridge in these molecules is in fact influenced.[158]

The generation of reduced complexes and study of their dioxygen reactivity is of paramount interest and importance, as discussed. Preliminary observations showed that a reduced species $[(^{6}L)Fe^{II}\text{-}Cu^{I}]^{+}$ (**37**) can be generated, and that its subsequent reaction with O_2 leads to $[(^{6}L)Fe^{III}\text{-}O\text{-}Cu^{II}]^{+}$ (**35**), Scheme 6.[159] Complex **37** is prepared by reduction of $[(^{6}L)Fe\text{-}OH]$ with dithionite followed by addition of a copper(I) salt. It is characterized by λ_{max} = 424 nm, and by ^{1}H-NMR properties consistent

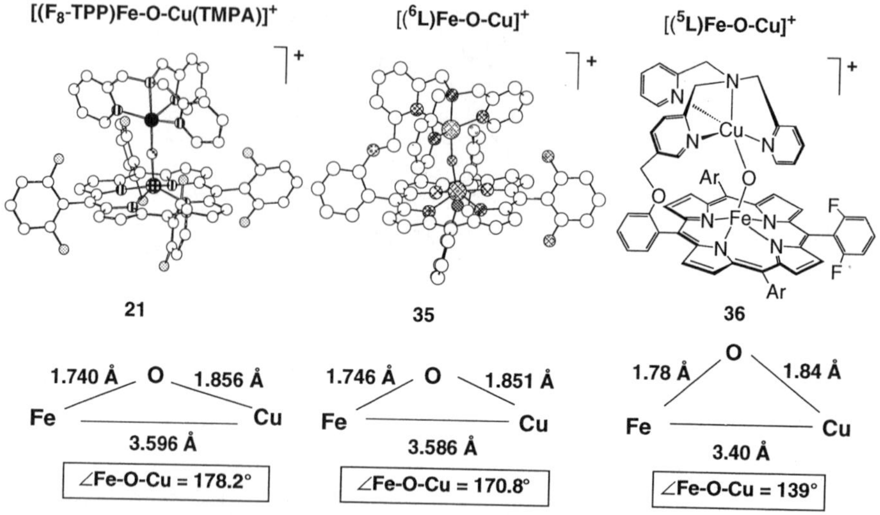

Fig. 7.17 X-ray diffraction or EXAFS-deduced structure comparisons of three iron(III)-oxo-copper(II) complexes.

with high-spin Fe(II) in tetrahydrofuran (THF) as solvent. Interestingly, when the oxygenation of **37** is carried out at –80 °C in THF or in MeCN, a new spectroscopically distinct species is observed;[160] this cleanly converts to μ-oxo complex **35** after warming to room temperature.[159] In fact, with a closely related ligand, Naruta and co-workers have recently obtained evidence (i.e. resonance Raman, electrospray mass spectrometry) for an Fe-peroxo-Cu intermediate.[151] These observations bode well for future investigations where we wish to probe the possibility of forming Fe-peroxo-Cu or other interesting intermediates which may have relevance to the processes of O_2-binding, reduction, O–O cleavage and proton translocation by the Fe-Cu center in cytochrome *c* oxidases.

3. Concluding Remarks

In this review, we have surveyed a number of recent examples of distinct, yet related, dioxygen binding modes and activation and/or reduction processes observed in inorganic complex models relevant to copper proteins and heme-copper oxidases. The recent progress in developing bioinorganic models with potential application as new reagents or catalysts for oxidative transformations has been remarkable. Yet, there will be continued extensive research involving structural and mechanistic interrogation of proteins, along with the design, synthesis and investigation of chemical model systems. New insights pertaining to biochemical mechanisms, details deduced from high resolution X-ray structures (e.g. PHM, C*c*O, etc. ...), as well as the appearance of novel organic cofactors (such as those found in galactose oxidase or amine oxidase), continue to bring fascinating new challenges for bioinorganic chemists.

Acknowledgments

We are very grateful to the National Institutes of Health for financial support of the research described in this report. We also thank Dr. Sean Prigge (Johns Hopkins University.) for providing us partial X-ray coordinates in order to prepare Fig. 7.2.

Note Added In Proof

Since the writing and submission of this chapter, a number of important and relevant papers have been published. Highlights are given here according to section in the article, and references are placed at the end. *Section 1.1:* A key new paper[161] and review[162] on copper chaperone proteins have appeared. Several notable reviews on biological NO_x chemistry and denitrification have been published.[163,164]

Section 1.2: A new oxyhemocyanin molluskan (octopus) protein X-ray structure has an active site Cu_2-O_2 arrangement which is very similar to the horseshoe crab structure; one of the six histidine ligands is also involved in a thioether bridge, i.e. a Cys-His link.[165] A newer purification of the particular Cu-dependent methane monooxygenase indicates the absence of iron; pMMO is a copper protein only.[166] In the amine oxidase topa quinone biogenesis, the stoichiometry of O_2 required for the six-electron oxidation of tyrosine to topa quinone (TPQ) has been determined; 2 mol of O_2 produces 1 mol of H_2O_2, i.e. E-Tyr + $2O_2 \rightarrow$ E-TPQ + H_2O_2.[7] Further elaboration and discussion of a Cu-mediated reaction mechanism is presented.[167]

Sections 2.1 & 2.2: Further description of structural aspects and novel reactivity with exogenous substrates has been described for $[\{(MePY2)Cu\}_2(O_2)]^{2+}$.[73] A detailed spectroscopic and theoretical analysis has been carried out for the series of complexes $[Cu_2(Nn)(O_2)]^{2+}$, showing how the length of the dinucleating ligand methylene linker n results in a series of complexes with systematically varying μ-η^2:η^2-peroxodicopper(II) structure; smaller n results in a more distorted "butterfly" Cu_2-O_2 core, with red-shifted 400 nm to 500 nm charge-transfer band and a higher O–O stretching vibration, the latter as observed by resonance Raman spectroscopy.[168] The full structural description and accompanying discussion of the μ_4-peroxo complex described by Krebs and co-workers has been published.[169] Utilizing molecular modeling (i.e. molecular mechanics) design aids, Comba and co-workers have synthesized and characterized new end-on μ-peroxo-dicopper (i.e. copper-dioxygen) complexes with considerable room-temperature stability.[170] Tolman and co-workers have clarified some aspects of the interconversion chemistry involving μ-η^2:η^2-peroxo and bis(μ-oxo)dicopper complexes,[171] and they have generated ligands which result in the apparent hydroxylation of an arene from a bis(μ-oxo)dicopper(III) core.[172] Considerably more work in the field is needed to resolve questions concerning the inherent reactivity of substrates with μ-η^2:η^2-peroxo versus bis(μ-oxo)dicopper complexes. Related relevant theoretical aspects of copper-dioxygen complex formation, preferred structures and reactivity are dealt with in a number of recent reports.[168,173–177]

Section 2.6: A number of more recent papers also describe copper-phenolate complexes,[178,179] and/or new Cu(II)-phenoxyl radical chemistry,[180,181] with possible relevance to galactose oxidase.

Section 3.1: An important review on cytochrome c oxidase emphasizing structural and spectroscopy has been published.[182] The His-Tyr link found at the Cu_B of the heme-copper active site is derived from a post-translational protein modification.[183] The O_2-reduction mechanism is now being discussed in terms of this important active-site tyrosine, with tyrosyl (phenoxyl) radical intermediates suggested to occur; this tyrosine thus may act as a key electron and proton donor.[184–187]

Section 3.2: The reaction of a mononuclear copper(I) complex, $[(MePY2)Cu^{I}(CH_3CN)]^+$ (see Sec. 2.1) with $(F_8\text{-TPP})Fe^{II}$ (20) (Fig. 7.11) and O_2 results in the formation of a μ-oxo Fe(III)-O-Cu(II) complex; iron-superoxo and heme-Cu peroxo intermediates are detected or implicated.[188]

Section 3.3: J. P. Collman and co-workers have developed still newer generations of heme-Cu complexes, which as models for cytochrome *c* oxidase, effect the electrocatalytic reduction of dioxygen.[189–191] Resonance Raman spectroscopic measurements reveal that the O_2-intermediate formed by reaction of $[(^6L)Fe^{II}\text{-}Cu^{I}]^+$ (37; Scheme 6) is indeed a peroxo complex $\{\nu_{(O\text{-}O)} = 787\ cm^{-1}\}$, while $^1H,^2H$-NMR spectroscopy indicates this $[(^6L)Fe^{III}\text{-}(O_2^{2-})\text{-}Cu^{II}]^+$ species is high-spin, the first example of such a heme-copper dioxygen (peroxo) complex.[192] An Fe(II)-Cu(I) complex with binucleating ligand employing a tethered heme axial pyridyl donor with pyridyl-amine tridentate chelate (for Cu) forms a peroxo level O_2-adduct.[193]

References

1. Solomon, E.I., Sundaram, U.M. and Machonkin, T.E. *Chem. Rev.* 1996, **96:** 2563–2605.
2. Holm, R.H., Kennepohl, P. and Solomon, E.I. *Chem. Rev.* 1996, **96:** 2239–2314.
3. Adman, E.T. *Adv. Protein Chem.* 1991, **42:** 145–197.
4. Karlin, K.D., Tyeklár, Z. and Zuberbühler, A.D. in *Bioinorganic Catalysis* (ed.) Reedijk, J. Marcel Dekker, Inc., New York, 1993, pp 261–315.
5. *Bioinorganic Chemistry of Copper* (ed.) Karlin, K.D. and Tyeklár, Z. Chapman & Hall, New York, 1993.
6. Karlin, K.D. *Science* 1993, **261:** 701–708.
7. Karlin, K.D. and Tyeklár, Z. *Adv. Inorg. Biochem.* 1994, **9:** 123–172.
8. Karlin, K.D. and Zuberbühler, A.D. in *Bioinorganic Catalysis,* 2nd Ed., Revised and Expanded (ed.) Reedijk, E.B.J. Marcel Dekker, Inc., New York, 1999, pp. 469–534.

9. Kaim, W. and Rall, J. *Angew. Chem. Int. Ed. Engl.* 1996, **35:** 43–60.
10. Fox, S. and Karlin, K.D. in *Active Oxygen in Biochemistry* (eds.) Valentine, J.S., Foote, C.S., Greenberg, A. and Liebman, J.F. Blackie Academic and Professional, Chapman & Hall, Glasgow, 1995, pp. 188–231.
11. Koch, K.A., Pena, M.O. and Thiele, D.J. *Chemistry and Biology* 1997, **4:** 549–560.
12. Pufahl, R.A. *et al. Science* 1997, **278:** 853–856.
13. Valentine, J.S. and Gralla, E.B. *Science* 1997, **278:** 817–818.
14. Valentine, J.S. *et al. Curr. Op. Chem. Biol.* 1998, **2:** 253–262.
15. Berlett, B.S. and Stadtman, E.R. *J. Biol. Chem.* 1997, **272:** 20313–20316.
16. Stadtman, E.R. *Annu. Rev. Biochem.* 1993, **62:** 797–821.
17. Mukhopadhyay, C.K., Mazumder, B., Lindley, P.F. and Fox, P.L. *Proc. Natl. Acad. Sci. USA* 1997, **94:** 11546–11551.
18. McNamara, J.O. and Fridovich, I. *Nature* 1993, **362:** 20–21.
19. Multhaup, G. *et al. Science* 1996, **271:** 1406–1409.
20. Brown, D.R. *et al. Nature* 1997, **390:** 684–687.
21. Averill, B.A. *Chem. Rev.* 1996, **96:** 2951–2964.
22. Zumft, W.G. and Kroneck, P.M.H. *Adv. Inorg. Biochem.* 1996, **11:** 193–221.
23. Magnus, K.A. *et al. Proteins: Struct., Funct., Genet.* 1994, **19:** 302–309.
24. Hazes, B. *et al. Protein Sci.* 1993, **2:** 597–619.
25. Klinman, J.P. *Chem. Rev.* 1996, **96:** 2541–2561.
26. Prigge, S.T., Kolhekar, A., Eipper, B.A., Mains, R.E. and Amzel, M. *Science* 1997, **278:** 1300–1305.
27. Lippai, I., Speier, G., Huttner, G. and Zsolnai, L. *Chem. Commun.* 1997, 741–742.
28. Messerschmidt, A. *Adv. Inorg. Chem.* 1993, **40:** 121–185.
29. Lindley, P. *et al. JBIC* 1997, **2:** 454–463.
30. Ito, N. *et al. Nature* 1991, **350:** 87–90.
31. Ferguson-Miller, S. and Babcock, G.T. *Chem. Rev.* 1996, **96:** 2889–2907.
32. Bertini, I., Mangani, S. and Viezzoli, M.S. *Adv. Inorg. Chem.* 1998, **45:** 127–250.
33. Fridovich, I. *J. Biol. Chem.* 1997, **272:** 18515–18517.
34. Sayre, L.M. and Nadkarni, D. *J. Am. Chem. Soc.* 1994, **116:** 3157–3158.
35. Cooksey, C.J. *et al. J. Biol. Chem.* 1997, **272:** 26226–26235.
36. Clews, J. *et al. Chem. Commun.* 1998, 77–78.
37. Boswell, J.S., Reedy, B.J., Kulathila, R., Merkler, D. and Blackburn, N.J. *Biochemistry* 1996, **35:** 12241–12250.

38. Elliott, S.J. *et al.* *J. Am. Chem. Soc.* 1997, **119:** 9949–9955.
39. Zahn, J.A. and DiSpirito, A.A. *J. Bacteriol.* 1996, **178:** 1018–1029.
40. Yuan, H., Perille Collins, M.L. and Antholine, W.E. *J. Am. Chem. Soc.* 1997, **119:** 5073–5074.
41. Elliott, S.J., Randall, D.W., Britt, R.D. and Chan, S.I. *J. Am. Chem. Soc.* 1998, **120:** 3247–3248.
42. Zaitseva, I. *et al.* *JBIC* 1996, **1:** 15–23.
43. Ducros, V. *et al.* *Nature Struct. Biol.* 1998, **5:** 310–316.
44. Parsons, M.R. *et al.* *Structure* 1995, **3:** 1171–1184.
45. Kumar, V. *et al.* *Structure* 1996, **4:** 943–955.
46. Wilce, M.C.J. *et al.* *Biochemistry* 1997, **36:** 16116–16133.
47. Li, R., Klinman, J.P. and Mathews, F.S. *Structure* 1998, **6:** 293–307.
48. Ruggiero, C.E., Smith, J.A., Tanizawa, K. and Dooley, D.M. *Biochemistry* 1997, **36:** 1953–1959.
49. Whittaker, J.M. in *Bioinorganic Chemistry of Copper* (eds.) Karlin, K.D. and Tyeklár, Z. Chapman & Hall, New York, 1993, pp. 447–458.
50. Whittaker, M.M. and Whittaker, J.W. *J. Biol. Chem.* 1988, **263:** 6074–6080.
51. Whittaker, M.M. and Whittaker, J.W. *Biophys. J.* 1993, **64:** 762–772.
52. Kitajima, N. and Moro-oka, Y. *Chem. Rev.* 1994, **94:** 737–757.
53. Tyeklár, Z. *et al.* *J. Am. Chem. Soc.* 1993, **115:** 2677–2689.
54. Jacobson, R.R., Tyeklár, Z., Karlin, K.D., Liu, S. and Zubieta, J. *J. Am. Chem. Soc.* 1988, **110:** 3690–3692.
55. Baldwin, M.J. *et al.* *J. Am. Chem. Soc.* 1991, **113:** 8671–8679.
56. Karlin, K.D. *et al.* *J. Am. Chem. Soc.* 1993, **115:** 9506–9514.
57. Karlin, K.D., Lee, D.-H., Kaderli, S. and Zuberbühler, A.D. *Chem. Commun.* 1997, 475–476.
58. Karlin, K.D., Kaderli, S. and Zuberbühler, A.D. *Acc. Chem. Res.* 1997, **30:** 139–147.
59. Lee, D.-H. *et al.* *J. Am. Chem. Soc.* 1995, **117:** 12498–12513.
60. Bol, J.E. *et al.* *Angew. Chem. Int. Ed. Engl.* 1997, **36:** 998–1000.
61. Fujusawa, K., Tanaka, M., Moro-oka, Y. and Kitajima, N. *J. Am. Chem. Soc.* 1994, **116:** 12079–12080.
62. Sanyal, I. *et al.* *Inorg. Chem.* 1992, **31:** 4322–4332.
63. Mahapatra, S., Halfen, J.A., Wilkinson, E.C., Que, L., Jr and Tolman, W.B. *J. Am. Chem. Soc.* 1994, **116:** 9785–9786.
64. Lynch, W.E., Kurtz, D.M., Jr., Wang, S. and Scott, R.A. *J. Am. Chem. Soc.* 1994, **116:** 11030–11038.

65. Sorrell, T.N., Allen, W.E. and White, P.S. *Inorg. Chem.* 1995, **34:** 952–960.
66. Halfen, J.A. *et al. Science* 1996, **271:** 1397–1400.
67. Blackburn, N.J., Strange, R.W., Farooq, A., Haka, M.S. and Karlin, K.D. *J. Am. Chem. Soc.* 1988, **110:** 4263–4272.
68. Karlin, K.D. *et al. J. Am. Chem. Soc.* 1988, **110:** 1196–1207.
69. Karlin, K.D. *et al. Inorg. Chem.* 1992, **31:** 1436–1451.
70. Pidcock, E., *et al. J. Am. Chem. Soc.* 1999, **121:** 1299–1308.
71. Itoh, S. *et al. J. Am. Chem. Soc.* 1998, **120:** 2890–2899.
72. Itoh, S. *et al. J. Am. Chem. Soc.* 1995, **117:** 4714–4715.
73. Obias, H.V., *et al. J. Am. Chem. Soc.* 1998, **120:** 12960–12961; Pidcock, E., *et al. J. Am. Chem. Soc.* 1999, **121:** 1870–1878.
74. Allen, W.E. and Sorrell, T.N. *Inorg. Chem.* 1997, **36:** 1732–1734.
75. Mahapatra, S. *et al. J. Am. Chem. Soc.* 1996, **118:** 11555–11574.
76. Mahadevan, V. *et al. J. Am. Chem. Soc.* 1997, **119:** 11996–11997.
77. Cole, A.P., Root, D.E., Mukherjee, P., Solomon, E.I. and Stack, T.D.P. *Science* 1996, **273:** 1848–1850.
78. DuBois, J.L. *et al. J. Am. Chem. Soc.* 1997, **119:** 8578–8579.
79. Reim, J. and Krebs, B. *Angew. Chem. Int. Ed. Engl.* 1994, **33:** 1969–1971.
80. Wada, A. *et al. Angew. Chem. Int. Ed.* 1998, **37:** 798–799.
81. Karlin, K.D. *et al. J. Am. Chem. Soc.* 1984, **106:** 2121–2128.
82. Nasir, M.S., Cohen, B.I. and Karlin, K.D. *J. Am. Chem. Soc.* 1992, **114:** 2482–2494 and references cited therein.
83. Karlin, K.D. *et al. J. Am. Chem. Soc.* 1994, **116:** 1324–1336.
84. Alzuet, G., Casella, L., Villa, M.L., Carugo, O. and Gullotti, M. *J. Chem. Soc., Dalton Trans.* 1997, 4789–4794.
85. Shen, C.-Y., Hu, M.-F., Luo, Q.-H. and Shen, M.-C. *Transition Met. Chem.* **20:** 1995, 634–635.
86. Mahapatra, S. *et al. Inorg. Chem.* 1997, **36:** 6343–6356.
87. Pidcock, E., Obias, H.V., Zhang, C.X., Karlin, K.D. and Solomon, E.I. *J. Am. Chem. Soc.* 1998, **120:** 7841–7847.
88. Karlin, K.D. *et al. J. Am. Chem. Soc.* 1987, **109:** 2668–2679.
89. Paul, P.P., Tyeklár, Z., Jacobson, R.R. and Karlin, K.D. *J. Am. Chem. Soc.* 1991, **113:** 5322–5332.
90. Brackman, W. and Havinga, E. *Rec. Trav. Chim. Pays-Bas* 1955, **74:** 937, 1021, 1070, 1103, 1107.
91. Karlin, K.D. and Gultneh, Y. *Prog. Inorg. Chem.* 1987, **35:** 219–327.
92. Casella, L., Gullotti, M., Radaelli, R. and Di Gennaro, P. *J. Chem. Soc., Chem. Commun.* 1991, 1611–1612.

93. Casella L. *et al. Inorg. Chem.* 1996, **35:** 7516–7525.
94. Maumy, M. and Capdevielle, P. *M. Mol. Catal. A: Chemical* 1996, **113:** 159–166.
95. Mandal, S., Nadkarni, D.V. and Sayre, L.M. in *ACS Cleveland Section Meeting-in-Miniature,* May 26, 1997, Cleveland State University, 1997; Sayre, L.M., personal communication.
96. Blain, I. *et al. Europ. Inorg. Chem.* 1998, 1297–1304.
97. Reddy, K.V. *et al. J. Am. Chem. Soc.* 1990, **112:** 2332–2340.
98. Wang, F.-J. and Sayre, L.M. *J. Am. Chem. Soc.* 1992, **114:** 248–255.
99. Patch, M.G., McKee, V. and Reed, C.A. *Inorg. Chem.* 1987, **26:** 776–778.
100. Capdevielle, P. and Maumy, M. *Tetrahedron Lett.* 1991, **32:** 3831–3834.
101. Mahapatra, S., Halfen, J.A. and Tolman, W.B. *J. Am. Chem. Soc.* 1996, **118:** 11575–11586.
102. Mahapatra, S., Young, J.V.G., Kaderli, S., Zuberbühler, A.D. and Tolman, W.B. *Angew. Chem. Int. Ed. Engl.* 1997, **36:** 130–133.
103. Gelling, O.J. and Feringa, B.L. *J. Am. Chem. Soc.* 1990, **112:** 7599–7604.
104. Halfen, J.A., Young, J.V.G. and Tolman, W.B. *J. Am. Chem. Soc.* 1996, **118:** 10920–10921.
105. Markó, I.E., Giles, P.R., Tsukazaki, M., Brown, S.M. and Urch, C.J. *Science* 1996, **274:** 2044–2046.
106. Markó, I.E., Tsukazaki, M., Giles, P.R., Brown, S.M. and Urch, C.J. *Angew. Chem. Int. Ed. Engl.* 1997, **36:** 2208–2210.
107. Adams, H., Bailey, N.A., Campbell, I.K., Fenton, D.E. and He, Q.-Y. *J. Chem. Soc., Dalton Trans.* 1996, 2233–2237.
108. Wang, Y. and Stack, T.D.P. *J. Am. Chem. Soc.* 1996, **118:** 13097–13098.
109. Halfen, J.A., Young, J.V.G. and Tolman, W.B. *Angew. Chem. Int. Ed. Engl.* 1996, **35:** 1687–1689.
110. Whittaker, M.M., Duncan, W.R. and Whittaker, J.W. *Inorg. Chem.* 1996, **35:** 382–386.
111. Zurita, D., Gautier-Luneau, I., Ménage, S., Pierre, J.-L. and Saint-Aman, E. *JBIC* 1997, **2:** 46–55
112. Itoh, S. *et al. Inorg. Chem.* 1997, **36:** 1407–1416.
113. Sokolowski, A. *et al. JBIC* 1997, **2:** 444–453.
114. Halfen, J.A. *et al. J. Am. Chem. Soc.* 1997, **119:** 8217–8227.
115. Sokolowski, A. *et al. J. Am. Chem. Soc.* 1997, **119:** 8889–8900.
116. Müller, J. *et al. Angew. Chem. Int. Ed.* 1998, **37:** 616–619.
117. Wang, Y., DuBois, J.L., Hedman, B., Hodgson, K.O. and Stack, T.D.P. *Science* 1998, **279:** 537–540.

118. Tsukihara, T. *et al. Science* 1996, **272:** 1136–1144.
119. Tsukihara, T. *et al. Science* 1995, **269:** 1069–1074.
120. Iwata, S., Ostermeier, C., Ludwig, B. and Michel, H. *Nature* 1995, **376:** 660–669.
121. Ostermeier, C., Harrenga, A., Ermler, U. and Michel, H. *Proc. Natl. Acad. Sci. USA* 1997, **94:** 10547–10553.
122. Yoshikawa, S. *et al. Science* 1998, **280:** 1723–1729.
123. Kitagawa, T. and Ogura, T. *Prog. Inorg. Chem.* 1997, **45:** 431–479.
124. Blackmore, R.S., Greenwood, C. and Gibson, Q.H. *J. Biol. Chem.* 1991, **266:** 19245–19249.
125. Oliveberg, M. and Malmström, B.G. *Biochemistry* 1992, **31:** 3560–3563.
126. Kitajima, N. *Adv. Inorg. Chem.* 1992, **39:** 1–77.
127. Nanthakumar, A., Nasir, M.S., Karlin, K.D., Ravi, N. and Huynh, B.H. *J. Am. Chem. Soc.* 1992, **114:** 6564–6566 and references cited therein.
128. Berry, K.J., Gunter, M.J. and Murray, K.S. in *Oxygen and Life,* The Royal Society of Chemistry, London, UK, 1980, pp. 170–179.
129. Gunter, M.J., Mander, L.N. and Murray, K.S. *J. Chem. Soc., Chem. Commun.* 1981, 799–801.
130. Karlin, K.D. *et al. J. Am. Chem. Soc.* 1994, **116:** 4753–4763.
131. Nanthakumar, A., Fox, S., Murthy, N.N. and Karlin, K.D. *J. Am. Chem. Soc.* 1997, **119:** 3898–3906 and references cited therein.
132. Fox, S., Nanthakumar, A., Wikström, M., Karlin, K.D. and Blackburn, N.J. *J. Am. Chem. Soc.* 1996, **118:** 24–34.
133. Nanthakumar, A., Fox, S. and Karlin, K.D. *J. Chem. Soc., Chem. Commun.* 1995, 499–501.
134. Scott, M.J. and Holm, R.H. *J. Am. Chem. Soc.* 1994, **116:** 11357–11367.
135. Holm, R.H. *Pure and Appl. Chem.* 1995, **67:** 217–224.
136. Kauffmann, K.E., Goddard, C.A., Zang, Y., Holm, R.H. and Münck, E. *Inorg. Chem.* 1997, **36:** 985–993.
137. Lee, S.C. and Holm, R.H. *J. Am. Chem. Soc.* 1993, **115:** 11789–11798.
138. Scott, M.J., Goddard, C.A. and Holm, R.H. *Inorg. Chem.* 1996, **35:** 2558–2567.
139. Scott, M.J. *et al. J. Am. Chem. Soc.* 1995, **117:** 568–569.
140. Yoshikawa, S., Mochizuki, M., Zhao, X.-J. and Caughey, W.S. *J. Biol. Chem.* 1996, **270:** 4270–4279.
141. Tsubaki, M., Mogi, T., Hori, H., Sato-Watanabe, M. and Anraku Y. *J. Biol. Chem.* 1996, **271:** 4017–4022.

142. Palmer, G. *J. Bioenerg. Biomembr.* 1993, **25:** 145–151.
143. Fann, Y.C. *et al. Biochemistry* 1995, **34:** 10245–10255.
144. Monzani, E. *et al. J. Mol. Catal. A: Chemical* 1997, **117:** 199–204.
145. Fuji, H., Yoshimura, T. and Kamada, H. *Chem. Lett.* 1996, 581–582.
146. Sasaki, T. and Naruta, Y. *Chem. Lett.* 1995, 663–664.
147. Naruta, Y., Sasaki, T. and Nakamura, N. *J. Inorg. Biochem.* 1997, **67:** 108.
148. Bulach, V., Mandon, D. and Weiss, R. *Angew. Chem. Int. Ed. Engl.* 1991, **30:** 572–575.
149. Bag, N., Chern, S.-S., Peng, S.-M. and Chang, C.K. *Inorg. Chem.* 1995, **34:** 753–756.
150. Baeg, J.-O. and Holm, R.H. *Chem. Commun.* 1998, 571–572.
151. Sasaki, T., Nakamura, N. and Naruta Y. *Chem. Lett.* 1998, 351–352.
152. Collman, J.P. *Inorg. Chem.* 1997, **36:** 5145–5155.
153. Collman, J.P. *et al. J. Am. Chem. Soc.* 1994, **116:** 9783–9784.
154. Collman, J.P. *et al. J. Am. Chem. Soc.* 1997, **119:** 3481–3489.
155. Collman, J.P., Fu, L., Herrmann, P.C. and Zhang, X. *Science* 1997, **275:** 949–951.
156. Karlin, K.D. *et al. Pure and Appl. Chem.* 1995, **67:** 289–296.
157. Martens, C.F., Murthy, N.N., Obias, H.V. and Karlin, K.D. *Chem. Commun.* 1996, 629–630.
158. Obias, H.V., *et al. J. Am. Chem. Soc.* 1998, **120:** 9696–9697.
159. Ju, T.D., *et al. J. Am. Chem.* 1999, **38:** 2244–2245.
160. Ghiladi, R.A. *et al.* (1999), submitted for publication.
161. Rae, T.D., Schmidt, P.J., Pufahl, R.A., Culotta, V.C. and O'Halloran, T.V. *Science* 1999, **284:** 805–808.
162. Harrison, M.D., Jones, C.E. and Dameron, C.T. *JBIC* 1999, **4:** 145–153.
163. Zumft, W.G. *Microbiol. Mol. Biol. Revs.* 1997, **61:** 533–616.
164. Berks, B.C., Ferguson, S.J., Moir, J.W.B. and Richardson, D.J. *Biochim. Biophys. Acta — Bioenergetics* 1995, **1232:** 97–173.
165. Cuff, M.E., Miller, K.I., van Holde, K.E. and Hendrickson, W.A. *J. Mol. Biol.* 1998, **278:** 855–870.
166. Nguyen, H.-H.T., Elliott, S.J., Yip, J.H.-K. and Chan, S.I. *J. Biol. Chem.* 1998, **273:** 7957–7966.
167. Ruggiero, C.E. and Dooley, D.M. *Biochemistry* 1999, **38:** 2892–2898.
168. Pidcock, E., *et al. J. Am. Chem. Soc.* 1999, **121:** 1299–1308.
169. Reim, J., Werner, R., Haase, W. and Krebs, B. *Chem. Eur. J.* 1998, **4:** 289–298.
170. Börzel, H., *et al. Chem. Eur. J.* 1999, **5:** 1716–1721.

171. Cahoy, J., Holland, P.L. and Tolman, W.B. *Inorg. Chem.* 1999, **38:** 2161–2168.
172. Holland, P.L., Rodgers, K.R. and Tolman, W.B. *Angew. Chem. Int. Ed.* 1999, **38:** 1139–1142.
173. Cramer, C.J., Smith, B.A. and Tolman, W.B. *J. Am. Chem. Soc.* 1996, **118:** 11283–11287.
174. Bérces, A. *Inorg. Chem.* 1997, **36:** 4831–4837.
175. Liu, X.-Y., Palacios, A.A., Novoa, J.J. and Alvarez S. *Inorg. Chem.* 1998, **37:** 1202–1212.
176. Pidcock, E., Obias, H.V., Zhang, C.X., Karlin, K.D. and Solomon, E.I. *J. Am. Chem. Soc.* 1998, **120:** 7841–7847.
177. Bérces, A. *Chem. Eur. J.* 1998, **4:** 1297–1304.
178. Jazdzewski, B.A., Young, J.V.G. and Tolman, W.B. *Chem. Commun.* 1998, 2521–2522.
179. Vaidyanathan, M., *et al. Inorg. Chem.* 1998, **37:** 6418–6427.
180. Chaudhuri, P., Hess, M., Flörke, U. and Wieghardt, K. *Angew. Chem. Int. Ed.* 1998, **37:** 2217–2220.
181. Halcrow, M.A., *et al. J. Chem. Soc. Dalton Trans.* 1999, 1753–1762.
182. Michel, H., Behr, J., Harrenga, A. and Kannt, A. *Annu. Rev. Biophys. Biomol. Struct.* 1998, **27:** 329–356.
183. Kanti Das, T., Pecoraro, C., Tomson, F.L., Gennis, R.B. and Rousseau, D.L. *Biochemistry* 1998, **37:** 14471–14476.
184. Gennis, R.B. *Biochim. Biophys. Acta* 1998, **1365:** 241–248.
185. Proshlyakov, D.A., Pressler, M.A. and Babcock, G.T. *Proc. Natl. Acad. Sci. USA* 1998, **95:** 8020–8025.
186. Sucheta, A., Istvan Szundi, I. and Einarsdóttir, O. *Biochemistry* 1998, **37:** 17905–17914.
187. Poulos, T.L., Li, H. and Raman, C.S. *Curr. Opin. Chem. Biol.* 1999, **3:** 131–137.
188. Kopf, M.-A., Neuhold, Y.-M., Zuberbühler, A.D. and Karlin, K.D. *Inorg. Chem.* 1999, **38:** 3093–3102.
189. Collman, J.P., *et al. Angew. Chem. Int. Ed.* 1998, 37: 3397–3400.
190. Collman, J.P., Schwenninger, R., Rapta, M., Bröring, M. and Fu, L. *Chem. Commun.* 1999, 137–138.
191. Collman, J.P., *et al. J. Am. Chem. Soc.* 1999, **121:** 1387–1388.
192. Ghiladi, R.A., *et al.* 1999, submitted for publication.
193. Kopf, M.-A. and Karlin, K.D. *Inorg. Chem.* 1999, **38:** in press.

Chapter 8

Iron-Containing Models of Catechol Dioxygenases

Hans-Jörg Krüger

Institut für Anorganische und Angewandte Chemie der Universität Hamburg, Martin-Luther-King-Platz 6, D-20146 Hamburg, Germany

1. Introduction

The conversion of hydroxylated aromatic into acyclic compounds plays a central role within the degradation of aromates by bacteria. The enzymes engaged in this key reaction are primarily non-heme iron-containing catechol dioxygenases, which catalyze the oxidative cleavage of an aromatic C–C bond in catechol derivatives by molecular oxygen, incorporating both oxygen atoms into the products. Depending on the position of the reactive C–C bond with respect to the hydroxyl groups of the catechol substrate, these enzymes are classified into *intradiol*-cleaving and *extradiol*-cleaving catechol dioxygenases (Reaction 8.1). Both types of enzymes further distinguish themselves by the fact that the former utilizes iron(III) and the latter, iron(II) ions in their active sites.

After providing an overview of the structural and mechanistic properties of the enzymes, this review will concentrate almost exclusively

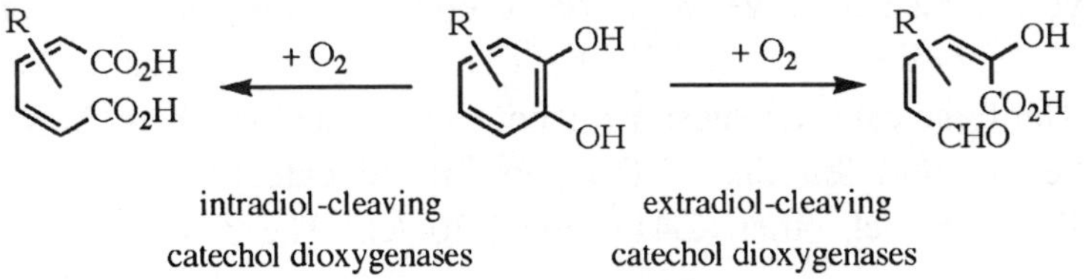

(8.1)

on the biomimetic reactivity of iron-containing model complexes. (For previous reviews on this topic see Ref. 1.)

2. Survey of the Structural and Mechanistic Properties of Intradiol- and Extradiol-Cleaving Catechol Dioxygenases

2.1 Intradiol-cleaving catechol dioxygenases

Among the intradiol-cleaving catechol dioxygenases, protocatechuate 3,4-dioxygenase (3,4-PCD), isolated from the soil bacterium *Pseudomonas putida* (formerly classified as *P. aeruginosa*), is probably the best characterized, with respect to the structure and reactivity of the active site. The basic structural and mechanistic features determined for this enzyme are equally valid for 3,4-PCD from other bacterial sources, for catechol 1,2-dioxygenase (1,2-CTD) and for all other enzymes of this type.

Structure determinations are now available for 3,4-PCD in the resting form,[2] for 3,4-PCD enzyme complexes with substrates,[3,4] substrate analogues[3] and inhibitors,[5] and for 3,4-PCD altered by site-directed mutagenesis.[6] Thus, in its resting state,[2] the active site of 3,4-PCD consists of a mononuclear iron(III) ion in a trigonal-bipyramidal coordination geometry (Fig. 8.1). Four of the ligand donor atoms derive from two tyrosine (Tyr 408 and Tyr 447) and two histidine residues (His 460 and His 462) of the peptide chain. Five-coordination around the iron ion is completed by a hydroxide ion in the equatorial plane. The axial coordination positions of the trigonal bipyramid are occupied by the Tyr 447 and His 462 residues. The structural changes that occur upon substrate-binding are summarized in Fig. 8.1.[3,4] First, the coordination geometry of the iron site changes from a trigonal bipyramid to a square pyramid. Second, both the hydroxide ion and the Tyr 447 residue serve as proton acceptors for the catechol protons and leave the coordination sphere of the iron ion after deprotonation of the substrate. Third, the catechol substrate is bound as a dianion where, however, the two Fe–O bond lengths of the coordinated catecholate unit vary substantially from each other (2.41Å and 2.16 Å).[3] The elongation of the

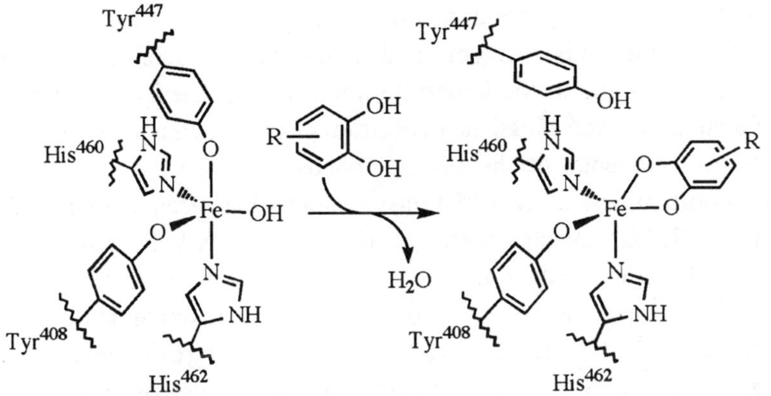

Fig. 8.1 Schematic views of the active sites in protocatechuate 3,4-dioxygenase in the resting form and in the enzyme-substrate complex.

Fe–O_{cat} bond *trans* to the Tyr 408 residue has been attributed to a *trans* influence exerted by the Fe–O_{tyr} bond. It has been proposed that the asymmetry is essential for the activation of the substrate and for the catalysis — in the sense that the weakening of the Fe–O_{cat} bond promotes ketonization of the corresponding C–O bond of the catecholate ligand. The last important structural change consists of the formation of a cavity with an open iron coordination site at approximately the position where formerly the hydroxide ion was bound to the iron ion. This cavity is large enough to be accessed by an oxygen molecule.

The spectroscopic investigation of the active site in intradiol-cleaving catechol dioxygenases serves as an impressive example of how powerful the sophisticated application of a combination of various spectroscopic methods can be in determining the structural properties of an unknown metal site in an enzyme. Thus, essentially, all the important structural features of the active site in intradiol-cleaving catechol dioxygenases were correctly predicted by spectroscopic results before they could be confirmed by recent structure determinations. In addition, the similarity between spectroscopic results from studies of various catechol dioxygenases emphasizes that the results obtained by the structure analysis seem to be equally applicable to all other intradiol-cleaving enzymes.

Mössbauer and EPR spectroscopy assign a high-spin iron(III) ion to the active site in the native enzyme and in the enzyme-substrate complex.[7,8] The rather low value of the hyperfine interaction of the iron electrons with the ^{57}Fe nucleus in the Mössbauer spectrum of the enzyme-substrate complex reflects the covalency of the Fe–O_{cat} bonds.

The coordination of two histidine residues has been demonstrated by Resonance Raman spectroscopy[9] as well as by EXAFS results.[10,11] The intense red color of the enzymes originates from moderately intense LMCT transitions due to the coordination of tyrosinate residues to an iron(III) ion.[12,13] Definite proof for the presence of tyrosine residues in the ligand sphere is given by Resonance Raman spectroscopy.[9,14,15] The observation of two different frequencies for the ν(C–O) vibration provides evidence for the presence of two distinct tyrosinate residues.[9,14] The absence of the ν(C–O) features corresponding to the Tyr 447 residue agrees with the crystallographic data showing the release of the Tyr 447 residue from the ligand sphere upon substrate-binding.[4]

The presence of a water-derived species, a water molecule itself or a hydroxide ion was shown by the hyperfine line broadening of the EPR signal of the native enzyme in the presence of ^{17}O-labeled water.[16] Interestingly, this line broadening disappeared in the spectrum of the enzyme-substrate complex, indicating that the solvent molecule is released from the coordination sphere when substrate is added. Based on the short Fe–O bond length of 1.9 Å supplied from EXAFS data, the solvent-derived species is attributed to a hydroxide ion.[11] Interestingly, retention of the five-coordination of the iron site upon substrate-binding was also predicted by XANES and EXAFS data prior to the determination of the crystal structure of the enzyme-substrate complex.[11]

Catecholate binding results in the appearance of an additional broad absorption at long wavelengths in the electronic absorption spectrum.[12] Based on Resonance Raman studies, these bands are attributed to catecholato-to-iron(III) CT transitions.[13,17] Furthermore, the frequencies for the vibrations observed in the Resonance Raman spectrum indicate a bidentate chelating coordination mode for the catecholate ligand.[13] An EPR study investigating the hyperfine line broadening effects of specifically ^{17}O-labeled substrate analogues on the EPR signals of enzyme-substrate complexes reached the same conclusion.[18] Bidentate coordination of catecholate derivatives was also detected by NMR spectroscopy with 3,4-PCD.[19] In contrast to the result obtained for 3,4-PCD, the NMR-spectroscopic investigation carried out with the enzyme 1,2-CTD reveals that, here, the catechol substrate is bound as a monodentate ligand.

Steady-state kinetic studies indicate an ordered bimolecular mechanism where the binding of the substrate begins the reaction sequence.[20] Stopped flow kinetic studies have provided evidence for the occurrence of two and three intermediates after the formation of the enzyme-substrate complex with 1,2-CTD and 3,4-PCD, respectively.[21] With genetically altered 3,4-PCD in which Tyr 447 is changed to a histidine by site-directed mutagenesis, the binding of the substrate can be resolved into four consecutive reaction steps.[6] Based on the observance of phenolato-to-iron(III) CT bands in the electronic absorption spectra of all intermediates, it was concluded that the iron ion does not change its trivalent oxidation state during the catalytic turnover.[20] This conclusion is supported by Mössbauer and EPR studies of steady–state intermediates of the reaction.[8] Considering that the binding of dioxygen to an iron(III) site is unlikely to occur, this finding indicates that the intradiol-cleavage reaction does not involve an oxygen activation mechanism; instead a substrate activation mechanism was proposed here.[22] The mechanistic details of the substrate activation have only been elucidated with the help of biomimetic model studies and will, therefore, be discussed at a later point in connection with these studies.

2.2 *Extradiol-cleaving catechol dioxygenases*

In contrast to the intradiol-cleaving catechol dioxygenases, extradiol-cleaving dioxygenases require high-spin iron(II) ions for their reactivity.[23,24] Due to the spectroscopic limitations of iron(II) ions, characterization of the active site was long restricted to Mössbauer studies. As a result, the knowledge of the structural and mechanistic details of the active site in extradiol-cleaving catechol dioxygenases was somewhat scarce. Within the last ten years, this situation has changed due to the employment of more sophisticated spectroscopic approaches and due to the availability of crystallographic information.

The understanding of the structural features of the active site in its resting state has been considerably advanced by recent structure determinations of the 2,3-dihydroxybiphenyl 1,2-dioxygenase (DHBD).[25] A schematic view of the active site is depicted in Fig. 8.2; the structure analysis assigns a square-pyramidal coordination geometry to the iron(II) ion. The metal ion is coordinated to two histidine residues, a glutamate residue, and two water molecules. The same structure is also found for the iron(III) site in the inactivated oxidized form of DHBD.[26] Steady-state kinetic studies indicate an ordered bimolecular reaction mechanism with catechol binding to the iron(II) center before the reaction with the oxygen molecule takes place.[27] The structure determination of the enzyme-substrate complex reveals that, upon coordination of the substrate 2,3-dihydroxybiphenyl, the two water molecules are expelled from the ligand sphere of the iron ion and that the catecholate derivative occupies those coordination sites which are *trans* to the two histidine residues, thereby leaving the coordination number and geometry at the iron center unchanged. In comparison to the structural parameters of a hydrogencatecholato iron(II) complex,[28] the two substantially different Fe–O bond lengths (2.10 Å and 1.93 Å) indicate that the substrate is bound in a similar way to the iron(II) ion in the enzyme (Fig. 8.2).

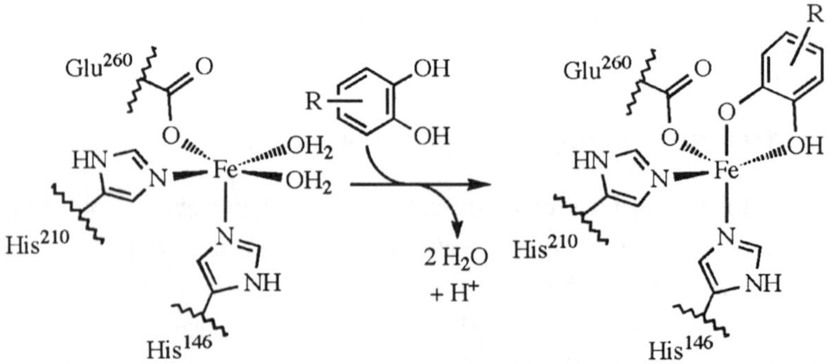

Fig. 8.2 Schematic views of the active sites in 2,3-dihydroxybiphenyl 1,2-dioxygenase before and after substrate-binding.

Based on several spectroscopic methods, the same coordination geometry and ligand sphere is proposed for the active sites in other extradiol-cleaving catechol dioxygenases (e.g. catechol 2,3-dioxygenase (2,3-CTD) and 4,5-protocatechuate dioxygenase (4,5-PCD)). Thus a square-pyramidal geometry is also suggested by the *d-d* transitions in the electronic absorption spectra of the resting form and the enzyme-substrate complex of 2,3-CTD, which were investigated by a combination of near-IR optical absorption, circular dichroism and magnetic circular dichroism spectroscopy.[29] X-ray absorption spectroscopy carried out on 2,3-CTD before and after addition of a substrate confirms that the coordination number and geometry are not altered upon binding of the substrate.[30,31] However, while the interpretation of the data by Que and co-workers is consistent with the square-pyramidal coordination geometry established by the structure determinations,[30] Bertini *et al.* suggest a six-coordination for the iron(II) ion.[31] It was proposed that this discrepancy in results arises from different methods of analyzing the raw data.[30]

The EXAFS data of the enzyme-substrate complex reveal that one O/N scatterer is located at the rather short distance of 1.93 Å, while the other four N/O ligand atoms are 2.10 Å away from the metal ion.[30] These bond lengths are in excellent agreement with those in a mononuclear hydrogencatecholato iron(II) model complex[28] and are consistent with the different bond lengths observed in the structure analysis for the DHBD-substrate complex. The histidine coordination is implied from EXAFS data[31] as well as from NMR spectroscopic evidence.[32]

Nitric oxide serves as an oxygen surrogate in the study of non-heme iron(II)-containing oxygen-activating enzymes. But more importantly, employment of this reagent provides the advantage that, upon binding, the spectroscopically rather inaccessible high-spin iron(II) site ($S = 2$ spin state) is converted to an Fe^{II}–NO species with a characteristic $S = 3/2$ spin state that can be investigated by EPR spectroscopy. The EPR spectra of the enzyme-NO complexes from 2,3-CTD and 4,5-PCD change upon reaction with the substrate, indicating that nitric oxide and the substrate are both bound to the iron ion in these enzyme-NO-substrate complexes.[23,33] When the aqueous solvent is enriched with ^{17}O-labeled water, the hyperfine line broadening of the EPR signals emanating from the enzyme-NO complex and certain enzyme-NO-inhibitor complexes versus no detectable line broadening of the EPR signal of the enzyme-NO-substrate complex indicates that at least one, possibly two water molecules are bound to the iron site in the native enzyme and that these water molecules are replaced by the catechol substrate.[33] The chelating binding mode of the catechol substrate has been further inferred from the hyperfine line broadening effects arising from ^{17}O-labeled substrate and inhibitor molecules.[34] It is interesting to note that the presence of a specific binding site for nitric oxide suggests that dioxygen is bound to the same iron coordination site during catalytic turnover.[33] Considering that the X-ray absorption spectrum of the enzyme-NO-substrate complex of 2,3-CTD clearly indicates six-coordination,[30] this observation suggests that the reaction of the enzyme-substrate complex with molecular oxygen might also result in the formation of a ternary complex in which the substrate and the dioxygen species together occupy three of the six coordination sites of the iron ion. In addition, substrate-binding increases the binding constant for nitric oxide by a factor of 1000; thus, if NO does react as a real O_2 analogue, this finding provides an attractive explanation for the ordering of the reaction mechanism implied by the steady-state kinetic studies.[27,33]

Taking into account all spectroscopic and crystallographic results, the mechanism depicted in Fig. 8.3 was proposed by Que *et al.* for the catalysis of extradiol-cleaving catechol dioxygenases.[30] In contrast to the intradiol-cleaving enzymes, all three protein-based ligands remain coordinated throughout the entire catalytic cycle. In the first step, the catechol substrate probably replaces both water molecules coordinated to the iron(II) ion in the resting state. The monoanionic catecholate unit binds to the metal ion in a chelating fashion. Substrate-binding increases the affinity of the iron(II) site for molecular oxygen, probably by lowering the Fe^{III}/Fe^{II} redox potential. Thus, in the reaction with dioxygen, an intermediary six-coordinate monohydrogencatecholato-superoxo iron(III) species is formed, which is subsequently converted to a semiquinonato-superoxo iron(II) complex by an intramolecular electron transfer reaction. The isolation of a similar semiquinonato-superoxo complex containing a rhodium(III) ion has been

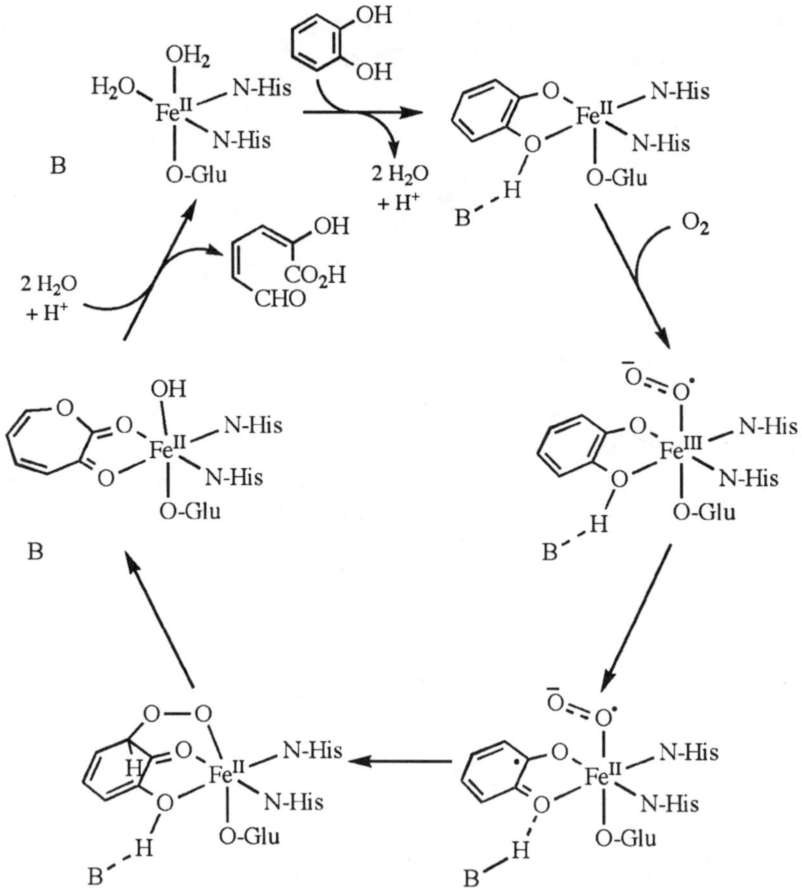

Fig. 8.3 Proposed mechanism for the reactivity of extradiol-cleaving catechol dioxygenases.[30] A basic site deriving from the peptide chain is denoted as "B".

reported.[35] The formation of a coordinated semiquinonate moiety alleviates the nucleophilic attack of the superoxide ligand onto the carbon atom adjacent to the enediol unit. By this means, an alkylperoxo complex is produced, which undergoes a Criegee rearrangement in the subsequent reaction step. As a result, a seven-membered 2,3-dioxolactone is formed, which is still coordinated to an iron(II) hydroxide species. Hydrolysis of the lactone and aquation affords the product α-hydroxymuconic semialdehyde and the original iron(II) site of the enzyme in the resting form.

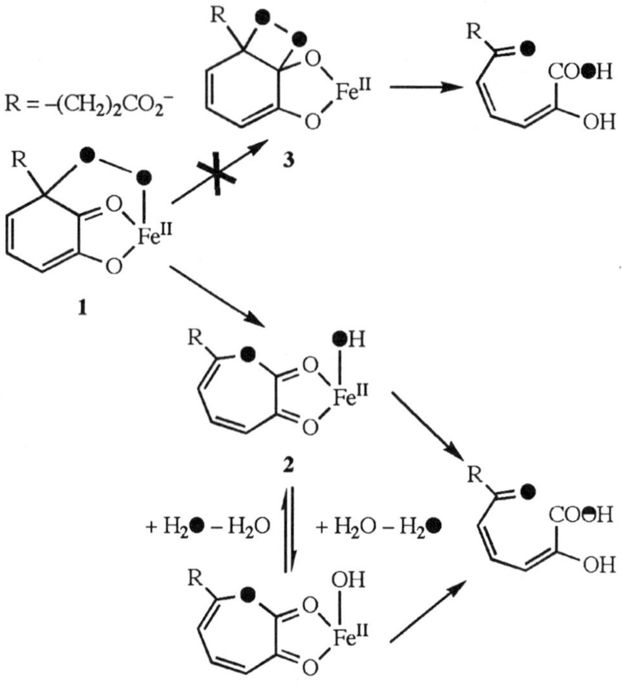

(8.2)

Two important mechanistic studies carried out with the extradiol-cleaving enzyme (2,3-dihydroxyphenyl)propionate 1,2-dioxygenase by Bugg *et al.* support the validity of the reaction mechanism described above. First, $^{18}O_2$ and $H_2{}^{18}O$ labeling experiments provide clear evidence that the alkylperoxo intermediate **1** reacts via a Criegee rearrangement to the unsaturated lactone **2** rather than via an intramolecular ring closure to the dioxetane intermediate **3**.[36] Thus, partial loss of ^{18}O-labeled oxygen atoms in the product is observed in the reaction with $^{18}O_2$ in $H_2{}^{16}O$ (Reaction 8.2), while partial incorporation of ^{18}O-labeled oxygen atoms in the product is found when the reaction with $^{16}O_2$ takes place in ^{18}O-enriched water. These results can only be explained by a Criegee rearrangement that is accompanied by the formation of either an iron(II)-bound hydroxide ion or water molecule, both of which could undergo solvent exchange reactions.

In a second experiment Bugg *et al.*[37] examined the extradiol-cleavage products of a catechol substrate that contains a cyclopropyl group attached to the aromatic ring (Reaction 8.3). Cyclopropyl groups are known to be capable of isomerizing via a radical-induced ring opening process. The observation of such an isomerization of the cyclopropyl group during catalytic turnovers suggests that a semiquinonate radical is indeed formed in the catalytic cycle.

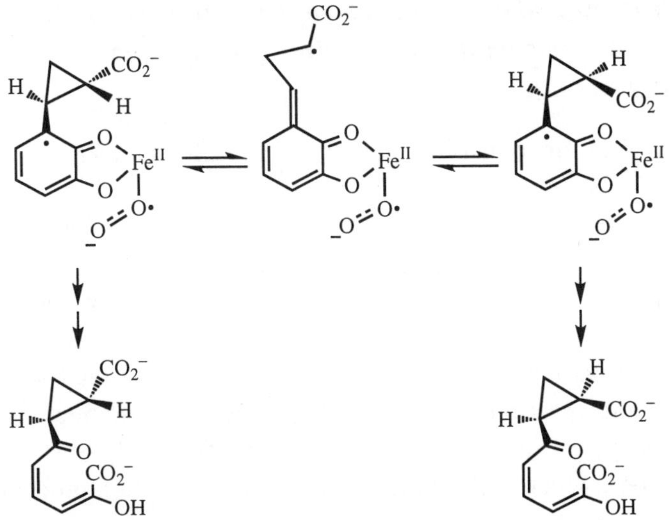

(8.3)

Finally it should be mentioned that an extradiol-cleaving catechol dioxygenase has been reported that requires manganese(II) instead of iron(II) ions for its reactivity.[38] However, knowledge about structural or mechanistic details for this type of enzyme is still very limited and model compounds are not yet available.

3. Functional Biomimetic Model Compounds

The occurrence of intradiol- and/or extradiol-cleavage reactivity of iron catecholate complexes is determined to the greatest extent by the type of coligand(s) present in the ligand environment of the iron ion. The composition of the donor atom set, the steric complexation properties, and especially the denticity of the coligand(s) are prominent factors that regulate the reaction of the coordinated catecholate ligand with dioxygen. This section reviews the biomimetic reactivity of mononuclear iron-containing model compounds for catechol dioxygenases with respect to the employed type of coligand(s).

3.1 Intradiol- and extradiol cleavage reactions with iron catecholate complexes containing mono- and bidentate ligands

One of the first functional iron-containing models mimicking the reactivity of intradiol-cleaving catechol dioxygenases was reported by Funabiki and co-workers.[39] Within a series of publications,[39–43] they investigated the reaction of 3,5-di-*tert*-butylcatechol H_2dbc with molecular oxygen in various anhydrous solvents (THF, DMF, MeCN) that contain $FeCl_2{\cdot}4H_2O$ or $FeCl_3$ and an excess of a pyridine derivative. They observed a diversity of products in various yields, depending on the specific reaction conditions applied in the reactions (Reaction 8.4). The main product, with yields generally ranging from 42% to 73%, was identified as 3,5-di-*tert*-butyl-1,2-quinone (dbq), which is the simple two-electron oxidation product of dbc^{2-}. Intradiol- (**4** and **5**) as well as extradiol-cleavage products (**6** and **7**) were detected together with some minor amounts of other oxygenation products (**8–10**). Utilization of ^{18}O-labeled oxygen, resulting in the incorporation of labeled oxygen atoms into the products, unambiguously establishes that products **4–10** arise from oxygenation reactions rather than from electron transfer reactions coupled with hydrolysis reactions.[40] Thus, in anhydrous tetrahydrofuran, the highest yields of intradiol-cleavage products (21%) are obtained for reaction mixtures containing H_2dbc (1 equivalent), $FeCl_3$

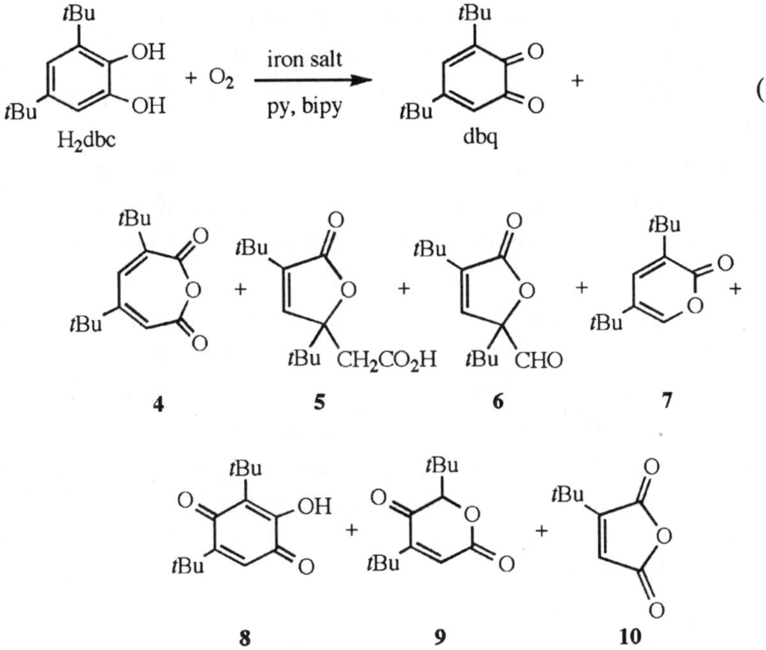

(8.4)

(0.125 equivalent), pyridine (11.2 equivalents) and bipyridine (0.125 equivalent); under these conditions the yields of extradiol-cleavage products and dbq were 9% and 52%, respectively.[40] Although rather low, turnover numbers of up to 2.2 still provide evidence for the catalytic properties of the iron salt. Pyridine acting as a ligand and as a deprotonation reagent is judged to be indispensable to these oxygenation reactions. Addition of bidentate π-acceptor ligands such as 2,2'-bipyridine and 1,10-phenanthroline is not essential for the reactions, but their presence increases the yields of the oxygenation reactions.[39] Because similar yields are achieved with either the iron(II) or iron(III) chloride salts[41] and because the same electronic absorption spectrum as that for the iron(III) solution was detected after exposure of the iron(II)-containing reaction mixture to an oxygen atmosphere,[40] it was concluded that the reaction proceeds through the same intermediate irrespective of the oxidation state of the utilized iron salt.

The yields for stoichiometric reactions could be improved to 50% by employing, as solvent, DMF containing additionally a borate buffer that served as a protecting reagent against oxidation of unbound catechol by oxygen.[42]

In the absence of pyridine, oxygenation reactions also occur if water is added as ligand to the reaction mixture;[43] the yields of intradiol-cleavage products are, however, significantly lower than in the case where pyridine is added. It is interesting to note that the presence of water instead of pyridine changes the selectivity for extradiol versus intradiol oxygenation; thus, in aqueous THF, especially in the iron(II)-containing reaction mixtures, extradiol oxygenation is favored over intradiol oxygenation. In addition, the amount of water added to the reaction solution exerts a profound influence on the degree of selectivity, indicating that different reaction intermediates are involved in the two oxygenation processes.[43]

Based on UV-vis spectroscopic evidence, the catecholate unit is bound to the iron(III) ion in a 1:1 ratio.[40] The electronic absorption spectrum of the iron(III) salt/H_2dbc mixture depends on whether pyridine or bipyridine is present in solution, indicating coordination of these ligands. Thus a pyridine-containing solution exhibits two intense absorption peaks at 550 nm and 975 nm, which are attributed to LMCT transitions. The disappearance of the absorption band upon reaction with oxygen indicates the involvement of this species in the catalytic process.

Solid compounds have been isolated with various catecholate derivatives, which render the same absorption spectra as their solutions prepared *in situ*.[44] Depending on the catecholate derivatives employed in the synthesis, complexes of different compositions were formed. Anionic complexes of the general formula $[FeCl_2(Cat)(py)]^-$ were obtained with catechol, 4-methylcatechol and 4-chlorocatechol, while 3,5-di-*tert*-butylcatechol and 3-methylcatechol produce neutral complexes of the type $[FeCl(Cat)(py)_2]$ where Cat and py denote a general unspecified catecholate derivative and a pyridine ligand, respectively. Based on EXAFS and XANES data, these complexes are described as containing five-coordinate iron ions and being dimeric in the solid states and monomeric in THF solutions.[44,45] In contrast to these compounds, a monomeric six-coordinate species of the composition $[FeCl_2(Cl_4\text{-cat})(py)_2]^-$ is formed with tetrachlorocatechol (H_2Cl_4-cat).

The reaction solution appears to be an intricate equilibrium involving at least two iron species.[44,46] Thus, under an argon atmosphere, the Mössbauer spectrum of a solution of iron(III) chloride with an approximate tenfold excess H_2dbc in pyridine reveals the presence of a high-spin iron(II) and a high-spin iron(III) species. Exposure of the solution to oxygen results in the disappearance of the iron(II) species. Funabiki *et al.* assign the high-spin iron(III) component to an iron(III) catecholate complex and the high-spin iron(II) component to an iron(II) semiquinonate complex. The NMR spectra of $H[FeCl_2(4\text{-Me-cat})(py)]$ and $[FeCl(3\text{-Me-cat})(py)_2]$ in pyridine/THF solution at different temperatures show that three and two iron species, respectively, are present in an equilibrium. The possible ligand exchange (Reaction 8.5) was proposed in order to explain, at least to some extent, the NMR data.

$$[FeCl_2(Cat)(py)_2]^- \xrightleftharpoons{+\,py\,-\,Cl^-} [FeCl(Cat)(py)_3] \xrightleftharpoons{+\,py\,-\,Cl^-} [Fe(Cat)(py)_4]^+ \qquad (8.5)$$

Funabiki *et al.* like to consider the Mössbauer and NMR data as evidence for the existence of an equilibrium between the mono- and bidentate coordination mode of the catecholate ligand to an iron(III) ion and, further, as evidence for the existence of an equilibrium between an iron(III) catecholate and an iron(II) semiquinonate species (Reaction 8.6). However, considering the experimental data available at the moment, this interpretation should be considered highly speculative.

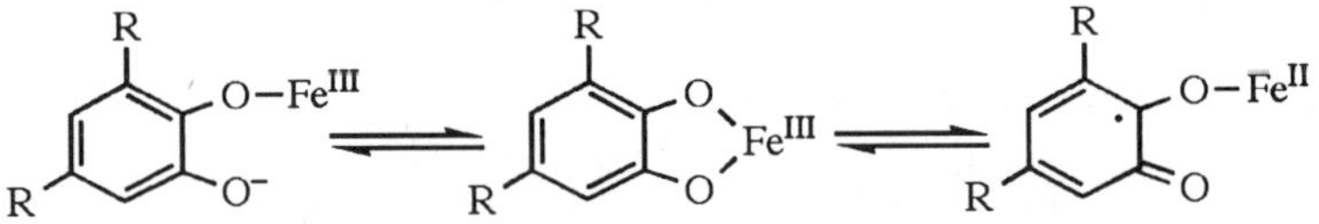

(8.6)

Recently it was shown that solutions of the isolated complex $[FeCl(dbc)(py)_2]$ do not support any oxygenation reaction with dioxygen.[45] Oxygenation reaction is observed only after an excess of H_2dbc is added.

This result was interpreted to mean that the catalytically active species contains two catecholate moieties. A reaction mechanism proposed by Funabiki for the intradiol ring cleavage reaction is depicted in Fig. 8.4.

In summary, simple iron(III) salts in the presence of pyridine and bidentate π-acceptor ligands have been shown to display modest catalytic properties for the conversion of catechol into intradiol-cleavage products by molecular oxygen. However, the reaction is not very selective. Thus, compared to the oxygenation reaction affording the cleavage of the intradiol C–C bond of the catechol, extradiol-cleavage reaction and other oxygenation reactions proceed to a noticeable extent. But more importantly, the sum of all oxygenation products is significantly outweighed by the product

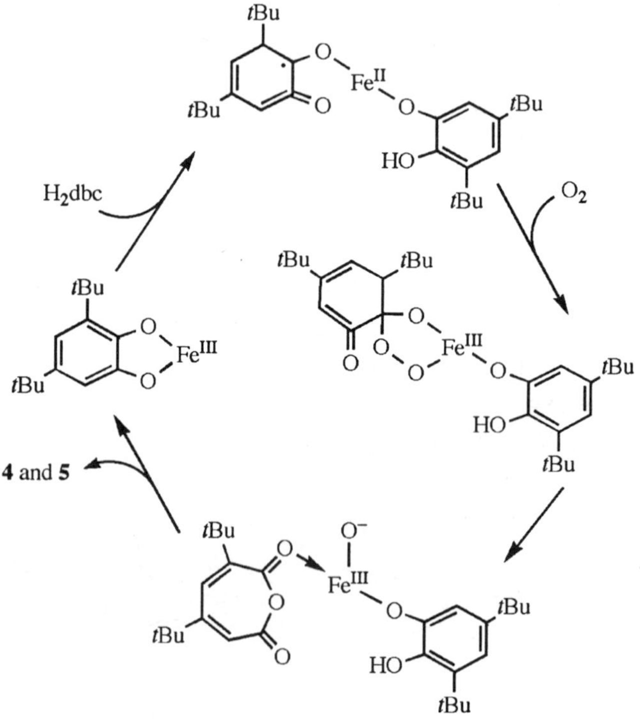

Fig. 8.4 Schematic mechanism postulated by Funabiki *et al.*[45] for the intradiol ring cleavage of 3,5-di-*tert*-butylcatechol H_2dbc catalyzed by iron(III) salts in the presence of pyridine ligand.

originating from the simple oxidation of the catechol derivative by two electrons without incorporation of any oxygen atoms. Last but not least, the identity of the oxygenation catalyst is by no means unambiguously established. The complexity of the reaction and the multitude of iron complexes present in solution make it very difficult to attribute any reliable reaction mechanism to the reaction or to extract any useful mechanistic aspects from these reactions for the elucidation of the catalytic mechanism of intradiol-cleaving catechol dioxygenases.

3.2 *Intradiol-cleavage reaction with iron catecholate complexes containing tetradentate ligands that enforce cis-octahedral coordination geometries*

A major breakthrough in the development of functional model compounds, which resulted in a better understanding of the reaction mechanism of the enzyme was achieved when it was realized that two easily accessible *cis*-coordination sites are needed for the occurrence of the intradiol ring cleavage reaction. First, two adjacent coordination sites support the bidentate coordination mode of catecholate ligand, while second, they provide the needed spacial proximity for the interaction between the coordinated catecholate moiety and the dioxygen molecule that allows the formation of a putative peroxoalkyl intermediate (complex **16** in Fig. 8.8). With tetradentate ligands that prefer equatorial coordination, the oxygen molecule and the monodentate bound catecholate unit are bound at two opposing axial coordination sites and therefore can only interact through electron transfer reactions via the metal ion, resulting in the simple oxidation of the coordinate catecholate to *o*-quinone by two electrons. Tripodal ligands are well-suited for establishing *cis*-octahedral coordination geometries around metal ions. A variety of ligands employed in model chemistry for intradiol-cleaving catechol dioxygenases is depicted in Fig. 8.5. The macrocycle L–N_4Me_2 is another type of tetradentate ligand that prefers *cis*-octahedral coordination geometry. Due to the small cavity of the twelve-membered ring and the enforced rigidity arising from the two pyridine rings, the macrocycle cannot coordinate to a metal ion in an equatorial fashion.

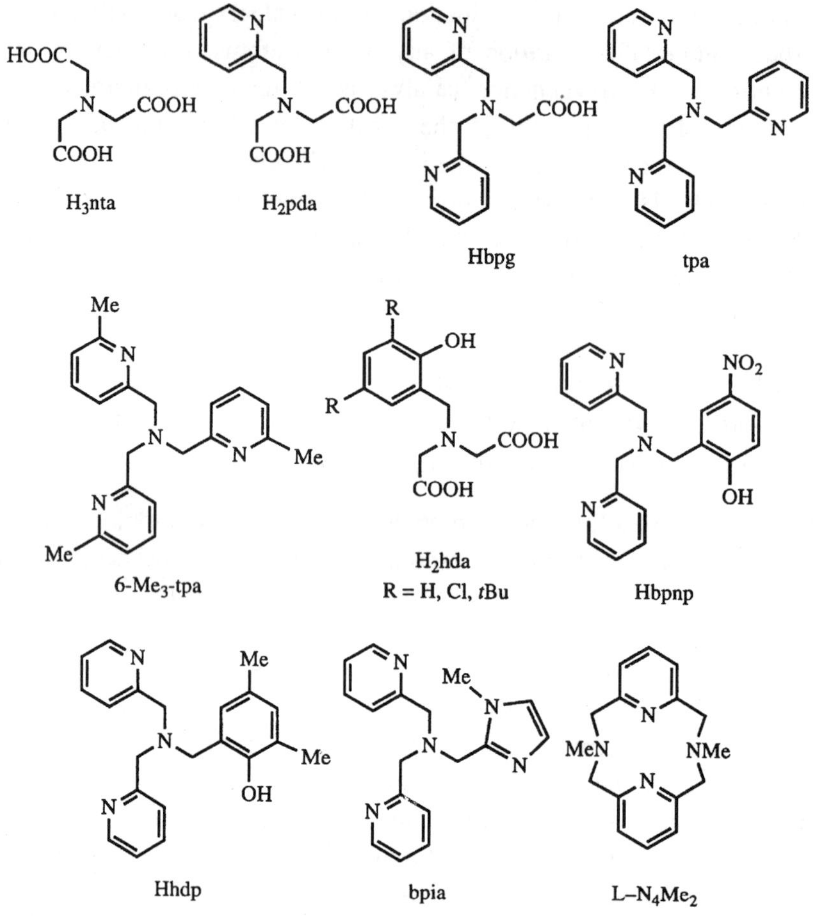

Fig. 8.5 Tetradentate ligands used in biomimetic complexes modeling the intradiol-cleavage reactivity.

One of the major advantages to investigate the reactivity of molecular oxygen with iron(III) catecholate complexes containing tetradentate instead of the previously discussed mono- and bidentate ligands is the fact that most complexes can be isolated under anaerobic conditions and characterized by various physical methods. By this means, the starting material for the investigation of the reactivity is well-defined.

3.2.1 Structural and electronic properties of the iron(III) catecholate complexes containing tetradentate coligands

In all these complexes, the catecholate unit is bound to the iron(III) ion in a bidentate chelating fashion. X-ray structure determinations were carried out on a number of model complexes.[47–53] For example, the perspective view of the complex cation $[Fe(L\text{–}N_4Me_2)(dbc)]^+$ is shown in Fig. 8.6. A comparison of Fe–O and C–O bond lengths is contained in Table 8.1. By virtue of belonging to the class of dioxolenes, coordinated catecholates are prone to behaving like non-innocent ligands. Thus, using the concept of formal oxidation states — which one has to keep in mind may not always be appropriate — the electron distribution in a given metal dioxolene complex can be described by any one of the three depicted redox isomers **11–13**. The C–O bond length can provide clear evidence for the correct assignment of the redox isomer. Thus, on the basis of the C–O bond lengths, the bound dioxolene moiety in all these complexes is identified as a catecholate ligand. No manifestation of a partial semiquinonate character

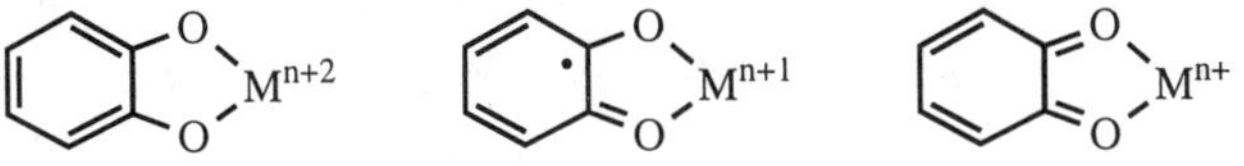

11 **12** **13**

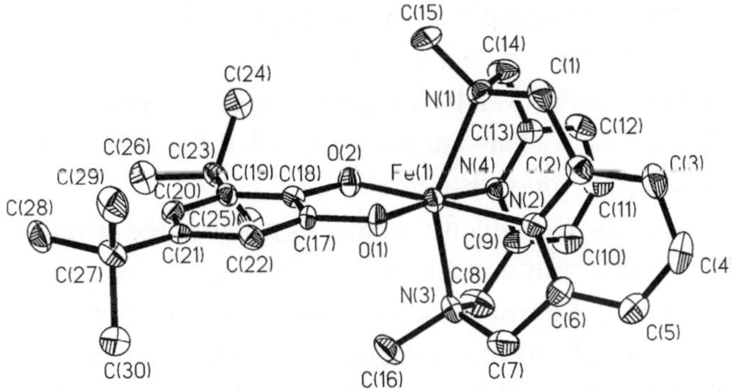

Fig. 8.6 Perspective view of the structure of the complex cation $[Fe(L\text{–}N_4Me_2)(dbc)]^+$.[52]

Table 8.1 Comparison of selected bond lengths in iron(III) catecholate complexes.

Compound	Bond Lengths (Å)		Ref.
	Fe–O	C–O	
$[Fe(nta)(dbc)]^{2-}$	1.887, 1.979	1.349, 1.338	47, 48
[Fe(bpg)(dbc)]	1.889, 1.989	1.356, 1.355	49
$[Fe(tpa)(dbc)]^{+}$	1.898, 1.917	1.347, 1.349	50
$[Fe(L\text{–}N_4Me_2)(dbc)]^{+}$	1.895, 1.911	1.344, 1.368	52
$[Fe(L\text{–}N_4Me_2)(cat)]^{+}$	1.903, 1.915	1.329, 1.353	51
$[Fe(L\text{–}N_4Me_2)(Cl_4\text{-}cat)]^{+}$	1.940	1.344	52
$[Fe(bpia)(Cl_4\text{-}cat)]^{+}$	1.919, 1.948	not reported	53

can be deduced from the structure analyses. A large difference in length between the two Fe–O_{cat} bonds has been observed for the complexes $[Fe(nta)(dbc)]^{2-}$ and [Fe(bpg)(dbc)], which has been attributed to the *trans* influence of the tetradentate ligand.[47–49] This asymmetry of Fe–O bonds of 0.09 Å–0.10 Å has been interpreted as a sign for the incipient formation of a monodentate coordination mode of the catecholate ligand in the solid state and was initially claimed to be a contributing factor in the reactivity of a coordinated catecholate unit with oxygen. A rather similar interpretation ensued from the observation of substantially different Fe–O_{cat} bond lengths in the enzyme-substrate complex.[3] The finding of rather similar Fe–O bond lengths in other complexes displaying even higher reactivities,[50–52] however, casts considerable doubts on the importance of asymmetric Fe–O bond lengths for the intradiol-cleavage reactivity of the biomimetic complexes as well as of the enzyme-substrate complex.

All complexes show an intense color. In general, the visible regions of their electronic absorption spectra are dominated by two absorption bands with molar extinction coefficients between 1000 $M^{-1}cm^{-1}$ and 3000 $M^{-1}cm^{-1}$. The absorption maxima of a selection of complexes are listed in Table 8.2 and in Table 8.3 (in Sec. 3.2.3). Based on the intensities, the dependence of the energies of the absorption maxima on the substituents at the aromatic ring of the catecholate ligand, the Resonance Raman spectra, and the redox properties of the complexes, the absorption bands are attributed to catecholato-to-iron(III) charge transfer transitions.[54,55] While the origin of the lower-energy LMCT band was proposed to involve a charge transfer from the π-orbitals of the oxygen atoms and the aromatic ring of the catecholate unit, the second LMCT band was suggested to arise from a charge transfer of the in-plane lone pair of the oxygen atoms to the σ-antibonding orbitals of the metal. From the data presented in Tables 8.2 and 8.3, it is evident that the more electron-withdrawing the substituents at the catecholate ring are and the less Lewis-acidic the iron(III) ion is, the greater the shift of absorption maxima to higher energies. A comparison of the electronic absorption spectra of the solids and the solutions of $[Fe(L{-}N_4Me_2)(dbc)]^+$ and $[Fe(L{-}N_4Me_2)(cat)]^+$ implies that the bidentate coordination mode of the catecholate ligands remains preserved to the most extent when the complexes are dissolved.[51]

When reported, the EPR spectra of the complexes clearly indicate the presence of high-spin iron(III) ions.[28,54,56,57] On the other hand, the solutions of the iron(III) catecholate complexes with $L{-}N_4Me_2$ as coligand render spectra that are comprised of two signals arising from a high-spin ($S = 5/2$) and a low-spin ($S = 1/2$) iron(III) component.[55] The reason why there are two different spin states observed when this tetradentate ligand is employed is still uncertain and requires further investigation. However, for all the iron(III) catecholate complexes thus far examined by EPR spectroscopy, the spectra provide no indication for the presence of a coordinated or unbound semiquinone radical.

Que and co-workers investigated a series of iron(III) 3,5-di-*tert*-butylcatecholate complexes with the tripodal ligands nta^{3-}, pda^{2-}, bpg^-, and tpa, where, starting with the ligand nta^{3-}, the individual carboxymethyl arms are successively replaced by pyridylmethyl arms.[49,50] They were able to demonstrate that the amounts of spin density transferred onto the coordinated catecholate ligand increases with the Lewis-acidity of the iron(III) center. Thus, the ^{1}H-NMR signals corresponding to the 3,5-di-*tert*-butylcatecholate

Table 8.2 Comparison of some properties of iron(III) 3,5-di-*tert*-butylcatecholate complexes.

Compound	λ_{max} (nm) [solvent]	kO_2 ($M^{-1}s^{-1}$) [solvent]	δ(dbc 6- and 4-H) (ppm)	Ref.
$[Fe(nta)(dbc)]^{2-}$	408, 622 [MeCN]	0.01 [MeOH]	19, 14	50
$[Fe(pda)(dbc)]^{-}$	444, 688 [MeCN]	0.05 [MeOH]	16, 9	50
[Fe(bpg)(dbc)]	488, 764 [MeCN]	0.43 [MeOH]	8, −18	50
$[Fe(tpa)(dbc)]^{+}$	568, 883 [MeCN]	10 [MeOH]	−5, −57	50
$[Fe(6\text{-}Me_3\text{-}tpa)(dbc)]^{+}$	600, 1020 [MeCN]	0.0095 [MeCN]	not reported	28
$[Fe(hdp)(dbc)]^{-}$	476, 726 [DMF]	0.0033 [DMF]	30, 16	49
$[Fe(bpia)(dbc)]^{+}$	558, 865 [MeOH]	4.3 [MeOH]	not reported	53
$[Fe(L\text{-}N_4Me_2)(dbc)]^{+}$	553, 784 [MeCN]	0.38 [MeCN]	12, −30	51

ligand experience considerable contact shifts depending on the nature of the tripodal ligand (Table 8.2). In contrast, the positions of ^{1}H-NMR signals arising from *m*-pyridine protons of the tripodal ligand are hardly affected by the change in Lewis-acidity and are characteristic of high-spin iron(III) complexes with this type of tripodal ligands. The large variation in the contact shifts for the individual resonances of the catecholate moiety was judged to be too large to only be attributed to an increase of unpaired spin density at the catecholate moiety via spin delocalization from the paramagnetic center. Instead, the shift pattern of the catecholate unit is considered to originate from the enhanced semiquinone character of the coordinated dioxolene unit as the Lewis-acidity of the iron(III) center is increased. To be perfectly clear, the iron compounds are still best described as high-spin iron(III) catecholate complexes, as evidenced by the structural data, the UV-vis and EPR spectra, and the NMR shifts of the pyridine protons. But as the Lewis-acidity and therefore the covalency of the Fe–O_{cat} bonds increase, mixing of a higher energy iron(II) semiquinonate state into the iron(III) catecholate ground state occurs to an extent large enough to be observed in the NMR spectra because of the high sensitivity of the paramagnetic shift to unpaired spin density and distribution, but too small to affect the structural or other electronic features of the complexes.

3.2.2 *Stoichiometric reactions of the iron(III) 3,5-di-tert-butylcatecholate complexes with molecular oxygen*

All iron(III) 3,5-di-*tert*-butylcatecholate complexes with the ligands depicted in Fig. 8.5 react with molecular oxygen to afford the intradiol-cleaving products **4** and **14** in varying yields. In general, less reactive complexes (e.g. [Fe(bpnp)(dbc)] and $[Fe(hda)(dbc)]^{2-}$)[48,58] display rather modest yields of intradiol-cleavage products and generate noticeable amounts of *o*-benzoquinone as by-product, while fast reacting complexes render nearly quantitative yields of intradiol-cleavage products (Reaction 8.7) in contrast to those iron complexes with mono- and bidentate nitrogen-containing ligands previously discussed.

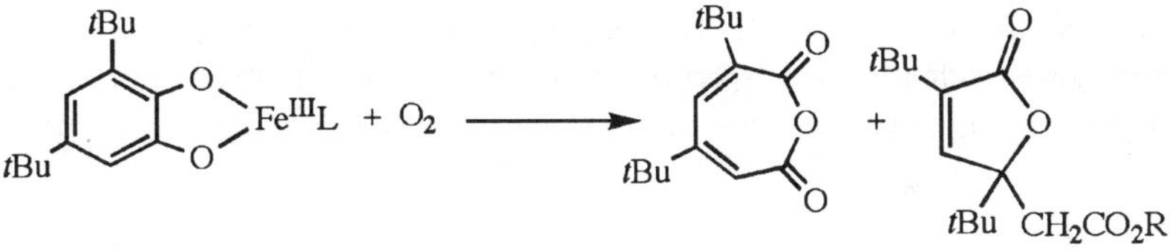

4 **14**

(8.7)

The progress of the stoichiometric reaction of the complexes with dioxygen can be conveniently monitored by the disappearance of the intense CT bands in the visible region of the electronic absorption spectrum (Fig. 8.7). The exponential decrease of the intensity of the absorption maximum of the lowest energy CT band over time shows that the reaction follows pseudo-first order kinetics (see inlet of Fig. 8.7). A comparison of the reaction velocity constants determined for some iron(III) 3,5-di-*tert*-butylcatecholate complexes is presented in Table 8.2. The finding that, upon exposure of solutions of the iron(III) catecholate complexes[49,50,51,53] to an atmosphere of dioxygen, no intermediary formation of a different chromophore but only the exponential decay of the absorption bands corresponding to the iron(III) starting material are noticed in the electronic

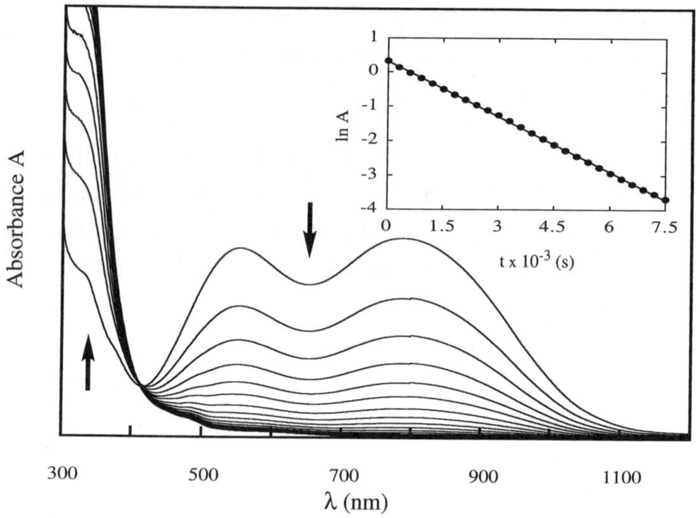

Fig. 8.7 Progress of the reaction of $[Fe(L\text{–}N_4Me_2)(dbc)]^+$ with air in MeCN at room temperature as monitored by electronic absorption spectroscopy in 10-min intervals. The arrows indicate how the intensities of the absorption bands change over time. Inlet: plot of ln *A* versus time *t*.[51]

absorption spectra recorded over time, allows the conclusion that the iron(III) 3,5-di-*tert*-butylcatecholate complexes themselves react with the oxygen molecules.

The first systematic study of the factors that governs the reactivity of iron(III) 3,5-di-*tert*-butylcatecholate complexes was initiated by Que and co-workers using the series of tripodal ligands nta^{3-}, pda^{2-}, bpg^- and tpa.[49,50] The results of this study prepared the basis for the current understanding of the reaction mechanism of intradiol-cleaving dioxygenases. It was observed that the rate of the cleavage reaction increases by three orders of magnitude and the yields of intradiol-cleavage products are progressively improved from 84% to 98% as the Lewis-acidity of the iron(III) center is increased. Since a wealth of spectroscopic information on these complexes is available, it is possible to correlate the reactivity with the electronic properties of the complexes.

A more Lewis-acidic metal ion will form stronger covalent metal–oxygen bonds with the catecholate ligand. The observation of no reactivity with a gallium(III) catecholate complex,[49] however, demonstrates that Lewis-acidity alone is not the important factor. The metal ion must be also redox-active; that means in the present context that the reduced metal oxidation state must be somewhat accessible. Thus, a higher covalency of the bonds in the redox-active iron(III) ion results in a higher degree of partial transfer of electron density from the coordinated catecholate ligand to the metal ion, thereby furnishing the coordinated catecholate moiety with some partial semiquinone character and partially reducing the trivalent metal ion. As described above, partial iron(II) semiquinonate character in the electronic structure of the iron(III) catecholate complexes has been detected by the NMR-shifts of the catecholate protons (Table 8.2). Thus, intradiol-cleavage reactivity is linked to the amount of iron(II) semiquinonate character of the starting complexes rather than the Lewis-acidity of the metal ion.[49,50]

Therefore, the role of the iron(III) ion in intradiol-cleaving dioxygenases consists of the activation of catecholate substrate rather than that of dioxygen. Enhancement of the semiquinone radical character on the coordinated catecholate moiety makes the enediol carbon atoms more susceptible to an electrophilic attack by a triplet oxygen molecule. The hypothesis of a mechanism based on substrate activation serves nowadays as a generally accepted explanation for the unusual reactivity of molecular oxygen with iron(III) complexes.

In addition to the NMR shifts of the catecholate ligand, a correlation of reactivity with the optical properties of the iron(III) catecholate complexes has been found. In general, the more reactive the complexes are, the lower in energy the two LMCT bands are. This relationship can be explained by the fact that these LMCT bands describe optical transitions from an iron(III) catecholate ground state to an iron(II) semiquinonate excited state. Therefore, the less the two states are separated by energy, the more they can mix, allowing the iron(III) catecholate ground state to obtain more iron(II) semiquinonate character. However, one has to keep in mind that this correlation between the reactivity and the optical properties of the complexes is not a strict one. For example, the lowest energy LMCT band for the complex $[Fe(6\text{-}Me_3\text{-}tpa)(dbc)]^+$ (1020 nm)[28] is found at a higher wavelength than that of $[Fe(tpa)(dbc)]^+$ (883 nm)[50], nevertheless the former is considerably less reactive than the latter.

It is interesting to note that the reaction velocity constants measured for four iron(III) 3,5-di-*tert*-butylcatecholate complexes with the same FeN_4O_2 coordination environment, ($[Fe(tpa)(dbc)]^+$,[50] $[Fe(bpia)dbc]^+$,[53] $[Fe(L\text{–}N_4Me_2)dbc]^+$,[51] and $[Fe(6\text{-}Me_3\text{-}tpa)(dbc)]^+$),[28] span a range of four orders of magnitude, indicating that slight changes in the σ- and π-bonding properties of the nitrogen donor atoms and in the steric factors of the ligand can exert a tremendous influence on the reactivity. In this connection, it is worthwhile to mention that the iron(III) 3,5-di-*tert*-butylcatecholate complex containing the tetraazamacrocycle SS-CTH as coligand exhibits no intradiol-cleavage reactivity at all although it displays LMCT bands at rather high wavelengths.[59]

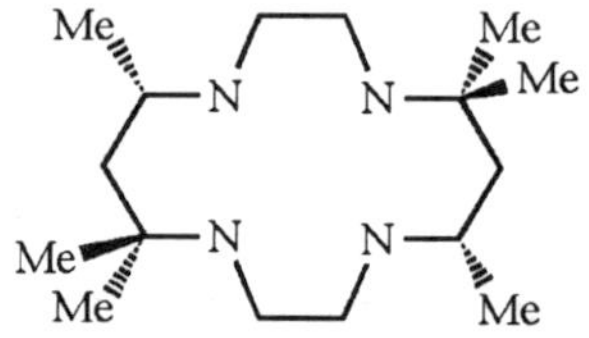

SS–CTH

A comparison of the reaction velocity constants listed in Table 8.2 reveals that, in addition to the catecholate ligand, only nitrogen donors are coordinated to the six-coordinate iron ion in the most reactive model complexes, while in the enzyme-substrate complex the five-coordinate iron ion is surrounded by two histidine residues, one tyrosinate residue and the catecholate unit. Thus, in contrast to the trend observed in the biomimetic model studies, the enzyme is considerably more reactive despite the less Lewis-acidic metal site.

Based on the biomimetic model studies, a mechanism has been proposed for the reactivity of intradiol-cleaving catechol dioxygenases (Fig. 8.8).[1a]

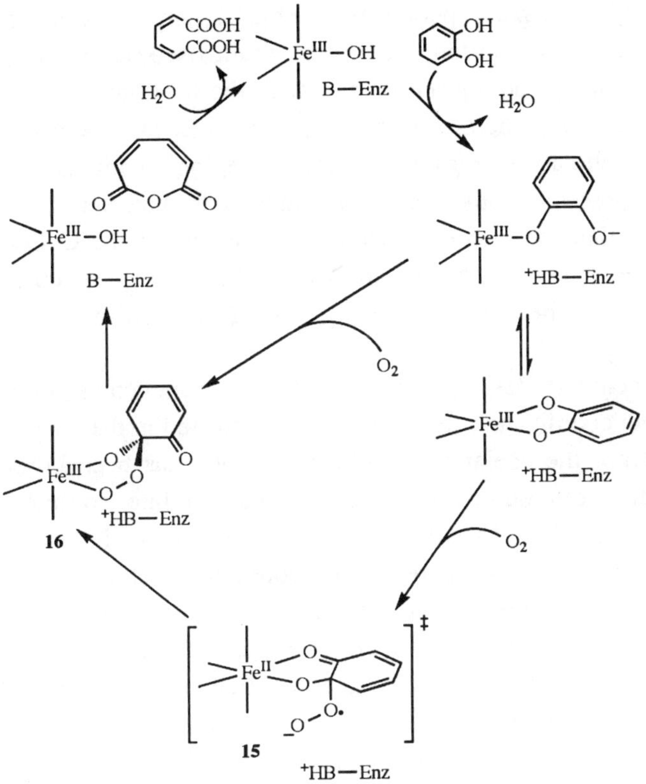

Fig. 8.8 Proposed mechanism for the iron(III) ion-catalyzed intradiol ring cleavage by dioxygen.

In the first step, the catecholate ligand coordinates to the iron(III) ion. It has already been commented on that in the enzyme the coordinated hydroxide and one of the tyrosinate residues function as proton acceptors in this step. In their proposed reaction mechanisms, some authors[49,50,59] prefer to depict the occurrence of an equilibrium reaction between the valence tautomeric iron(III) catecholate and iron(II) semiquinonate complexes where the equilibrium lies very much on the side of the former species. An estimate of the equilibrium constant for the equilibrium between the valence tautomers of $[Fe(SS\text{-}CTH)(dbc)]^+$ has been reported to be

$6 \cdot 10^{-6}$.[59] While the participation of an iron(II) semiquinonate species may therefore indeed be given, no direct spectroscopic evidence for the involvement of this equilibrium reaction in the dioxygenation reaction has been provided yet. Therefore, in the present mechanistic cycle, only the iron(III) catecholate species is depicted; keeping in mind, however, that this species possesses some iron(II) semiquinonate character. It is worthwhile to point out that the incipient ketonization of one of the C–O catecholate bonds inferred from the structure analyses of the enzyme-substrate complexes has been described as promoting the formation of a coordinated semiquinonate.[3,4]

The catecholate may then be bound in a monodentate or a bidentate fashion; both coordination modes have been detected in the enzyme-substrate complex. Only the chelating coordination mode has been observed in all these model compounds. For the iron catecholate complex with the macrocycle L–N_4Me_2 as coligand, it has been noted that, because of its rigidity, the macrocyclic ligand, once coordinated, remains bound through all nitrogen donor atoms to the metal ion at all times. In addition, the electronic absorption spectrum of the complex in the solid state resembles that of the dissolved complex, suggesting that the chelating coordination mode of the catecholate unit, demonstrated in the solid state, is to the most extent preserved in solution. Thus, a direct attack of the oxygen molecule onto the bidentate catecholate ligand (which displays, as previously discussed, some semiquinone radical character) seems very likely. The resulting, very short-lived alkylperoxy radical coordinated to an iron(II) ion (**15**) immediately reacts further to an alkylperoxo iron(III) intermediate (**16**). Alternatively, monodentate coordination of the catecholate unit may create an open coordination site, allowing the oxygen molecule to react in a concerted fashion with the catecholate ligand *and* the iron ion to the alkylperoxo iron(III) intermediate. The possibility of an alkylperoxo iron(III) intermediate is supported by the structural characterization of a similar alkylperoxo complex, which was formed in the reaction of dioxygen with an iridium(III) catecholate complex containing triphos as coligand.[60] In the enzyme, it was suggested that the alkylperoxide oxygen atom coordinates to the open coordination site generated upon substrate-binding.[3]

A Criegee rearrangement of the alkylperoxo complex affords muconic anhydride and an iron(III) hydroxide species, which may participate in the subsequent hydrolysis of the acid anhydride. Indications that this reaction step does not proceed via a dioxetane intermediate were provided by the reaction of $[Fe(nta)(dbc)]^{2-}$ with $^{18}O_2$, where a single ^{18}O label is incorporated in the 3,5-di-*tert*-butyl-5-carboxymethyl-2-furanone derivative.[47] Similar labeling experiments with the enzyme 1,2-CTD showed the loss of some ^{18}O-labeled atoms in the product due to the solvent exchange reaction of the iron(III) hydroxide species with $H_2{}^{16}O$.[61] In addition, with faster reacting model complexes in less nucleophilic solvents, 3,5-di-*tert*-butylmuconic anhydride (**4**) can be obtained as the main product, providing definite evidence for the occurrence of a Criegee rearrangement in the reaction mechanism.

3.2.3 Intradiol-cleavage reaction of other coordinated catechol derivatives

The substrate 3,5-di-*tert*-butylcatechol is the most widely employed catecholate derivative in investigations of the reactivity of biomimetic model compounds for catechol dioxygenases because the bulky *tert*-butyl groups prevent oxidative polymerization of the catechol and because rather stable intradiol-cleavage products can be obtained. In addition, the electron-donating alkyl groups attached to the aromatic ring increase the reactivity of the catecholate ligand. Until very recently, only one study employed another catechol derivative other than 3,5-di-*tert*-butylcatechol;[57] nowadays the investigation of biomimetic reaction has been extended to include a wide variety of catechol derivatives (Table 8.3).[51–53,55,62]

Koch and Krüger were originally able to demonstrate a ring-opening for the reaction of oxygen with coordinated catechol itself — a true substrate of the intradiol-cleaving enzyme 1,2-CTD. The reaction is, however, considerably slower than that of the enzyme ($2.5 \cdot 10^5$ M^{-1} s^{-1}). In a more recent study, the oxidative cleavage of chlorocatechol derivatives, which are also authentic substrates for chlorocatechol dioxygenases, was reported by Funabiki and co-workers.[62]

Table 8.3 Comparison of some properties of iron(III) complexes with various catecholate ligands.

Compound	λ_{max} (nm) [solvent]	k_{o2} (M^{-1} s^{-1}) [solvent]	Ref.
$[Fe(L\text{-}N_4Me_2)(dbc)]^+$	553, 784 [MeCN]	$3.8 \cdot 10^{-1}$ [MeCN]	51
$[Fe(L\text{-}N_4Me_2)(3\text{-}MeO\text{-}cat)]^+$	607, 743 (sh) [MeCN]	$8.6 \cdot 10^{-2}$ [MeCN]	55
$[Fe(L\text{-}N_4Me_2)(4\text{-}t\text{Bu-}cat)]^+$	519, 766 [MeCN]	$1.4 \cdot 10^{-2}$ [MeCN]	55
$[Fe(L\text{-}N_4Me_2)(cat)]^+$	501, 734 [MeCN]	$2.8 \cdot 10^{-3}$ [MeCN]	51
$[Fe(L\text{-}N_4Me_2)(3\text{-}EtO_2C\text{-}cat)]^+$	484, 693 [MeCN]	$<1.2 \cdot 10^{-5}$ [MeCN]	55
$[Fe(L\text{-}N_4Me_2)(Cl_4\text{-}cat)]^+$	555 (sh), 677 [MeCN]	$3.5 \cdot 10^{-5}$ [MeCN]	52
$[Fe(bpia)(dbc)]^+$	558, 865 [MeOH]	4.3 [MeOH]	53
$[Fe(bpia)(4\text{-}Me\text{-}cat)]^+$	515, 842 [MeOH]	2.6 [MeOH]	53
$Fe(bpia)(4\text{-}t\text{Bu-}cat)]^+$	519, 836 [MeOH]	2.3 [MeOH]	53
$[Fe(bpia)(cat)]^+$	510, 800 [MeOH]	$8.3 \cdot 10^{-2}$ [MeOH]	53
$[Fe(bpia)(Cl_4\text{-}cat)]^+$	528, 729 [MeOH]	$>10^{-5}$ [MeOH]	53
$[Fe(tpa)(dbc)]^+$	568, 883 [MeCN]	10 [MeOH]	50
$[Fe(tpa)(3\text{-}Cl\text{-}cat)]^+$	532, 766 [MeCN]	$6.4 \cdot 10^{-2}$ [MeCH]	62
$[Fe(tpa)(4\text{-}Cl\text{-}cat)]^+$	479, 786 [MeCN]	$8.4 \cdot 10^{-2}$ [MeCH]	62

In Table 8.3, the velocity constants of iron(III) complexes with various catecholate derivatives are correlated to the optical properties. It can be seen that the reactivity decreases as more electron-withdrawing substituents are attached to the aromatic ring. This is consistent with the hypothesis of an electrophilic attack of the dioxygen onto the enediol carbon. In general, the LMCT bands shift to higher energies and the redox potentials for the iron(III) catecholate/iron(II) catecholate and iron(III) semiquinonate/iron(III) catecholate redox couples (values not shown)[52,55] increase as the substituents at the aromatic ring become more electron-withdrawing.

3.2.4 Catalytic intradiol-cleavage reaction

The first catalytic reaction with an iron complex coordinated to a tetradentate ligand was reported with $[Fe(nta)(dbc)]^{2-}$ anions.[56,57] However, the reaction proceeds very slowly, requiring several days for completion. In addition, a borate buffer is needed to prevent the oxidation of the catechol derivative by air itself to *o*-benzoquinone. Further and most noteworthy, the product distribution (84% of **14**) has been questioned and, instead of the oxygenation products, *o*-benzoquinone was found to be the main product in this reaction (58% of dbq, 14% of (**4** + **14**)).[42] A further catalytic reaction was reported with the complex [Fe(bpnp)(dbc)].[58] But here again the reaction is very slow and *o*-benzoquinone was obtained as the main product (40% of dbq, 10% of **4**).

The first well-defined stoichiometrically functioning *and* highly catalytic biomimetic model system for the reactivity of *intradiol*-cleaving catechol dioxygenases was introduced by Koch and Krüger.[51] They employed the complex $[Fe(L\text{–}N_4Me_2)(dbc)]^+$ as an efficient catalyst in the conversion of H_2dbc to 3,5-di-*tert*-butylmuconic anhydride (**4**). With 10% of the iron(III) catalyst, yields of 70% for **4** and 13% for the corresponding *o*-benzoquinone

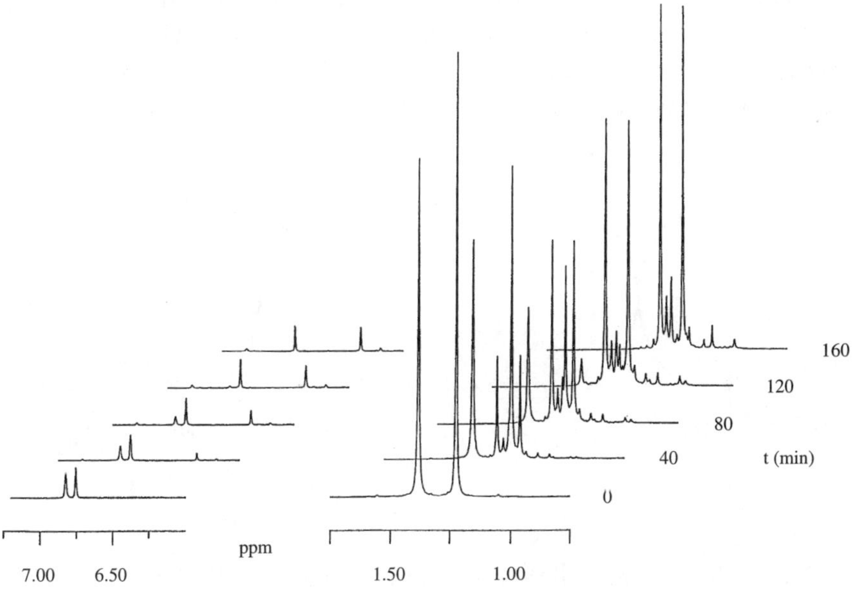

Fig. 8.9 Catalytic oxidation of H_2dbc as monitored by ^{1}H NMR spectroscopy (with 10% of the catalyst $[Fe(L\text{–}N_4Me_2)(dbc)]^+$; NMR signals for unbound H_2dbc: $\delta = 6.83$, 6.76, 1.38, and 1.22 ppm, for **4**: $\delta = 6.74$, 6.29, 1.28, and 1.18 ppm, and for dbq: $\delta = 7.08$, 6.15, 1.25, and 1.23 ppm).[51]

are obtained after 3 h of reaction time under an atmosphere of pure oxygen. Turnover numbers of 54 are achieved with 1% of the catalyst. The progress of the reaction can be monitored by ^1H NMR spectroscopy as shown in Fig. 8.9. The direct observation of the products composition over time during catalytic turnovers allows the conclusion that muconic anhydride is indeed the direct product of the oxygenation reaction. The development of reaction products over time definitely excludes any involvement of *o*-benzoquinone as an intermediate in the oxygenation reaction. Since the amounts of the minor product, *o*-benzoquinone, have increased in the catalytic reaction compared to the stoichiometric reaction and, since catechol itself is susceptible to oxidation reaction by dioxygen in basic solutions, the formation of *o*-benzoquinone is not thought to be mediated by the iron complex. For the catalytic reaction, only two equivalents of a base are needed per iron(III) ion. In contrast to the previous study with $[Fe(nta)(dbc)]^{2-}$, no borate buffer is needed to protect H_2dbc from oxygen because the reaction proceeds in considerably less time. This very efficient catalytic cycle is summarized by Reaction 8.8. The exact identity of the putative intermediate "$[(L–N_4Me_2)Fe^{3+}–O^{2-}]^{+}$" is not yet known.

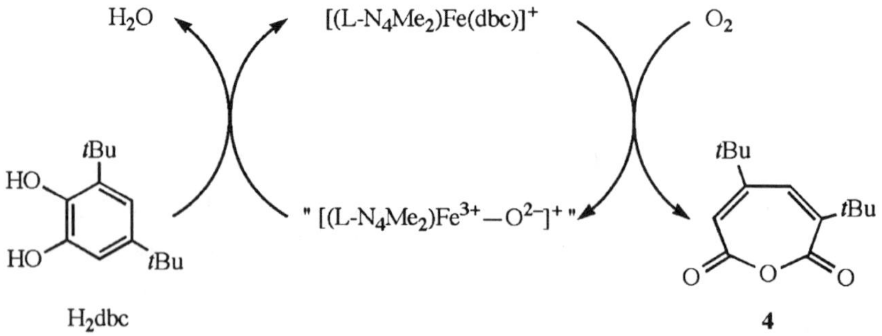

(8.8)

Since then, other highly catalytic systems, $[Fe(bpia)(dbc)]^{+}$ and $[Fe(tpa)(dbc)]^{+}$,[53,63] have appeared, with even better yields of 3,5-di-*tert*-butylmuconic anhydride. From this, a general trend can be discerned: namely as the stoichiometric reaction of the corresponding complex with oxygen becomes faster, the yields of 3,5-di-*tert*-butylmuconic anhydride **4** and the turnover number increase and the amounts of the side-product 3,5-di-*tert*-butyl-*o*-benzoquinone (dbq) decrease.

Following the same argument, the best strategy for observing catalytic reactions with even less reactive catechol derivatives involves utilization of the ligand system that yields the fastest reaction for a 3,5-di-*tert*-butylcatecholate complex. This strategy was indeed successfully employed for the catalytic oxygenation of chlorocatechol derivatives by an iron(III) catalyst that contains tpa as coligand.[62]

3.2.5 *Reactivity of iron(II) 3,5-di-*tert*-butylcatecholate and iron(III) 3,5-di-tert-butylbenzosemiquinonate complexes*

By virtue of belonging to the class of dioxolene complexes, a rich redox chemistry is expected to be associated with iron(III) 3,5-di-*tert*-butylcatecholate complexes, since redox reactions can involve the metal ion as well as the ligand. Thus, the iron(III) complex $[Fe(L\text{–}N_4Me_2)(dbc)]^+$ (**17**) can be electrochemically reduced to the high-spin iron(II) catecholate complex $[Fe(L\text{–}N_4Me_2)(dbc)]$ (**18**) and oxidized to the iron(III) semiquinonate complex $[Fe(L\text{–}N_4Me_2)(dbsq)]^{2+}$ (**19**, $dbsq^-$ = 3,5-di-*tert*-butylbenzosemiquinonate(1–)) (Fig. 8.10). The iron(III) semiquinonate complex **19** has been isolated and fully characterized by structure analysis, Mössbauer spectroscopy and magnetochemical methods.[64] All data are consistent with the description of the compound as a low-spin iron(III) complex in which a strong antiferromagnetic coupling occurs between the $S = 1/2$ spin of the metal ion and the $S = 1/2$ spin of the coordinated ligand radical, resulting in diamagnetism for the complex at room temperature. The appearance of the oxidative response in the cyclic voltammogram as a seemingly irreversible oxidation is caused by the existence of equilibrium (Reaction 8.9). This equilibrium reaction has been unambiguously established by NMR, Mössbauer and electronic absorption spectroscopy as well as by electrochemical methods.[64]

As previously discussed, the intradiol-cleavage reactivity of the iron(III) 3,5-di-*tert*-butylcatecholate complex **17** has been linked to a partial iron(II) semiquinonate character that this complex possesses. Thus, with the

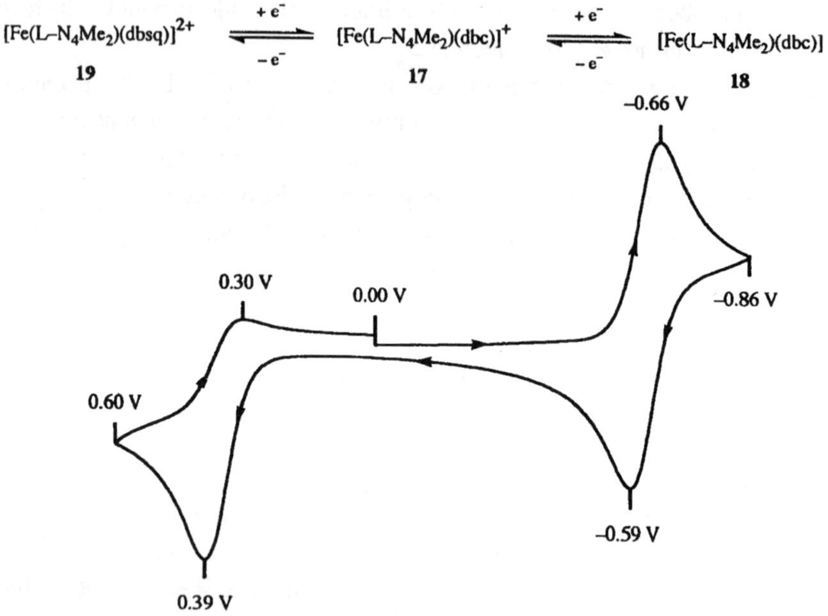

Fig. 8.10 Cyclic voltammogram of $[Fe(L–N_4Me_2)(dbc)]^+$ at a Pt-foil electrode in acetonitrile at 25°C (peak potentials versus SCE are indicated).

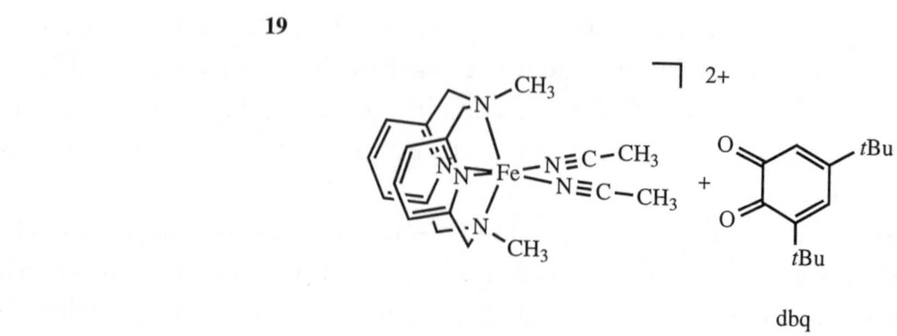

(8.9)

availability of the reduction and oxidation products of $[Fe(L–N_4Me_2)(dbc)]^+$, a unique opportunity exists for investigating the influence that each component (the iron(II) ion and the semiquinonate radical ligand) exerts on the reaction with molecular oxygen. Upon exposing a solution of **19** in acetonitrile to air, the intensity of the absorption band attributed to the iron(III) semiquinonate species decreases to about 90% of its original intensity within the first 24 h and then stays approximately constant, indicating that the semiquinonate coordinated to an iron(III) ion, by itself, does not react with molecular oxygen, whereas the reaction of the iron(III) 3,5-di-*tert*-butylcatecholate complex **17** with air is essentially completed within 3 h. On the other hand, the electrochemically generated iron(II) catecholate complex **18** reacts immediately with air. The electronic absorption spectrum of the solution resulting from the reaction of the iron(II) complex **18** with oxygen reveals the immediate formation of the iron(III) catecholate complex **17** in a yield of 47%. It is noteworthy that, in contrast to the iron(II) monohydrogencatecholate complex $[Fe(6\text{-}Me_3\text{-tpa})(Hdbc)]^+$ reported by Que *et al.*,[28] the iron(III) catecholate complex $[Fe(L–N_4Me_2)(dbc)]^+$ is *not* formed in a 100% yield. These results demonstrate that the right oxidation states of *both* the metal ion *and* the coordinated dioxolene unit are crucial for the occurrence of any cleavage of the intradiol C–C bond by molecular oxygen.

The addition of equivalent amounts of potassium superoxide to a solution of the iron(III) semiquinonate complex **19** in acetonitrile, through which a stream of nitrogen is vigorously bubbled to expel any generated oxygen in solution, affords the iron(III) catecholate complex **17** in quantitative yields (based on the electronic absorption spectrum), as expected from the redox potentials ($E_{1/2}(O_2/O_2^-) = -0.87$ V and $E_{1/2}$(**19**/**17**) = 0.35 V). This finding provides evidence that the well-established intradiol ring cleavage reaction of **17** with molecular oxygen does not occur *via* an initial electron transfer step, resulting in the formation of the iron(III) semiquinonate or a ternary semiquinonato-superoxo iron(III) complex as intermediate (Reaction 8.10), but instead *via* the direct attack of the oxygen molecule onto the catecholate moiety. This conclusion is also supported by the finding that the spectrophotometric investigation of the reaction of the iron(III)

3,5-di-*tert*-butylcatecholate complex **17** with molecular oxygen does not provide any indication for an absorption band originating from an intermediary formation of the iron(III) 3,5-di-*tert*-butylbenzosemiquinonate complex **19**.

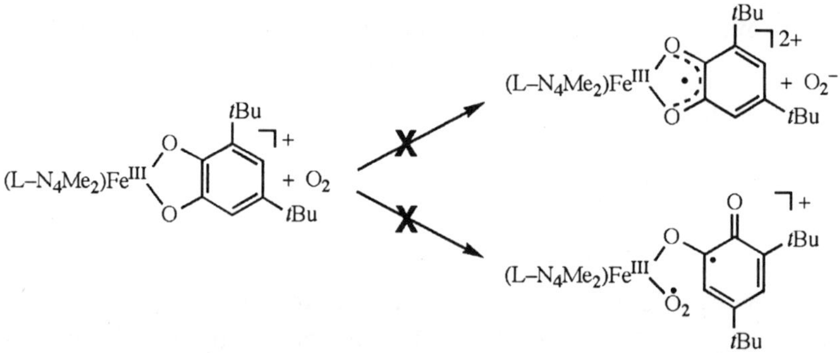

(8.10)

With the equilibrium constant determined for Reaction 8.9 at 25 °C, the equilibrium constant for the dissociation of $[Fe(L\text{–}N_4Me_2)(dbc)]^+$ into $[Fe(L\text{–}N_4Me_2)(MeCN)_2]^{3+}$ and dbc^{2-} (Reaction 8.11) or $[Fe(L\text{–}N_4Me_2)(MeCN)_2]^{2+}$ and $dbsq^-$ (Reaction 8.12) can be calculated. The equilibrium constants for these reactions portend that the catecholate ligand in complex **17** will stay coordinated in solution. This finding suggests that neither uncoordinated $dbsq^{-1}$ nor uncoordinated dbc^{2-} are very likely to play any significant role in the intradiol-cleavage reaction. The thermodynamic data presented here, however, provide no insight into the question of how easily the monodentate coordination mode of the dioxolene moiety can be achieved in solution, i.e. whether an open coordination site at the octahedral iron ion is accessible

$$[Fe(L\text{–}N_4Me_2)(dbc)]^+ + 2\ MeCN \xrightleftharpoons{K = 3.29\cdot10^{-57}\ M^{-1}} [Fe(L\text{–}N_4Me_2)(MeCN)_2]^{3+} + dbc^{2-} \qquad (8.11)$$

$$[Fe(L\text{–}N_4Me_2)(dbc)]^+ + 2\ MeCN \xrightleftharpoons{K = 2.29\cdot10^{-20}\ M^{-1}} [Fe(L\text{–}N_4Me_2)(MeCN)_2]^{2+} + dbsq^- \qquad (8.12)$$

so that a concerted attack of an oxygen molecule on the iron ion *and* the catecholate moiety is possible.

3.3 Intradiol-cleavage activity of iron complexes with miscellaneous phenolate-containing ligands

The ligand environment in the most reactive biomimetic complexes for intradiol-cleaving enzymes does not resemble that of the active site in the enzyme since the ligands in these complexes contain no phenolate donor group. There have been some research efforts to investigate the reactivity of iron(III) catecholate with phenolate-containing ligands. Some of the ligands employed in these studies are depicted in Figs. 8.5 and 8.11.

In the iron(III) catecholate complexes with the ligands salen^{2-},[48,65] hben^{2-},[66] and ehgs^{3-},[67] the dianionic catecholate moiety is bound to the iron(III) ion in a bidentate fashion. All three complexes have the same *cis*-octahedral FeN_2O_4 coordination sphere. Upon exposure to dioxygen, the complex $[Fe(hben)(dbc)]^-$ exhibits no reactivity,[66] while the complex

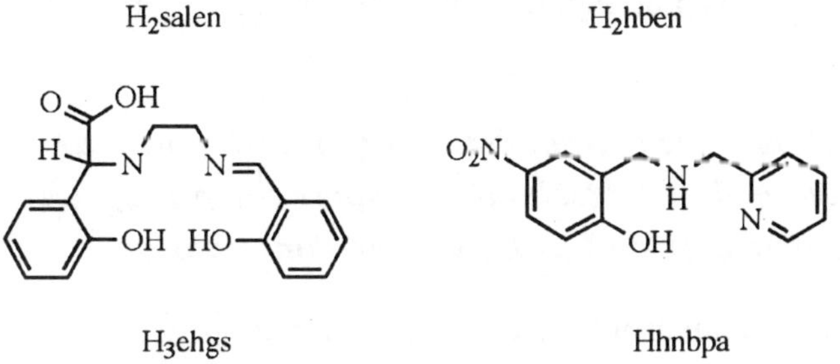

Fig. 8.11 Some selected phenolate-containing ligands used in biomimetic complexes modeling intradiol-cleavage reactivity.

$[Fe(salen)(dbc)]^-$ yields 35% intradiol-cleavage products after 4–6 d of reaction time.[48] This finding is in accordance with the general trend showing that the reactivity increases with the Lewis-acidity of the iron ion. In addition to the main product *o*-benzoquinone, intradiol-cleavage products were also reported to be formed in "significant" (but unfortunately, unspecified) amounts after 4 d in the reaction of air with an iron catecholate complex containing $ehgs^{3-}$ as coligand.[67] Among the iron(III) 3,5-di-*tert*-butylcatecholate complexes with the tripodal ligands hda^{3-} and hdp^-, the latter complex displays the higher reactivity, affording intradiol-cleavage products in a 91% yield, albeit in a rather slow reaction (Table 8.2).[48,49]

In a recent study,[68] a series of tridentate ligands were utilized to prepare iron(III) chloride complexes. All of these ligands possess one oxygen donor atom from a phenolate group and two nitrogen donor atoms deriving from either imine, amine, pyridyl, imidazole or benzimidazole groups. These iron(III) chloride complexes were tested as precursors to catalysts for the oxygenation reaction of H_2dbc. With 2% of the employed iron(III) chloride complex, turnover numbers between 9.9 to 21.2 were achieved after 4 d of reaction time. The highest yields of intradiol-cleavage products were obtained with the precursor compound containing the ligand $hnbpa^-$. Unfortunately, the corresponding catecholate complexes were not isolated. Therefore no structural or kinetic information is available at the moment.

In summary, intradiol-cleavage reactivity with varying yields has been demonstrated with phenolate-containing iron(III) complexes. However, it is noteworthy that the reaction is considerably slower than with complexes that contain π-acceptor ligands with only nitrogen donor atoms.

3.4 Stoichiometric extradiol-cleavage reaction with an iron(III) 3,5-di-tert-butylcatecholate complex containing a tridentate coligand with nitrogen donor atoms

Appropriate functional model compounds for the reactivity of the iron(II) site in extradiol-cleaving catechol dioxygenase are still lacking. As previously mentioned, Funabiki *et al.* were able to demonstrate that, in addition to intradiol-cleavage products and the simple two-electron oxidation product dbq, extradiol-cleavage products are obtained in the reaction of H_2dbc with dioxygen in the presence of catalytic amounts of iron(II) as well as of iron(III) salts.[39–43] Mechanistic details are, however, still not resolved and the identity of the active species is unknown.

Chiou and Que prepared the hydrogencatecholato iron(II) complex $[Fe(6\text{-}Me_3\text{-}tpa)(Hdbc)]^+$, which models the coordination mode of the catecholate in the enzyme-substrate complex of extradiol-cleaving catechol dioxygenases.[28,30] In the reaction with molecular oxygen, however, they observed no formation of extradiol-cleavage products; instead the iron(III) catecholate complex $[Fe(6\text{-}Me_3\text{-}tpa)(dbc)]^+$ was produced after the consumption of 0.25 equivalent of dioxygen, indicating that only an electron transfer reaction takes place, without the involvement of any extradiol-oxygenation reaction.[28]

In contrast, extradiol ring cleavage of coordinated catecholate occurs in iron(III)-containing complexes under certain conditions. This reaction has biological relevance because extradiol-oxygenation is detected in the enzymatic cleavage reaction of some catechol derivatives catalyzed by iron(III)-containing intradiol-cleaving catechol dioxygenases.[69] The first successful extradiol-cleavage reaction with a well-defined model compound was discovered by Dei *et al.* who observed that the stoichiometric reaction of the iron(III) complex [Fe(tacn)(dbc)Cl] (**20**) with molecular oxygen in acetonitrile yields extradiol-cleavage products in a 29% yield (3,5-di-*tert*-butyl-2-pyrone (**7**) and 4,6-di-*tert*-butyl-2-pyrone (**21**)) in addition to the two-electron oxidation product dbq (50% yield) (Reaction 8.13).[59]

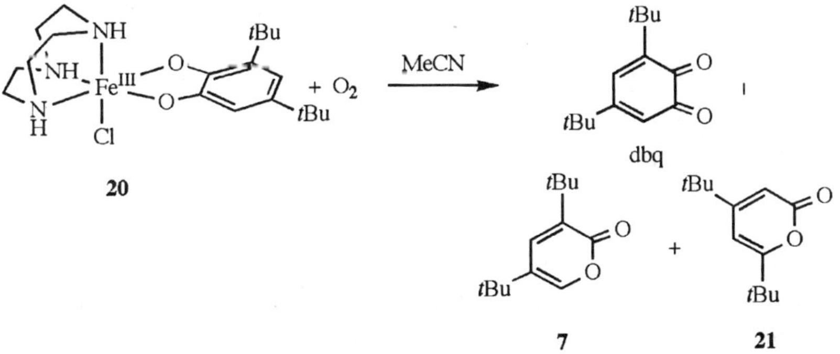

(8.13)

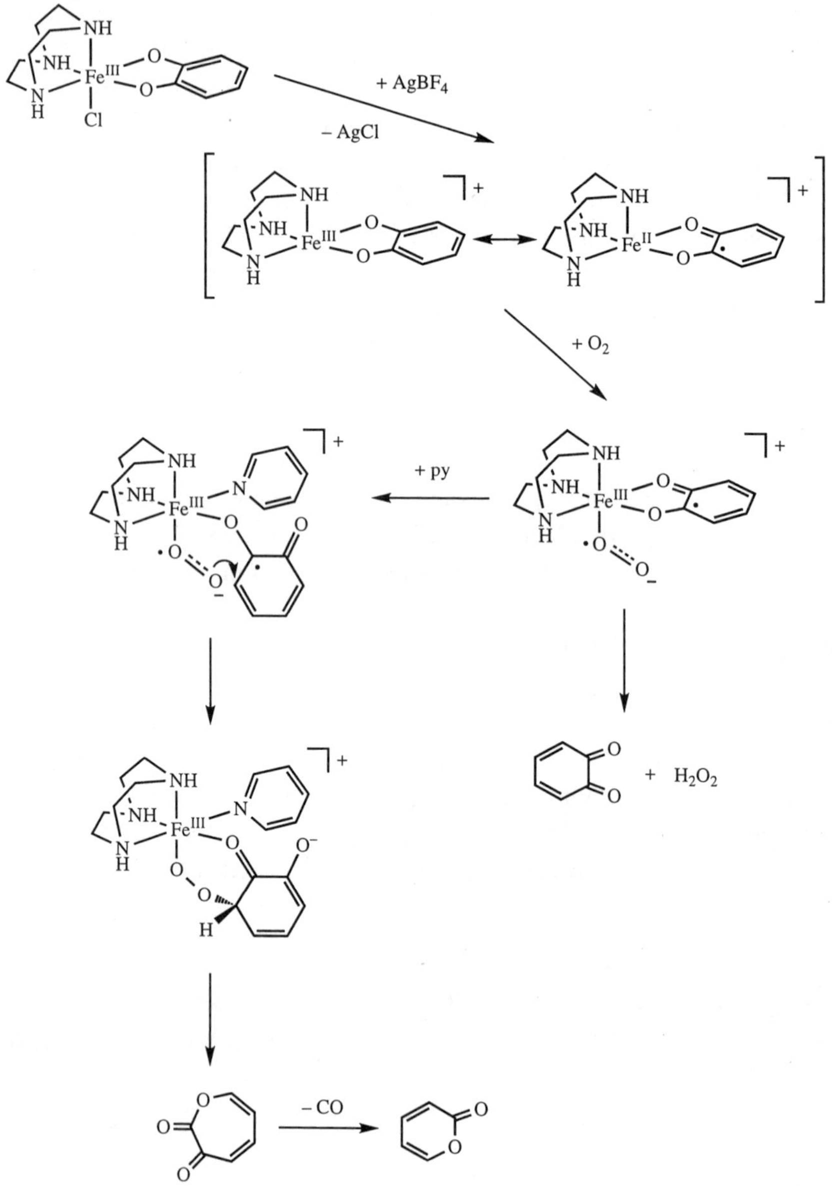

Fig. 8.12 Postulated mechanism for the extradiol-ring cleavage of the coordinated catecholate ligand in the stoichiometric reaction of **20** with molecular oxygen.[70]

This reaction mechanism was further investigated by Ito and Que, who could improve the yields of the extradiol-cleavage products up to 98% by adding 1 equivalent of silver tetrafluoroborate *and* an excess of a pyridine derivative (e.g. 2-methylpyridine) to the reaction solution.[70] In the absence of a potentially coordinating pyridine derivative, the addition of 1 equivalent silver tetrafluoroborate to a non-coordinating solvent renders almost exclusively dbq as the product for the reaction of complex **20** with dioxygen.

The experimental findings were explained by the postulated reaction mechanism depicted in Fig. 8.12. For the occurrence of a reaction with molecular oxygen, a readily accessible coordination site is needed at the metal ion. The reaction of complex **20** with silver tetrafluoroborate creates a readily available coordination site at the metal ion and thereby increases the iron(II) semiquinonate character of the complex, both of which are needed for the occurrence of any reaction with molecular oxygen. After the formation of a superoxo iron(III) complex, the importance of the Lewis-base comes into play. In the absence of a coordinating pyridine derivative, a simple electron transfer reaction from the coordinated superoxide to the coordinated catecholate ligand occurs, resulting in the formation of dbq and hydrogen peroxide. In contrast, the coordination of a pyridine ligand induces a change in the coordination of the catecholate ligand from a bidentate to a monodentate coordination mode. This entails a spacial proximity between the superoxide and the carbon atom next to the enediol moiety of the catecholate ligand. Formation of an alkylperoxo intermediate affords the 2-pyrone derivatives after the loss of a carbon monoxide. The difference between intradiol- and extradiol-cleavage of catecholate ligand coordinated to an iron(III) ion can essentially be ascribed to the difference in the type of reactivity of the dioxygen species involved in the reactions: nucleophilicity of a superoxide anion versus electrophilicity of an oxygen molecule.

It is interesting to note that iron(III) complexes with tetradentate ligands, where no additional coordination site for the reaction with the oxygen molecule is available, afford exclusively intradiol-cleavage but no extradiol-cleavage products. In contrast, no intradiol-cleavage products were detected in the reaction of [Fe(tacn)(dbc)Cl] with dioxygen. Within this context, it is worthwhile to mention that employment of the tridentate ligand bis(2-pyridylmethyl)amine instead of tacn results in intradiol- as well as extradiol-cleavage products.[71]

Finally, it may be pointed out that the mechanism postulated for the reactivity of [Fe(tacn)(dbc)Cl] provides an attractive explanation for the kinds of products observed in the reaction of H_2dbc with oxygen in the presence of simple iron(III) and iron(II) salts, as reported by Funabiki and co-workers;[39–43] utilization of iron(II) salts as the iron source requires only the amendment that the iron(II) ions are immediately oxidized to the iron(III) state upon exposure to oxygen. Thus, intradiol-cleavage might be linked to an octahedral iron(III) catecholate complex with four pyridine ligands in the ligand sphere whereas extradiol-cleavage and simple two-electron oxidation might proceed via a five-coordinate iron(III) catecholate species containing three pyridine molecules as coligands.

4. Conclusion

Studies modeling the reactivity of intradiol-cleaving catechol dioxygenases by small inorganic complexes have progressed considerably within the last ten years and have supplemented the knowledge gleaned through work on enzymes. Several highly catalytic model systems have been developed. It is noteworthy that, in the most reactive biomimetic complexes, the coligand consists only of nitrogen donor atoms, preferably from π-acceptor ligands. In contrast, the iron ion in the enzyme-substrate complex is surrounded by one tyrosinate and two histidine residues in addition to the bidentate catecholate substrate Despite the lesser Lewis-acidity and therefore lesser iron(II) semiquinonate character of the active site, the enzymes are considerably more reactive than the model complexes. In addition, exclusive intradiol ring cleavage has only been observed with octahedral iron complexes, whereas the active site in the intradiol-cleaving enzymes remains five-coordinate. Future efforts will most likely be directed towards the development of biomimetic complexes modeling the structural as well as the functional aspects of the active site in the enzymes. Furthermore, some mechanistic details of the reaction of the iron(III) catecholate complexes with dioxygen (for example the actual attack of the oxygen molecule onto the iron(III) catecholate complex) are still not unambiguously resolved and require further investigation.

In contrast, the progress in developing functional model complexes for the iron(II) site in extradiol-cleaving catechol dioxygenases lags considerably behind that of the intradiol-cleaving catechol dioxygenases. Extradiol ring cleavage has been demonstrated in the reaction of a five-coordinate iron(III) catecholate complex with dioxygen. This study can be regarded as the first successful attempt to model the catalytic mechanism of extradiol-cleaving catechol dioxygenases; however, the iron ion in this complex is present in the trivalent and not in the divalent oxidation state as required by the enzyme. At this time, no well-defined iron(II) catecholate complex exists which displays a functional model character with respect to the extradiol-cleaving enzyme in a stoichiometric reaction with molecular oxygen. This lack of functional biomimetic model complexes will certainly stimulate further research in the next few years and the results of these endeavors should advance our knowledge of the catalytic mechanism of the extradiol-cleaving catechol dioxygenases and of oxygen-activating enzymes in general.

5. Notes Added in Proof

1. Recent X-ray absorption studies[72] confirmed that for other intradiol-cleaving catechol dioxygenases with different substrate specificities, namely for hydroxyquinol and chlorocatechol dioxygenases, the active sites in the resting states and the enzyme-substrate complexes are very similar to those of the well-characterized protocatechuate 3,4-dioxygenase.

2. The crystal structure analysis[73] of the extradiol-cleaving catechol 2,3-dioxygenase (2,3-CTD) at 2.8 Å resolution revealed that the active site (22) consists of an iron(II) ion in a highly distorted tetrahedral coordination geometry coordinated by two nitrogen and two oxygen donor atoms, deriving from two histidine residues (His 153 and His 214), one glutamate residue

(Glu 265) and an acetone molecule. The acetone molecule originates from the buffer solutions employed in the purification and crystallization of the enzyme. No water molecule is bound to the iron ion. Although the protein-derived ligands in 2,3-CTD are the same as those in the previously investigated DHBD, the coordination geometry and coordination number of the iron site differ from each other. In addition, the entrance to the substrate-binding pocket in 2,3-CTD is considerably narrower than that of DHBD, thus allowing the binding of catechol but preventing the entrance of 2,3-dihydroxybiphenyl to the active site of the 2,3-CTD.

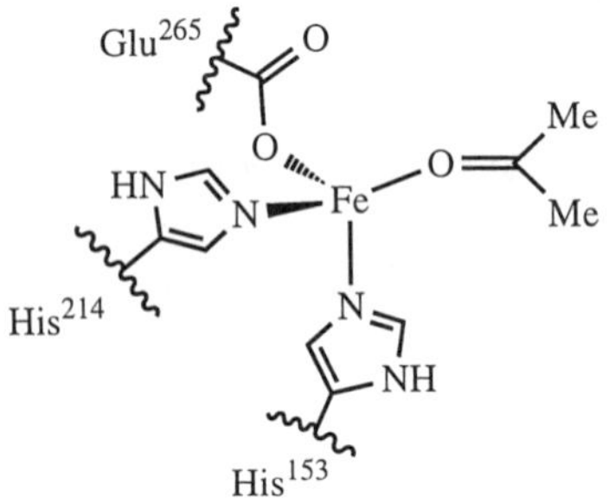

22

3. The importance of an easily accessible coordination site for the extradiol-cleaving reactivity of iron(III) 3,5-di-*tert*-butylcatecholate complexes with oxygen molecules has also been demonstrated with the tridentate ligand hydridotris(3,5-di-*iso*-propyl-1-pyrazolyl)borate.[74] Thus the pseudo-octahedral iron(III) complex **23** releases the coordinated acetonitrile molecule in solution, resulting in the formation of a square-pyramidal iron(III) catecholate complex which yields, in a stoichiometric reaction with oxygen, the extradiol-cleavage products **7** and **21** and the intradiol-cleavage product **4** in a 28, 39 and 33% yield, respectively (Reaction 8.14).

In contrast, the introduction of sterically more demanding groups into the tridentate ligand, as demonstrated with the ligand derivative hydridotris(3-*tert*-butyl-5-*iso*-propyl-1-pyrazolyl)borate, results in the formation of the

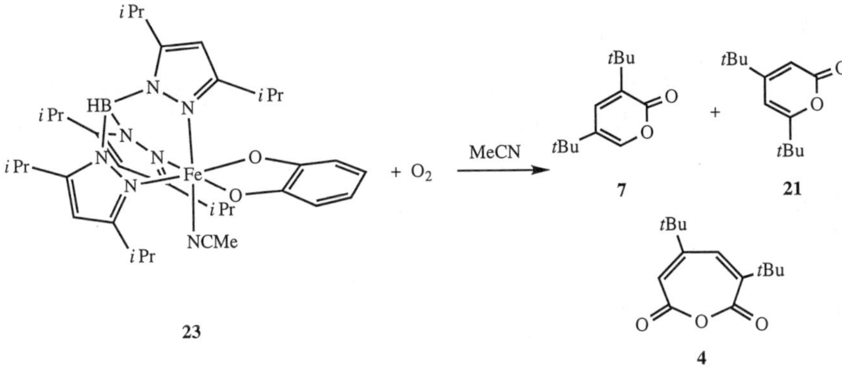

(8.14)

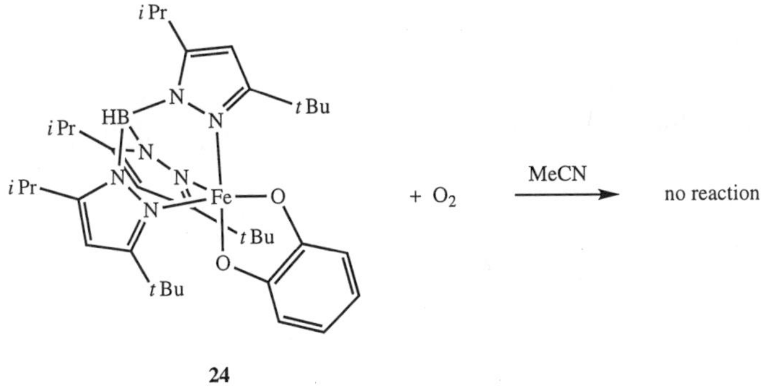

(8.15)

trigonal-bipyramidal iron(III) catecholate complex **24**. The increased steric hindrance of the tridentate ligand causes complex **24** to be stable towards oxygen (Reaction 8.15).

4. Iron(III) chloride complexes with a variety of tripodal co-ligands containing one or two *p*-nitrophenolate units were investigated with respect to their catalytic behavior in the intradiol-cleavage reaction of H_2dbc with oxygen.[75] Of interest is the fact that, after a reaction time of 4 d, the yields of intradiol-cleavage product **4** observed in the presence of phenolate-containing iron(III) chloride complexes were comparable to those in the presence of the iron(III) chloride complex containing tpa as co-ligand. However, an isolation and a kinetic investigation of the involved iron(III) catecholate species are needed in order to explain these observations.

6. Abbreviations

bpia	*N*,*N*-bis(2-pyridylmethyl)-*N*-(1-methylimidazol-2-yl-methyl)amine
CT	charge transfer
1,2-CTD	catechol 1,2-dioxygenase
2,3-CTD	catechol 2,3-dioxygenase
dbq	3,5-di-*tert*-butyl-1,2-benzoquinone
DHBD	2,3-dihydroxybiphenyl 1,2-dioxygenase
EPR	electron paramagnetic resonance
EXAFS	extended X-ray absorption fine structure
$H_2$4-*t*Bu-cat	4-*tert*-butylcatechol
Hbpg	*N*, *N*-bis(2-pyridylmethyl)glycine
Hbpnp	*N*-(2-hydroxy-5-nitrobenzyl)-*N*, *N*-bis(2-pyridylmethyl) amine
H_2cat	catechol
$H_2$3-Cl-cat	3-chlorocatechol
$H_2$4-Cl-cat	4-chlorocatechol
H_2Cl_4-cat	tetrachlorocatechol
H_2dbc	3,5-di-*tert*-butylcatechol
Hdbsq	3,5-di-*tert*-butyl-1,2-benzosemiquinone
H_3ehgs	*N*-[2-((2-hydroxyphenyl)glycino)ethyl]salicylideneamine
$H_2$3-EtO_2C-cat	3-(ethylcarboxymethyl)catechol
H_2hben	*N*, *N'*-bis((3-*tert*-butyl-2-hydroxy-5-methylphenyl)methyl)-*N*, *N'*-bismethyl-1,2-diaminoethane
H_2hda	*N*-(2-hydroxybenzyl)-*N*-carboxymethylglycine
Hhdp	*N*-(2-hydroxybenzyl)-*N*, *N*-bis(2-pyridylmethyl)amine
Hhnbpa	*N*-(2-hydroxy-5-nitrobenzyl)-2-pyridylmethylamine
$H_2$4-Me-cat	4-methylcatechol
$H_2$3-MeO-cat	3-methoxycatechol
H_3nta	*N*, *N*-bis(carboxymethyl)glycine
H_2pda	*N*-(carboxymethyl)-*N*-(2-pyridylmethyl)glycine
H_2salen	*N*, *N'*-ethylenebis(salicylideneamine)
LMCT	ligand-to-metal charge transfer
L-N_4Me_2	*N*, *N'*-dimethyl-2,11-diaza[3.3](2,6)pyridinophane

NMR	nuclear paramagnetic resonance
6-Me_3-tpa	tris[(6-methyl-2-pyridyl)methyl]amine
3,4-PCD	protocatechuate 3,4-dioxygenase
4,5-PCD	protocatechuate 4,5-dioxygenase
SCE	saturated calomel electrode
SS-CTH	(*SS*)-5,5,7,12,12,14-hexamethyl-1,4,8,11-tetraazacyclo-tetradecane.
tacn	1,4,7-triazacyclononane
tpa	tris(2-pyridylmethyl)amine
triphos	2,2-bis(diphenylphosphinomethyl)-1-diphenylphosphino-propane
XANES	X-ray absorption near-edge spectroscopy

7. References

1. (a) Que, L., Jr. in *Iron Carriers and Iron Proteins* (ed.) Loehr, T.M. VCH, New York, 1989, pp. 467–524; (b) Lipscomb, J.D. and Orville, A.M. in *Metal Ions in Biological Systems* (eds.) Sigel, H. and Sigel, A. Marcel Dekker Inc., New York, 1992, Vol. 28, pp. 243–298; (c) Que, L., Jr. in *Bioinorganic Catalysis* (ed.) Reedijk, J. Marcel Dekker Inc., New York, 1993, 347–393; (d) Bertini, I., Briganti, F., Mangani, S., Nolting, H.F. and Scozzafava, A. *Coord. Chem. Rev.* 1995, **144:** 321–345; (e) Que, L., Jr. and Ho, R.Y.N. *Chem. Rev.* 1996, **96:** 2607–2624; (f) Funabiki, T. in *Oxygenases and Model Systems* (ed.) Funabiki, T. Kluwer Academic Pub., Dordrecht, 1997, pp. 19–104; (g) Funabiki, T. in *Oxygenases and Model Systems* (ed.) Funabiki T. Kluwer Academic Pub., Dordrecht, 1997, 105–155.
2. (a) Ohlendorf, D.H., Lipscomb, J.D. and Weber, P.C. *Nature* 1988, **336:** 403–405; (b) Ohlendorf, D.H., Orville, A.M. and Lipscomb, J.D. *J. Mol. Biol.* 1994, **244:** 586–608.
3. Orville, A.M., Lipscomb, J.D. and Ohlendorf, D.H. *Biochemistry* 1997, **36:** 10052–10066.
4. Elgren, T.E., Orville, A.M., Kelly, K.A., Lipscomb, J.D., Ohlendorf, D.H. and Que, L., Jr. *Biochemistry*, 1997, **36:** 11504–11513.

5. Orville, A.M., Elango, N., Lipscomb, J.D. and Ohlendorf, D.H. *Biochemistry* 1997, **36:** 10039–10051.
6. Frazee, R.W., Orville, A.M., Dolbeare, K.B., Yu, H., Ohlendorf, D.H. and Lipscomb, J.D. *Biochemistry* 1998, **37:** 2131–2144.
7. (a) Fujisawa, H., Uyeda, M., Kojima, Y., Nozaki, M. and Hayaishi, O. *J. Biol. Chem.* 1972, **247:** 4414–4421; (b) Whittaker, J.W., Lipscomb, J.D., Kent, T.A. and Münck, E. *J. Biol. Chem.* 1984, **259:** 4466–4475.
8. (a) Que, J., Jr., Lipscomb, J.D., Zimmermann, R., Münck, E., Orme-Johnson, N.R. and Orme-Johnson, W.H. *Biochim. Biophys. Acta* 1976, **452:** 320–334; (b) Kent, T.A., Münck, E., Pyrz, J.W., Widom, J. and Que, L., Jr. *Inorg. Chem.* 1987, **26:** 1402–1408.
9. Siu, D.C.-T., Orville, A.M., Lipscomb, J.D., Ohlendorf, D.H. and Que, L. Jr. *Biochemistry* 1992, **31:** 10443–10448.
10. Felton, R.H., Barrow, W.L., May, S.W., Sowell, A.L. and Goel, S. *J. Am. Chem. Soc.* 1982, **104:** 6132–6134.
11. True, A.E., Orville, A.M., Pearce, L.L., Lipscomb, J.D. and Que, L., Jr. *Biochemistry* 1990, **29:** 10847–10854.
12. (a) Kojima, Y., Fujisawa, H., Nakazawa, A., Nakazawa, T., Kanetsuna, F., Taniuchi, H., Nozaki, M. and Hayaishi, O. *J. Biol. Chem.* 1967, **242:** 3270–3278; (b) Nakazawa, A., Nakazawa, T., Kotani, S., Nozaki, M. and Hayaishi, O. *J. Biol. Chem.* 1969, **244:** 1527–1532.
13. Que, L., Jr. *Coord. Chem. Rev.* 1983, **50:** 73–108.
14. (a) Que, L., Jr. Heistand, R.H., II, Mayer, R. and Roe, A.L. *Biochemistry* 1980, **19:** 2588–2593; (b) Que, L., Jr. and Epstein, R.M. *Biochemistry* 1981, **20:** 2545–2549.
15. (a) Tatsuno, Y., Saeki, Y., Iwaki, M., Yagi, T. and Nozaki, M. *J. Am. Chem. Soc.* 1978, **100:** 4614–4615; (b) Keyes, W.E., Loehr, T.M. and Taylor, M.L. *Biochem. Biophys. Res. Commun.* 1978, **83:** 941–945; (c) Bull, C., Ballou, D.P. and Salmeen, I. *Biochem. Biophys. Res. Commun.* 1979, **87:** 836–841.
16. Whittaker, J.W. and Lipscomb, J.D. *J. Biol. Chem.* 1984, **259:** 4487–4495.
17. (a) Felton, R.H., Cheung, L.D., Phillips, R.S. and May, S.W. *Biochem. Biophys. Res. Commun.* 1978, **85:** 844–850; (b) Que, L., Jr. and Heistand, R.H., II *J. Am. Chem. Soc.* 1979, **101:** 2219–2221.
18. Orville, A.M. and Lipscomb, J.D. *J. Biol. Chem.* 1989, **264:** 8791–8801.
19. (a) Lauffer, R.B. and Que, L., Jr. *J. Am. Chem. Soc.* 1982, **104:** 7324–7325; (b) Que, L., Jr., Lauffer, R.B., Lynch, J.B., Murch, B.P. and Pyrz, J.W. *J. Am. Chem. Soc.* 1987, **109:** 5381–5385.

20. Nozaki, M. in *Molecular Mechanism of Oxygen Activation* (ed.) Hayaishi, O. Academic Press, New York, 1974, pp. 135–165.
21. (a) Fujisawa, H., Hiromi, K., Uyeda, M., Nozaki, M. and Hayaishi, O. *J. Biol. Chem.* 1971, **246:** 2320–2321; (b) Fujisawa, H., Hiromi, K., Uyeda, M., Okuno, S., Nozaki, M. and Hayaishi, O. *J. Biol. Chem.* 1972, **247:** 4422–4428; (c) Bull, C., Ballou, D.P. and Otsuka, S. *J. Biol. Chem.* 1981, **256:** 12681–12686; (d) Walsh, T.A., Ballou, D.P., Mayer, R. and Que, L., Jr. *J. Biol. Chem.* 1983, **258:** 14422–14427; (e) Walsh, T.A. and Ballou, D.P. *J. Biol. Chem.* 1983, **258:** 14413–14421.
22. Que, L., Jr., Lipscomb, J.D., Münck, E. and Wood, J.M. *Biochim. Biophys. Acta* 1977, **485:** 60–74.
23. Arciero, D.M., Lipscomb, J.D., Huynh, B.H., Kent, T.A. and Münck, E. *J. Biol. Chem.* 1983, **258:** 14981–14991.
24. Tatsuno, Y., Saeki, Y., Nozaki, M., Otsuka, S. and Maeda, Y. *FEBS Lett.* 1980, **112:** 83–85.
25. Han, S., Eltis, L.D., Timmis, K.N., Muchmore, S.W. and Bolin, J.T. *Science* 1995, **270:** 976–980.
26. Senda, T., Sugiyama, K., Narita, H., Yamamoto, T., Kimbara, K., Fukuda, M., Sato, M., Yano, K. and Mitsui, Y. *J. Mol. Biol.* 1996, **255:** 735–752.
27. Hori, K., Hashimoto, T. and Nozaki, M. *J. Biochem.* 1973, **74:** 375–384.
28. Chiou, Y.-M. and Que, L., Jr. *Inorg. Chem.* 1995, **34:** 3577–3578.
29. Mabrouk, P.A., Orville, A.M., Lipscomb, J.D. and Solomon, E.I. *J. Am. Chem. Soc.* 1991, **113:** 4053–4061.
30. Shu, L., Chiou, Y.-M., Orville, A.M., Miller, M.A., Lipscomb, J.D. and Que, L., Jr. *Biochemistry* 1995, **34:** 6649–6659.
31. (a) Bertini, I., Briganti, F., Mangani, S., Nolting, H.F. and Scozzafava, A. *FEBS Lett.* 1994, **350:** 207–212; (b) Bertini, I., Briganti, F., Mangani, S., Nolting, H.F. and Scozzafava, A. *Biochemistry* 1994, **33:** 10777–10784.
32. Bertini, I., Capozzi, F., Dikiy, A., Happe, B., Luchinat, C. and Timmis, K.N. *Biochem. Biophys. Res. Commun.* 1995, **215:** 855–860.
33. Arciero, D.M., Orville, A.M. and Lipscomb, J.D. *J. Biol. Chem.* 1985, **260:** 14035–14044.
34. Arciero, D.M. and Lipscomb, J.D. *J. Biol. Chem.* 1986, **261:** 2170–2178.
35. Bianchini, C., Frediani, F., Laschi, F., Meli, A., Vizza, F. and Zanello, P. *Inorg. Chem.* 1990, **29:** 3402–3409.
36. Sanvoisin, J., Langley, G.J. and Bugg, T.D.H. *J. Am. Chem. Soc.* 1995, **117:** 7836–7837.

37. Spence, E.L., Langley, G.J. and Bugg, T.D.H. *J. Am. Chem. Soc.* 1996, **118:** 8336–8343.
38. Whiting, A.K., Boldt, Y.R., Hendrich, M.P., Wackett, L.P. and Que, L., Jr. *Biochemistry* 1996, **35:** 160–170.
39. Funabiki, T., Sakamoto, H., Yoshida, S. and Tamara, K. *J. Chem. Soc., Chem. Commun.* 1979, 754–755.
40. Funabiki, T., Mizoguchi, A., Sugimoto, T., Tada, S., Tsuji, M., Sakamoto, H. and Yoshida, S. *J. Am. Chem. Soc.* 1986, **108:** 2921–2932.
41. Funabiki, T., Mizoguchi, A., Sugimoto, T. and Yoshida, S. *Chem. Lett.* 1983, 917–920.
42. Funabiki, T., Konishi, T., Kobayashi, S., Mizoguchi, A., Takano, M. and Yoshida, S. *Chem. Lett.* 1987, 719–722.
43. Funabiki, T., Yoneda, I., Ishikawa, M., Ujiie, M., Nagai, Y. and Yoshida, S. *J. Chem. Soc., Chem. Commun.* 1994, 1453–1454.
44. Funabiki, T., Kojima, H., Kaneko, M., Inoue, T., Yoshioka, T., Tanaka, T. and Yoshida, S. *Chem. Lett.* 1991, 2143–2146.
45. Funabiki, T., Ishikawa, M., Nagai, Y., Yorita, J. and Yoshida, S. *J. Chem. Soc., Chem. Commun.* 1994, 1951–1952.
46. Funabiki, T., Tada, S., Yoshioka, T., Takano, M. and Yoshida, S. *J. Chem. Soc., Chem. Commun.* 1986, 1699–1700.
47. White, L.S., Nilsson, P.V., Pignolet, L.H. and Que, L., Jr. *J. Am. Chem. Soc.* 1984, **106:** 8312–8313.
48. Que, L., Jr., Kolanczyk, R.C. and White, L.S. *J. Am. Chem. Soc.* 1987, **109:** 5373–5380.
49. Cox, D.D. and Que, L., Jr. *J. Am. Chem. Soc.* 1988, **110:** 8085–8092.
50. Jang, H.G., Cox, D.D. and Que, L., Jr. *J. Am. Chem. Soc.* 1991, **113:** 9200–9204.
51. Koch, W.O. and Krüger, H.-J. *Angew. Chem. Int. Ed. Engl.* 1995, **34:** 2671–2674.
52. Krüger, H.-J. and Koch, W.O. in *Bioinorganic Chemistry: Transition Metals in Biology and Their Coordination Chemistry* (ed.) Trautwein, A.X. Wiley-VCH, Weinheim, 1997, pp. 632–648.
53. Duda, M., Pascaly, M. and Krebs, B. *J. Chem. Soc., Chem. Commun.* 1997, 835–836.
54. Cox, D.D., Benkovic, S.J., Bloom, L.M., Bradley, F.C., Nelson, M.J., Que, L., Jr. and Wallick, D.E. *J. Am. Chem. Soc.* 1988, **110:** 2026–2032.
55. Krüger, H.-J., unpublished results.
56. Weller, M.G. and Weser, U., *J. Am. Chem. Soc.* 1982, **104:** 3752–3754.
57. Weller, M.G. and Weser, U. *Inorg. Chim. Acta* 1985, **107:** 243–245.

58. Nishida, Y., Shimo, H. and Kida, S. *J. Chem. Soc., Chem. Commun.* 1984, 1611–1612.
59. Dei, A., Gatteschi, D. and Pardi, L. *Inorg. Chem.* 1993, **32:** 1389–1395.
60. Barbaro, P., Bianchini, C., Mealli, C. and Meli, A. *J. Am. Chem. Soc.* 1991, **113:** 3181–3183.
61. Mayer, R.J. and Que, L., Jr. *J. Biol. Chem.* 1984, **259:** 13056–13060.
62. Funabiki, T., Yamazaki, T., Fukui, A., Tanaka, T. and Yoshida, S. *Angew. Chem. Int. Ed. Engl.* 1998, **37:** 513–515.
63. Que, L., Jr., private communication.
64. Koch, W.O., Schünemann, V., Gerdan, M., Trautwein, A.X. and Krüger, H.-J. *Chem. Eur. J.* 1998, **4:** 1255–1265.
65. Heistand, R.H., II, Lauffer, R.B., Fikrig, E. and Que, L., Jr. *J. Am. Chem. Soc.* 1982, **104:** 2789–2796.
66. Mialane, P., Anxolabéhère-Mallart, E., Blondin, G., Nivorojkine, A., Guilhem, J., Tchertanova, L., Cesario, M., Ravi, N., Bominaar, E., Girerd, J.-J. and Münck, E. *Inorg. Chim. Acta* 1997, **263:** 367–378.
67. Spartalian, K. and Carrano, C.J. *Inorg. Chem.* 1989, **28:** 19–24.
68. Viswanathan, R. and Palaniandavar, M. *J. Chem. Soc., Dalton Trans.* 1995, 1259–1266.
69. (a) Fujiwara, M., Golovleva, L.A., Saecki, Y., Nozaki, M. and Hayaishi, O. *J. Biol. Chem.* 1975, **250:** 4848–4855; (b) Que, L., Jr. *Biochem. Biophys. Res. Commun.* 1978, **84:** 123–129.
70. Ito, M. and Que, L., Jr. *Angew. Chem. Int. Ed. Engl.* 1997, **36:** 1342–1344.
71. Lim, J.H., Kim, J.J., Lee, K.-B. and Jang, H.G. *J. Bioinorg. Chem.* 1997, **61:** 326.
72. Briganti, F., Mangani, S., Podocchi, L., Scozzafava, A., Golovleva, L.A., Jadan, A.P. and Solyanikova, I.P. *FEBS Letts.* 1998, **433:** 58–62.
73. Kita, A., Kita, S., Fujisawa, I., Inaka, K., Ishida, T., Horiike, K., Nozaki, M. and Miki, K. *Structure* 1999, **7:** 25–34.
74. Ogihara, T., Hikichi, S., Akita, M. and Moro-oka, Y. *Inorg. Chem.* 1998, **37:** 2614–2615.
75. Viswanathan, R., Palaniandavar, M., Balasubramanian, T. and Muthiah, T.P. *Inorg. Chem.*, **37:** 2943–2951.

Chapter 9

Biomimetic Chemistry of Molybdenum

Charles G. Young

School of Chemistry, University of Melbourne, Parkville, Victoria 3052, Australia

1. Introduction

Molybdenum enzymes catalyze many important biological reactions, are widely distributed in nature, and are vital components of Earth's carbon, nitrogen, and sulfur cycles. Most of these enzymes contain a mononuclear, pterin-containing molybdenum active site and catalyze net oxygen atom transfer reactions; examples include sulfite oxidase, xanthine oxidase, aldehyde oxidoreductase, nitrate reductase, dimethyl sulfoxide reductase, and D-biotin *S*-oxide reductase. The biomimetic chemistry of these oxomolybdenum enzymes is the subject of this chapter.

In contrast to the oxygenases, and many oxidases and hydroxylases, the molybdenum-containing oxidases utilize water rather than dioxygen as the source of oxygen and generate rather than consume reducing equivalents during substrate (X) turnover (Eq. 9.1). These reactions are not driven by the favorable reduction of dioxygen nor are they impeded by the spin-forbidden nature of triplet dioxygen reactions. As a result, for some substrates, XO, the reverse of Eq. 9.1 is thermodynamically favorable and is catalyzed by molybdenum-containing reductases.

$$X + H_2O \rightarrow XO + 2H^+ + 2e^- \qquad (9.1)$$

Consequently, a treatment of the biomimetic chemistry of molybdenum requires a discussion of both oxidation *and* reduction reactions. This chapter will focus on biomimetically important oxygen atom transfer (OAT) and coupled electron-proton transfer (CEPT) reactions. The emphasis is on the reactivity of model systems, rather than on the structure and spectroscopic properties of their components, and the literature is selectively not exhaustively reviewed. The aim is to update and complement the many excellent reviews in this area, especially with regard to molybdenum's OAT and CEPT chemistries.[1–8] Some important aspects of the modeling endeavor, e.g. emerging synthetic strategies,[9,10] advances in understanding electronic structure,[11,12] and Mo-sulfur redox chemistry,[13–15] etc. will not be addressed in any detail here.

2. Molybdenum Enzymes

The molybdenum enzymes are generally large, multi-subunit enzymes containing a number of prosthetic groups and a mononuclear molybdenum active site.[6] Our current understanding of these active sites is based largely on extended X-ray absorption fine structure (EXAFS),[16] electron paramagnetic resonance (EPR),[17] resonance Raman,[18] and recent X-ray crystallographic studies.[19–26] The presence of molybdopterin, MPT, or a dinucleotide derivative, e.g. molybdopterin cytosine dinucleotide, MCD, which acts as ene-1,2-dithiolate (dithiolene) ligand to molybdenum is a defining feature of these enzymes.[1–8,27,28]

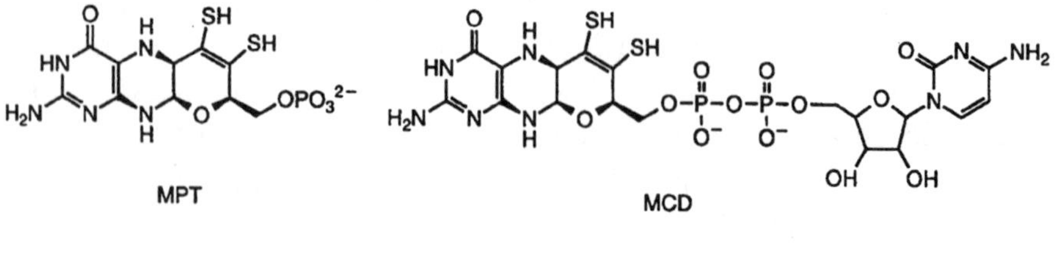

Various water-based (oxo, hydroxo, and/or aqua), thio, serinate and cysteinate (or selenocysteinate) coligands complete the Mo coordination spheres. These ligands are intimately involved in the essential reaction chemistry of the enzymes.[1–8] The active sites operate at the Mo(VI), Mo(V) and Mo(IV) oxidation levels and catalyze redox reactions which involve the net transfer of an oxygen atom between a substrate and water (Eq. 9.1). This net two-electron process inter-converts the Mo(VI) and Mo(IV) states (cf. OAT reactions, Sec. 3.1). The active sites are then regenerated by a sequence of one-electron transfers involving the prosthetic groups and external electron acceptor/donor molecules; EPR-active Mo(V) species are observed in some cases (cf. CEPT reactions, Sec. 3.2). The detailed molecular mechanism employed by each enzyme depends on the active site and the nature of the substrate (Secs. 2.1–2.3).

There is strong evidence that molybdenum enzymes can participate in OAT reactions, although OAT is not necessarily involved in *in vivo* reactions.[1] Murray *et al.*[29] demonstrated that partial (0.67 atom fraction) incorporation of ^{18}O from labeled nicotinamide *N*-oxide into uric acid takes place during the oxidation of xanthine by xanthine oxidase. More recently, Schultz *et al.*[30] demonstrated that the reaction shown in Eq. 9.2 (cf. Eq. 9.9) is catalyzed by the dimethyl sulfoxide reductase from *Rhodobacter sphaeroides*. Also, Hille and Sprecher[31] have shown that ^{16}O-labeled xanthine oxidase reacts with sub-stoichiometric quantities of xanthine in $H_2{}^{18}O$ to form uric acid containing ^{16}O at the 8-position of the purine ring. Conversely, turnover of sub-stoichiometric amounts of xanthine by ^{18}O-labeled enzyme in $H_2{}^{16}O$ resulted in ^{18}O incorporation into the product. These results can be interpreted in terms of either OAT or hydroxo ligand transfer *(vide infra)*.

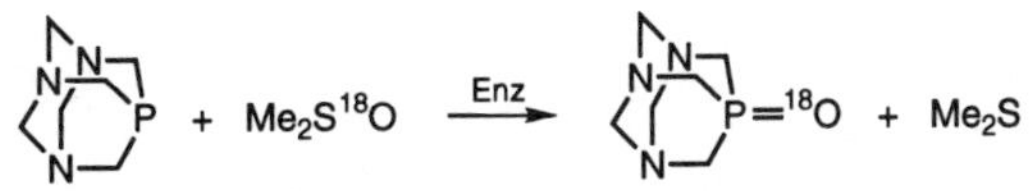

(9.2)

Hille[6] has classified the oxomolybdenum enzymes into three families. Brief descriptions of each of these families are given in the following sections.

2.1 *The sulfite oxidase family*

This family includes the sulfite oxidases and assimilatory nitrate reductases found in bacteria, plants, birds and animals.[1–8] Chicken liver sulfite oxidase is a 110 kDa homodimer containing one Mo center and one b_5-type heme center per subunit; it catalyzes the oxidation of sulfite to sulfate, the terminal step in the oxidation of cysteine and methionine, employing cytochrome *c* as physiological oxidant (Eq. 9.3). Deficiency of the enzyme in humans leads to major neurological abnormalities and early death.[19,32] Assimilatory nitrate reductases are involved in the first step of nitrate assimilation by fungi, algae, and plants; these 110 kDa homodimers contain one *b*-type cytochrome, one FAD and one Mo center per subunit; they catalyze the reaction shown in Eq. 9.4 (Reductant: NADH in plants and most algae, NADPH in fungi).

$$SO_3^{2-} + H_2O + 2\text{cyt}\ c^{III} \rightarrow SO_4^{2-} + 2H^+ + 2\text{cyt}\ c^{II} \quad (9.3)$$

$$NO_3^- + NAD(P)H + H^+ \rightarrow NO_2^- + H_2O + NAD(P)^+ \quad (9.4)$$

EPR and EXAFS studies of chicken liver sulfite oxidase and the nitrate reductase from *Chlorella vulgaris* indicate that both enzymes possess similar active sites. The protein structure and details of the stereochemistry and ligand set at the active site have been provided by the recent crystal structure of chicken liver sulfite oxidase.[19] The types of centers formed under various conditions are shown in Fig. 9.1.

With respect to the molybdenum center, it appears that a spectator oxo ligand[33,34] controls the electronic structure allowing the second oxo ligand to participate in OAT reactions with the substrate, X = SO_3^{2-} for sulfite

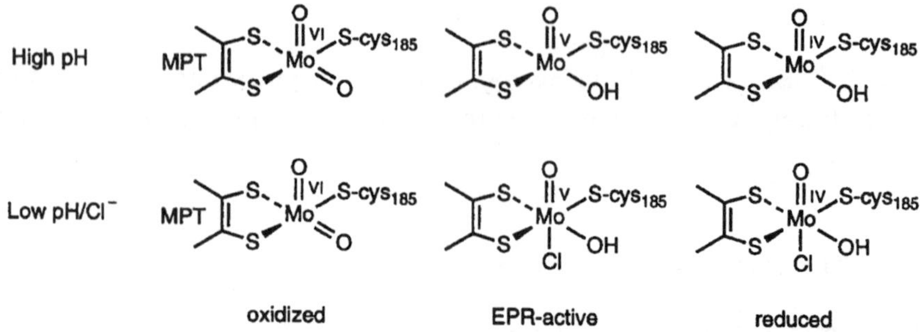

Fig. 9.1 Partial active site structures for chicken liver sulfite oxidase (and assimilatory nitrate reductases, charges not shown).

oxidase (Scheme 9.1, **a→b**). Egress of the oxidized substrate, XO = SO_4^{2-}, permits the generation of an aqua-Mo(IV) species (**b→c**), which undergoes CEPT reactions leading to the regeneration of the active site (**c→d→a**). Nitrate reductases would cycle in the opposite direction (cf. Scheme 9.2 for dimethyl sulfoxide reductase). The OAT chemistry is represented as a concerted two-electron event, the CEPT chemistry as two consecutive (e^-, H^+) transfers.[5,7]

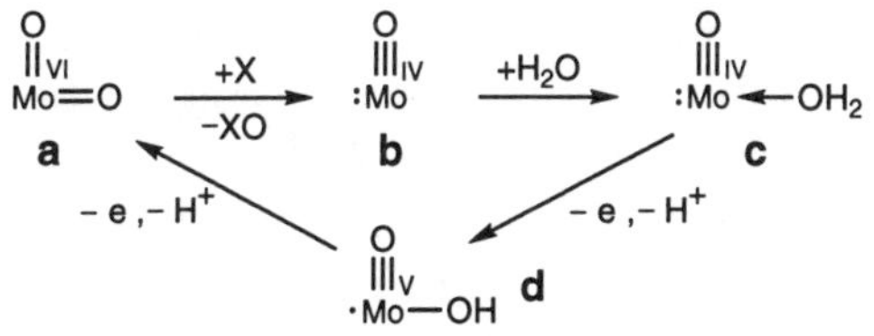

Scheme 9.1

2.2 *The dimethyl sulfoxide reductase family*

These enzymes, all from bacterial or archaeal sources, act as terminal respiratory oxidases during anaerobic growth in the presence of dimethyl sulfoxide; they catalyze the reaction shown in Eq. 9.5. The bacterial production of dimethyl sulfide in the world's oceans has a profound effect on marine life (especially feeding behavior) and, through its role in cloud nucleation, global albedo (reflectance) and climate.[35]

$$Me_2SO + 2H^+ + 2e^- \rightleftharpoons Me_2S + H_2O \tag{9.5}$$

The enzyme has been isolated and crystallized under a variety of conditions from the bacterial sources *R. sphaeroides* and *R. capsulatus* and a number of crystal structures are now available.[20–23] These reveal the presence of two molybdopterin guanine dinucleotide ligands, P and Q, one of which (Q) displays a range of coordination modes and redox levels (Fig. 9.2).

The structures in Figs. 9.2(a) and 9.2(c), and those of corresponding reduced states, have been corroborated or established by EXAFS

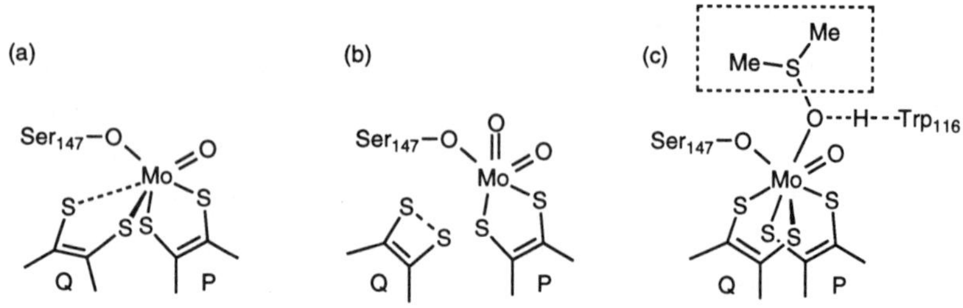

Fig. 9.2 Structures of dimethyl sulfoxide reductases. (**a**) Oxidized enzyme from *R. sphaeroides*.[20] (**b**) Oxidized enzyme from *R. capsulatus*.[21] (**c**) Most recent structure determined for oxidized enzyme from *R. capsulatus*, showing the native form (boxed region absent) and the Me_2S adduct (boxed region present).[22,23]

spectroscopy.[23,36] As Rees[28] observes, "An important objective of future research is to establish the relationship between these multiple forms .... and to determine the mechanistic significance, if any, of these states." The structure of another member of this family, the selenocysteine-containing formate dehydrogenase H from *Escherichia coli*, has also been determined.[24]

A mechanism for dimethyl sulfoxide reductase is given in Scheme 9.2; the relationship of the mechanisms in Schemes 9.1 and 9.2, one essentially the reverse of the other, should be noted. OAT and CEPT reactions feature once again.

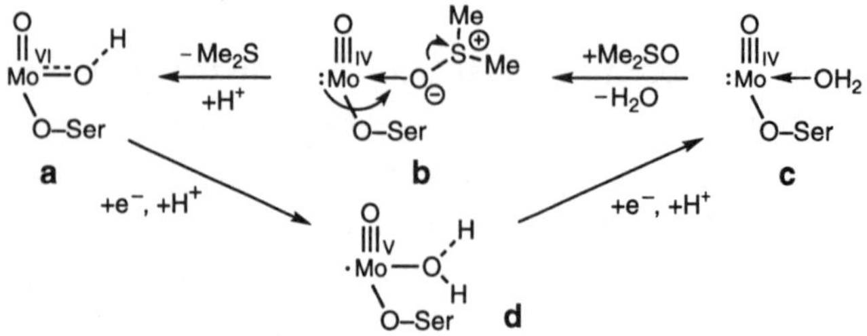

Scheme 9.2

2.3 *The xanthine oxidase family*

These enzymes catalyze the hydroxylation of a wide range of aldehydes and aromatic heterocycles, in reactions involving the net insertion of an oxygen atom into a C–H bond, e.g. xanthine oxidase converts xanthine to uric acid (Eq. 9.6).[6]

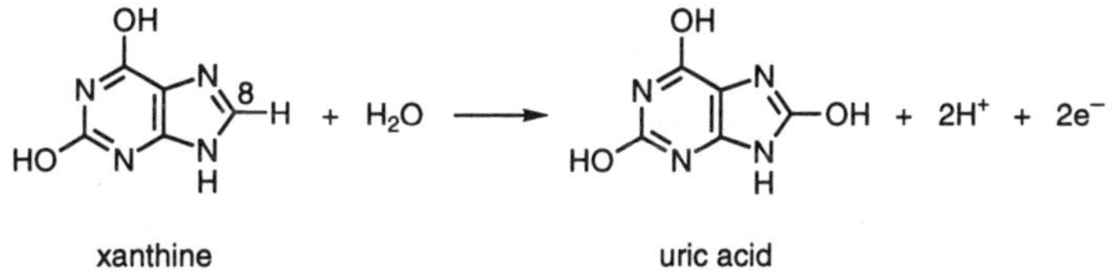

(9.6)

The presence of a catalytically essential active site sulfur atom, believed to be a terminal thio ligand, is supported by cyanolysis, EPR and EXAFS studies.[1–8] A five-coordinate $[MoOS(OH_n)(dithiolene)]^{0/-}$ oxidized active site is consistent with recent crystal structures of several forms of *Desulfovibrio gigas* aldehyde oxidoreductase[25,26] and density functional theory calculations.[37,38] The latter strongly support a five-coordinate $[Mo^{VI}OS(OH)(dithiolene)]^-$ oxidized active site and a mechanism involving hydroxo group, rather than oxo group transfer. There is some evidence[37,39] for an Mo-C interaction during turnover and in some inhibited forms of the enzyme but such an interaction is yet to be firmly established. Proposed structures for several states of xanthine oxidase are shown in Fig. 9.3.

The mechanism shown in Scheme 9.3[6] is supported by many lines of evidence but is not the only one possible.[4,26,40]

Modeling of the substrate reactions of these oxo-thio-enzymes remains a difficult and unmet challenge.[15] Oxo-thio-Mo(VI) complexes are known but few have been reported to effect biomimetic reactions and none catalyze the oxidation of xanthine. Six-coordinate oxo-thio-Mo(VI) complexes are subject to S···S interactions and partial redox involving the thio group and coligands, but are known to mimic the enzymes' cyanolysis/sulfidation reactions (Sec. 7.1). The state and directions of the modeling endeavor have been appraised[15] and will not be treated in detail here.

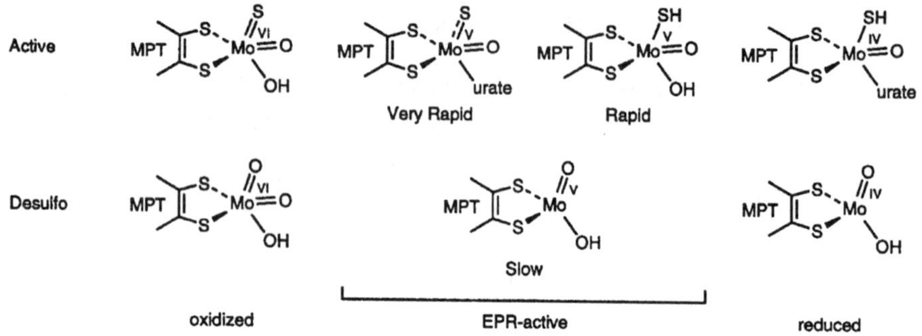

Fig. 9.3 Active sites of active and desulfo (cyanolyzed) xanthine oxidase.

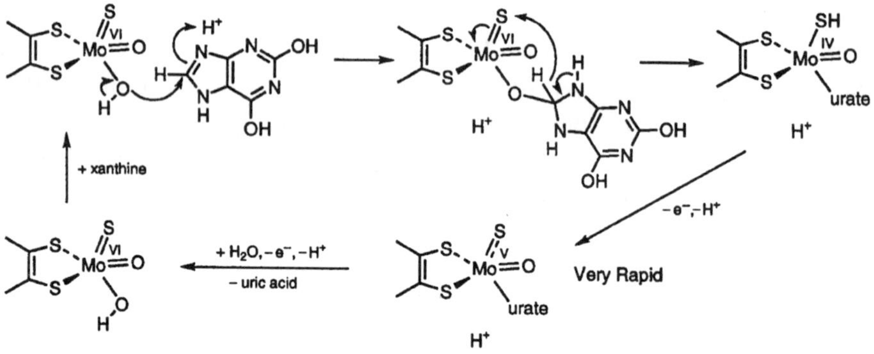

Scheme 9.3

3. Overview of Biomimetic Systems

3.1 *Oxygen atom transfer reactions*

The "oxo transfer" hypothesis is supported by enzyme studies *(vide supra)* and molybdenum's predilection for OAT reactions, as reflected in the later sections of this chapter. Reactions of this type have been comprehensively reviewed by Holm.[1,41]

Oxygen atom transfer is a concerted two-electron process which results in reduction of the oxygen atom donor and oxidation of the acceptor. For molybdenum, the most common OAT reactions are represented by Eqs. 9.7 and 9.8. The coupling of these reactions results in a catalytic system for the oxidation of X by X′O (Eq. 9.9, compare Eq. 9.2 catalyzed by dimethyl sulfoxide reductase).

$$Mo^{VI}O_2L_n + X \rightarrow Mo^{IV}OL_n + XO \tag{9.7}$$

$$Mo^{IV}OL_n + X'O \rightarrow Mo^{VI}O_2L_n + X' \tag{9.8}$$

$$X + X'O \rightarrow XO + X' \tag{9.9}$$

The formation of dinuclear μ-oxo-Mo(V) complexes via comproportionation (Eq. 9.10) is a prevalent complication in molybdenum's OAT chemistry. In some cases, but by no means all, the dinuclear complex is in equilibrium with its mononuclear constituents.[1,2,5,7,41]

$$Mo^{VI}O_2L_n + Mo^{IV}OL_n \leftrightharpoons Mo^{V}{}_2O_2(\mu\text{-}O)L_{2n} \tag{9.10}$$

Catalysis of Eq. 9.9 remains feasible when dinucleation is reversible, or when the dinuclear species is itself capable of participating in OAT reactions. These two cases are exemplified in the dithiocarbamate (L) systems shown in Schemes 9.4[42–44] and 9.5,[45,46] respectively; both catalyze the oxidation of phosphines by oxidants such as *S*- and *N*-oxides and dioxygen. The substrate oxidations depend crucially on the generation of a vacant site for binding XO, at $MoOL_n$ via disproportionation (Eq. 9.10) or through loss of labile solvent (solv) ligands, respectively.

Many other inorganic systems are complicated by comproportionation and dinucleation reactions. These include systems based on cysteine ester complexes, which mimic the behavior shown in Scheme 9.4,[47,48] and

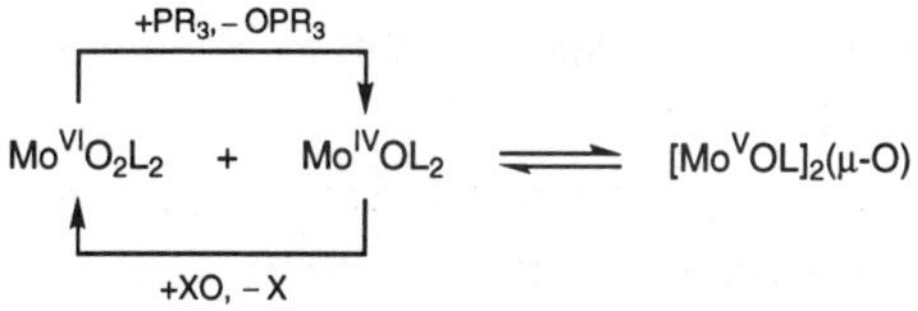

Scheme 9.4

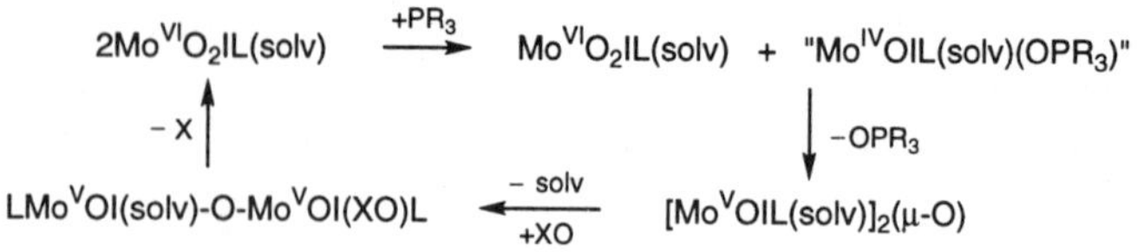

Scheme 9.5

tridentate Schiff base complexes, which mimic the type of behavior in Scheme 9.5 (Sec. 4).[49]

Comproportionation can also lead to the formation of mononuclear Mo(V) species which adversely effect catalytic systems based on OAT (Eq. 9.11) — Mo(V) is separated from Mo(VI) and Mo(IV) by one-electron rather than two-electron processes. This complication can be avoided by immobilizing the Mo(VI) and Mo(IV) complexes in various ways (see Sec. 8).

$$Mo^{VI}O_2L_n + Mo^{IV}OL_n + 2H^+ \rightarrow 2[Mo^{V}OL_n]^+ + H_2O \qquad (9.11)$$

Dinucleation and (to a lesser extent) comproportionation reactions are undesirable in models of the *mononuclear* enzyme active sites. Accordingly, strategies have been developed to limit dinucleation, which is a particular problem at the Mo(V) and Mo(IV) oxidation levels. The use of sterically bulky ligands that prevent the close approach of the metal centers is one such strategy (adopted by the enzymes, and the systems in Secs. 5–8). Another is the use of charged complexes which remain mononuclear by virtue of electrostatic repulsions (see Secs. 7–9).

The kinetics and thermodynamics of OAT reactions have been extensively studied.[1,41] In substrate oxidation reactions (Eq. 9.7), the oxygen atom acceptor is commonly a phosphine. More recent models involve the biologically relevant substrates sulfite and benzoin (Sec. 9). When $X = PR_3$, the reactions are first order with respect to both complex and phosphine, giving the overall second order rate law in Eq. 9.12. A kinetics treatment for systems complicated by the reversible equilibrium in Eq. 9.10 is also available.[44] Kinetics data (Table 9.1) are indicative of associative mechanisms. These reactions are proposed to take place by nucleophilic attack of the phosphine lone pair on the $\pi*$ Mo=O orbital and are facilitated, both electronically and energetically, by the presence of a "spectator oxo group."[33,34] A theoretical study by Pietsch and Hall[50] has provided insights into these reactions.

$$-\frac{d[Mo(VI)]}{dt} = k[Mo(VI)][PR_3] \qquad (9.12)$$

A wide variety of complexes and substrates, including enzyme substrates such as *S*- and *N*-oxides and nitrate, participate in the OAT reaction in Eq. 9.8. Where mononuclear oxo-Mo(IV) complexes are involved, associative second order reactions defined by the rate law in Eq. 9.13 are observed. When dinuclear Mo(IV) or Mo(V) species are involved the kinetics behavior is complicated, usually exhibiting substrate saturation kinetics.

$$-\frac{d[Mo(IV)]}{dt} = k[Mo(IV)][XO] \qquad (9.13)$$

Finally, Holm and co-workers have provided valuable insights into the thermodynamic fitness of molybdenum centers to promote OAT reactions involving enzyme substrates.[1,51] The activation for oxygen atom transfer appears to depend heavily on factors which stabilize or destabilize the Mo(IV) and Mo(VI) oxidation states. The general belief that sulfur-donor ligands are necessary for efficient OAT has been challenged by recent results (see Sec. 10).

3.2 Coupled electron-proton transfer reactions

Regeneration of the active sites following OAT is generally proposed to take place by two one-electron CEPT reactions (see Schemes 9.1–9.3).[2,52] These reactions are favored on electrostatic grounds and are facilitated by the pivotal relationship of the water-based aqua, hydroxo, and oxo ligands, exemplified in Eq. 9.14.

$$Mo^{VI}{=}O \underset{-e^-,-H^+}{\overset{+e^-,+H^+}{\rightleftharpoons}} Mo^{V}{-}OH \underset{-e^-,-H^+}{\overset{+e^-,+H^+}{\rightleftharpoons}} Mo^{IV}{-}OH_2 \tag{9.14}$$

π-bonding in Mo=E (E = O, S) systems permits dramatic modulation of the pK_a of the E atom in Mo=E and Mo-EH units (and coligands) when redox occurs at the metal center. Formal reduction of the metal center increases the basicity of the Mo=E ligands through the population of π^* orbitals with significant ligand character. This induces irreversible chemical processes, including formation and loss of ligand water; these reactions can be prevented by exploiting ligand design (Secs. 5 and 7) or high dilution.[53] Conversely, oxidation of aqua-Mo(IV) species results in the formation of hydroxo and oxo species. CEPT reactions feature in the electrochemical inter-conversion of amino-thiol and amido-thiol complexes (Sec. 5), the modeling of Mo(V) states, and reactions of Mo(VI) complexes with thiols and benzoin (Secs. 5–9).

While water-based CEPT reactions (Eq. 9.14) appear to play a central role in enzymatic processes, there are few examples where water plays a recognized role in model systems. In a few systems,[48,54] water is known to affect the rate of OAT reactions, but the mechanisms involved remain obscure. In one case, the role of water and CEPT reactions in OAT catalysis has been delineated; this system, discussed in Sec. 7, constitutes one of the closest models for the overall behavior of enzymes featuring the dioxo-Mo(VI) unit.

4. Schiff Base Systems

Many molybdenum Schiff base complexes have been reported[55] but only tridentate, dianionic ligands, such as sap, ssp, and hpd, feature in "biomimetic" studies.

Work in this area was initiated by Rajan and Chakravorty,[56] who reported oligomeric MoO_2L and monomeric $MoO_2L(D)$ complexes containing 3-X- and 5-X-sap and -ssp ligands (L), and N-, O- and P-donor ligands and solvents (D). Shortly thereafter, Boyd and Spence[57] reported that $MoO_2L(dmf)$ (L = sap, ssp; dmf = *N,N*-dimethylformamide) reacted with

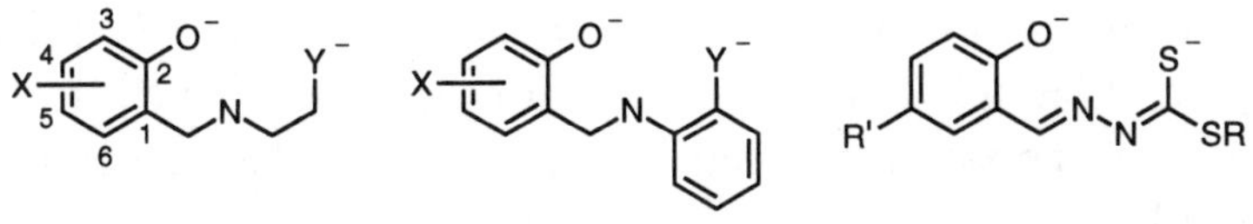

PPh_2Et in dmf to form oxo-Mo(IV) complexes, MoOL(dmf). However, Craig *et al.*[49] proved that the products of these and related reactions were dinuclear μ-oxo-Mo(V) complexes, $Mo_2O_3L_2(solv)_2$, reducing the biomimetic relevance of these systems and defining their chemistry in terms of Eqs. 9.15 and 9.16 (solv = solvent).

$$Mo^{VI}O_2L(solv) + PR_3 \rightarrow Mo^{IV}OL(solv) + OPR_3 \tag{9.15}$$

$$Mo^{VI}O_2L(solv) + Mo^{IV}OL(solv) \rightarrow Mo^{V}{}_2O_3L_2(solv)_2 \tag{9.16}$$

The reactions of MoO_2(5-X-sse) and MoO_2(5-X-ssp) complexes with PPh_2Et are first order with respect to complex and phosphine (Eq. 9.12) and selected kinetics data are summarized in Table 9.1.[58,59] The formation of the dinuclear complex must be fast and complete, making OAT the rate-determining step of the reaction. The rates and activation parameters correlate with the Mo(VI)/Mo(V) irreversible reduction potentials, the frequency of the antisymmetric $\nu(MoO_2)$ IR band, and the Hammett σ_p functions of the X groups. The correlation of k with $\nu_{as}(MoO_2)$ is consistent with nucleophilic attack of the Mo=O π^* antibonding orbital.

Certain of these dinuclear Mo(V) complexes react cleanly and completely with Ph_2SO and NO_3^- to give Ph_2S and NO_2^-, respectively, along with the corresponding Mo(VI) complex. A pathway paralleling the oxidation shown in Scheme 9.5 is likely to describe these reactions.[49]

Related dioxo-Mo(VI) thiocarbazate and dithiocarbazate complexes, MoO_2L and $MoO_2L(D)$, have been the subject of more recent studies.[60–63] The crystal structure of orange MoO_2(Me-hpd)(H_2O) reveals a six-coordinate, distorted octahedral structure with a mer Me-hpd ligand.[63] Reactions of MoO_2L or $MoO_2L(D)$ (L = Me-hpd and Bz-hpd) with PPh_3 are proposed to produce dinuclear oxo-Mo(IV) complexes (e.g. Eq. 9.17) but this has not been confirmed, e.g. by ^{31}P NMR.[61,62] The reactions obey the rate law in Eq 9.12 and various data are summarized in Table 9.1.

$$Mo^{VI}O_2(Me\text{–}hpd) + PPh_3 \rightarrow 1/2[Mo^{IV}O(Me\text{–}hpd)]_2 + OPPh_3 \quad (9.17)$$

$$Mo^{VI}O_2(Me\text{–}hpd)(MeOH) + 2H_2tga \rightarrow 1/2[Mo^{IV}O(Me\text{–}hpd)]_2 + dtga + H_2O + MeOH \quad (9.18)$$

Table 9.1 Selected kinetics data for reactions of Mo(VI) complexes with phosphines.[a]

Complex, T in °C	Phosphine	$k \times 10^3$ $M^{-1}s^{-1}$	$\Delta H^\ddagger$ kJ mol^{-1}	$\Delta S^\ddagger$ J K^{-1} mol^{-1}	Ref.
$MoO_2(S_2CNEt_2)_2$, 25[b]	PPh_3	71(3)			44
MoO_2(L-Cys-OEt)$_2$, 35[c]	PPh_3	0.29	46	−153	47
MoO_2(ssp), 30	$PEtPh_2$	1.02(9)	65.2	−86.5	58
MoO_2(sse), 30	$PEtPh_2$	0.21(2)	70.3	−82.6	58
MoO_2(Bz-hpd)(dmf), 25	PPh_3	8.9(2)	55(1)	−102(3)	61
MoO_2(Me-hpd)(dmf), 30	PPh_3	14.9(2)			62
(L-NS_2)MoO_2, 23	PPh_3	7(1)			76
(L-NS_2)MoO_2, 23	$P(C_6H_4F)_3$	9.7(4)	49(3)	−119(7)	79
MoO_2(L-NS)$_2$,	PEt_3	5.6(1)	40(2)	−155(8)	83
(L-N_3)MoO_2L					
L = SPh^-, 25	PPh_3	0.59(2)			85
L = OPh^-, 25	PPh_3	0.002(1)			85
L = $S_2P(OEt)_2^-$, 25[d]	PPh_3	3.4(1)	64(1)	−121(8)	84
L = $S_2PPr^i{}_2^-$, 30[d]	PPh_3	0.25(3)			87
$(NEt_4)_2[MoO_2(bdt)_2]$, 30[e]	PPh_3	3.9			109
$MoO_2Cl_2(dmf)_2$, 25	PPh_3	43(1)			115
$(PPh_4)_2[MoO_2(NCS)_4]$[f]	PPh_3	2.2×10^4			117
$MoO_2(NCS)_2(Bu^t{}_2bpy)$[f]	PPh_3	1.7×10^3			117

[a] Reactions in dmf unless indicated. [b] In 1,2-dichloroethane. [c] In benzene, Cys-OEt = cysteine ethyl ester. [d] In toluene. [e] In acetonitrile. [f] In dichloromethane at 25 °C.

The methanol complex MoO_2(Me-hpd)(MeOH) reacts with thioglycolic acid, H_2tga, according to Eq. 9.18 (dtga = disulfide of H_2tga) to form the same brown, insoluble Mo(IV) complex. This reaction was proposed to proceed via two consecutive steps: (i) displacement of MeOH by H_2tga, which is proposed to bind via the sulfur atom, and (ii) induction of redox by a second H_2tga molecule, to produce disulfide and MoO(Me-hpd); the latter rapidly dimerizes to form $[MoO(Me\text{-}hpd)]_2$. The involvement of CEPT processes is likely.

The behavior of these dithiocarbazate systems, and the spectral properties of the various complexes involved, bears an uncanny resemblance to that reported for the sap and ssp systems, known to involve dinuclear Mo(V) complexes. The recent structural characterization of a salt of $[\{VO(Me\text{-}hpd)\}_2(\mu\text{-}O)]^-$ shows that, on steric grounds at least, the formation of a dinuclear Mo(V) complex is possible.[64]

5. Tetradentate Amino(thiolate) and Related Systems

The development of models for mononuclear, EPR-active enzyme states and the elucidation of CEPT chemistry have been major thrusts of the work in this area. The structures of the principal ligands employed are given below.

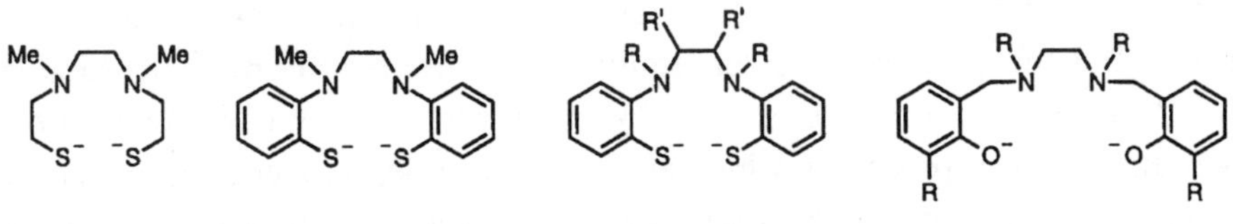

L^1-N_2S_2 L^2-N_2S_2

L^3-$N_2H_2S_2$ (R = R' = H)
L^3-N_2S_2 (N-deprotonated)
L^4-$N_2H_2S_2$ (R = H, R' = Me)
L^4-N_2S_2 (N-deprotonated)

L-N_2O_2

A number of dioxo-Mo(VI) complexes have been prepared and characterized; [65–69] the structure of *cis,trans*-(L^2-N_2S_2)MoO_2 [Fig. 9.4(a)] is typical of those that have been structurally characterized (the

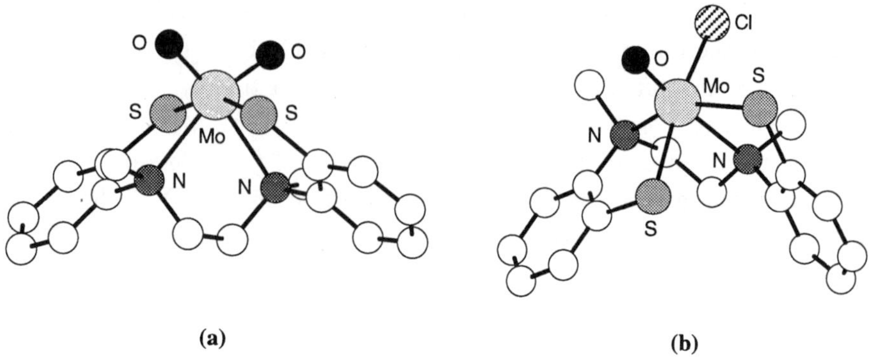

Fig. 9.4 (**a**) Structure of *cis,trans*-(L^2-N_2S_2)MoO_2.[67] (**b**) Structure of *cis,cis*-(L^2-N_2S_2)MoOCl (the *cis,trans* isomer has also been structurally characterized).[70]

stereochemical prefix refers firstly to the relative positions of the oxo groups and secondly to the relative positions of the thiolate groups).[66–69] The d(Mo=O), d(Mo-S) and ∠(O-Mo-O) values are typically ca. 1.70 Å, 2.43 Å, and 108°, respectively.

Molybdenum(V) and Mo(IV) complexes are accessed by CEPT and OAT reactions involving the dioxo-Mo(VI) species. The electrochemical reduction of (L^3-$N_2H_2S_2$)MoO_2 to [(L^3-N_2S_2)MoO]$^-$ is proposed to take place through an initial two-electron reduction of the amino-Mo(VI) complex, followed by comproportionation to produce the amido-Mo(V) complex. Loss of an oxo group upon reduction is facilitated by deprotonation of the amine N atoms and the production of an equivalent of water.[71] Further, Küsthardt *et al.*[69] have shown that (L^4-$N_2H_2S_2$)MoO_2 is electrochemically reduced by two one-electron processes which are coupled to deprotonation/protonation of the amine N atoms. The irreversible reduction of (L^4-$N_2H_2S_2$)MoO_2 to (L^4-$N_2H_2S_2$)MoO(OH) is followed by rapid loss of water from the Mo(V) complex, whereupon a reversible reduction to [(L^4-N_2S_2)MoO]$^{2-}$ becomes possible. An examination of Scheme 9.6 (NMI = *N*-methylimidazole) reveals the dominant role of CEPT reactions in this system.

In contrast, studies of *N*-protected L^n-N_2S_2 and L-N_2O_2 complexes revealed CEPT processes involving the terminal chalcogenide groups rather than the amine N atoms. Initial studies by Zubieta and co-workers[65]

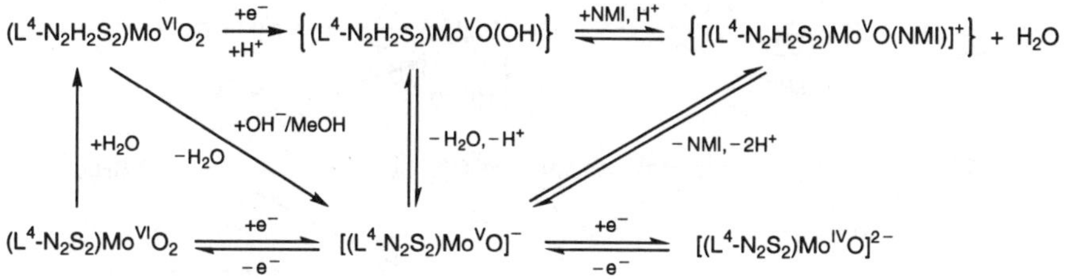

Scheme 9.6

were developed by Wedd, Spence and co-workers.[67,68,70,72–74] They showed that the reduction of *cis,trans*-(L^2-N_2S_2)MoO_2 produced [(L^2-N_2S_2)MoO_2]$^-$, which was characterized by a highly anisotropic EPR signal with <g> = 1.9007 and large A_{Mo} values. The conjugate acid, *cis*-(L^2-N_2S_2)MoO(OH), which was stable at -42 °C, was characterized by a doublet EPR signal (<g> = 1.957, $A_H = 15.1 \times 10^{-4}$ cm^{-1}). At higher temperatures, the (L^2-N_2S_2)MoO(OH) signal was replaced by a singlet signal assigned to [(L^2-N_2S_2)MoO]$^+$; this complex reacts with water to form dinuclear complexes.[70] Reaction of (L^2-N_2S_2)MoO_2 with excess SH^- produces [(L^2-N_2S_2)MoOS]$^-$, which can be protonated at low temperature to form *cis*-(L^2-N_2S_2)MoO(SH). The oxo-thio-Mo(V) complex is characterized by EPR spectra with a higher <g> = 1.9435, and a reduced rhombicity compared to the dioxo-Mo(V) analogue. The complex *cis*-(L^2-N_2S_2)MoO(SH) decomposes at room temperature, and the formation of *trans*-(L^2-N_2S_2)MoO(SH), then [(L^2-N_2S_2)MoO]$^+$, was proposed. The complex *trans*-(L^2-N_2S_2)MoO(SH) was also reported to form from *trans*-(L^2-N_2S_2)MoOCl, now known to be *cis,cis*-(L^2-N_2S_2)MoOCl [Fig. 9.4(b)]. Maybe the species previously assigned *trans*-MoOX geometries are in fact *cis,cis* complexes; Scheme 9.7, drawn from original schemes,[67,73] new information,[70] and relevant enzyme states, reflects this possibility. Note the prevalence of CEPT chemistry in Scheme 9.7.

The oxo-thio- and oxo(hydrosulfido)-Mo(V) complexes of L^2-N_2S_2 are important models for the Mo(V) states of the hydroxylases. Coordination of sulfur was demonstrated using a compound doubly-labeled with ^{98}Mo

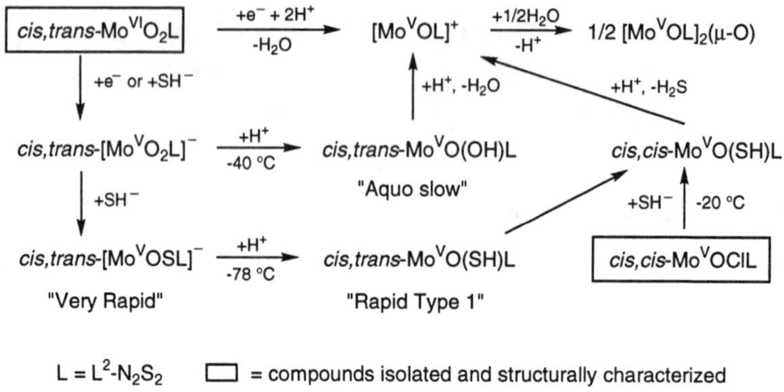

Scheme 9.7

($I = 0$) and ^{33}S ($I = 3/2$). The isotopomer $[(L^2\text{-}N_2S_2)^{98}MoO^{33}S]^-$ shows a remarkably large and anisotropic coupling to ^{33}S ($A_S = 10.8 \times 10^{-4}$ cm^{-1}), whereas $(L^2\text{-}N_2S_2)^{98}MoO(^{33}SH)$ has a greatly reduced, less anisotropic coupling to ^{33}S ($A_S = 1.4 \times 10^{-4}$ cm^{-1}).[74] EPR studies of ^{17}O-labeled complexes were also reported.[73] A comparison of the EPR coupling parameters of xanthine oxidase and these models provided strong support for the assignment of the Very Rapid and Rapid 1 and 2 signals to $[Mo^VOS(OR)]^+$ and $[Mo^VO(OH)(SH)]$ centers, respectively.[74]

Few OAT reactions involving $L^n\text{-}N_2S_2$ complexes have been reported. Reaction of $(L^4\text{-}N_2H_2S_2)MoO_2$ with PPh_2Me in dmf/NMI is proposed to form unstable $(L^4\text{-}N_2H_2S_2)MoO(NMI)$, which comproportionates with starting material to produce $(NMIH)[(L^4\text{-}N_2S_2)MoO]$ and water (cf. Scheme 9.6, lower).[69] Reaction of *cis,trans*-$(L^2\text{-}N_2S_2)MoO_2$ with PPh_3 forms C_i-$[(L^2\text{-}N_2S_2)MoO]_2(\mu\text{-}O)$.[70]

6. (Thiolato)pyridine Systems

6.1 Bis(thiolato)pyridine system

This system, developed by Holm and co-workers,[75–80] involves complexes of the 2,6-bis(2,2-diphenyl-2-thioethyl)pyridinate(2-) ligand, L-NS_2. The

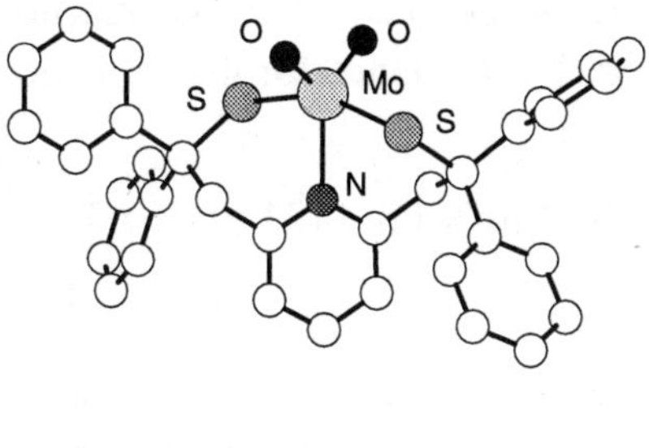

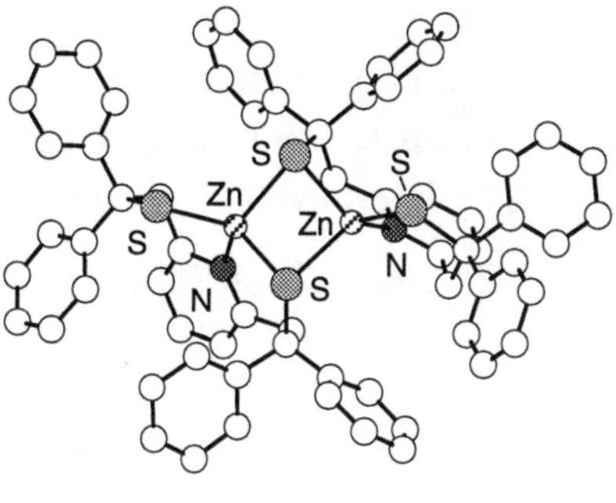

Fig. 9.5 **(a)** Structure of (L-NS_2)MoO_2.[75] **(b)** Structure of Zn_2(L-NS_2)$_2$.[81]

structure of trigonal-bipyramidal (L-NS_2)MoO_2 is shown in Fig. 9.5(a); the sulfur donor atoms occupy the two *trans* axial sites and the equatorial plane contains the two oxo groups (mean d(Mo=O) = 1.694 Å, ∠(O-Mo-O) = 110.5°) and the nitrogen donor atom. Reaction of (L-NS_2)MoO_2 with PPh_3 in dmf results in the formation of purple (L-NS_2)MoO(dmf), which was characterized by microanalysis, electrochemistry, and infra-red and UV-vis spectroscopy. The stoichiometry of the reaction of (L-NS_2)MoO_2 with PPh_3 was established by ^{31}P NMR experiments, which were consistent with the reaction in Eq. 9.19. This Mo(IV) complex has not been structurally characterized and there has been some speculation[1] about its structure. The L-NS_2 ligand may act as a bridging thiolate ligand, as observed in the structure of Zn_2(L-NS_2) [Fig. 9.5(b)],[81] in a dinuclear, six-coordinate oxo-Mo(IV) species. The kinetics of reactions involving (L-NS_2)MoO(dmf) (see below) would also be explained by a dinuclear formulation. For simplicity, we will adopt thc monomer formulation in this review.

$$(L\text{-}NS_2)MoO_2 + PPh_3 + dmf \rightarrow (L\text{-}NS_2)MoO(dmf) + OPPh_3 \qquad (9.19)$$

Both OAT and CEPT reactions have been observed in this system. As indicated above, phosphines reduce (L-NS_2)MoO_2 to (L-NS_2)MoO(dmf) with concomitant formation of phosphine oxide. The reactions follow the rate law in Eq. 9.12 and available kinetics data are summarized in Table 9.1.[76]

As well, $(L\text{-}NS_2)MoO_2$ is reduced to $(L\text{-}NS_2)MoO(dmf)$ by thiols, which become oxidized to the corresponding disulfides (Eq. 9.20).[77–79] This reaction would appear to fall into the CEPT category but evidence for the formation of Mo(V) intermediates is not available.

$$(L\text{-}NS_2)MoO_2 + 2RSH + dmf \rightarrow (L\text{-}NS_2)MoO(dmf) + H_2O + RSSR \quad (9.20)$$

The complex $(L\text{-}NS_2)MoO(dmf)$ reduces a wide range of compounds, XO, including biological substrates such as Me_2SO, D-biotin *S*-oxide, nicotinamide *N*-oxide, amine *N*-oxides, and nitrate.[76,79,80] The reactions of $(L\text{-}NS_2)MoO(dmf)$ with sulfoxides, $XO = R_2SO$, have been studied in detail and are proposed to proceed according to Scheme 9.8. The rate of the reaction is first order in $(L\text{-}NS_2)MoO(dmf)$ but at high sulfoxide concentrations substrate saturation kinetics pertain. The rate law is given by Eq. 9.21, where ligand Y = dmf + Me_2SO.

$$\begin{aligned}\frac{d[(L\text{-}NS_2)MoOY]}{dt} &= -k[(L\text{-}NS_2)MoO(R_2SO)] \\ &= -k\left(\frac{K[R_2SO]}{K[R_2SO]+[dmf]}\right)[(L\text{-}NS_2)MoOY]\end{aligned} \quad (9.21)$$

Data for a number of reactions are given in Table 9.2. The equilibrium and rate constants for the reactions fall into narrow ranges, $(4–16) \times 10^3$ and $(1.4–1.7) \times 10^{-3}\ s^{-1}$, respectively; this behavior is not typical of systems involving mononuclear Mo(IV) complexes (Secs. 6.2 and 7.1). The rates of sulfoxide reduction compare favorably with those

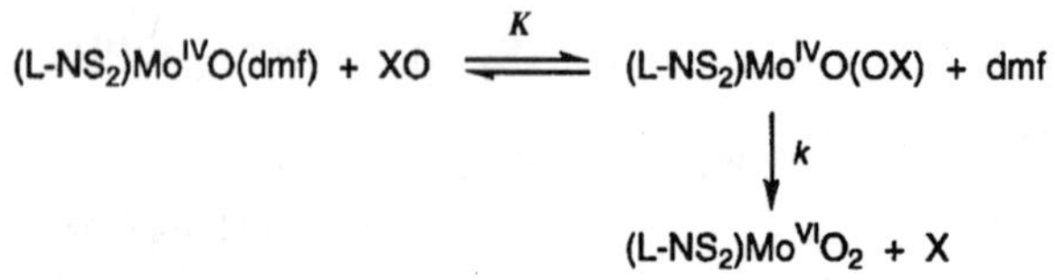

Scheme 9.8

Table 9.2 Kinetics data for the oxidation of (L-NS_2)MoO(dmf) by XO in dmf.

XO	T (K)	10^{-3} K	$10^3 k$, s^{-1}	$\Delta H^{\ddagger}$ kJ mol^{-1}	$\Delta S^{\ddagger}$ J deg^{-1} Mol^{-1}	Ref
Me_2SO	296	4.2(3)	1.50(3)			76
Ph_2SO	296	11(10)	1.43(3)			76
(p-$FC_6H_4)_2SO$	297.5		1.40(7)	92.5(5.4)	10.9(6.7)	79
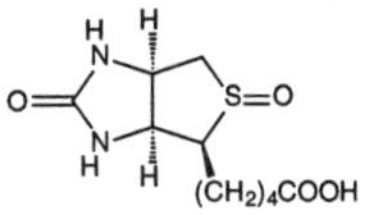	296	16(10)	1.36(3)			76
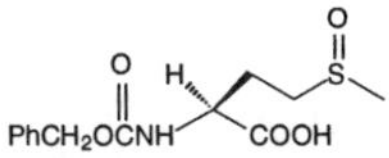	296	5.5(4)	1.63(3)			76
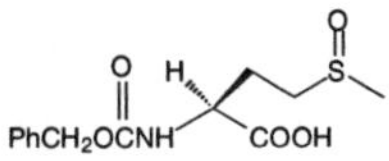	296	5.0(4)	1.70(3)			76
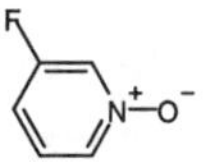	297.5		1.60(8)	97.9(5.9)	30.1(8.4)	79
NO_3^-	295.5		1.49(5)	99.2(2.5)	33.5(8.4)	80

of enzyme systems, the methionine S-oxide reductase from *E. coli* having a maximum specific activity equivalent to $k = 1.6 \times 10^{-4}$ s^{-1}.[76]

The reactions of (L-NS_2)MoO(dmf) with N-oxides and nitrate are also described by Scheme 9.8 and the kinetics treatment given by Eq. 9.21. The results of kinetics studies are summarized in Table 9.2. For nitrate reduction over the temperature range 282–317 K, K varies from 4.4×10^3 to 21×10^3. The moderate entropy of activation of 33.5(8.4) J deg^{-1} mol^{-1} suggests that the structure in the transition state resembles that of the ground state substrate complex and that the Mo-ONO_2 bond is retained. Comparison with other studies allowed Craig and Holm[80] to estimate that the N–O bond in nitrate is ca. 20 kJ mol^{-1} and 80 kJ mol^{-1} stronger than the X–O bond in Me_2SO and pyridine *N*-oxide, respectively.

Finally, catalytic systems which couple forward and reverse OAT reactions have been developed (Scheme 9.9). Initially, phosphines were employed as the oxygen atom acceptor but systems utilizing thiols in a combined OAT/CEPT cycle were later realized.[78,79] The catalytic rate is limited by the rate of reduction of (L-NS_2)MoO_2 by phosphine or thiol, the former being faster than the latter (see Note Added In Proof).

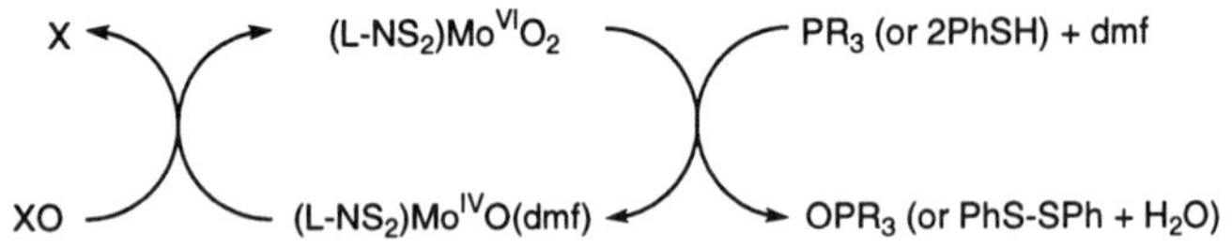

Scheme 9.9

6.2 (Thiolato)pyridine system

In a refinement of the bis(thiolato)pyridine system above, Holm and co-workers developed a second biomimetic system based on the complexes MoO_2(L-NS)$_2$ and MoO(L-NS)$_2$, where L-NS = bis(4-*t*-butylphenyl)-2-pyridylmethanethiolate(1-).[82,83] In this system, it was possible to structurally characterize both the Mo(VI) and Mo(IV) complexes and to demonstrate that both were mononuclear. Six-coordinate MoO_2(L-NS)$_2$ exhibits a distorted octahedral structure with *cis*-oxo- and *trans*-thiolate groups; d(Mo=O) and $\angle$(O-Mo-O) are 1.696(4) Å and 107.7(3)°, respectively [Fig. 9.6(a)]. Five-coordinate MoO(L-NS)$_2$ possesses a distorted trigonal bipyramidal structure with an $MoOS_2$ equatorial plane, axial nitrogen ligands, and d(Mo=O) = 1.681(5) Å [Fig. 9.6(b)].[82]

The kinetics of the reaction of MoO_2(L-NS)$_2$ with PEt_3, which produces MoO(L-NS)$_2$ and $OPEt_3$ (Eq. 9.7) have been investigated; the reaction follows the rate law in Eq. 9.12 and data are summarized in Table 9.1.[83]

In dmf, MoO(L-NS)$_2$ cleanly reduces a variety of oxygen atom donors, including the biological substrates nicotinamide *N*-oxide, adenine *N*-oxide,

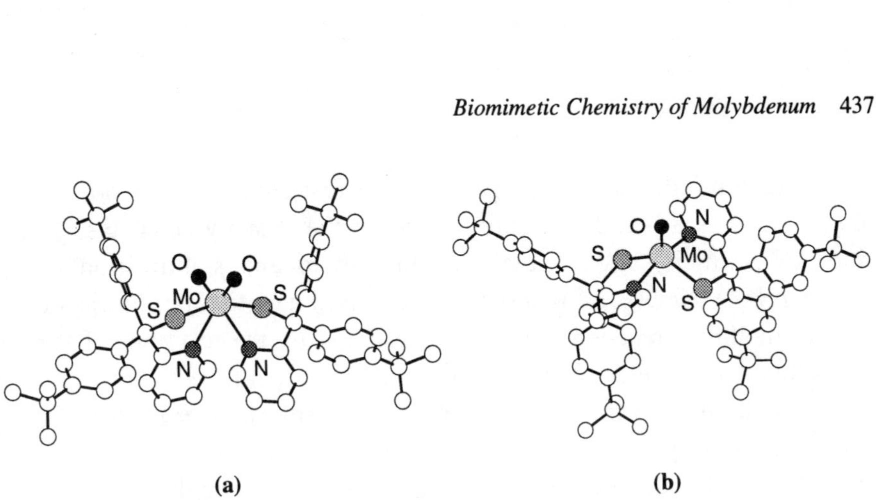

Fig. 9.6 Structures of **(a)** $MoO_2(L\text{-}NS)_2$ and **(b)** $MoO(L\text{-}NS)_2$.

Me_3NO and Me_2SO, and non-biological *N*-, *S*-, *Se*- and *As*-oxide substrates. The rates of XO reduction are substrate-dependent and follow a second order rate law; available rate constants and activation parameters are collected in Table 9.3.[83] For reactions involving sterically unencumbered substrates, rate constants range from 2.2 to 10^{-4} $M^{-1}s^{-1}$; an associative mechanism is indicated by the large, negative entropies of activation. As shown in Table 9.3, the values of k and $\Delta H^{\ddagger}$ do not consistently

Table 9.3 Kinetics data for substrate reduction by $MoO(L\text{-}NS)_2$ in dmf at 298 K.[83]

Substrate	$k(M^{-1}\,s^{-1})$	$\Delta H^{\ddagger}$ kJ mol^{-1}	$\Delta S^{\ddagger}$ J K^{-1} mol^{-1}	$D_{X\text{-}O}$ kJ mol^{-1}	pK_{BH^+}
Me_2SO	$1.01(2) \times 10^{-4}$	56.9(8)	−130(4)	364	−1.80
Ph_2SO	$3.14(4) \times 10^{-4}$	59(2)	−113(8)	372	−2.54
$(CH_2)_4SO$	$1.44(2) \times 10^{-3}$	48.1(8)	−138(4)	(~364)	−1.34
Ph_3AsO	$5.6(2) \times 10^{-2}$	44(2)	−117(8)	430	0.99
pyNO	$3.04(5) \times 10^{-1}$				0.79
3-F-pyNO	$6.5(3) \times 10^{-2}$	41.4(8)	−130(4)	(~301)	(0.79)
2,6-Me_2pyNO	$4.3(2) \times 10^{-3}$				1.02
2,6-Ph_2pyNO	$4.3(3) \times 10^{-5}$				
Ph_2SeO	2.16(6)	44(2)	−88(8)	(~180)	0.35

correlate with the bond energy of the substrate's X–O bond energy, $D_{X\text{-}O}$. There is a broad correlation between the basicity constants, pK_{BH^+}, and $\Delta H^\ddagger$, which suggests that the binding to and stabilization of the (reorganized) $MoO(L\text{-}NS)_2$ intermediate contributes to $\Delta H^\ddagger$. Steric factors clearly modulate reaction rate as demonstrated by the reactions of various substituted pyridine *N*-oxides.

An initial pseudo-rotation to a trigonal-bipyramidal *cis,trans*-$Mo(O)N_2S_2$ intermediate is proposed to precede substrate-binding by an associative process. An intramolecular oxygen atom transfer through an $[Mo\cdots OX]^\ddagger$ transition state then follows. This is consistent with ^{18}O-labeling studies which have shown that OAT takes place directly between substrate and Mo complex.[82] Significantly, this system is thermodynamically capable of oxidizing and reducing all enzymatic substrates except those requiring the $[MoOS]^{2+}$ center as oxidant.

7. Tris(pyrazolyl)borate Systems

Tris(pyrazolyl)borate and related tripodal, N-donor ligands, $L^n\text{-}N_3$, bind to Mo in a facial manner, thereby restricting most chemical and stereochemical changes to the coligand sites. Substitution at the 3-position of the heterocyclic rings generally ensures mononuclearity. These ligands feature in one of the broadest model systems, featuring forward and reverse OAT and CEPT reactions, and many of the oxo-Mo(VI,V,IV) states of biological importance. The OAT and CEPT chemistry developed by Young and co-workers is the main focus of this section.

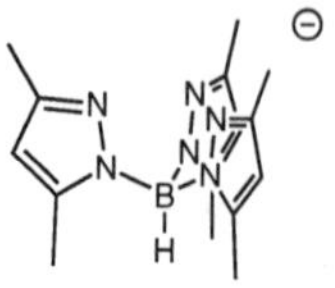

L-N_3

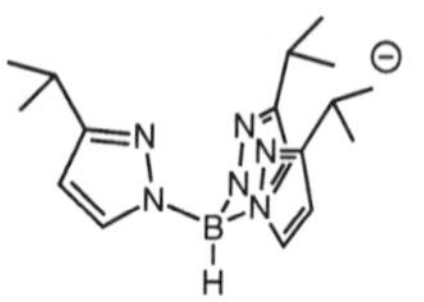

L^2-N_3

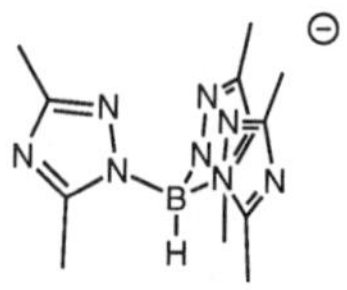

L^3-N_3

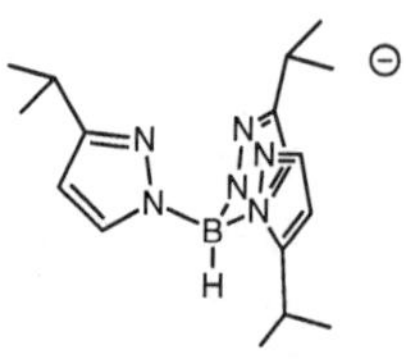

L^4-N_3

Many dioxo-Mo(VI) complexes containing L-N_3 ligands have been prepared and characterized.[84–87] The halo complexes (L$^{1\text{-}3}$-N_3)MoO_2X (X = F, Cl, Br) are readily prepared and complexes containing monodentate, monoanionic, hard (O-, N-) or soft (S) donor coligands are accessible via metathesis reactions (e.g. Eq. 9.22).[85,86]

$$(\text{L-N}_3)\text{MoO}_2\text{Br} + \text{NEt}_3 + \text{HX} \rightarrow (\text{L-N}_3)\text{MoO}_2\text{X} + \text{NHEt}_3\text{Br} \quad (9.22)$$

These Mo(VI) complexes react with PPh_3 to produce $OPPh_3$ and a number of Mo-containing products depending on the conditions of the reaction (Scheme 9.10).[85,86,88,89] The initial product, coordinatively unsaturated "(L-N_3)Mo^{IV}OX" can be trapped at the Mo(IV) level as (L-N_3)MoOX(py) with pyridine (py). In the absence of stabilizing ligands, Mo(V) species are formed via solvent reactions or comproportionation (cf. Eq. 9.10). The putative aqua-Mo(IV) species undergoes further reactions leading to Mo(V) species *(vide infra).*

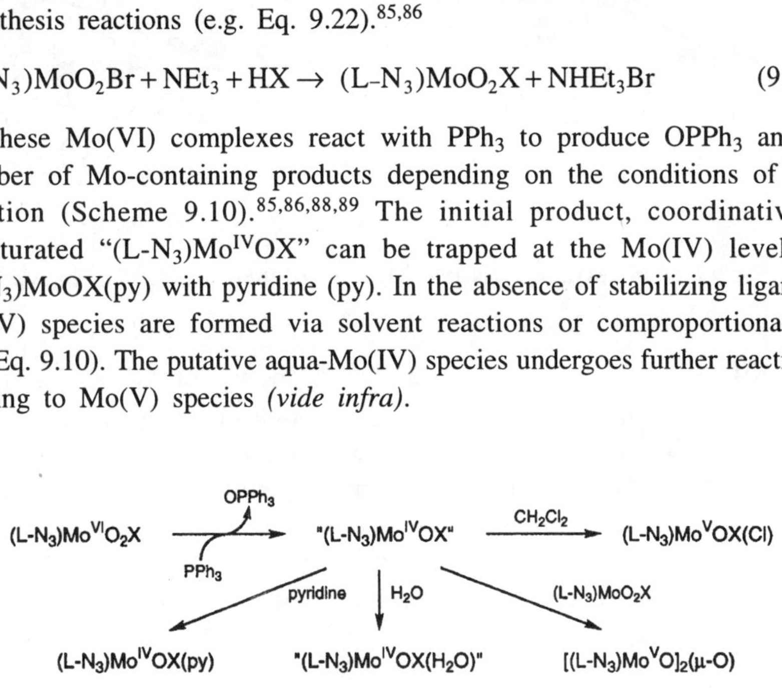

Scheme 9.10

The molecular structures of (L-N_3)MoO_2(SPh)[86] and (L-N_3)MoO (SPh)(py)[89] are shown in Fig. 9.7. These, and all related complexes, feature six-coordinate, distorted octahedral coordination spheres, with d(Mo=O) = 1.64 Å–1.72 Å and ∠(O-Mo-O) ca. 103°. The L-N_3 ligand adopts its usual tridentate binding mode and the d(Mo-N) values invariably reflect the relative *trans* influences of the coligands.

Complexes containing ambidentate, monoanionic coligands, L, such as dithiocarbamates, dithiophosphates, dithiophosphinates, and pyridine thiolate (pyS) can be prepared by reacting the precursors $MoOL_2$ or $Mo_2O_3L_4$[42,43]

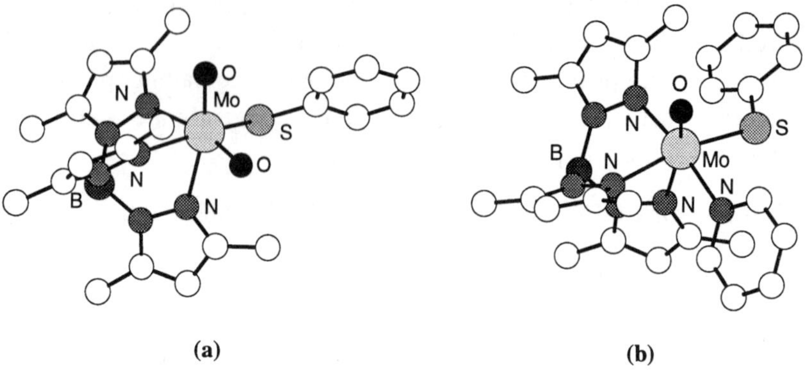

Fig. 9.7 Structures of **(a)** (L-N_3)$Mo^{VI}O_2$(SPh) and **(b)** (L-N_3)Mo^{IV}O(SPh)(py).

with K(L-N_3), for example, Eq. 9.23.[84,87,90] The use of K(L^2-N_3) in place of K(L-N_3) in Eq. 9.23 results in the formation of (L^4-N_3)MoOL, which reacts with boron sulfide to yield the analogous thio complexes, e.g. (L^4-N_3)MoS(S_2PR_2). In turn, these are susceptible to OAT reactions leading to rare oxo-thio species of relevance to xanthine oxidase.[91] The (L-N_3)Mo^{IV}OL complexes can be converted to Mo(VI) species by OAT, with a concomitant change in the denticity of L, as shown in Eq. 9.24.

$$Mo^{IV}OL_2 + K(L\text{-}N_3) \rightarrow (L\text{-}N_3)Mo^{IV}OL + KL \quad (9.23)$$

$$(L\text{-}N_3)Mo^{IV}OL + XO \rightarrow (L\text{-}N_3)Mo^{VI}O_2L + X \quad (9.24)$$

The structures of $LMoO_2\{S_2P(OEt)_2\}$ and $LMoO\{S_2P(OEt)_2\}$[84] are shown in Fig. 9.8. The dioxo-Mo(VI) complex features a monodentate dithio ligand and *cis-* arrangement of oxo ligands. The monodentate dithio ligand adopts an extended conformation which positions the uncoordinated sulfur atom distal to and at a distance of 5.42 Å from the Mo center. The oxo-Mo(IV) complex contains a single terminal oxo ligand and a symmetrically bidentate dithio ligand.

Some feeling for the very extensive OAT and CEPT chemistries of the (L-N_3)MoO_2X- and (L-N_3)MoO_2L-based systems can be gleaned from Scheme 9.11, which summarizes the reactions of the L = dithiophosphinate

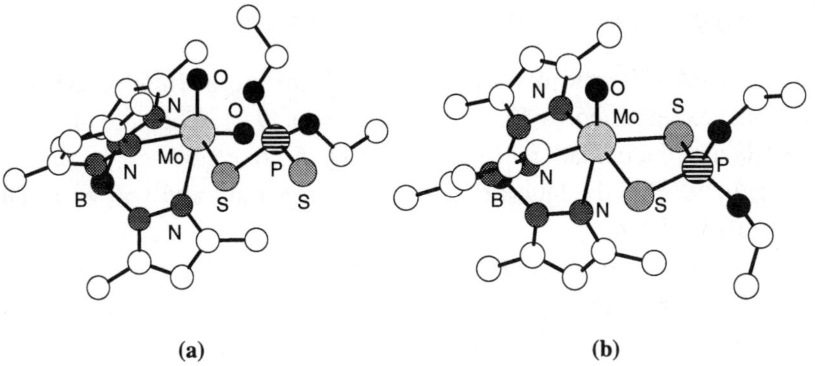

Fig. 9.8 Molecular structures of **(a)** $(L\text{-}N_3)Mo^{VI}O_2\{S_2P(OEt)_2\}$ and **(b)** $(L\text{-}N_3)Mo^{IV}O\{S_2P(OEt)_2\}$.

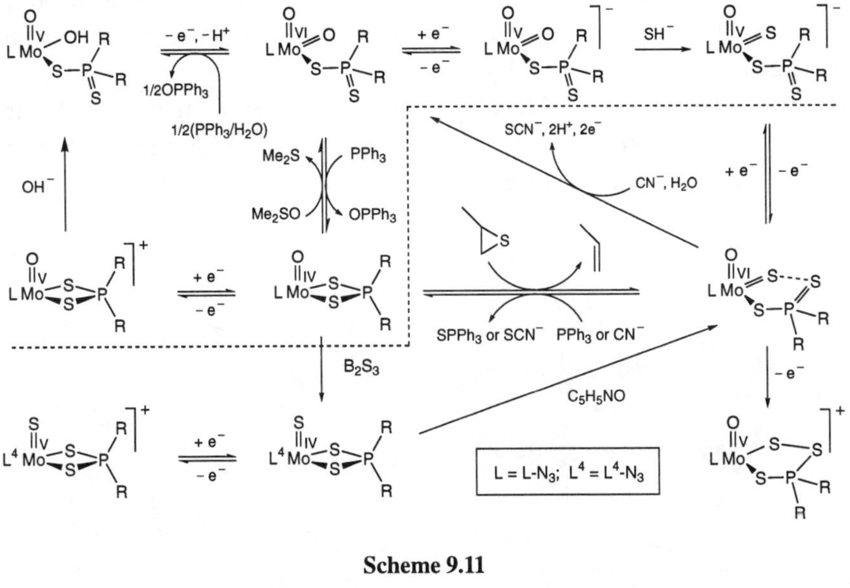

Scheme 9.11

system.[87,90–92] The types of reactions with a direct parallel in the $(L\text{-}N_3)MoO_2X$ systems are shown in the area above the dashed line; above the line, the monodentate and bidentate L ligand can read as X and X(py), respectively.

7.1 Oxygen atom transfer and related reactions

The dioxo-Mo(VI) and oxo-Mo(IV) complexes are interconverted by OAT reactions, exemplified by Eq. 9.24 and Scheme 9.11. They are also capable of catalyzing the oxidation of phosphines by sulfoxides, *N*-oxides and water/dioxygen;[84,85,88,89] the latter is particularly important and will be discussed in more detail later.

Sulfur atom transfer (SAT) reactions have also been demonstrated in some cases. For example, $(L^{1,4}\text{-}N_3)MoO(S_2PR_2)$ and $(L\text{-}N_3)MoO(pyS)$ react with propylene sulfide to form $(L^{1,4}\text{-}N_3)MoOS(S_2PR_2)$[92] and $(L\text{-}N_3)MoO(pyS_2)$,[90] respectively. The spectroscopic and structural [Fig. 9.9(a) and (b)] properties of these complexes, particularly the observed Mo-S(1) and S(1)-S(2) distances, are consistent with their formulation as oxo-thio-Mo(VI) and (dithio)oxo-Mo(IV) species, respectively. Interestingly, facets of internal Mo/S redox chemistry permit the reduction *and* oxidation of both complexes to Mo(V) species,[90] as shown in Fig. 9.9(c). The facile redox interplay of Mo and S may have important but undervalued roles in biological systems.[13–15] The $(L\text{-}N_3)MoOS(S_2PR_2)$ complexes are the only ones known to undergo cyanolysis and regeneration reactions

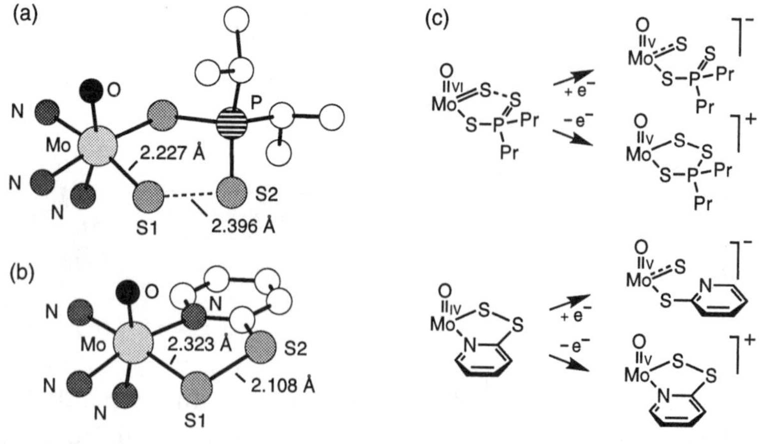

Fig. 9.9 **(a)** Partial structure of $(L\text{-}N_3)MoOS(S_2PPr^i{}_2)$. **(b)** Partial structure of $(L\text{-}N_3)MoO(pyS_2)$. **(c)** Redox reactions of these complexes, $L\text{-}N_3$ not shown.

mimicking the cyanolysis/reactivation reactions of xanthine oxidase etc. (top right Scheme 9.11).

The reactions of $(L\text{-}N_3)MoO_2X$ complexes with PPh_3 obey the rate law in Eq. 9.12; see Table 9.1 for data.[85,87] The reactions of $(L\text{-}N_3)MoO(S_2PR_2)$ with Me_2SO and pyridine *N*-oxide (Eq. 9.24) are first order with respect to both the Mo complex and substrate (Eq. 9.13); second order rate constants and activation parameters are collected in Table 9.4. The rate constants depend on R in the order Me < Et < Pr. The corresponding trends for $\Delta H^{\ddagger}$ (Me > Et > Pr) and $\Delta S^{\ddagger}$ (Me > Et > Pr) lead to a similar dependence on R for $\Delta G^{\ddagger}$.[84,87] An associative mechanism is indicated.

Table 9.4 Kinetics data for reactions of $(L\text{-}N_3)MoO(S_2PR_2)$ with Me_2SO (and C_5H_5NO) in toluene at 25 °C.

	$k \times 10^5$ $M^{-1}s^{-1}$	$\Delta H^{\ddagger}$ kJ mol^{-1}	$\Delta S^{\ddagger}$ J K^{-1}mol^{-1}	$\Delta G^{\ddagger\,b}$ kJ mol^{-1}
$(L\text{-}N_3)MoO(S_2PMe_2)$	2.1(1)	73(1)	−88(1)	100(2)
$(L\text{-}N_3)MoO(S_2PEt_2)$	3.6(1)	68(1)	−99(2)	99(2)
$(L\text{-}N_3)MoO(S_2PPr^i_2)$	5.5(3)	63(1)	−111(1)	98(2)
$(L\text{-}N_3)MoO(S_2PPr^i_2)^a$	48.0(3)	62(1)	−90(1)	90(1)
$(L\text{-}N_3)MoO\{S_2P(OEt)_2\}$	1.51(6)	64(2)	−121(3)	102.8

[a] Data for reaction of $(L\text{-}N_3)MoO(S_2PPr^i_2)$ and C_5H_5NO. [b] Calculated at 40 °C.

7.2 *Coupled electron-proton transfer and related reactions*

The complexes $(L^{1\text{-}3}\text{-}N_3)MoO_2(SPh)$ (and some others) undergo a reversible one-electron electrochemical reduction to produce EPR-active, dioxo-Mo(V) species characterized by broad, anisotropic EPR spectra with g ca. 1.92, $\Delta g > 0.14$, and $W_{1/2}$ ca. 20 G.[86] The $L^{1,2}\text{-}N_3$ complexes are readily protonated to form the conjugate acids, $(L^{1,2}\text{-}N_3)MoO(OH)X$, which exhibit less anisotropic, doublet EPR signals with g ca. 1.94 and A_H ca. 11×10^{-4} cm^{-1}. The analogous $(L^{1,2}\text{-}N_3)MoO(OD)X$ species, generated in the presence of D_2O, exhibit apparent singlet signals, confirming the presence of the hydroxo ligand. These species can also be generated by chemical reducing agents such as cobaltocene and hydrosulfide. Cobaltocene reduction of $(L^{1-3}\text{-}N_3)Mo^{VI}O_2(SPh)$ results in the kinetic syntheses of the novel, air-sensitive compounds, $[CoCp_2][(L^{1-3}\text{-}N_3)Mo^{V}O_2(SPh)]$, one of which has been structurally characterized.[93] The complex $(L\text{-}N_3)MoO(OH)(SPh)$ can be isolated as a co-precipitate with the cobaltocenium salt of its conjugate base.[93] The highly anisotropic EPR signals characteristic of dioxo-Mo(V) species have never been observed in any enzyme EPR spectrum but the conjugate acid has been implicated in the turnover of sulfite oxidase and nitrate reductase. The dioxo-Mo(V) species are not observed to participate in two-electron OAT reactions.

Reactions of $(L^{1-3}\text{-}N_3)MoO_2X$ with SH^- result in the initial formation of $[(L^{1-3}\text{-}N_3)MoO_2X]^-$, then $[(L^{1-3}\text{-}N_3)MoOSX]^-$, as demonstrated by EPR spectroscopy. The EPR spectra of the oxo-thio complexes are characterized by a larger g and a reduced anisotropy compared to those of the oxo analogues. Hydrosulfido complexes, $(L^{1-3}\text{-}N_3)MoO(SH)X$ have not been detected. The accessibility and stability of the various Mo(V) species as a function of X are detailed elsewhere. Analogous $[(L\text{-}N_3)MoOE(S_2PR_2)]^-$ species are also known.[87]

The complexes $(L\text{-}N_3)MoO_2X$ are converted to air-stable complexes $(L\text{-}N_3)MoO(OSiMe_3)X$ by one-electron coupled electron-electrophile transfer (CEET) reactions involving cobaltocene and the electrophilic reagent Me_3SiCl.[94] These complexes may also be obtained by reacting $[(L\text{-}N_3)MoO_2X]^-$ with Me_3SiCl or $(L\text{-}N_3)MoO(OH)X$ with Me_3SiCl/NEt_3. The crystal structure of $(L\text{-}N_3)MoO(OSiMe_3)(SCH_2Ph)$ reveals a six-coordinate octahedral complex with a *cis*-oxo(silyloxo) fragment and $d(Mo{=}O) = 1.675(4)$ Å, $d(Mo\text{-}OSi) = 1.932(4)$ Å and $\angle(O\text{-}Mo\text{-}O) = 102.0(2)°$. The EPR spectra of $(L\text{-}N_3)MoO(OSiMe_3)X$ are comparable to those of $(L\text{-}N_3)MoO(OD)X$, which are free of the ^1H coupling observed for $(L\text{-}N_3)MoO(OH)X$. The CEPT reactions important in enzymes are a specific example of the broader CEET reaction type.

The role of water in the OAT/CEPT reactions is particularly interesting.[86–89] Water appears to compete effectively for the binding site of "$(L\text{-}N_3)MoOX$" or "$(L\text{-}N_3)MoOL$" formed by OAT to phosphines. The redox potential of the aqua complex is such that the reaction of Eq. 9.25, for example, is possible in wet solvents. Here, comproportionation mimics oxidation of reduced enzyme by external oxidant.

$$(L\text{-}N_3)Mo^{IV}OX(OH_2) + (L\text{-}N_3)Mo^{VI}O_2X \rightarrow 2(L\text{-}N_3)Mo^{V}O(OH)X \quad (9.25)$$

Moreover, a catalytic system can be maintained despite comproportionation since the resultant (*hydroxo*)oxo complex is readily converted back to the dioxo-Mo(VI) species by dioxygen. ^{18}O- and ^{17}O-labeling studies have confirmed that the oxygen atom transferred to PPh_3 comes from $(L\text{-}N_3)MoO_2X$ and that water not dioxygen is the original source of the oxygen atom. The overall reaction of $(L\text{-}N_3)MoO_2X$ with PPh_3 in wet aerobic solutions can be written as shown in Eq. 9.26. The lack of substantial geometrical rearrangements in the sequence $(L\text{-}N_3)MoOX(OH_2)$ $\rightarrow (L\text{-}N_3)MoO(OH)X \rightarrow (L\text{-}N_3)MoO_2X$ is likely to facilitate the chemistry observed. The OAT and CEPT capabilities of the enzymes, especially those exposed by the Schultz/Hille/Holm[30] and Hille/Sprecher[31] experiments (Sec. 2), are closely mirrored by this system.

$$PPh_3 + H_2{*}O + 2O_2 \rightarrow {*}OPPh_3 + 2O_2H \quad (9.26)$$

8. Bidentate Mercaptoethanoate Systems

Cervilla and co-workers have described biomimetic and catalytic systems based on complexes of 2,2-diphenyl-2-mercaptoethanoate(2-), L-SO. The dioxo-Mo(VI) anion of $(NH_4)_2[MoO_2(L\text{-}SO)_2]$, exhibits the distorted octahedral structure shown in Fig. 9.10(a).[95] The $[MoO_2]^{2+}$ core adopts the usual *cis* structure with d(Mo=O) = 1.712 Å and ∠(O-Mo-O) = 104.4(1)°. The reaction of $[MoO_2(L\text{-}SO)_2]^{2-}$ with PPh_3 in refluxing methanol is reported to form $[MoO(L\text{-}SO)_2]^{2-}$ and $OPPh_3$.[95]

In dry ethanol, $[MoO_2(L\text{-}SO)_2]^{2-}$ reacts with $H_2(L\text{-}SO)$ and other thiols to form the oxo-Mo(V) complex, $[MoO(L\text{-}SO)_2]^{-}$, according to Eq. 9.27. The EPR-active oxo-Mo(V) complex possesses a square-pyramidal structure [Fig. 10(b)] with Mo=O and Mo-S distances of 1.633(8) Å and

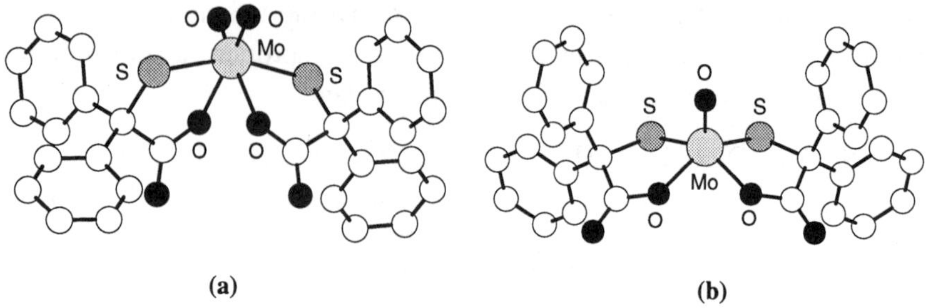

Fig. 9.10 (**a**) Structure of the anion in $NH_4[MoO_2(L\text{-}SO)_2]\cdot 2H_2O$. (**b**) Structure of the anion in $NBu^n_4[MoO(L\text{-}SO)_2]$.

2.342(3) Å, respectively.[96] The formation of the Mo(V) species effectively prevents the catalytic oxidation of thiols but can be prevented by intercalating $[MoO_2(L\text{-}SO)_2]^{2-}$ into a synthetic layered double hydroxide (LDH, e.g. hydrotalcite).

$$[MoO_2(L\text{-}SO)_2]^{2-} + RSH + H^+ \rightarrow [MoO(L\text{-}SO)_2]^- + 1/2RSSR + H_2O \tag{9.27}$$

Exchange of the interlamellar nitrate counteranions in cationic Zn(II)-Al(III) LDHs of composition $[Zn_{3-x}Al_x(OH)_6](NO_3)_x \cdot H_2O$ results in the formation of an air-stable material, originally formulated as $[Zn_{3-x}Al_x(OH)_6][MoO_2(L\text{-}SO)_2]_{y/2}(NO_3)_{x-y} \cdot H_2O$ (x = 0.75);[97,98] we refer to this material as Mo-LDH. Recent XANES and EXAFS experiments support the presence of an (aqua)trioxo species of the type $[MoO_3(L\text{-}SO\text{-}S)_2(OH_2)]^{4-}$, within the LDH layers.[99] The structural changes upon intercalation are reversible as extraction of the complex from the host results in the isolation of $[MoO_2(L\text{-}SO)_2]^{2-}$. Restricted migration of the complexes within the LDH layers is believed to prevent comproportionation and formation of catalytically inactive Mo(V) species.

Consequently, Mo-LDH is a catalyst for the oxidation of thiols to the corresponding disulfides by dioxygen,[97,98] dimethyl sulfoxide,[97] or aromatic nitro compounds.[100] The rate of catalysis is enhanced by the addition of small amounts (ca. 0.1%) of protonic zeolite or amberlite IR-120(H). A

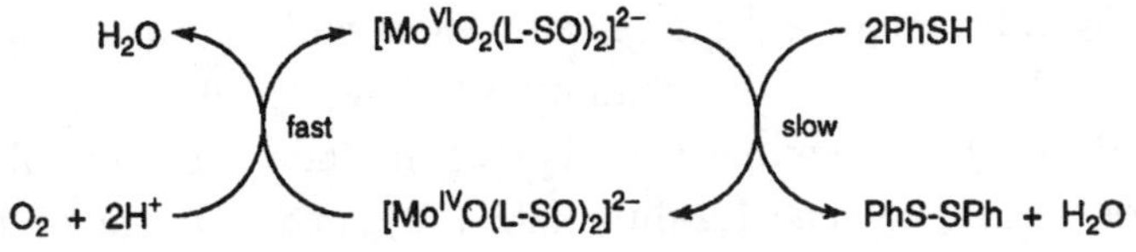

Scheme 9.12

plausible catalytic cycle for the thiol/O_2 system (O_2 acts as a two-electron oxidant to Mo(IV)) is given in Scheme 9.12.

The reduction of nitro compounds is of biomimetic interest, being catalyzed by nitroreductases such as xanthine dehydrogenase, quinone reductase and the nitroreductases from enteric bacteria.[101] The reaction of $PhNO_2$ (5.3 mmol) with HSPh in the presence of Mo-LDH (0.065 mmol) and amberlite proceeds cleanly, effecting full conversion according to Eq. 9.28.[101]

$$PhNO_2 + 6HSPh \rightarrow PhNH_2 + 3PhSSPh + 2H_2O \qquad (9.28)$$

A mechanism involving CEPT and OAT reactions at molybdenum was proposed for this reaction and a kinetics treatment was developed.[100] Initially, thiol oxidation is postulated to be coupled to reduction of Mo(VI) through two CEPT reactions. Molybdenum(VI) is regenerated by OAT from $PhNO_2$ to Mo(IV), this being the first step in the reduction of the nitro compound. The thiol reduction of PhNO to $PhNH_2$ via ArNHOH then takes place. The details of this mechanism may need to be re-assessed in the light a trioxo-Mo(VI) formulation for Mo-LDH.[99]

9. Bis(dithiolene) Systems

Mononuclear mono- and bis(dithiolene) complexes are key models for the enzyme systems. The former are extremely rare,[9,102] the latter are more common and are of two principal classes. Complexes of the first class feature true dithiolene ligands, the second features arene-dithiolates wherein the ene function is part of an aromatic ring system. Symmetrical dithiolene ligands include 1,2-dicyanoethylenedithiolate (mnt), benzene-1,2-dithiolate (bdt), and 1,2-dicarbamoylethylene-1,2-dithiolate (dced = $S_2C_2(CONH_2)_2^{2-}$). Unsymmetrical ligands include the sterically bulky 3-Ph_3Si-bdt ligand, toluene-1,2-dithiolate (tdt), and a range of pterinyl-, quinoxalinyl and other derivatives now accessible via protected dithiolene derivatives.[10]

Three main types of complex, viz. *cis*-$[Mo^{VI}O_2L_2]^{2-}$, $[Mo^{V}OL_2]^{-}$, and $[Mo^{IV}OL_2]^{2-}$, comprise these systems; a number of synthetic routes are summarized in Scheme 9.13. The success of Route A is dependent on the careful control of pH and the use of a stable dithiolene.[103] Route B, employing protected dithiolene precursors, provides for considerable synthetic flexibility.[10] The exploitation of BH_4^- to access oxo-Mo(IV) species from available Mo(V) complexes was crucial in the development of route C.[104,105] Thiol ligand exchange on $[MoO(SPh)_4]^-$, or its derivatives, is a common route to the $[Mo^{V}OL_2]^-$ complexes (not shown in Scheme 9.13).[106]

Many compounds, at all three oxidation levels, have been structurally characterized. As shown in Fig. 9.11, the dioxo-Mo(VI) complexes exhibit distorted octahedral structures, with *cis* oxo groups (d(Mo=O) = 1.70–1.73 Å) and two sets of Mo-S distances, the larger of the two (ca. 2.6 Å versus 2.4 Å) being *trans* to the oxo ligands.[103,108,109] The oxo-Mo(V) and -Mo(IV) complexes are invariably square-pyramidal with an apical oxo group.[104–107,110]

Nakamura and co-workers have been pre-eminent in the development of the OAT chemistry in these systems.[105,107–112] Dinuclear Mo(V) species are

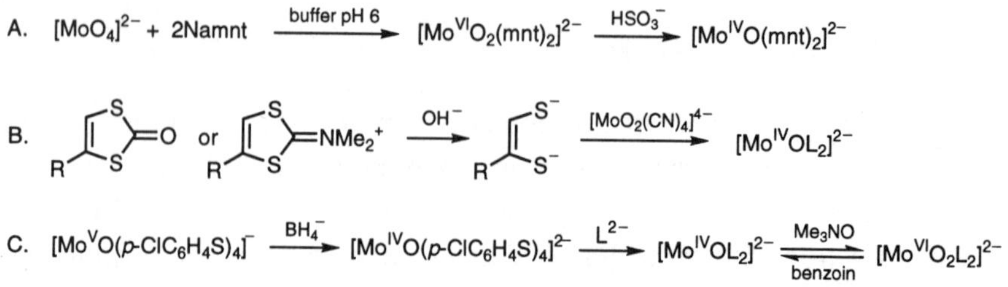

Scheme 9.13

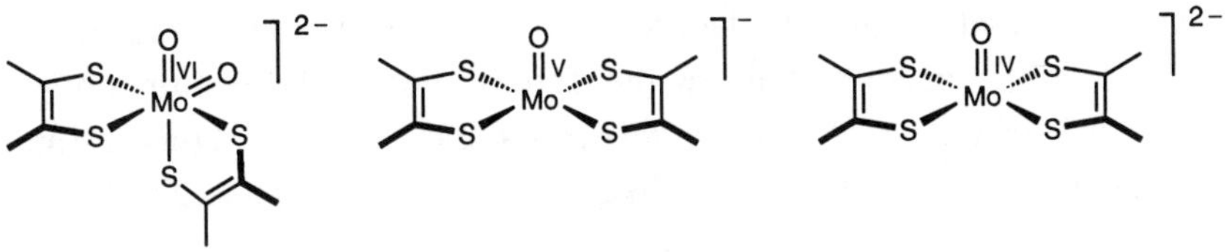

Fig. 9.11 Structural frameworks for *cis*-$[Mo^{VI}O_2L_2]^{2-}$, $[Mo^{V}OL_2]^{-}$, and $[Mo^{IV}OL_2]^{2-}$ complexes containing dithiolene or pseudo-dithiolene ligands.

not observed to form via comproportionation (Eq. 9.10), presumably due to unfavorable electrostatic interactions. Careful reaction of $[MoOL_2]^{2-}$ complexes with Me_3NO results in the formation of $[MoO_2L_2]^{2-}$, which is the final step in the synthetic route shown in Scheme 9.13, route C. The reactions are pseudo-first order with k_{obs} ($[Me_3NO]$ = 10 mM) values in the range 2.6(5)–5.7(4) × 10^{-3} s^{-1} for the complexes with L = bdt, tdt and their 3-Ph_3Si derivatives.[105] The reactions are proposed to occur via binding of Me_3NO to the vacant *trans* site in $[MoOL_2]^{2-}$, followed by a *trans-cis* rearrangement and cleavage of the N–O bond. The stability of the dioxo-Mo(VI) product depends on the electronic structure of the dithiolene,[110] and resonance Raman studies[111] suggest the dithiolene ligand may exert a *trans* effect on the oxo ligands.

The influence of NH···S(dithiolene) hydrogen bonding on structure and reactivity has been explored in a recent study.[107] The crystal structure of $(NPr^n_4)_2[MoO(dced)_2]$ reveals a number of NH···S and NH···O hydrogen bonds in the anion. Reaction of this complex with Me_3NO proceeds according to Eq. 9.8 but with an enormously enhanced (× 10^4–10^5) rate, with k_{obs} ~ 1 s^{-1}. The rate enhancement is ascribed to the stabilization of the dioxo-Mo(VI) complex by NH···S hydrogen bonds, and their influence in facilitating the *cis-trans* rearrangement along the reaction coordinate. A less dramatic effect on the rate of OAT is observed when potential hydrogen bonding diamides are added to reaction mixtures.[107]

The oxidation of benzoin by $[MoO_2L_2]^{2-}$ proceeds according to Eq. 9.29.[109,112] Under pseudo-first order conditions, k_{obs} = 2.2(4) × 10^{-5} s^{-1} ([benzoin] = 10 mM), with a deuterium isotope effect, k_H/k_D of 4.0 for α-deuterobenzoin.[112] This is consistent with α-hydrogen atom abstraction by an oxo ligand in the rate-determining step. The biological relevance of this reaction is accentuated by the report that molybdenum enzymes from *Proteus vulgaris* and *P. mirabilis* catalyze the conversion of $RCH(OH)CO_2^-$, a substrate very like benzoin, to $RCOCO_2^-$.[113]

$$[MoO_2L_2]^{2-} + PhCH(OH)COPh \rightarrow [MoOL_2]^{2-} + PhCOCOPh + H_2O \tag{9.29}$$

The complex $[MoO_2(bdt)_2]^{2-}$ reacts with PPh_3 according to Eq. 9.7 (see Table 9.1)[109] and the oxidation of $PPh_{3-x}Et_x$ (x = 0–3) by $(NBu^n{}_4)_2[MoO_2(mnt)_2]$ in acetonitrile has been investigated by Lorber *et al.*[54] Unlike other reactions of this type, the observed rate constants are linearly dependent on the phosphine concentrations, according to Eq. 9.30 with an intercept k_1 of ca. 4.4×10^{-4} s^{-1}. Saturation kinetics were not observed at high phosphine concentrations. Data are given in Table 9.5. The trend in k_2 shows higher rates of OAT with increasing basicity of phosphine. A dependence on the relative σ donor capability of the phosphines (χ) but not the cone angle of the phosphine (θ) has been established.

$$k_{obs} = k_1 + k_2[PR_3] \tag{9.30}$$

Finally, and perhaps most importantly, a unique system mimicking sulfite oxidase has been studied by Sarkar and co-workers[103,114] and Lorber *et al.*[54] The reaction of $(NBu^n{}_4)_2[MoO_2(mnt)_2]$ with $NaHSO_3$ in MeCN-H_2O (3.3% water) proceeds cleanly according to Eq. 9.31.

Table 9.5 Kinetics data and phosphine electronic and steric parameters for the reactions of $(NBu^n{}_4)_2[MoO_2(mnt)_2]$ with phosphines in acetonitrile at 25 °C.[54]

Phosphine	10^4k_1 s^{-1}	k_2 $M^{-1}s^{-1}$	$\Delta H^\ddagger$ kJ mol^{-1}	$\Delta S^\ddagger$ J K^{-1}mol^{-1}	χ cm^{-1}	θ deg.
PEt_3	4.84 ± 2	0.109(5)	62(6)	−56(20)	6.30	132
$PPhEt_2$	1.23 ± 0.3	0.0784(5)			9.30	136
PPh_2Et	7.42 ± 1	0.0614(30)			11.30	140
PPh_3	4.20 ± 0.6	0.0559(10)	45(4)	−119(10)	13.25	145

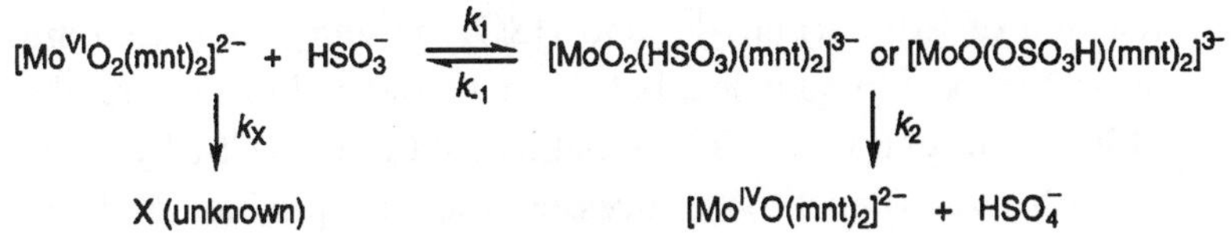

Scheme 9.14

$$[Mo^{VI}O_2(mnt)_2]^{2-} + HSO_3^- \rightarrow [Mo^{IV}O(mnt)_2]^{2-} + HSO_4^- \qquad (9.31)$$

The reaction is first order in the Mo(VI) complex but the [$NaHSO_3$] dependence is consistent with a pre-equilibrium between $[MoO_2(mnt)_2]^{2-}$ and HSO_3^-: Scheme 9.14 combines two proposed schemes.[54,103]

For a reaction of the type shown in Scheme 9.14, k_{obs} is given by Eq. 9.32, where $K_M = (k_2 + k_{-1})/k_1$. In this case, $k_2 << k_{-1}$ and K_M becomes k_{-1}/k_1. A double reciprocal plot of $1/k_{obs}$ versus $1/[HSO_3^-]$ yielded[103] values of 0.87(4) s^{-1} and 1.0(1) × 10^{-2} M, respectively, for k_2 and K_M at 20 °C. Formation of the catalyst-substrate complex may follow the breaking of one of the Mo–S bonds to generate an active site very like that of sulfite oxidase (see Fig. 9.1). Alternatively, binding of HSO_3^- to form a seven-coordinate species may induce conversion of the $[MoO_2]^{2+}$ center to an $[Mo(\equiv O)O^{\delta+}]^{2+}$ center which subsequently transforms to product through OAT. The inhibition of the sulfite oxidation reaction by SO_4^{2-}, $H_2PO_4^-$ and $H_2PO_3^-$ has also been investigated.[114]

$$\frac{1}{k_{obs}} = \frac{1}{k_2} + \frac{K_M}{k_2[HSO_3^-]} \qquad (9.32)$$

More recently, Lorber *et al.*[54] examined the kinetics of this reaction in a range of MeCN/H_2O mixtures ([H_2O] = 13.9 M – 27.8 M) by stopped-flow techniques. They confirmed the saturation kinetics reported by Das *et al.*, determining that k_2 = 1.02(7) s^{-1} and K_M = 3.9(5) × 10^{-2} M for 1:1 MeCN:H_2O solutions, and observed that the rate constants decrease as the concentration of water in the solvent increases. The water-dependence was rationalized by the hydrogen bonding of water to $[MoO_2(mnt)_2]^{2-}$ or formation of a seven-coordinate $[MoO_2(H_2O)(mnt)_2]^{2-}$ complex.

The reaction of $[MoO_2(mnt)_2]^{2-}$ and HSO_3^- in anhydrous acetonitrile is first order with respect to complex, but again saturation kinetics are observed at high HSO_3^- concentrations. The kinetics data are best fit by a rate law consistent with two competitive processes; the first producing the species which undergoes OAT to give $[MoO(mnt)_2]^{2-}$ and HSO_4^-, the second involving either the reversible formation of an inactive species (X) (with a rate constant k_X) or the irreversible formation of a decomposition product. Here, $K_1 = 1.6(1) \times 10^{-2}$ M, $k_2 = 3.2(1) \times 10^{-2}$ s^{-1}, and $k_X = 6.7(4) \times 10^{-3}$ s^{-1}. Addition of water increases the rate of the reaction, possibly as a result of pH changes generating small amounts of more reactive SO_3^{2-}. The role of water in these and other systems warrants further study.

The oxidation of $[MoO(mnt)_2]^{2-}$ using $[Fe(CN)_6]^{3-}/H_2O$ results in partial reclamation (ca. 10%) of $[MoO_2(mnt)_2]^{2-}$, possibly via CEPT reactions mimicking the regeneration the active sulfite oxidase.[103]

Recent advances in this area are described in the Note Added In Proof.

10. Other Systems

Recently, Arnaiz *et al.*,[115] Arzoumanian *et al.*,[116,117] and Lorber *et al.*[118] have challenged the need for bulky S-donor ligands for efficient OAT reactions. Arnaiz *et al.*[115] reported that the complexes $MoO_2X_2(Me_2SO)_2$ (X = Cl, Br) were effective catalysts for the oxidation of PPh_3 by Me_2SO in neat Me_2SO. It appears that only mononuclear Mo(VI) and Mo(IV) compounds are involved in the catalytic reaction and k (Eq. 9.12) for the Mo(VI)/PPh_3 reaction is especially large (Table 9.1).

Arzoumanian *et al.*[116,117] reported that $(PPh_4)_2[MoO_2(NCS)_4]$ and various neutral derivatives, such as $MoO_2(NCS)_2(Bu^t{}_2bpy)$, catalyze the oxidation of PPh_3 by Me_2SO. The reaction of $(PPh_4)_2[MoO_2(NCS)_4]$ with PPh_3, which results in the formation of $(PPh_4)_2[MoO(NCS)_4]$, is exceptionally fast (Table 9.1, Eq. 9.12) and oxidation of the Mo(IV) complex is the rate-limiting step in the oxidation of PPh_3 by Me_2SO.[117]

In the system described by Lorber *et al.*,[118] $MoO_2(acac)_2$ (acac = acetylacetonate) is employed as a catalyst precursor for the oxidation of primary, benzylic and allylic alcohols by Me_2SO or various *N*-oxides.

Evidence suggests that the reaction proceeds via a hydride transfer from a coordinated alkoxide to an oxo ligand on Mo(VI), thereby forming aldehyde and a Mo(IV) species, which is oxidized by OAT from Me_2SO to reform the active catalyst.

11. Conclusion

This chapter highlights the quintessential biomimetic chemistry of molybdenum. OAT and CEPT reactions, which pervade high-valent molybdenum chemistry and provide a sensible and satisfying nexus between the inorganic and biological chemistries of this crucial trace element, lie at the heart of the many investigations described. These biomimetic studies have made a vital contribution to our understanding of the mononuclear oxomolybdenum enzymes and will underpin future work in this area.

Note Added In Proof

Important new developments are summarized below (in order of Secs. 2–10).

Enzymes. The crystal structure of the periplasmic trimethylamine *N*-oxide reductase from the marine bacterium *Shewanella massilia* has been reported.[119] The 2.5 Å resolution structure reveals an Mo center coordinated by two MGD ligands, serine-149 and, it is believed, two oxo ligands.

Mo K–edge EXAFS and EPR studies of wild and mutant *R. sphaeroides* dimethyl sulfoxide reductase have confirmed the presence of mono–oxo (oxidized) and des–oxo (reduced) Mo active sites.[120] The findings of other workers[20,22,23] were critically examined and a case was made for bidentate MPT ligation at all stages of catalysis (cf. Fig. 9.2). The interaction of human sulfite oxidase with arsenite (via an Mo–O–As link) has been probed by EXAFS and EPR spectroscopies.[121] Interestingly, the presence of an Se–S interaction in the Mo formate dehydrogenases from *E. coli* and *D. desulfuricans* ATCC 27774 has been revealed by Se K–edge EXAFS.[122,123]

Presumably, the interaction involves selenocysteine and a sulfur atom from MPT. Its biological role remains unclear.

The mechanism of xanthine oxidase and related enzymes has been probed in computational studies by Voityuk *et al.*[124]

Models. The suggestion (Sec. 4) that purported oxo-Mo(IV) dithiocarbazate complexes may be dinuclear μ-oxo-Mo(V) species is supported by studies of NS_2-donor thiosemicarbazide complexes of the type $Mo_2O_3L_2$.[125] A number of dioxo-Mo and -W complexes with N_2O_2- and N_2S_2-donor ligands closely related to L^1-N_2S_2 and L-N_2O_2 (Sec. 4) have been described by Ng and co-workers;[126,127] these catalyze the oxidation of benzoin by dimethyl sulfoxide.

Re-examination of the bis(thiolato)pyridine model developed by Holm and co-workers (Sec. 6.1) has revealed that purple MoO(L-NS_2)(dmf) is in fact the dinuclear μ-oxo-Mo(V) compound $Mo_2O_3(L\text{-}NS_2)_2 \cdot 2dmf$.[128] Microanalytical, IR and mass spectrometric data were consistent with the new formulation and careful ^{31}P NMR experiments were used to establish the correct reaction stoichiometry (Eq. 9.33, rather than Eq. 9.19; compound crystallizes as disolvate). The dinuclear formulation was confirmed by the crystal structures of $Mo_2O_3(L\text{-}NS_2)_2 \cdot 2sol$ (sol = dmf, $CHCl_3$). The new insights demand a re-evaluation of the chemistry described in Sec. 6.1.[75–80] Significantly, this system does not support OAT without comproportionation; the trispyrazolylborate systems (Sec. 7) were the first to achieve this aim. Scheme 9.5 can be adapted for catalysis by the bis(thiolato)pyridine system.

$$2(L\text{–}NS_2)MoO_2 + PPh_3 \rightarrow Mo_2O_3(L\text{–}NS_2)_2 + OPPh_3 \qquad (9.33)$$

Holm and co-workers have made significant advances in the modeling of the dimethyl sulfoxide reductases (Fig. 9.2). Initial studies involved the synthesis of $NEt_4[MoO(OSiPh_2Bu^t)(bdt)_2]$ and $NEt_4[Mo(OSiPh_2Bu^t)(bdt)_2]$ via direct silylation of $(NEt_4)_2[MoO_2(bdt)_2]$ and $(NEt_4)_2[MoO(bdt)_2]$, respectively.[129] A subsequent report described complete characterization data and the crystal structures of the two compounds (and various mnt analogues).[130] The des-oxo complexes react with strong oxygen atom donors (but not sulfoxides) to give the mono-oxo species. These reactions are neither clean nor complete. Dimethyl sulfide did not react with the mono-oxo-Mo(VI) complexes to produce Me_2SO adducts (cf. enzyme, in Fig. 9.2(c)). Complexes such as $Mo(bdt)_2(PMePh_2)_2$ are produced in reactions of the mono-oxo complexes with phosphines. Studies[129–131] of the tungsten analogues and a comparative study[132] of OAT from $[MO_2(mnt)_2]^{2-}$ (M = Mo, W) to various $(RO)_{3-n}R'_nP$ (Eq. 9.7) have also been reported. Typically, $k_{Mo}/k_W \sim 10^2$–10^3 (Eq. 9.12) and $\Delta H^{\ddagger}_W - \Delta H^{\ddagger}_{Mo} \sim 12$–$17$ kJ mol^{-1}. The reducibility of the metal center and the strengths of the M=O bonds are important factors in determining reactivity. Computational studies of the reaction of $[MoO_2(mnt)_2]^{2-}$ with HSO_3^- (Sec. 9) are consistent with direct attack of HSO_3^- at an oxo ligand and a straightforward OAT mechanism.[133]

Aspects of the chemistry of thiocyanato systems based on $[MoO_2(NCS)_4]^{2-}$ (Sec. 10) have been reviewed by Arzoumanian.[134]

Acknowledgments

I am grateful to the following colleagues for helpful discussions, or for providing information or reprints that have been used in the preparation of this article: Profs./Drs. M. R. Bray, R. C. Bray, S. Burgmayer, A. Cervilla, R. J. Deeth, J. H. Enemark, C. D. Garner, G. N. George, M. B. Hall, L. J. Laughlin, A. Nakamura, M. Onishi, D. C. Rees, F. A. Schultz, E. I. Stiefel, and Z. Xiao. The assistance of Mr. C. Doonan in the preparation of the figures and the financial support of the Australian Research Council are gratefully acknowledged.

12. References

1. Holm, R.H. *Coord. Chem. Rev.* 1990, **100:** 183–221.
2. Enemark, J.H. and Young, C.G. *Adv. Inorg. Chem.* 1993, **40:** 1–88.
3. Pilato, R.S. and Stiefel, E.I. in *Bioinorganic Catalysis* (ed.) Reedijk, J. Dekker, New York, 2nd Ed., 1999, pp. 81–152.
4. Hille, R. *Biochim. Biophys. Acta* 1994, **1184:** 143–169.
5. Young, C.G. and Wedd, A.G. in *Encyclopedia of Inorganic Chemistry* (ed.) King, R.B. Wiley, Chichester, 1994, pp. 2330–2346.

6. Hille, R. *Chem. Rev.* 1996, **96:** 2757–2816.
7. Young, C.G. and Wedd, A.G. *J. Chem. Soc., Chem. Commun.* 1997, 1251–1257.
8. Kisker, C. *et al. Annu. Rev. Biochem.* 1997, **66:** 233–267.
9. Pilato, R.S. *et al. J. Am. Chem. Soc.* 1991, **113:** 9372–9374.
10. Collison, D., Garner, C.D. and Joule, J.A. *Chem. Soc. Rev.* 1996, 25–32.
11. Balagopalakrishna, C. *et al. Inorg. Chem.* 1996, **35:** 7758–7768
12. Inscore, F. *et al. Inorg. Chem.* 1999, **38:** 1401–1410.
13. Wedd, A.G. in *Sulfur, Its Significance for Chemistry, for the Geo-, Bio- and Cosmosphere and Technology* (eds.) Müller, A. and Krebs, B. Elsevier, Amsterdam, 1984, pp. 181–193.
14. Stiefel, E.I. *J. Chem. Soc., Dalton Trans.* 1997, 3915–3923.
15. Young, C.G. *J. Biol. Inorg. Chem.* 1997, **2:** 810–816.
16. Cramer, S.P. *Adv. Inorg. Bioinorg. Mech.* 1983, **2:** 259–310.
17. Bray, R.C. *Quart. Rev. Biophys.* 1988, **21:** 299–329.
18. Garton, S.D. *et al. J. Am. Chem. Soc.* 1997, **119:** 2590–2591.
19. Kisker, C. *et al. Cell* 1997, **91:** 973–983.
20. Schindelin, H. *et al. Science* 1996, **272:** 1615–1621.
21. Schneider, F. *et al. J. Mol. Biol.* 1996, **263:** 53–69.
22. Baugh, P.E. *et al. J. Biol. Inorg. Chem.* 1997, **2:** 634–643.
23. McAlpine, A.S. *et al. J. Biol. Inorg. Chem.* 1997, **2:** 690–701.
24. Boyington, J.C. *et al. Science* 1997, **275:** 1305–1308.
25. Romão, M.J. *et al. Science* 1995, **270:** 1170–1176.
26. Huber, R. *et al. Proc. Nat. Acad. Sci. USA.* 1996, **93:** 8846–8851.
27. Rajagopalan, K.V. *Adv. Enzy. Relat. Areas Mol. Biol.* 1991, **64:** 215–289.
28. Rees, D.C. *et al. J. Chem. Soc., Dalton Trans.* 1997, 3909–3914.
29. Murray, K.N. *et al. J. Biol. Chem.* 1966, **241:** 4798–4801.
30. Schultz, B.E. *et al. J. Am. Chem. Soc.* 1995, **117:** 827–828.
31. Hille, R. and Sprecher, H. *J. Biol. Chem.* 1987, **262:** 10914–10917.
32. Johnson, J.L. and Wadman, S.K. in *The Metabolic Basis of Inherited Disease* (eds.) Scriver, C.R., Beaudet, A.L., Sly, W.S. and Valle, D. McGraw-Hill, New York, 1989, pp. 1463–1475.
33. Rappé, A.K. and Goddard, W.A. III *J. Am. Chem. Soc.* 1982, **104:** 448–456.
34. Rappé, A.K. and Goddard, W.A. III *J. Am. Chem. Soc.* 1982, **104:** 3287–3294.
35. Stiefel, E.I. *Science* 1996, **272:** 1599–1600.
36. George, G.N. *et al. J. Am. Chem. Soc.* 1996, **118:** 1113–1117.
37. Bray, M.R. and Deeth, R.J. *J. Chem. Soc., Dalton Trans.* 1997, 4005–4009.
38. Voityuk, A.A. *et al. J. Am. Chem. Soc.* 1997, **119:** 3159–3160.

39. Howes, B.D. *et al. J. Am. Chem. Soc.* 1994, **116:** 11624–11625.
40. Xia, M. *et al. Biochem. Soc. Trans.* 1997, 768–773.
41. Holm, R.H. *Chem. Rev.* 1987, **87:** 1401–1449.
42. Barral, R. *et al. Kinet. Catal. (Engl. Trans.)* 1973, **14:** 130–133.
43. Chen, G.J.–J. *et al. Inorg. Chem.* 1976, **15:** 2612–2615.
44. Reynolds, M.S. *et al. Inorg. Chem.* 1984, **23:** 3057–3062.
45. Baird, D.M., Falzone, S. and Haky, J.E. *Inorg. Chem.* 1989, **28:** 4561–4562.
46. Baird, D.M. *et al. Inorg. Chim. Acta* 1995, **237:** 117–122.
47. Deli, J. and Speier G. *Trans. Met. Chem.* 1981, **6:** 227–229.
48. Ueyama, N. *et al. J. Chem. Soc., Dalton Trans.* 1990, 387–394.
49. Craig, J.A. *et al. Inorg. Chem.* 1989, **28:** 2082–2091.
50. Pietsch, M.A. and Hall, M.B. *Inorg. Chem.* 1996, **35:** 1273–1278.
51. Holm, R.H. and Donahue, J.P. *Polyhedron* 1993, **12:** 571–589.
52. Stiefel, E.I. *Proc. Nat. Acad. Sci. USA* 1973, **70:** 988–992.
53. Paffett, M.T. and Anson, F.C. *Inorg. Chem.* 1981, **20:** 3967–3972.
54. Lorber, C. *et al. J. Chem. Soc., Dalton Trans.* 1997, 3997–4003.
55. Syamal, A. and Mayrya, M.R. *Coord. Chem. Rev.* 1989, **95:** 183–238.
56. Rajan, O.A. and Chakravorty, A. *Inorg. Chem.* 1981, **20:** 660–664.
57. Boyd, I.W. and Spence, J.T. *Inorg. Chem.* 1982, **21:** 1602–1606.
58. Topich, J. and Lyon, J.T. III. *Inorg. Chem.* 1984, **23:** 3202–3206.
59. Topich, J. and Bachert, J.O. III *Inorg. Chem.* 1992, **31:** 511–515.
60. Purohit, S. *et al. Inorg. Chem.* 1989, **28:** 3735–3742.
61. Bhattacharjee, S. and Bhattacharyya, R. *J. Chem. Soc., Dalton Trans.* 1992, 1357–1364.
62. Bhattacharjee, S. and Bhattacharyya, R. *J. Chem. Soc., Dalton Trans.* 1993, 1151–1158. Correction: *ibid* (1993), 1631.
63. Dutta, S.K. *et al. Inorg. Chem.* 1997, **36:** 2517–2522.
64. Dutta, S.K. *et al. Inorg. Chem.* 1997, **36:** 4954–4960.
65. Pickett, C. *et al. Inorg. Chem.* 1982, **21:** 908–916.
66. Bruce, A. *et al. Inorg. Chem.* 1982, **21:** 917–926.
67. Dowerah, D. *et al. J. Am. Chem. Soc.* 1987, **109:** 5655–5665.
68. Hinshaw, C.J. *et al. Inorg. Chem.* 1989, **28:** 4483–4491.
69. Küsthardt, U. *et al. Inorg. Chem.* 1993, **32:** 1838–1843.
70. Barnard, K.R. *et al. Inorg. Chem.* 1997, **36:** 637–649.
71. Rajan, O.A. *et al. Inorg. Chem.* 1983, **22:** 3065–3072.
72. Wilson, G.L. *et al. J. Am. Chem. Soc.* 1988, **110:** 6923–6925.
73. Wilson, G.L. *et al. J. Am. Chem. Soc.* 1991, **113:** 6803–6812.

74. Greenwood, R.J. *et al. J. Am. Chem. Soc.* 1993, **115:** 5385–5392.
75. Berg, J.M. and Holm, R.H. *J. Am. Chem. Soc.* 1985, **107:** 917–925.
76. Berg, J.M. and Holm, R.H. *J. Am. Chem. Soc.* 1985, **107:** 925–932.
77. Harlan, E.W. *et al. J. Am. Chem. Soc.* 1986, **108:** 6992–7000.
78. Caradonna, J.P. *et al. J. Am. Chem. Soc.* 1986, **108:** 7856–7858.
79. Caradonna, J.P. *et al. J. Am. Chem. Soc.* 1988, **110:** 2139–2144.
80. Craig, J.A. and Holm, R.H. *J. Am. Chem. Soc.* 1989, **111:** 2111–2115.
81. Kaptein, B. *et al. J. Chem. Soc., Chem. Commun.* 1987, 1457–1459.
82. Schultz, B.E. *et al. J. Am. Chem. Soc.* 1993, **115:** 2714–2722.
83. Schultz, B.E. and Holm, R.H. *Inorg. Chem.* 1993, **32:** 4244–4248.
84. Roberts, S.A. *et al. Inorg. Chem.* 1988, **27:** 3044–3051.
85. Roberts, S.A. *et al. Inorg. Chem.* 1990, **29:** 3650–3656.
86. Xiao, Z. *et al. Inorg. Chem.* 1995, **34:** 5950–5962.
87. Laughlin, L.J. and Young, C.G. *Inorg. Chem.* 1996, **35:** 1050–1058.
88. Xiao, Z. *et al. J. Am. Chem. Soc.* 1992, **114:** 9194–9195.
89. Xiao, Z. *et al. Inorg. Chem.* 1996, **35:** 7508–7515.
90. Hill, J.P. *et al. Inorg. Chem.* 1996, **35:** 3447–3448.
91. Young, C.G. *et al. Inorg. Chem.* 1996, **35:** 5368–5377.
92. Eagle, A.A. *et al. J. Am. Chem. Soc.* 1992, **114:** 9195–9197.
93. Xiao, Z. *et al. J. Am. Chem. Soc.* 1996, **118:** 2912–2921.
94. Xiao, Z. *et al. J. Biol. Inorg. Chem.* 1996, **1:** 415–423.
95. Palanca, P. *et al. J. Chem. Soc., Chem. Commun.* 1990, 531–533.
96. Sanz, V. *et al. Inorg. Chem.* 1991, **30:** 3113–3115.
97. Cervilla, A. *et al. J. Am. Chem. Soc.* 1994, **116:** 1595–1596.
98. Cervilla, A. *et al. J. Chem. Soc., Dalton Trans.* 1994, 2953–2957.
99. Corma, A. *et al. J. Chem. Soc., Chem. Commun.* 1996, 1613–1614.
100. Cervilla, A. *et al. J. Am. Chem. Soc.* 1995, **117:** 6781–6782.
101. Bryant, C. and DeLuca, M. *J. Biol. Chem.* 1991, **266:** 4119–4125.
102. Nicholas, K.M. and Khan, M.A. *Inorg. Chem.* 1987, **26:** 1633–1636.
103. Das, S.K. *et al. J. Am. Chem. Soc.* 1994, **116:** 9061–9070.
104. Kondo, M. *et al. Bull. Chem. Soc. Jpn.* 1993, **66:** 1391–1396.
105. Oku, H. *et al. Inorg. Chem.* 1994, **33:** 209–216.
106. Ellis, S.R. *et al. J. Chem. Soc., Dalton Trans.* 1989, 413–417.
107. Oku, H., Ueyama, N. and Nakamura, A. *Inorg. Chem.* 1997, **36:** 1504–1516.
108. Yoshinaga, N. *et al. Chem. Lett.* 1990, 1655–1656.
109. Ueyama, N. *et al. Inorg. Chem.* 1996, **35:** 643–650.
110. Oku, H. *et al. Chem. Lett.* 1994, 607–610.

111. Oku, H., Ueyama, N. and Nakamura, A. *Inorg. Chem.* 1995, **34:** 3667–3676.
112. Oku, H., Ueyama, N. and Nakamura, A. *Chem. Lett.* 1996, 1131–1132.
113. Trautwein, T. *et al. Eur. J. Biochem.* 1994, **222:** 1025–1032.
114. Chaudhury, P.K., Das, S.K. and Sarkar S. *Biochem. J.* 1996, **319:** 953–959.
115. Arnaiz, F.J. *et al. Polyhedron* 1994, **13:** 3257–3259.
116. Arzoumanian, H. *et al. Inorg. Chem.* 1994, **33:** 3177–3179.
117. Arzoumanian, H. *et al. New J. Chem.* 1996, **20:** 699–705.
118. Lorber, C.Y. *et al. Bull. Soc. Chim. Fr.* 1996, **133:** 755–758, 1156.
119. Czjzek, M. *et al. J. Mol. Biol.* 1998, **284:** 435–447.
120. George, G.N. *et al. J. Am. Chem. Soc.* 1999, **121:** 1256–1266.
121. George, G.N. *et al. J. Am. Chem. Soc.* 1998, **120:** 4522–4523.
122. George, G.N. *et al. J. Am. Chem. Soc.* 1998, **120:** 1267–1273.
123. George, G.N. *et al. J. Am. Chem. Soc.* 1999, **121:** 2625–2626.
124. Voityuk, A.A. *et al. Inorg. Chem.* 1998, **37:** 176–180.
125. Stelzig, L. *et al. J. Chem. Soc., Dalton Trans.* 1998, 2921–2926.
126. Mong, Y.–L. *et al. J. Chem. Soc., Dalton Trans.* 1998, 3057–3064.
127. Wong, Y.–L. *et al. Eur. J. Inorg. Chem.* 1999, 313–321.
128. Doonan, C.J. *et al. J. Am. Chem. Soc.* 1999, **121:** 6430–6436.
129. Donahue, J.P. *et al. J. Am. Chem. Soc.* 1998, **120:** 3259–3260.
130. Donahue, J.P. *et al. J. Am. Chem. Soc.* 1998, **120:** 12869–12881.
131. Lorber, C. *et al. J. Am. Chem. Soc.* 1998, **120:** 8102–8112.
132. Tucci, G.C. *et al. Inorg. Chem.* 1998, **37:** 1602–1608.
133. Thapper, A. *et al. Inorg. Chem.* 1999, **38:** 1015–1018.
134. Arzoumanian, H. *Coord. Chem. Rev.* 1998, **178–180:** 191–202.

Chapter 10

Models of Superoxide Dismutases

Diane E. Cabelli*, Dennis Riley,† Jorge A. Rodriguez,‡
Joan Selverstone Valentine‡ & Haining Zhu‡

**Department of Chemistry, Brookhaven National Laboratory, Upton, NY 11973*

†MetaPhore Pharmaceuticals, Inc., 3655 Vista Avenue, St. Louis, MO 63110

‡Department of Chemistry and Biochemistry, UCLA, Los Angeles, CA 90095-1569

1. Introduction

1.1 Superoxide dismutases

Superoxide dismutase enzymes catalyze the disproportionation of superoxide to give dioxygen and hydrogen peroxide[1] (Eq. 10.1).

$$2O_2^- + 2H^+ \rightarrow O_2 + H_2O_2 \qquad (10.1)$$

Three different, structurally unrelated types of superoxide dismutase (SOD) enzymes have been identified: the copper-zinc SODs, the manganese or iron SODs, and the nickel SODs. The manganese and iron SODs are clearly related to one another based on similarities in their amino acid sequences and their three-dimensional structures.

1.2 Superoxide toxicity and biological function of SODs

Superoxide anion, O_2^-, is formed as a normal by product of aerobic metabolism from a large number of sources, such as cellular respiration, activated polymorphonuclear leukocytes, endothelial cells, and mitochondrial electron flux; and it has been demonstrated to be a mediator of ischemia-reperfusion injury, inflammatory diseases and vascular diseases.[2–4] SOD enzymes are believed to play important roles as antioxidants in accelerating the conversion of superoxide to peroxide, thereby preventing direct reactions of superoxide that may do damage to sensitive targets in the cell.[1]

The identities of several individual chemical reactions that contribute to the toxicity of superoxide have been defined. One particularly important type of biological reaction of superoxide is with solvent-exposed iron-sulfur clusters in enzymes such as aconitase. The reaction of superoxide with these centers has been demonstrated to inactivate such enzymes both *in vitro* and *in vivo* and to increase levels of intracellular free reduced iron, which itself can catalyze the oxidative damage of other cellular components by hydrogen peroxide and other oxidants (Fenton-type chemistry).[5] Similarly, it has been shown that superoxide can induce the reductive release of iron from ferritin, the major intracellular iron storage protein.[6] The resultant Fenton chemistry which is unleashed has been proposed to be the source of cardiotoxicity of anthracycline antitumor agents.[7]

Recent literature has also documented the high reactivity of the protonated conjugate acid derived from superoxide, perhydroxyl radical, with important biomolecules. This potent oxidizing agent has been shown to be a competent and selective initiator of the autoxidation of lipid membrane components with its site of attack being *only* at the allylic positions in unsaturated fatty acids.[8,9] The products from the resultant highly selective chain autoxidations are leukotrienes (e.g. LTB_4) and prostanoids, which are responsible for the upregulation of proinflammatory cytokines (e.g. TNF_a) implicated in autoimmune diseases. The perhydroxyl radical has also been demonstrated to nick DNA under biomimetic conditions and, unlike some other highly reactive oxidants such as hydroxyl radical, with high sequence specificity.[10]

Our understanding of the physiological roles of SODs is as yet far from complete. A particularly surprising recent result was the demonstration that mice lacking the gene for CuZnSOD develop relatively normally,[11] although fibroblasts from these CuZnSOD-deficient mice are markedly more sensitive to redox cycling drugs than are the wild type cells.[12] By contrast, yeast lacking the gene for CuZnSOD grow poorly in air and die rapidly in stationary phase,[13] and both yeast and mice lacking the genes for the mitochondrial MnSOD have drastically reduced life spans.[13–15]

Superoxide and other reactive oxygen species have also been shown to function as secondary messengers in signaling pathways in higher organisms.[3,16] Thus, a cell may respond to a stimulus that is not due to changes in oxidative stress by generating reactive oxygen species that diffuse to a target, react, and thereby transduce the signal. Thus the possibility exists that SODs could play roles in signaling pathways in addition to performing their antioxidant functions, suggesting that modulation of the redox state of the cell may provide an explanation for the protective effects against cancer exerted by some antioxidants, as well as why ROS (reactive oxygen species) activate transcription factors such as NFκB.[16]

1.3 Synthetic and engineered SODs

Attempts to make synthetic SODs began shortly after the initial discovery of the enzymes and have continued to this day. However, investigators have only recently succeeded in making complexes that not only have high SOD activity but also show activity *in vivo* and are potentially suitable for use as drugs.[17] Initial studies focused on copper complexes as mimetic compounds for CuZnSOD, but it was quickly realized that synthetic copper complexes usually do not withstand the high affinities of complexing species that are found *in vivo*, such as bovine serum albumin and the myriad of proteins and amino acids that are found in cells and sera. The copper complexes of stronger binding ligands, such as copper porphyrin or EDTA complexes, have generally been found to be inactive to superoxide radicals.[18]

This endeavor has been more successful in the case of manganese complexes as SOD mimics. Two types of manganese-containing synthetic SODs, one with porphyrin ligands[18–20] and the other with a variety of macrocyclic ligands[21–23] have been shown to have significant SOD activity *in vivo*. There has been much less information about the potential of iron and nickel complexes as SODs, although recent studies suggest that SODs based on macrocyclic complexes of iron are feasible.[24] A synthetic FeSOD, prepared by reengineering the protein thioredoxin to contain an iron-binding site, has also been reported to have high SOD activity *in vitro*.[25]

1.4 Focus of this review

Our knowledge of the mechanistic details of the reaction of superoxide with transition metal complexes has increased significantly in recent years, and dramatic successes have been achieved recently in the development of synthetic SODs with great potential as pharmaceuticals. At the same time, our understanding of the mechanisms of the reactions of the naturally occurring SOD enzymes has also deepened. In this review, we first describe the mechanistic pathways that appear to be operating in reactions of some of the coordination complexes that have been studied as SOD mimics. We then attempt to apply this knowledge in analyzing the enzymatic reactions of the superoxide dismutase enzymes.

2. Superoxide Chemistry in Aqueous Solution

2.1 Background

The physical and chemical behavior of superoxide, O_2^-, and its protonated form, the perhydroxyl radical, HO_2, in both aqueous and non-aqueous solutions is described in a two-volume series by Afanas'ev,[26] and several reviews covering aspects of HO_2 and O_2^- chemistry in aqueous[27] or non-aqueous[28] chemistry have appeared recently. This section will briefly review the physical properties of superoxide and the perhydroxyl radical in aqueous solution as a background to understanding its reactivity with metal ions and complexes as well as with the superoxide dismutase (SOD) enzymes.

2.2 *Physical properties*

The equilibrium between HO_2 and its conjugate base, O_2^-, occurs with a pK of 4.8 in aqueous solution (Eq. 10.2).[29,30]

$$HO_2 \rightleftharpoons O_2^- + H^+$$
$$K_1 = 1.6 \times 10^{-5} \text{M} \qquad (10.2)$$

Both oxy-radicals absorb light in the ultraviolet, with maxima at 245 nm and 225 nm for O_2^- and HO_2 respectively. Their extinction coefficients are substantially different: $\varepsilon_{245\ \text{nm}} = 2350\ \text{M}^{-1}\text{cm}^{-1}$ for O_2^- and $\varepsilon_{225\ \text{nm}} = 1400\ \text{M}^{-1}\text{cm}^{-1}$ for HO_2.

Reduction potentials for O_2, O_2^-, and HO_2 in aqueous solution ε versus NHE, with unit concentration used as the standard state for all reactants and products, are $E^0(O_2 + e^- \rightarrow O_2^-) = -0.16$ V at pH 7, $E^0(O_2 + H^+ + e^- \rightarrow HO_2) = +0.12$ V at pH 0, $E^0(O_2^- + 2H^+ + e^- \rightarrow H_2O_2) = +0.89$ V at pH 7, and $E^0(HO_2 + H^+ + e^- \rightarrow H_2O_2) = +1.44$ V at pH 0. (The first two E^0 values are shifted by −0.17 V if unit pressure is used for the standard state of dioxygen (O_2). Compilations of thermodynamic parameters for HO_2 and O_2^- under many different conditions are available elsewhere.[31,32]

2.3 *Chemical reactivity of superoxide and perhydroxyl radicals*

The disproportionation of superoxide and perhydroxyl radicals to yield hydrogen peroxide and dioxygen occurs via a pH-dependent mechanism involving Reactions 10.3–10.5. Reaction 10.5 does not occur in the pH range of 0.2–13.

$$HO_2 + HO_2 \rightarrow H_2O_2 + O_2$$
$$k = 8.3 \times 10^5 \text{M}^{-1}\text{s}^{-1} \qquad (10.3)$$

$$HO_2 + O_2^- \xrightarrow{H^+} H_2O_2 + O_2$$
$$k = 9.7 \times 10^7\, M^{-1}s^{-1} \quad (10.4)$$

$$O_2^- + O_2^- \rightarrow \text{no reaction} \quad (10.5)$$

The disparity in the rates of these three reactions illustrates two important aspects of superoxide reactivity: (a) Reductions by superoxide are generally faster than reductions by perhydroxyl (compare Reactions 10.3 and 10.4). (b) Perhydroxyl is capable of fast one-electron oxidation of substrates while superoxide anion is usually not (compare Reactions 10.4 and 10.5). Those exceptional cases in which superoxide is observed to oxidize substrates rapidly occur only when a proton is available to transfer to the O_2^- moiety simultaneously with acceptance of an electron, i.e. proton-coupled electron transfer, resulting in formation of hydroperoxide anion, HO_2^- (Reaction 10.6), rather than simple electron transfer to give peroxide dianion, O_2^{2-}.

$$X \cdots O_2^- \cdots H - Y \rightarrow X^+ + HO_2^- + Y^- \quad (10.6)$$

An example of a fast oxidation by superoxide in which proton-coupled electron transfer to superoxide is likely to occur, is the rapid oxidation of hydroquinones by superoxide[33] (Reaction 10.7).

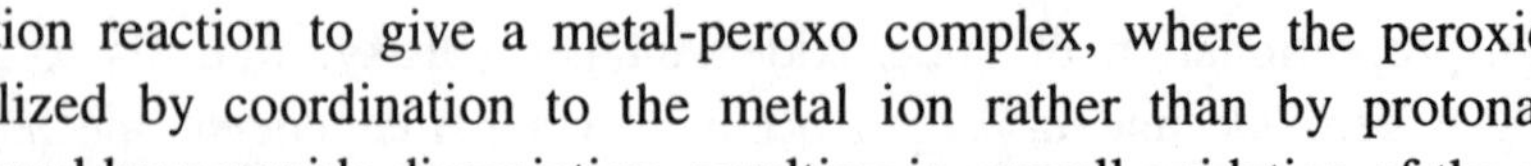

$$k = 1.7 \times 10^7\, M^{-1}s^{-1} \quad (10.7)$$

Alternatively, a metal ion may be oxidized by superoxide in an oxidative addition reaction to give a metal-peroxo complex, where the peroxide is stabilized by coordination to the metal ion rather than by protonation, followed by peroxide dissociation, resulting in overall oxidation of the metal ion. In this case, the electron transfer to form a metal bound peroxide can precede the protonation step because the metal ion stabilizes the O_2^{2-} ligand as it is formed (Reaction 10.8).

$$M^{n+} + O_2^- \rightarrow M^{(n+1)+}(OO^{2-}) \xrightarrow{H^+} M^{(n+1)+} + H_2O_2 \quad (10.8)$$

3. Mechanism of the Superoxide Dismutase Reaction

3.1 The superoxide dismutase reaction

Superoxide dismutases are catalysts of the superoxide disproportionation reaction (Eq. 10.1). The superoxide dismutase enzymes known to date are relatively small metalloproteins containing either Cu and Zn, Mn, Fe, or Ni. The ubiquitous presence of these enzymes in living organisms has led to great interest in the mechanism by which these enzymes so efficiently catalyze Reaction 10.1, and in the synthesis of model compounds that will function as efficiently as the enzymes.

The known superoxide dismutase mechanisms are conveniently divided into two separate steps, the first involving reduction by superoxide to afford dioxygen and the second involving oxidation by superoxide yielding hydrogen peroxide. In the following discussion, these two steps are discussed and analyzed separately.

3.2 Reduction by superoxide to give dioxygen

The most characteristic reaction of superoxide is as a one-electron reducing agent, and many examples of relatively fast reactions of this type are known. Two general mechanistic pathways involving redox changes at a metal center should be considered (Reaction 10.9): coordination (inner-sphere) of superoxide followed by electron transfer to a metal ion and outer-sphere electron transfer to a metal ion without prior coordination of superoxide.

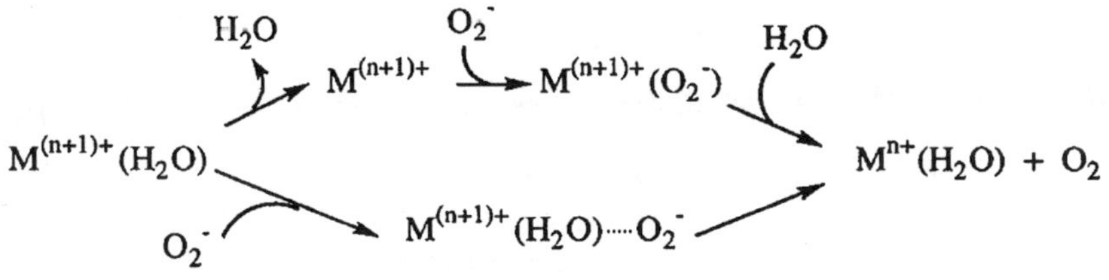

(10.9)

Variants of these reactions are reactions in which the starting complex is a metal-peroxo complex and hydrogen peroxide is released as the metal ion is reduced (Reaction 10.10; see Reaction 10.44 below for an example).

$$M^{(n+1)+}(OOH^-) \xrightarrow[O_2^-,\ H^+]{} M^{n+} + O_2 + H_2O_2 \qquad (10.10)$$

Another possible mechanistic pathway in which superoxide might act as a reducing agent in SOD reactions is via reduction of a metal bound superoxide affording a metal bound peroxide without change of oxidation state of the metal ion[34] (Reaction 10.11).

$$M^{(n+1)+}(O_2^-) + O_2^- \rightarrow M^{(n+1)+}(O_2^{2-}) + O_2 \qquad (10.11)$$

3.3 *Oxidation by superoxide to give hydrogen peroxide*

As discussed in Sec. 2.3 above, superoxide is observed to be reduced rapidly by substrates to the peroxide level only when proton-coupled electron transfer can occur, i.e. when a proton transfers to the O_2^- moiety simultaneously with transfer of an electron (Reaction 10.6). For the case where the reducing agent is a redox metal center, proton-coupled electron transfer to superoxide could conceivably occur with superoxide in the first coordination sphere of the metal ion (Reaction 10.12). Such an inner sphere pathway requires a vacant coordination site on the metal. So it is probable that such a pathway is limited to the rate at which ligand exchange occurs on the metal ion.

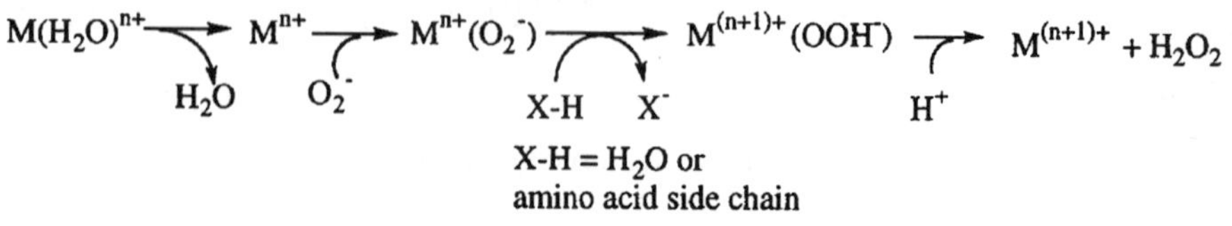

(10.12)

Two different types of outer sphere mechanisms are also possible. Such mechanistic pathways might be advantageous since they could bypass formation of stable metal hydroperoxide complexes, $M^{(n+1)+}(OOH^-)$, whose rates of dissociation could slow down SOD reactions. The first of these outer sphere mechanisms would require that superoxide be hydrogen bonded to two proton donors and not be bonded to the metal ion prior to electron transfer (Reaction 10.13). In the case of synthetic SOD mimics, it is difficult to imagine avoiding superoxide complexation as well as achieving the required hydrogen bonding configuration prior to electron transfer, but it is nevertheless a feasible mechanism for SOD enzymes. This mechanism is proposed as a possible mechanism for CuZnSOD in Sec. 5.1.3.3 below.

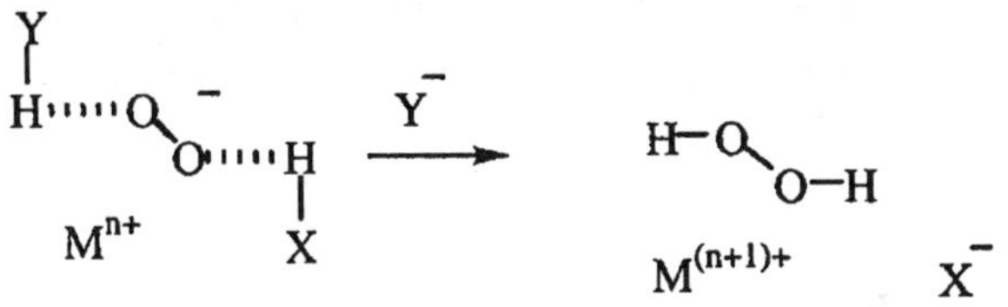

(10.13)

The second outer sphere mechanism involves a coordinated water molecule that is the hydrogen atom donor for a proton-coupled electron transfer (Reaction 10.14).

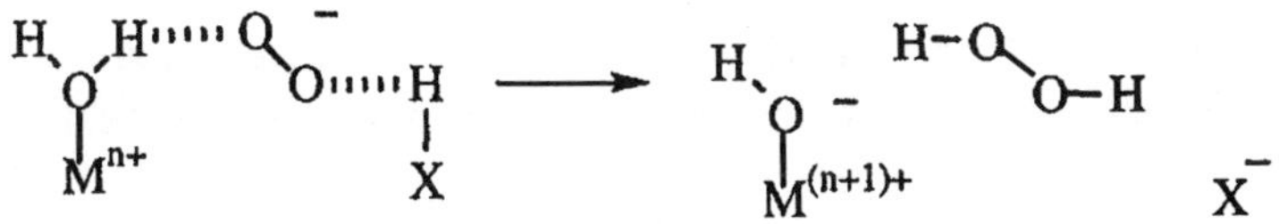

(10.14)

Such a mechanism is reminiscent of the proton-coupled electron transfer pathway postulated for the self-exchange reaction for $Fe(H_2O)_6^{2+/3+}$, i.e. hydrogen atom transfer from $[Fe^{II}(H_2O)_6]^{2+}$ to $[Fe^{III}(H_2O)_5(OH)]^{2+}$ (Reaction 10.15).[35,36]

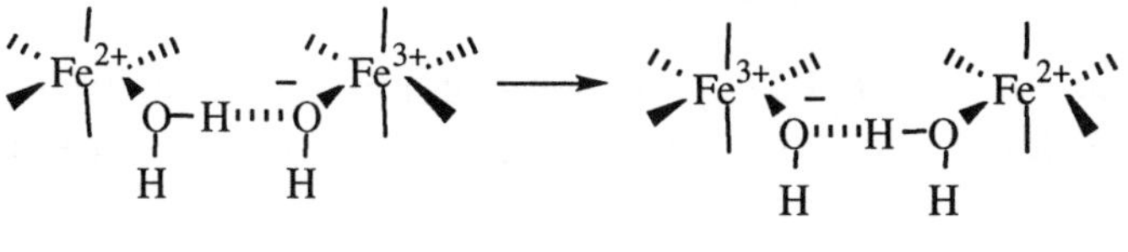

(10.15)

Reaction 10.14 probably cannot occur at a fast rate if the proton donor H-X is not present and oriented correctly for proton transfer simultaneous with electron transfer. In such cases, a pH-dependent pathway may be observed in which the oxidant is actually HO_2 (Reaction 10.16). Such a pathway would also fulfill a very important criteria for fast electron transfer; namely, that charge separation cannot develop in the transition state for the electron transfer step if the process is to possess extremely fast rates; i.e. near the diffusion limit, as is observed with SOD enzymes.

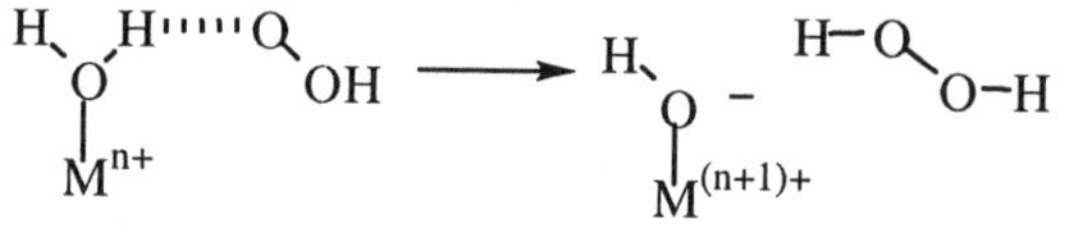

(10.16)

4. Reactions of Superoxide with Copper, Manganese, and Iron Ions and Low Molecular Weight Complexes

4.1 Copper

4.1.1 Aqueous copper ions

The search for low molecular weight copper complexes as model complexes for CuZnSOD has historically received the most attention, in large part due to the high prevalence of CuZnSOD *in vivo* and the fact that it was the first SOD identified. The reactions of copper ions and many small copper complexes with superoxide have consequently been studied in great detail using pulse radiolysis techniques.

Cupric ions (Cu^{2+}) are reduced by superoxide and by perhydroxyl in aqueous solution with rate constants that are very large: $k_{17} = 10^{10}$ $M^{-1}s^{-1}$ and $k_{18} = 10^9$ $M^{-1}s^{-1}$.[37]

$$O_2^{-} + Cu^{2+} \rightarrow Cu^{+} + O_2 \tag{10.17}$$

$$HO_2 + Cu^{2+} \rightarrow Cu^{+} + O_2 + H^{+} \tag{10.18}$$

Re-oxidation of the formed cuprous ion (Cu^+) by superoxide or perhydroxyl occurs with somewhat slower rate constants: $k_{19} = (5-8) \times 10^9$ $M^{-1}s^{-1}$ and $k_{20} = 10^8$ $M^{-1}s^{-1}$.[37]

$$O_2^- + Cu^+ \xrightarrow{2H^+} Cu^{2+} + H_2O_2 \quad (10.19)$$

$$HO_2 + Cu^+ \xrightarrow{H^+} Cu^{2+} + H_2O_2 \quad (10.20)$$

The high rate constant for Reaction 10.19 may suggest that the reaction proceeds by proton-coupled electron transfer as in Reaction 10.14 where HX = H_2O. An inner sphere pathway for this reaction is also consistent with the known very fast (> 10^9) water exchange rates on the aquo copper(II) ion which arise because Cu^{2+} is a Jahn-Teller d^9 spin system.

As the catalytic process is limited by the slowest rate constant in the cycle, it is clear that at physiological pH (pH = 7.4), the catalytic process involves predominantly O_2^- (Reactions 10.17 and 10.19) and is very rapid. The limiting feature in this system is that copper hydroxides form as the pH is increased and the insolubility of these copper hydroxides in water, limits the observed rate at higher pH.

The rate constants for reactions of superoxide with cuprous or cupric ion are so large at ambient pH that contamination by free copper ions can be a significant problem in determining O_2^- lifetime or reactivity in aqueous solution.[38] The use of copper chelators can obviate this problem because, as discussed below, the rate constants and reaction mechanisms for copper ions change markedly when they are incorporated in a coordination complex or metalloenzyme.

4.1.2 Copper amino acid and other coordination complexes

The original studies of the reactions of superoxide with copper amino acid complexes demonstrated that the reactions occur with quite large rate constants, but these studies were carried out at a single pH or over a narrow pH range.[39–41] Subsequent studies over a wide pH range were carried out for Cu(II) histidine complexes[42] and Cu(II) arginine complexes[43] which established that, at least in these two systems, the large rate constants occur over the narrow pH range in which the copper(II) amino acid complex had either an open coordination site or an exchangeable water. This observation suggests that the reaction occurs via an inner sphere mechanism, as in Reaction 10.9 above. In the case of Cu(II) arginine complexes, the formation of a transient copper(II) superoxo complex was detected in pulse radiolysis experiments (Reaction 10.21).

$$[Cu(II)ArgH_2]^+ + O_2^- \rightarrow [Cu(II)(ArgH_2^+)(O_2^-)] \rightarrow Cu(I) - Arginine + O_2 \qquad (10.21)$$

A particularly interesting series of copper complexes whose reactions with superoxide were studied contain 1,10-phenanthroline or substituted 1,10-phenanthrolines as ligands. The reaction of such complexes with superoxide, hydrogen peroxide, and dioxygen holds particular interest because the copper bis-phenanthroline complexes are known to cleave DNA in the presence of hydrogen peroxide.[44–47] This reaction is probably the result of a site, specific Fenton-type reaction of $Cu(phen)_2^+$ with H_2O_2. (Reaction 10.22 is the generalized description of this reaction.)

$$LCu(I) + H_2O_2 \longrightarrow \begin{cases} LCu(II)(HO\cdot) + OH^- \\ LCu(II)(OH^-) + HO\cdot \\ Cu(III)L + 2OH^- \end{cases} \qquad (10.22)$$

All that is needed to make a catalytic cycle of the $Cu(phen)_2^{2+}$ plus H_2O_2 DNA-cleavage system is the presence of a reductant; e.g. superoxide anion. The rate constants for reactions such as Reaction 10.22 are not fast (see Table 1), but the facile and efficient binding of $Cu(phen)_2^{2+}$ to DNA ensures a high rate of reaction of the oxidant produced in Reaction 10.22 with DNA.

The overall mechanism for the reaction of $Cu(phen)_2^{2+}$[48] with O_2^- proceeds via the formation of a transient copper(II)-superoxo complex (Reactions 10.23–10.26).

$$Cu(II)(phen)_2^{2+} + O_2^- \rightarrow Cu(I)(phen)_2^+ + O_2 \qquad (10.23)$$

$$Cu(I)(phen)_2^+ + O_2^- \xrightarrow{2H^+} Cu(II)(phen)_2^{2+} + H_2O_2 \qquad (10.24)$$

$$Cu(I)(phen)_2^+ + O_2 \rightarrow [Cu(II)(phen)_2(O_2^-)]^+ \qquad (10.25)$$

$$[Cu(II)(phen)_2(O_2^-)]^+ + Cu(I)(phen)_2^+ \xrightarrow{2H^+} 2Cu(II)(phen)_2^{2+} + H_2O_2 \qquad (10.26)$$

Table 1 Rate constants for oxidation of various copper(I) complexes by O_2^-, O_2 or H_2O_2 (pH = 7.0, unless specified).

Complex	O_2^-	O_2	H_2O_2
Cu^+_{aq}	10^{10}[a]	4.6×10^5[b]	4.7×10^3[c]
Cu(4,11-dieneN4)$^+$[d]	–	2.6×10^7	–
$Cu(bpy)_2^+$[e]	1.8×10^8	5.8×10^4	1.5×10^3
$Cu(phen)_2^+$[f]	3.0×10^8	5.0×10^4	9.4×10^2
$Cu(5\text{-}Clphen)_2^+$[g]	2.1×10^8	5.0×10^3	–
$Cu(5\text{-}NO_2phen)_2^+$[h]	8.3×10^8	5.8×10^2	4.4×10^2
$Cu(5\text{-}Mephen)_2^+$[i]	2.3×10^8	6.6×10^4	1.6×10^3
$Cu(2,9\text{-}Me_2phen)_2^+$[j]	2.4×10^8	–	–
CuZnSOD[k]	2×10^9	very slow	0.04

[a]pH = 3–6.5 (Ref. 49).
[b]pH = 2.1 (Ref. 50).
[c]pH = 2.3 (Ref. 51).
[d]Cu(4,11-dieneN4)$^+$ = 5,7,7,12,14,14-hexamethyl-1,2,8, 11- tetraazocyclo-tetradeca-4,11-dienecopper(I) (Ref. 52).
[e]$Cu(bpy)_2^+$ = bis(2,2'-bipyridine)copper(I) (Ref. 53).
[f]$Cu(phen)_2^+$ = bis(1,10-phenanthroline)copper(I) (Ref. 54).
[g]$Cu(5\text{-}Clphen)_2^+$ = bis(5-chloro-1,10-phenanthroline)copper(I) (Ref. 53).
[h]$Cu(5\text{-}NO_2phen)_2^+$ = bis(5-nitro-1,10-phenanthroline)copper(I) (Ref. 53).
[i]$Cu(5\text{-}Mephen)_2^+$ = bis(5-methyl-1,10-phenanthroline)copper(I) (Ref. 53).
[j]$Cu(2,9\text{-}Me_2phen)_2^+$ = bis(2,9-dimethyl-1,10-phenanthroline)copper(I) (Ref. 53).
[k](Ref. 55)

4.2 Manganese

4.2.1 Manganese ions and complexes

Reactions of manganese ions and complexes with superoxide are complicated by the simultaneous occurrence of inner and outer sphere pathways for oxidation of manganese(II). Understanding these reactions in detail is important since similar mechanistic pathways are likely to occur in reactions of MnSOD with superoxide.

Manganous ion is only soluble and stable at low to neutral pH, and the manganic ion is only stable to disproportionation in acidic solution. In contrast, upon complexation, manganese(II) can be studied over a broad pH range. Pulse radiolysis studies have demonstrated that both Mn^{2+} ions and manganese(II) polyaminocarboxylate complexes react with superoxide to give transient manganous superoxide complexes[56–63] (Reactions 10.27–10.29).

$$LMn(II) + O_2^- \rightleftharpoons LMn(II)(O_2^-) \quad (10.27)$$

$$LMn(II)(O_2^-) \xrightarrow{2H^+} LMn(III) + H_2O_2 \quad (10.28)$$

$$LMn(II)(O_2^-) + O_2^- \xrightarrow{2H^+} LMn(II) + O_2 + H_2O_2 \quad (10.29)$$

The protonated adduct, $LMn(HO_2)$, which could be either $LMn(II)(HO_2)$ or $LMn(III)(HO_2^-)$, is formed at low pH, either by protonation of $LMn(II)(O_2^-)$ (Reaction 10.30) or by direct reaction of HO_2 with LMn(II) (Reaction 10.31). The mechanism by which the protonated adduct disappears can involve the formation of a bridged binuclear adduct, leading to the formation of LMn(III) (Reactions 10.26–10.28). A recent publication[64] suggests that the binuclear adduct is not necessarily formed with manganous ions at low pH. (It should be noted that the species $LMn(III)(HO_2^-)$ is written here as a Mn(III) complex of the hydroperoxide anion HO_2^-, but it could instead be $LMn(II)(HO_2)$, a Mn(II) complex of the perhydroxy radical HO_2.)

$$LMn(II)(O_2^-) + H^+ \rightleftharpoons LMn(III)(HO_2^-) \quad (10.30)$$

$$LMn(II) + HO_2 \rightarrow LMn(III)(HO_2^-) \quad (10.31)$$

$$LMn(III)(HO_2^-) + LMn(II) \rightleftharpoons [LMn(III)(HO_2^-)Mn(II)L] \quad (10.32)$$

$$[LMn(III)(HO_2^-)Mn(II)L] \rightarrow LMn(III) + LMn(II) + HO_2^- \quad (10.33)$$

$$[LMn(III)(HO_2^-) \rightarrow LMn(III) + HO_2^- \quad (10.34)$$

Protonation of the transient intermediate $LMn(II)(O_2^-)$ to yield peroxide has been observed to be very ligand sensitive. For example, the rate of formation of the Mn(III) complex of pyrophosphate ($P_2O_7^{4-}$) has been shown to be concomitant with the rate of disappearance of O_2^-; no $Mn(II)(O_2^-)$ adduct was observed. An inner sphere mechanism with adduct formation is not precluded in that reaction, but it is clearly impossible to accelerate the formation of the adduct sufficiently enough for it to have an observable lifetime. In contrast, for a manganese(II) complex of TTHA[65] (TTHA = triethylenehexaaminehexaacetate), release of free peroxide occurred via reaction of the Mn(II) superoxo complex with a proton and not water such that, at alkaline pH, the $Mn(II)(O_2^-)$TTHA complex disappeared by a second-order process that can be attributed to dissociation of superoxide (reverse of equilibrium 10.35), followed by spontaneous dismutation of O_2^- via Reaction 10.1.

$$LMn(III) + O_2^- \rightleftharpoons LMn(II) + O_2 \tag{10.35}$$

In addition to the facile oxidation of Mn(II) by superoxide, a superoxide dismutase catalytic cycle also, of course, requires the rapid reduction of Mn(III) by O_2^- (the forward step of the equilibrium in Reaction 35). Rate constants for this reaction are not easily measured because manganic complexes are generally not very stable at ambient pH. An interesting example of such a determination is described in a pulse radiolysis study of reactions of superoxide with manganese complexes of EDDA (EDDA = ethylenediaminediacetate).[63] The oxidation of Mn(II)EDDA by O_2^- (Reactions 10.36 and 10.37) again proceeded via a transient adduct (Reaction 10.36), as has been seen for several complexes. In this system however, it was also possible to form significant concentrations of Mn(III)EDDA by reaction of Mn(II)EDDA with a large burst of O_2^-. The subsequent reduction (Reactions 10.38 and 10.39) of Mn(III)EDDA by superoxide could then be observed using a second, smaller burst of O_2^- and was found to display saturation kinetics, suggesting a mechanism that includes formation of a $Mn(III)(O_2^-)$ transient adduct (Reaction 10.38). This experiment was only possible because Reaction 10.37 was sufficiently fast at neutral pH to allow the rapid formation of Mn(III)EDDA and because equilibrium 10.38 and Reaction 10.39 were sufficiently slow relative to Reaction 10.37. This system is clearly not an effective biomimetic catalyst for MnSOD because of these slow rates.

$$Mn(II)EDDA + O_2^- \rightarrow Mn(II)EDDA(O_2^-) \quad (10.36)$$

$$Mn(II)EDDA(O_2^-) + H_2O \xrightarrow{H+} Mn(III)EDDA + H_2O_2 + OH^- \quad (10.37)$$

$$Mn(III)EDDA + O_2^- \rightleftharpoons Mn(III)EDDA(O_2^-) \quad (10.38)$$

$$Mn(III)EDDA(O_2^-) \rightarrow Mn(II)EDDA + O_2 \quad (10.39)$$

It should be noted that the rate constants for reduction of model Mn(III) complexes by O_2^- given in Table 2 are significantly lower than those measured for MnSOD. The reduction of Mn(III)SOD by O_2^- to yield Mn(II)SOD and O_2 is virtually diffusion-controlled (ca. 10^9 $M^{-1}s^{-1}$) with no detectable formation of a transient adduct (see Sec. 5.2 below).

The dismutation of O_2^- by various Mn(II)-containing model complexes is generally slow (k_{cat} ca. $10^6 - 10^7$ $M^{-1}s^{-1}$) relative to the enzyme. This catalysis is often limited by reduction of Mn(III) by O_2^-; oxidation of the Mn(II) complexes by superoxide is usually fast (see Table 2). This pattern is also seen with manganous and manganic porphyrins. Oxidation of Mn(II) porphyrins (k ca. $10^8 - 10^{10}$ $M^{-1}s^{-1}$) and reduction of Mn(III) porphyrins (k ca. $10^5 - 10^8$ $M^{-1}s^{-1}$) by O_2^- have been shown to occur via outer sphere mechanisms.[20] The latter rate constants are the ones that control the catalytic efficiency of manganese porphyrins as SOD mimics.

A number of more effective MnSOD synthetic mimics have been synthesized and studied recently. These include the Mn(III)desferrioxamine B green complex,[66–69] Mn(II)salen complexes,[70] and Mn(II) pentaazamacrocycle complexes.[17,21,22]

Kinetic studies of the reactions of superoxide with a large series of pentaazamacrocyclic complexes of manganese using stopped flow techniques have been particularly informative concerning the details of the mechanistic pathways involved, especially with regard to the role of protons in

Table 2 Rate constants for reactions of Mn(II) and Mn(III) complexes with superoxide.

Ligand	$k(LMn(II) + O_2^-)$, $M^{-1}s^{-1}$ (pH) (Ref.)	$k(LMn(III) + O_2^-)$, $M^{-1}s^{-1}$ (pH) (Ref.)
NTA[a]	4×10^8 (4.5) (61) 1.2×10^8 (5.5) (61)	1.2×10^7 (6.0) (59)
EDTA[b]	3×10^7 (4.5) (61) 7.5×10^6 (5.5) (61)	5×10^4 (10.0) (62)
CyDTA[c]	–	7.2×10^5 (9.2) (62)
TTHA[d]	1.75×10^5 (6–8.3) (65)	
$P_2O_7^{4-}$[e]	2.6×10^7 (6.5) (57) 1.3×10^7 (7.3) (71)	
Formate	4.6×10^7 (4–7) (57)	
Sulfate	5.4×10^7 (2.8–5.6) (57)	
Phosphate	5×10^7 (2–6) (58)	
EDDA[f]	3×10^7 (7.1) (63)	
DFB[g]	3×10^6 (66)	1×10^6 (66)
TMPyP[h]	$(2.5–3.3) \times 10^9$ (6.7–9.3) (20)	$(3.5–5.1) \times 10^7$ (5.6–9.3) (20,72)
TAPP[i]	$(5.6–20) \times 10^8$ (6.7–9.3) (20)	$(1.5–13) \times 10^6$ (5.6–9.3) (20,72)
PFP[j]	$(5–9) \times 10^9$ (6.7–9.3) (20)	1.7×10^7 (6.7) (20) 4.0×10^5 (9.3) (20)
MnSOD[k]	1.5×10^9 (9.3) (72)	1.5×10^9 (9.3) (72)

[a]NTA = nitriloacetate
[b]EDTA = ethylenediaminetetraacetate
[c]CyDTA = diaminocyclohexane-N,N,N',N'-tetraacetate
[d]TTHA = triethylenetetraaminehexaacetate
[e]$P_2O_7^{4-}$ = pyrophosphate
[f]EDDA = ethylenediaminediacetate
[g]DFB = desferrioxamine B
[h]TMPyP = tetrakis(4-N-methylpyridyl)porphine
[i]TAPP = tetrakis-4-(N,N,N,-trimethylamino)phenylporphine
[j]PFP = tetrakis(N-methylisonicotinomidophenyl)porphine
[k]from *Thermus thermophilus*

determining the relative rates of reduction of superoxide when it is bound to a metal ion during catalysis.[22]

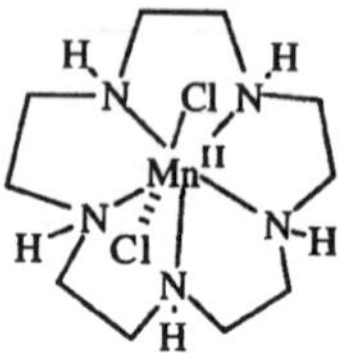

1

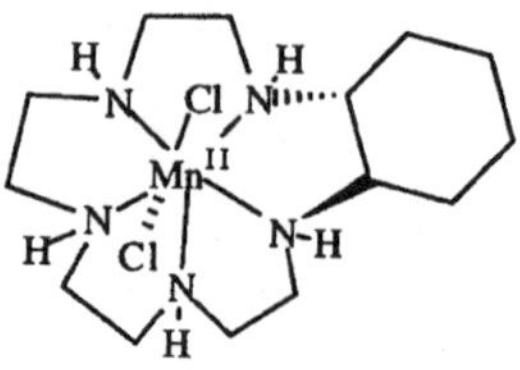

2

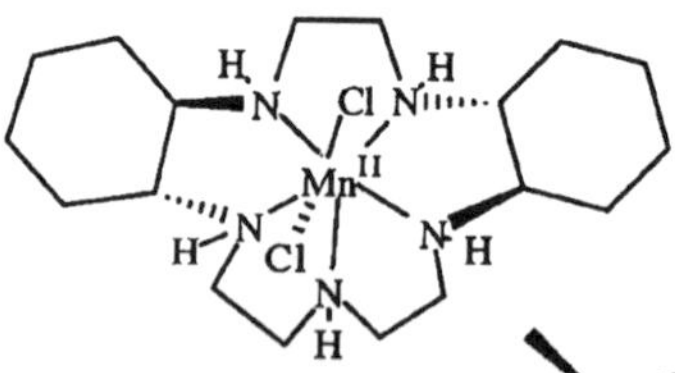

3

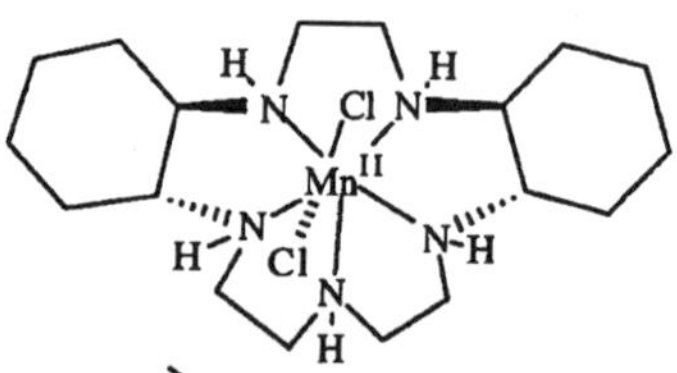

4

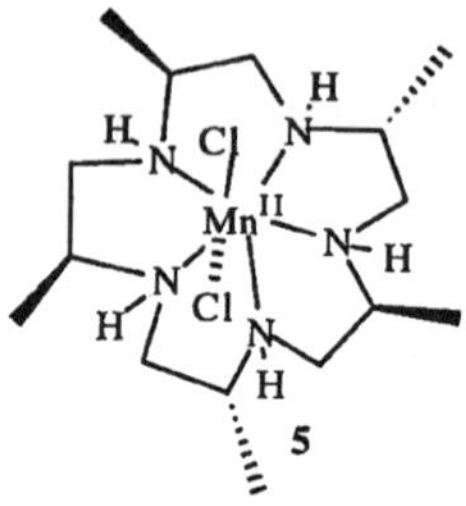

5

For each of the complexes **1–5**, the catalytic rate of disproportionation of superoxide, k_{cat}, was found to be first order in $[O_2^-]$ and [Mn complex] added. For **1–4**, a plot of the observed k_{cat} at a given pH versus the proton concentration gives a straight line with a non-zero intercept. Thus both a pH-dependent and a pH-independent pathway were observed. In the case of **5**, by contrast, the rate is pH-independent.

The observed rate law is given in Eq. 10.40.

$$-\mathrm{d}[O_2^-]/\mathrm{d}t = k_{cat}[\text{Mn complex}][O_2^-] = [\text{Mn complex}][O_2^-]\left\{2k_1/K_a[H^+] + 2k_2\right\} \tag{10.40}$$

where K_a is the acidity constant for HO_2 ($K_a = 2.04 \times 10^{-5}$). These results are consistent with the following mechanism (Reactions 10.41–10.44).

$$LMn(II)(OH_2) + HO_2 \xrightarrow{k_{41}} LMn(III)(OH^-) + H_2O_2 \qquad (10.41)$$

$$LMn(II)(H_2O) + O_2^- \xrightarrow{k_{42}} LMn(III)(OO^{2-}) + H_2O \text{ (inner sphere)} \qquad (10.42)$$

$$LMn(III)(OO^{2-}) + H^+ \xrightarrow{\text{fast}} LMn(III)(OOH^-) \qquad (10.43)$$

$$LMn(III)(OOH^-) + O_2^- + H^+ \xrightarrow{\text{fast}} LMn(II) + H_2O_2 \qquad (10.44)$$

Several interesting comparisons can be made between catalysts **1–5**. Comparison of the k_{cat} observed for the two bis (*trans*-fused) cyclohexano derivatives, **3** and **4**, is particularly interesting (see Table 3). Cyclic voltammetry of **3** and **4** showed identical behavior, with a well-defined reversible one-electron oxidation at $E_{1/2}$ at 0.74 V versus. SHE in anhydrous methanol.[22] Nevertheless, their ability to function as SOD catalysts is strikingly different. Complex **3** was found to function at a rate equivalent to native mitochondrial MnSOD at pH = 7.4, while complex **4** exhibited no detectable SOD activity. This striking difference in the catalytic rates of superoxide dismutation for these two stereoisomeric complexes is indicative of a powerful influence of the ligand substituents on the rates of reaction with superoxide.

Table 3 Representative values for the proton-dependent and proton-independent catalytic rate constants for SOD activity of Mn complexes 1–5 measured over the pH range 7.4–8.3.

Complex	k_{41}, 10^{10} M^{-1} s^{-1}	k_{42}, 10^7 s^{-1}	$k_{cat} \times 10^{-7}$ $M^{-1}s^{-1}$ (pH = 7.4)
1	0.29	0.91	
2	0.78	1.44	
3	2.25	1.58	12.1
	(1.07 in D_2O)	(1.83 in D_2O)	
4	0	0	0
5	0	1.91	3.82
		(1.01 in D_2O)	

Studying the kinetics of the SOD reaction for **3** and **5** in D_2O provided further insight into the details of the rate-limiting electron transfer steps. Complex **5** exhibits no $[H^+]$ dependence in its SOD activity. The catalytic rates for **3** and **5** in D_2O, exhibited kinetic behavior similar to that observed in H_2O. The pH-independent rate constant k_2 for **5** was decreased ~45%. For **3**, k_1 decreased by ~50%, but k_2 increased by ~10% (see Table 3). These results, combined with the observations that **4** is not a catalyst, that **5** has no pH-dependent component to its k_{cat}, and that k_1 for **1**, **2**, and **3** are approaching diffusion-controlled limits, has led to some important conclusions about the mechanism.

The observation that k_2, the rate constant for the pH-independent pathway, is approximately 10^7 $M^{-1}s^{-1}$ for each of these complexes suggests that a dissociative process may be rate-limiting since the rate of loss of water from the inner-coordination sphere of aquo Mn(II) is 0.8×10^7 s^{-1}. The modest deuterium isotope effect for k_2 for both **3** and **5**, suggest that this is indeed the case. Rates of solvent exchange on metal ions can be both faster and slower in D_2O than in H_2O, but the fact that there is an effect strongly implies that aquo ligand exchange is involved in the rate-determining step.

Another conclusion that can be drawn from the kinetic experiments is that the near diffusion-controlled rate constant k_1 for the pH-dependent electron transfer process requires that the barrier to electron transfer to HO_2 be very small. Thus, we can conclude that little charge separation develops in the transition state and that the Mn(III) product complex has a structure very similar to the starting Mn(II) complexes so that the reorganization energy is low. The pH dependence on the rate in D_2O supports these conclusions. The decrease in rate for **3** in D_2O resembles that seen for the Fe^{2+}/Fe^{3+} self-exchange reaction in water which is believed to occur via a hydrogen atom transfer from a bound water on Fe(II) (Reaction 10.15).[35,36]

Recent studies provide support for proton-coupled electron transfer as an important pathway for fast electron transfer in biological systems, as suggested recently for photosystem II,[73] thereby avoiding the build up of charge separation in the transition state for electron transfer. Additionally, a recent report indicates that the thermodynamics of an H atom transfer from a Mn bound water to HO_2 will be favorable; i.e. the bond dissociation energy of an O–H bond of water bound to a Mn(III)(Salen) derivative is 86 kcal/mol. Thus H atom transfer to HO_2 is energetically allowed since the bond dissociation energy of the O–H bond of hydrogen peroxide is reported to be 87.2 kcal/mol.[74]

The ability of **1**, **2**, and **3** (but not **4**) to adopt pseudo-octahedral conformations by folding the macrocyclic ring so that one of the secondary amines occupies an axial site also provides a logical explanation for the dramatic loss of SOD reactivity in the case of **4**. Thus, if the macrocyclic [15]aneN_5 ligand possesses C-substituents which, due to intramolecular steric repulsions and angle strains, enable the ligand to readily adopt a folded pseudo-octahedral geometry about the spherically symmetrical Mn(II) ion (as is the case for **1**, **2**, and **3**), then the Mn(II) complex would be poised to undergo facile electron transfer. Thus the outer sphere path is expected to proceed through an intermediate pseudo-octahedral $[LMn(II)(H_2O)]^{2+}$ complex with a folded macrocyclic ligand (Reaction 10.45), while the inner sphere substitution pathway would involve a superoxo $[LMn(II)(O_2^-)]^+$ six-coordinate intermediate complex, containing the folded macrocyclic ligand, which would be protonated to generate the $[Mn(III)(L)(HO_2)]^{2+}$ pseudo-octahedral complex (Reaction 10.46).

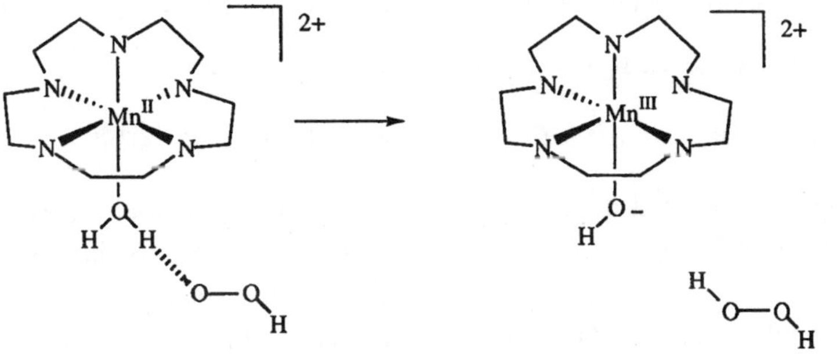

(10.45)

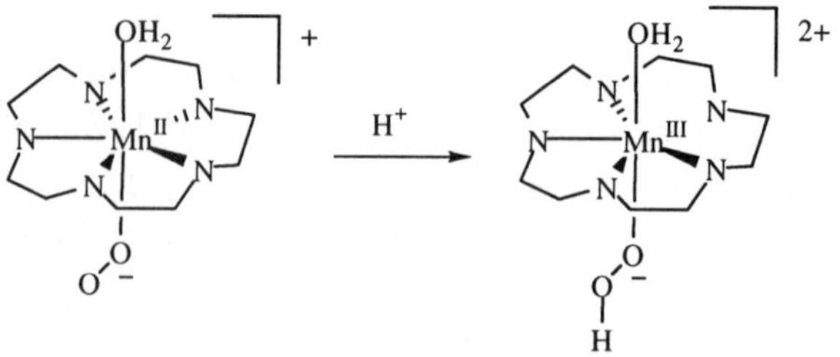

(10.46)

This need to fold the ligand into a conformation that will stabilize a pseudo-octahedral geometry on Mn(II) may be the reason why **5** shows no rate for this pH-dependent pathway. Thus, the presence of a methyl substituent symmetrically positioned on one carbon of each chelate ring may inhibit folding, and thereby block this pathway for fast electron transfer to HO_2.

4.3 *Iron*

The aquo-iron system is not an effective superoxide dismutase, as evidenced by the rate constants for the individual reactions in the overall mechanism[75–77] (Reactions 10.47–10.50):

$$Fe^{2+} + HO_2 \rightarrow Fe^{3+} + HO_2^- \qquad k_{47} = 1.2\times10^6\,M^{-1}s^{-1} \tag{10.47}$$

$$Fe^{2+} + O_2^- \rightarrow Fe^{3+} + HO_2^- \qquad k_{48} = 1\times10^7\,M^{-1}s^{-1} \tag{10.48}$$

$$Fe^{3+} + HO_2 \rightarrow Fe^{2+} + H^+ + O_2 \qquad k_{49} < 10^3\,M^{-1}s^{-1} \tag{10.49}$$

$$Fe^{3+} + O_2^- \rightarrow Fe^{2+} + O_2 \qquad k_{50} = 1.5\times10^8\,M^{-1}s^{-1} \tag{10.50}$$

The kinetic values are relatively independent of formate and sulfate, which complex only weakly to iron at low pH. The oxidation of Fe^{2+} by HO_2 occurs by a multistep process at low pH (Reactions 10.51–10.54) that is analogous to that described in Reactions 10.31–10.34 in the manganese system.[78–80]

$$Fe^{2+} + HO_2 \rightarrow Fe^{3+}(HO_2^-) \tag{10.51}$$

$$Fe^{3+}(HO_2^-) + Fe^{2+} \rightleftharpoons [Fe^{3+}(HO_2^-)Fe^{2+}] \tag{10.52}$$

$$Fe^{3+}(HO_2^-) \rightarrow Fe^{3+} + HO_2^- \tag{10.53}$$

$$[Fe^{3+}(HO_2^-)Fe^{2+}] \rightarrow Fe^{3+} + HO_2^- + Fe^{2+} \tag{10.54}$$

A number of synthetic iron complexes that catalyze the dismutation of superoxide to oxygen and hydrogen peroxide have been reported, although far fewer examples exist than for either copper or manganese. The earliest of these is the $Fe^{III}(EDTA)(H_2O)$ complex.[81] While this complex is a poor catalyst, with catalytic rate $< 10^6\ M^{-1}s^{-1}$, the mechanism is nevertheless instructive in view of the preceding discussion. Reduction of the ferric EDTA complex with superoxide anion is accomplished with a rate constant of $2 \times 10^6\ M^{-1}s^{-1}$, producing oxygen and the ferrous complex. Reoxidation occurs via a pH-independent process with a rate constant of $\sim 10^6\ M^{-1}s^{-1}$. This pH independence and the magnitude of the rate constant suggest that the reoxidation is accomplished via an inner sphere binding of superoxide to the coordination site generated by the loss of a bound water (see Reaction 10.13 above).

The iron(II) complex of the synthetic porphyrin ligand, tetrakis (4-N-methylpyridyl)porphine has been reported to catalyze this dismutation,[82] but mechanistic details are limited owing to the complex's propensity to form oxo-bridged dimers and as a consequence lose activity.[83] Other Fe-based synthetic SOD catalysts have been reported with amino pyridine ligands such as N,N,N',N'-tetrakis(2-pyridylmethyl)ethylenediamine and tris[N-(2-pyridylmethyl)-2-aminoethyl]amine,[84] but mechanistic details are lacking.

Recently, a series of Fe(III) pentaazamacrocyclic ligands were reported to be active SOD catalysts and to exhibit reaction kinctics that reveal that reduction is the rate-limiting process in the catalytic cycle.[24] These same macrocyclic ligands were reported to provide highly active SOD catalysts with Mn(II) (see Sec. 4.2.1 above), but with the major difference that the rate-limiting step is oxidation. Thus, the resting state of the catalyst for each metal is the high-spin d^5 complex. These Fe(III) complexes exist in water in the physiological pH regime as the equilibrium mixture of the *trans*-$[Fe^{III}(L)(H_2O)(OH)]^{2+}$ and the *trans*-$[Fe^{III}(L)(OH)_2]^+$ complexes. From kinetic and mechanistic studies, the authors establish that the aquo, hydroxo complex, *trans*-$[Fe^{III}(L)(H_2O)(OH)]^{2+}$ is the active form. The reaction proceeds via a dissociative inner sphere pathway in which the loss of the axial water ligand from Fe(III) generating a vacant coordination site is the rate-limiting process. Subsequent binding of superoxide anion to Fe(III) provides the pathway for reduction of Fe(III) to Fe(II). This mechanism is analogous to the general mechanism represented by Eq. 10.9 for an inner sphere reduction step.

5. Superoxide Dismutase Enzymes

5.1 Copper-zinc superoxide dismutases

5.1.1 Structural characteristics

CuZnSOD is found in the cytosol and extracellular compartments of virtually all eukaryotic cells and in the periplasm of gram-negative bacteria.[1] (One notable exception is a large group of marine arthropods with no CuZnSOD but with a cytosolic MnSOD in its place.[85]) Most CuZnSODs are homodimeric proteins, with each of the two equivalent subunits containing a copper and a zinc ion and a molecular weight of approximately 16 000. All of the characterized CuZnSODs are remarkably stable relative to other intracellular proteins. Two monomeric CuZnSODs (from rice[86] and from the periplasmic space of *E. coli*[87]) have also been isolated. With the exception of the CuZnSODs from *E. coli*[88] and *P. leiognathi*,[89] the amino acid sequences, protein fold, substrate binding-channel, and active site configurations of CuZnSOD have been highly conserved throughout evolution.[90]

The metal binding site of CuZnSOD consists of six histidines and an aspartate.[91] Three of the histidines (His 46, His 48, and His 120, using yeast or human CuZnSOD numbering) bind exclusively to copper in either the reduced, copper(I)-containing or the oxidized, copper(II)-containing forms of the protein. Two of the histidines (His 71 and His 80) and the aspartate

(Asp 83) bind exclusively to the zinc(II) ion. The imidazole ring of the "bridging histidine", His 63, binds exclusively to the zinc ion in reduced CuZnSOD and the copper(I) ion sits in a nearly trigonal plane of three histidines.[91,92] In the oxidized form of the enzyme, the copper ion has changed position, moving closer to His 63 and binding to it so that His 63 acts as a bridging ligand between the Cu^{2+} and the Zn^{2+} ions.[91] A water molecule also enters the first coordination sphere of the copper(II) ion (Fig. 1).

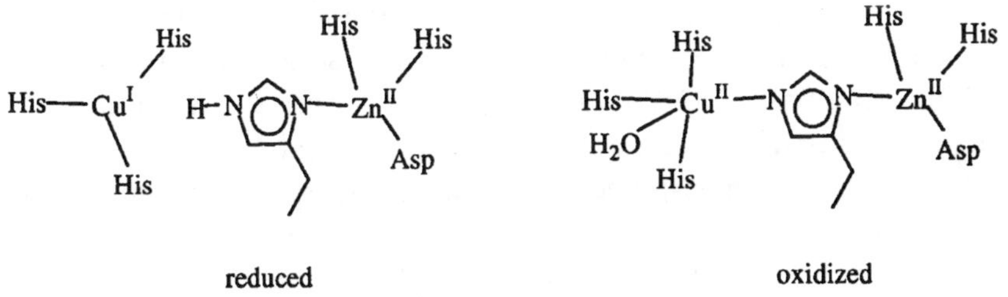

Fig. 1 Schematic representation of the metal-binding sites in CuZnSOD.

Aside from the structural differences in the metal binding region, which consist mainly of a large displacement of the copper ion in reduced versus oxidized CuZnSOD, and some small changes in liganding histidines and the water ligand at the copper site, the structures of the oxidized and reduced enzymes are extremely similar. One particularly important structural feature that is virtually identical in both the oxidized and reduced enzymes is the funnel-shaped channel that provides access to the copper ion. This channel is 24 Å wide at the surface of the protein but becomes progressively narrower until it is only 4 Å across directly above the copper ion.[93–95] This narrowing of the bottom of the channel has the effect of limiting ligand binding to Cu^{2+}, in the oxidized form of the enzyme, to small anions similar in size and charge to superoxide, e.g. cyanide, azide, and fluoride, and making the enzyme highly specific for reaction with superoxide. One anion, particularly prevalent *in vivo* and apparently excluded from approaching the copper(II) center closely, is ascorbate anion, whose rate of reaction with the oxidized enzyme is extremely slow, despite very favorable redox potentials.[96]

5.1.2 What makes CuZnSOD a good SOD?

5.1.2.1 Substrate recognition

A wealth of information strongly supports substrate recognition occurring via electrostatic interactions between O_2^- and either oxidized or reduced CuZnSOD. Superoxide is drawn into and docked in the active site by an electrostatic field, created by positively charged amino acid side chains near and in the funnel-like channel that lies directly above the catalytic copper ion. A series of theoretical calculations[97–99] of the electrostatic attraction of the enzyme for O_2^- and experimental determinations[100–102] of the SOD activities of CuZnSODs (containing chemical modification or mutagenic substitution of conserved amino acid residues that form the channel) have demonstrated the essential role of electrostatic guidance in the enzymatic activity. Recent NMR studies on monomeric mutant CuZnSOD further suggest that changes and/or fluctuations in the channel structure may be important in determining the catalytic efficiency of those enzymes.[103–105] On the basis of all of these experiments and calculations, it seems clear (1) that the selectivity of the enzyme for reaction with superoxide and the efficiency of the disproportionation reaction catalyzed by the enzyme depends critically on the ability of the enzyme to maintain the undisturbed precisely engineered channel and (2) that the electrostatic steering of O_2^- into the active site of enzyme is one particularly important feature of CuZnSOD that makes it an efficient biological catalyst of O_2^- disproportionation.

5.1.2.2 The role of Arginine 143

Arg 143 is located in the active site channel, approximately 6 Å above the copper(II) ion in the oxidized structure, and it plays an important role in generating the electrostatic field that draws superoxide to the mouth of the active site channel.[98,100] In addition, the positively charged guanidinium group of that residue is well-positioned to assist in guiding the superoxide anion down the channel, hydrogen bonding with it as it approaches the copper center in either the oxidized or reduced enzyme.

This hydrogen bonding interaction localizes O_2^- directly above the copper ion in a position from which a redox reaction could occur by either an outer or inner sphere electron transfer mechanism (see discussion of possible mechanisms below). Recently, the pH dependence of the catalytic rate of wild type single and double mutants of *Xenopus leavis* CuZnSOD was determined using pulse radiolysis.[106] The results, which are in agreement with those from site-directed mutagenesis experiments on the human enzyme,[98] indicated that 25% of the electrostatic steering of superoxide is due to Arg 143 and that this residue plays a key role in orienting O_2^- as it approaches the copper ion.

5.1.2.3 Coordination of superoxide and other anions to copper(I) and copper (II)

One particularly important issue is whether the superoxide substrate enters the first coordination sphere of either the copper(II) or copper(I) ions in the oxidized or reduced enzyme, respectively, in the course of the catalytic mechanism. In the case of oxidized CuZnSOD, anions such as azide (N_3^-), cyanide (CN^-), and fluoride (F^-) bind directly to Cu^{2+} in the enzyme, and the latter two ligands have been shown to exchange rapidly. Therefore, it is likely that superoxide enters into the first coordination sphere of the Cu^{2+} ion, forming a $Cu^{2+}(O_2^-)$ complex, prior to the electron transfer that forms the products Cu^+ and O_2.[91] For the re-oxidation of reduced CuZnSOD by superoxide however, the situation is less clear. The trigonal planar $Cu(I)(His)_3$ center seems unlikely to have a high affinity for binding a fourth ligand, based on comparisons with model cuprous complexes. Nevertheless, clear evidence of anion binding to the reduced protein has been obtained by NMR.[107]

A recent study of azide binding to both oxidized and reduced CuZnSOD using Fourier transform infrared spectroscopy appears to have resolved this apparent dilemma.[108a] The antisymmetric stretching band of N_3^- was observed to shift to higher energy upon coordination to oxidized CuZnSOD, consistent with coordination to copper(II). By contrast, interaction of N_3^- with reduced CuZnSOD or with the copper-free enzyme resulted in a band shift to lower energy similar to that observed upon reaction of N_3^- with free lysine or arginine. It was concluded that N_3^- does not directly coordinate to copper(I) in reduced CuZnSOD but instead binds to the guanidinium group of Arg 143. It is likely that superoxide binds in a similar fashion in both cases, i.e. to copper(II) in oxidized CuZnSOD but to Arg 143 in reduced CuZnSOD. (A recent X-ray crystal structure of the reduced CuZnSOD-azide complex has confirmed that the distance between copper(I) and azide is long.)[108b] The implications of this finding for the mechanism are discussed further in Sec. 5.1.3.

5.1.2.4 The role of Histidine 63

The imidazole ring of His 63 binds solely to Zn^{2+} in the reduced enzyme, whereas it is deprotonated to form imidazolate which bridges Cu^{2+} and Zn^{2+} when the enzyme is oxidized. The His 63-Zn^{2+} moiety clearly plays an important role in the SOD mechanism since removal of zinc from the enzyme makes the SOD activity pH-dependent.[109] A similar pH-dependent decrease in activity was observed in studies of the site-directed mutant His63Ala that lacks the bridging histidine.[110] Studies of the SOD activity of the zinc-deficient enzyme by pulse radiolysis established that the reduction of the copper(II) form of the enzyme by superoxide remained fast and pH-independent. Therefore, it could be concluded that oxidation of the copper(I) form of the enzyme by superoxide had become rate-determining and pH-dependent. These observations suggest that the His 63-Zn^{2+} moiety plays an important role in determining the rate of the oxidation of reduced CuZnSOD by superoxide and its lack of dependence on pH. It was postulated that the coordination of the His 63-Zn^{2+} moiety to copper(II) creates the imidazolate bridge, which prevents strong binding of the hydroperoxy anion to copper and thus facilitates rapid loss of the peroxide product.

5.1.2.5 Thermal stability and pH independence

CuZnSODs share the characteristic of remarkably high thermal stability relative to other intracellular proteins; irreversible deactivation has an onset of greater than 70 °C. Differential scanning calorimetry (DSC) studies of both bovine and human CuZnSOD indicate that unfolding occurs at ca. 80–90 °C.[111,112] An enzyme with increased thermal stability was constructed using site-directed mutagenesis by replacing Cys 6 by Ala and Cys 111 by Ser in the human enzyme.[113] This thermal stability seems to result in large part from reducing the extent of irreversible unfolding. The copper and zinc ions are known to confer thermal stability to the enzyme as the apoenzyme undergoes denaturation approximately 40 °C lower than the holoenzyme.[114]

The rate constant for superoxide dismutation is independent of pH over the pH range of 5–9 while the activity drops at higher pH. A suggestion for this decease in activity has been electrostatic effects resulting from deprotonation of amino acid residue(s) at or near the active site channel.[115,116] The pH independence of the activity is in marked contrast with the pH dependence observed for small amino acid complexes of copper (see Sec. 4.1). It is thus quite striking that the pH independence of CuZnSOD is lost when zinc is removed from the active site (see Sec. 5.1.3.3).

5.1.3 Mechanism of superoxide disproportionation catalyzed by CuZnSOD

5.1.3.1 Overall reaction

The SOD reaction catalyzed by CuZnSOD occurs by a two-step mechanism, just as in the model complexes, with reduction of Cu^{2+} to Cu^{+} by the first O_2^{-} and reoxidation of Cu^{+} to Cu^{2+} by a second O_2^{-}. During the reduction step, the bond between His 63 and the copper ion is broken, and one equivalent of H^{+} per enzyme subunit is required to protonate the imidazolate ring of His 63. Upon reoxidation, two equivalents of H^{+} are eliminated in the form of H_2O_2 and the Cu^{2+}-His63-Zn^{2+} bridge reforms. Under conditions of saturating superoxide concentrations, there is apparently not enough time for the imidazolate bridge to break and reform between successive reactions with superoxide, suggesting that the breaking of the bridge in reduced CuZnSOD is not absolutely required prior to reoxidation by superoxide in the catalytic mechanism.[117] However, it is unlikely that such high concentrations of superoxide are reached in a cell *in vivo* under normal circumstances.

5.1.3.2 Reduction of oxidized CuZnSOD by superoxide

Reduction of oxidized CuZnSOD by O_2^- is believed to occur through an inner sphere electron transfer mechanism, as discussed above in Sec. 5.1.2.3. A likely mechanistic pathway is summarized in Fig. 2. In this mechanism,

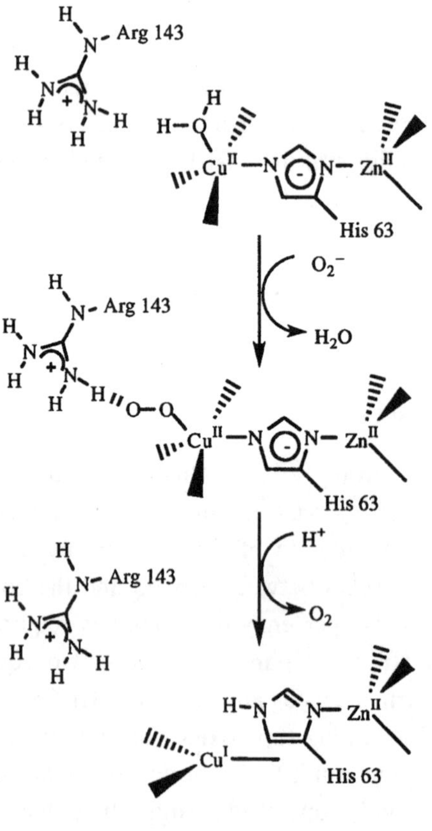

Fig. 2 Inner sphere mechanism for reduction of oxidized CuZnSOD by superoxide.

superoxide is drawn into and maneuvered through the funnel-like channel by a positive electrostatic field. Once at the bottom of the funnel, it binds to the Cu^{2+} ion, hydrogen bonds with Arg 143, and displaces the water that was bound to the copper ion. X-ray crystal structures of either cyanide[118] or azide[119] bound to copper(II) in CuZnSOD support such a configuration for superoxide if it binds to the copper ion prior to electron transfer. Recent results from Quian *et al.*[120] suggest that protonation of the bridging imidazolate of His 63 is very rapid and coupled to the electron transfer, eliminating the interaction between His 63 and copper and thus breaking the Cu^{2+}-Im-Zn^{2+} bridge. Once reduced to Cu^+, the copper ion moves 1 Å away from its initial position toward the remaining three histidine ligands, making room for a proton to attach to His 63, and adopting a trigonal planar ligand geometry. In contrast to the copper site, the zinc site appears to remain invariant during the reaction and His 63 is held in place by coordination to Zn^{2+}, ready to reform the imidazolate bridge upon reoxidation.

5.1.3.3 Oxidation of reduced CuZnSOD by superoxide

The rate of reaction of superoxide with CuZnSOD is strongly dependent on ionic strength,[121] consistent with the dominance of the electrostatic field that attracts superoxide to the metalloenzyme, but is independent of the oxidation state of the copper ion, as evidenced by the identical rate constants for reaction of superoxide with the oxidized and the reduced forms of CuZnSOD, i.e. 2×10^9 $M^{-1}s^{-1}$.[91] Thus, superoxide enters the substrate channel in the reduced enzyme just as it did in the case of the oxidized enzyme and presumably docks in a similar fashion to Arg 143 but without complexing to the copper (see Sec. 5.1.2.3). For superoxide to oxidize the Cu^+ ion, there must be at least one proton intimately associated with the O_2^- as it accepts an electron to become peroxide (see Sec. 2.3). Presumably, the proton on His 63, which is very close according to the crystal structure,[92] plays that role.

If only one proton transfers to the peroxide as it is formed, the product is the hydroperoxide anion, HOO^-. This oxy anion is likely to complex to Cu^{2+}, if it is formed at this point (Reaction 10.55).

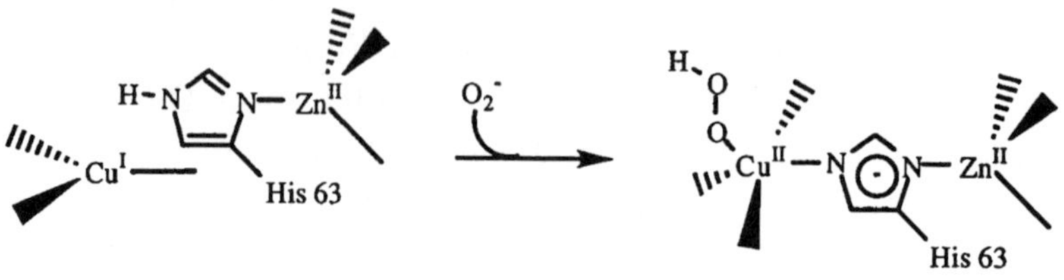

(10.55)

Protonation and loss of hydrogen peroxide would then follow. An alternative possibility is that two protons, one from a water molecule and the other from His 63, transfer simultaneously with a through-space electron transfer from Cu^+, forming hydrogen peroxide directly. This possibility is described in Fig. 3.

When zinc is removed from the zinc binding site of CuZnSOD, the oxidation of the reduced enzyme by superoxide becomes pH-dependent (see Sec. 5.1.2.4 above). One possible explanation is that the strong binding of the zinc-imidazolate moiety to copper(II) in oxidized CuZnSOD weakens the bond to the hydroperoxide anion so that it binds only weakly or not at all (Reaction 10.56). In the absence of the zinc-imidazolate ligand on copper(II), the loss of peroxide may be fast only at low pH.[109]

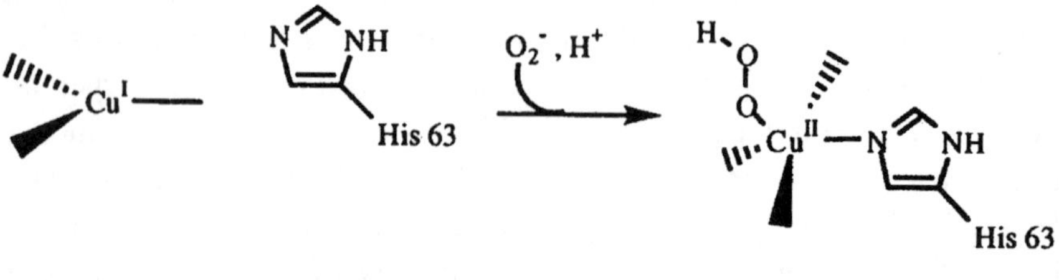

(10.56)

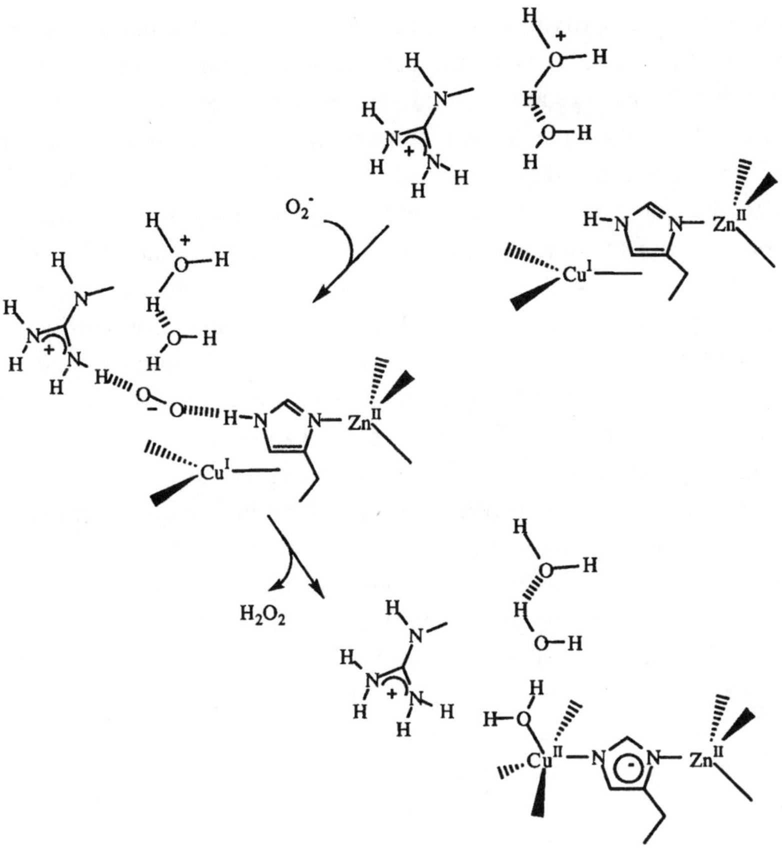

Fig. 3 Outer sphere mechanism for oxidation of reduced CuZnSOD by superoxide. A chain of conserved water molecules has been observed in the active site channel in all of the crystal structures.[91,92]

5.2 Manganese superoxide dismutase

5.2.1 Protein structure: active site and substrate access channel

Manganese superoxide dismutase (MnSOD) occurs with great frequency in both prokaryotic and eukaryotic cells where it exists as monomers, dimers, or tetramers of approximately 26 000 Da subunits. Human mitochondrial MnSOD, for example, is a homotetramer and the *E. coli* enzyme is a homodimer. X-ray crystal structural studies of both oxidized $Mn^{III}SOD$, and reduced $Mn^{II}SOD$ have shown conserved structures and metal binding sites.[122–124] Three histidines, one aspartate, and one solvent water molecule bind to the manganese ion in a nearly trigonal bipyramidal geometry.

The MnSOD subunits fold into two domains, N-terminal and C-terminal, which consist mainly of α-helix. The subunits are held together by polar interactions and hydrogen bonds. Within each subunit, four residues, two histidines from the N-terminal domain (His 28, His 81; numbering according to the *E. coli* enzyme) and a histidine and an aspartate from the C-terminal domain (His 171 and Asp 167), plus a solvent molecule (water or hydroxide) are bound to the manganese ion forming a trigonal bipyramidal geometry. The metal binding site is further stabilized by a highly conserved extended hydrogen bonding network involving Tyr 34 and Gln 146, which are hydrogen bonded to the metal binding ligands H_2O and Asp 167.[125] This proton linkage network is involved in the binding of the ligands to the metal center as well as in proton transfer, both of which are critical to the efficient turnover of the catalytic reaction.

The pathway from bulk solvent to the manganese ion is a channel lined with residues from both subunits. The residues in the channel, many of

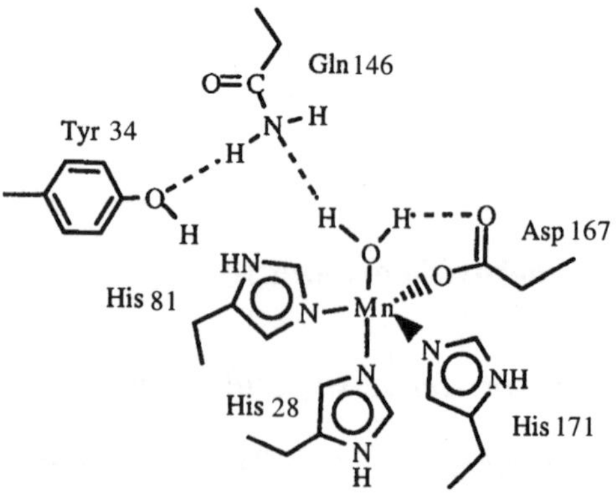

Fig. 4 Active site configuration of MnSOD.

which are either conserved or semi-conserved, may electrostatically facilitate entry of the substrate.

5.2.2 *Mobile metal center in the anion adduct*

Distorted trigonal bipyramid geometry and anion (F^-, N_3^-) binding to the metal center for both the oxidized and reduced forms of the enzyme are indicated from a variety of spectral measurements, including optical absorption, circular dichroism, NMR and EPR.[126] While the X-ray crystal structure of the Mn(III)SOD azide complex shows six atoms around the metal center,[123] the ligand field optical spectra of Mn(III)SOD are characteristic of five-coordination at room temperature (295 K) with a clean conversion to a six-coordinate geometry at low temperature.[127] An X-ray crystal structure of the Mn(III)SOD azide complex revealed that the water ligand was retained and that the largest structural change was the elongation of the metal carboxylate bond from 1.80 Å to 2.25 Å,[123] suggesting that the carboxylate group rather than water may be displaced upon anion binding (see Fig. 5). This mobile five-coordinate metal center may serve to facilitate the release of peroxide through rebinding of the displaced ligand to the metal ion, a key kinetic element in the redox catalysis of the enzyme.[125,127]

5.2.3 *MnSOD mechanism*

5.2.3.1 Reduction of oxidized enzyme by superoxide

The reduction of Mn(III)SOD to Mn(II)SOD by superoxide and the release of O_2 is fast (ca. 10^9 $M^{-1}s^{-1}$). According to the proposed mobile five-coordinate metal center model discussed earlier, superoxide approaches the metal center and binds to the Mn^{3+} ion, and a ligand detaches from the metal ion (see Fig. 5). The detached ligand remains in the outer coordination sphere until it rebinds to the Mn^{2+} ion after dioxygen is released.

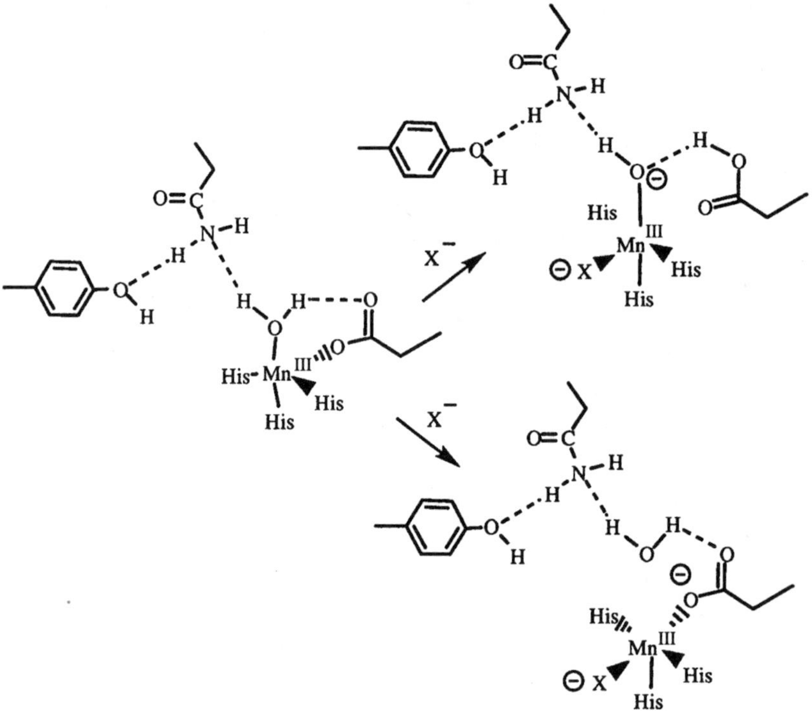

Fig. 5 Anion binding to Mn(III)SOD induces dissociation of either the Asp (upper) or water (lower) ligand from the mobile five-coordinate metal center.

$$(H_2O)Mn^{3+}SOD \xrightarrow{-H_2O} Mn^{3+}SOD \xrightarrow{O_2^-} (O_2^-)Mn^{3+}SOD \xrightarrow{\text{fast}} Mn^{2+}SOD + O_2 \quad (10.57)$$

5.2.3.2 Oxidation of the reduced enzyme by superoxide

A general assumption is that the oxidation of $Mn^{2+}SOD$ by superoxide occurs by an inner sphere pathway, but recent studies of manganese-containing synthetic SODs suggest that an outer sphere pathway should be considered as well. The mechanism widely assumed for the oxidation of

Mn^{2+}SOD to Mn^{3+}SOD by superoxide involves inner sphere electron transfer from the metal ion to the superoxide ligand coupled with proton transfer to the peroxide product as it is formed. In accordance with this mechanism is the observation that Mn^{2+}SOD is known to be capable of binding simple anions similar to superoxide. However, FeSOD, which is extremely similar to MnSOD in its structure and many of its other properties, does not bind such anions in its reduced Fe^{2+} state (see discussion below). If the SOD mechanism is the same for both MnSOD and FeSOD, then an outer sphere mechanism should be considered for both. Both inner and outer sphere mechanisms for oxidation of Mn^{2+}SOD to Mn^{3+}SOD by superoxide are therefore considered in this section.

Inner sphere electron transfer from Mn^{2+} to superoxide and transfer of a proton to the peroxo group would form a Mn^{3+}-hydroperoxo complex. An additional proton is needed to release H_2O_2 from the Mn^{3+}-hydroperoxo intermediate, reforming the oxidized enzyme. Among these processes, O_2^- binding to Mn^{2+}SOD is believed to be diffusion controlled, and inner sphere electron transfer within $Mn^{2+}(O_2^-)$ would be expected to be very rapid. Therefore, if this mechanism is correct, the rate-determining step is probably release of H_2O_2 from the Mn^{3+}-hydroperoxo intermediate (Reaction 10.58).

$$Mn^{2+}SOD + O_2^- \rightarrow (O_2^-)Mn^{2+}SOD \xrightarrow{H^+} (HOO^-)Mn^{3+}SOD \xrightarrow{H^+} Mn^{3+}SOD + H_2O_2 \tag{10.58}$$

Efficient proton transfer is critical to the release of the peroxo group and regeneration of the active oxidized enzyme. The rate-determining step in the overall catalytic reaction will therefore depend on the efficiency of delivery of protons from the enzyme. The protons necessary for oxidation of Mn^{2+}SOD by O_2^- may come from the hydrogen bond network enveloping the metal binding site, involving Tyr 34, Gln 146 and two metal bound ligands (solvent H_2O and Asp 167) (see Fig. 6). This mechanism involves both the mobile five-coordinate metal center and the proton network at the active site (see Fig. 6). Tyr 34 is proposed to play an important role in this proton network by maintaining the appropriate protonation state of the

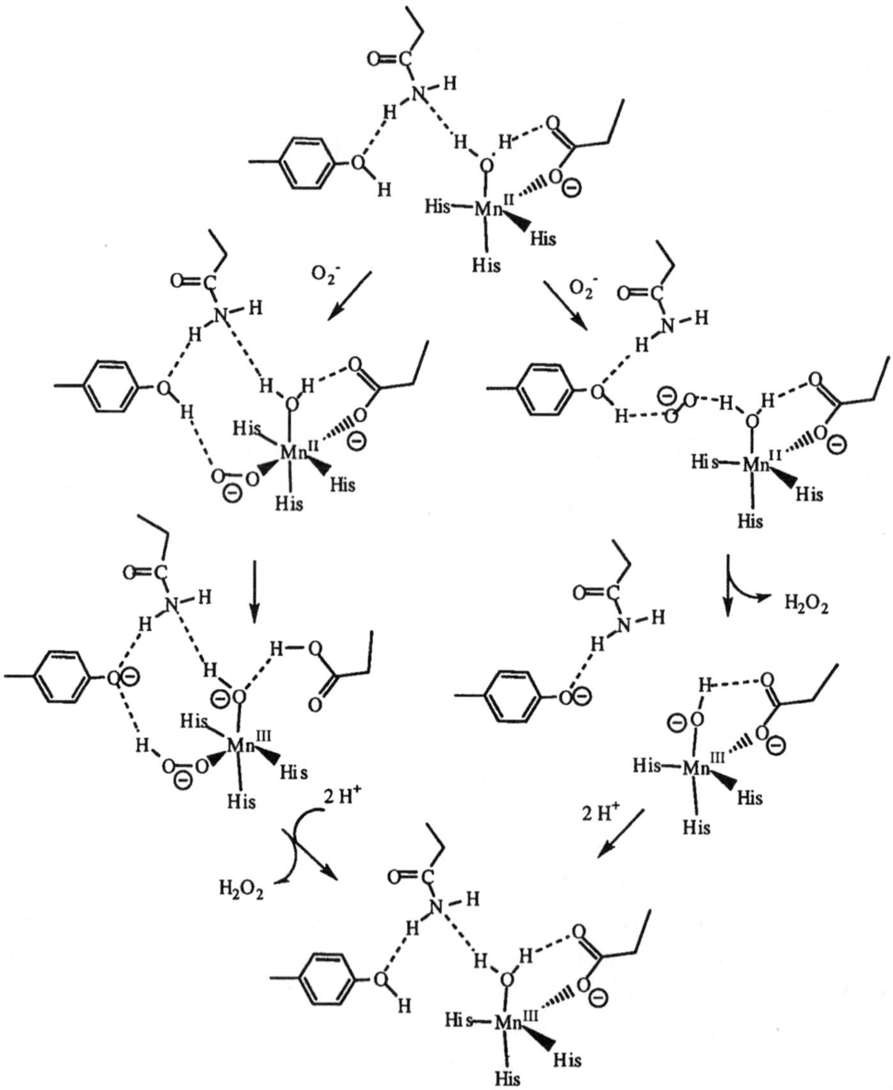

Fig. 6 Inner (left) and outer (right) sphere mechanism for oxidation of Mn(II)SOD by superoxide.

network and the proper coordination state and geometry of the active metal center. Tyr 34 is strictly conserved in all Mn and FeSODs. Mutation of Tyr to Phe greatly reduces the activity of MnSOD.[123,125,127]

Outer sphere electron transfer from superoxide to Mn^{2+} through the metal bound H_2O ligand (see Fig 5) would avoid formation of a Mn^{3+}-hydroperoxide. This alternative mechanism looks particularly attractive in light of the recent demonstration of such a mechanism in some pentaazamacrocyclic complexes of manganese[22] (see Sec. 4.2.1).

5.2.3.4 Inactive form of MnSOD

The disappearance of superoxide radical in the presence of MnSOD is first order at relatively low superoxide-to-enzyme concentrations while, at sufficiently high concentrations of superoxide-to-enzyme, a multiphase process is observed. This process involves a rapid initial disappearance of some percentage of O_2^-, followed by a zero order decay of the substrate and then a final abrupt depletion of the remaining O_2^-. A mechanism involving conversion of MnSOD to an inactive form is consistent with the observed kinetic behavior, where the zero order decay represents buildup of the inactive enzyme. Under conditions where the concentration of O_2^- is sufficiently low to avoid buildup of the inactive form, catalytic disproportionation of superoxide by the active MnSOD is diffusion-controlled ($k \sim 10^9$ $M^{-1}s^{-1}$).

Two different mechanisms have been suggested previously for the formation and nature of the inactive form of the enzyme. One mechanism postulates that a conformational change of the metal binding site leads to inactive MnSOD.[128,129] Another suggestion is that the bound peroxy ligand, O_2^{2-}, isomerizes from an end-on bound peroxy species to a triangular side-on bound peroxy.[72,130] We suggest here a third possibility: If the reoxidation of Mn^{2+}SOD occurs normally by an outer sphere oxidation via coordinated water (see Fig. 6), the inactive complex could be one in which a hydroperoxide ligand has entered the first coordination sphere of the Mn^{3+} ion and from which protonation and hydrogen peroxide dissociation occur only slowly. Thus, if proton delivery from amino acid residues or water in the active site to the bound hydroperoxy is not in the proper steric alignment, the proton transfer could be slow giving rise to a less active or inactive form of the enzyme. In this regard, it is interesting to note that the pentaazamacrocyclic complexes of manganese that are SOD mimics release hydrogen peroxide only when they are reduced by superoxide (see Reaction 10.44).

Of interest is a recent study[130] of the dismutation mechanism of human MnSOD, where isomerization from the inhibited form/dissociation of the bound peroxide was reported to be 30 times faster than the corresponding process in *T. thermophilus* MnSOD. It is tempting to speculate that in MnSODs, the entrance of hydroperoxide into the first coordination sphere and subsequent dissociation of peroxide may be processes that are very sensitive to subtle changes in the environment. Whether the enzyme enters into the inactive cycle will depend significantly on the enzyme: substrate ratio.

5.3 *Iron superoxide dismutase*

5.3.1 *Active metal center and metal binding specificity*

MnSOD and FeSOD share highly homologous sequences and almost identical protein structure and metal binding sites. However, with a few notable exceptions, the catalytic activity is greatly decreased if Mn is substituted into FeSOD or the reverse. Close examination of the high resolution crystal structures reveal important differences in the hydrogen bond network around the metal centers in these enzymes. The only significant change within 10 Å of the active site of FeSOD occurs at the position corresponding to Glu 146 in MnSOD, where histidine or alanine is substituted for the highly conserved glutamine. In FeSOD, the histidine or alanine residue in that position does not interact with Fe bound solvent while in MnSOD, a glutamine residue at this position forms a hydrogen bond with Mn bound solvent. This rather subtle difference may be critical to the metal selection.[131,132]

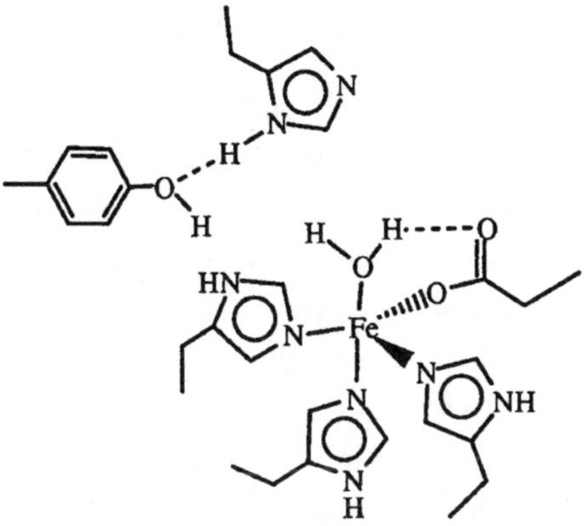

Fig. 7 Active site configuration of FeSOD. (Note that some FeSODs have Ala in place of the non-liganding His.)

Although most MnSODs and FeSODs are metal ion-specific, some SODs from bacteria such as *P. shermanii, B. thetaiotaomicron and S. mutans,* are active with either Mn or Fe. These SODs have been termed "cambialistic". They can be isolated from cell culture containing either Mn or Fe, or both, depending on the availability of the metal ions in the medium.[133–135]

5.3.2 FeSOD mechanism

MnSOD and FeSOD are very similar in many respects, but they differ in one important property which may provide useful insights into the mechanism of reaction of the reduced enzymes with superoxide. Unlike Mn^{2+}SOD, Fe^{2+}SOD does not appear to bind anions in its inner coordination sphere[126] and therefore is unlikely to react by the inner sphere mechanism described in Fig. 6. This failure of Fe^{2+}SOD to bind anions, combined with the slow rate of dissociation expected for hydroperoxide complexes of Mn^{3+} or Fe^{3+}, make the outer sphere pathway in Fig. 6 a strong possibility.

5.4 Reactions of SODs with hydrogen peroxide

Reduced CuZnSOD and FeSOD are oxidized slowly by hydrogen peroxide in a Fenton-type reaction and they are inactivated in the process by the hydroxyl radical that is generated at the active site. In the case of reduced CuZnSOD, the reaction goes exclusively via HO_2^-.[55,136] In contrast, MnSOD does not undergo Fenton-type chemistry with hydrogen peroxide. The deleterious effects of OH radical production makes it tempting to speculate that the presence of MnSOD rather than CuZnSOD or FeSOD in mitochondria of most eukaryotic cells may be advantageous since that organelle is likely to produce the highest concentrations of hydrogen peroxide in the cell as a consequence of respiration. The unreactive behavior with peroxide is certainly an essential part of any synthetic SOD that is to be used as a drug and is a property of the pentaazamacrocylic complexes of manganese that are being developed as SOD mimics.[23]

6. Conclusion

In this review, we have focused much of our discussion on the mechanistic details of how the native enzymes function, and how mechanistic developments/insights with synthetic small molecule complexes possessing SOD activity have influenced our understanding of the electron transfer processes involved with the natural enzymes. A few overriding themes have emerged. Clearly, the SOD enzymes operate at near diffusion-controlled rates and to achieve such catalytic turnover activity, several important physical principles must be operative. Such fast electron transfer processes require a role for protons; i.e. proton-coupled electron transfer ("H atom transfer") solves the dilemma of charge separation developing in the transition state for the electron transfer step. Additionally, outer sphere electron transfer is most likely an important pathway for manganese and iron dismutases. This situation arises because the ligand exchange rates on these two ions in water never exceed $\sim 10^7\ s^{-1}$; consequently catalytic rate constants of 10^9 require more subtle mechanistic insights. In contrast, copper complexes can achieve diffusion-controlled ($> 10^9$) exchange rates in water;

thus inner sphere electron transfer processes are more likely to be operative in the copper-zinc enzymes.

Recent studies have continued to expand our understanding of the mechanism of action of this important class of redox active enzymes — the superoxide dismutases — which have been critical in the successful adaptation of life on this planet to an oxygen-based metabolism. The design of SOD mimic drugs (synthetic model compounds) that incorporate this superoxide dismutase catalytic activity, and are capable of functioning *in vivo*, offers clear potential benefits in the control of diseases; ranging from the control of neurodegenerative conditions, such as Parkinson's or Alzheimer's disease, to cancer.

References

1. Fridovich, I., *J Biol Chem* 1997, **272:** 18515–18517.
2. Scandalios, J.G. (ed.) *Oxidative Stress and the Molecular Biology of Antioxidant Defenses.* Cold Spring Harbor Laboratory Press, Plainview, NY, 1997.
3. Valentine, J.S. *et al. Curr. Op. Chem. Biol.* 1998, **2:** 253–262.
4. Halliwell, B. and Gutteridge, J.M.C. *Free Radicals in Biology and Medicine,* 2nd Ed. Oxford University Press, New York, 1989.
5. Keyer, K. and Imlay, J.A., *Proc. Natl. Acad. Sci. USA* 1996, **93:** 13635–13640.
6. Aust, S.D. *Toxicol. Lett.* 1995, **82:** 941–944.
7. Thomas, C.E. and Aust, S.D. *Arch. Biochem. Biophys.* 1986, **248:** 684–689.
8. Aikens, J. and Dix, T.A. *Arch. Biochem. Biophys.* 1993, **305:** 516–525.
9. Aikens, J. and Dix, T.A. *J. Biol. Chem.* 1991, **266:** 15091–15098.
10. Dix, T.A. *et al. Biochemistry* 1996, **35:** 4578–4583.
11. Reaume, A.G. *et al. Nat. Genet.* 1996, **13:** 43–47.
12. Huang, T.T. *et al. Arch. Biochem. Biophys.* 1997, **344:** 424–432.
13. Longo, V.D., Gralla, E.B. and Valentine, J.S. *J. Biol. Chem.* 1996, **271:** 12275–12280.
14. Li, Y. *et al. Nat. Genet.* 1995, **11:** 376–381.
15. Lebovitz, R.M. *et al. Proc. Natl. Acad. Sci. USA* 1996, **93:** 9782–9787.
16. Irani, K. *et al. Science* 1997, **275:** 1649–1652.
17. Weiss, R.H. *et al. J. Biol. Chem.* 1996, **271:** 26149–26156.

18. Peretz, D. Solomon, D., Weinraub, D. and Faraggi, M. *Int. J. Radiat. Biol.* 1982, **42:** 449–456.
19. Batinić-Haberle, I., Liochev, S.I., Spasojević, I. and Fridovich, I. *Arch. Biochem. Biophys.* 1997, **343:** 225–233.
20. Weinraub, D., Levy, P. and Faraggi, M. *Int. J. Radiat. Biol.* 1986, **50:** 649–658.
21. Riley, D.P. and Weiss, R.H. *J. Am. Chem. Soc.* 1994, **116:** 387–388.
22. Riley, D.P., Lennon, P.J., Neumann, W.L. and Weiss, R.H. *J. Am. Chem. Soc.* 1997, **119:** 6522–6528.
23. Riley, D.P. and Weiss, R.H. *CATTECH* 1997, **1:** 41–49.
24. Zhang, D.-L. *et al. Inorg. Chem.* 1998, **37:** 956–963.
25. Pinto, A.L., Hellinga, H.W. and Caradonna, J.P., *Proc. Natl. Acad. Sci. USA* 1997, **94:** 5562–5567.
26. Afanas'ev, I.B. *Superoxide Ion: Chemistry and Biological Implications.* CRC Press Inc., Boca Raton, FL, 1991 Vol. I 1989, Vol. II 1991.
27. Bielski, B.H.J. and Cabelli, D.E. *Int. J. Radiat. Biol.* 1991, **59:** 291–319.
28. Sawyer, D.T. *Oxygen Chemistry.* Oxford University Press, New York, 1991.
29. Bielski, B.H.J. *Photochem. Photobiol.* 1978, **28:** 645–649.
30. Bielski, B.H.J., Cabelli, D.E., Arudi, R.L. and Ross, A.B. *J. Phys. Chem. Ref. Data* 1985, **14:** 1041–1100.
31. Wardman, P.J. *Phys. Chem. Ref. Data* 1989, **18:** 1637–1755.
32. Stanbury, D.M. *Adv. Inorg. Chem.* 1989, **33:** 69–138.
33. Rao, P.S. and Hayon, E. *J. Phys. Chem.* 1975, **79:** 109–115.
34. Sawyer, D.T. and Valentine, J.S. *Acc. Chem. Res.* 1981, **14:** 393–400.
35. Friedman, H.L. and Newton, M.D.J. *Electroanal. Chem.* 1986, **204:** 21–29.
36. Hudis, J. and Dodson, R.W. *J. Am. Chem. Soc.* 1956, **78:** 911–913.
37. Buxton, G.V., Greenstock, C.L., Helman, W.P. and Ross, A.B. *J. Phys. Chem. Ref. Data* 1988, **17:** 513–588.
38. Riley, D.P., Rivers, W.J. and Weiss, R.H. *Analyt. Biochemistry* 1991, **196:** 344–349.
39. Klug-Roth, D. and Rabani, J. *J. Phys. Chem.* 1976, **80:** 588–591.
40. Brigelius, R. *et al. Hoppe-Seyler's Z. Physiol. Chem.* 1975, **356:** 739–745.
41. Brigelius, R. *et al. FEBS Lett.* 1974, **47**: 72–75.
42. Weinstein, J. and Bielski, B.H.J. *J. Am. Chem. Soc.* 1980, **102:** 4916–4919.
43. Cabelli, D.E., Bielski, B.H.J. and Holcman, J. *J. Am. Chem. Soc.* 1987, **109:** 3665–3669.
44. Graham, D.R., Marshall, L.E., Reich, K.A. and Sigman, D.A. *J. Am. Chem. Soc.* 1980, **102:** 5419–5421.

45. Que, B.G., Downey, K.M. and So, A.G. *Biochemistry* 1980, **19:** 5987–5991.
46. Marshall, L.E., Graham, D.R., Reich, K.A. and Sigman, D.A. *Biochemistry* 1981, **20:** 244–250.
47. Sigman, D.S. *Acc. Chem. Res.* 1986, **19:** 180–186.
48. Goldstein, S. and Czapski, G. in *Superoxide and Superoxide Dismutase in Chemistry, Biology and Medicine* (ed.) Rotilio, G. Elsevier Science Pub., New York, 1985, pp. 64–66.
49. Rabani, J., Klug-Roth, D. and Lilie, J. *J. Phys. Chem.* 1973, **77:** 1169–1175.
50. Bjergebakke, E., Sehested, K. and Rasmussen, O.L. *Radiat. Res.* 1976, **66:** 433–442.
51. Kozlov, Y.N. and Berdnikov, V.M. *Russ. J. Phys. Chem.* 1973, **47:** 338–340.
52. Tait, A.M., Hoffman, M.Z. and Hayon, E. *Inorg. Chem.* 1976, **15:** 934–939.
53. Goldstein, S. and Czapski, G. *Inorg. Chem.* 1985, **24:** 1087–1092.
54. Goldstein, S. and Czapski, G. *J. Am. Chem. Soc.* 1983, **105:** 7276–7280.
55. Cabelli, D.E., Allen, D., Bielski, B.H.J. and Holcman, J. *J. Biol. Chem.* 1989, **264:** 9967–9971.
56. Baral, S., Lume-Pereira, C., Janata, E. and Henglein, A. *J. Phys. Chem.* 1986, **90:** 6025–6028.
57. Cabelli, D.E. and Bielski, B.H.J. *J. Phys. Chem.* 1984, **88:** 3111–3115.
58. Cabelli, D.E. and Bielski, B.H.J. *J. Phys. Chem.* 1984, **88:** 6291–6294.
59. Koppenol, W.H. *et al. Arch. Biochem. Biophys.* 1986, **251:** 594–599.
60. Pick-Kaplan, M. and Rabani, J. *J. Phys. Chem.* 1976, **80:** 1840–1843.
61. Lati, J. and Meyerstein, D. *J. Chem. Soc., Dalton Trans.* 1978, 1105–1118.
62. Stein, J. *et al. Inorg. Chem.* 1979, **18:** 3511–3519.
63. Rush, J.D. and Maskos, Z. *Inorg. Chem.* 1990, **29:** 897–905.
64. Jacobsen, F., Holcman, J. and Sehested, K. *J. Phys. Chem. A* 1997, **101:** 1324–1328.
65. Cabelli, D.E. in *Photochemistry and Radiation Chemistry: Complementary Methods for Electron Transfer Studies* (eds.) Wishart, J.F. and Nocera, D.G. American Chemical Society, Washington D.C., 1998, pp. 247–262.
66. Faulkner, K.M., Stevens, R.D. and Fridovich, I. *Arch. Biochem. Biophys.* 1994, **310:** 341–346.
67. Darr, D., Zarilla, K.A. and Fridovich, I. *Arch. Biochem. Biophys.* 1987, **258:** 351–355.
68. Rush, J.D., Maskos, Z. and Koppenol, W.H. *Arch. Biochem. Biophys.* 1991, **289:** 97–102.
69. Gray, B. and Carmichael, A.J. *Biochem. J.* 1992, **281:** 795–802.

70. Baudry, M. *et al. Biochem. Biophys. Res. Comm.*, 1993, **192:** 964–968.
71. Götz, F. and Lengfelder, E. in *Oxy-Radicals and Their Scavenger Systems, Molecular Aspects* (eds.) Cohen, G. and Greenwald, R.A., Elsevier Biomedical, New York, 1983, Vol. 1, pp. 228–233.
72. Bull, C., Niederhoffer, E.C., Yoshida, T. and Fee, J.A. *J. Am. Chem. Soc.* 1991, **113:** 4069–4076.
73. Hoganson, C.W. and Babcock, G.T. *Science* 1997, **277:** 1953–1956.
74. Caudle, M.T. and Pecoraro, V.L. *J. Am. Chem. Soc.* 1997, **119:** 3415–3416.
75. Barb, W.G., Baxendale, J.H., George P. and Hargrave, K.R. *Trans. Faraday Soc.* 1951, **47:** 462–500.
76. Barb, W.G., Baxendale, J.H., George, P. and Hargrave, K.R. *Trans. Faraday Soc.* 1951, **47:** 591–616.
77. Rush, J.D. and Bielski, B.H.J. *J. Phys. Chem.* 1985, **89:** 5062–5066.
78. Jayson, G.G., Parsons, B.J. and Swallow, A.J. *J. Chem. Soc., Faraday Trans. I* 1972, **68:** 2053–2058.
79. Jayson, G.G., Parsons, B.J. and Swallow, A.J. *J. Chem. Soc., Faraday Trans. I* 1973, **69:** 1079–1089.
80. Jayson, G.G. Parsons, B.J. and Swallow, A.J. *J. Chem. Soc., Faraday Trans. I* 1973, **69:** 236–242.
81. Bull, C., McClune, G.J. and Fee, J.A. *J. Am. Chem. Soc.* 1983, **105:** 5290–5300.
82. Pasternack, R.F. and Halliwell, B. *J. Am. Chem. Soc.* 1979, **101:** 1026–1033.
83. Weiss, R.H. *et al. J. Biol. Chem.* 1993, **268:** 23049–23054.
84. Nagano, T., Hirano, T. and Hirobe, M. *J. Biol. Chem.* 1989, **264:** 9243–9249.
85. Brouwer, M. *et al. Biochemistry* 1997, **36:** 13381–13388.
86. Kanematsu, S. and Asada, K. *Plant Cell Physiol.* 1989, **30:** 381–391.
87. Battistoni, A. and Rotilio, G. *FEBS Lett.* 1995, **374:** 199–202.
88. Pesce, A. *et al. J. Mol. Biol.* 1997, **274:** 408–420.
89. Bourne Y. *et al. Proc. Natl. Acad. Sci. USA* 1996, **93:** 12774–12779.
90. Bordo, D., Djinovic, K. and Bolognesi, M. *J. Mol. Biol.* 1994, **238:** 366–386.
91. Bertini, I., Mangani, S. and Viezzoli, M.S. *Adv. Inorg. Chem.* 1998, **45:** 127– 250.
92. (a) Ogihara, N.L. *et al. Biochemistry* 1996, **35:** 2316–2321; (b) Hart, P.J., Balbirnie, M.M., Ogihara, N.L., Nersissian, A.M., Weiss, M.S., Valentine, J.S., Eisenberg, D. *Biochemistry* 1999, **38:** 2167–2178.

93. Tainer, J.A. *et al. J. Mol. Biol.* 1982, **160:** 181–217.
94. Getzoff, E.D. *et al. Nature* 1983, **306:** 287–290.
95. Tainer, J.A., Getzoff, E.D., Richardson, J.S. and Richardson, D.C. *Nature* 1983, **306:** 284–287.
96. Lyons, T.J. *et al. Proc. Natl. Acad. Sci. USA* 1996, **93:** 12240–12244.
97. Sines, J.J., Allison, S.A. and McCammon, J.A. *Biochemistry* 1990, **29:** 9403–9412.
98. Fisher, C.L. *et al. Proteins — Structure Function and Genetics* 1994, **19:** 24–34.
99. Desideri, A. *et al. J. Mol. Biol* 1992, **223:** 337–342.
100. Beyer, W.F., Jr., Fridovich, I., Mullenbach, G.T. and Hallewell, R. *J. Biol. Chem.* 1987, **262**: 11182–11187.
101. Cocco, D. *et al. FEBS Lett.* 1982, **150:** 303–306.
102. Getzoff, E.D. *et al. Nature* 1992, **358:** 347–351.
103. Banci, L. *et al. Magnetic Resonance in Chemistry* 1997, **35:** 845–853.
104. Banci, L. *et al. Eur. J. Biochem.* 1995, **234:** 855–860.
105. Banci, L. *et al. J. Biol. Inorg. Chem.* 1997, **2:** 295–301.
106. Polticelli, F. *et al. Protein Sci.* 1996, **5:** 248–253.
107. Mota de Freitas, D., Ming, L.J., Ramasamy, R. and Valentine, J.S. *Inorg. Chem.* 1990, **29:** 3512–3518.
108. (a) Leone, M. *et al. Biochemistry* 1998, **37:** 4459–4464; (b) Ferraroni, M. *et al. J. Biol. Inorg. Chem.* 1998, **3:** 411–422.
109. Ellerby, L.M., Cabelli, D.E., Graden, J.A. and Valentine, J.S. *J. Am. Chem. Soc.* 1996, **118:** 6556–6561.
110. Graden, J.A., Ellerby, L.M., Roe, J.A. and Valentine, J.S. *J. Am. Chem. Soc.* 1994, **116:** 9743–9744.
111. Roe, J.A. *et al. Biochemistry* 1988, **27:** 950–958.
112. Lepock, J.R. *et al. Arch. Biochem. Biophys.* 1985, **241:** 243–251.
113. McRee, D.E. *et al. J. Biol. Chem.* 1990, **265:** 14234–14241.
114. Lepock, J.R., Frey, H.E. and Hallewell, R.A. *J. Biol. Chem.* 1990, **265:** 21612–21618.
115. Banci, L., Bertini, I., Luchinat, C. and Viezolli, M.S. *Inorg. Chem.* 1993, **32:** 1403–1406.
116. Polticelli, F. *et al. Arch. Biochem. Biophys.* 1995, **321:** 123–126.
117. Fee, J.A. and Bull, C. *J. Biol. Chem.* 1986, **261:** 13000–13005.
118. Djinovic-Carugo, K. *et al. FEBS Lett.* 1994, **349:** 93–98.
119. Djinovic-Carugo, K. *et al. J. Mol. Biol.* 1994, **240:** 179–183.

120. Qian, W., Luo, Q.H. and Shen, M.C. *Bioelectrochem. Bioenerg.* 1996, **39:** 291–294.
121. Argese, E. *et al. Biochemistry* 1987, **26:** 3224–3228.
122. Ludwig, M.L., Metzger, A.L., Pattridge, K.A. and Stallings, W.C. *J. Mol. Biol.* 1991, **219:** 335–358.
123. Lah, M.S. *et al. Biochemistry* 1995, **34:** 1646–1660.
124. Stallings, W.C. *et al. Free Radical Res. Commun.* 1991, **12:** 259–268.
125. Whittaker, M.M. and Whittaker, J.W. *Biochemistry* 1997, **36:** 8923–8931.
126. Whittaker, J.W. and Whittaker, M.M. *J. Am. Chem. Soc.* 1991, **113:** 5528–5540.
127. Whittaker, M.M. and Whittaker, J.W. *Biochemistry* 1996, **35:** 6762–6770.
128. Pick, M., Rabani, J., Yost, F. and Fridovich, I. *J. Am. Chem. Soc.* 1974, **96:** 7329–7333.
129. McAdam, M.E., Fox, R.A., Lavelle, F. and Fielden, E.M. *Biochem. J.* 1977, **165:** 71–79.
130. Hsu, J.-L. *et al. J. Biol. Chem.* 1996, **271:** 17687–17691.
131. Lim, J.-H. *et al. J. Mol. Biol.* 1997, **270:** 259–274.
132. Cooper, J.B. *et al. J. Mol. Biol.* 1995, **246:** 531–544.
133. Sehn, A.P. and Meier, B. *Biochem. J.* 1994, **304:** 803–808.
134. Meier, B. *et al. Biochem. J.* 1995, **310:** 945–950.
135. Matsumoto, T. *et al. Biochemistry* 1991, **30:** 3210–3216.
136. Fuchs, H.J.R. and Borders, C.L., Jr. *Biochem. Biophys. Res. Comm.* 1983, **116:** 1107–1113.

Chapter 11

Modeling the Oxygen-Evolving Complex in Photosystem II

Julian Limburg, Gary W. Brudvig & Robert H. Crabtree

Department of Chemistry, Yale University,
P.O. Box 208107, New Haven,
CT 06520-8107

1. Introduction

The oxygen-evolving complex (OEC) of photosystem II (PS II) has been the subject of intense study for many years. Even so, both the mechanism of photosynthetic water oxidation, and the precise structure of the OEC remain unknown.

In this chapter, we compare the reactivity of the OEC with that of the relevant inorganic chemistry. Rüttinger and Dismukes[1] have published a very thorough review of water oxidation by synthetic manganese and ruthenium complexes, so we will not include a detailed survey of the field, including only a discussion of the catalytic systems with an emphasis on proposed active species. The chapter is divided into four sections. Section 1 contains a brief discussion of the protein. Section 2 summarizes spectroscopic and chemical reactivity studies that have been carried out on the OEC. Section 3 covers the major proposals on the mechanism of water oxidation. Finally, section 4 contains a discussion of functional modeling, and the current strategies for the design of a functional model.

PS II (Fig. 11.1) is a multicomponent membrane bound protein complex that uses light energy to oxidize water and reduce plastoquinone. This process also creates a transmembrane proton gradient, which is then

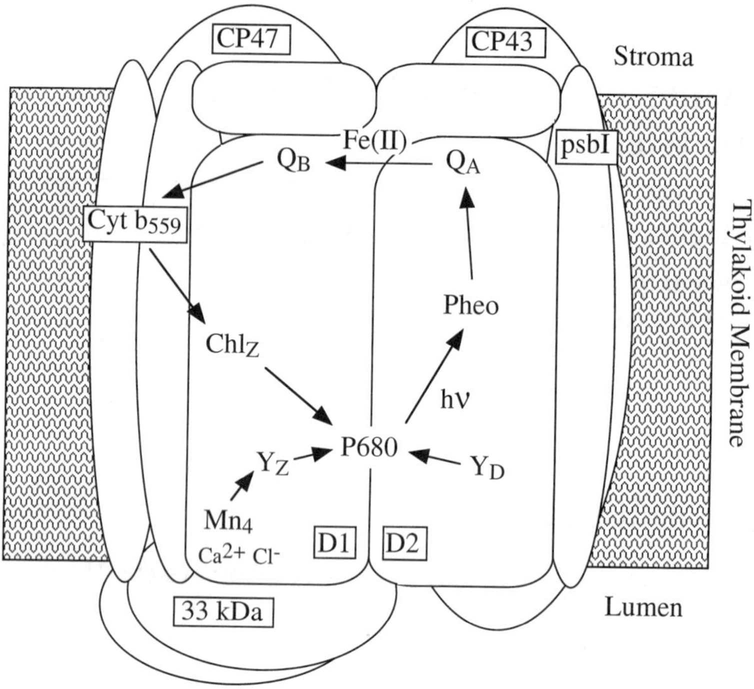

Fig. 11.1 A schematic representation of photosystem II. The structure is based on analogy to the bacterial reaction center whose crystal structure is known. The arrows denote pathways for electron transfer. Abbreviations: Y, Tyrosine; P680, special chlorophyll that functions as the primary electron donor; Q, plastoquinone; Cyt , cytochrome; Chl, chlorophyll; Pheo, pheophytin.

harnessed in the synthesis of the ATP subsequently used in carbon fixation. As a direct result of this light-driven reaction, water is oxidized to dioxygen[2–5] at the OEC which consists of a tetranuclear manganese cluster and an associated redox-active tyrosine, Y_Z. PS II contains a chlorophyll complex, P680, that can undergo a light-induced charge separation, reducing the plastoquinone Q_A, which in turn reduces Q_B. $P680^+$ is then reduced by Y_Z, to form a neutral tyrosyl radical $Y_Z\bullet$, and finally $Y_Z\bullet$ is reduced by the manganese cluster. After four such excitations, dioxygen is released. The four-photon cycle was first observed by Joliot[6] in flash experiments, which led to Kok's[7] proposal of the currently accepted store-, or S-state cycle (Fig. 11.2). Each advancement in S-state has since been correlated with an oxidation state change in the OEC.

In 1937, Pirson[8] was first to show, in nutritional studies on algae, that manganese was required for oxygen evolution, but it was not until 1970 that the stoichiometry of manganese per PS II was determined by Cheniae and Martin[9] who studied the release of manganese ions from thylakoid membranes. Major advances in our understanding of this reaction followed the isolation of PS II-enriched membranes from chloroplasts and, later, more highly purified oxygen-evolving PS II preparations,[10,11] allowing for the detailed spectroscopic and biochemical studies that we have today.

Ca^{2+} and Cl^- ions are both required for oxygen-evolving activity.[3] Cl^- can be replaced by Br^-, giving samples that have lower oxygen-evolving activity.[12] Calcium[13] can be functionally replaced by strontium[14–16] and vanadyl,[17] again with reduced activity, and non-functionally by a number of other cations, such as dysprosium.[18] Although the precise function of each of these ions is as yet unknown, there are many proposals. Cl^- has been proposed to modify the electrostatics of the OEC, thereby tuning the redox potential. Alternatively, it may be an exchangeable ligand which is replaced by water in one of the precursor states to water oxidation. The second case requires that Cl^- be ligated to the manganese cluster but, while this seems likely, it has yet to be established. Ca^{2+} might be required to ligate a substrate water or it may have a structural role.

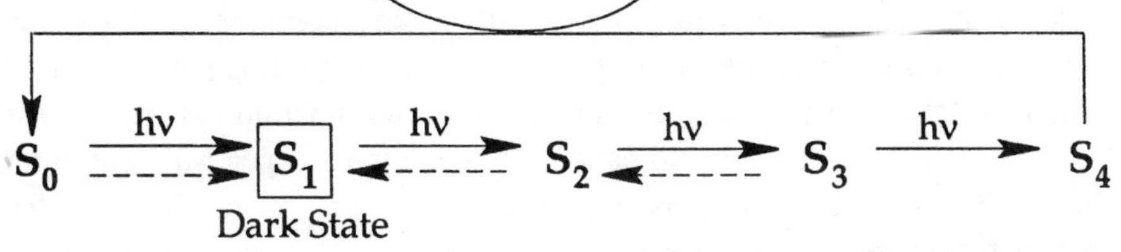

Fig. 11.2 The S-state cycle proposed, Kok *et al.* (Ref. 7). The solid arrows represent light-driven reactions and the dashed arrows represent dark reactions.

If PS II is allowed to come to equilibrium in the dark, the OEC is mainly in the S_1 state, the dark-stable state.[7] Under physiological conditions, the S_2 and S_3 states are unstable and are reduced to S_1 in the dark. The S_4 state releases dioxygen and is rapidly reduced to S_0, which itself is oxidized to S_1 in the dark.[19] At low temperatures (< 200 K), the reduction of Q_B is inhibited, and under such conditions the S_2 state is stable to reduction and can be formed in high yields by illumination.[20] This method has been used extensively for the spectroscopic studies of the S_2 state. The S_0 and S_3 states can also be prepared in good yield by using a combination of chemical treatments and/or flash illumination.[21] Once generated, these states can be stabilized by cooling the sample below 200 K.

2. The OEC

While the precise structure of the manganese tetramer is not known, the X-ray absorption and EPR studies have led to some fairly detailed proposals.[21–25] Figure 11.3 shows examples that are consistent with X-ray data.[2] The dimer-of-dimers structure shown in Fig. 11.4 is favored as best accounting for X-ray data.[23]

2.1 X-ray absorption studies[2]

The Mn-extended x-ray absorption fine structure (Mn-EXAFS) data from the S_1 state exhibit peaks from three shells of back scatterers at 1.7–1.9 Å, 2.7 Å and 3.3 Å.[23,26–29] The 2.7 Å peak is attributed to a Mn back scatterer, and plausibly modeled with anywhere between 1 to 1.5 Mn back scatterers per Mn. Comparison with model complexes shows that this distance is characteristic of di-μ-oxo-bridged manganese pairs. The first shell peak at 1.7–1.9 Å is fitted to an inner coordination sphere of O/N ligands, with an average coordination number of 6 ± 1 per Mn. The third shell at 3.3 Å is thought to have contributions from Mn or Ca^{2+}, but the precise fitting remains controversial. Latimer *et al.*[30] found that replacing Ca^{2+} with Sr^{2+} caused the intensity of the 3.3 Å peak to increase. Penner-Hahn *et al.*[31,32] were unable to reproduce this result with substitution of Ca^{2+} by either Sr^{2+} or Dy^{3+}.

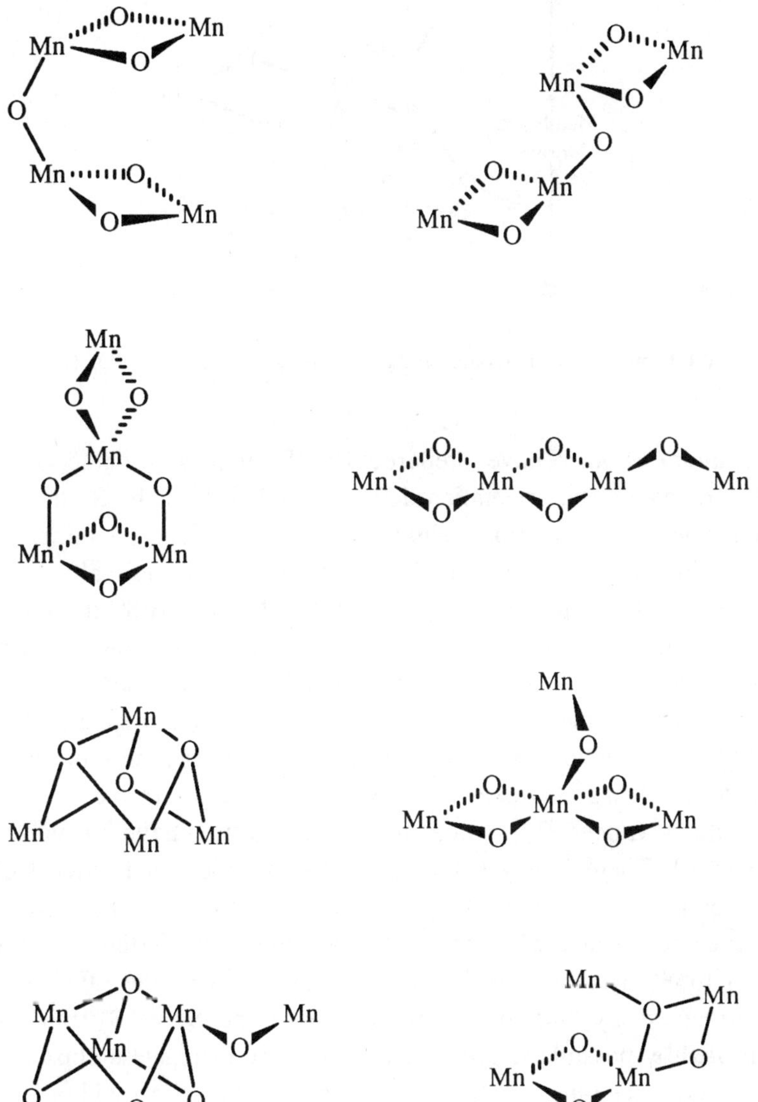

Fig. 11.3 Proposals for the structure of the manganese cluster of the OEC based on X-ray data (adapted from Ref. 2).

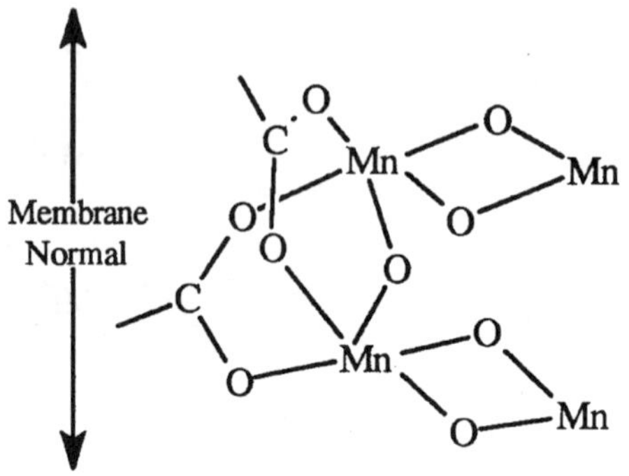

Fig. 11.4 Dimer-of-dimers model for the manganese cluster in the OEC (Ref. 23).

Yachandra *et al.*[23] have prepared PS II samples oriented on mylar sheets and studied the orientation-dependent EXAFS of the S_1 state. Their data have led to the proposal shown in Fig. 11.4. Here, a dimer of Mn di-μ-oxo dimers is oriented parallel to the membrane normal.

Mn-X-ray absorption near edge structure (Mn-XANES) studies of the S_1 state are best fitted using a combination of two Mn(III) and two Mn(IV), but no Mn(II).[2,33,34] As the fits are obtained by comparisons with model complexes, and as the exact make-up of the coordination sphere of Mn in the OEC is not known, an assignment of four Mn(III) cannot be ruled out.

Mn-XANES has also been used to follow changes in the OEC linked to S state advances.[35,36] Once again, the results have proven to be controversial. There is a general consensus that the Mn K-edge shifts to higher energies in going from S_0 to S_1, and from S_1 to S_2. However, there is disagreement about whether the Mn K-edge shifts further when going from S_2 to S_3. It seems likely that the oxidations of the OEC are Mn-centered up to S_2, but the interpretation of the S_2 to S_3 transition is complicated by possible structural changes to the manganese cluster in this step,[23] and has been interpreted to involve either Mn, amino acid or substrate oxidation.

To summarize, X-ray absorption data on the OEC have shown that it contains di-μ-oxo-bridged manganese pairs, and that there are probably Mn-centered oxidation state changes associated with advances from S_0 to S_2.

There is also evidence that a structural rearrangement occurs between the S_2 and S_3 states.

2.2 *EPR*[21]

The S_2 state gives a multiline EPR signal centered at $g = 2$, arising from a strongly exchange coupled S = 1/2 ^{55}Mn cluster, where the exchange interactions arise from μ-oxo-bridged manganese ions.[20,37] This multiline signal can be simulated using various models, of which three have emerged as the most likely candidates. Most studies have fitted the signal to a $Mn^{III}Mn^{IV}_3$ complex.[38] Zheng *et al.*[39,40] conclude from simulations using anisotropic hyperfine couplings that the S_2 state is a $Mn^{III}_3Mn^{IV}$ complex, containing five-coordinate manganese. Åhrling and Pace[41] have modeled the S_2 multiline as arising from a $Mn^{III}Mn^{IV}$ dimer that has large quadrapole couplings from five-coordinate manganese. Randall *et al.* found that their ^{55}Mn-ENDOR spectra were best fitted using a tetranuclear Mn cluster.[42]

The S_2 state can also exist in a form that produces a signal centered at $g = 4.1$ which comes from an S = 3/2 or 5/2 Mn cluster. This can be generated in samples illuminated at T = 140 K, and warming causes the signal to decay with the simultaneous formation of the $g = 2$ multiline signal. Ammonia-treated samples exhibit a $g = 4.1$ signal, which has resolved ^{55}Mn hyperfine structure in oriented samples,[43] and the observation of these hyperfine splittings shows that the $g = 4.1$ EPR signal also arises from a Mn cluster. The multiline and $g = 4.1$ EPR signals in untreated samples are thought to come from two different comformers of the same Mn cluster.[21]

The S_1 state of the OEC in spinach thylakoids is an even electron configuration, and so it is an integer spin system having no conventional mode EPR signals. It can exist in a diamagnetic ground state, the resting state, and a state that is either paramagnetic or has a low lying exited state, the active state.[44] These two forms are thought to be conformers with different exchange couplings between the Mn, analogous to the two forms of the S_2 state. Recently, dark-adapted samples of *Synechocystis* sp. PS II were found to have a parallel mode EPR signal characteristic of an integer spin Mn cluster.[45]

As the S_1 resting state is diamagnetic, the S_0 state was expected to be an odd-electron, and so EPR active, species, but until recently no signal was observed. However, Messinger *et al.*[46] and Åhrling *et al.*[47] independently reported a multiline signal centered at $g = 2$ from methanol-treated samples. Initial indications are that the S_0 state contains an exchanged coupled Mn(II)-Mn(III) dimer with either two Mn(III) or two Mn(IV).

The S_3 state can be generated in good yield using PNPDU,[48] a redox active herbicide that binds in place of Q_B. This state of the OEC has no conventional mode EPR signal, and so is either integer spin or diamagnetic.

There are a number of treatments to produce samples that give an EPR signal arising from the interaction of the S_2 state with an organic radical, identified as and termed the $S_2Y_Z^{\bullet}$ state.[49–52] Although these samples are inactive toward dioxygen evolution, suggesting that they contain an altered form of the OEC, they have shown that Y_Z is located near the manganese cluster[49–52] (distances have been estimated between 3.5 Å and 8 Å[50–53]). Y_Z is thought to be involved in proton-coupled electron transfer (PCET) with the manganese cluster (see section 3).

Electron spin echo envelope modulation (ESEEM) studies[54] suggest that the inner coordination sphere of the manganese cluster is primarily O with at most two N ligands. This conclusion agrees with that from the EXAFS data. Using ^{15}N-labeled PS II, Tang *et al.*[55] were able to show that there is one histidine ligated to the manganese cluster. The remaining ligands are oxo bridges, carboxylates, and perhaps water and/or chloride.

In summary, EPR spectroscopy has shown that the net oxidation state of the Mn in the OEC increases on going from S_0 to S_1 and S_2, in agreement with the XANES data.

2.3 *Chemically reduced samples*

In addition to the states first demonstrated by Kok *et al.*, the OEC can exist in states more reduced than S_0, termed S_{-1}, S_{-2} and S_{-3}, prepared by chemical reduction using either hydroxylamine or hydrazine. All were first identified by an increase in the number of flashes required for the maximal yield of oxygen evolution on illumination of treated, dark-adapted samples. Clearly, all these states are functionally active, and so relevant to physiological water oxidation.

The best studied of these reduced states is S_{-1},[3,56,57] formed by hydroxylamine treatment of dark adapted PS II which shifts the Mn K-edge to lower energies,[31,34] suggesting that the S_{-1} state is either a structurally altered or a reduced form of the manganese cluster. Removing hydroxylamine from such reduced samples does not restore the S_1 state, further supporting the idea that the OEC is reduced.

The S_{-2}[58–61] and S_{-3} states,[46] produced by similar methods to S_{-1}, are less characterized. Although there have been no direct chemical or spectroscopic studies on centers in these states, it is a reasonable assumption based on analogy to S_{-1} that they are further reduced forms of the OEC. However, there is clearly a need for further studies before such assignments can be made with confidence.

2.4 *Reactions of the OEC*

We now discuss the reactivities of the S_0, S_1, S_2 and S_3 states towards chemical reduction and ligand/water exchange.

2.4a *Exchange with ^{18}O-labeled water*

$H_2{}^{18}O$ was mixed with centers in states from S_0 to S_3, and then sufficient flashes were given to the samples to allow one turnover for oxygen evolution. The oxygen so produced was then analyzed by mass spectrometry. These experiments were used to ascertain when water binds to the OEC. They were first attempted by Radmer[62] and then Bader *et al.*[63,64] who found only rapidly exchanging waters, with the $^{36}O_2$:$^{34}O_2$:$^{32}O_2$ ratios reflecting the isotopic distribution of the water. However, the experimental setup used did not have a short enough time resolution to detect "fast" ($k < 1\ s^{-1}$)

reactions. More recently, Messinger *et al.*[65,66] developed an apparatus with a resolution of 20 ms, allowing for far more detailed studies. They have found that there are two substrate waters with different exchange rates, $k_f = 40\ s^{-1}$ and $k_s = 2.2\ s^{-1}$ respectively, that exchange with the S_2 and S_3 states. States below S_2 have no slowly exchanging water molecules. The slow exchanger is thought to be an oxo or hydroxo ligated to Mn, while the fast exchanger either enters in the S_4 state or is possibly ligated to something other than Mn, Ca^{2+} for instance.

2.4b *Reactivity of the S states towards reduction*

Messinger *et al.*[67] studied the reactivity of the OEC with hydroxylamine and hydrazine as a function of S state, finding the following trends in rates:

$$k_{NH_2R}(S_0) > k_{NH_2R}(S_1) << k_{NH_2R}(S_2) >> k_{NH_2R}(S_3)$$

The marked difference in reactivity between the S_2 and S_3 states could be caused by a change in structure occurring during the S_2 to S_3 transition. This change would either reduce the accessibility of the manganese cluster to the external reductants (a change to the environment of the OEC) or slow the rate of electron transfer between the bound reductant and the acceptor (a change in the reactivity of the OEC itself). To support the latter hypothesis, Messinger *et al.*[67] note that larger molecules that shorten the lifetime of the S_2 and S_3 states, such as bulky secondary amines, do not exhibit this reactivity difference. However, the OEC, even in the S_2 state, is only accessible to small molecules so the significance of this result is unclear. Ammonia, on the other hand, is less reactive with the S_3 state than the S_2 state,[68,69] but is small enough for differences in site accessibility to have only minimal effects on the rate. This suggests that the slower rate of reduction of S_3 is an intrinsic property of the manganese cluster, and not due to the protein structure around it. A difference in oxidation states would be a clear candidate to explain these differences in reactivity, as will be discussed in the next section.

3. Proposed Mechanisms of the OEC

We have seen that a large body of both spectroscopic and chemical information exists about the OEC. In this section, we will attempt to draw some conclusions about the structure and oxidation states of the OEC, including a discussion of the relevant inorganic model chemistry.

The earliest attempts to model the mechanism of the OEC were based on a proposed predominance of Mn(II) and Mn(III) in the active states.[70,71] There was also much debate as to the number of Mn at the active site with both dimers and tetramers being proposed, but with so little spectroscopic information available on the protein, no attempts were made to propose a detailed structure.

Although no one has synthesized a manganese tetramer that exactly mimics the spectroscopic data of all the S states, a large number of complexes are known that do show similar spectra,[24,25,72] and these have allowed for several mechanistic proposals.

As previously mentioned, model complexes have provided strong support for the interpretation that the OEC contains di-μ-oxo bridged manganese ions, and that these oxo bridges can account for the strong Mn-Mn exchange couplings present in all the S states. A majority of these oxomanganese complexes are dimers, but there are a number of tetramers. The synthesis of such tetramers allowed for the first detailed proposals of both the structure of the OEC and of the mechanism of water oxidation.

The first of these, which has been termed the Brudvig/Crabtree model[73] (Fig. 11.5), was inspired by the isolation of an adamantane complex by Weighardt *et al.*[74] The structural rearrangement thought to occur between the S_2 and S_3 states was explained by a conversion of the manganese cluster from a cubane to an adamantane form. O–O bond formation was proposed to occur between two of the bridging oxos when the S_4 state is reached, initiated by the formation of a Mn(V) center. The S_4 state was, therefore, postulated to contain three Mn(IV) and a Mn(V).

Another proposal by Vincent and Christou[75] was based upon a butterfly Mn tetramer that converted to a cubane on the formation of S_3 and subsequently released O_2 by a double-pivot mechanism (Fig. 11.6). Again, O–O bond formation was proposed to occur in S_4 between two bridging

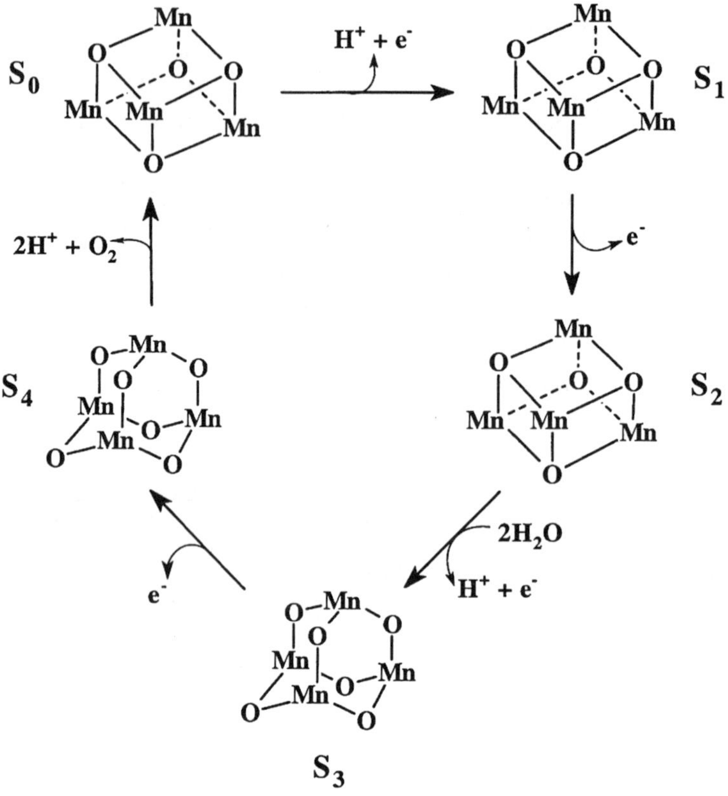

Fig. 11.5 The Brudvig/Crabtree model for water oxidation (adapted from Ref. 73).

oxo ligands to form a side-on bound μ^2-η^2:η^2-peroxo complex, which then released dioxygen. Here, the S_4 state was formulated as four Mn(IV). These models reflected advances in inorganic model chemistry, and were the first to propose detailed mechanisms for O–O bond formation.

More recent proposals involve the participation of a terminal Mn=O group in the O–O bond-forming step, with the Mn being in a +4 or +5 oxidation state. Analogous species are key intermediates in metalloporphyrin based organic oxidations in biological systems, such as cytochrome P450.[76,77] Mononuclear compounds with a terminal Mn(V)=O group have

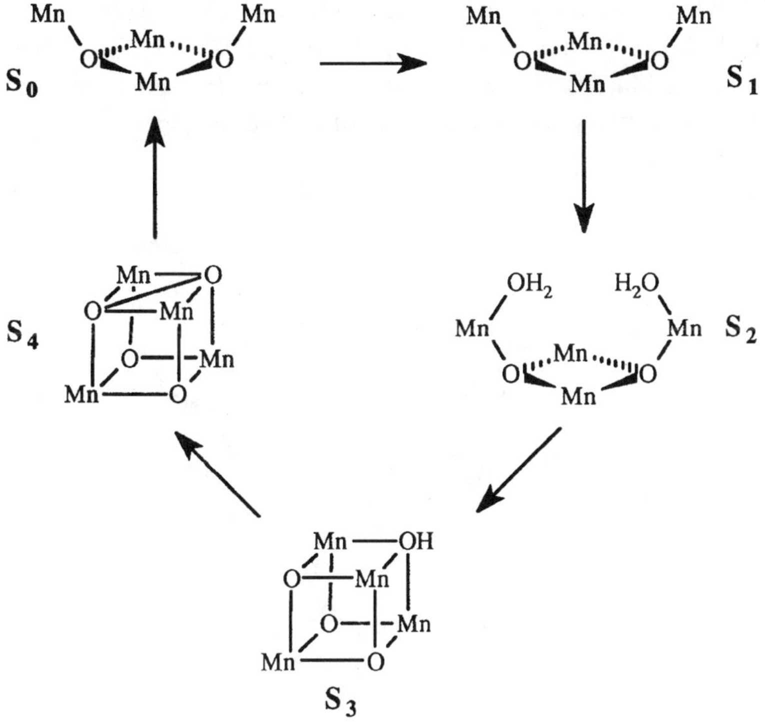

Fig. 11.6 The Vincent/Christou model for water oxidation (adapted from Ref. 75).

been isolated using highly stabilizing amide ligands,[78–80] but no stable clusters containing such a species were ever synthesized.

On the basis of Hückel calculations[81,82] Proserpio *et al.* proposed that the O–O bond forms between two terminal Mn=O groups bridged by a proton (Fig. 11.7). In a revision of their double pivot mechanism that was presented in Ref. 1, Vincent and Christou proposed O–O bond formation occurring between a bridging and a terminal oxo that forms in an intermediate between S_3 and S_4 (Fig. 11.8).

The proposal that Y_Z is located close (≤ 8 Å) to the manganese cluster has led to speculation on its involvement in H atom abstraction from water bound to the manganese cluster. Hoganson *et al.*[83,84] have proposed that a

Mn(IV)=O species forms at S_3 as a result of H atom abstraction from a terminally ligated water molecule by $Y_Z^{\bullet}$ (Fig. 11.9). The Mn(IV)=O then reacts with a Mn(IV)-OH in concert to H atom abstraction, shown as S_4' in Fig. 11.9, resulting in the formation of a μ-peroxo species which would be the S_4 state.

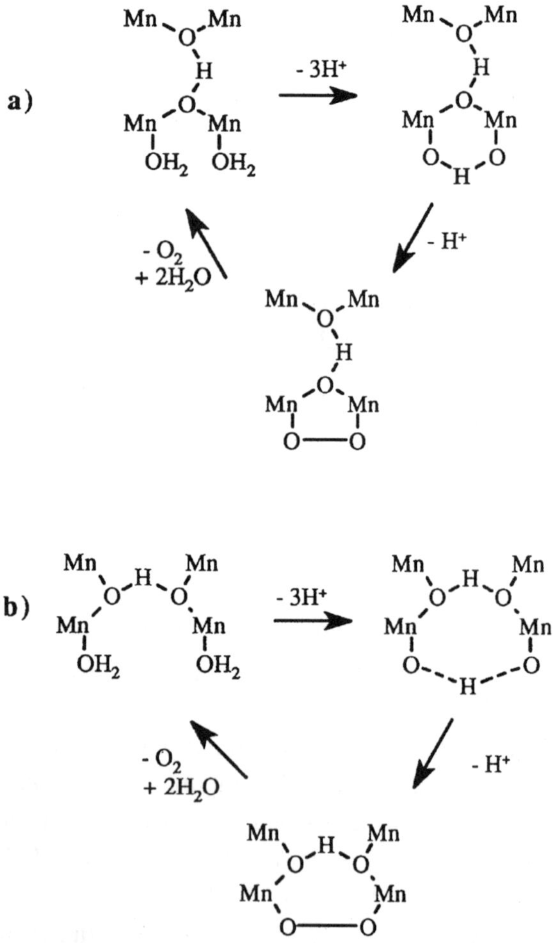

Fig. 11.7 Models proposed for photosynthetic water oxidation by Proserpio *et al.* (adapted from Ref. 82).

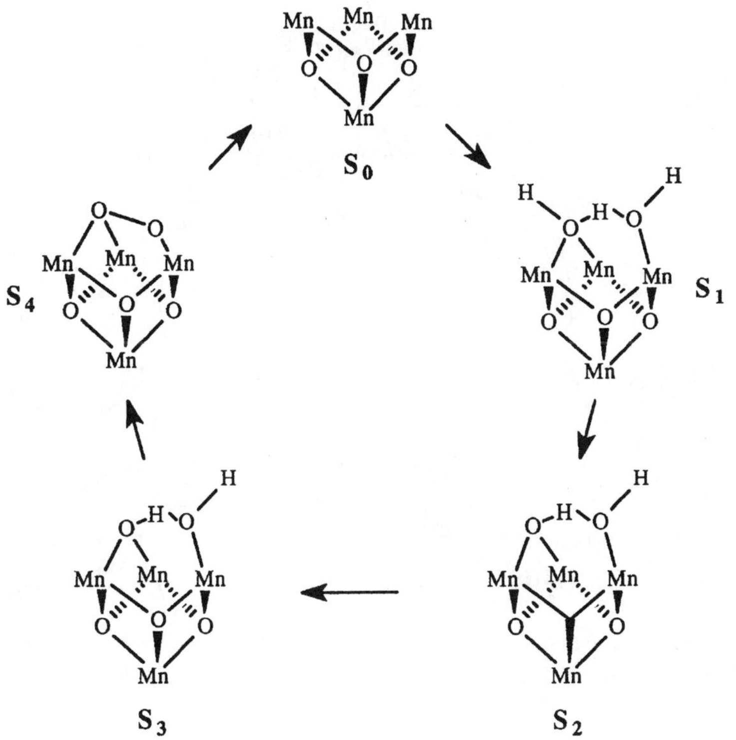

Fig. 11.8 Revised version of the Vincent/Christou model (adapted from Ref. 1).

Gardner and Mayer[85] have demonstrated the importance of PCET in determining the pathway of organic oxidations. Their methodology was applied to manganese complexes by Caudle and Pecoraro,[86] who showed that ligation to a high-valent manganese significantly lowers the O–H bond dissociation energy of water, and so H abstraction by $Y_Z^{\bullet}$ is feasible. Blomberg, Siegbahn and co-workers[87] obtained a similar result using *ab initio* calculations. The O–H bond strength is 119 kcal/mol in uncoordinated water, but a water terminally ligated to Mn(III) or Mn(IV) is estimated to have an O–H bond strength between 78 and 89 kcal/mol. This compares with a value of 86.5 kcal/mol in tyrosine,[88,89] allowing for H atom abstraction from a bound water to be exothermic and so provide extra driving force relative to a purely outer sphere oxidation of the manganese cluster by $Y_Z^{\bullet}$.

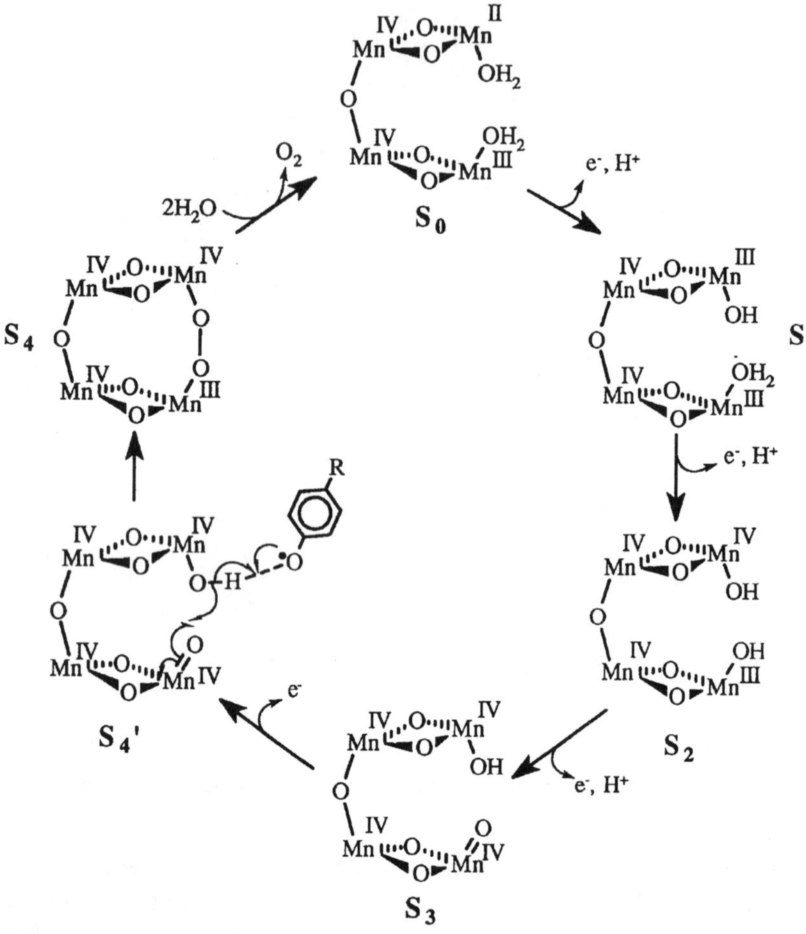

Fig. 11.9 Proposed mechanism for water oxidation by the OEC using H atom abstraction by $Y_Z^{\bullet}$ (adapted from Ref. 83).

In section 2, we noted that states as low as S_{-3} have been prepared using external reductants. If all the Mn ions are reduced to a +2 oxidation state, they dissociate from the protein,[57] suggesting that this highly reduced S_{-3} state contains at least one Mn(III). Assuming Mn-centered oxidation occurs from S_{-3} to S_1, this favors an assignment of S_1 as $Mn^{III}_2Mn^{IV}_2$.

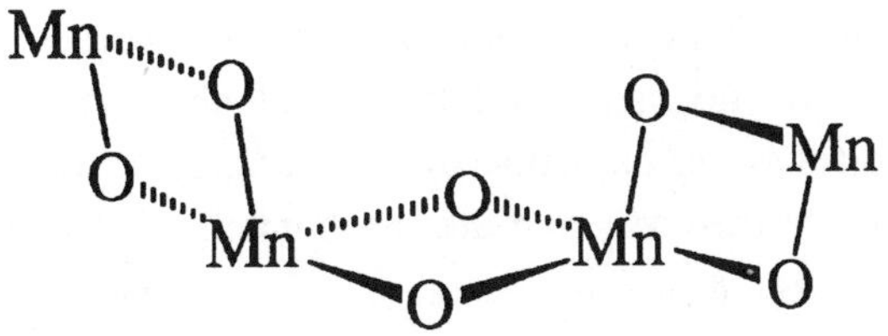

Fig. 11.10 Structure of the $[Mn_4O_6]^{4+}$ core from the complex studied by Blondin *et al.* (Ref . 72).

We also noted that greater reactivity of the S_2 state towards reduction as compared to the S_3 state can be explained by an oxidation state change. Manganese clusters containing only Mn(IV) are less reactive than their mixed-valence counterparts, as Mn(IV) has slower ligand exchange rates than Mn(III) (high-spin d^3 versus high-spin d^4). If the S_3 state is Mn^{IV}_4 and the S_2 is $Mn^{III}Mn^{IV}_3$, then the slower exchange rate of the ligands in the S_3 state would cause it to be less susceptible to reduction, assuming that the rate-determining step is binding of NH_2R. The observation that there is little difference in water exchange rates between these states does not support this hypothesis, but it is unclear whether the rate-determining step in water exchange involves binding to Mn.

Blondin *et al.*[72] have reported the *in situ* generation of an S = 1/2 ground state $Mn^{III}Mn^{IV}_3$ 2,2'-bipyridine (bpy) complex (Fig. 11.10) whose EPR spectrum is broadly similar to the S_2 multiline signal of PS II. The complex was obtained by γ-irradiation of a solution of the Mn^{IV}_4 complex. The topology of the complex is the same as that proposed by Yachandra *et al.*[23] based on EXAFS studies (Fig. 11.4), and this result provides support for the dimer-of-dimers model.

4. Model Chemistry

4.1 Functional models for the OEC

Despite the large number of complexes proposed as structural models for the manganese cluster in PS II that have been synthesized, there are few analogous complexes that can act as homogeneous water oxidation catalysts. Here we discuss the relevant catalytic model chemistry in terms of the models proposed for water oxidation by the OEC, concentrating in particular on the proposed intermediates and the mechanism of O–O bond formation, and also suggest strategies for the design of a functional model, with specific emphasis on a system we developed.

There are a few reports of heterogeneous catalytic water oxidation by manganese complexes. Shilov and co-workers[90] have found that addition of catalytic amounts of MnO_2 to solutions of strong one-electron oxidants, such as $Ru(bpy)_3^{3+}$, leads to water oxidation. Shilov and co-workers[91] also reported that Mn(IV) incorporated into phospholipid membranes releases dioxygen under similar reaction conditions, and proposed that both these reactions proceed via a Mn(V) intermediate.

$[(bpy)_2Mn(O)_2Mn(bpy)_2]^{3+}$ acts as a heterogeneous water oxidation catalyst, in the presence of excess $(NH_4)_6Ce(NO_3)_2$.[92,93] There is also a report of analogous chemistry with asymmetric di-μ-oxo dimanganese complexes formulated as $[Mn_2(O)_2L_3A_2]$, where L = bpy or 2,2'-phenanthroline (phen) and A = dimethylformamide or pyridine.[94] The di-μ-oxo-dimers ultimately decompose to permanganate. In similar studies, Gobi *et al.*[95] found that $[Mn(salen)(H_2O)]^+$ (salenH_2 = bis(salicylaldehyde) ethylenediimine), either as a saturated solution or in a heterogeneous adsorbed state on kaolin clay, evolves O_2 and N_2, with turnover numbers of between 2 and 3 for water oxidation. The N_2 was proposed to result from NH_4^+ oxidation.

None of these heterogeneous systems is well-characterized, limiting the amount of mechanistic information that they can provide. However, they have unambiguously demonstrated that Mn can act as a catalyst for water oxidation in model systems.

There are only two structurally characterized metal cluster complexes that can act as homogeneous water oxidation catalysts. One is an o-phenylene-bridged Mn(III)-porphyrin dimer that can electrochemically oxidize water.[96a] No hydrogen peroxide is released during the reaction, the complex has no catalase activity, and the electrochemistry suggests a four-electron process. This led the authors to propose a four-electron mechanism proceeding via a Mn(V)=O dimer intermediate. This assignment has been supported by studies involving olefin epoxidation.[96b] O–O bond formation may occur either by a direct reaction of the two oxo groups or by attack of a water on one of the oxo groups.

The second, and mechanistically more important system, $[(bpy)_2(H_2O)RuORu(H_2O)(bpy)_2]^{4+}$ was first reported in 1982 by Gersten *et al.*[97] and has a catalytic wave in its cyclic voltammagram that was assigned to water oxidation. Perhaps the most significant mechanistic result was the detection of the O=Ru(V)ORu(V)=O intermediate[98] as the precursor for oxygen evolution, a result that led to speculation as to exactly how such a species may react to form O_2.

Geselowitz and Meyer prepared (III/III) and (III/IV) dimers using isotopically-labeled water,[99] and then allowed one turnover to occur in unlabeled water. Their ratio of $^{36}O_2$:$^{34}O_2$:$^{32}O_2$ of 13 : 64 : 23 apparently disagrees with similar experiments carried out by Hurst *et al.*,[98] who labeled the oxo-bridge as well as the water ligands, and found $^{34}O_2$:$^{32}O_2$ ratios of approximately 50 : 50 with no fully labeled dioxygen. It should be noted that the two experiments were run at different pH (pH 1 versus pH 0) and used different oxidants (Ce(IV) versus Co(III)). Hurst *et al.* found that the isotopic distribution of the products was invariant with time, ruling out any slow exchange processes between water ligated to Ru and solvent water, although some rapid exchange process may operate during the initial stages of the reaction. No single mechanism can account for both of these results, but some conclusions can be drawn.

Multiple pathways must be invoked to account for all the products, assuming that exchange processes are slow compared to the rate of dioxygen evolution. Figure 11.11 shows some proposals for the formation of unlabeled, singly-labeled and fully-labeled dioxygen starting from a labeled Ru-bpy di-oxo intermediate. Formation of $^{34}O_2$ [Fig. 11.11(c)] requires an attack of solvent water on an Ru=O group, and a similar mechanism has also been proposed to account for $^{32}O_2$ [Figs. 11.11(a) and (b)].[98,99] $^{36}O_2$ must form from the reaction of two Ru=O groups, and both intramolecular [Fig. 11.11(d)] and intermolecular [Fig. 11.11(e)] mechanisms are possible.[99]

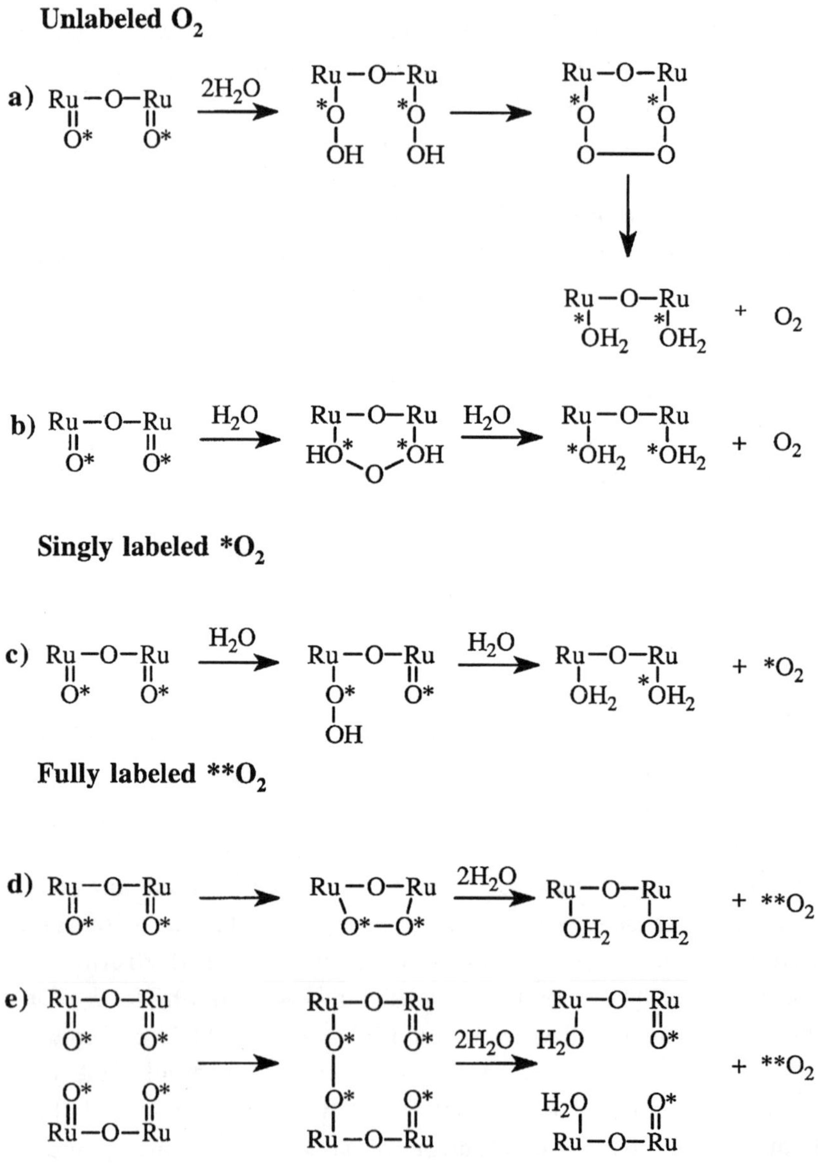

Fig. 11.11 Proposed mechanisms for O–O bond formation by $[(bpy)_2(H_2O)RuORu(H_2O)(bpy)_2]^{4+}$ based upon the 18-O labeling studies (see Refs. 98 and 99).

The Ru-bpy dimer can evolve dioxygen from either the IV/V or the V/V state. If bpy-5,5'-dicarboxylic acid[100] or the 4,4' analog[101] are used in place of bpy, the III/IV dimer can evolve dioxygen with a stoichiometry of four dimers per dioxygen produced.[102] The initial step of the reaction is thought to be the formation of the IV/IV dimer via a disproportionation reaction. This shows that the basicity of the ligands on the metal can greatly alter the characteristics of the active species.

Manganese complexes tend to exchange ligands rapidly and have an inherently complicated solution chemistry. A good example of this can be found in the chemistry of $[(bpy)_2Mn(O)_2Mn(bpy)_2]^{3+}$, which in aqueous solution exists in dynamic equilibrium with at least six other complexes of varying nuclearity.[103–106] As a result of this, although there has been some success in oxidizing water with manganese complexes, it has been difficult to characterize the active species. One parallel approach, avoiding this problem, has been to try and design a catalyst.

Initial attempts involved the use of hexadentate ligands, such as amide homologues of EDTA, to make a *cis*-aqua di-μ-oxo dimanganese complex analogous to the Ru-bpy dimer. Analogous complexes were isolated with the *cis* sites coordinated with a bridging acetate,[107,108] but no activity toward water oxidation was reported.

In 1997, we reported a reaction between Mn(II)-dpa/terpy (dpa = dipicolinate and terpy = 2,2':6,2"-terpyridine) complexes and oxygen atom transfer reagents, K-oxone (potassium peroxomonosulfate) and dimethlydioxirane, that produced dioxygen catalytically.[109a] We chose oxone as it was previously used with manganese catalysts in organic oxidations which were postulated to involve Mn(V)=O intermediates.[110,111] In addition to dioxygen, permanganate was produced in a side reaction that eventually stopped the catalysis by consuming all the manganese. EPR and UV/vis spectroscopic studies on these solutions showed that the major species in solution are III/IV and IV/IV di-μ-oxo dimanganese complexes, and when the ligand is dpa, there are isosbestic points in the UV/vis spectrum that indicate a clean conversion to permanganate. No reactions occur with Lewis acids (Ni(II), Zn(II), and Al(III)), and only non-catalytic reactivity is found for Mn when bpy, phen and pic (picolinate) complexes are used.

The successful ligands favor meridional coordination, leading to the structure shown in Fig. 11.12. The dimers in solution contain *trans*-aqua ligands, and we proposed that these exchangeable coordination sites are responsible for the observed catalytic activity. Complexes of this formulation avoid the tendency to form bridging oxos, such as is

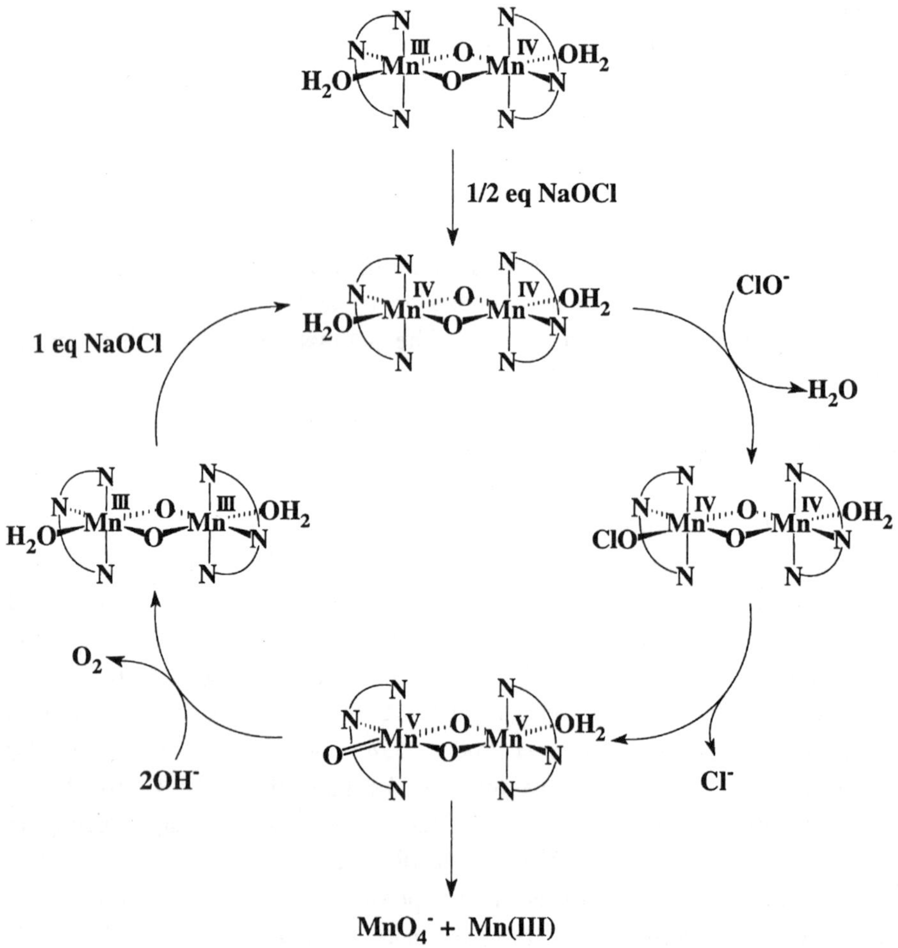

Fig. 11.12 Proposed mechanism for dioxygen evolution and permanganate formation by $[(H_2O)(terpy)Mn(O)_2Mn(terpy)(H_2O)]^{3+}$ (Ref. 109b).

found with triazazcyclononane, which makes tri-μ-oxo[112] and di-μ-oxo μ-peroxo-bridged complexes.[113] Further, the chelate effect means that tridentate ligands are less likely to dissociate then their bidentate analogs, so the dimers tend to maintain their structure during turnover, allowing for a well-defined system.

The complex $[(terpy)(H_2O)Mn(O)_2Mn(OH_2)(terpy)](NO_3)_3$ was isolated from the reaction between the Mn(II) acetate, terpy and oxone.[109b] This mixed valence dimer catalyzes the production of O_2 from NaOCl (sodium hypochlorite), again with permanganate formation leading to a loss of catalytic activity. This complex is the first example of a di-μ–oxo dimanganese complex that evolves dioxygen in a non-catalase fashion, making $[(terpy)(H_2O)Mn(O)_2Mn(OH_2)(terpy)]^{3+}$ a functional model for the OEC.[109b].

The proposed mechanism for this reaction (Fig. 11.12) involves oxygen atom transfer to a IV/IV dimer[109b] resulting in an active species that contains a Mn(V)=O. The terminal oxo ligand can then undergo exchange with the solvent water in accordance with the mechanism proposed by Bernadou and Meunier.[109c] The O–O bond forming step could involve the reaction between a solvent water and the terminal oxo. Permanganate would form by disproportionation of the Mn(V)=O containing intermediate.

4.2 *Comparisons between the model chemistry and the OEC*

4.2a O–O bond-forming reactions

By far the best characterized system that evolves dioxygen, the Ru-bpy dimer, almost conclusively requires the formation of a terminal oxo species. Such species are involved in many mechanistic proposals put forward for photosynthetic water oxidation. These proposals are further supported by our model system, which appears to make dioxygen via a Mn(V)=O intermediate, and such a species seems to be a viable candidate for the S_4 state. By analogy with the Ru-bpy dimer, O–O bond formation could occur either inter-or intramolecularly. The first would involve attack of an external water on the Mn=O, with the water either being "free" or ligated to another metal, such as Ca^{2+} [Fig. 11.13(a)]. This kind of nucleophilic attack would be facilitated if the terminal oxo species on a high-valent, and so electron poor metal, has a ∂^+ character. The second would involve a reaction between a Mn=O and either a second Mn=O, a Mn-OH [Fig. 11.13(b)] or a bridging oxo ligand [Fig. 11.13(c)]. A Mn=O can be thought of as either an

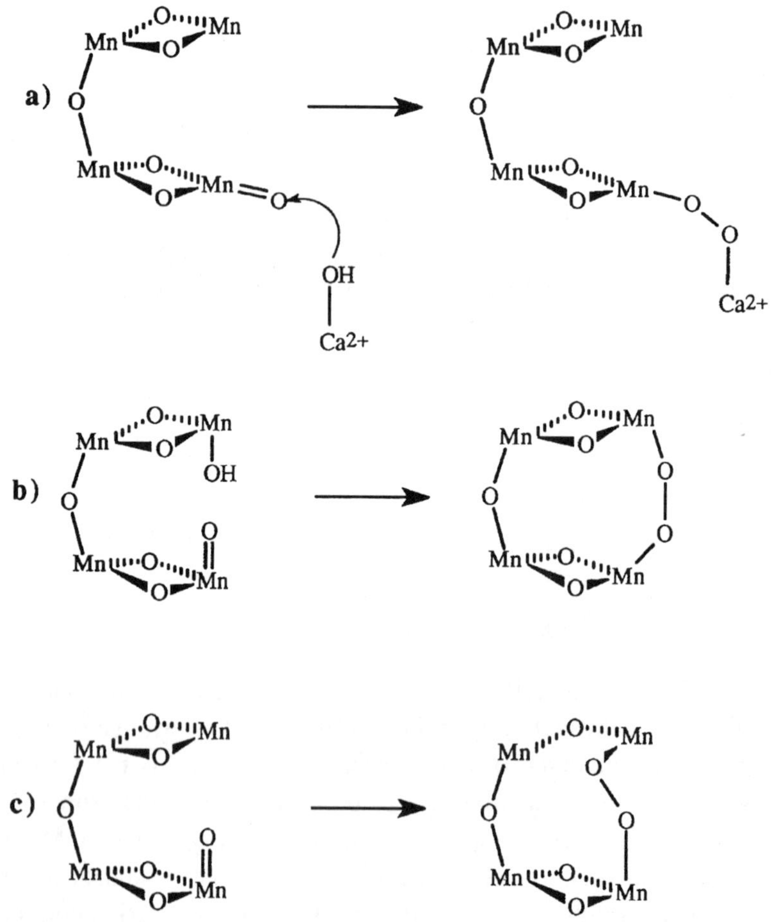

Fig. 11.13 Possible reaction pathways for O–O bond formation from a Mn(V)=O containing active state of the OEC.

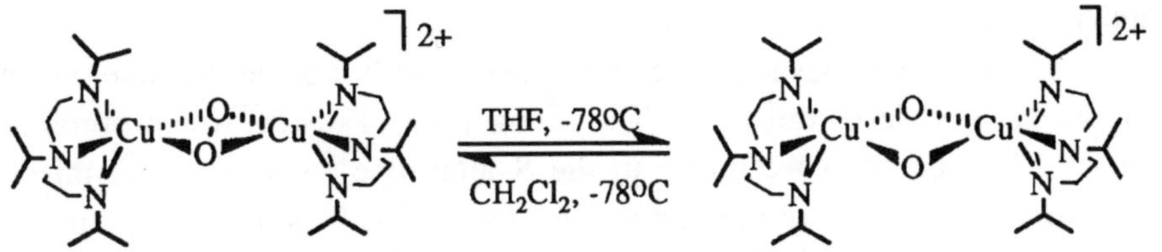

Fig. 11.14 The interconversion of the two forms of $[LCu(O)_2CuL]^{2+}$, where L = 1,4,7-trisopropyl-1,4,7-triazacyclononane (Ref. 118).

electrophile[114–116] or as a radical species.[117]

Halfen *et al.*[118] reported a di-μ-oxo dicopper complex capable of reversible O–O bond formation at –78 °C involving the formation of a μ-η^2:η^2–O_2 core (Fig. 11.14). Although this complex does not release dioxygen, it does emphasize the possibility that this mechanism may be employed by the OEC, as originally proposed by Vincent and Christou (Fig. 11.6) and later by Pecoraro *et al.*[119] While no analogous chemistry has ever been observed in manganese complexes, introducing a Mn(V) may destabilize the bridging oxos toward reactivity,[73] and such a proposal cannot be excluded. The O–O bond-forming step may alternatively involve reaction between a bridging and a terminal oxo, as proposed by Vincent and Christou (Fig. 11.8).

4.2b Oxidation states for the active state of a Mn-based water oxidation catalyst

In section 3, four sets of oxidation states were proposed as possibilities for the S_4 state of the OEC:

$$Mn^{III}{}_2Mn^{IV}{}_2Y_Z^{\bullet},$$

$$Mn^{III}Mn^{IV}{}_3,$$

$$Mn^{IV}{}_4Y_Z^{\bullet},$$

and $Mn^{IV}{}_3Mn^{V}$

and none of these can be ruled out.

Any viable hypothesis must explain the instability of the S_4 state which causes the release of dioxygen. The preparation of a $Mn^{III}Mn^{IV}_3$ complex whose EPR spectrum is similar to the S_2 multiline[72] suggests that this is the correct assignment of oxidation states for the S_2. This proposal is further supported by the simulations of the S_2 state multiline EPR spectrum and by XANES studies. The observation of an active S_{-3} state, which suggests that the S_1 state is $Mn^{III}_2Mn^{IV}_2$, provides further support for this assignment of the oxidation states of Mn in S_2. Assuming that the OEC is oxidized past S_2, the latter two proposals for S_4 are, therefore, the most likely. Mn(V) has been proposed as an intermediate in heterogeneous water oxidation catalysts,[90,91] as well as in our dioxygen evolving system,[109a,b] and we believe that Mn(V) is required to produce a terminal oxo species. The only characterized Mn=O complexes, other than MnO_4^{n-} species, contain Mn(V), and these species are proposed as models for the active states of Mn-porphyrin and salen oxidation catalysts.[78–80] The variability of the oxidation states of the active species of the Ru-bpy water oxidation catalyst shows that the ligating environment plays an important role. It may be that in the protein the ligation is such that Mn(IV) can support a terminal oxo, so although we favor formation of a terminal oxo only upon formation of a Mn(V), formation of a Mn(IV)=O in the OEC is possible. This assignment was postulated for the S_3 state by Hoganson and Babcock (see Fig. 11.9), where O–O bond formation occurs during the transition to the S_4 state.

5. Conclusions

While the precise nature of the OEC is still unknown, the number of possibilities for both structural and mechanistic models have been steadily narrowing. There is a growing body of evidence that the active species contains a high-valent Mn=O intermediate, which we assign as Mn(V), and that formation of this species is the key step in water oxidation. We have developed a model system that may have analogous chemistry, and further characterization should provide insight into the behavior of a Mn(V)=O-containing cluster, and its possible role in water oxidation.

6. Acknowledgments

This work was supported by grants from the National Institutes of Health.

7. References

1. Rüttinger, W. and Dismukes, G.C. *Chem. Rev.* 1997, **97:** 1–24.
2. Yachandra, V.K., Sauer, K. and Klein, M.P. *Chem. Rev.* 1996, **96:** 2927–2950.
3. Debus, R.J. *Biochim. Biophys. Acta* 1992, **1102:** 269–352.
4. Ghanotakis, D.F. and Yocum, C.F. *Annu. Rev. Plant Physiol. Plant Mol. Biol.* 1990, **41:** 255–276.
5. Rutherford, A.W., Zimmermann, J.-L. and Boussac, A. in *The Photosystems: Structure, Function and Molecular Biology* (ed.) Barber, J. Elsevier, Amsterdam, The Netherlands, 1992, pp. 179–229.
6. Joliot, P., Barbieri, G. and Chabaud, R. *Photochem. Photobiol.* 1969, **10:** 309–329.
7. Kok, B., Forbush, B. and McGloin, M. *Photochem. Photobiol.* 1970, **11:** 457–475.
8. Pirson, A. *Z. Botan.* 1937, **31:** 193–267.
9. Cheniae, G.M. and Martin, I.F. *Biochim. Biophys. Acta* 1970, **197:** 219–239.
10. Berthold, D.A., Babcock, G.T. and Yocum, C.F. *FEBS Lett.* 1981, **134:** 231–234.
11. Murata, N., Miyao, M., Omata, T., Matsunami, H. and Kuwabara, T. *Biochim. Biophys. Acta.* 1984, **765:** 363–369.
12. Sandusky, P.O. and Yocum, C.F. *Biochim. Biophys. Acta* 1986, **849:** 85–93.
13. Yocum, C.F. *Biochim. Biophys. Acta* 1991, **1059:** 1–15.
14. Ghanotakis, D.F., Babcock, G.T. and Yocum, C.F. *FEBS Lett.* 1984, **167:** 127–130.
15. Pistorius, E.K. *Eur. J. Biochem.* 1983, **135:** 217–222.
16. Boussac, A. and Rutherford, A.W. *Biochemistry* 1988, **27:** 3476–3483.
17. Lockett, C.J., Demetriou, C., Bowden, S.J. and Nugent, J.H.A. *Biochim. Biophys. Acta* 1990, **1016:** 213–218.
18. Ghanotakis, D.F., Babcock, G.T. and Yocum, C.F. *Biochim. Biophys Acta* 1985, **809:** 173–180.

19. Vermaas, W.J.F., Renger, G. and Dohnt, G. *Biochim. Biophys. Acta* 1984, **704:** 194–202.
20. Brudvig, G.W., Casey, J.L. and Sauer, K. *Biochim. Biophys. Acta* 1983, **723:** 366–371.
21. Brudvig, G.W. *Adv. Chem. Ser.* 1995, **246:** 249–263.
22. Wieghardt, K. *Angew. Chem., Int. Ed. Engl.* 1989, **28:** 1153–1172.
23. Yachandra, V.K., DeRose, V.J., Latimer, M.J., Mukerji, I., Sauer, K. and Klein, M.P. *Science* 1993, **260:** 675–679.
24. Thorp H.H. and Brudvig, G.W. *New J. Chem.* 1991, **15:** 479–490.
25. Manchanda, R., Brudvig, G.W. and Crabtree, R.H. *Coord. Chem . Rev.* 1995, **144:** 1–38.
26. George, G.N. Prince, R.C. and Cramer, S.P. *Science* 1989, **243:** 789–791.
27. Penner-Hahn, J.E., Fronko, R.M., Pecoraro, V.L., Yocum, C.F., Betts, S.D. and Bowlby, N.R. *J. Am. Chem. Soc.* 1990, **112:** 2549–2557.
28. Kusunoki, M. Ono, T.-a., Matsushita, T., Oyanagi, H. and Inoue, Y. *J. Biochem.* 1990, **108:** 560–567.
29. MacLachlan, D.J., Hallahan, B.J., Ruffle, S.V., Nugent, J.H.A., Evans, M.C.W., Strange, R.W. and Hasnain, S.S. *J. Biochem.* 1992, **285:** 569–576.
30. Latimer, M.J., DeRose, V.J., Mukerji, I., Yachandra, V.K., Sauer, K. and Klein, M.P. *Biochemistry* 1995, **34:** 10898–10909.
31. Riggs-Gelasco, P.J., Mei, R., Yocum, C.F. and Penner-Hahn, J.E. *J. Am. Chem. Soc.* 1996, **118:** 2387–2399.
32. Penner-Hahn, J.E., Riggs-Gelasco, P.J., Yu, E., DeMarois, P. and Yocum, C.F. in *Photosynthesis: From Light to Biosphere* (ed.) Mathis, P. Kluwer, Dordrecht, The Netherlands, 1995, Vol. 2, pp. 241–246.
33. Yachandra, V.K., DeRose, V.J., Latimer, M.J., Mukerji, I., Sauer, K. and Klein, M.P. in *Research in Photosynthesis*, (ed.) Murata, N. Kluwer, Dordrecht, The Netherlands, 1992, Vol. 2, pp. 281–287.
34. Riggs, P.J., Mei, R., Yocum, C.F. and Penner-Hahn, J.E. *J. Am. Chem. Soc.* 1992, **114:** 10650–10651.
35. Ono, T.-a., Noguchi, T., Inoue, Y., Kusunoki, M., Matsushita, T. and Oyanagi, H. *Science* 1992, **258:** 1335–1337.
36. Reolofs, T.A., Liang, W., Latimer, M.J., Cinco, R.M., Rompel, A., Andrews, J.C., Sauer, K., Yachandra, V.K. and Klein, M.P. *Proc. Natl. Acad. Sci. USA.* 1996, **93:** 3335–3340.
37. Dismukes, G.C. and Siderer, Y. *Proc. Natl. Acad. Sci. USA.* 1981, **78:** 274–278.

38. Bonvoisin, J., Blondin, G., Girerd, J.-J. and Zimmermann, J.-L. *Biophys. J.* 1992, **61:** 1076–1086.
39. Zheng, M. and Dismukes, G.C. *Inorg. Chem.* 1996, **35:** 3307–3319.
40. Zheng, M. and Dismukes, G.C. in *Research in Photosynthesis* (ed.) Murata, N. Kluwer, Dordrecht, The Netherlands, 1992, Vol. 2, pp. 305–308.
41. Åhrling, K.A. and Pace, R.J. *Biophys. J.* 1995, **68:** 2081–2090.
42. Randall, D.W., Sturgeon, B.E., Ball, J.A., Lorigan, G.A., Chan, M.K., Klein, M.P., Armstrong, W.H. and Britt, R.D. *J. Am. Chem. Soc.* 1995, **117:** 11780–11789.
43. Kim, D.H., Britt, R.D., Klein, M.P. and Sauer, K.J. *J. Am. Chem. Soc.* 1990, **112:** 9389–9391.
44. Koulougliotis, D., Hirsh, D.J. and Brudvig, G.W. *J. Am. Chem. Soc.* 1992, **114:** 8322–8323.
45. Campbell, K.A., Peloquin, J.M., Pham, D.P., Debus, R.J. and Britt, R.D. *J. Am. Chem. Soc.* 1998, **120:** 447–448.
46. Messinger, J., Seaton, G, Wydrzynski, T., Wacker, U. and Renger, G. *Biochemistry* 1997, **36:** 6862–6873.
47. Åhrling, K.A., Peterson, S. and Styring, S. *Biochemistry* 1997, **36:** 13148–13152.
48. Bocarsly, J.R. and Brudvig, G.W. *J. Am. Chem. Soc.* 1992, **114:** 9762–9767.
49. Gilchrist, M.L., Jr., Ball, J.A., Randall, D.W. and Britt, R.D. *Proc. Natl. Acad. Sci. USA* 1995, **92:** 9545–9549.
50. Szalai, V.A. and Brudvig, G.W. *Biochemistry* 1996, **35:** 1946–1953.
51. Lakshmi, K.V., Eaton, S.S., Eaton, G.R., Frank, H.A. and Brudvig, G.W. *J. Phys. Chem. B* 1998, **102:** 8327–8335.
52. Boussac, A., Zimmermann, J.-L., Rutherford, A.W. and Lavergne, J. *Nature* 1990, **347:** 303–306.
53. Force, D.A., Randall, D.W. and Britt, R.D. *Biochemistry* 1997, **36:** 12062–12070.
54. DeRose, V.J., Yachandra, V.K., McDermott, A.E., Britt, R.D., Sauer, K. and Klein, M.P. *Biochemistry* 1991, **30:** 1335–1341.
55. Tang, X.-S., Diner, B.A., Larsen, B.S., Gilchrist, M.L., Jr., Lorigan, G.A. and Britt, R.D. *Proc. Natl. Acad. Sci. USA* 1994, **91:** 704–708.
56. Guiles, R.D., Yachandra, V.K., McDermott, A.E., Cole, J.L., Dexheimer, S.L., Britt, R.D., Sauer, K. and Klein, M.P. *Biochemistry* 1990, **29:** 486–496.

57. Brudvig, G.W. and Beck, W.F. in *Manganese Redox Enzymes* (ed.) Pecoraro, V.L. VCH Publishers, New York, 1992, 119–140.
58. Renger, G., Messinger, J. and Hansumm, B. in *Current Research in Photosynthesis* (ed.) Baltscheffsky, M. Kluwer, Dordrecht, The Netherlands, 1990, Vol. 1, pp. 845–848.
59. Messinger J. and Renger, G. *FEBS Lett.* 1990, **277:** 141–146.
60. Messinger, J., Pauly, S. and Witt, H.T., *Z. Naturforsch* 1991, **46c:** 1033–1038.
61. Kretschmann, H., Pauly, S. and Witt, H.T. *Biochim. Biophys. Acta* 1991, **1059:** 208–214.
62. Radmer, R. and Ollinger, O. *FEBS Lett.* 1980, **110:** 57–61.
63. Bader, K.P. and Schmid, G.H. *Biochim. Biophys. Acta* 1988, **936:** 179–186.
64. Bader, K.P., Renger, G. and Schmid, G.H. *Photosynthesis Res.* 1993, **38:** 355–361.
65. Messinger, J., Badger, M. and Wydrzynski, T. *Proc. Natl. Acad. Sci. USA* 1995, **92:** 3209–3213.
66. Messinger, J., Hiller, W., Badger, M. and Wydrzynski, T. in *Photosynthesis: From Light to Biosphere* (ed.) Mathes, P. Kluwer, Dordrecht, The Netherlands, 1995, Vol. 2, pp. 283–286.
67. Messinger, J., Wacker, U. and Renger, G. *Biochemistry* 1991, **30:** 7852–7862.
68. Velthuys, B.R. *Biochem. Biophys. Acta* 1975, **396:** 392–401.
69. Boussac, A., Rutherford, A.W. and Styring, S. *Biochemistry* 1990, **29:** 24–32.
70. Govindjee, Kambara, T. and Coleman, W. *Photochem. Photobiol.* 1985, **42:** 187–210.
71. Dismukes, G.C. *Photochem. Photobiol.* 1986, **43:** 99–115.
72. Blondin, G., Davydov, R., Philouze, C., Charlot, M.-F., Styring, S., Åkermark, B., Girerd, J.-J. and Boussac, A. *J. Chem. Soc. Dalton Trans.* 1997, 4069–4074.
73. Brudvig, G.W. and Crabtree, R.H. *Proc. Natl. Acad. Sci. USA* 1986, **83:** 4586–4588.
74. Wieghardt, K., Bossek, U. and Gebert, W. *Angew. Chem. Int. Ed. Engl.* 1983, **22:** 328–329.
75. Vincent, J.B. and Christou, G. *Inorg. Chim. Acta* 1987, **136:** L41–L43.
76. Ferguson-Miller, S. and Babcock, G.T. *Chem. Rev.* 1996, **96:** 2889–2909.
77. Sono, M., Roach, M.P., Coulter, E.D. and Dawson, J.H. *Chem. Rev.* 1996, **96:** 2841–2888.

78. Collins, T.J. and Gordon-Wylie, S.W. *J. Am. Chem. Soc.* 1989, **111:** 4511–4513.
79. (a) Collins, T.J., Powell, R.D., Slebodnick, C. and Uffelman, E.S. *J. Am. Chem. Soc.* 1990, **112:** 899–901; (b) Miller, C.G., Gordon-Wylie, S.W., Horwitz, C.P., Strazisar, S.A., Peraino, D.K., Clark, G.R., Weintraub, S.T. and Collins, T.J. *J. Am. Chem. Soc.* 1998, **120:** 11540–11541.
80. MacDonnell, F.M., Fackler, N.L.P., Stern, C. and O'Halloran, T.V. *J. Am. Chem. Soc.* 1994, **116:** 7431–7432.
81. Proserpio, D.M., Hoffmann, R. and Dismukes, G.C. *J. Am. Chem. Soc.* 1992, **114:** 4374–4382.
82. Proserpio, D.M., Rappe, A.K. and Gorun, S.M. *Inorg. Chim. Acta* 1993, **213:** 319–324.
83. Hoganson, C.W. and Babcock, G.T. *Science* 1997, **277:** 1953–1956.
84. Hoganson, C.W., Lydakis-Simantiris, N., Tang, X.-S., Tommos, C., Warncke, K., Babcock, G.T., Diner, B.A., McCracken, J. and Styring, S. *Photosynthesis Res.* 1995, **46:** 177–184.
85. Gardener, K.A. and Mayer, J.M. *Science* 1995, **269:** 1849–1851.
86. Caudle, M.T. and Pecoraro, V.L. *J. Am. Chem. Soc.* 1997, **119:** 3415–3416.
87. Blomberg, M.R.A., Siegbahn, P.E.M., Styring, S., Babcock, G.T., Åkermark, B. and Korall, P. *J. Am. Chem. Soc.* 1997, **119:** 8285–8292.
88. Bordwell F.G. and Cheng, J.-P. *J. Am. Chem. Soc.* 1991, **113:** 1736–1743.
89. Lind, J., Shen, X., Eriksen, T.E. and Merenyi, G.J. *J. Am. Chem. Soc.* 1990, **112:** 479–482.
90. Shafirovich, V.Y., Khannanov, N.K. and Shilov, A.E. *J. Inorg. Biochem.* 1981, **15:** 113–129.
91. Luneva, N.P., Knerelman, E.I., Shafirovich, V.Y. and Shilov, A.E. *J. Chem. Soc., Chem. Commun.* 1987, 1504–1505.
92. Ramaraj, R., Kura, A. and Kaneko, M. *Chem. Lett.* 1987, **264:** 261–264.
93. Ramaraj, R., Kira, A. and Kaneko, M. *Angew. Chem. Int. Ed. Engl.* 1986, **25:** 825–827.
94. Swarnabala, G. and Rajasekharan, M.V. *Proc. Indian. Acad. Sci. (Chem. Sci.)* 1990, **102:** 87–98.
95. Gobi, K.V., Ramaraj, R. and Kaneko, M. *J. Mol. Catal.* 1993, **81:** L7–L11.
96. (a) Naruta, Y., Sasayama, M. and Sasaki, T. *Angew. Chem. Int. Ed. Engl* 1994, **33:** 1839–1841; (b) Ichihara, K. and Naruta, Y. *Chem. Lett.* 1998, 185–186.
97. Gersten, S.W., Samuels, G.J. and Meyer, T.J. *J. Am. Chem. Soc.* 1982, **104:** 4029–4030.

98. Hurst, J.K., Zhou, J. and Lei, Y. *Inorg. Chem.* 1992, **31:** 1010–1017.
99. Geselowitz, D. and Meyer, T.J. *Inorg. Chem.* 1990, **29:** 3894–3896.
100. Rotzinger, F.P., Munavalli, S., Comte, P., Hurst, J.K., Gratzel, M., Den, F.-J. and Frank, A.J. *J. Am. Chem. Soc.* 1987, **109:** 6619–6626.
101. Comte, P., Nazeerudin, M.K., Rotzinger, F.P., Frank, A.J. and Gratzel, M. *J. Mol. Catal.* 1989, **52:** 63–84.
102. Nazeerudin, M.K., Rotzinger, F.P., Comte, P. and Gratzel, M. *J. Chem. Soc., Chem. Commun.* 1988, 872–874.
103. Cooper S.R. and Calvin, M. *J. Am. Chem. Soc.* 1977, **99:** 6623–6630.
104. Dave, B.C., Czernuszewicz, R.S., Bond, M.R., and Carrano, C.J. *Inorg. Chem.* 1993, **32:** 3593–3594.
105. Manchanda, R., Brudvig, G.W., Crabtree, R.H., Sarneski, J.E. and Didiuk, M. *Inorg. Chim. Acta* 1993, **212:** 135–137.
106. Philouze, C., Blondin, G., Ménage, S., Auger, N., Girerd, J.-J., Vigner, D., Lance, M. and Nierlich, M. *Angew. Chem. Int. Ed. Engl.* 1992, **31:** 1629–1631.
107. Pal, S. and Armstrong, W.H. *Inorg. Chem.* 1992, **31:** 5417–5423.
108. Pal, S., Chan, M.K. and Armstrong, W.H. *J. Am. Chem. Soc.* 1992, **114:** 6398–6406.
109. (a) Limburg, J., Brudvig, G.W. and Crabtree, R.H. *J. Am. Chem. Soc.* 1997, **119:** 2761–2762; (b) Limburg, J., Vrettos, J.S., Liable-Sands, L.M., Rheingold, A.L., Crabtree, R.H. and Brudvig, G.W. *Science* 1999, **283:** 1524–1527; (c) Bernadou, J. and Meunier, B. *J. Chem. Soc., Chem. Comm.* 1998, 2167–2173.
110. Robert, A. and Meunier, B. *New J. Chem.* 1988, **12:** 885–896.
111. Wessel, J. and Crabtree, R.H. *J. Mol. Catal.* 1996, **133A:** 13–22.
112. Wieghardt, K., Bossek, U., Nuber, B., Weiss, J., Bonvoisin, J., Corbella, M., Vitols, S.E. and Girerd, J.-J. *J. Am. Chem. Soc.* 1988, **110:** 7398–7411.
113. Bossek, U., Weyhermüller, T., Wieghardt, K., Nuber, B. and Weiss, J. *J. Am. Chem. Soc.* 1990, **112:** 6387–6388.
114. Limburg, J., Szalai, V.A. and Brudvig, G.W. *J. Chem. Soc., Dalton Trans.* 1999, 1353–1361.
115. Szalai, V.A. and Brudvig, G.W. *Amer. Sci.* 1998 **86:** 542–551.
116. Pecoraro, V.L., Baldwin, M.J., Caudle, M.T., Hsieh, W. and Law, N.A. *Pure and Appl. Chem.* 1998, **70:** 925–929.
117. Siegbahn, P.E.M. and Crabtree, R.H. *J. Am. Chem. Soc.* 1999, **121:** 117–127.

118. Halfen, J.A., Mahapatra, S., Wilkinson, E.C., Kaderli, S., Young, V.G., Que, L., Zuberbuhler, A.D. and Tolman, W.B. *Science* 1996, **271:** 1397–1400.
119. Pecoraro, V.L., Baldwin, M.J. and Gelasco, A. *Chem. Rev.* 1994, **94:** 807–826.

Chapter 12
Asymmetric Biomimetic Oxidations

Anne Robert & Bernard Meunier

Laboratoire de Chimie de Coordination du CNRS,
205 route de Narbonne, 31077 Toulouse cedex 4, France

1. Introduction

The increasing demand for enantiomerically pure materials is the constant driving force for the development of new methods of asymmetric oxidations allowing the preparation of highly useful synthons from low cost starting material e.g. olefins, ketones, etc.

Among all the different asymmetric reactions leading to the incorporation of one (or two) oxygen atom(s) within a substrate, the most desirable and the most challenging is the catalytic asymmetric oxidation, since only one chiral catalyst molecule can create from few to thousand chiral product molecules.[1,2] Following the development of asymmetric catalytic hydrogenation (due to the pioneering work of Kagan and Knowles in the early 1970s), the field of catalytic asymmetric oxidation was opened in 1980 by Sharpless and Katsuki on the asymmetric epoxidation of allylic alcohols by combining $Ti(iOPr)_4$ an optically active tartrate ester and *t*-butyl hydroperoxide ("Sharpless I").[3,4] This reaction has been successfully developed at the industrial scale for the preparation of many different allylic epoxides. Furthermore, this catalytic system has been adapted to the asymmetric oxidation of sulfides to chiral sulfoxides with high enantiomeric purity.[5,6] These optically active sulfoxides are useful chiral auxiliaries in asymmetric organic synthesis. A second major asymmetric reaction concerns the asymmetric dihydroxylation of olefins ("Sharpless II") by using catalytic amounts of osmium tetroxide with bis(cinchona) alkaloid ligands based on quinidine or quinine.[7–9] In the present chapter, the field of asymmetric oxidations will be limited to catalytic systems bio-inspired by the work performed in the field of biomimetic oxidations as described in the preceding chapters of this book. The field of biomimetic asymmetric oxidations started initially with chiral metalloporphyrins, but was found to be applicable in organic synthesis when chiral Schiff base complexes were developed independently by Jacobsen[10,11] and Katsuki.[12–14] Both aspects of these biomimetic asymmetric oxidations will be reported in the present chapter as well as few examples of asymmetric Baeyer-Villiger reactions catalyzed by chiral complexes. The complexes used for this latter reaction are not directly related to biomimetic chemistry since this reaction is catalyzed by flavin-based enzymes without metal ions at the active site, but it will be highly useful to develop such a powerful asymmetric oxidation reaction with efficient transition metal catalysts.

2. Asymmetric Oxidations Catalyzed by Chiral Metalloporphyrins

Cytochrome P450 enzymes are able to perform asymmetric oxygenation reactions (epoxidation and hydroxylation) because of the chiral environment created by the amino acids of the active site. Both reactions, asymmetric epoxidation of non-activated olefins and asymmetric hydroxylation of aliphatic C–H bonds, are two highly useful reactions allowing the preparation of high value chiral products from cheap raw materials. However, both reactions were just considered as dreams in the late 1970s since no chemical catalysts were able to perform such reactions (the classical autoxidation reactions with intermediate free organic radicals are obviously not adapted for asymmetric oxidations). The discovery of metalloporphyrins as biomimetic catalysts able to epoxidize olefins and hydroxylate alkanes without free radical intermediates, triggered in 1983–1988 the initial reports on asymmetric epoxidations of non-activated olefins catalyzed with chiral metalloporphyrins (Scheme 1 for reactions and Scheme 2 for catalyst structures).

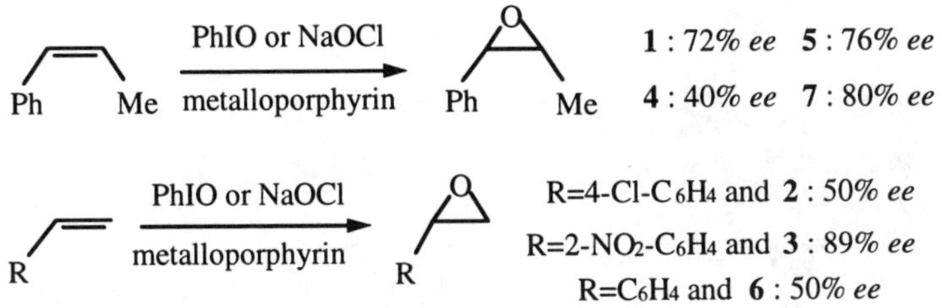

Scheme 1 Examples of asymmetric epoxidations catalyzed by chiral metalloporphyrins (see Schemes 2(a) and (b) for catalyst structures).

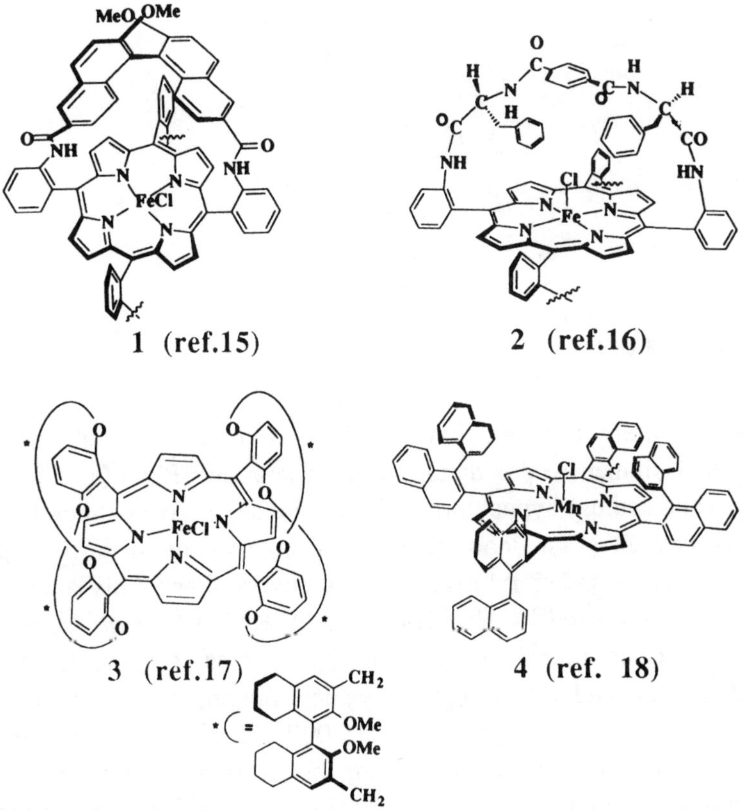

Scheme 2(a) Structures of chiral metalloporphyrin catalysts used in asymmetric epoxidations (modified from Ref. 14).

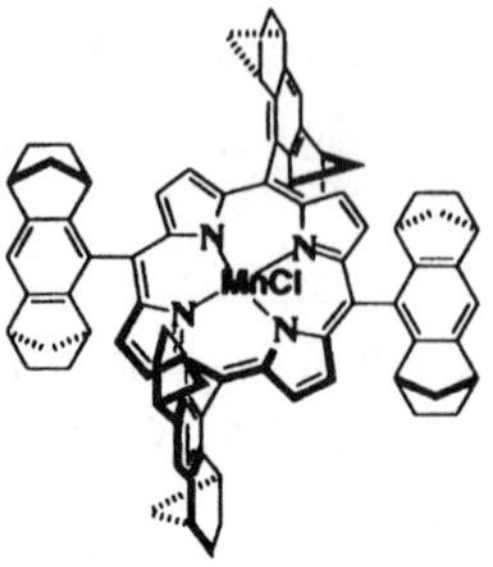

5 (ref.19)

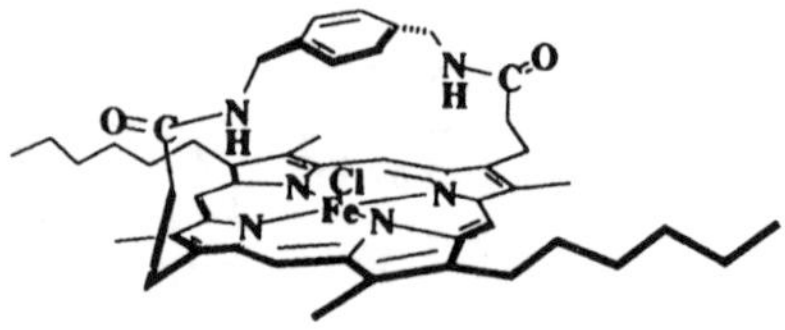

6 (ref.20)

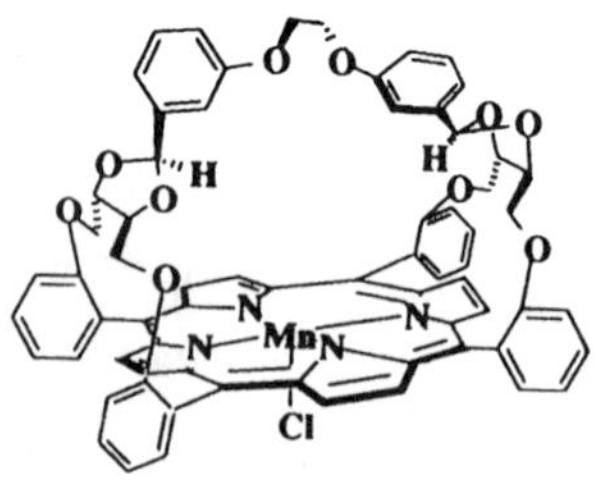

7 (ref.21)

Scheme 2(b) Structures of chiral metalloporphyrin catalysts used in asymmetric epoxidations (modified from Ref. 14).

Two different oxidants have been used in these metalloporphyrin-catalyzed asymmetric epoxidations: iodosylbenzene[22] PhIO (a rather smooth oxidant) and sodium hypochlorite[23,24] (a more aggressive oxidant but a very efficient oxygen atom donor to generate high-valent metal-oxo species). With this oxidant and chiral manganese porphyrins **4** or **5** (this latter complex was initially prepared by Halterman *et al.*), Kodadek *et al.* reported very high turnover numbers with styrene-type olefins, ranging from 950 to 18 720 cycles, and rather high *ee* values, ranging from 40% to 70%.[25] With aliphatic olefins and catalyst **5,** 15 000 cycles with a remarkable 85% *ee* value can be obtained. So far with this class of olefins, these are the best results reported in the literature using any type of chiral epoxidation catalyst. *N*-oxides have also been recently used as oxidant in ruthenium porphyrin-catalyzed epoxidations.[26]

Collman *et al.* provided some rational aspects to help in the design of chiral metalloporphyrin catalysts.[27] Among all different chiral structures studied, the D_4 symmetry seems to provide the most efficient metalloporphyrin catalysts in asymmetric epoxidations. Glyco-conjugated manganese porphyrins gave rather low enantiomeric excesses (10%–30% *ee*).[28]

Prochiral alkyl sulfides gave sulfoxides with 14%–48% *ee* with a vaulted iron porphyrin.[29] Higher *ee* values have been obtained with prochiral aryl sulfides.[30]

The first asymmetric hydroxylations were reported by Groves and Viski in 1990 with a vaulted manganese porphyrin catalyst with *ee* values ranging from 40% to 72%.[29] A detailed study with this chiral catalyst indicated that (*R*)-(1-deuterioethyl)benzene provided the *S* alcohol in excess (16% *ee*) while (*S*)-(1-deuterioethyl)benzene provided the *R* alcohol but with a much higher, isotopically enhanced stereoselectivity (77% *ee*).[31] Stereoselective hydroxylation at single unactivated carbon atom of chiral substrates was recently performed with a manganese porphyrin carrying four β-cyclodextrin substituents.[32] It must be noted that the rate of this sophisticated hydroxylation reaction is close to that observed for cytochrome P450 i.e. 260 catalytic cycles obtained within 10 min.

The major limitation of these metalloporphyrin-catalyzed asymmetric oxidations is the rather tedious multistep syntheses of the chiral porphyrin ligands. This point has been solved by using chiral salen-type catalysts.

3. Asymmetric Oxidations Catalyzed by Chiral Schiff Base Complexes

In 1980, it was reported that catalytic amounts of manganese Schiff bases associated with sodium hypochlorite as oxygen atom donor were able to epoxidize styrene (8% epoxide yield after 3 h, corresponding to 13 catalytic cycles).[23,24] Some years later, Kochi *et al.* used chromium[33] and manganese[34] Schiff base complexes as catalysts for olefin epoxidation in the presence of iodosylbenzene. Burrows C. also reported the epoxidation of non-functionalized olefins catalyzed by nickel Schiff base complexes associated with sodium hypochlorite.[35] Although the obtained yields were moderates with these non-chiral catalysts, these preliminary works evidenced that Schiff base complexes were able to mimic the monooxygenase activity of heme proteins, in a similar way as synthetic metalloporphyrins. It was therefore reasonable to use chiral Schiff base ligands in order to perform asymmetric oxygenations of prochiral substrates.

3.1 *First enantioselective oxidations catalyzed by chiral Schiff base complexes*

Brunner *et al.* described the enantioselective cyclopropanation of 1,1-diphenylethylene (*ee* = 24%) in the presence of a copper complex obtained by condensation of salicylaldehyde and 1,1'-binaphthyl-2,2'-diamine as chiral moiety.[36] The catalytic oxidation of styrene into 1-phenylethanol was carried on with cobalt complexes of Schiff bases bearing 1,2-diaminocyclohexyl, 2,2'-diamino-6,6'-dimethyl-1,1'-biphenyl, or 2,2'-diamino-1,1'-binaphthyl moieties. This latter complex exhibited the highest enantiomeric excesses, up to 38%.[37]

Katsuki *et al.* synthesized Schiff base ligands and manganese complexes starting from chiral derivatives of ethylene diamine (especially 1,2-diphenylethylene-1,2-diamine) and of salicylaldehydes bearing also chiral

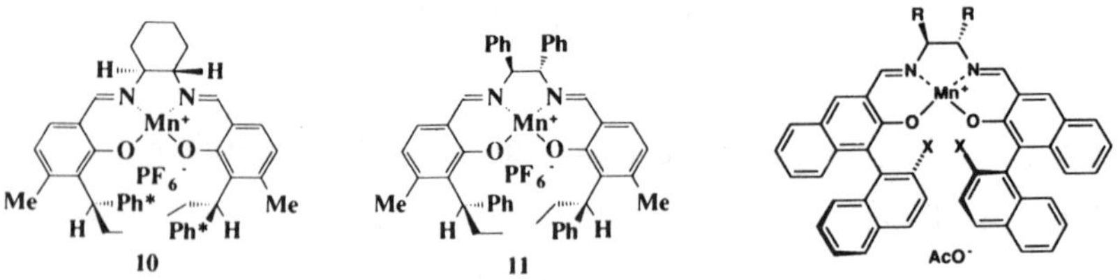

8 (Ref. 40)

9 (Ref. 38)

10a R=Ph, X=Me
10b R=Ph, X=Ph
10c R=3,5-$(CH_3)_2C_6H_3$, X=Ph
10d R=-$(CH_2)_4$-, X=Ph
(Ref. 41)

Scheme 3 Structure of chiral Schiff base complexes used in asymmetric epoxidations.

substituents in *ortho* position of the -OH function.[38–41] These manganese complexes (Scheme 3), bearing four chiral centers, have been used for the epoxidation of non-functionalized olefins, with iodosylbenzene as oxidant. For the epoxidation of *cis*-olefins, the highest enantiomeric excesses have been obtained with catalyst **10c** (96% *ee* for dihydronaphthalene and 92% for indene[41]) and catalyst **8** (91% *ee* for dihydronaphthalene[40]). For the epoxidation of *trans*-stilbene, 62% *ee* could be achieved with **8**,[40] 48% *ee* with catalyst **9**,[39] and only 9% *ee* with catalyst **10c**. For a recent review on this family of Schiff bases with four chiral centers, see Ref. 14.

3.2 Development of the Jacobsen's catalytic system

During the same period, Jacobsen *et al.* developed a series of manganese Schiff base complexes bearing only two asymmetric centers on the diamine moiety and *tert*-butyl substituents in *ortho* position to the phenol function.[42] With iodosylmesitylene as the oxygen atom source, the epoxidation of 1,2-dihydronaphthalene was obtained with 72% yield and 78% *ee*.

Then Jacobsen *et al.* described the synthesis and use of about one hundred metal complexes based on C_2 symmetric salen chiral ligands.[11,43] Among all these different catalysts, the highest catalytic activities and the best enantioselectivities were obtained with the manganese complex of the Schiff base obtained by condensation of chiral *trans*-1,2-diaminocyclohexane with 3,5-di-*tert*-butylsalicylaldehyde, which is easily prepared by formylation of 2,4-di-*tert*-butylphenol (Scheme 4).[44] Two reasons explain the importance of these bulky substituents on the salicylidene moieties: (i) they avoid "west, south and east approaches," and enforce the prochiral olefin approach near the dissymmetric diimine bridge, and (ii) they create a cage effect around the generated manganese-oxo species, thereby reducing the formation of catalytically inactive μ-oxo dimers and ligand bleaching. The latter phenomenon was found to be the main limitation of the NaOCl/Mn(salen)[24] or PhIO/Mn(salen)[34] oxidation systems.

In some cases, the presence of bases as axial ligands (especially 4-phenylpyridine *N*-oxide) increases both catalytic activity and enantioselectivity.[45] The presence of electron-donating groups on the catalyst

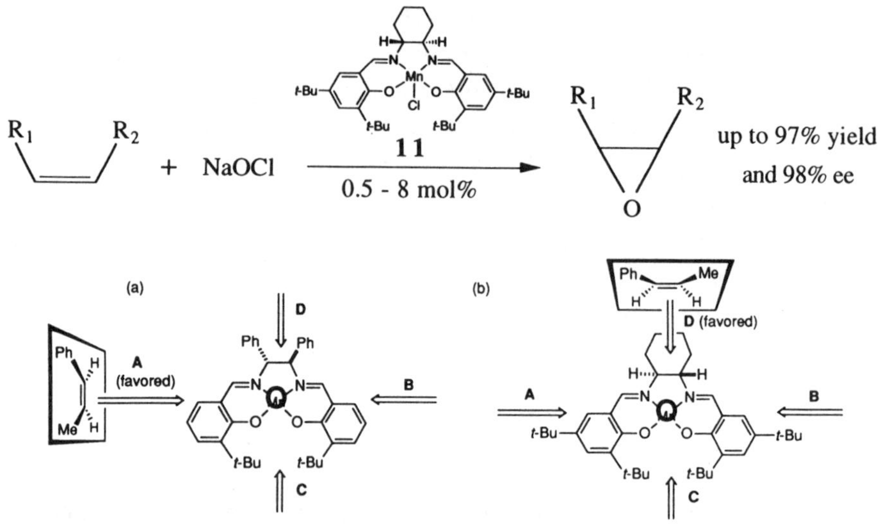

Scheme 4 Asymmetric epoxidation using chiral manganese Schiff base catalysts (modified from Ref. 11).

provides higher enantioselectivities in epoxidation reactions.[46] For recent investigations on the mechanism of oxygen-atom transfer from the high-valent metal-oxo intermediate to the substrate, see Refs. 47 and 48.

The best "Jacobsen's catalyst" **11** can be prepared on a large scale[49] and is now commercially available. When associated with the simple and cheap oxidant sodium hypochlorite, this catalyst was found to be efficient and widely developed for the asymmetric epoxidation of a variety of interesting and important substrates; some examples are given below.

Remarkably high selectivities are observed in the epoxidation of 2,2-dimethyl-3-chromene derivatives (*ee* = 97%–98%).[50] The enantioselective epoxidation of 6-cyano-2,2-dimethyl-3-chromene was a key step for the synthesis of the antihypertensive agent chromakalim. The epoxidation of *cis*-ethyl cinnamate ester (95%–97% *ee*) was applied to the synthesis of the taxol C-13 side chain.[45] This method of direct epoxidation of an α, β-unsaturated ester via an oxo transfer was also used in the asymmetric synthesis of the antihypertensive diltiazem.[51] Monoepoxides of 1,3-dienes[52] and conjugated enynes[53] are valuable intermediates in organic synthesis.

The facts that *cis*-olefins are generally more reactive in the Mn(III)salen-catalyzed epoxidation, and that the epoxidation of conjugated *cis*-olefins give *trans*-epoxide as the major product,[53] suggest a strategy for the regio- and enantioselective synthesis of *trans,trans*-diene monoepoxides by epoxidation of *cis,trans*-conjugated dienes. This strategy was successfully applied to the synthesis of a variety of optically active vinyl epoxides, including a key intermediate in the synthesis of leucotriene A_4 (*ee* = 80%–83%).[54] The scope of the manganese-salen-catalyzed epoxidation was extended by epoxidation of one of the chemical industry's most important prochiral olefins — the styrene. At –78 °C, enantiomeric excess as high as 83%–86% could be achieved with *m*-chloroperbenzoic acid associated with catalyst **11** and *N*-methylmorpholine *N*-oxide as co-catalyst. It was observed that this additive is critical in order to reach high enantioselectivities.[55]

During the catalytic epoxidation of 1,2-dihydronaphthalene with a manganese Schiff base associated with hypochlorite, the asymmetric hydroxylation at the benzylic position of the produced epoxide occurs as a side reaction (Scheme 5). As the major enantiomer of the dihydronaphthalene oxide is also the less reactive in this hydroxylation side reaction, the reaction mixture is progressively enriched in that epoxide.[56] In fact, the stereoselective oxidation of benzylic C–H bonds using chiral manganese(III) Schiff base complexes allows a kinetic resolution process of the produced epoxides. In this manner, dihydronaphthalene-derived epoxides with *ee*'s > 98% can be produced.

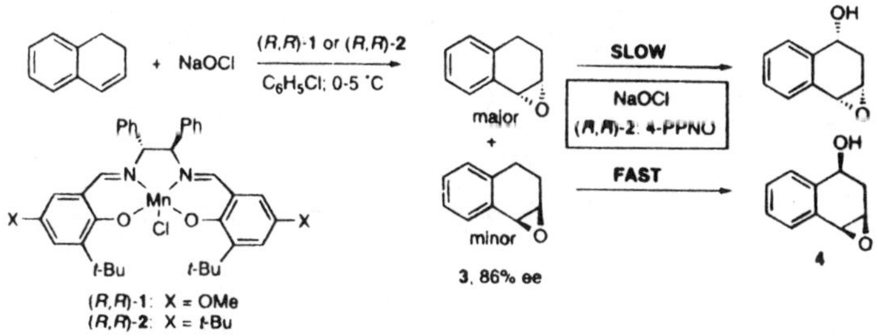

Scheme 5 Kinetic resolution of dihydronaphthalene oxide by enantioselective hydroxylation of a benzylic C–H bond (reproduced with permission from Ref. 56).

Enhancement of the selectivity for the formation of *trans*-epoxides from aryl- and alkyl-disubstituted *cis*-olefins was achieved by using a cinchona alkaloid-derived salt in the epoxidation medium.[57]

The same Schiff base ligands as previously described, but metallated with chromium(III) were used for the enantioselective ring-opening of epoxides with trimethylsilyl azide. This reaction was first used for the kinetic resolution of *meso* epoxides,[58] but it can also provide an access to optically pure α-azido trimethylsiloxy intermediates (Scheme 6).

From this intermediate, a desilylation/reduction sequence yielded synthetically important α-amino alcohols.[59,60] The mechanism of the asymmetric ring-opening of epoxides by trimethylsilyl azide catalyzed by chromium(III) salen was recently investigated.[61] This asymmetric ring-opening of *meso* and racemic epoxides by $TMSN_3$ displayed a particularly high enantioselectivity for the opening of epoxides fused to five-membered rings.[62] This method has been applied for (i) the synthesis of carbocyclic nucleoside analogues used as potential anti-HIV agents,[63] and (ii) the preparation of an aryl oxazolidinone antibacterial agent,[64] and of the protein kinase C inhibitor balanol.[65] This reaction was extended using cobalt(II) Schiff bases associated with carboxylic acids as nucleophiles,[66] or water itself, thus yielding α-diols.[67]

It is noteworthy that the optically active manganese(III) Schiff base complex associated with hydrogen peroxide as oxidant is able to catalyze

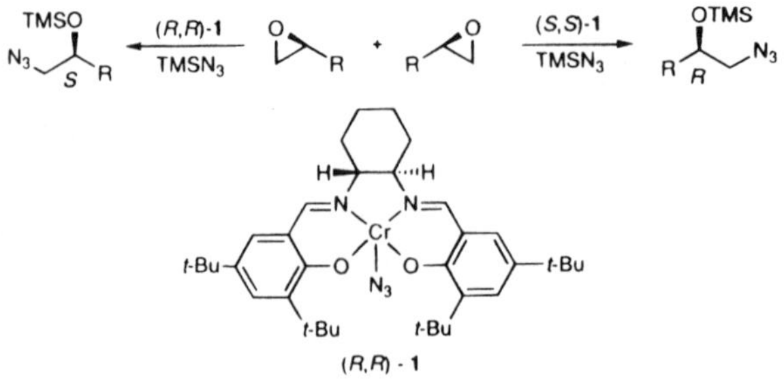

Scheme 6 Asymmetric ring-opening of epoxides by trimethylsilyl azide (reproduced with permission from Ref. 59).

the oxidation of aryl sulfides to sulfoxides, even if synthetically useful enantioselectivities have not yet been obtained with catalysts bearing only diaminocyclohexyl as chiral motif.[68] On the other hand, with complexes derived from **10b** and **10d**, bearing four chiral centers, associated with iodosylbenzene, enantioselectivity up to 94% was achieved.[69]

By analogy to cyclopropanation and epoxidation, diimine-copper(I) complexes were shown to catalyze the aziridination reaction of olefins with PhI = NTs.[70]

3.3 Other developments of chiral Schiff base catalysts

In order to improve the activity and selectivity, and to broaden the scope of substrates of chiral manganese Schiff base complexes (especially in the epoxidation of terminal olefins),[71] a new series of ligands and metal complexes based on the optically active binaphthyl moiety was developed.[72,73] These compounds were obtained by condensation of 1,1'-binaphtyl-2,2'-diamine with various substituted achiral salicylaldehydes and consecutive metallation (Scheme 7). In order to enlarge the availability of chiral binaphthyl-based complexes, the corresponding Mn, Fe, Co, Ni, Cu and Zn compounds have been prepared.[73] In spite of a chiral

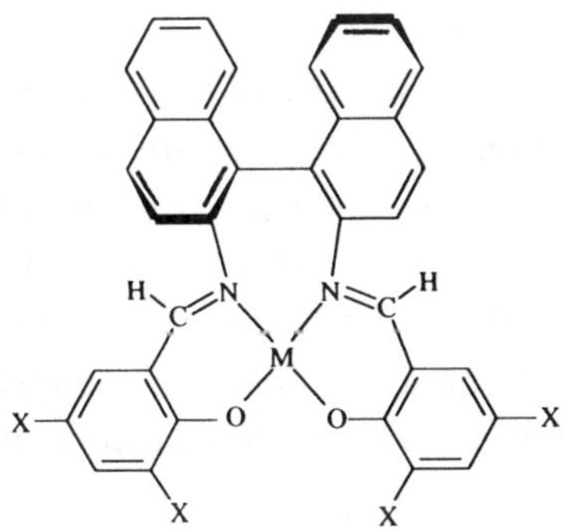

12 X = H, Cl, Br, *t*-Bu or NO_2
M = Mn, Ni, Fe, Co, Zn, Cu

Scheme 7 "Salbinapht" complexes.

environment, the manganese "salbinapht" complexes **12-Mn** displayed poor catalytic activity and enantioselectivity, even in the favorable case of a *cis*-disubstituted olefin like 1,2-dihydronaphthalene, in the presence of sodium hypochlorite, hydrogen peroxide or *m*-chloroperbenzoic acid/*N*-methylmorpholine *N*-oxide as oxidant.

This low activity is probably due to extensive degradation of the manganese binaphthyl-based complexes under the oxidation reaction conditions. A possible explanation for the instability of these manganese complexes is the highly distorted geometry of the salbinapht complexes, more compatible with tetrahedral rather than octahedral or square-pyramidal coordination. The distorted environment around the central metal ion was confirmed by X-ray crystallography of binaphthyl-based Schiff bases metallated with manganese or iron.[74] These authors reported a 76% enantiomeric excess in the epoxidation of the highly reactive *cis*-β-methylstyrene with PhIO at –20 °C with the manganese catalyst; no reaction was obtained with the iron analogue which undergoes rapid decomposition under the oxidative reaction conditions.

Hydrogen peroxide is an attractive oxidant, giving only water as a by-product. It has been used (30% aqueous solution) in the presence of imidazole for the salen-catalyzed asymmetric epoxidation of 1,2-dihydronaphthalene and *trans*-β-methylstyrene.[75]

In 1990, A. Berkessel *et al.* developed the synthesis of Nickel(II) Schiff base complexes bearing a chiral side chain with a thioether group acting as axial ligand to the metal. As expected, this ligand was degraded rapidly with dioxygen in solution.[76] More recently, several chiral pentadentate dihydrosalen ligands bearing an imidazole group as a fifth axial donor ligand were synthesized and metallated with manganese(III) (Scheme 8).

With catalysts **13** and **14** activated with diluted aqueous hydrogen peroxide (1%) and without any added co-catalyst, the epoxidation of 1,2-dihydronaphthalene was achieved with enantiomeric excesses up to 66% respectively, the result being ascribed to an efficient enantioface selection by the chiral catalyst.[77]

Ruthenium(III) complexes with a dissymmetric Schiff base containing the 1,2-diaminocyclohexane moiety have been evaluated for the epoxidation

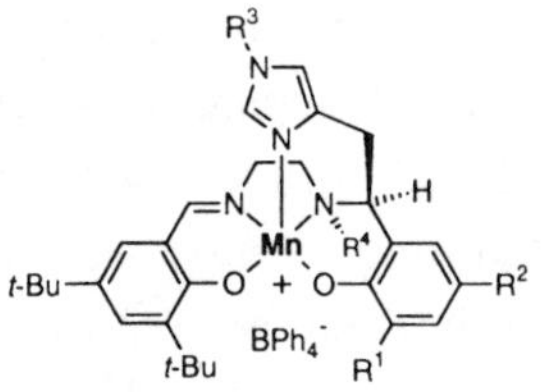

13 $R_1=R_2=R_3=H$
$R_4=CH_3$

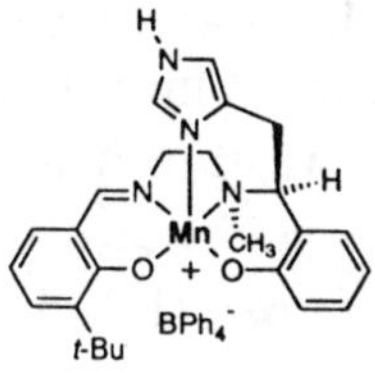

14

Scheme 8 Dihydrosalen manganese(III) complexes with a covalently linked imidazole as axial ligand (modified from Ref. 77).

of *cis*-, *trans*- and terminal olefins (styrene) by iodosylbenzene.[78,79] Only moderate enantiomeric excesses were obtained, in the same range as those observed with the manganese(III) analogues under similar conditions.

Chromium(III) salen complexes associated with PhIO, which are less efficient for the epoxidation of conjugated olefins, were found to exhibit interesting solvent-dependent stereochemisty (better *ee*'s with toluene than with acetonitrile).[80]

Among the other possible oxidants, dimethyldioxirane was used as an oxygen atom donor for the epoxidation of 2,2-dimethyl-3-chromene by the catalyst **11**. The enantioselectivities varied within the range of 63% to 93% *ee* under neutral conditions.[81] This substrate was also epoxidized with molecular oxygen and an aldehyde as reductant in the presence of manganese(III) salen complexes and *N*-alkylimidazoles.[82] But this latter reaction system probably involves a peracid intermediate generated by autoxidation of the aldehyde which can act as an oxygen atom donor or directly as an epoxidizing agent.

Recently, the first asymmetric epoxidation system using an achiral manganese(III) salen complex and (-)-sparteine as chiral amine was reported by Katsuki *et al.*[83] With iodosylbenzene as terminal oxidant and chromene derivatives *ee*'s ranging from 60% to 73% were obtained, but epoxide yields were low (5%–10%).

The possibility of carrying out reactions in media other than pure organic solvents has also been investigated. In order to prepare Schiff base complexes being soluble in water or which can be fixed onto a support by electrostatic interactions, a ligand bearing dimethylamino groups was synthesized by condensation of *trans*-1,2-diaminocyclohexane with 4-dimethylamino-2-hydroxybenzaldehyde.[73] This ligand was metallated with manganese. Unfortunately, the assays of epoxidation of several olefins and sulfoxidation of alkyl arylsulfides by ClO^- in a homogeneous mixture water/acetonitrile were unsuccessful. To make the recovery and reuse of the catalyst easier, the Jacobsen's complex **11** was occluded in an elastomeric polydimethylsiloxane membrane.[84] Enantiomerically pure manganese(III) salen complexes based on 1,2-diaminocyclohexane and bearing styrene-type vinyl groups were submitted to radical co-polymerization with styrene and divinylbenzene.[85] In both cases, the heterogeneous catalysts were found to be easily recycled, to have mainly unmodified catalytic efficiencies for several runs, and to be more resistant to inactivation with respect to the corresponding homogeneous catalyst.

4. Asymmetric Baeyer-Villiger Oxidations Catalyzed by Metal Complexes

The Baeyer-Villiger reaction corresponds to the addition of a peracid onto a ketone to produce an ester. Enzymes are able to catalyze the asymmetric Baeyer-Villiger reaction with active sites involving a flavin-peroxide surrounded by chiral amino acids.[86,87] Only very few enantioselective chemical catalysts are able to perform an asymmetric Baeyer-Villiger reaction. Strukul *et al.* reported such a reaction with a chiral Pt catalyst, but weak *ee*'s and turnovers were obtained.[88,89] The second example of a catalyzed asymmetric Baeyer-Villiger reaction involving a chiral copper complex and dioxygen in the presence of pivalaldehyde as oxygen atom source[90–92] was reported by Bolm *et al.* With a low catalyst/cyclic ketone ratio (1/100), the expected lactone with 41% yield and an *ee* value of 65% was obtained. Cyclobutanones are oxidized to lactones via enantiodivergent pathways as observed with enzymes.[92] The "Sharpless I" catalyst has been used for the asymmetric Baeyer-Villiger oxidation of similar cyclobutanones with *ee*'s ranging from 34% to 75% depending on the structure of the starting ketone.[93]

5. Concluding Remarks

We should consider this chapter as a starting point containing the key references of this rapidly expanding area of the biomimetic oxidation domain, and not as an exhaustive review on biomimetic asymmetric oxidations. In fact, this area will not suffer from parameters which will be considered as drawbacks for ton-scale preparation of oxidation products with non-chiral catalysts, like salt-producing oxidants (NaOCl, $KHSO_5$, etc.), limited turnover numbers (50 to 200 catalytic cycles) or rather elaborated ligands. The increasing need of new chiral synthons and drugs will be a constant driving force to develop new efficient bio-inspired asymmetric oxidation reactions. The creation of a new C–O bond in prochiral substrates using an optically active catalyst will continue to be a challenging area in the future. In particular, there is a real need for efficient asymmetric hydroxylation reactions catalyzed by cheap and easily accessible chiral catalysts.

The present contribution may have suffered due to the fact that the colleague who initially agreed to be in charge of this chapter decided to cancel his participation at the final stage of the elaboration of this book. Consequently, the writing of this chapter was performed within a short period of time, but we hope that the present survey allows the readers to find sufficient key references for entry into this quickly expanding area concerning biomimetic asymmetric oxidations.

Abbreviation

Salen: salicylaldehyde ethylenediimine, the progenitor of a successful series of Schiff base ligands.

References

1. *Catalytic Asymmetric Synthesis* (ed.) Ojima, I. VCH, Weinheim, 1993.
2. Corey, E.J. and Guzman-Perez, A. *Angew. Chem. Int. Ed.* 1998, **37:** 388–401.
3. Katsuki, T. and Sharpless, K.B. *J. Am. Chem. Soc.* 1980, **102:** 5974–5976.
4. Johnson, R.A. and Sharpless, K.B., in Ref. 1, Chap. 4.1, pp. 103–158.
5. Kagan, H.B., in Ref. 1, Chap. 4.3, pp. 203–226.
6. Brunel, J.M. and Kagan, H.B. *Bull. Soc. Chim. Fr.* 1996, **133:** 1109–1115.
7. Sharpless, K.B., Amberg, W., Bennani, Y.L., Crispino, G.A., Hartung, J., Jeong, K.S., Kwong, H.L., Morikawa, K., Wang, Z.M., Xu, D. and Zhang, X.L. *J. Org. Chem.* 1992, **57:** 2768–2771.
8. Johnson, R.A. and Sharpless, K.B., in Ref. 1, Chap. 4.4, pp. 227–272.
9. Guzman-Perez, A. and Corey, E.J. *Tetrahedron Lett.* 1997, **38:** 5941–5944.
10. Zhang, W., Loebach, J.L., Wilson, S.R. and Jacobsen, E.N. *J. Am. Chem. Soc.* 1990, **112:** 2801–2803.
11. Jacobsen, E.N., in Ref. 1, Chap. 4.2, pp. 159–202.
12. Irie, R., Noda, K., Ito, Y., Matsumoto, N. and Katsuki, T. *Tetrahedron Lett.* 1990, **31:** 7345–7348.
13. Katsuki, T. *J. Mol. Cat. A: Chem.* 1996, **113:** 87–107.
14. Katsuki, T. *Coord. Chem. Rev.* 1995, **140:** 189–214.
15. Groves, J.T. and Myers, R.S. *J. Am. Chem. Soc.* 1983, **105:** 5791–5796.
16. Mansuy, D., Battioni, P., Renaud, J.P. and Guerin, P. *J. Chem. Soc., Chem. Commun.* 1985, 155–156.
17. Naruta, Y., Tani, F., Ishihara, N. and Maruyama, K. *J. Am. Chem. Soc.* 1991, **113:** 6865–6872.
18. O'Malley, S. and Kodadek, T. *J. Am. Chem. Soc.* 1989, **111:** 9116–9117.
19. Halterman, R.L. and Jan, S.T. *J. Org. Chem.* 1991, **56:** 5253–5254.
20. Konishi, K., Oda, K., Nishida, K., Aida, T. and Inoue, S. *J. Am. Chem. Soc.* 1992, **114:** 1313–1317.
21. Collman, J.P., Lee, V.J., Kellen-Yuen, C.L., Zhang, X., Ibers, J.A. and Braumann, J.I. *J. Am. Chem. Soc.* 1993, **117:** 692–703.
22. Groves, J.T., Nemo, T.E. and Myers, R.S. *J. Am. Chem. Soc.* 1979, **101:** 1032–1033.
23. Guilmet, E. and Meunier, B. *Tetrahedron Lett.* 1980, **21:** 4449–4450.
24. Meunier, B., Guilmet, E., De Carvalho, M.E. and Poilblanc, R. *J. Am. Chem. Soc.* 1984, **106:** 6668–6676.

25. Barry, J.F., Campbell, L., Smith, D.W. and Kodadek, T. *Tetrahedron* 1997, **53:** 7753–7776.
26. Berkessel, A. and Frauenkron, M. *J. Chem. Soc., Perkin Trans 1* 1997, 2265–2266.
27. Collman, J.P., Zhang, X., Lee, V.J., Uffelman, E.S. and Braumann, J.I. *Science* 1993, **261:** 1404–1411.
28. Vilain-Deshayes, S., Robert, A., Maillard, P., Meunier, B. and Momenteau, M. *J. Mol. Cat. A: Chem.* 1996, **113:** 23–34.
29. Groves, J.T. and Viski, P. *J. Org. Chem.* 1990, **55:** 3628–3634.
30. Chiang, L.C., Konishi, K., Aida, T. and Inoue, S. *J. Chem. Soc., Chem. Commun.* 1992, 254–256.
31. Groves, J.T. and Viski, P. *J. Am. Chem. Soc.* 1989, **111:** 8537–8538.
32. Breslow, R., Huang, Y., Zhang, X. and Yang, J. *Proc. Natl. Acad. Sci. USA* 1997, **94:** 11156–11158.
33. Samsel, E.G., Srinivasan, K. and Kochi, J.K. *J. Am. Chem. Soc.* 1985, **107:** 7606–7617.
34. Srinivasan, K., Michaud, P. and Kochi, J.K. *J. Am. Chem. Soc.* 1986, **108:** 2309–2320.
35. Yoon, H. and Burrows, C.J. *J. Am. Chem. Soc.* 1988, **110:** 4087–4089.
36. Brunner, H. and Miehling, W. *Monatsh. Chem.* 1984, **115:** 1237–1254.
37. Nishinaga, A., Yamato, H., Abe, T., Maruyama, K. and Matsuura, T. *Tetrahedron Lett.* 1988, **29:** 6309–6312.
38. Irie, R., Noda, K., Ito, Y., Matsumoto, M. and Katsuki, T. *Tetrahedron Lett.* 1990, **31:** 7345–7348.
39. Irie, R., Noda, K., Ito, Y. and Katsuki, T. *Tetrahedron Lett.* 1991, **32:** 1055–1058.
40. Hosoya, N., Irie, R. and Katsuki, T. *Synlett.* 1993, 261–263.
41. Sasaki, H., Irie, R. and Katsuki, T. *Synlett.* 1994, 356–358.
42. Zhang, W., Loebach, J.L., Wilson, S.R. and Jacobsen, E.N. *J. Am. Chem. Soc.* 1990, **112:** 2801–2803.
43. Pospisil, P.J., Carsten, D.H. and Jacobsen, E.N. *Chem. Eur. J.* 1996, **2:** 974–980.
44. Jacobsen, E.N., Zhang, W., Muci, A.R., Ecker, J.R. and Deng, L. *J. Am. Chem. Soc.* 1991, **113:** 7063–7064.
45. Deng, L. and Jacobsen, E.N. *J. Org. Chem.* 1992, **57:** 4320–4323.
46. Jacobsen, E.N., Zhang, W. and Güler, M.L. *J. Am. Chem. Soc.* 1991, **113:** 6703–6704.

47. Finney, N.S., Pospisil, P.J., Chang, S., Palucki, M., Konsler, R.G., Hansen, K.B. and Jacobsen, E.N., *Angew. Chem. Int. Ed. Engl.* 1997, **36:** 1720–1723.
48. Linde, C., Arnold, M., Norrby, P.-O. and Akermark, B. *Angew. Chem. Int. Ed. Engl.* 1997, **36:** 1723–1725.
49. Larrow, J.F., Jacobsen, E.N., Gao, Y., Hong, Y., Nie, X. and Zepp, C.M. *J. Org. Chem.* 1994, **59:** 1939–1942.
50. Lee, N.H., Muci, A.R. and Jacobsen, E.N. *Tetrahedron Lett.* 1991, **32:** 5055–5058.
51. Jacobsen, E.N., Deng, L., Furukawa, Y. and Martinez, L.E. *Tetrahedron* 1994, **50:** 4323–4334.
52. Chang, S., Heid, R.M. and Jacobsen, E.N. *Tetrahedron Lett.* 1994, **35:** 669–672.
53. Lee, N.H. and Jacobsen, E.N. *Tetrahedron Lett.* 1991, **32:** 6533–6536.
54. Chang, S., Lee, N.H. and Jacobsen, E.N. *J. Org. Chem.* 1993, **58:** 6939–6941.
55. Palucki, M., Pospisil, P.J., Zhang, W. and Jacobsen, E.N. *J. Am. Chem. Soc.* 1994, **116:** 9333–9334.
56. Larrow, J.F. and Jacobsen, E.N. *J. Am. Chem. Soc.* 1994, **116:** 12129–12130.
57. Chang, S., Galvin, J.M. and Jacobsen, E.N. *J. Am. Chem. Soc.* 1994, **116:** 6937–6938.
58. Martinez, L.E., Leighton, J.L., Carsten, D.H. and Jacobsen, E.N. *J. Am. Chem. Soc.* 1995, **117:** 5897–5898.
59. Larrow, J.F., Schaus, S.E. and Jacobsen, E.N. *J. Am. Chem. Soc.* 1996, **118:** 7420–7421.
60. Schaus, S.E., Larrow, J.F. and Jacobsen, E.N. *J. Org. Chem.* 1997, **62:** 4197–4199.
61. Hansen, K.B., Leighton, J.L. and Jacobsen, E.N. *J. Am. Chem. Soc.* 1996, **118:** 10924–10925.
62. Leighton, J.L. and Jacobsen, E.N. *J. Org. Chem.* 1996, **61:** 389–390.
63. Martinez, L.E., Nugent, W.A. and Jacobsen, E.N. *J. Org. Chem.* 1996, **61:** 7963–7966.
64. Schaus, S.E. and Jacobsen, E.N. *Tetrahedron Lett.* 1996, **37:** 7937–7940.
65. Wu, M.H. and Jacobsen, E.N. *Tetrahedron Lett.* 1997, **38:** 1693–1696.
66. Jacobsen, E.N., Kakiuchi, F., Konsler, R.G., Larrow, J.F. and Tokunaga, M. *Tetrahedron Lett.* 1997, **38:** 773–776.
67. Tokunaga, M., Larrow, J.F., Kakiuchi, F. and Jacobsen, E.N. *Science* 1997, **277:** 936–938.

68. Palucki, M., Hanson, P. and Jacobsen, E.N. *Tetrahedron Lett.* 1992, **33:** 7111–7114.
69. Kokubo, C. and Katsuki, T. *Tetrahedron* 1996, **52:** 13895–13900.
70. Li, Z., Quan, R.W. and Jacobsen, E.N. *J. Am. Chem. Soc.* 1995, **117:** 5889–5890.
71. Robert, A., Tsapara, A. and Meunier, B. *J. Mol. Cat.* 1993, **85:** 13–19.
72. Bernardo, K., Robert, A., Dahan, F. and Meunier, B. *New J. Chem.* 1995, **19:** 129–131.
73. Bernardo, K., Leppard, S., Robert, A., Commenges, G., Dahan, F. and Meunier, B. *Inorg. Chem.* 1996, **35:** 387–396.
74. Cheng, M.-C., Chan, M.C.-W., Peng, S.-M., Cheung, K.-K. and Che, C.-M. *J. Chem. Soc., Dalton Trans* 1997, 3479–3482.
75. Pietikäinen, P. *Tetrahedron Lett.* 1994, **35:** 941–944.
76. Berkessel, A., Bats, J.W. and Schwarz, C. *Angew. Chem. Int. Ed. Engl.* 1990, **29:** 106–108.
77. Berkessel, A., Frauenkron, M., Schwenkreis, T., Steinmetz, A., Baum, G. and Fenske, D. *J. Mol. Cat. A: Chem.* 1996, **113:** 321–342.
78. Kureshy, R.I., Khan, N.H., Abdi, S.H.R. and Bhatt, A.K. *J. Mol. Cat. A: Chem.* 1996, **110:** 33–40.
79. Kureshy, R.I., Khan, N.H., Abdi, S.H.R., Iyer, P. and Bhatt, A.K. *J. Mol. Cat. A: Chem.* 1997, **120:** 101–108.
80. Imanishi, H. and Katsuki, T. *Tetrahedron Lett.* 1997, **38:** 251–254.
81. Adam, W., Jekö, J., Lévai, A., Nemes, C., Patonay, T. and Sebök, P. *Tetrahedron Lett.* 1995, **36:** 3669–3672.
82. Imagawa, K., Nagano, T., Yamada, T. and Mukaiyama, T. *Chem. Lett.* 1994, 527–530.
83. Hashihayata, T., Ito, Y. and Katsuki, T. *Tetrahedron* 1997, **53:** 9541–9552.
84. Vankelecom, I.F.J., Tas, D., Parton, R.F., Van de Vyver, V. and Jacobs, P.A. *Angew. Chem. Int. Ed. Engl.* 1996, **35:** 1346–1348.
85. Minutolo, F., Pini, D., Petri, A. and Salvadori, P. *Tetrahedron: Asym.* 1996, **7:** 2293–2302.
86. Alphand, V. and Furstoss, R. in *Enzyme Catalysis in Organic Synthesis* (eds.) Drauz, K. and Waldmann, H. VCH, Weinheim, 1995, pp. 745–772.
87. Azerad, R. *Bull. Soc. Chim. Fr.* 1995, **132:** 17–51.
88. Gusso, A., Baccin, C., Pinna, F. and Strukul, G. *Organometallics* 1994, **13:** 3442–3451.
89. Strukul, G., Varagnolo, A. and Pinna, F. *J. Mol. Cat. A: Chem.* 1997, **117:** 413–423.

90. Bolm, C., Schlingloff, G. and Weickhardt, K. *Angew. Chem. Int. Ed. Engl.* 1994, **33:** 1848–1849.
91. Bolm, C. and Schlingloff, G. *J. Chem. Soc., Chem. Commun.* 1995, 1247–1248.
92. Bolm, C., Schlingloff, G. and Bienewald, F. *J. Mol. Cat. A: Chem.* 1997, **117:** 347–353.
93. Lopp, M., Paju, A., Kanger, T. and Pehk, T. *Tetrahedron Lett.* 1996, **37:** 7583–7586.

Chapter 13

Bioinspired Oxidations Catalyzed by Ruthenium Complexes

Shun-Ichi Murahashi & Naruyoshi Komiya

Osaka University, Department of Chemistry,
Graduate School of Engineering Science, Toyonaka,
Osaka 560–8531, Japan

1. Introduction

Oxidation using transition metal complexes is one of the most important and basic reactions in organic synthesis and is being developed as a topic of much current interest.[1] Stoichiometric oxidations with traditional oxidants such as permanganate and dichromate usually produce vast amounts of inorganic salts which are difficult to dispose of. Owing to the current need to develop forward-looking technologies, environmentally acceptable reactions must be constructed. Catalytic oxidations may be considered to be one of the more environmentally-friendly reactions. Organic oxidizing agents such as hydrogen peroxide and *t*-butyl hydroperoxide have been used for oxidations catalyzed by trace amounts of transition metal catalysts.

Ruthenium complexes have specific features, since they exhibit octavalency and have a wide range of oxidation states from VIII to –II, corresponding to electron configurations of d^0 to d^{10}. Accordingly, catalytic oxidation reactions can be explored using many oxidation states of ruthenium.[2] The most high-valent ruthenium complex is ruthenium(VIII) tetraoxide (RuO_4) which is generally known as a strong oxidant.[3a,b] RuO_4 can be used catalytically in the presence of a suitable co-oxidant such as sodium periodate or hypochlorite, but it is a fairly unselective oxidant, and carbon-carbon bond cleavage is often a competitive process.[3] Recently, lower-valent ruthenium complexes have been used in conjunction with a variety of oxidants to effect specific oxidations. Oxo-ruthenium complexes attract particular attention in the oxidation chemistry. Novel oxidation chemistry of oxo-ruthenium complexes have been reported by various groups. An important feature of these complexes is their capability to oxidize various substrates under mild conditions.

Porphyrin oxo-ruthenium[4,6–8,10] and non-porphyrin oxo-ruthenium[5,21c] complexes were synthesized and characterized, and stoichiometric oxidations of organic substrates have been reported. Furthermore, it is revealed that catalytic formation of these species is possible, and catalytic oxidations can be operated. Dioxo(porphyrinato)ruthenium(VI) complexes,[6–12] which are generated from $Ru(TMP)(O)_2$ (TMP = *meso*-tetramesitylporphyrinato) (**1**) with molecular oxygen[6a,7,8] or 2,6-dichloropyridine *N*-oxide,[10] $Ru(OEP)(PPh_3)Br$ (OEP = octaethylporphyrinato) with iodosylbenzene,[9] and Ru(TPFPP)(CO) (TPFPP = *meso*-tetrakis(pentafluorophenyl)porphyrinato) with 2,6-dichloropyridine *N*-oxide[6b] or acetaldehyde/molecular oxygen[11] are able to catalyze the oxidations of alkenes, alkanes, or primary amines. Non-porphyrin oxo-ruthenium complexes have also been generated[13–21] from $[Ru(dmp)(MeCN)_2]^{2+}$ (dmp = 2,9-dimethyl-1,10-phenanthroline) with molecular oxygen[15a,c] or hydrogen peroxide,[15a,b,d] and $[Ru(Me_3tacn)(O)_2(CF_3CO_2)]^+$ (Me_3tacn = *N,N',N"*-trimethyl-1,4,7-triazacyclononane) (**2**) with *t*-butylhydroperoxide,[21c] and their catalytic activities have been studied.

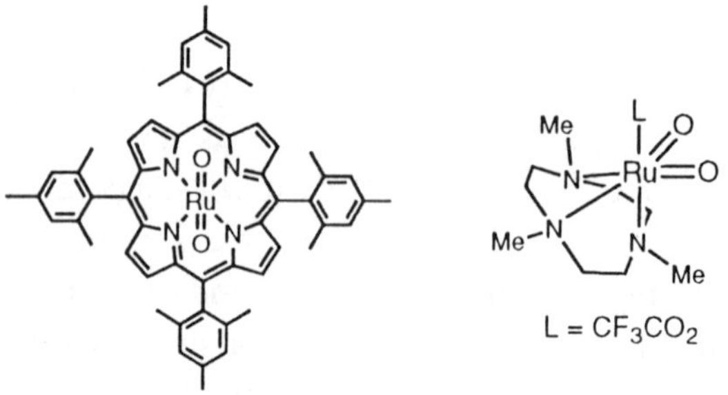

$Ru(TMP)(O)_2$
(1)

cis-$[Ru(Me_3tacn)(O)_2(CF_3CO_2)]^+$
(2)

Scheme 13.1

For organic synthesis, oxo-ruthenium species bearing simple moieties have been generated by treatment of low-valent ruthenium complexes (Ru^n) such as $RuCl_2(PPh_3)_3$, $RuCl_3$, and even Ru/C with oxidants such as *t*-butyl hydroperoxide, hydrogen peroxide, peracetic acid, and molecular oxygen / acetaldehyde [Eq. (13.1)].[22]

$$Ru^n + \begin{cases} t\text{-BuOOH} \\ H_2O_2 \\ CH_3CO_3H \\ CH_3CHO - O_2 \end{cases} \longrightarrow \underset{(R = t\text{-Bu}, H, CH_3CO)}{Ru^n\text{-OOR}} \longrightarrow Ru^{n+2}=O \qquad (13.1)$$

Suitable combinations of simple ruthenium catalysts with oxidants provide novel cytochrome P450-type catalytic oxidations of a wide variety of substrates, which include amines, amides, β-lactams, alkenes, alcohols, phenols, and unactivated alkanes, without using porphyrins. In this chapter, we describe the exploitation of novel ruthenium-catalyzed oxidations and their application to organic synthesis.

2. Ruthenium-Catalyzed Oxidation of Tertiary Amines

Selective demethylation of tertiary methyl amines is one of the most specific and important metabolic functions of cytochrome P450.[23] Novel cytochrome P450-type oxidation behavior with tertiary amines has been found in the reaction using ruthenium complex catalyst and *t*-BuOOH to give α-*t*-butyldioxyamines. From this clue, some novel biomimetic catalytic reactions have been explored.

2.1 Oxidation with tert-butyl hydroperoxide

The oxidative *N*-demethylation of *N*-methylamines is one of the important reactions specific to cytochrome P450,[23] and several model reactions using iron-porphyrins have been reported.[24] We found that low-valent ruthenium complex-catalyzed oxidation of tertiary amine **3** with *t*-BuOOH gave the corresponding α-(*t*-butyldioxy)alkylamine **4** efficiently [Eq. (13.2)].[25] This is in contrast to the usual catalytic oxidation of tertiary amines with

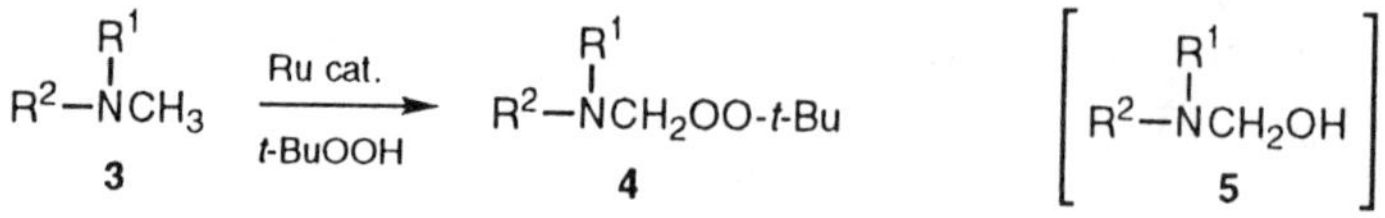

(13.2)

peroxides to give *N*-oxides. Hemiaminal-type product **4** has a similar structure to α-hydroxymethylamine **5** derived from the oxidation with cytochrome P450.

The catalytic activity of various metal complexes was examined for the oxidation of *N*,*N*-dimethylaniline (**6**) with *t*-BuOOH [Eq. (13.3)]. $RuCl_2(PPh_3)_3$ has proven to be the most effective catalyst for the formation

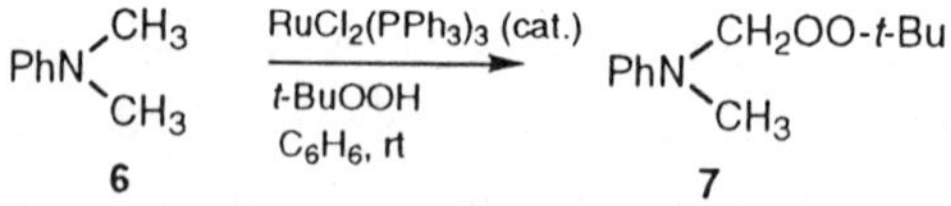

(13.3)

of *N*-methyl-*N*-(*tert*-butyldioxymethyl)aniline (**7**) (93%). The *N*-methyl group is oxidized chemoselectively in the presence of other alkyl and alkenyl groups. Thus, $RuCl_2(PPh_3)_3$-catalyzed oxidation of *N*-ethyl-*N*-(methyl)aniline with *t*-BuOOH gives *N*-ethyl-*N*-(*tert*-butyldioxymethyl)aniline exclusively (78%).

The oxidation of 4-(*N*,*N*-dimethylamino)pyridine **8** gave the corresponding α-*t*-butyldioxyamine **9** in 72% yield [Eq. (13.4)]. Benzylic and allylic positions and carbon-carbon double bonds tolerate the oxidation.

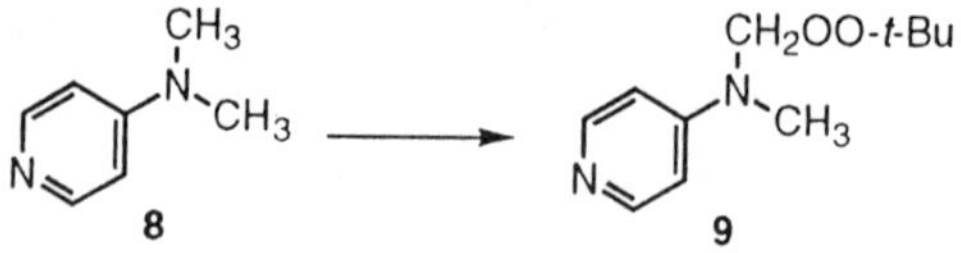

(13.4)

(13.5)

For example, *N*-methyl-*N*-(3-butenyl)aniline **10** can be converted into the corresponding peroxide **11** in 87% yield [Eq. (13.5)].

To gain insight into the mechanism, the relative reaction rates of the oxidation of five substituted *N,N*-dimethylanilines [X-$C_6H_4N(CH_3)_2$, X = *p*-CH_3O, *p*-CH_3, H, *p*-Cl, and *m*-Cl] with *t*-BuOOH in benzene were determined. The rate data correlate well with the Hammett linear free energy relationship with use of σ values The ρ value is –0.84, which is similar to the value (–0.74) with P450,[26] indicating the formation of a cationic intermediate. The intramolecular deuterium isotope effect of the ruthenium-catalyzed oxidation of *N*-methyl-*N*-(trideuteriomethyl)aniline was determined to be 3.53. Furthermore, the intermolecular isotope effect of the oxidation of *N,N*-dimethylanilines was determined to be 1.64, on the competitive reaction of *N,N*-dimethylaniline and *N,N*-bis(trideuteriomethyl)aniline. The observed intra- and intermolecular isotope effects (3.53 and 1.64) are larger than the values observed for *N*-demethylation with cytochrome P450 (1.6–3.1[27] and 1.0–1.3[28]), suggesting that the C–H bond breaking in the present reaction proceeds with more radical character.

The mechanism of this oxidation can be rationalized by assuming Scheme 13.2, which is similar to the oxidation by cytochrome P450.[23] The key intermediate of the reaction may be *t*-butyldioxy-ruthenium **12** or oxo-ruthenium **13** species. The reaction of low-valent ruthenium complex with *t*-BuOOH would form Ru(II)-OO-*t*-Bu (**12**), which is transformed to the Ru(IV) = O complex (**13**) by cleavage of the O–O bond. Electron transfer and subsequent hydrogen transfer or hydrogen abstraction and subsequent electron transfer results in forming the iminium complex **14**. Nucleophilic attack of *t*-BuOOH on **14** gives the product, water, and Ru(II) species to

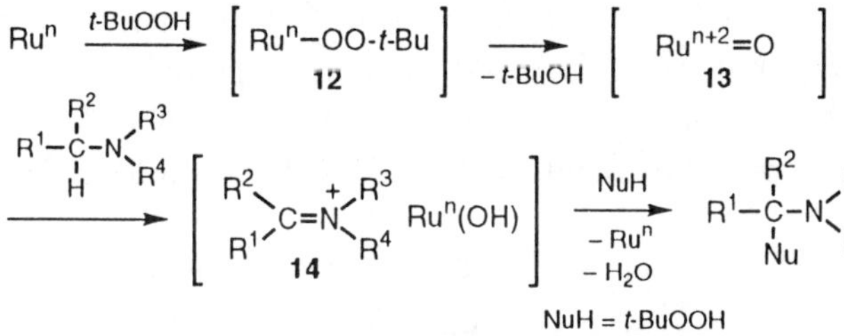

Scheme 13.2

complete the catalytic cycle. The possibility that Ru(II)-OO-*t*-Bu (**12**) directly reacts with amines cannot be ruled out.

The selective oxidation of *N*-methylamines provides various useful methods for organic synthesis. First is the selective demethylation of tertiary amines by the ruthenium-catalyzed oxidation and subsequent hydrolysis reaction.[25] This is the first synthetically practical method for the *N*-demethylation of tertiary amines, although several metalloporphyrin-catalyzed reactions have been reported.[24] The methyl group is removed chemoselectively in the presence of various alkyl groups. For example, catalytic oxidation of *N*-ethyl-*N*-methylaniline followed by treatment of the reaction mixture with 2M HCl solution gives *N*-ethylaniline in 61% isolated yield [Eq. (13.6)]. The reaction seems to involve protonation, removal of *t*-butylhydroperoxide, and hydrolysis of the iminium ion to give the secondary amine and formaldehyde.

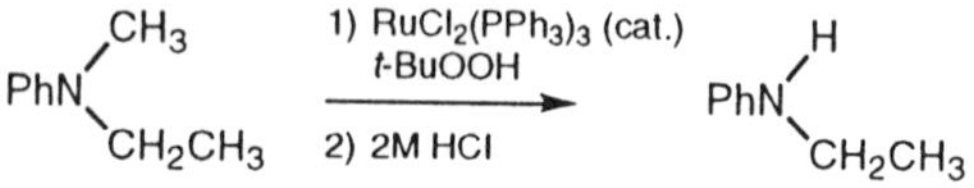

(13.6)

Second application of the oxidation is a novel, biomimetic method for construction of the piperidine skeleton from *N*-methylhomoallylamine by means of an olefin-iminium ion cyclization reaction. Although the biological system is able to introduce a carbon atom of the *N*-methyl group into a piperidine ring to synthesize alkaloids such as (–)-scoulerine from (+)-reticuline, such a synthetic method has never been reported.[29] The first biomimetic method in which the *N*-methyl group of amines is incorporated into piperidine skeletons is as follows. Cyclization of the peroxide bearing a substituted 3-butenyl group gives *trans*-3,4-disubstituted piperidines stereoselectively. For example, *trans*-1-phenyl-3-propyl-4-chloropiperidine **16** has been obtained stereoselectively from *N*-methyl-*N*-(3-heptenyl)aniline via **15** (55%) upon treatment with a 2 N HCl solution [Eq. (13.7)].

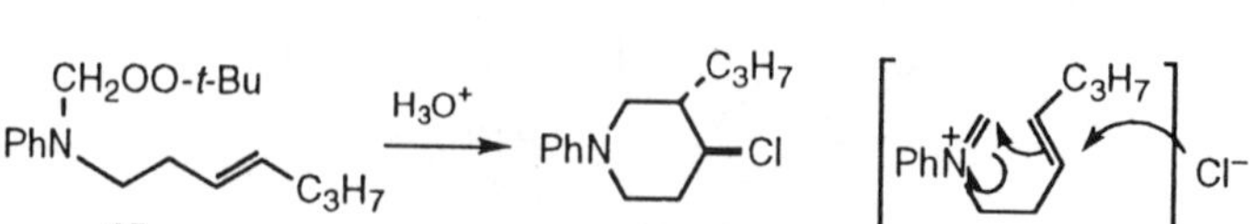

15 **16** **17**

(13.7)

This cyclization can be rationalized by assuming the formation of iminium ion **17** by protonation of **15** and subsequent elimination of *t*-BuOOH, nucleophilic attack of an alkene, giving a carbonium ion, which is trapped with Cl^- nucleophile from the less hindered side

Similar treatment using CF_3CO_2H in place of HCl gave hydroxy derivatives. Thus, *cis*-4a-hydroxy-2-phenyldecahydroisoquinoline (**19**) was obtained selectively from *N*-methyl-*N*-2-(1-cyclohexenyl)ethylaniline (**18**) upon treatment of the corresponding peroxide with an aqueous CF_3CO_2H solution [Eq. (13.8)].

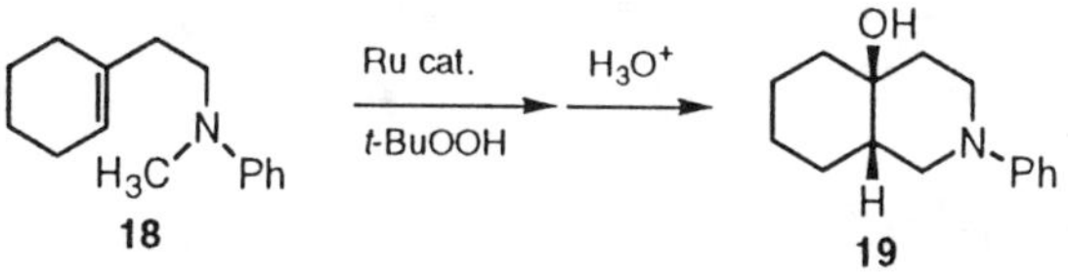

(13.8)

Construction of tetrahydroquinoline skeletons can be performed by the ruthenium-catalyzed oxidation of *N*-methyl-*N*-alkylanilines with *t*-BuOOH followed by treatment with olefins.[30] The reactions of the peroxide **7** with allylsilane in the presence of $TiCl_4$ in CH_2Cl_2 at −78 °C gave the corresponding tetrahydroquinolines **20** in 70% yield [Eq. (13.9)].

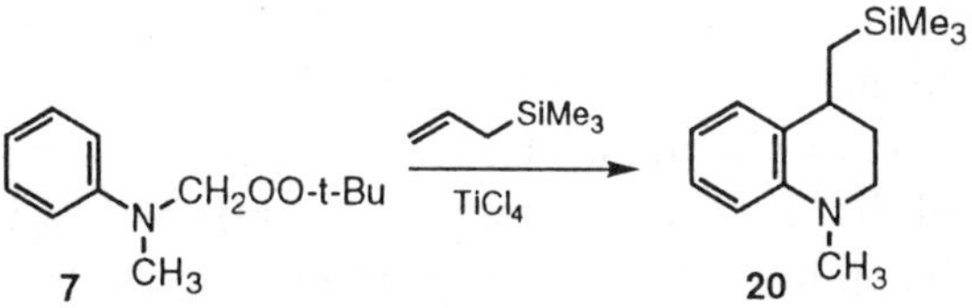

(13.9)

2.2 *Oxidation with hydrogen peroxide*

Based on Scheme 13.2, both generation of the oxo-ruthenium species using other oxidizing reagents and trapping the iminium ion intermediate **14** with other nucleophiles can be expected. Indeed, ruthenium-catalyzed oxidation of tertiary amines with hydrogen peroxide in methanol can be performed to give the corresponding α-methoxyamines with high efficiency [Eq. (13.10)].[31]

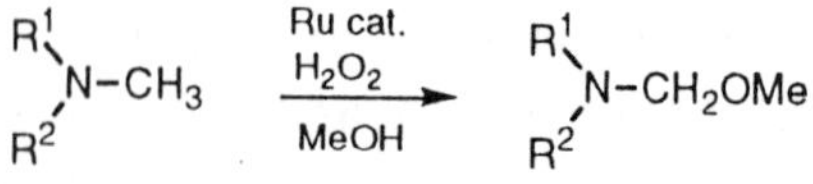

(13.10)

$RuCl_3{\bullet}nH_2O$ has proven to be the most effective catalyst among metal catalysts examined, and $RuCl_2(PPh_3)_3$, which is a good catalyst for the oxidation of amines with *t*-BuOOH, shows moderate activity. Tertiary amines **21** and **23** gave the corresponding α-methoxyamines **22** and **24** in 87% and 60% yield, respectively [Eqs. (13.11) and (13.12)].

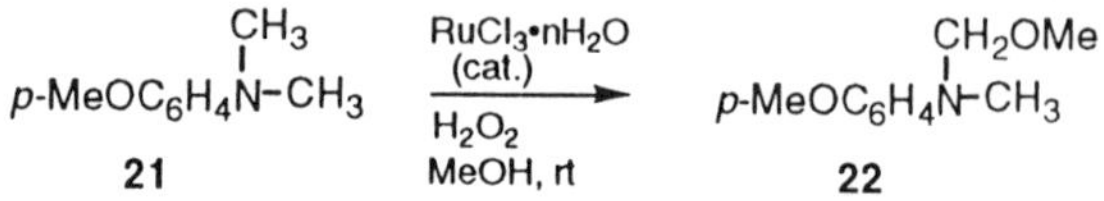

21 **22**

(13.11)

23 **24**

(13.12)

The relative reaction rates of the oxidation of four substituted *N,N*-dimethylanilines gave the ρ value of –3.26. The intra- and intermolecular isotope effects for the oxidation of *N,N*-dimethylanilines were determined to be 3.47 and 3.72, respectively. Considering the observed kinetic data and isotope effects, the present reaction can be rationalized by assuming the following mechanism. The ruthenium(III) complex reacts with H_2O_2 to give an oxo-ruthenium(V) species, which undergoes electron transfer followed by hydrogen transfer to afford an iminium ion complex **25**. Nucleophilic attack of methanol then gives the methoxylated amines (Scheme 13.3). Compared to the $RuCl_2(PPh_3)_3$ / *t*-BuOOH system mentioned in Sec. 2.1, the slope of ρ value in $RuCl_3$ / H_2O_2 system is sharp, indicating that a more cationic intermediate may form at the rate-determining step.

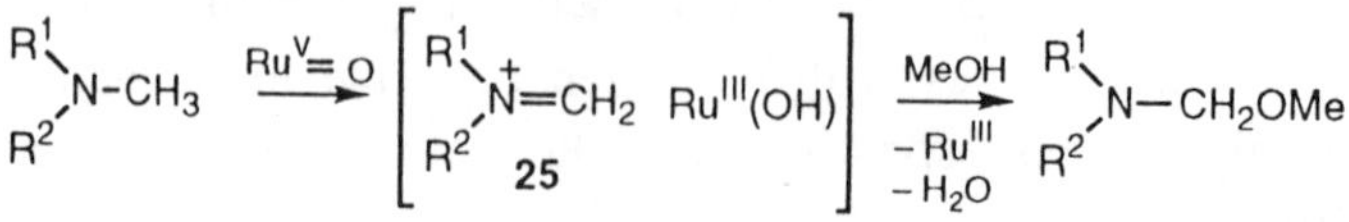

Scheme 13.3

Selective demethylation of tertiary methylamines can be achieved by the hydrolysis of α-methoxylated amines as well as α-*t*-butyldioxyamines.[31] For example, the methoxylation of *N*,*N*-dimethyl-*p*-toluidine followed by treatment with 2N HCl aqueous solution gave *N*-methyl-*p*-toluidine in 75% yield [Eq. (13.13)].

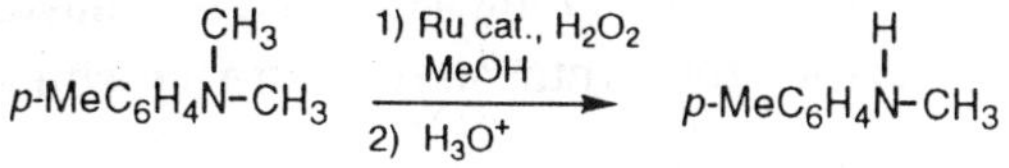

(13.13)

The present oxidation reaction provides an efficient method for construction of the quinoline skeleton. *cis*-Fused tricyclic amine **27** can be synthesized from *N*,*N*-dimethylaniline via **26** in two steps (91%) [Eq. 13.14)].

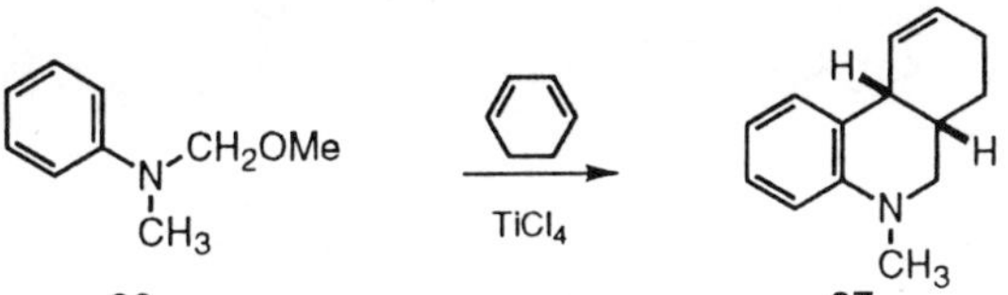

26 **27** (13.14)

3. Ruthenium-Catalyzed Oxidations of Secondary and Primary Amines

Imines can be synthesized by either the condensation of carbonyl compounds such as ketones and aldehydes with amines, accompanied by the elimination of water, or the reaction of Grignard reagents with nitriles; however, these methods are limited to the synthesis of non-cyclic imines.[32] We have found that the ruthenium-catalyzed reaction of secondary amines gives imines [Eq. (13.15)].[33] This is the first catalytic oxidative transformation of secondary amines into imines.

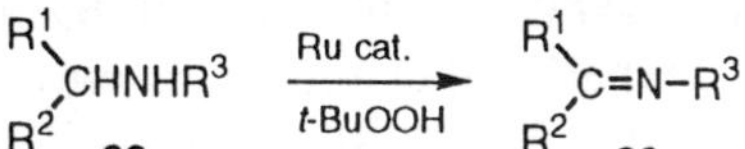

28 **29** (13.15)

Treatment of a secondary amine with two equivalents of *t*-BuOOH in benzene, in the presence of $RuCl_2(PPh_3)_3$ at room temperature gives the corresponding imines in a single step efficiently. 4Å molecular sieves are needed to prevent the hydrolysis of product imines in some cases. For example, the oxidations of dibenzylamine (**30**), allylamine **32**, and tetrahydroisoquinoline **34** give the corresponding imine **31**, azadiene **33**, and cyclic imine **35** in 80%, 69%, and 82% yields, respectively [Eqs. (13.16), (13.17) and (13.18)]. The obtained 1-azadiene **33** and 3,4-dihydroisoquinoline **35** are important precursors for synthesis of nitrogen-containing natural

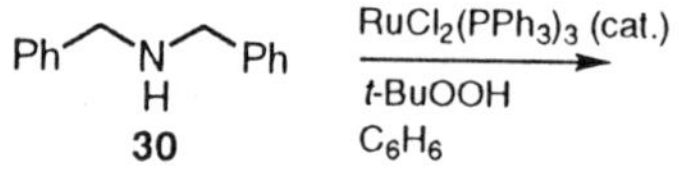

(13.16)

(13.17)

(13.18)

products. Generally, this oxidation gives thermodynamically stable imines. Aromatization takes place when an excess amount of *t*-BuOOH oxidant is used. For example, tetrahydroquinoline **36** can be converted into quinoline (**37**) (73%) [Eq. (13.19)].

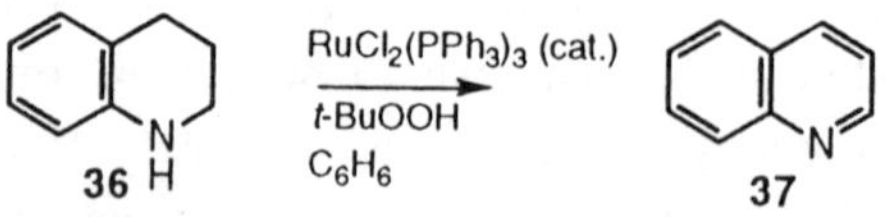

(13.19)

We have already reported that secondary amines can be converted into nitrones by using tungsten catalyst and hydrogen peroxide.[34] Therefore, we are able to prepare either imines or nitrones from secondary amines by selecting the catalyst. Thus, the ruthenium-catalyzed oxidation of amine **28** gives imine **29** [Eq. (13.15)]. In contrast, the tungstate-catalyzed oxidation of the same amine **28** with hydrogen peroxide gives nitrone **38** [Eq. (13.20)]. These two methods are particularly useful to introduce a

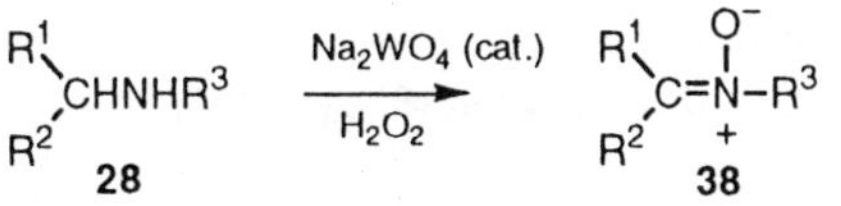

(13.20)

substituent at the α-position of amines, because either imines or nitrones react readily with nucleophiles to give the α-substituted amine derivatives (see Sec. 5).

The catalytic reaction can be extended to the oxidation of primary amines. The ruthenium-catalyzed oxidation of diphenylmethylamine with *t*-BuOOH gave benzophenone (88%), which was formed by hydrolysis of the imine intermediate [Eq. (13.21)].

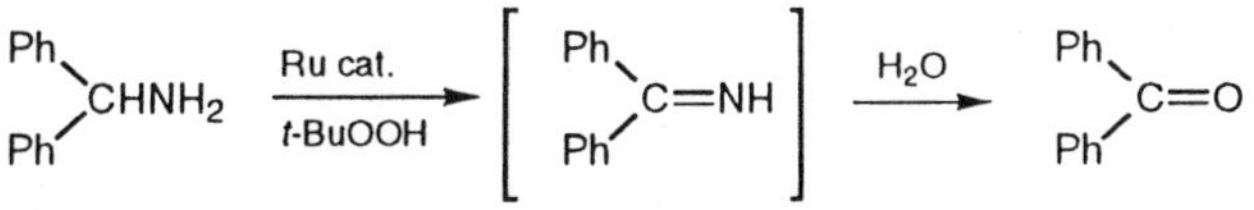

(13.21)

The mechanism of the ruthenium-catalyzed oxidation of secondary amines with *t*-BuOOH can be rationalized by assuming the mechanism which is similar to the oxidation of tertiary amines (Scheme 13.2). As shown in Scheme 13.4, electron transfer from secondary amines to oxo-ruthenium (IV) species followed by hydrogen abstraction leads to the iminium ion complex **39**, which decomposes to give the product imines **29**, Ru^{II} species, and water to complete the catalytic cycle.

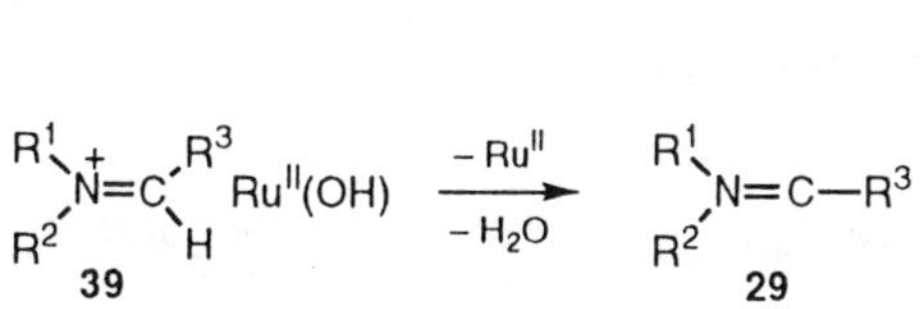

Scheme 13.4

Goti *et al.* reported that tetra-*n*-propylammonium perruthenate (TPAP) /*N*-methylmorpholine *N*-oxide (NMO) system, which was first utilized by Ley and Griffith, showed good efficiency for the catalytic oxidation of secondary amines [Eq. (13.22)].[35]

(13.22)

Aerobic oxidation of amines was reported for a non-porphyrin system; however, forcing conditions were required and the selectivity was low. For example, the oxidation of benzylamine results in the formation of benzonitrile and benzamide which is formed by hydrolysis of nitrile [Eq. (13.23)].[14]

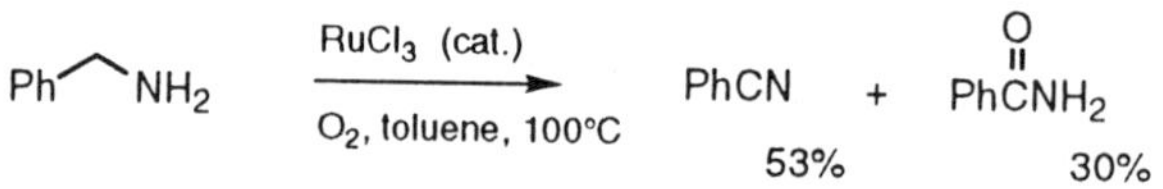

(13.23)

Recently, James *et al.* reported that efficient aerobic oxidation of primary and secondary amines can be performed by using ruthenium porphyrin complex. The complex Ru(TMP)$(O)_2$ (**1**) undergoes dehydrogenation of primary and secondary amines catalytically in the presence of air.[7a] For example, primary amines such as *n*-butylamine were oxidized to nitriles in 100% yields; the intermediate imines (R-CH = NH) are readily dehydrogenated [Eq. (13.24)]. The secondary benzyl amines, such as dibenzylamine, can be transformed into the corresponding imines (70%) [Eq. (13.25)], although secondary alkyl amines are difficult to dehydrogenate.

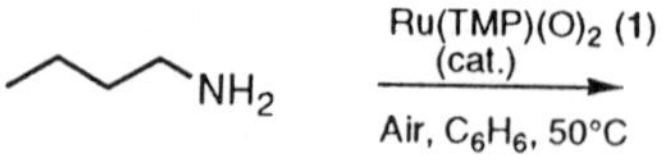

(13.24)

(13.25)

During the oxidative deamination of amino acids, imino-acids are often postulated as intermediates. Imino ester Ru(II) complexes were isolated from the reaction of α-amino esters with dioxo-Ru(IV) picket fence complex.[4b]

4. Ruthenium-Catalyzed Oxidation of Amides

The oxidation of the α-C–H bond of amides is of importance in view of the xenobiotic amino compounds[23] and is one of the most attractive strategies for the synthesis of biologically active nitrogen compounds.[36] Cytochrome P450 enzymes catalyze specific oxidation of amides; however, biomimetic method for selective oxidation of amides are limited to electrochemical processes.[37] The cytochrome P450-type oxidation of tertiary amines can also be applied to the oxidation of amides.[38] Thus, ruthenium-catalyzed oxidation of amides with *t*-BuOOH under mild conditions gives the corresponding *t*-butyldioxyamide **40**, which is a useful synthetic intermediate [Eq. (13.26)].

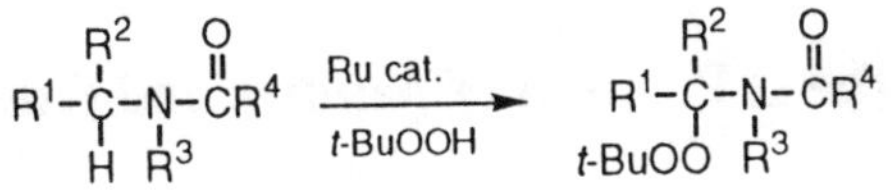

40 (13.26)

Various *N*-protected secondary amines can be converted into the corresponding α-(*t*-butyldioxy)amides. Particularly, the oxidation of cyclic amine derivatives, which have some resistance to the oxidation, proceed efficiently. For example, the oxidation of 1-(methoxycarbonyl)pyrrolidine

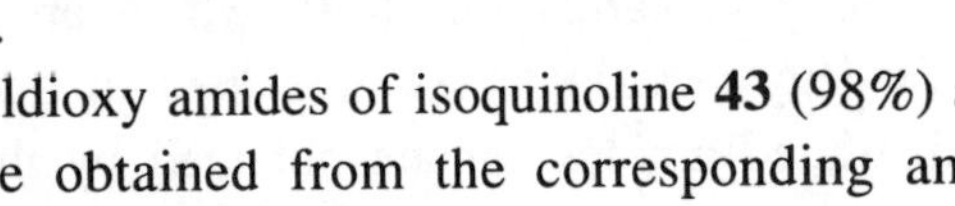

41 **42** (13.27)

(**41**) gives 2-(*t*-butyldioxy)-1-(methoxycarbonyl)pyrrolidine (**42**) in 60% yield [Eq. (13.27)].

The *t*-butyldioxy amides of isoquinoline **43** (98%) and indole **44** (92%), which can be obtained from the corresponding amides, are important synthetic intermediates of natural products [Eqs. (13.28) and (13.29)].

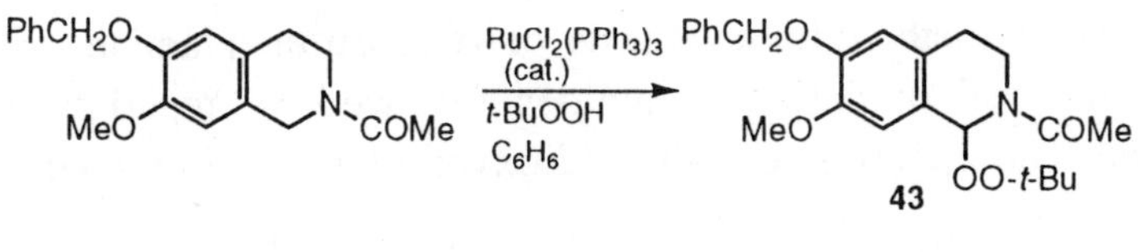

43 (13.28)

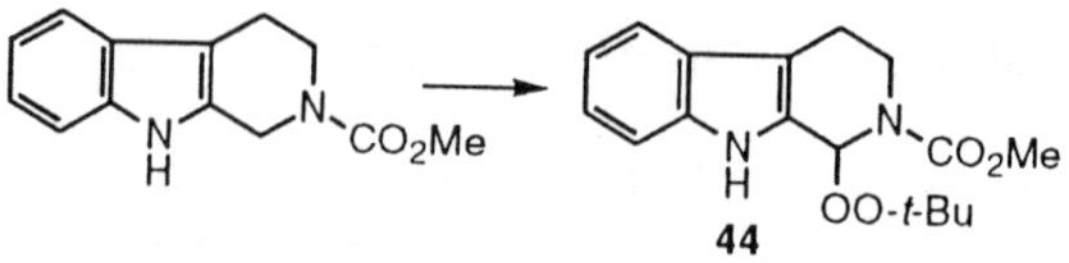

(13.29)

The oxidation of γ-substituted-γ-lactams **45** proceeds selectively to give γ-(*t*-butyldioxy)lactams **46** in 80% yield [Eq. (13.30)], while manganese-catalyzed oxidation of **45** gave a complex mixture of products.[39]

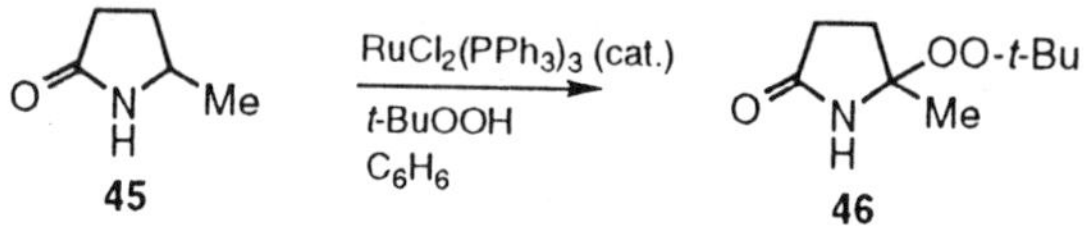

(13.30)

5. Substitution at the α Position of Amines and Amides

The substitution at the α-carbon atom of amines is of importance in view of synthesis of nitrogen-containing natural products and biologically active compounds. Introduction of a substituent has been performed using electrophilic reagents. Thus, *N*-protected amines (**48**) with an electron-withdrawing group Z such as formamidines (Z = CH = NR), amides (Z = C(O)R), and nitrosoamines (Z = NO) undergo lithiation with

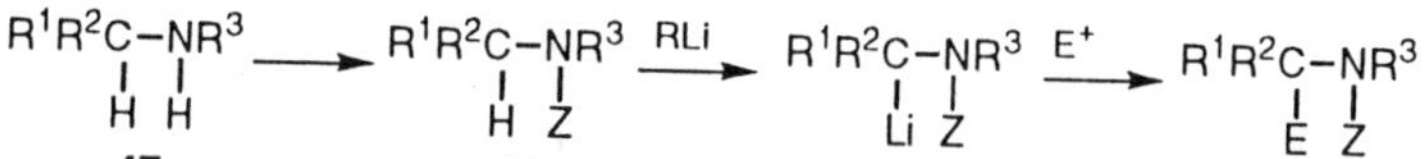

(13.31)

organolithium reagents to give carbanions (**49**), which are trapped with electrophiles to give **50** [Eq. (13.31)].[40] As described in Eq. (13.15) and Eq. (13.20), the α-substituted amines can be obtained by the addition of nucleophiles to either imines or nitrones derived by catalytic oxidations of amines.

An alternative method for synthesis of α-substituted amines is the ruthenium-catalyzed oxidation of amides followed by reactions with nucleophiles. The reaction of *t*-butyldioxyamides **51,** obtained with nucleophiles in the presence of a Lewis acid, gives *N*-acyl-α-substituted amides **52** very efficiently.[41] Thus, the selective carbon-carbon bond formation at the α position of amines can be performed according to Eq. (13.32).

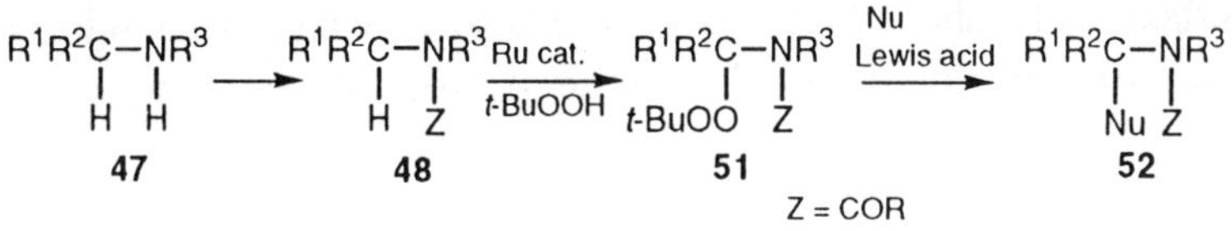

(13.32)

For example, $TiCl_4$-promoted reaction of *t*-butyldioxy derivatives **53** with allylsilane gives the 1-allyl derivative **54** in 89% yield [Eq. (13.33)].

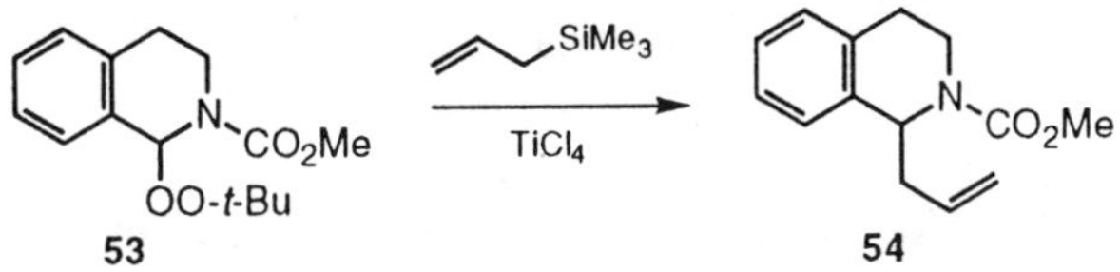

(13.33)

The titanium-promoted reactions of α-*t*-butyldioxypyrrolidine **42** with silyl enol ether gave keto amide **55** (81%) [Eq. (13.34)]. Less reactive 1,3-diene can be also applied to $TiCl_4$-promoted reaction of *t*-butyldioxyamide to give α-substituted amides [Eq. (13.35)]. This is the first example of nucleophilic attack of 1,3-diene to an iminium ion. These reactions clearly show the efficiency of the present oxidation reaction for the synthesis of various key synthetic intermediates.

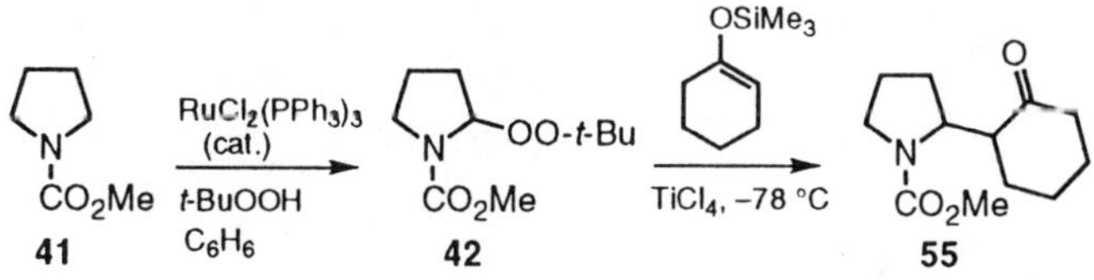

(13.34)

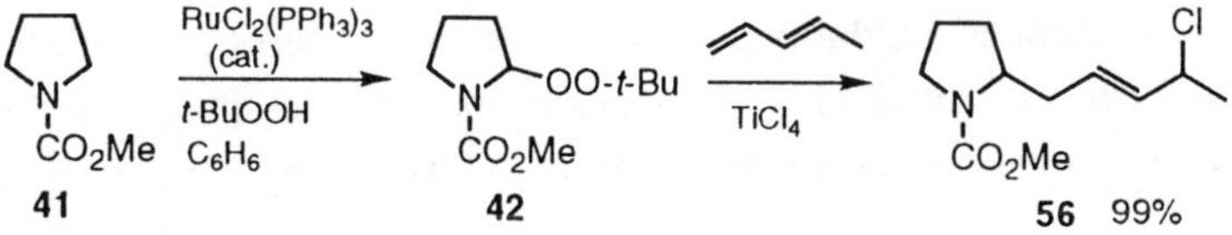

(13.35)

6. Oxidative Cyanation of Amines and Amides

As shown in Sec. 5, one can consider that the C–C bond formation of amides at the α position of nitrogen has been achieved in two steps; oxidation of amines to give *t*-butyldioxyamides, and subsequent reaction with carbon nucleophiles (Scheme 13.5). However, considering Scheme 13.2, iminium ion intermediate **14** can be trapped directly with external

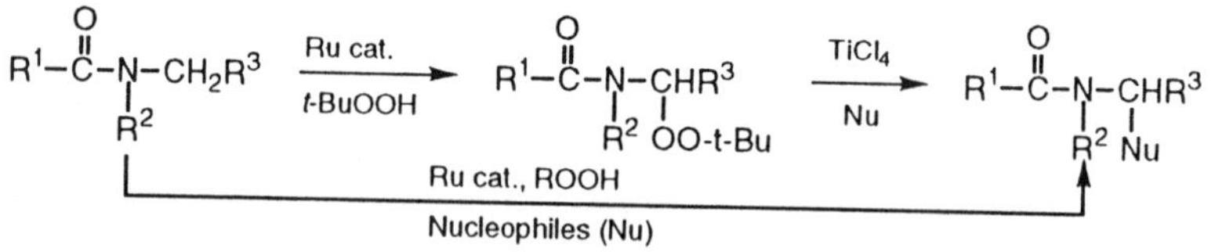

Scheme 13.5

nucleophiles before trapping with the second molecule of *t*-BuOOH. Indeed, ruthenium-catalyzed direct oxidative cyanations of tertiary amines and amides can be performed upon treatment with trimethylsilyl cyanide (TMSCN), giving α-cyanated amines such as **57** (75%) and **58** (78%) [Eqs. (13.36) and (13.37)].[42]

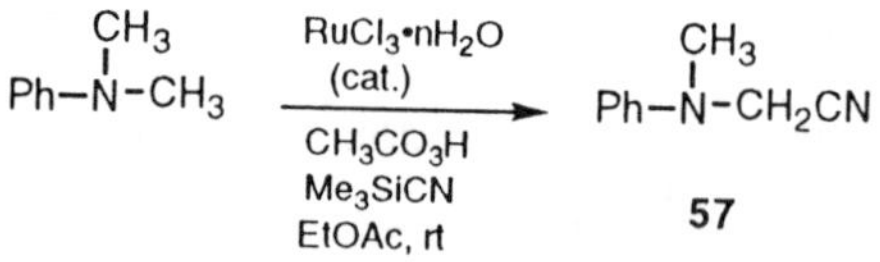

57 (13.36)

58 (13.37)

Recently, cyanation reactions of nitrogen-containing compounds by anodic oxidation,[43] photoinduced oxidation,[44] ClO_2 oxidation,[45] aerobic oxidation,[46] and oxidation with 1-cyano-3-(1*H*)-1,2-benziodoxol[47] or cyanogen iodide[48] were reported; however, these reactions occur with limited amines. So, the

Ru cat./CH_3CO_3H/TMSCN system is convenient for oxidative cyanations of both amines and amides.

This oxidation reaction provides a novel and convenient method for synthesis of α-amino acid derivatives from amines, because α-cyano amines are similar to the intermediates of Strecker amino acid synthesis.[49] For example, *N*-cyanomethyl-*p*-methoxy-*N*-methylaniline (**59**), which is readily obtained by the oxidative cyanation of *N,N*-dimethyl-*p*-methoxyaniline, can be converted into *N*-methyl-*N*-(*p*-methoxyphenyl)glycine (**60**) in 82% yield upon hydrolysis [Eq. (13.38)].

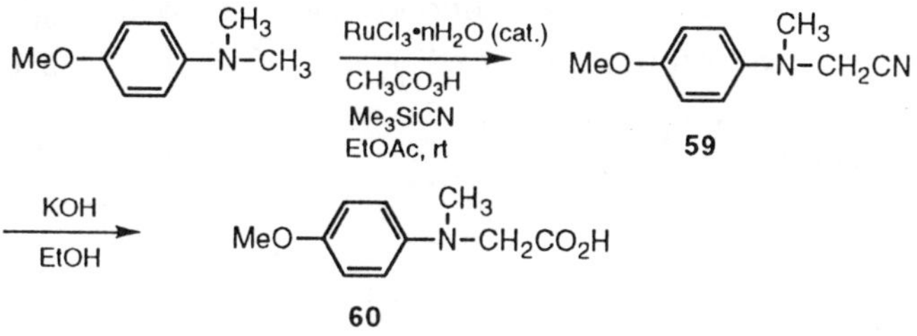

(13.38)

The azide group can also be introduced directly at the α position to nitrogen of amides. Thus, the ruthenium-catalyzed oxidation of *N*-*t*-butoxycarbonylpyrrolidine (**61**) with peracetic acid in the presence of sodium azide gave α-azido-*N*-*t*-butoxycarbonylpyrrolidine (**62**), which is an important precursor of α-amino amides, in 57% yield [Eq. (13.39)].[50]

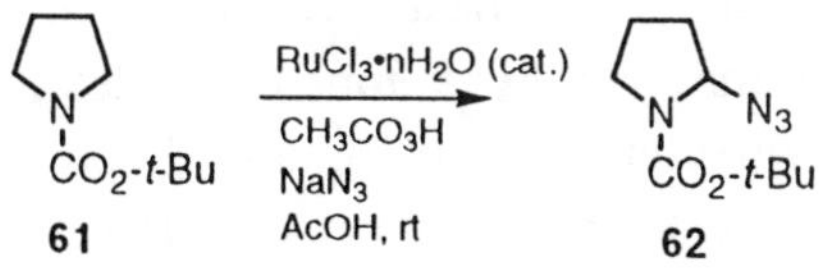

(13.39)

7. Ruthenium-Catalyzed Oxidation of β-Lactams

One of the most challenging topics among the oxidation of amides is the catalytic oxidation of β-lactams, since they are usually unstable under the normal oxidation conditions. Selective oxidation of β-lactams can be performed using cytochrome P450-type catalytic reactions.

7.1 Oxidation with peracetic acid

Oxidation of β-lactams requires specific reaction conditions because of the higher strain of the four-membered acyliminium ion intermediates. Direct oxidative acyloxylation of β-lactams can be performed by ruthenium-catalyzed oxidation with peracetic acid in acetic acid under mild conditions. The products obtained are highly versatile and are key intermediates for the synthesis of antibiotics. Thus, ruthenium-catalyzed oxidations of 2-azetidinones **63** and **65** with peracetic acid in acetic acid in the presence of sodium acetate at room temperature give the corresponding 4-acetoxy-2-azetidinones **64** and **66** in high yields (94% and 86% respectively) [Eqs. (13.40) and (13.41)].[38] One can use $RuCl_2(PPh_3)_3$ and $RuCl_3$, but for practical synthesis ruthenium on charcoal is used conveniently.

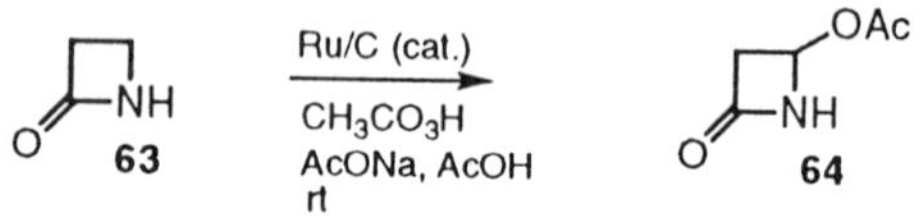

63 → **64** (13.40)

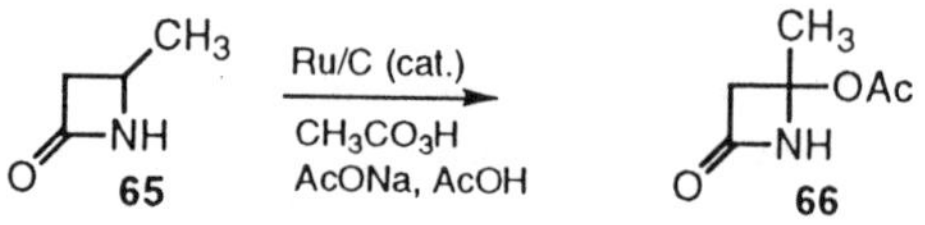

65 → **66** (13.41)

Although peracetic acid is the best oxidant, other oxidants such as *m*-chloroperbenzoic acid, methyl ethyl ketone peroxide, and iodosylbenzene can be used for the acyloxylation of β-lactams. Importantly, (1'*R*, 3*S*)-3-[1'-(*tert*-butyldimethylsilyloxy)ethyl]azetidin-2-one (**67**) can be converted into the corresponding 4-acetoxyazetidinone **68** with extremely high diastereoselectivity (94%, > 99% de) [Eq. (13.42)].

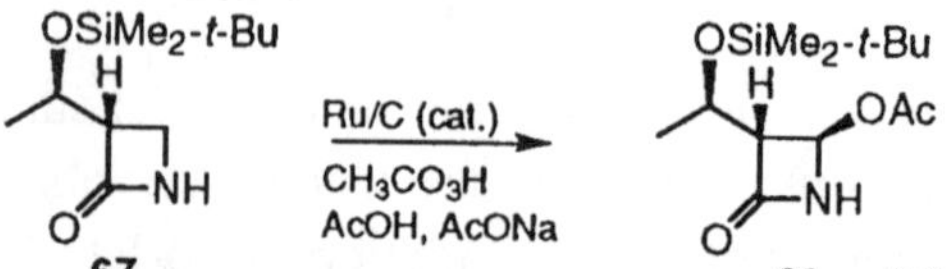

67 → **68** 94%, >99% de (13.42)

The product **68** is a versatile and key intermediate for the synthesis of carbapenem **69**, which is an important, commercially available antibiotic [Eq. (13.43)].[51]

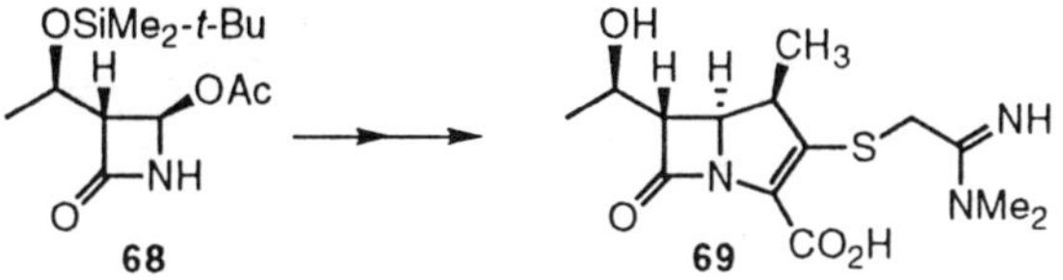

68 69 (13.43)

Recently, this method was applied to the synthesis of 3-amino-4-acetoxy-azetidinones **71** in 85% yield [Eq. (13.44)].[52] The diastereoselectivity of the reaction was in favor of the more stable *trans* isomer.

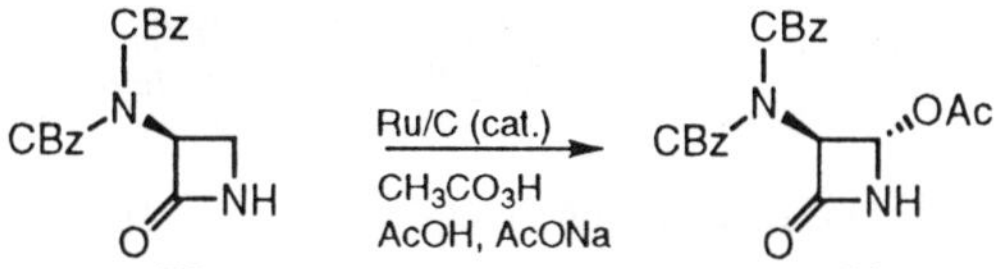

70 71 (13.44)

7.2 Oxidation with molecular oxygen

The ruthenium-catalyzed oxidation of β-lactams requires peroxides such as peracetic acid and methyl ethyl ketone peroxide. However, these peroxides are cumbersome to utilize and sometimes contain undesirable materials. Therefore, the formation of peroxides *in situ* from aldehydes and molecular oxygen in the presence of catalysts was examined. It is well-known that peracetic acid can be prepared upon treatment of acetaldehyde with molecular oxygen in the presence of a catalyst such as cobalt salts (Scheme 13.6).[53] Ruthenium-catalyzed oxidation of β-lactams with molecular oxygen can be performed very efficiently in the presence of acetaldehyde, an acid, and sodium acetate.[54] Typically, the $RuCl_3$-catalyzed oxidation of β-lactam **67** with molecular oxygen (1 atm) in the presence of acetaldehyde and sodium carboxylate gave the corresponding 4-acyloxy β-lactam **68** in 91% and 81% yields, respectively [Eq. (13.45)]. This aerobic oxidation has the same efficiency as the ruthenium-catalyzed oxidation with peracetic acid.

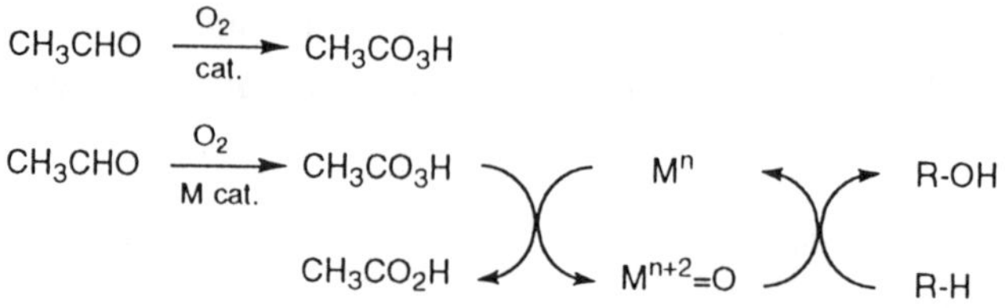

Scheme 13.6

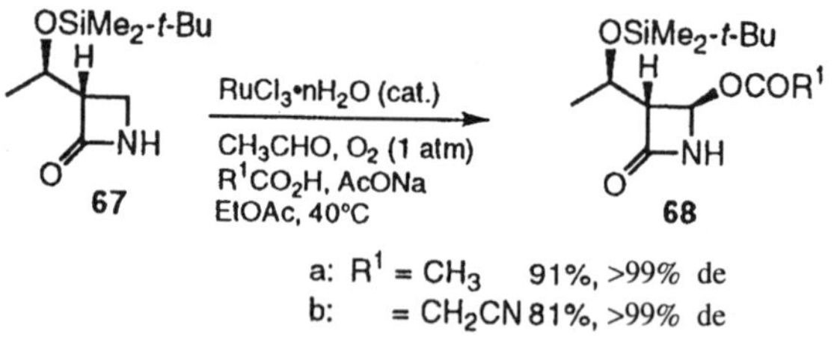

(13.45)

The trimethylsilyl triflate (TMSOTf)-promoted reaction of 4-cyanoacetoxy β-lactam (**68b**) with ketene silyl acetal in CH_2Cl_2 at −78 °C gave 4-methoxycarbonylmethyl β-lactam (**72**) stereoselectively in 92% yield [Eq. (13.46)].

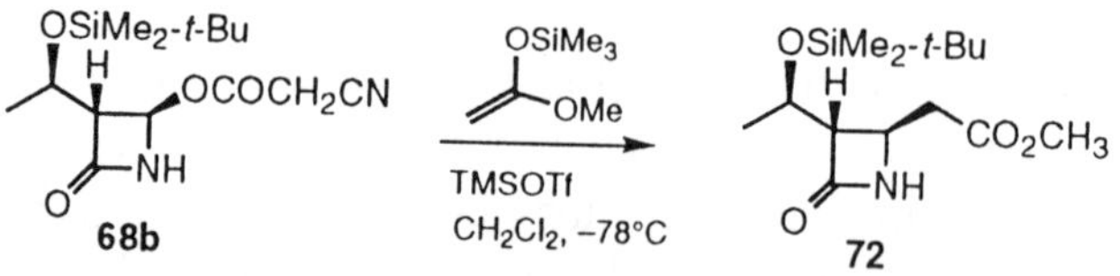

(13.46)

8. Ruthenium-Catalyzed Epoxidation of Alkenes

One of the typical functions of cytochrome P450 is the epoxidation of alkenes. In 1979, Groves and co-workers first reported the metalloporphyrin-catalyzed oxidation of alkenes with iodosylbenzene as a simulation of the catalytic cycle of cytochrome P450.[55] For example, the reaction of *cis*-stilbene with iodosylbenzene in the presence of Fe(TPP)Cl catalyst gave the corresponding *cis*-stilbene oxide in 82% yield. After this success, metalloporphyrin-catalyzed epoxidations of alkenes with various oxidizing agents have been investigated extensively.[1a,b]

Ruthenium complexes are applicable to the epoxidation of alkenes. Octavalent RuO_4 generated from $RuCl_3$/hypochlorite or periodate system is usually too reactive, and C = C bond cleavage is often a major reaction; however, the addition of a bipyridine ligand enables the epoxidation of alkenes, because it works as an electron-donating ligand to enhance the electron density on the metal and to modulate the reactivity of RuO_4.[56–59] Balavoine *et al.* first reported that $RuCl_3$ associated with bipyridyl[56] and phenanthrolines,[57] catalyzes the epoxidation of alkenes with sodium periodate [Eq. (13.47)]. The reactions are stereospecific for both *cis* and *trans* alkenes. Griffith and Williams *et al.* synthesized $[RuO_2(bpy)\{IO_3(OH)_3\}]\cdot 1.5H_2O$ (**73**) which proved to be an effective epoxidation catalyst.[58] Complex **73** in a biphasic water-dichloromethane mixture is an efficient epoxidation catalyst using sodium periodate as a co-oxidant [Eq. (13.47)].

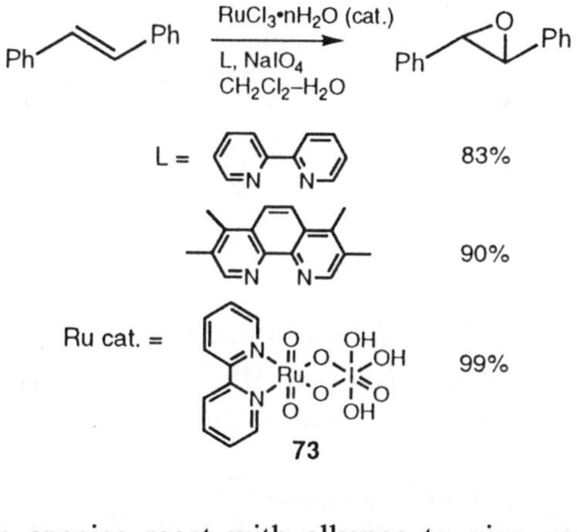

(13.47)

Oxo-ruthenium species react with alkenes to give epoxides. Ru(OEP)(PPh_3)Br[9] and $[RuCl(dpp)_2]$[17a] catalyze oxidations of alkenes with PhIO to give epoxides in moderate yield.

Hirobe[10d,f] and Groves[6b] *et al.* reported that oxygen transfer reactions from 2,6-dichloropyridine *N*-oxide to alkenes were efficiently catalyzed by ruthenium porphyrins under mild conditions, and various alkenes were converted into the corresponding epoxides in high yield [Eqs. (13.48) and (13.49)]. Substituents at the 2 and 6 positions on the pyridine ring of the oxidant are necessary for high efficiency. This is because simple pyridine, which results from the oxygen transfer of pyridine *N*-oxide, coordinates more strongly to the ruthenium than 2,6-dichloropyridine, causing it to inhibit the catalytic activity of ruthenium porphyrin.

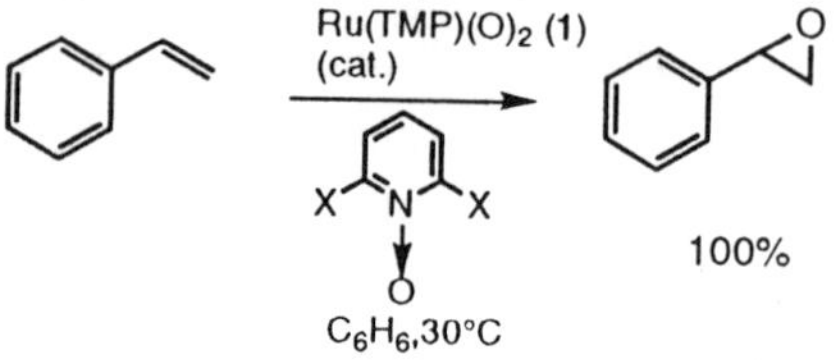

(13.48)

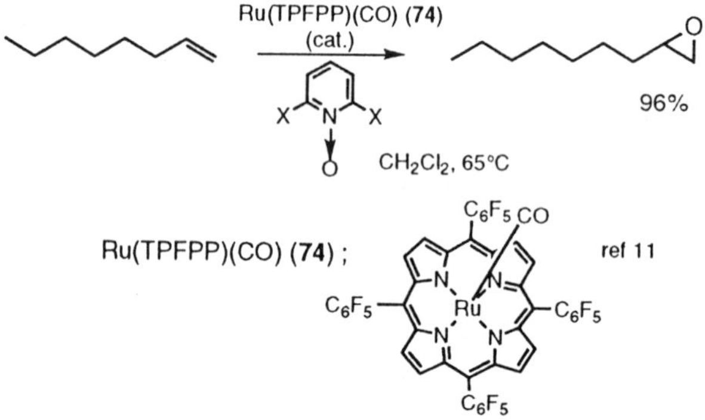

(13.49)

Che[21c,d] developed the *cis*-dioxo-ruthenium(VI) complexes **2** for the oxidation of alkenes with *t*-BuOOH [Eq. (13.50)]. The complex **2** was isolated and characterized by X-ray analysis.

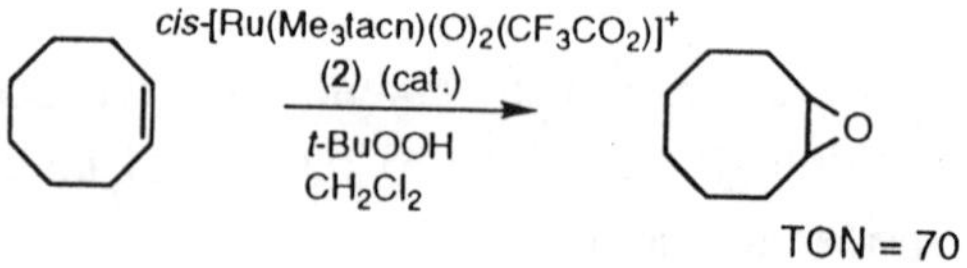

(13.50)

Aerobic oxidation has been explored by several groups.[6a,8,13,15a,60] Groves *et al.* reported Ru(TMP)(O)$_2$ (**1**)-catalyzed aerobic epoxidation of olefins using 1 atm of molecular oxygen.[6a] The fact that two moles of epoxide were produced for each mole of dioxygen shows this reaction proceeds without reductant. The epoxidation is nearly stereospecific. It has been observed that *cis*-olefin is much more reactive than the *trans*-olefin. The mechanism in Scheme 13.7 has been suggested. This oxidation can be applied to the oxidation of unsaturated steroids.[8]

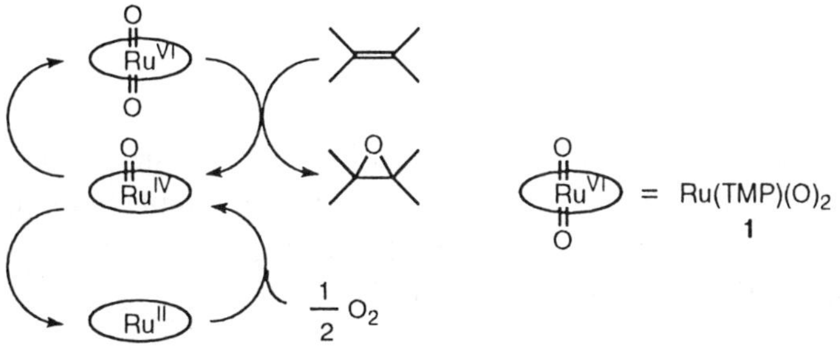

Scheme 13.7

Drago[15a] *et al.* reported that *cis*-dioxo-ruthenium(VI) complex, which is generated from *cis*-$[Ru(dmp)(CH_3CN)_2]^{2+}$ and O_2, catalyzed the epoxidation of norbornene to give the corresponding epoxide efficiently [Eq. (13.51)]. The bulky and rigid ligand is necessary to make a comparatively unstable *cis*-dioxo species.

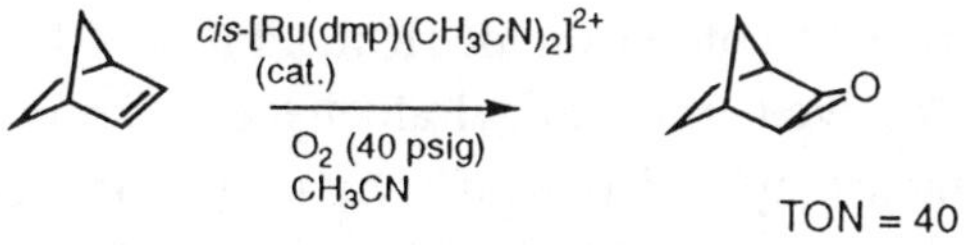

(13.51)

Recently, Knochel *et al.* reported that the ruthenium catalyst bearing perfluorinated 1,3-diketone ligands **75** catalyzes the aerobic epoxidation of alkenes in perfluorinated solvent under Mukaiyama's[61] epoxidation conditions [Eq. (13.52)].[62] Perfluorinated hydrocarbons for the solvent are suitable for the aerobic oxidation since they are chemically inert and solubilize a range of gases such as molecular oxygen well.

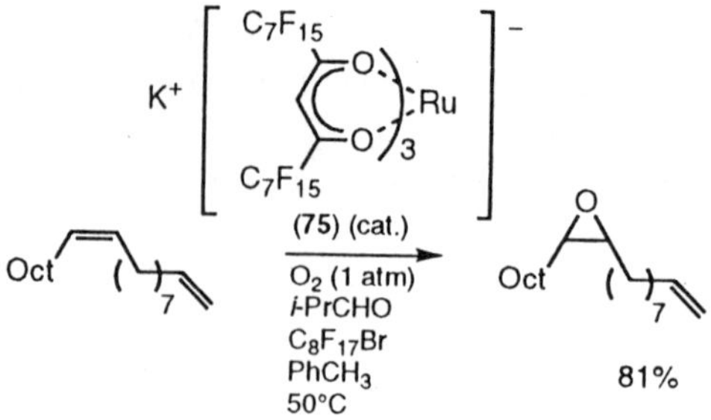

(13.52)

Asymmetric epoxidations were reported for several ruthenium complexes. The asymmetric oxidation of styrene catalyzed by homochiral ruthenium porphyrin **76** with 2,6-dichloropyridine *N*-oxide gives (*R*)-styrene oxide in 57%*ee* with 16% conversion [Eq. (13.53)].[63]

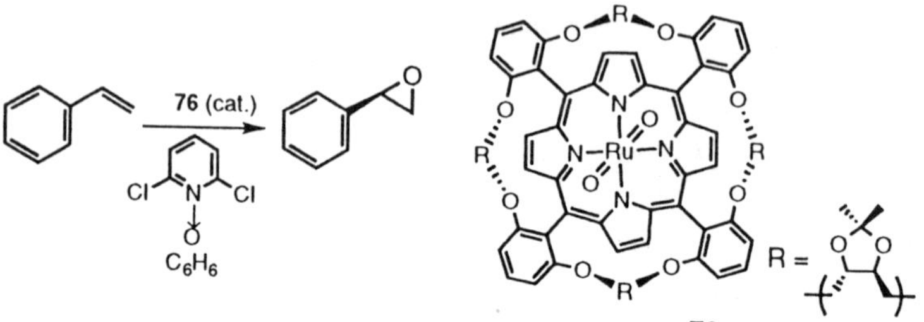

76

(13.53)

Chiral Ru(II) Schiff base complexes of $[RuL(PPh_3)_3(H_2O)_2]$ (**77**) where L = chiral Schiff base derived from salicyl aldehyde and amino acids have been reported. Asymmetric epoxidation of styrene by complex **77** resulted in the formation of (*S*)-styrene oxide in 80%*ee* [Eq. (13.54)].[64]

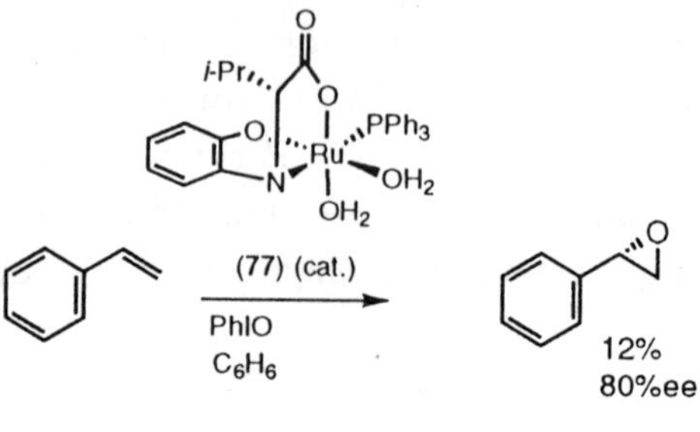

(13.54)

Recently, catalytic asymmetric epoxidation of *trans*-stilbene by chiral ruthenium bis(oxazolinyl)pyridine complex **78** with iodosylbenzene diacetate was reported to give (1*S*,2*S*)-*trans*-stilbene oxide in 74%*ee* [Eq. (13.55)].[65]

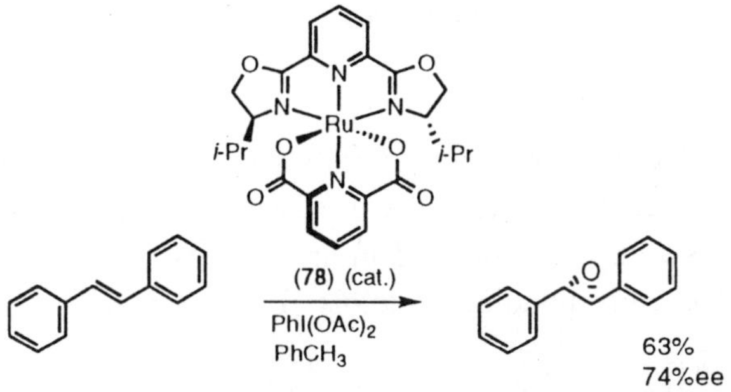

(13.55)

Ruthenium complexes are able to effect another type of oxidation. Che *et al.* synthesized bis(tosyl)imidoruthenium(VI) porphyrin complex [Ru(TPP)(NTs)$_2$]. The imidoruthenium(VI) complex is stable in the solid state at −20 °C for days. This complex can undergo stoichiometric imido group transfer reactions with alkenes to afford aziridines [Eq (13.56)].[66]

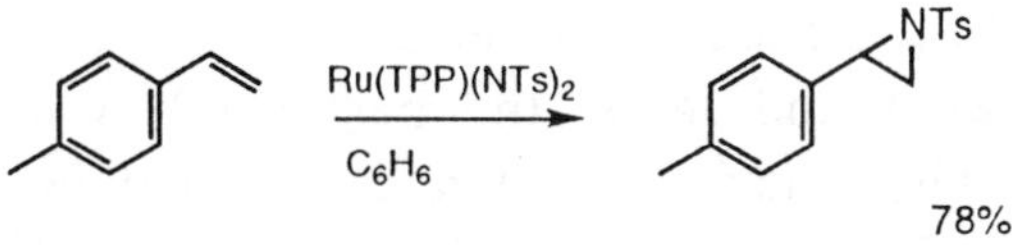

(13.56)

9. Ruthenium-Catalyzed Oxidative Transformation of Alkenes to α-Ketols

As one of the possible pathway of epoxidations using chromium tetraphenylporphyrin, cationic intermediate **79** has been postulated as a key intermediate (Scheme 13.8).[67]

If one could trap the intermediate **79** with external nucleophiles such as water, a new type of catalytic oxidation of alkenes can be performed.

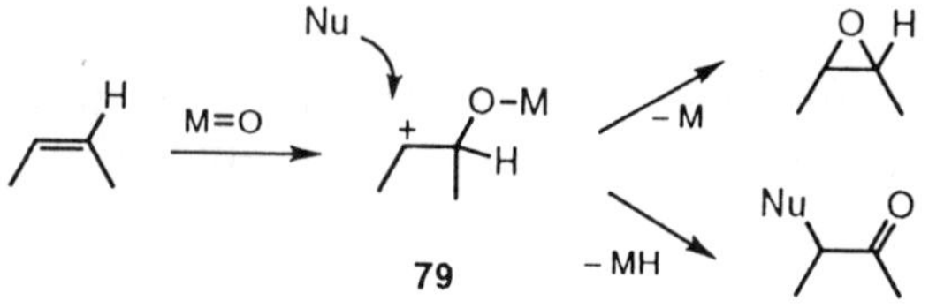

Scheme 13.8

Indeed, novel oxidative transformation of olefins to α-ketols was discovered to proceed very efficiently. Thus, the low-valent ruthenium-catalyzed oxidation of alkenes with peracetic acid in an aqueous solution under mild conditions gives the corresponding α-ketols [Eq. (13.57)].[68]

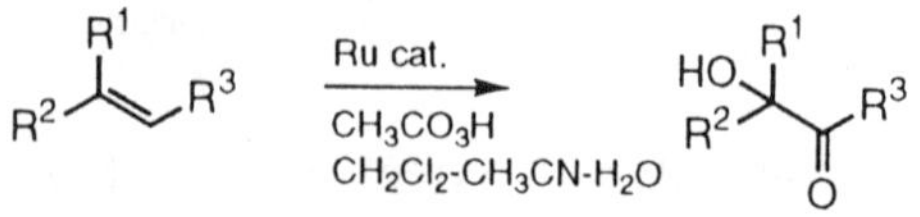

(13.57)

α-Ketols are important synthetic intermediates and partial structures of various biologically active compounds such as cortisone acetate. The reported methods for the synthesis of α-ketols from alkenes are limited to the oxidations with stoichiometric amount of $KMnO_4$/$CuSO_4$•$5H_2O$[69] and with isobutylaldehyde/O_2 in the presence of OsO_4 and bis(3-methyl-2,4-pentanedionato)nickel(II) catalyst.[70] Ruthenium trichloride has been found to be the best catalyst among those examined. The two-phase aqueous system is required for the present reaction. Typically, $RuCl_3$-catalyzed oxidation of 3-acetoxy-1-cyclohexene (**80**) with peracetic acid in H_2O-CH_3CN-CH_2Cl_2 (1:1:1) affords (2R*, 3S*)-3-acetoxy-2-hydroxycyclohexanone (**81**) chemo- and stereoselectively in 78% yield [Eq. (13.58)].

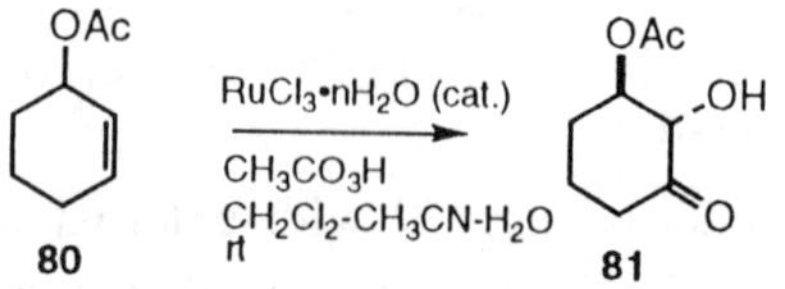

(13.58)

The oxidation is quite different from that promoted by ruthenium tetraoxide.[3] Indeed, the oxidation of 1-methylcyclohexane **82** under these conditions gives 2-hydroxy-2-methylcyclohexanone (**83**) (67%), while the oxidation of the same substrate **82** under the conditions in which ruthenium tetraoxide is generated catalytically gives 6-oxoheptanoic acid (**84**) (91%) [Eq. (13.59)].

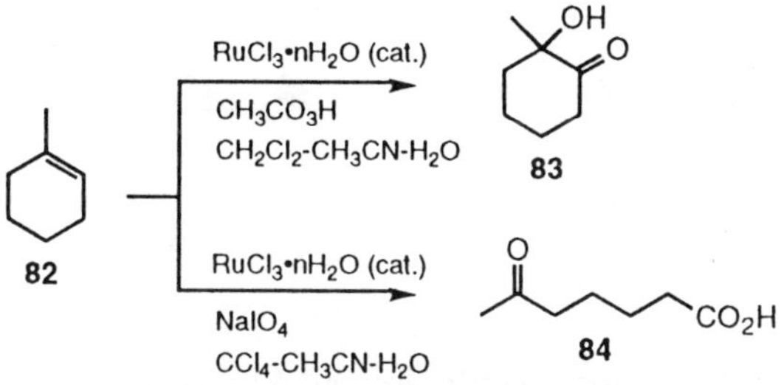

(13.59)

The present oxidation can be applied to the oxidation of substituted alkenes having functional groups such as acetoxy, methoxycarbonyl, and azide groups to give the corresponding α-ketols in good to excellent yields. The oxidation of 3-azide-1-cyclohexene (**85**) gave (2S*, 3R*)-3-azide-2-hydroxycyclohexanone (**86**) chemo- and stereoselectively (65%) [Eq. (13.60)].

(13.60)

The efficiency of the present reaction has been demonstrated by the synthesis of cortisone acetate **90**,[71] which is a valuable antiinflammatory agent. The strategy for the synthesis of **90** is shown in [Eq. (13.61)]. The oxidation of 3β, 21-diacetoxy-5α-pregn-17-ene (**88**) proceeded stereoselectively to give 20-oxo-5α-pregnane-3β,17α,21-triol 3,21-acetate (**89**) [57%, Eq. (13.61)]. Conventional treatment of **89** followed by microbial oxidation with *Rhizopus nigricaus* gave **90**.

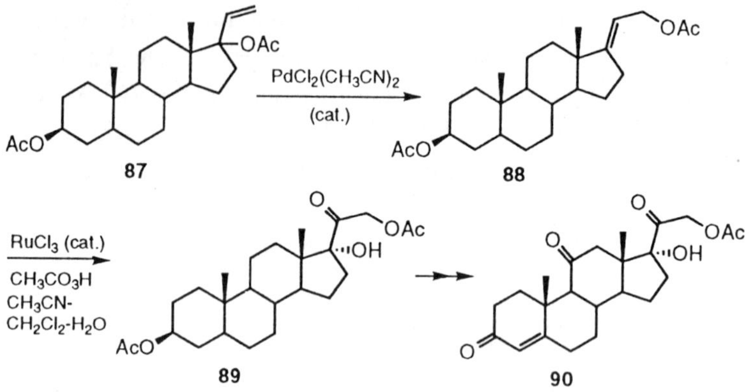

(13.61)

10. Ruthenium-Catalyzed Oxidation of Alcohols

10.1 Oxidation with peroxides

Several methods for oxidation of alcohols have been reported using various catalytic systems which include $RuCl_2(PPh_3)_3$ with oxidants such as *N*-methylmorpholine *N*-oxide (NMO),[72a] iodosylbenzene,[72b] TMSOOTMS,[72c] $RuCl_3$ with hydrogen peroxide,[72d] and K_2RuO_4 with peroxodisulfate.[72e] In particular, Ley *et al.* discovered that the oxidation of primary and secondary alcohols to carbonyl compounds by tetrapropylammonium perruthenate (TPAP)/NMO is an excellent method.[73] One of the key features of the TPAP system is its ability to tolerate other potentially reactive groups. For example, double bonds, polyenes, enones, halides, cyclopropanes, epoxides, and acetals all remain intact during TPAP oxidation. The oxidation of primary alcohols with TPAP gives the corresponding aldehyde [Eqs. (13.62)

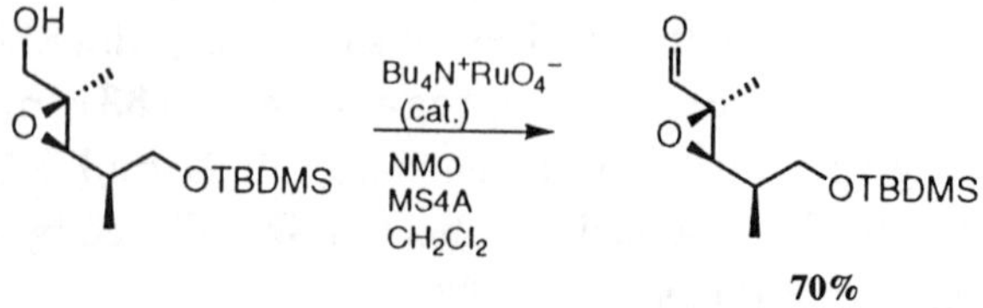

(13.62)

and (13.63)], whereas RuO_4 oxidation results in the formation of carboxylic acid.

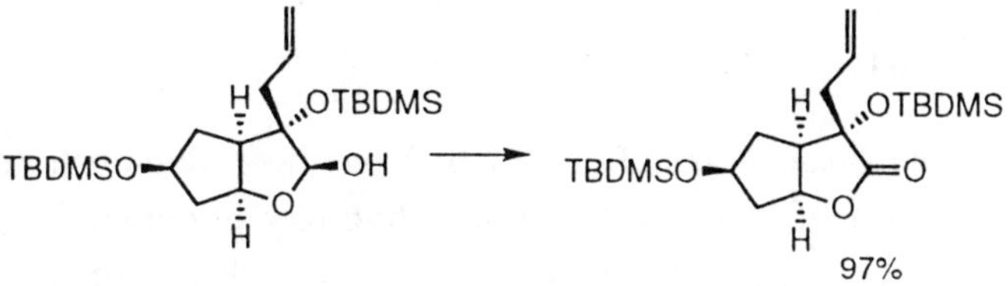

(13.63)

The $RuCl_2(PPh_3)_3$-catalyzed reaction of secondary alcohols with *t*-BuOOH gives ketones under mild conditions.[74,75] This oxidation can be applied to the transformation of cyanohydrins into acyl cyanides,[75] which are versatile synthetic intermediates such as acylating reagents. Typically, oxidation of 3,4-dimethoxybenzaldehyde cyanohydrin **91** with two equivalents of *t*-BuOOH in dry benzene at room temperature gives 3,4-dimethoxybenzoyl cyanide (**92**) in 89% yield [Eq. (13.64)]. It is worth noting that the acyl cyanides thus obtained are excellent reagents for the chemoselective acylation reaction. The reaction of amino alcohols with acyl cyanides gives selectively *N*-acylated amino alcohols. Furthermore, primary amines are selectively acylated in the presence of secondary amines.[76]

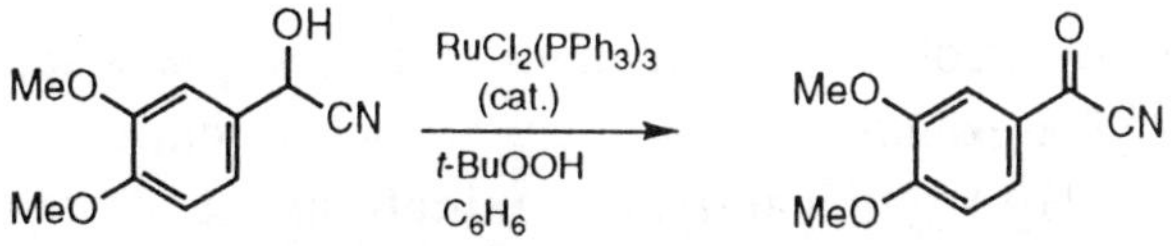

91 **92**

(13.64)

The reaction of alcohols with peracetic acid in the presence of $RuCl_3$ catalyst also gave the corresponding ketones with high efficiency [Eq. (13.65)].[77] It is noteworthy that primary alcohols are oxidized faster than the secondary alcohols in the present reactions. Various aliphatic and aromatic secondary alcohols can be oxidized at room temperature to give the corresponding ketones very efficiently. For example, decahydro-1-naphthol (**93**) is oxidized selectively to give 1-decalone (**94**) (95%) [Eq. (13.65)].

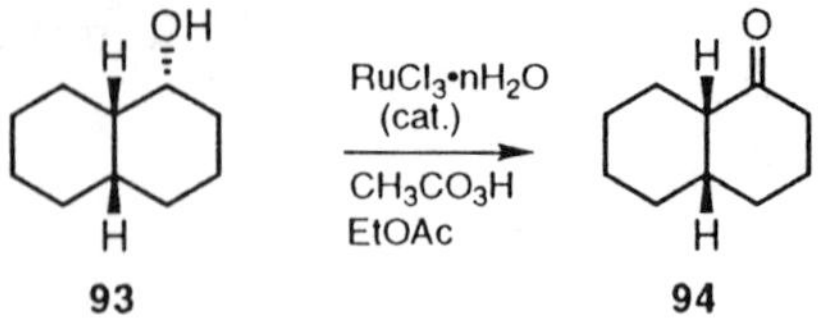

(13.65)

The utility of the present oxidation reaction has been demonstrated by the selective oxidation of nitrogen-containing hydroxy compounds. Thus, the Ru/C-catalyzed oxidation of 1-(4-pyridyl)ethanol (**95**) with peracetic acid gave 4-acetylpyridine (**96**) (85%) chemoselectively [Eq. (13.66)]. This is in contrast to the oxidation of **95** with peracetic acid itself, giving the corresponding *N*-oxide (**97**) preferentially [Eq. (13.66)]. Similar treatment of 1-(2-pyridyl)ethanol with Ru/C catalyst gave 2-acetylpyridine (83%) selectively.

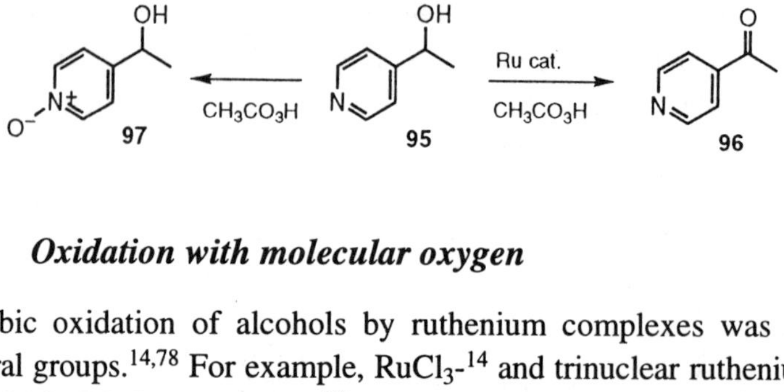

(13.66)

10.2 *Oxidation with molecular oxygen*

Aerobic oxidation of alcohols by ruthenium complexes was reported by several groups.[14,78] For example, $RuCl_3$-[14] and trinuclear ruthenium complex $[Ru_3O(O_2CC_2H_5)_3(H_2O)_3]$-[78a]catalyzed oxidations of 2-octanol and cyclohexanol give 2-octanone and cyclohexanone in moderate yields; however, these reactions require rigorous conditions such as high temperature and 3 atms of O_2 [Eqs. (13.67) and (13.68)]. The oxidation of alcohol **98** by $RuCl_2(PPh_3)_3$/4-BzOTEMPO-O_2[78b] system gives the corresponding ketone **99** [Eq. (13.69)].

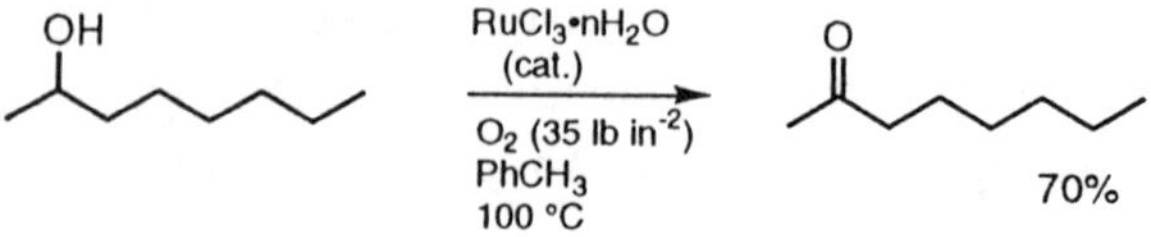

(13.67)

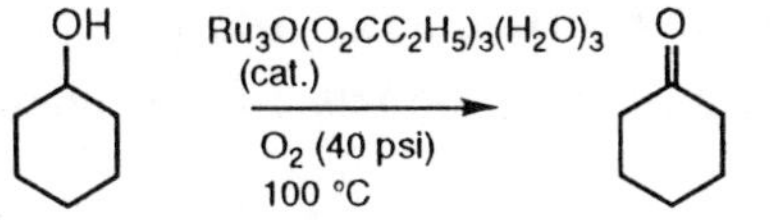

(13.68)

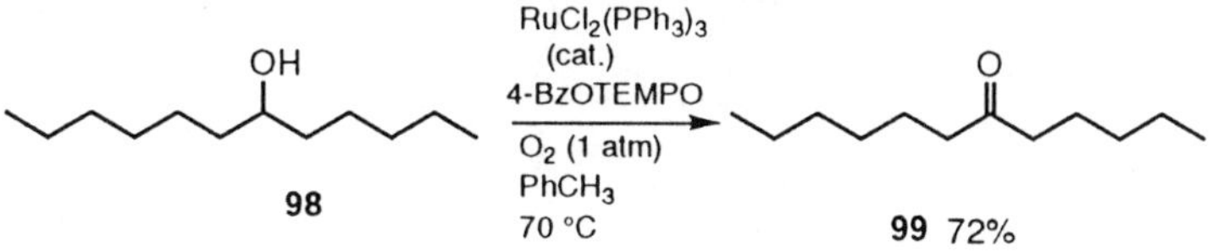

(13.59)

Allylic alcohols can be converted into α,β-unsaturated aldehydes with 1 atm of molecular oxygen in the presence of RuO_2[78c] catalyst [Eq. (13.70)].

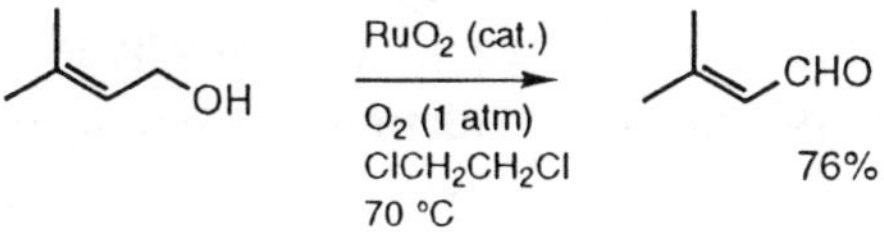

(13.70)

The aerobic oxidation of alcohols by oxo-ruthenium species has been studied. Oxo-ruthenium species can be generated by the reaction of ruthenium complexes with molecular oxygen in the presence of aldehydes as shown in Sec. 7.2; however, under these conditions, the oxidation of alcohols does not occur because the formation of peracids is prevented by strong coordination of alcohol to the ruthenium. Therefore, we have examined catalysts which are known to form peracetic acid from acetaldehyde and molecular oxygen and have no strong coordination ability toward alcohols. We have found a novel, general, and efficient method for the aerobic oxidation of alcohols by using a ruthenium-cobalt bimetallic catalyst. The oxidation of alcohols is performed at room temperature with molecular oxygen (1 atm) in the presence of an aldehyde and $RuCl_3$–$Co(OAc)_2$ bimetallic catalyst [Eq. (13.71)].[79]

$$R^1CH(OH)R^2 \xrightarrow[\text{O}_2\text{ (1 atm)},\ \text{CH}_3\text{CHO}]{\text{Ru/Co cat.}} R^1C(=O)R^2 \qquad (13.71)$$

Thus, aerobic oxidation of alcohols **100** with $RuCl_3$ and $Co(OAc)_2$ catalysts in the presence of acetaldehyde in ethyl acetate at room temperature gave the corresponding ketone **101** in 89% yield [Eq. (13.72)]. This method is highly convenient, because the products can be readily isolated simply

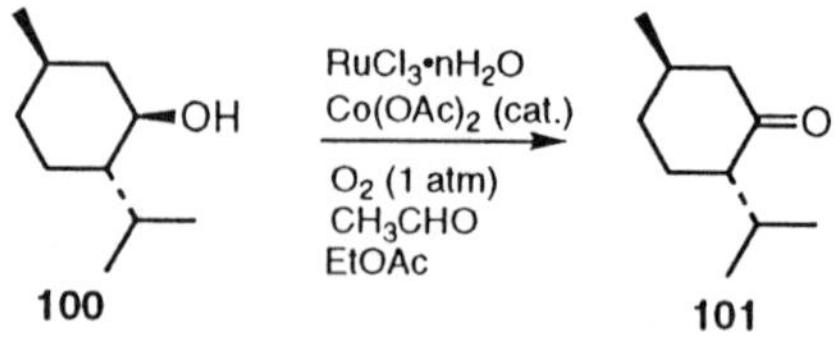

100 101

(13.72)

by removal of acetic acid and the catalyst through washing with small amount of water. Primary alcohols are oxidized to the corresponding carboxylic acids smoothly. The present aerobic oxidation can be rationalized by assuming the following two sequential pathways: (i) formation of peracids by a cobalt-mediated radical chain reaction of aldehydes with molecular oxygen and (ii) ruthenium-catalyzed oxidation of alcohols with the peracids thus formed.

Recently, Markó *et al.* discovered that TPAP can be used as an effective catalyst for the aerobic oxidation of alcohols to the corresponding carbonyl compounds [Eq. (13.73)]. It is noteworthy that even secondary alcohols undergo smooth oxidation into the corresponding ketones under mild conditions [Eq. (13.74)].[80]

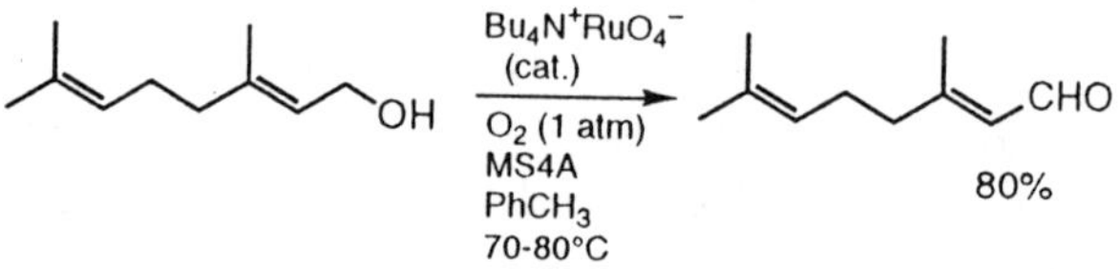

(13.73)

(13.74)

11. Ruthenium-Catalyzed Oxidation of Phenols

The oxidative transformation of phenols is of importance in view of its biological and synthetic aspects. However, the oxidation of phenols generally lacks selectivity because of coupling reactions caused by phenoxyl radicals,[81] and selective oxidation of phenols is limited to phenols bearing bulky substituents at 2- and 6-positions.[82] Using ruthenium catalysts, a biomimetic and selective oxidation of phenols can be performed. Thus, the oxidation of *p*-substituted phenols with *t*-BuOOH in the presence of $RuCl_2(PPh_3)_3$ catalyst gives 4-(*tert*-butyldioxy)-4-alkylcyclohexadienones [Eq. (13.75)].[83] The reaction provides an efficient method for selective oxidation of phenols

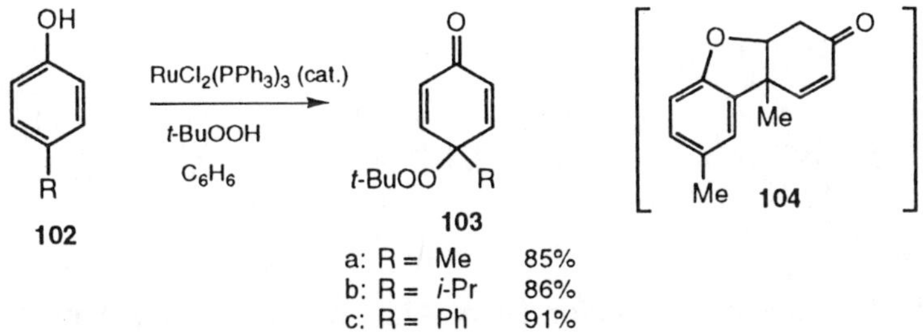

(13.75)

bearing no substituent at the 2- and 6-positions. This is because of the fast electron transfer of ruthenium complexes. The reaction can be rationalized by assuming the mechanism which involves oxo-ruthenium complex (Scheme 13.9). Hydrogen abstraction with oxo-ruthenium species gives phenoxyl radical **105**, which undergoes fast electron transfer to ruthenium to give a cationic intermediate **106**. Nucleophilic reaction with the second molecule of *t*-BuOOH gives **103**. It is noteworthy that iron porphyrin (Fe(TPP)Cl)-catalyzed oxidation of *p*-cresol (**102a**) with *t*-BuOOH

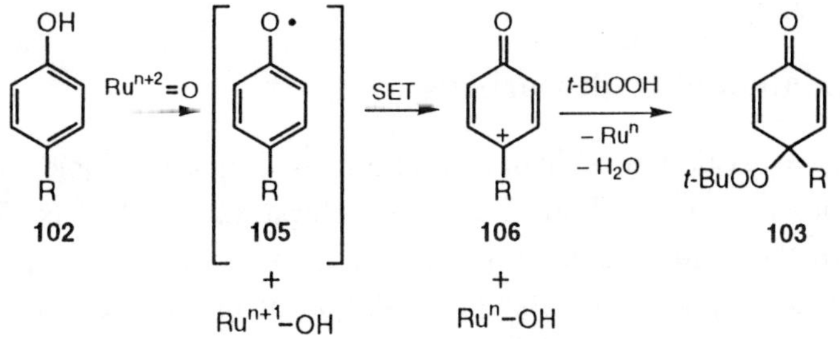

Scheme 13.9

gave radical coupling product **104** as a major product along with **103a** (**104**:**103a** = 1.2 : 1),[84] indicating that the nature of the active intermediate depends on the metal.

4-(*tert*-Butyldioxy)-4-alkylcyclohexadienones thus obtained are versatile synthetic intermediates. For example, $TiCl_4$-promoted transformation of **107** gives 2-substituted quinones **108** which are rearrangement products of alkyl groups (93%) [Eq. (13.76)].

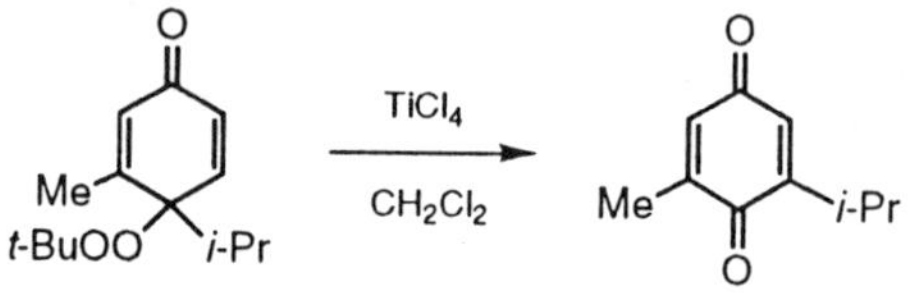

107 **108**

(13.76)

Interestingly, sequential migration-Diels–Alder reactions of *tert*-butyldioxy dienone **110** in the presence of *cis*-1,3-pentadiene gave *cis*-fused octahydroanthraquinone **111** stereoselectively (73%) [Eq. (13.77)]. The two-step synthesis of **111** from **109** shows great promise for further synthetic applications.

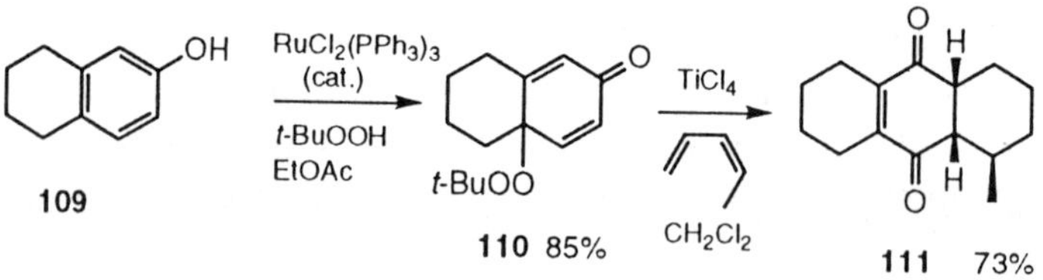

109 **110** 85% **111** 73%

(13.77)

12. Oxidation of Hydrocarbons

The selective oxidation of hydrocarbons is an important topic in view of the economical and ecological use of natural raw materials. However, practical methods are limited to the oxidations of reactive substrates bearing functional groups, and the catalytic oxidation of unactivated hydrocarbons remains a challenging topic. Direct oxidation of hydrocarbons is one of the typical reactions of cytochrome P450.[23] As a model of this enzyme, catalytic oxidations of alkanes with metalloporphyrins and non-porphyrin metals have been explored using various oxidants.[1] We have explored novel and practical systems for the oxidation of alkanes using ruthenium, iron, copper, and other metals.

12.1 Ruthenium-catalyzed oxidation with peroxide

Ruthenium porphyrins such as $Ru(OEP)(PPh_3)_3$ shows the catalytic activity for the oxidation of alkanes with PhIO.[9] As described in Eq.(13.48), Hirobe *et al.* studied the unique reactivity of $Ru(TMP)(O)_2$ complexes to catalyze the efficient epoxidation of olefins by 2,6-dichloropyridine *N*-oxide.[10d,f] Moreover, the catalytic ability of ruthenium porphyrins is enhanced by the presence of a small amount of HCl or HBr, and the oxidation of alkanes with pyridine *N*-oxide is also catalyzed by ruthenium porphyrins with high efficiency.[10a,c] For example, the oxidation of adamantane proceeds efficiently to give 1-adamantanol, 1,3-adamantanediol, and 2-adamantanone with high turnover numbers [Eq. (13.78)]. Lau *et al.* reported that the addition of Lewis acids such as $ZnCl_2$, $FeCl_3$, and BF_3•$MeCO_2H$ greatly accelerated the reaction rates in the stoichiometric oxidation of alkanes by $BaRu(O)_2(OH)_3$.[18b]

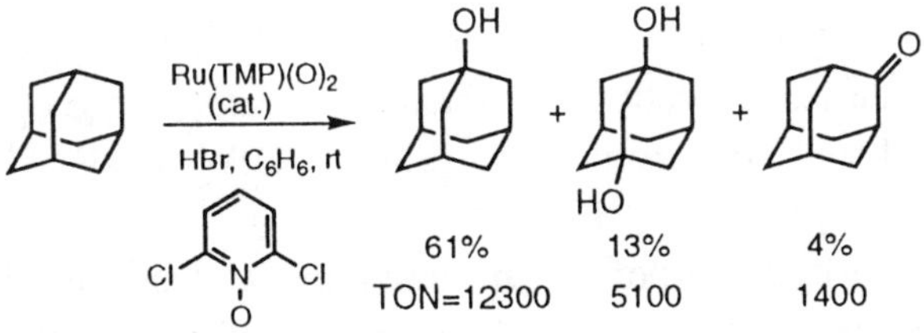

(13.78)

Groves *et al.* reported the very rapid oxygenation of a variety of substrates including alkanes, terminal olefins, and benzene, catalyzed by perfluorinated ruthenium complex Ru(TPFPP)(CO) (**74**) with 2,6-pyridine *N*-oxide.[6b] Hydroxylation of adamantane and *cis*-decalin were achieved with high selectivity, complete stereo-retention, high rates (up to 800 turnovers/min), and high efficiency (400–14600 turnovers) [Eqs. (13.79) and (13.80)].

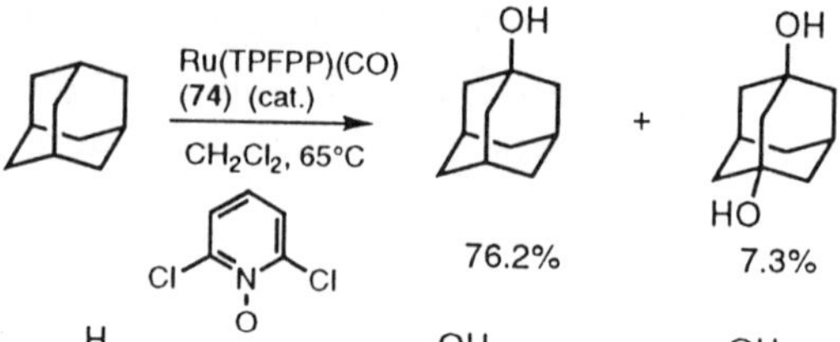

(13.79)

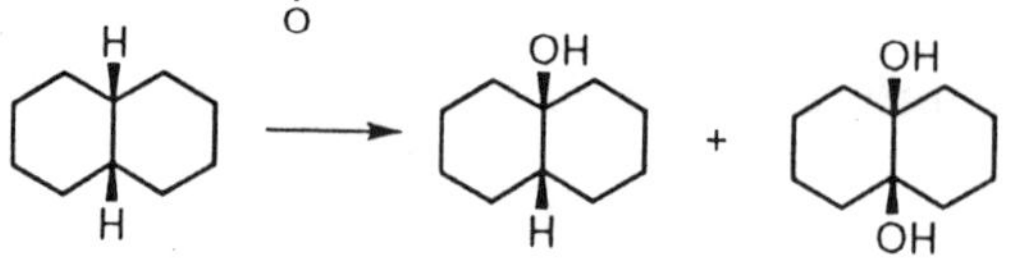

(13.80)

Recently, perfluorinated ruthenium phthalocyanine was reported as a oxidation catalyst with *t*-BuOOH.[85]

Non-porphyrin ruthenium complexes with peroxides can be used for the catalytic oxidation of alkanes. $BaRuO_3(OH)_2$-catalyzed oxidation of cyclohexane with PhIO gives oxidation products with high efficiency.[18a] $[RuCl(dpp)_2]^+$ with PhIO or LiClO system was reported for the alkane oxidation.[17] Che and Drago independently investigated the reactivities towards *cis*- and *trans*-dioxo-ruthenium species and found that *cis*-$[Ru(dmp)_2(MeCN)_2]^{2+}$ with H_2O_2,[15a] *cis*-$[Ru(Me_3tacn)(O)_2(CF_3CO_2)]^+$ (**2**) with *t*-BuOOH,[21c] and *cis*-$[Ru(6,6\text{-}Cl_2bpy)_2(OH_2)_2]^{2+}$ with *t*-BuOOH[21d] systems are more efficient for the oxidation of cyclohexane. Ruthenium(III) monomer complexes such as $[RuCl_2(TPA)]^+$ and $[RuCl(Me_2SO)(TPA)]^+$ bearing tripodal ligand TPA (TPA = tris(2-pyridylmethyl)amine) was synthesized, and catalytic oxidations of adamantane with *m*-chloroperbenzoic acid were reported.[19,20] Polyoxometalate $[SiRu(H_2O)W_{11}O_{39}]^{5-}$ also works as an oxidation catlayst using $KHSO_5$[86a] and H_2O_2.[86b]

The oxidations of hydrocarbons with ruthenium catalysts bearing simple moiety are highly effective. Actually, the oxidation of hydrocarbons in the presence of ruthenium catalysts such as $RuCl_2(PPh_3)_3$[87] or Ru/C[88] have been reported to give the corresponding ketones and alcohols in good yields [Eq. (13.81)]. Typically, ruthenium-catalyzed oxidations of cyclohexane with

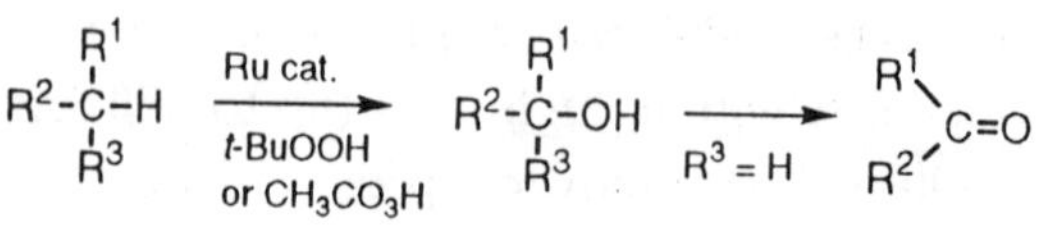

(13.81)

peroxides such as *t*-BuOOH and peracetic acid give cyclohexanone along with a small amount of cyclohexanol efficiently [Eq. (13.82)].

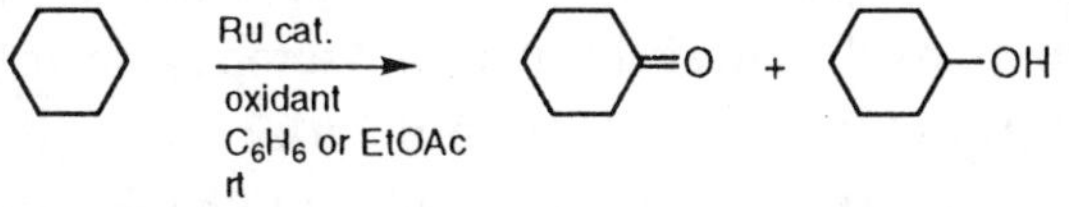

(13.82)

$RuCl_2(PPh_3)_3$ –*t*-BuOOH	conv. 41%	47%	7%
Ru/C–CH_3CO_3H	conv. 62%	67%	1%

When *t*-BuOOH is used as an oxidant for metal-catalyzed hydroxylation of alkanes, two modes of C–H activation can be observed: one resulting from RO• or ROO• formed from the homolytic cleavage of the peroxidic O–O bond, the other due to an oxo-metal (M = O) species produced by heterolytic cleavage of the O–O bond of metal peroxide. These two possible mechanisms have been evidenced by Mansuy and Momenteau in 1982 in a comparative study using several metalloporphyrins in association with ROOH or PhIO.[89] In order to investigate the character of active species in the ruthenium-catalyzed oxidation of alkanes with *t*-BuOOH, kinetic data were compared with other systems (Table 13.1). The ρ value obtained from $RuCl_2(PPh_3)_3$-catalyzed oxidation of substituted toluenes with *t*-BuOOH

Table 13.1 Kinetics and Intra- and Intermolecular Deuterium Isotope Effect for the Catalytic Oxidation of Hydrocarbons

System	ρ^a	$k_H/k_D{}^b$	$k_H/k_D{}^c$
cytochrome P450	-1.6^d	-11^g	
$RuCl_2(PPh_3)_3$ *t* BuOOH	−1.39	9.0	8.4
t-BuO•	-0.4^e		4.2^h
t-BuOO•	-0.6^f		

[a]Values from Hammet plots of log k_{rel} versus σ^+ for the competitive oxidation of substituted toluenes. [b]Intramolecular isotope effect for the oxidation of 1,1-dideuterio-1,3-diphenylpropane. [c]Intermolecular isotope effect for the oxidation of cyclohexane/cyclohexane-d_{12}. [d]Ref. 90. [e]Ref. 91. [f]Ref. 92. [g]Ref. 93. [h]Ref. 94.

is –1.39, which is similar to that (–1.6)[90] obtained for the oxidation by cytochrome P450 and is quite different from those with *t*-BuO• (–0.4)[91] and *t*-BuOO• (–0.6).[92] These facts support the existence of oxo-ruthenium species rather than alkoxyl radical. Deuterium isotope effects also support the existence of these species. Intramolecular and intermolecular deuterium isotope effects (k_H/k_D) were determined to be 9.0 and 8.4 for the oxidation of 1,1-dideuterio-1,3-diphenylpropane and cyclohexane/cyclohexane-d_{12}, respectively. These large k_H/k_D values are similar to that obtained for the oxidation with cytochrome P450 (11).[93] In contrast, smaller isotope effect (4.2) was obtained with *t*-BuO•.[94] This evidence shows the involvement of ruthenium-dependent species such as oxo-ruthenium and *t*-butylperoxy-ruthenium species.

Assuming the reaction of oxo-ruthenium species, it is expected that more reactive species will be generated in the presence of strong acids such as trifluoroacetic acid. Indeed, the $RuCl_3{\bullet}nH_2O$-catalyzed oxidation of cyclohexane in trifluoroacetic acid and dichloromethane (5:1) with peracetic acid gives cyclohexyl trifluoroacetate in 77% [Eq. (13.83)].[88]

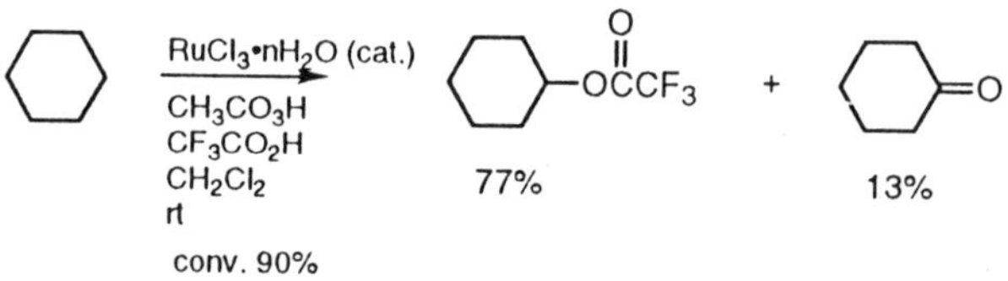

(13.83)

12.2 *Metal-catalyzed oxidation with molecular oxygen*

The catalytic oxidation of alkanes with molecular oxygen under mild conditions is an especially rewarding goal, since direct functionalization of unactivated C–H bonds in saturated hydrocarbons usually requires drastic conditions such as high pressure and high temperature. Aerobic oxidation of alkanes with [Ru^{III}-EDTA] complex and ruthenium-substituted polyoxometalate $[WZnRu_2(OH)(H_2O)(ZnW_9O_{34})_2]^{11-}$ catalysts are reported.[16,95]

As shown in Eq. (13.81), we have found that a low-valent ruthenium catalyst/peroxide system is efficient for the biomimetic oxidation of hydrocarbons. Furthermore, the oxo-metal species can be generated by the reaction of a low-valent transition metal complex with molecular oxygen in the presence of an aldehyde [Eqs. (13.45) and (13.71)]. As the consequence of continuous study on the development of efficient aerobic oxidation reactions, we have found that ruthenium catalysts are effective for the oxidation of alkanes with molecular oxygen, in the presence of acetaldehyde [Eq. (13.84)].[96] Furthermore, we found that even iron[96] and copper[97] catalysts are efficient for the oxidation of non-activated

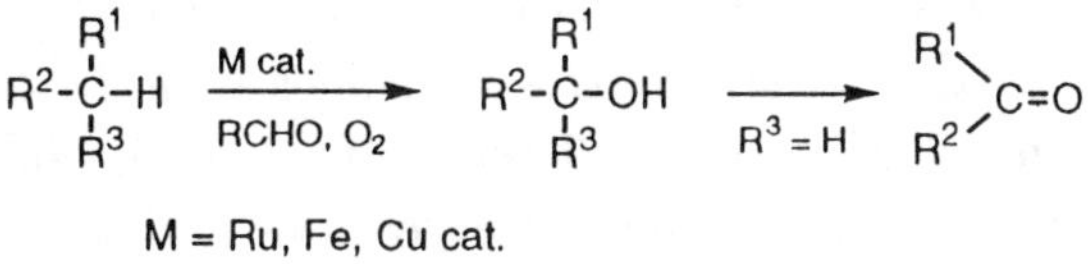

(13.84)

hydrocarbons at room temperature under 1 atm of molecular oxygen [Eq. (13.84)]. Actually, $RuCl_3{\bullet}nH_2O$, iron powder, and $Cu(OH)_2$ have proven to be effective catalysts, although $Fe(OAc)_3$, $FeCl_3$, $RuCl_2(PPh_3)_3$, and $Cu(OAc)_2$ can also be used. Aldehydes such as acetaldehyde and heptanal are effective. Typically, $RuCl_3{\bullet}nH_2O$ and Fe-catalyzed oxidations of cyclooctane and adamantane with molecular oxygen in the presence of an aldehyde give the corresponding alcohols and ketones selectively [Eqs. (13.85) and (13.86)]. Practically, the oxidation of cyclohexane can be performed without solvent under the conditions of 8 atms of O_2 and 70 °C [Eq. (13.87)].

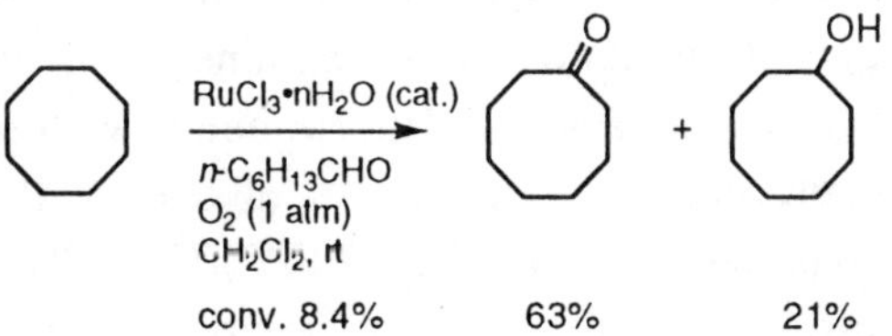

(13.85)

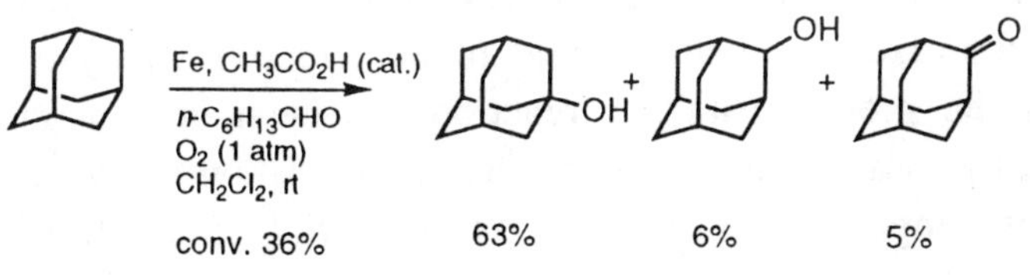

(13.86)

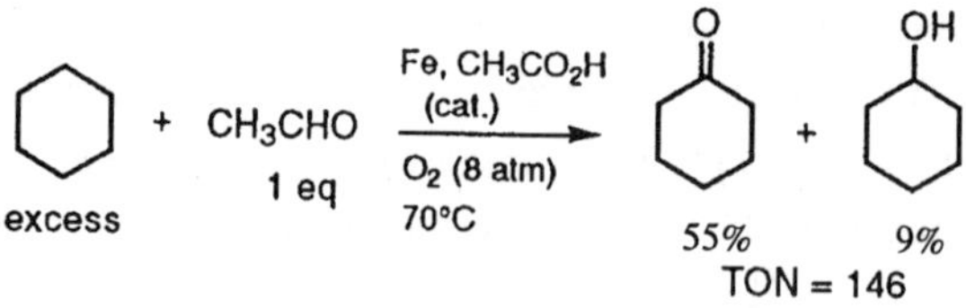

(13.87)

These aerobic oxidations can be rationalized by assuming the sequence shown in Scheme 13.10. The metal-catalyzed radical chain reaction of an aldehyde with molecular oxygen affords the corresponding peracid. The reaction of metal catalyst with peracids thus formed would give an oxo-metal intermediate. Hydrogen abstraction of alkanes with the oxo-metal is followed by oxygen transfer to afford the corresponding alcohols. The alcohols are further oxidized to the corresponding ketones under these conditions. Recently, similar iron[98a,b] and cobalt[98c,d]-catalyzed aerobic oxidations of alkanes with aldehydes were reported.

$$RCHO + O_2 \xrightarrow{\text{M cat.}} RCO_3H$$

$$M^n + RCO_3H \longrightarrow M^{n+2}{=}O + RCO_2H$$

$$M^{n+2}{=}O + RH \longrightarrow M^n + ROH$$

Scheme 13.10

Generation of the oxo-metal porphyrin using an aldehyde and O_2 in place of single oxygen atom donors was examined, and we found that metalloporphyrins bearing meso-pentafluorophenyl groups are highly efficient catalysts for the aerobic oxidation of alkanes with acetaldehyde [Eq. (13.88)].[11] We prepared the novel Ru(TPFPP)(CO) (**74**) complex and found that **74** is an efficient catalyst for the aerobic oxidation of alkanes using acetaldehyde [Eq. (13.88)]. Furthermore, Mn(TPFPP)Cl (**112**) and Co(TPFPP) (**113**) were also found to be effective catalysts.[11] The efficiency of the present catalytic reaction has been demonstrated by extremely high turnover numbers in the aerobic oxidation of cyclohexane, which is most important in view of its industrial application. Thus,

74, **112**, and **113**-catalyzed oxidation of cyclohexane with molecular oxygen in the presence of acetaldehyde gave cyclohexanone and cyclohexanol in 62–75% yields based on acetaldehyde and turnover numbers of 14 100–17 000 [Eq. (13.88)].

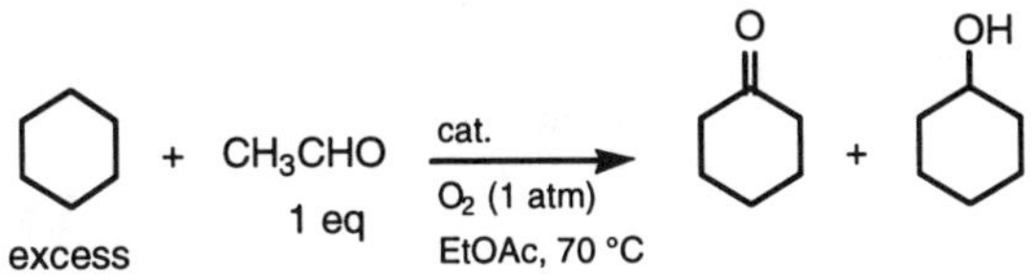

(13.88)

cat. M =					turnover number
Ru^{II}(CO)	: Ru(TPFPP)(CO)(**74**)		54%	8.1%	1.41×10^4
Mn^{III}(Cl)	: Mn(TPFPP)Cl	(**112**)	61%	9.4%	1.59×10^4
Co^{II}	: Co(TPFPP)	(**113**)	64%	11%	1.70×10^4
			(based on acetaldehyde)		

In the copper-catalyzed oxidation of alkanes, the catalytic activity of copper salts is in the order of $Cu(OH)_2 > Cu(OAc)_2 >> Cu(OCOCF_3)_2 > CuCl_2$, indicating that a copper salt bearing a more basic or an electron-donating ligand is suitable for the present oxidation. Divalent copper ions are known to form the most stable complexes with ligands amongst all the metals of the second half of the first transition series. So we thought that crown ethers could be used as an electron-donating ligand for the copper catalysts, although transition metals usually do not make a stable complex with crown ethers. As a consequence of this study we found that the combination of a copper complex and a crown ether shows high catalytic activity for the present aerobic oxidation of alkanes with acetaldehyde.[97b,c]

The efficiency of the present catalysts is highlighted by extremely high turnover numbers in the aerobic oxidation of cyclohexane. Actually, the oxidation of cyclohexane in the presence of $CuCl_2$ (2.50×10^{-3} mol%) and 18-crown-6 (2.50×10^{-3} mol%) and acetaldehyde at 70 °C under O_2 atmosphere (1 atm) gave cyclohexanone (61% yield based on acetaldehyde) and cyclohexanol (10%) with turnover number of 16 200 [Eq. (13.89)]. Although stable transition metal complexes of crown ethers are rarely isolated, we succeeded in the isolation of copper crown ether complex $[(CuCl_2)_4(C_{12}H_{24}O_6)_2(H_2O)_2]$ (**114**). Importantly, the complex **114**

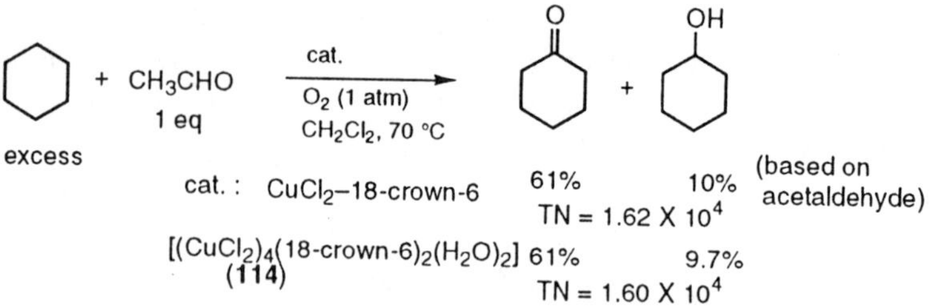

(13.89)

is as efficient a catalyst as the combined use of $CuCl_2$ and 18-crown-6, indicating that ligation of crown ether is significant for the catalytic activity.

Cobalt-catalyzed aerobic oxidation of cyclohexane has been used commercially, however, the total yield of cyclohexanone and cyclohexanol is quite low.[99] Our oxidation reaction provides a powerful industrial strategy for the synthesis of cyclohexanone by combination of Wacker oxidation of ethylene with the present metal-catalyzed oxidation of cyclohexane (Scheme 13.11).

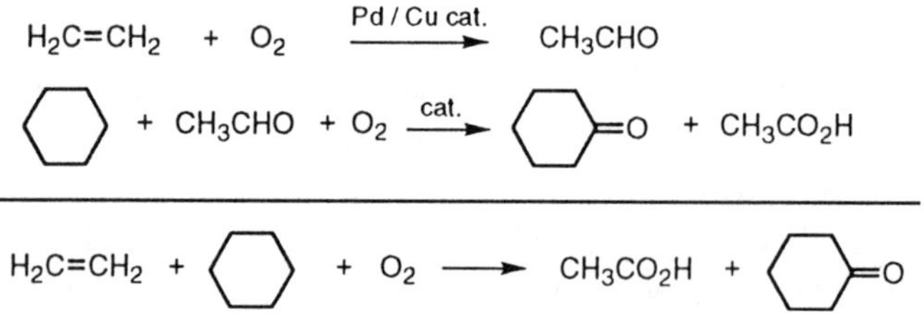

Scheme 13.11

References

1. (a) *Metalloporphyrins Catalyzed Oxidations* (eds.) Montanari, F. and Casella, L. Kluwer Academic Publishers, Dordrecht, 1994; (b) *Metalloporphyrins in Catalytic Oxidations* (ed.) Sheldon, R.A. Marcel Dekker, New York, 1994; (c) Sheldon, R.A. and Kochi, J.K. *Metal-Catalyzed Oxidations of Organic Compounds.* Academic Press, New York, 1981; (d) *Activation and Functionalization of Alkanes* (ed.) Hill, C.L. Wiley, New York, 1989.

2. (a) Holm, R.H. *Chem. Rev.* 1987, **87:** 1401–1449; (b) Jørgensen, K.A. *Chem. Rev.* 1989, **89:** 431–458; (c) Griffith, W.P. *Chem. Soc. Rev.* 1992, **21:** 179–185.
3. (a) Lee, D.G. and van den Engh, M. *Oxidation in Organic Chemistry* (ed.) Trahanovski, W.S. Academic Press, New York, 1973, Part B, Chap. 4; (b) Courtney, J.T. *Organic Synthesis by Oxidation with Metal Compounds* (eds.) Mijs, W.J. and de Jonge, C.R.H.I. Plenum Press, New York, 1984, Chap. 8; (c) Carlsen, P.H.J., Katsuki, T., Martin, V.S. and Sharpless, K.B. *J. Org. Chem.* 1981, **46:** 3936–3938.
4. (a) Leung, W.H. and Che, C.M. *J. Am. Chem. Soc.* 1989, **111:** 8812–8818; (b) Morice, C., Maux, P.L. and Simonneaux, G.J. *Tetrahedron Lett.* 1996, **37:** 6701–6704; (c) Maux, P.L., Bahri, H. and Simonneaux, G. *Inorg. Chem.* 1995, **34:** 4691–4697.
5. (a) Cheng, W.C., Yu, W.Y., Li, C.K. and Che, C.M. *J. Org. Chem.* 1995, **60:** 6840–6846; (b) Cheng, W.C., Yu, W.Y., Cheung, K.K. and Che, C.M. *J. Chem. Soc., Dalton Trans.* 1994, 57–62; (c) Che, C.M., Tang, W.T., Wong, W.T. and Lai, T.F. *J. Am. Chem. Soc.* 1989, **111:** 9048–9056; (d) Fung, W.H., Cheng, W.C., Yu, W.Y., Che, C.M. and Mak, T.C.W. *J. Chem. Soc., Chem. Commun.* 1995, 2007–2008; (e) Stultz, L.K., Binstead, R.A., Reynolds, M.S. and Meyer, T.J. *J. Am. Chem. Soc.* 1995, **117:** 2520–2532; (f) Dovletoglou, A. and Meyer, T.J. *J. Am. Chem. Soc.* 1994, **116:** 215–223; (g) Thompson, M.S. and Meyer, T.J. *J. Am. Chem. Soc.* 1982, **104:** 4106–4115; (h) Thompson, M.S. and Meyer, T.J. *J. Am. Chem. Soc.* 1982, **104:** 5070–5076; (i) Muller, J.G., Acquaye, J.H., and Takeuchi, K.J. *Inorg. Chem.* 1992, **31:** 4552–4557; (j) Marmion, M.E. and Takeuchi, K.J. *J. Chem. Soc., Chem. Commun.* 1987, 1396–1397; (k) Marmion, M.E. and Takeuchi, K.J. *J. Am. Chem. Soc.* 1986, **108:** 510–511.
6. (a) Groves, J.T. and Quinn, R. *J. Am. Chem. Soc.* 1985, **107:** 5790–5792; (b) Groves, J.T., Bonchio, M., Carofiglio, T. and Shalyaev, K. *J. Am. Chem. Soc.* 1996, **118:** 8961–8962.
7. (a) Bailey, A.J. and James, B.R. *Chem. Commun.* 1996, 2343–2344, (b) Cheng, S.Y.S., Rajapakse, N., Rettig, S.J. and James, B.R. *J. Chem. Soc., Chem. Commun.* 1994, 2669–2670.
8. Tavarès, M., Ramasseul, R., Marchon, J.-C., Bachet, B., Brassy, C. and Mornon, J.-P. *J. Chem. Soc., Perkin Trans. 2,* 1992, 1321–1329.
9. (a) Leung, T., James, B.R. and Dolphin, D. *Inorg. Chim. Acta* 1983, **79:** 180–181; (b) Dolphin, D., James, B.R. and Leung, T. *Inorg. Chim. Acta* 1983, **79:** 25–27.

10. (a) Shingaki, T., Miura, K., Higuchi, T., Hirobe, M. and Nagano, T. *Chem. Commun.* 1997, 861–862; (b) Higuchi, T., Satake, C. and Hirobe, M. *J. Am. Chem. Soc.* 1995, **117:** 8879–8880; (c) Ohtake, H., Higuchi, T. and Hirobe, M. *J. Am. Chem. Soc.* 1992, **114:** 10660–10662; (d) Higuchi, T., Ohtake, H., and Hirobe, M. *Tetrahedron Lett.* 1989, **30:** 6545–6548; (e) Higuchi, T. and Hirobe, M. *Tetrahedron Lett.* 1991, **32:** 7435–7438; (f) Ohtake, H., Higuchi, T. and Hirobe, M. *Tetrahedron Lett.* 1992, **33:** 2521–2524.
11. Murahashi, S.-I., Naota, T. and Komiya, N. *Tetrahedron Lett.* 1995, **36:** 8059–8062.
12. Liu, C.J., Li, S.G., Pang, W.Q. and Che, C.M. *Chem. Commun.* 1997, 65–66.
13. Leising, R.A. and Takeuchi, K.J. *Inorg. Chem.* 1987, **26:** 4391–4393.
14. Tang, R., Diamond, S.E., Neary, N. and Mares, F. *J. Chem. Soc., Chem. Commun.* 1978, 562–562.
15. (a) Goldstein, A.S., Beer, R.H. and Drago, R.S. *J. Am. Chem. Soc.* 1994, **116:** 2424–2429; (b) Goldstein, A.S. and Drago, R.S. *J. Chem. Soc., Chem. Commun.* 1991, 21–22; (c) Davis, S. and Drago, R.S. *J. Chem. Soc., Chem. Commun.* 1990, 250–251; (d) Robbins, M.H. and Drago, R.S. *J. Chem. Soc., Dalton Trans.* 1996, 105–110.
16. (a) Taqui Khan, M.M. and Shukla, R.S. *J. Mol. Catal.* 1992, **77:** 221–228; (b) Taqui Khan, M.M., Bajaj, H.C., Shukla, R.S. and Mirza, S.A. *J. Mol. Catal.* 1988, **45:** 51–56.
17. (a) Bressan, M. and Morvillo, A. *J. Chem. Soc., Chem. Commun.* 1988, 650–651; (b) Bressan, M. and Morvillo, A. *J. Chem. Soc., Chem. Commun.* 1989, 421–423.
18. (a) Lau, T.C. and Mak, C.K. *J. Chem. Soc., Chem. Commun.* 1995, 943–944; (b) Lau, T.C. and Mak, C.K. *J. Chem. Soc., Chem. Commun.* 1993, 766–767.
19. Kojima, T. *Chem. Lett.* 1996, 121–122.
20. Yamaguchi, M. Kousaka, H. and Yamagishi, T. *Chem. Lett.* 1997, 769–770.
21. (a) Ho, P.K.K., Cheung, K.K. and Che, C.M. *Chem. Commun.* 1997, 1197–1198; (b) Cheng, W.C., Fung, W.H. and Che, C.M. *J. Mol. Catal. A: Chemical* 1996, **113:** 311–319; (c) Cheng, W.C., Yu, W.Y. Cheung, K.K. and Che, C.M. *J. Chem. Soc., Chem. Commun.* 1994, 1063–1064; (d) Lau, T.C., Che, C.M., Lee, W.O. and Poon, C.K. *J. Chem. Soc., Chem. Commun.* 1988, 1406–1407.
22. (a) Murahashi, S.-I. *Angew. Chem. Int. Ed. Engl.* 1995, **34:** 2443–2465; (b) Murahashi, S.-I. and Naota, T. in *Comprehensive Organic Chemistry II*, Vol. 12 (eds.) Abel, E.W., Stone, F.G.A. and Wilkinson, G. Pergamon, Oxford, 1995, pp. 1177–1192.

23. (a) *Cytochrome P450* (eds.) Sato, R. and Omura, T. Kodansha, Tokyo, 1978; (b) *Cytochrome P450, Structure, Mechanism and Biochemistry*, 2nd Ed. (ed.) Ortiz de Montellano, P.R. Plenum Press, New York, 1995; (c) Woggon, W.-D. in *Topics in Current Chemistry*, Vol 184, *Bioorganic Chemistry. Models and Applications* (ed.) Schmidtchen, F.P. Springer, Berlin, 1996, pp. 39–96.
24. (a) Shannon, P. and Bruice, T.C. *J. Am. Chem. Soc.* 1981, **103:** 4580–4582; (b) Dicken, C.M., Lu, F.-L., Nee, M.W. and Bruice, T.C. *J. Am. Chem. Soc.* 1985, **107:** 5776–5789; (c) Santa, T., Miyata, N., Hirobe, M. *Chem. Pharm. Bull.* 1984, **32:** 1252–1255; (d) Lindsay Smith, J.R. and Mortimer, D.N. *J. Chem. Soc., Chem. Commun.* 1985, 64–65; (e) Fujimori, K., Fujiwara, S., Takata, T. and Oae, S. *Tetrahedron Lett.* 1986, **27:** 581–584.
25. Murahashi, S.-I., Naota, T. and Yonemura, K. *J. Am. Chem. Soc.* 1988, **110:** 8256–8258.
26. Burka, L.T., Guengerich, F.P., Willard, R.J. and Macdonald, T.J. *J. Am. Chem. Soc.* 1985, **107:** 2549–2551.
27. (a) Shono, T., Toda, T. and Oshino, N. *J. Am. Chem. Soc.* 1982, **104:** 2639–2641; (b) Miwa, G.T., Garland, W.A., Hodshon, B.J., Lu, A.Y.H. and Northrop, D.B. *J. Biol. Chem.* 1980, **255:** 6049–6054; (c) Miwa, G.T., Walsh, J.S., Kedderis, G.L. and Hollenberg, P.F. *J. Biol. Chem.* 1983, **258:** 14445–14449.
28. (a) Thompson, J.A. and Holtzman, J.L. *Drug Metab. Dispos.* 1974, **2:** 577–582; (b) Heimbrook, D.C., Murray, R.I., Egeberg, K.D., Sligar, S.G., Nee, M.W. and Bruice, T.C. *J. Am. Chem. Soc.* 1984, **106:** 1514–1515.
29. Cordell, G.A. *Introduction to Alkaloids; A Biogenetic Approach* Wiley, New York, 1981, Chap. 8, pp. 275–573.
30. Murahashi, S.-I., Naota, T., Nakato, T. *Synlett.* 1992, 835–836.
31. Murahashi, S.-I., Naota, T., Miyaguchi, N. and Nakato, T. *Tetrahedron Lett.* 1992, **33:** 6991–6994.
32. (a) Sandler, S.R. and Karo, W. *Organic Functional Group Preparations,* Vol. II. Academic Press, New York, 1986, pp. 291–322; (b) Dayagi, S., in *The Chemistry of the Carbon-Nitrogen Double Bond* (ed.) Patai, S. Interscience Publishers, London, 1970, pp. 61–147.
33. Murahashi, S.-I., Naota, T. and Taki, H. *J. Chem. Soc., Chem. Commun.* 1985, 613–614.
34. (a) Mitsui, H., Zenki, S., Shiota, T. and Murahashi, S.-I. *J. Chem. Soc., Chem. Commun.* 1984, 874–875; (b) Murahashi, S.-I. and Shiota, T. *Tetrahedron Lett.* 1987, **28:** 2383–2386.
35. Goti, A. and Romani, M. *Tetrahedron Lett.* 1994, **35:** 6567–6570.

36. Dalton, D.R. *The Alkaloids.* Marcel Dekker, New York, 1979.
37. Shono, T. *Tetrahedron* 1984, **40:** 811–850.
38. Murahashi, S.-I., Naota, T., Kuwabara, T., Saito, T., Kumobayashi, H. and Akutagawa, S. *J. Am. Chem. Soc.* 1990, **112:** 7820–7822.
39. Doumaux, A.R., Jr. and Trecker, D.J. *J. Org. Chem.* 1970, **35:** 2121–2125.
40. (a) Meyers, A.I. *Aldrichimica Acta* 1985, **18:** 59–68; (b) Beak, P., Zajdel, W.J. and Reitz, D.B. *Chem. Rev.* 1984, **84:** 471–523; (c) Seebach, D. and Enders, D. *Angew. Chem. Int. Ed. Engl.* 1975, **14:** 15–32.
41. Naota, T., Nakato, T. and Murahashi, S.-I. *Tetrahedron Lett.* 1990, **31:** 7475–7478.
42. Murahashi, S.-I., Naota, T., Komiya, N. and Koshida S. in preparation.
43. (a) Andreades, S. and Zahnow, E.W. *J. Am. Chem. Soc.* 1969, **91:** 4181–4190; (b) Konno, A., Fuchigami, T., Fujita, Y. and Nonaka, T. *J. Org. Chem.* 1990, **55:** 1952–1954; (c) Chiba, T. and Takata, Y. *J. Org. Chem.* 1977, **42:** 2973–2977.
44. (a) Santamaria, J. *Pure Appl. Chem.* 1995, **67:** 141–147; (b) Santamaria, J., Kaddachi, M.T. and Rigaudy, J. *Tetrahedron Lett.* 1990, **31:** 4735–4738.
45. Chen, C.-K., Hortmann, A.G. and Marzabadi, M.R. *J. Am. Chem. Soc.* 1988, **110:** 4829–4831.
46. Murata, S., Teramoto, K., Miura, M. and Nomura, M. *Bull. Chem. Soc. Jpn.* 1993, **66:** 1297–1298.
47. Zhdankin, V.V., Kuehl, C.J., Krasutsky, A.P., Bolz, J.T., Mismash, B., Woodward, J.K. and Simonsen, A.J. *Tetrahedron Lett.* 1995, **44:** 7975–7978.
48. Lee, B.H., Clothier, M.F. and Pickering, D.A. *Tetrahedron Lett.* 1997, **38:** 6119–6122.
49. Barrett, G.C. in *Chemistry and Biochemistry of the Amino Acids* (ed.) Barrett, G.C. Chapman and Hall, London, 1985, pp. 246–296.
50. Murahashi, S.-I., Naota, T., and Hirai, N. in preparation.
51. (a) Nagao, Y., Abe, T., Shimizu, H., Kumagai, T. and Inoue, Y. *J. Chem. Soc., Chem. Commun.* 1989, 821–823; (b) Deziel, R. and Endo, M. *Tetrahedron Lett.* 1988, **29:** 61–64; (c) Shirai, F. and Nakai, T. *J. Org. Chem.* 1987, **52:** 5491–5492; (d) Ito, Y. and Terashima, S. *Tetrahedron Lett.* 1987, **28:** 6625–6628; (e) Meyers, A.I., Sowin, T.J., Scholz, S. and Ueda, Y. *Tetrahedron Lett.* 1987, **28:** 5103–5106; (f) Fuentes, L.M., Shinkai, I. and Salzmann, T.N. *J. Am. Chem. Soc.* 1986, **108:** 4675–4676; (g) Nagao, Y., Kumagai, T., Tamai, S., Abe, T., Kuramoto, Y., Taga, T., Aoyagi, S., Nagase, Y., Ochiai, M., Inoue, Y. and Fujita E. *J. Am. Chem. Soc.* 1986, **108:** 4673–4675; (h) Shibata, T., Iino, K., Tanaka, T., Hashimoto, T., Kameyama, Y. and Sugimura, Y. *Tetrahedron Lett.* 1985, **26:** 4739–4742.

52. Cainelli, G., DaCol, M., Galletti, P. and Giacomini, D. *Synlett.* 1997, 923–924.
53. Ref. 1c, p. 140.
54. Murahashi, S.-I., Saito, T., Naota, T., Kumobayashi, H. and Akutagawa, S. *Tetrahedron Lett.* 1991, **32:** 5991–5994.
55. Groves, J.T., Nemo, T.E. and Myers, R.S. *J. Am. Chem. Soc.* 1979, **101:** 1032–1033.
56. Balavoine, G., Eskenazi, C., Meunier, F. and Rivière, H. *Tetrahedron Lett.* 1984, **25:** 3187–3190.
57. Eskénazi, C., Balavoine, G., Meunier, F. and Rivière, H. *J. Chem. Soc., Chem. Commun.* 1985, 1111–1113.
58. (a) Bailey, A.J., Griffith, W.P., White, A.J.P. and Williams, D.J. *J. Chem. Soc., Chem. Commun.* 1994, 1833–1834; (b) Bailey, A.J., Griffith, W.P. and Savage, P.D. *J. Chem. Soc., Dalton Trans.* 1995, 3537–3542.
59. Augier, C., Malara, L., Lazzeri, V. and Waegell, B. *Tetrahedron Lett.* 1995, **36:** 8775–8778.
60. Bailey, C.L. and Drago, R.S. *J. Chem. Soc., Chem. Commun.* 1987, 179–180.
61. (a) Yamada, T., Takai, T., Rhode, O. and Mukaiyama, T. *Chem. Lett.* 1991, 1–4; (b) Mukaiyama, T. and Yamada, T. *Bull. Chem. Soc. Jpn.* 1995, **68:** 17–35.
62. Klement, I., Lütjens, H. and Knochel, P. *Angew. Chem. Int. Ed. Engl.* 1997, **36:** 1454–1456.
63. Gross, Z., Ini, S., Kapon, M. and Cohen, S. *Tetrahedron Lett.* 1996, **37:** 7325–7328.
64. Kureshy, R.I., Khan, N.H., Abdi, S.H.R. and Bhatt, K.N. *Tetrahedron: Asymmetry* 1993, **4:** 1693–1701.
65. Nishiyama, H., Shimada, T., Ito, H., Sugiyama, H. and Motoyama, Y. *Chem. Commun.* 1997, 1863–1864.
66. Au, S.M., Fung, W.H., Cheng, M.C., Che, C.M. and Peng, S.M. *Chem. Commun.* 1997, 1655–1656.
67. (a) Ostovic, D. and Bruice, T.C. *Acc. Chem. Res.* 1992, **25:** 314–320; (b) Castellino, A.J. and Bruice, T.C. *J. Am. Chem. Soc.* 1988, **110:** 158–162.
68. Murahashi, S.-I., Saito, T., Hanaoka, H., Murakami, Y., Naota, T., Kumobayashi, H. and Akutagawa, S. *J. Org. Chem.* 1993, **58:** 2929–2930.
69. Baskaran, S., Das, J. and Chandrasekaran, S. *J. Org. Chem.* 1989, **54:** 5182–5184.
70. Takai, T., Yamada, T. and Mukaiyama, T. *Chem. Lett.* 1991, 1499–1502.
71. Horiguchi, Y., Nakamura, E. and Kuwajima, I. *J. Am. Chem. Soc.* 1989, **111:** 6257–6265.

72. (a) Sharpless, K.B., Akashi, K. and Oshima, K. *Tetrahedron Lett.* 1976, 2503–2506; (b) Müller, P. and Godoy, J. *Tetrahedron Lett.* 1981, **22:** 2361–2364; (c) Kanemoto, S., Oshima, K., Matsubara, S., Takai, K. and Nozaki, H. *Tetrahedron Lett.* 1983, **24:** 2185–2088; (d) Barak, G., Dakka, J. and Sasson, Y. *J. Org. Chem.* 1988, **53:** 3553–3555; (e) Kim, K.S., Kim, S.J., Song, Y.H. and Hahn, C.S. *Synthesis* 1987, 1017–1018.
73. Ley, S.V., Norman, J., Griffith, W.P. and Marsden, S.P. *Synthesis,* 1994, 639–666.
74. (a) Tsuji, Y., Ohta, T., Ido, T., Minbu, H. and Watanabe, Y. *J. Organomet. Chem.* 1984, **270:** 333–341; (b) Tanaka, M., Kobayashi, T. and Sakakura, T. *Angew. Chem. Int. Ed. Engl.* 1984, **23:** 518–518.
75. (a) Murahashi, S.-I., Naota, T., and Nakajima, T. *Tetrahedron Lett.* 1985, **26:** 925–928; (b) Murahashi, S.-I. and Naota, T. *Synthesis,* 1993, 433–440.
76. Murahashi, S.-I., Naota, T., and Nakajima, N. *Chem. Lett.* 1987, 879–882.
77. Murahashi, S.-I., Naota, T., Oda, Y. and Hirai, N. *Synlett.* 1995, 733–734.
78. (a) Bilgrien, C., Davis, S. and Drago, R.S. *J. Am. Chem. Soc.* 1987, **109:** 3786–3787; (b) Inokuchi, T., Nakagawa, K. and Torii, S. *Tetrahedron Lett.* 1995, **36:** 3223–3226; (c) Matsumoto, M. and Watanabe, N. *J. Org. Chem.* 1984, **49:** 3435–3436; (d) Matsumoto, M. and Ito, S. *J. Chem. Soc., Chem. Commun.* 1981, 907–908.
79. Murahashi, S.-I., Naota, T., and Hirai, N. *J. Org. Chem.* 1993, **58:** 7318–7319.
80. Markó, I.E., Giles, P.R., Tsukazaki, M., Chellé-Regnaut, I., Urch, C.J. and Brown, S.M. *J. Am. Chem. Soc.* 1997, **119:** 12661–12662.
81. Ingold, K.U. in *Free Radicals,* Vol. 1 (ed.) Kochi, J.K. John Wiley & Sons, New York, 1973, pp. 37–112.
82. Shimizu, M., Orita, H., Hayakawa, T., Watanabe, Y. and Takehira, K. *Bull. Chem. Soc. Jpn.* 1991, **64:** 2583–2584.
83. Murahashi, S.-I., Naota, T., Miyaguchi, N. and Noda, S. *J. Am. Chem. Soc.* 1996, **118:** 2509–2510.
84. Brovo, A., Fontana, F. and Minisci, F. *Chem. Lett.* 1996, 401–402.
85. Balkus, K.J., Jr., Eissa, M. and Levado, R. *J. Am. Chem. Soc.* 1995, **117:** 10753–10754.
86. (a) Neumann, R. and Abu-Gnim, C. *J. Chem. Soc., Chem. Commun.* 1989, 1324–1325; (b) Matsumoto, Y., Asami, M., Hashimoto, M. and Misono, M. *J. Mol. Catal. A: Chemical* 1996, **114:** 161–168.
87. Murahashi, S.-I., Oda, Y., Naota, T. and Kuwabara, T. *Tetrahedron Lett.* 1993, **34:** 1299–1302.

88. Murahashi, S.-I., Oda, Y., Komiya, N. and Naota, T. *Tetrahedron Lett.* 1994, **35:** 7953–7956.
89 (a) Mansuy, D., Bartoli, J.-F., Chottard, J.-C. and Lange, M. *Angew. Chem. Int. Ed. Engl.* 1980, **19:** 909–910; (b) Mansuy, D., Bartoli, J.-F. and Momenteau, M. *Tetrahedron Lett.* 1982, **23:** 2781–2784.
90. Blake, R.C., II and Coon, M.J. *J. Biol. Chem.* 1981, **256:** 12127–12133.
91. (a) Sakurai, H. and Hosomi, A. *J. Am. Chem. Soc.* 1967, **89:** 458–460; (b) Walling, C. and McGuinness, J.A. *J. Am. Chem. Soc.* 1969, **91:** 2053–2058.
92. Russell, G.A. *Free Radicals*, Vol. 1 (ed.) Kochi, J.K. John Wiley & Sons, New York, 1977, pp. 275–332.
93. Hjelmeland, L.M., Aronow, L. and Trudell, J.R. *Biochem. Biophys. Res. Commun.* 1977, **76:** 541–549.
94. Kim, J., Harrison, R.G., Kim, C. and Que, L., Jr. *J. Am. Chem. Soc.* 1996, **118:** 4373–4379.
95. Neumann, R., Khenkin, A.M. and Dahan, M. *Angew. Chem. Int. Ed. Engl.* 1995, **34:** 1587–1589.
96. Murahashi S.-I., Oda Y., and Naota T., *J. Am. Chem. Soc.* 1992, **114:** 7913–7914.
97. (a) Murahashi, S.-I., Oda, Y., Naota, T. and Komiya, N. *J. Chem. Soc., Chem. Commun.* 1993, 139–140; (b) Komiya. N., Naota. T. and Murahashi, S.-I. *Tetrahedron Lett.* 1996, **37:** 1633–1636; (c) Komiya, N., Naota, T., Oda, Y. and Murahashi, S.-I. *J. Mol. Catal. A: Chemical* 1997, **117:** 21–37.
98. (a) Giannandrea, R., Mastrorilli, P., Nobile, C.F. and Suranna, G.P. *J. Mol. Catal.* 1994, **94:** 27–36; (b) Mastrorilli, P. and Nobile, C.F. *Tetrahedron Lett.* 1994, **35:** 4193–4196; (c) Punniyamurthy, T., Kalra, S.J.S. and Iqbal, J. *Tetrahedron Lett.* 1995, **36:** 8497–8500; (d) Mandal, A.K. and Iqbal, J. *Tetrahedron* 1997, **53:** 7641–7648.
99. Schuchardt, U., Carvalho, W.A. and Spinacé, E.V. *Synlett.* 1993, 713–718.

Chapter 14

Biocatalytic and Biomimetic Oxidations from an Industrial Perspective

Roger A. Sheldon

Delft University of Technology,
Faculty of Chemical Technology and Materials Science,
Laboratory for Organic Chemistry and Catalysis,
Julianalaan 136, 2628 BL Delft, The Netherlands

1. Introduction

Catalytic oxidations of petroleum hydrocarbons (alkanes, alkenes and aromatics) with molecular oxygen form the basis for the manufacture of a variety of commodity chemicals.[1] Nevertheless, there is still considerable room for improvement in the selectivity of many of these processes. Moreover, several highly desirable oxidative transformations such as the "dream reactions" shown below (Reactions 14.1–14.3) have not yet been realized.

$$CH_4 + {}^1/_2O_2 \longrightarrow CH_3OH \qquad (14.1)$$

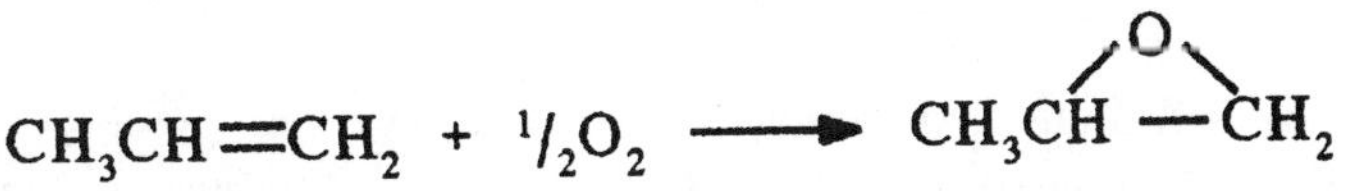

(14.2)

(14.3)

Oxidative transformations are also widely used in fine chemicals manufacture. In many cases this involves the use of stoichiometric amounts of inorganic oxidants, e.g. chromium(VI) reagents and manganese dioxide, resulting in the generation of copious amounts of inorganic waste. Increasingly stringent environmental regulations are making the use of such stoichiometric methodologies prohibitive and attention is being focused on the development of catalytic alternatives employing clean oxidants such as O_2 and H_2O_2.[2]

Oxidations mediated by redox enzymes (oxidoreductases) are key steps in many fundamental processes *in vivo*.[3] These transformations proceed at ambient temperature and pressure and are often highly chemo-, regio- and enantioselective.

A perfect example is the selective microbial hydroxylation of progesterone to 11α-hydroxy-progesterone (Reaction 14.4), which led to a commercially viable synthesis of cortisone that replaced a 31-step chemical synthesis from a bile acid[4].

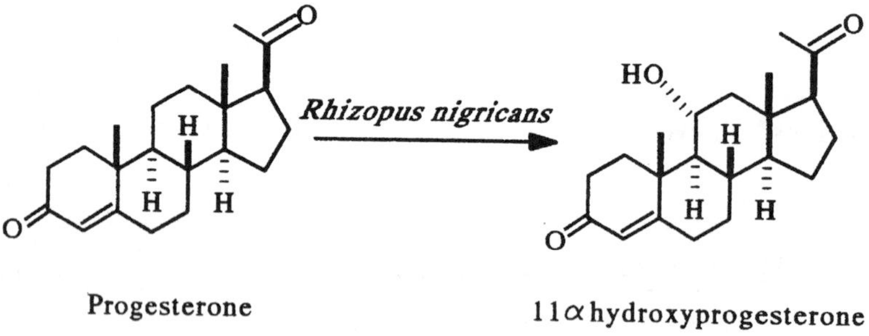

(14.4)

Reactions 14.1–14.3 are also mediated by monooxygenase enzymes, albeit by employing stoichiometric amounts of a sacrificial reductant (see later). The harnessing of this catalytic potential of enzymes for industrial synthesis is one of the most important challenges in modern chemistry. Indeed, the eminent Russian chemist and Nobel prizewinner, N.N. Semonov, speculated in 1974: "By applying the ideas of biochemistry, chemical science may solve the energy crisis, make industrial production infinitely more efficient, and provide mankind with wings".[5] In this chapter we shall examine, in the context of oxidation chemistry, what progress has been made towards realization of this ambitious goal. In the context of practical utility, high selectivities alone are not sufficient. For industrial applications the catalyst productivity (g product/g catalyst/h) and volumetric productivity (g product/ liter/h) are important criteria. The level of productivity required for a viable process is obviously related to the added value of the product. Table 1 compares the volumetric productivities of selected fermentation, enzymatic and chemical processes.

It is readily apparent that low volumetric productivities are viable for complex, expensive products such as Penicillin G and Vitamin B_{12} where chemical synthesis is not a feasible alternative. It is also clear that in order to compete with existing chemical processes, biocatalytic or biomimetic methods need to have high volumetric productivities. A rule of thumb for fine chemicals is that the output should be at least 100 kg of product per m^3 per day, or 4 g/ltr/hr.

Table 14.1 Volumetric productivities of various processes.

Process/Product	Productivity ($g.h^{-1}.h^{-1}$)	Price \$ kg^{-1}
Fermentations		
Citric acid	7	1.5
Lactic acid	4	2.5
Lysine	0.6	13
Penicillin G	0.17	20
Vitamin B_{12}	0.001	8000
Enzymatic		
Glucose → fructose	200	2.0
Fumaric acid → aspartic acid	20	3.0
Chemical		
Acetic acid (MeOH + CO)	500	0.75
Ethylene → ethanol	80	0.70

2. Oxygen Activation and Oxidation Mechanisms

The existence of life in an aerobic environment depends on the low kinetic reactivity of dioxygen with organic compounds which are highly exothermic. This low reactivity is due to the fact that reactions of the triplet ground state of dioxygen with singlet organic molecules are spin forbidden. Reaction of $^3O_2$ with a singlet molecule affording two free radicals (doublets) is spin allowed (Reaction 14.5) but highly endothermic (up to 50 kcal.mol^{-1}). Hence, it is observed at moderate temperatures only with very reactive substrates, that afford highly resonance stabilized radicals (Reaction 14.6). Nature takes advantage of this property of reduced flavins in the activation of $^3O_2$ by flavin-dependent oxygenases (see later). Alternatively, this spin conservation obstacle can be overcome by combining $^3O_2$ with a paramagnetic transition metal ion (Reaction 14.7).

$$RH + O_2 \longrightarrow R\bullet + HO_2\bullet \tag{14.5}$$

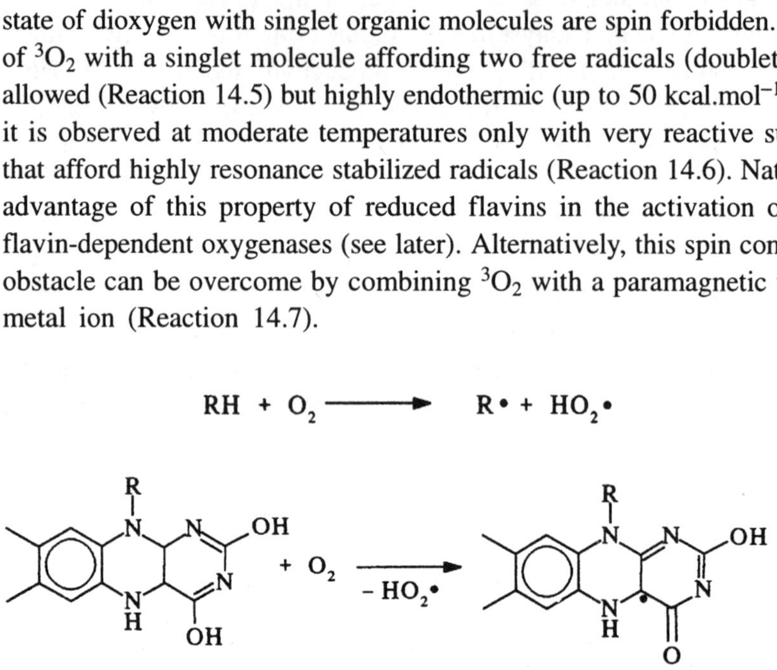

(14.6)

$$M^n + O_2 \longrightarrow M^{n+1}{-}O{-}O\bullet \tag{14.7}$$

Liquid phase oxidations with dioxygen generally involve a free radical chain autoxidation mechanism (Reactions 14.8 and 14.9). Variable valence metal ions, e.g. Co and Mn, catalyze this process by generating chain-initiating radicals, under mild conditions, in one-electron transfer reactions with traces of RO_2H (Reactions 14.10 and 14.11) or, in some cases, with the substrate, RH.[1] Gas phase oxidations, in contrast, often involve the so-called Mars-van Krevelen mechanism[6] in which the substrate undergoes oxidation by an oxometal species, e.g. V(V) = O or Mo(VI) = O, followed by regeneration with dioxygen (Reactions 14.12 and 14.13). In the liquid phase, with relatively high concentrations of RH, autoxidation chain lengths are often long and it is difficult to compete with the ubiquitous free radical autoxidation. In the gas phase, in contrast, concentrations of RH are much lower and a Mars-van Krevelen pathway is more feasible.

$$R\bullet + O_2 \longrightarrow RO_2\bullet \qquad (14.8)$$

$$RO_2\bullet + RH \longrightarrow RO_2H + R\bullet \qquad (14.9)$$

$$M^n + RO_2H \longrightarrow M^n + RO_2\bullet + H^+ \qquad (14.10)$$

$$M^{n-1} + RO_2H \longrightarrow M^n + RO\bullet + HO^- \qquad (14.11)$$

$$M^n = O + S \longrightarrow M^{n-2} + SO \qquad (14.12)$$

$$2M^{n-2} + O_2 \longrightarrow 2M^n = O \qquad (14.13)$$

Free radical autoxidation is a largely indiscriminate process giving high selectivities only with molecules containing one reactive position and giving a stable product, e.g. toluene to benzoic acid. There is a genuine need, therefore, for catalysts that are able to compete with free radical chain autoxidation, i.e. a Mars-van Krevelen type mechanism in the liquid phase. Monooxygenases which mediate the reaction of simple hydrocarbons (see later) with dioxygen are generally believed to involve a high-valent oxoiron species as the active oxidant. In this case, oxygen transfer to the substrate can compete with free radical autoxidation by virtue of the fact that quasi gas phase conditions are created within the active site of the enzyme. In contrast, such conditions do not pertain to biomimetic oxidations with simple metal complexes. Moreover, when hydrogen peroxide or alkyl hydroperoxides are used as the stoichiometric oxidant in biomimetic systems, this can lead to the formation of $RO_2\bullet$ radicals (R = H or alkyl) by one electron transfer (see earlier). Bimolecular termination of these radicals leads to the formation of dioxygen which can react further via a free radical autoxidation mode. This does not appear to have been fully appreciated by many authors (see later).

3. Biocatalytic Oxidations

3.1 Classification of oxidoreductases

Oxidoreductases can be conveniently divided into five categories according to the oxidant and stoichiometry involved (Fig. 14.1).

From a practical viewpoint, it is important to distinguish between enzymes that consume stoichiometric amounts of cofactor and those with no cofactor regeneration requirement. Dehydrogenases and monooxygenases (with the exception of internal monooxygenases) belong to the former, and dioxygenases, oxidases and peroxidases to the latter category. Dehydrogenases are usually employed as whole cell systems, mostly for the reverse reaction (e.g. Baker's yeast reduction of carbonyl compounds) and will not be discussed further in this chapter.

Oxygenases are widely distributed in plants, animals and microorganisms where they play essential metabolic roles. They are further classified into monooxygenases and dioxygenases, depending on whether one or both oxygen atoms of dioxygen are incorporated into the substrate.

Dehydrogenases

$$RH + D \rightleftarrows R(-H) + DH_2$$

Monooxygenases

$$RH + O_2 + D \longrightarrow ROH + D + H_2O$$

Dioxygenases

$$RH + O_2 \longrightarrow RO_2H$$

Oxidases

$$RH + O_2 \longrightarrow R(-H) + H_2O_2$$

Peroxidases

$$RH + H_2O_2 \longrightarrow ROH + H_2O$$

$D, DH_2 = \text{cofactor}$

Fig. 14.1 Types of oxidoreductases.

Monooxygenases require two reducing equivalents (two electrons and two protons) to reduce the second oxygen atom to water. These are derived from the cofactor ($NADH_2$ or $NADPH_2$) or the substrate itself and are designated external and internal monooxygenases, respectively (see Fig. 14.2). Similarly, dioxygenases are either inter- or intramolecular depending on whether both oxygen atoms are incorporated into the substrate or one into the substrate and the other into an additional cofactor. It should be emphasized that the terms mono- and dioxygenase say nothing about the mechanism of the transformation.

Monooxygenases

External: $S + DH_2 + O_2 \longrightarrow SOH + D + H_2O$

Internal: $SH_2 + O_2 \longrightarrow SOH + H_2O$

Dioxygenases

Intermolecular: $S + D + O_2 \longrightarrow SO + DO$

Intramolecular: $S + O_2 \longrightarrow SO_2$

Fig. 14.2 Mono- and dioxygenases.

Oxygenases can also be divided into four general classes on the basis of the cofactor or prosthetic group involved: flavin, copper, heme-iron and non-heme iron. Flavin-dependent monooxygenases catalyze the oxidation of relatively reactive substrates, e.g. cyclohexanone monooxygenase which catalyzes the Baeyer-Villiger oxidation of cyclohexanone (see later). Copper-dependent mono- and dioxygenases mediate the oxidation of electron-rich aromatics, e.g. Reactions 14.14 and 14.15.

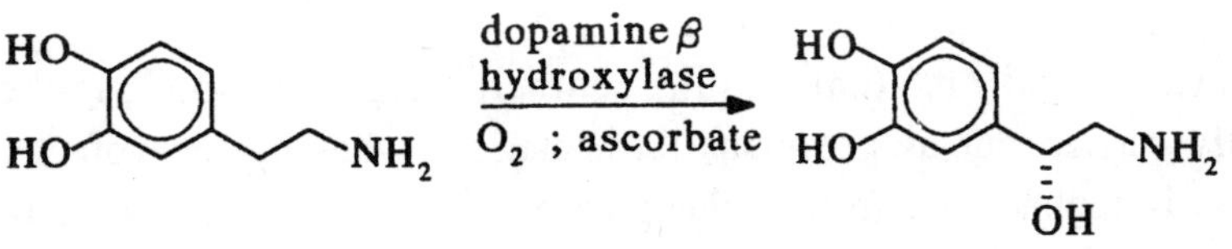

(14.14)

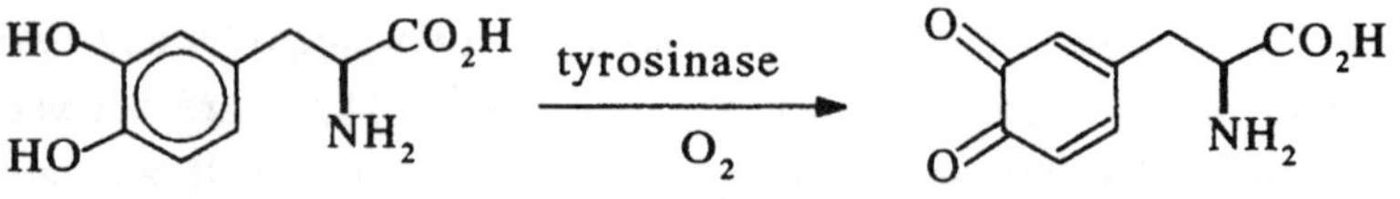 (14.15)

The oxidation of simple hydrocarbons is catalyzed only by iron-dependent oxygenases,[8] which can be divided into three types according to the nature of the iron-containing prosthetic group: heme, non-heme mononuclear and non-heme dinuclear iron. The best known and most well-studied are the heme-dependent oxygenases, typified by cytochrome P450.[9–11] More recently, attention has focused on dinuclear non-heme iron oxygenases.[12–14] This interest stems from the fact that one example of this group, methane monooxygenase[15] mediates the conversion of methane to methanol (Reaction 14.1).

The majority of peroxidases contain heme iron as the prosthetic group although vanadium-, manganese- and selenium-dependent peroxidases are also known.[16,17] Peroxidases catalyze the oxidation of substrates with H_2O_2. Oxidases catalyze the oxidative dehydrogenation of organic substrates with concomitant formation of H_2O_2 (or sometimes water). They are generally flavoproteins or copper- or molybdenum- (sometimes in combination with iron) dependent enzymes.

3.2 *Whole cells versus isolated enzymes*

Biocatalytic conversions can be carried out with whole cells or with isolated enzymes (Fig. 14.3). When the former involves the use of growing microbial cells, it is referred to as a fermentation. Fermentation can be further divided into *de novo* fermentation — the multistep conversion of a carbon and energy source, such as glucose, into primary or secondary metabolites — and microbial transformations of organic substrates, sometimes referred to as precursor fermentations (Fig. 14.3).

Enzymatic oxidations that require cofactor regeneration can be conveniently carried out as precursor fermentations whereby the cofactor is continuously regenerated from a cheap carbon source, e.g. an alkane or the substrate itself. Moreover, the enzyme is also continuously being generated by the growing cells. This is important since oxygenases are generally unstable towards oxidative degradation. Regeneration of cofactors and enzymes is not possible in conversions with dead (resting) cells. The use of whole (resting) cells rather than isolated enzymes is often preferred due to the limited stability of many oxidoreductases outside the cell and/ or the requirement for non-stoichiometric cofactors.

It is also worth noting that microorganisms constitute by far the best source of enzymes, production from plants and animals being much less efficient (see Table 14.2).

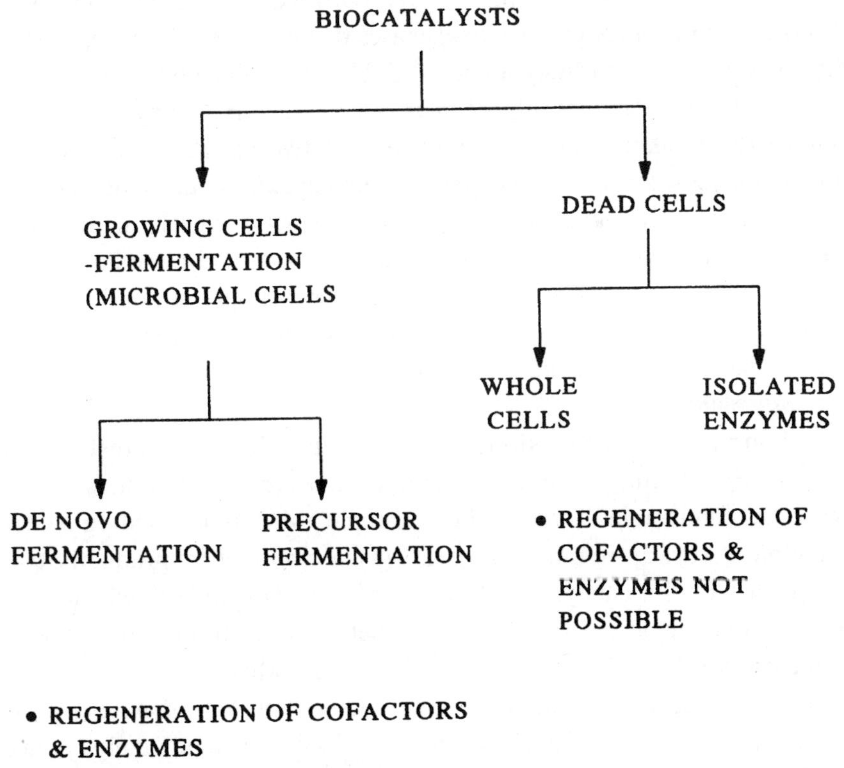

Fig. 14.3 Biocatalytic conversions.

Table 14.2 Productivity of enzyme production.

Source	kg biomass for 1 kg enzyme	Reactor volume (ltr)
Plants	100 000	2 000 000
Animals	10 000	200 000
Microorganisms	1 000	20 000
Recombinant microorganisms	10 – 20	200 – 400

3.3 Oxygenases

When microorganisms, particularly bacteria, are allowed to grow on hydrocarbons — alkanes, olefins or (alkyl)aromatics — as the sole carbon and energy source, they produce oxygenases that catalyze selective oxidations with dioxygen, e.g. hydroxylation of C–H bonds and olefin epoxidation. The cytochrome P450 family of heme-dependent monooxygenases, that catalyze the oxidation of a wide variety of hydrophobic substrates, are probably the best known.[9–11] Methanotrophic bacteria produce the non-heme iron-dependent methane monooxygenase (MMO) which mediates the conversion of methane to methanol (Reaction 14.1) and a variety of other selective oxidations (Table 14.3).[7,15,18] Another example of non-heme iron-dependent monooxygenases is provided by the ω-hydroxylases which catalyze the selective oxidation of a terminal methyl group in a linear alkane or a fatty acid.[19]

Although such conversions are, in principle, of interest for the manufacture of bulk chemicals, their volumetric productivities are too low to be commercially viable. For example, MMO-mediated conversion of methane (its optimum substrate) has a volumetric productivity of 1.1 $g.l^{-1}.h^{-1}$ (and a catalyst productivity of 0.9 g/g cells/h)[20] which is two orders of magnitude too low. Such volumetric productivities are, however, of interest for fine chemicals production (see earlier).

Lonza workers, for example, have shown that *Pseudomonas putida* species, previously grown on xylene, produce a xylene monoxygenase (XMO) which catalyzes the selective oxidation of methyl-substituted heteroaromatics to the corresponding monocarboxylic acids (see

Table 14.3 Selective oxidations mediated by methane monooxygenase MMO.[18]

Substrate	Product	Selectivity (%)	
		M. capsulatus	*M. trichosporium*
		39	44
	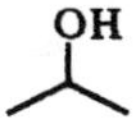	61	56
	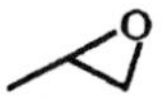	100	100
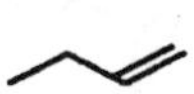		100	100
		30	26
		70	74
	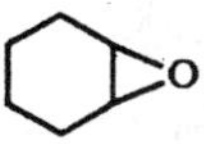	9	7
	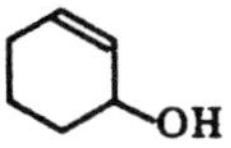	91	93
	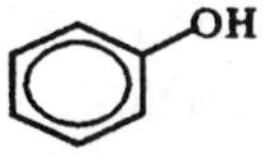	100	100
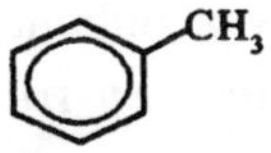	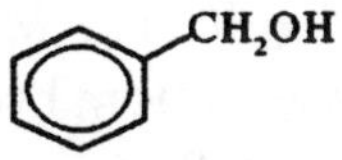	100	100

Table 14.4).[21] An obvious advantage compared to the fermentation of aromatic hydrocarbons is the increased water solubility of the substrate. Many of the products are of commercial interest as agrochemicals or pharmaceutical intermediates. For example, 5-methylpyrazinecarboxylic acid (MPCA) is an intermediate in the production of Acipimox (an antilypolitic) and Glipicide (an antidiabetic).

Table 14.4 *P. putida* ATC 33015-mediated oxidation of heteroaromatic side-chains with O_2 .[21]

Substrate	Product	Substrate	Product
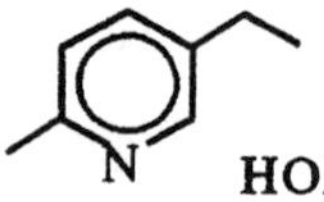	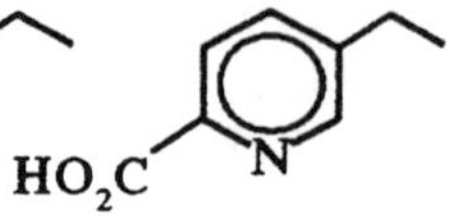	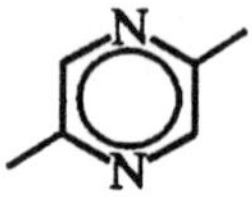	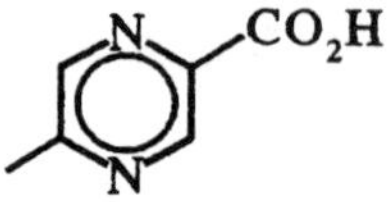
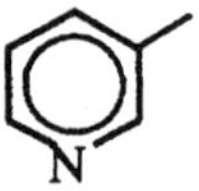	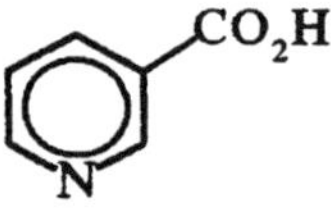	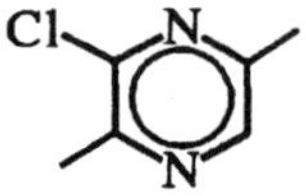	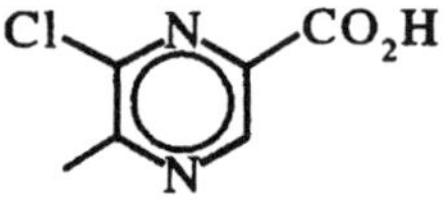
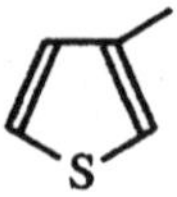	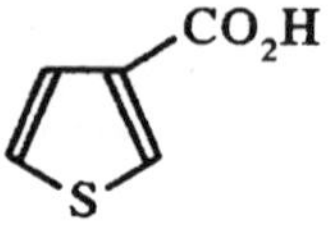	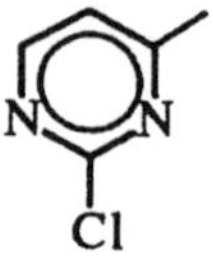	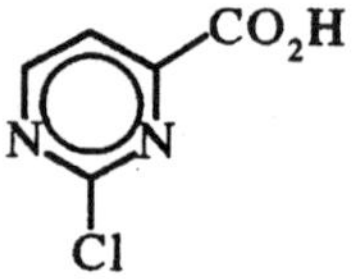

When the oxidation of 2,5-dimethylpyrazine (DMP) was performed with live cells growing on xylenes (Reaction 14.16), in a fed-batch operation to maintain the concentration of DMP below 2 g/l, MPCA was obtained in 95% yield and 24 g/l^{-1} in 54 h fermentation time, i.e. a volumetric productivity of ca. 0.5 $g.l^{-1}.h^{-1}$. MPCA was obtained by acidification of the cell-free fermentation broth after ten-fold concentration. Lonza is currently producing MPCA on a 15 m^3 scale using this procedure.

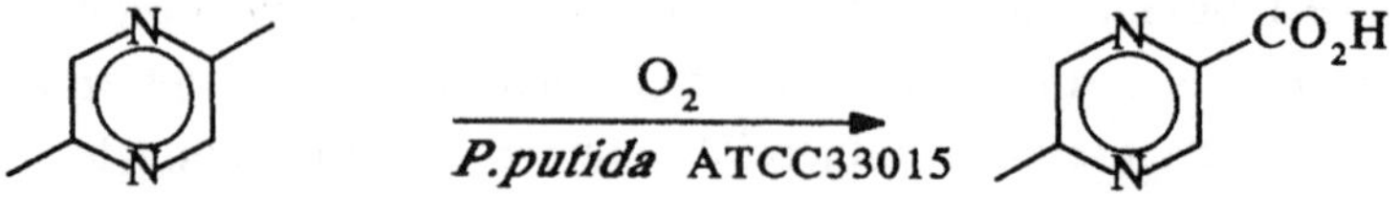

DMP

MPCA
95% yield
24 $g.l^{-1}$

(14.16)

Lonza workers also showed[21] that *P. oleovorans* ATCC 29347 previously grown on *n*-octane catalyzes the oxidation of ethyl substituents in heteroaromatic rings to the corresponding acetic acid derivatives (Table 14.5).

The product of 3-ethyl pyridine oxidation, pyridyl-3-acetic acid, is an intermediate in the synthesis of Risedronate, a drug for the treatment of

Table 14.5 *P. oleovorans* ATCC 29347-mediated oxidation of ethyl-substituted heteroaromatics.[21]

Substrate	Product
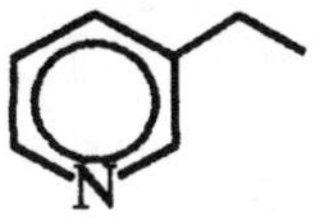	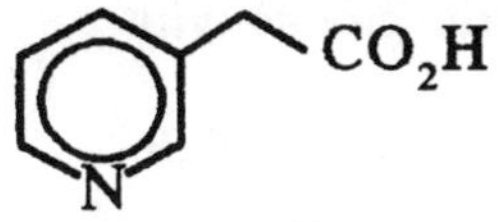
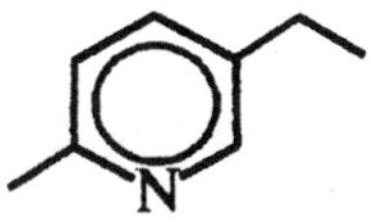	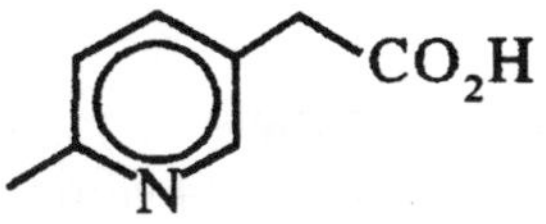
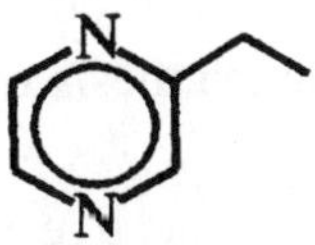	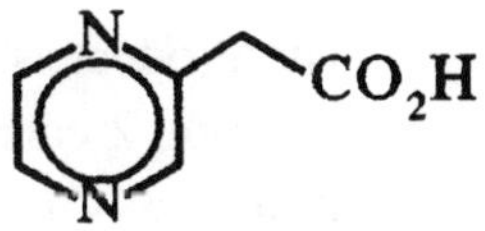
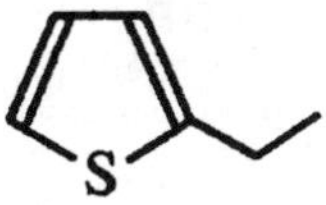	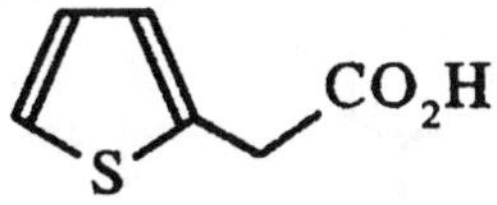

osteoporosis. The active enzyme in these conversions is a ω-hydroxylase (see earlier), which also converts ethylbenzene to phenylacetic acid. The remarkable selectivities that are possible in such bioconversions are well-illustrated by the oxidation of 5-ethyl-2-methylpyridine: XMO from *P. putida* mediates exclusive oxidation of the methyl group while the ω-hydroxylase from *P. oleovorans* oxidizes only the ethyl group (see Tables 14.4 and 14.5).

A third possible conversion of heteroaromatics is ring hydroxylation.[21,22] For example, Lonza has developed an industrial process for the manufacture of 6-hydroxynicotinic acid, an insecticide intermediate, by hydroxylation of nicotinic acid (Reaction 14.17) mediated by cells of *Achromobacter xylosoxidans* DSM 2783 grown on nicotinic acid.[21,22] More than 100 g.l^{-1} of product was obtained, in > 90% yield, within 12 h (i.e. a volumetric productivity of ca. 10 g.l^{-1}.h^{-1}).

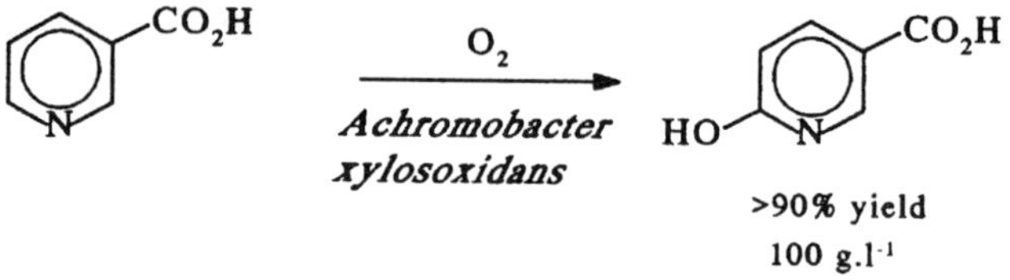

(14.17)

Similarly, a variety of pyridines and pyrazines were shown to undergo regiospecific microbial hydroxylation with bacterial cells grown on pyrazine or DMP (see Table 14.6). For example, pyrazine-2-carboxylic acid was converted to 5-hydroxypyrazine-2-carboxylic acid in 96% isolated yield at concentrations of 75 g.l^{-1} in 24 h (i.e. a volumetric productivity of 3 g.l^{-1}.h^{-1}). The product is a starting material for the synthesis of various 5-substituted pyrazine-2-carboxylic acids.[21]

Another commercially interesting example of a regioselective microbial hydroxylation is the XMO-mediated conversion of indole (Reaction 14.18) to 3-oxindole, the precursor of indigo, reported by Mitsubishi workers.[23]

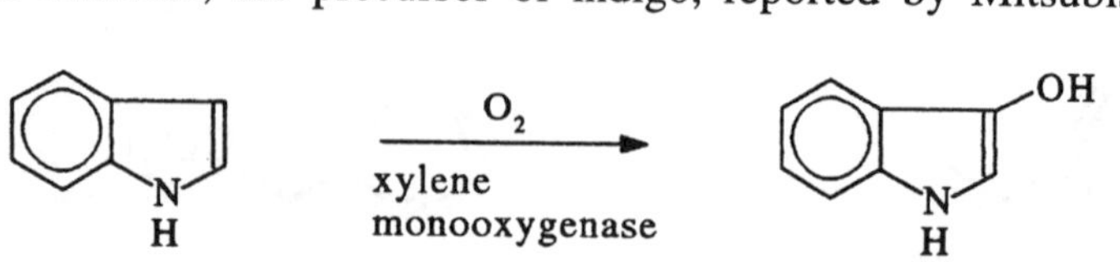

(14.18)

Table 14.6 Microbial hydroxylation of heteroaromatic rings.[21]

Substrate	Product

Agrobacterium sp. DSM 6136 :

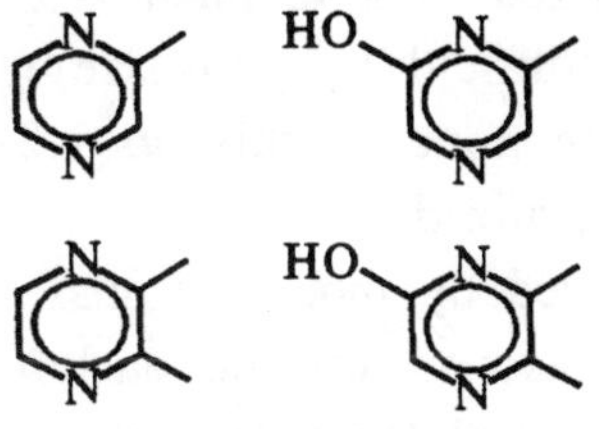

Rhodococcus erythropolis DSM 6138 :

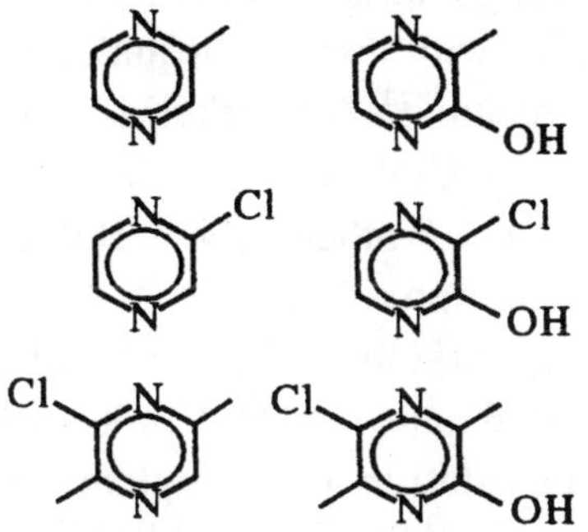

Alcaligenes faecalis DSM 6269 :

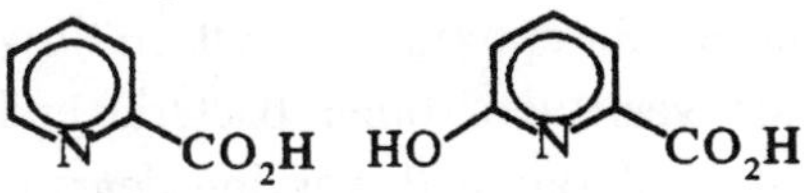

Pseudomonas acidovorans DSM 4746 :

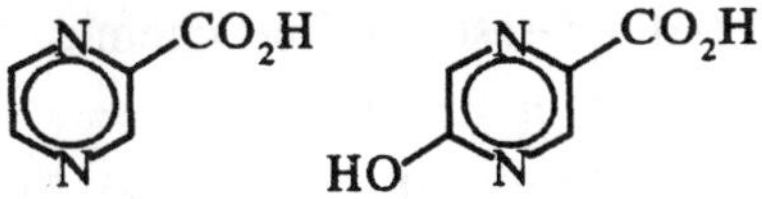

Arene dioxygenases catalyze the first step in the metabolism of unactivated aromatics, yielding the corresponding *cis*-1,2-dihydrocatechols in high regio- and enantioselectivity (Reaction 14.19).[24] The reaction requires NADH as a stoichiometric cofactor.

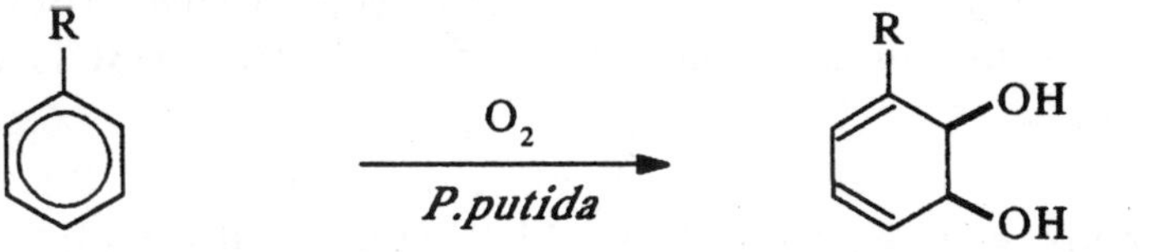

(14.19)

R = CH_3 , F , Cl , etc.

ICI workers originally developed a fermentation procedure, using a genetically manipulated *P. putida* species, for Reaction 14.19. The products could be converted to polyphenylenes by esterification and pyrolysis. The enantiomerically pure *cis*-1,2-dihydrocatechols are potentially interesting chiral synthons or can be rearomatized to give phenols or catechols with novel substitution patterns that are difficult to synthesize by other means.[24] Nevertheless, to our knowledge none of this *cis*-1,2-dihydrocatechol chemistry has yet been commercialized.

Another synthetically useful oxidation catalyzed by monooxygenases is olefin epoxidation.[25–29] Here again, for reasons outlined earlier, this conversion is best carried out as a precursor fermentation. As mentioned above MMO produced by methanotrophic bacteria catalyzes the epoxidation of propylene to give the (R)-epoxide, albeit in low enantioselectivity (*ee* = 10%).[27,30] A wide variety of other bacteria and fungi also mediate the aerobic epoxidation of olefins with varying degrees of enantioselectivity.[25-29] The low volumetric productivities of these conversions are prohibitive for the production of commodity chemicals, such as propylene oxide, but are of potential interest for fine chemicals production, particularly of enantiomerically pure epoxides.

For example, the ω-hydroxylase of *P. oleovorans* catalyzes the epoxidation of terminal olefins.[31–33] 1-octene, for example, afforded (R)-octene oxide in 70% *ee* and 10–20 $g.l^{-1}$ in 140 h (i.e. a volumetric productivity of ca. 0.1 $g.l^{-1}.h^{-1}$) by fermentation with *P. oleovorans* in a two-phase system.[32]

Nocardia corallina B-276 cells grown on propylene catalyze the asymmetric epoxidation of a broad range of olefins.[25] Selected examples are shown in Table 14.7. The technology has been developed to industrial scale by Nippon Mining for the small scale production of several (R)-epoxides for use in the synthesis of ferroelectric crystals and pharmaceuticals.[25]

Another interesting group of enzymes comprises the Baeyer-Villiger monooxygenases (BVMOs) which catalyze the regio- and enantioselective oxidation of acyclic and cyclic ketones to esters and lactones, respectively.[34–36] BVMOs are found in a wide variety of bacteria

Table 14.7 Optically active epoxides produced by *N. corallina* B-276.[25]

Substrate	Epoxide configuration	*ee* (%)
Propylene	R	83
1-Octene	R	91
1-Hexadecene	R	82
2-Methyl-1-hexene	R	76
Allyl chloride	S	81
Styrene	R	69
m-Chlorostyrene		86
Pentyl allyl ether	R	84

and fungi where they play key roles in metabolic pathways. They are flavin-dependent monooxygenases requiring stoichiometric amounts of NAD(P)H as a sacrificial reductant.

The most studied and best characterized BVMO is cyclohexanone monooxygenase (CHMO) from *Acinetobacter calcoaceticus*, an NADPH-dependent monomeric protein containing flavin adenine dinucleotide (FAD) as the tightly-bound prosthetic group.[37,38] The amino acid sequence of CHMO has been determined and the gene responsible for CHMO has been identified, cloned and expressed in *E. coli*.[39] The catalytic cycle of CHMO is illustrated in Fig. 14.4. FAD is reduced to $FADH_2$ by the NADPH cofactor. Reaction of $FADH_2$ with O_2 affords 4a-hydroperoxy flavin which is the active oxidant. Addition of the latter to cyclohexanone affords a tetrahedral intermediate which rearranges to caprolactone with concomitant formation of FAD-4a-OH which eliminates water to regenerate FAD.

BVMOs have been widely used in organic synthesis for regio- and enantioselective conversion of chiral or prochiral cyclic ketones to optically active lactones. These reactions have been carried out as precursor fermentations or with the isolated enzyme. In the latter case, some means of cofactor regeneration is necessary, and linked enzyme systems (e.g. the use of glucose dehydrogenase or formate dehydrogenase to regenerate NAD(P)H in a membrane bioreactor), have been devised for this purpose.[40]

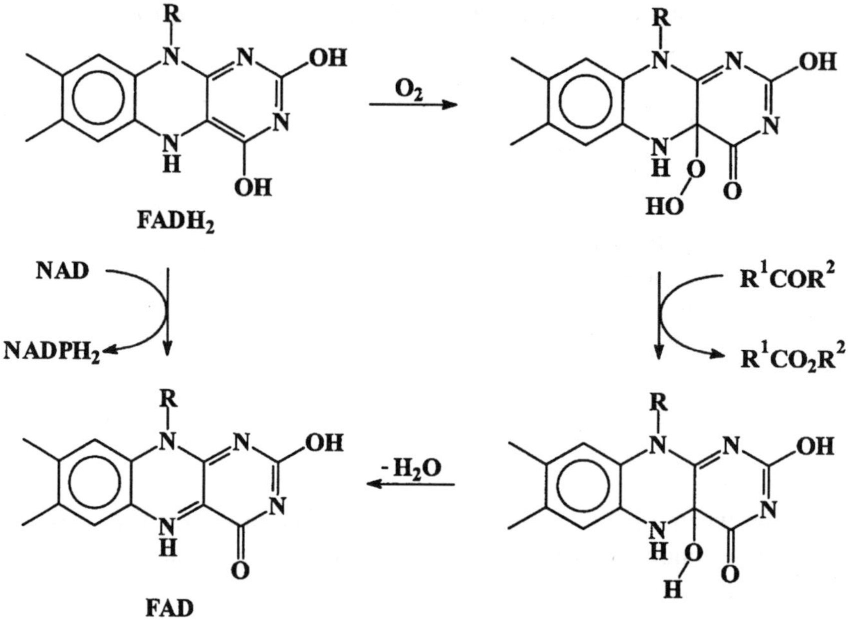

Fig. 14.4 Catalytic cycle of CHMO.

An example of a kinetic resolution of a chiral ketone and an asymmetric conversion of a prochiral ketone are shown in Reaction 14.20[41] and 14.21,[42] respectively. BVMOs have also been shown to catalyze other oxidative conversions, notably asymmetric sulfoxidation.[43] However, notwithstanding their enormous synthetic potential, no example of a BVMO-catalyzed oxidation has, to our knowledge, been commercialized.

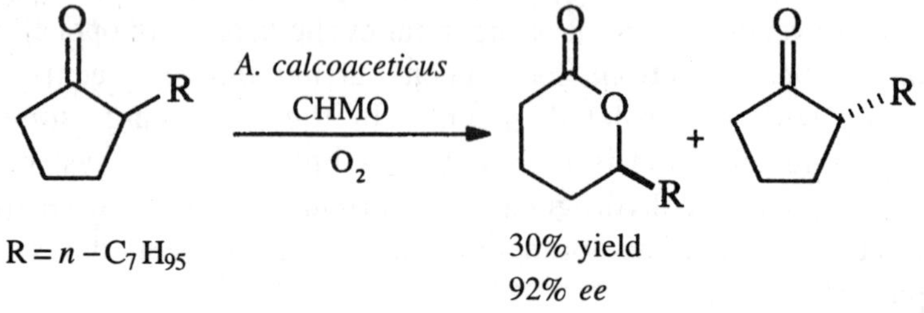

(14.20)

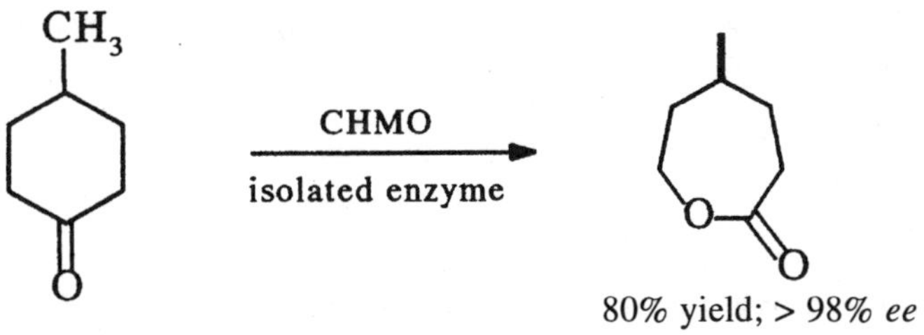

(14.21)

Lipoxygenases (LOX; E.C. 1.13.11.12) constitute a group of closely related non-heme iron-dependent dioxygenases which catalyze the regio- and enantioselective oxidation of polyunsaturated fatty acids containing a *cis,cis*-1,4-diene unit to the corresponding *cis,trans* conjugated allylic hydroperoxide.[44,45] Natural substrates include linoleic, γ-linolenic and arachadonic acids. They are widely distributed in plants and animals, the most well-known example being soybean lipoxygenase which catalyzes the regio- and enantioselective conversion of linoleic acid to the 13S-hydroperoxide (Reaction 14.22).

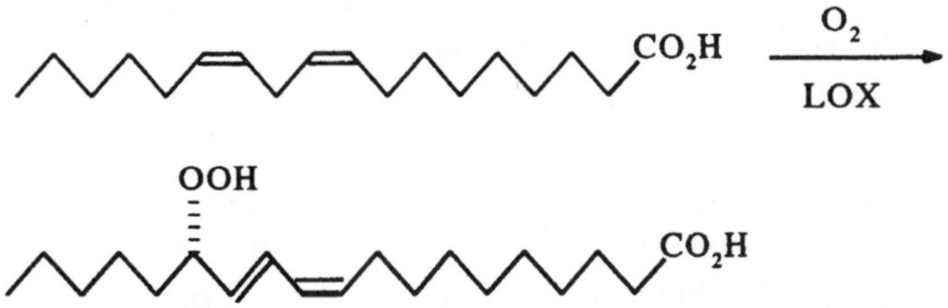

(14.22)

Lipoxygenases which afford the 9S-hydroperoxide[45] are also known. Lipoxygenases have no cofactor requirement, and hence, can be used as the free enzyme. They are non-heme iron-dependent dioxygenases, the resting state of which contains ferrous iron. The key step is a one-electron oxidation of the substrate by Fe(III), giving a highly resonance stabilized bis-allylic radical which suggests that the reaction is initiated by reaction of Fe(II) with traces of RO_2H derived from classical free radical autoxidation. LOX-catalyzed conversion of linoleic and linolenic acids are of commercial interest, since the resulting hydroperoxides can be cleaved by a second enzyme (hydroperoxide lyase) to (un)saturated aldehydes which are natural flavors.[46] Such products command a high price because they fulfil the requirements for labeling as "natural" flavors.

3.4 Oxidases

The oxidation of hydroxyl moieties is one of the fundamental transformations in organic synthesis. *In vivo,* such transformations are mediated by alcohol dehydrogenases and oxidases. The latter have the advantage that they have no stoichiometric cofactor requirement. Methyltrophic yeasts of, for example, the genera *Candida*, *Hansenula* and *Pichia* produce alcohol oxidases AOX when grown on methanol. These enzymes mediate the oxidation of aliphatic and benzylic alcohols (Reaction 14.23).[47] The reaction is preferably performed with whole cells as these contain an endogeneous catalase which decomposes the H_2O_2 formed which otherwise inhibits the enzyme.

$$RCH_2OH + O_2 \xrightarrow{\text{AOX}} RCHO + H_2O_2 \tag{14.23}$$

Alcohol oxidases are of commercial interest for the synthesis of natural aldehyde flavours (see above) by oxidation of naturally occurring alcohols. For example, whole (resting) cells of *P. pastoris* and *H. polymorpha* mediate the aerobic oxidation of 1-pentanol and 1-hexanol to the corresponding aldehydes in a biphasic system.[48]

Leafy green plants, notably spinach, produce the enzyme glycolate oxidase (E.C. 1.1.3.15) which mediates the aerobic oxidation of glycolic acid and L-lactic acid to glyoxylic and pyruvic acids, respectively. DuPont workers[49] employed permeabilized whole cells of recombinant *P. pastoris*, expressing both spinach glycolate oxidase and an endogeneous catalase, as catalysts for the oxidation of glycolic to glyoxylic acid (see Fig. 14.5). The latter was formed in 98% selectivity at 100% conversion with a volumetric yield of 55 $g.l^{-1}.h^{-1}$. The reaction is carried out in the presence of ethylene diamine which forms a Schiff's base with the product, thus preventing overoxidation. This constitutes a commercially attractive route to glyoxylic acid which is currently manufactured via environmentally unattractive nitric acid oxidation of ethylene glycol or acetaldehyde or the more expensive ozonolysis of maleic anhydride.

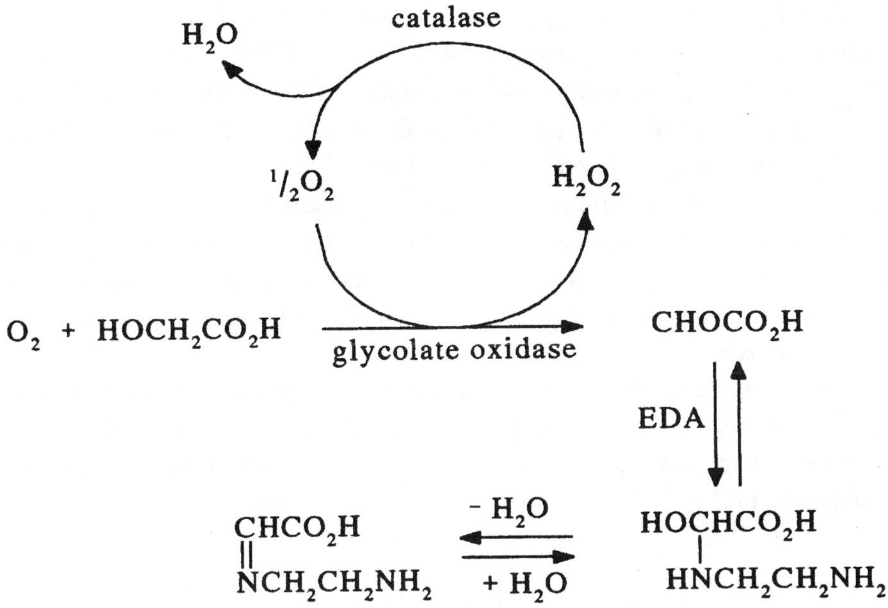

Fig. 14.5 Biocatalytic production of glyoxylic acid.

Similarly, the same group[50] developed a process for the oxidation of L-lactate to pyruvate (Reaction 14.24) using permeabilized whole cells of recombinant *H. polymorpha* or *P. pastoris* containing glycolate oxidase. The volumetric yield was 18 $g.l^{-1}.h^{-1}$.

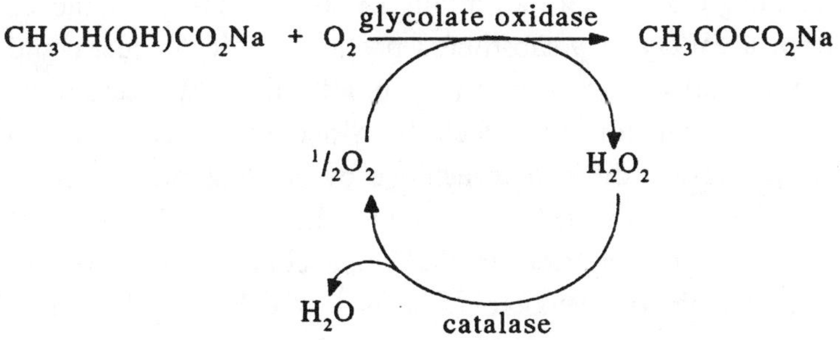

(14.24)

Carbohydrates are also substrates for various alcohol oxidases, notable examples of which are glucose oxidase (E.C. 1.1.3.4), galactose oxidase (E.C. 1.1.3.9) and pyranose-2-oxidase (E.C. 1.1.3.10). Glucose oxidase is the enzyme involved in the industrial production of ca. 4500 tons per annum of *D*-gluconic acid by fermentation of glucose.[51] Galactose oxidase has the broadest substrate specificity, and hence, the greatest potential for synthetic applications (e.g. various diols and triols are oxidized enantioselectively to the corresponding aldehydes[52]). However, no commercial applications of such oxidations have been forthcoming possibly due to low productivities and/or limited availability of the enzyme.

Another potentially interesting oxidase is polyphenol oxidase (laccase, tyrosinase; E.C. 1.10.3.2), a copper-dependent oxidase which catalyzes the oxidation of phenols or catechols to the corresponding 1,2-benzoquinones (Reaction 14.25).[53]

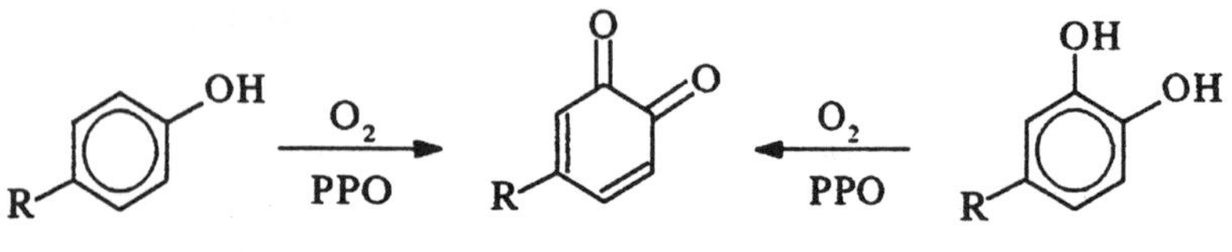

(14.25)

Polyphenol oxidase (PPO) occurs widely in plants and fungi and is probably involved in the biosynthesis of lignin. The *o*-quinone product is unstable in aqueous media where it undergoes facile polymerization. This problem was circumvented by employing mushroom polyphenol oxidase in chloroform (containing 0.5% water), affording the corresponding *o*-quinones in high yield from a variety of *p*-substituted phenols.[54] The product could be reduced non-enzymatically in a separate step, affording a two-step route for the conversion of phenols to catechols. Since regiospecific *ortho* hydroxylation of phenols is difficult to achieve by conventional synthetic methods, this provides an attractive biocatalytic alternative. However, to our knowledge, it has not yet been applied on a commercial scale. An interesting application of polyphenol oxidase is in the bioremediation of phenol-polluted water.[56] A number of phenols can be effectively removed, over a wide range of concentrations, using mushroom polyphenol oxidase. The resulting *o*-quinone polymers are adsorbed on chitosan.

Polyphenol oxidase has been used in conjunction with the electron transfer mediator, 2,2'-azino-bis(3-ethyl-benzothiazoline-6-sulfonic acid) (ABTS) diammonium salt for the catalytic aerobic oxidation of benzylic alcohols to the corresponding benzaldehydes (Reaction 14.26).[56] Yields were > 90% and volumetric productivities ca. 0.1 $g.l^{-1}.h^{-1}$.

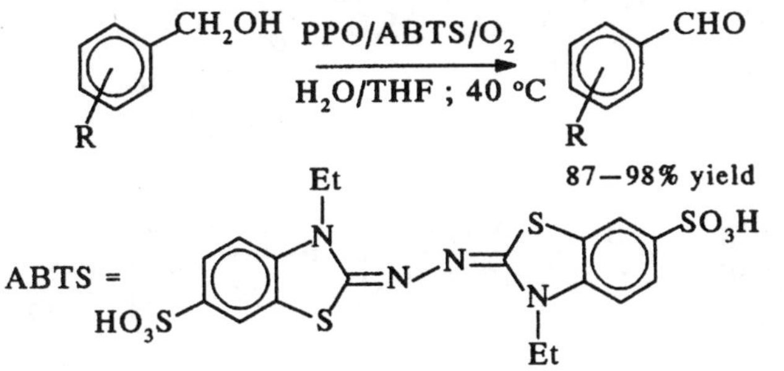

(14.26)

3.5 *Peroxidases*

Peroxidases are ubiquitous in nature, e.g. lignin peroxidase catalyzes the oxidative degradation of lignin. One of the most versatile and synthetically interesting peroxidases is the chloroperoxidase (CPO) produced by the marine fungus *Caldariomyces fumago.*[57] CPO is a heme-dependent enzyme, which *in vivo* catalyzes the oxidation of chloride ion by H_2O_2 to give hypochlorite. This forms the basis for the chlorination of organic compounds *in vivo*.

The active oxidant is a high-valent oxoiron species bearing a distinct resemblance to the active oxidant from cytochrome P450 (in both cases the axial ligand is a cysteinate residue). Hence, we reasoned that this oxoiron intermediate should, in the absence of chloride ion, be capable of undergoing (enantio)selective oxygen transfer reactions with organic substrates. Indeed, we[58] and others[59] found that CPO catalyzes the enantioselective oxidation of arylmethyl sulfides (Reaction 14.27), the enantioselective epoxidation of olefins (Reaction 14.28)[60] and the regioselective oxidation of indoles to the corresponding 2-oxoindoles (Reaction 14.29).[61]

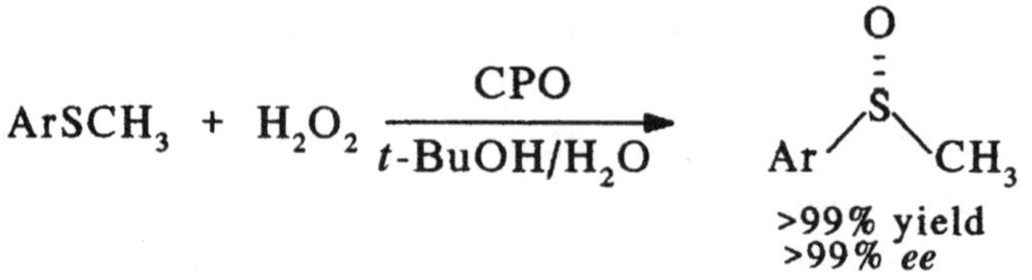

(14.27)

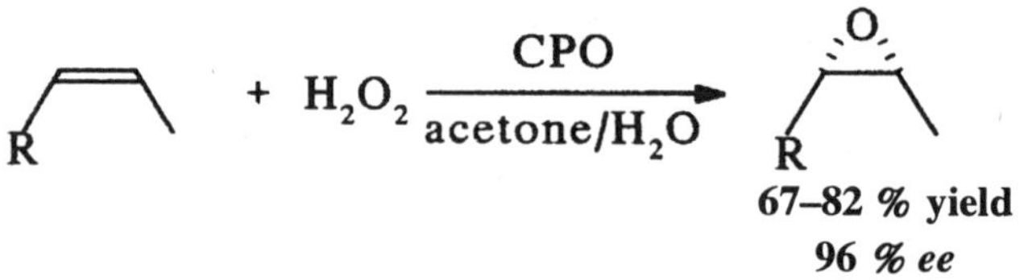

(14.28)

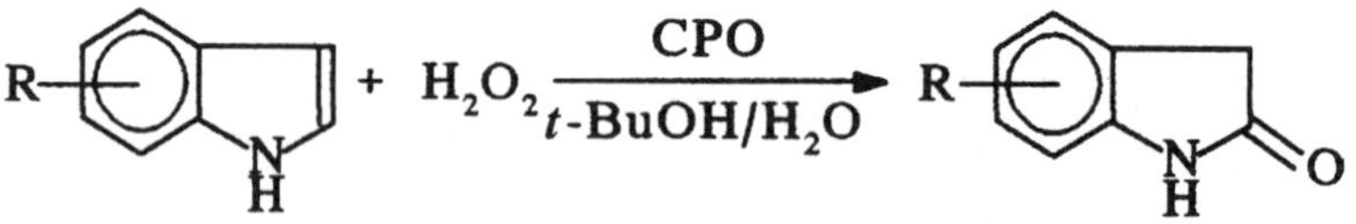

(14.29)

Since peroxidases have no cofactor requirement, such transformations are performed with the free enzyme. However, in common with other heme-dependent peroxidases, CPO has a major drawback: limited stability owing to rapid oxidative degradation of the porphyrin ring by H_2O_2. The stability can be increased by employing aqueous *tert*-butanol as the reaction medium and slow, feed-on-demand addition of H_2O_2 regulated by means of a H_2O_2 stat.[62]

Although this leads to substantial improvements,[63] the volumetric and catalyst productivities are still economically unattractive. Catalyst costs are directly related to catalyst productivities, as shown in Table 14.8 (which compares enantioselective epoxidation using CPO with corresponding reactions using the Mn Salen/NaOCl system developed by Jacobsen.[64]) With the former productivities are at least two orders of magnitude too low to be interesting. In contrast, indole oxidation mediated by CPO has productivities which are "in the ballpark" at least.

Since limited stability is an inherent feature of heme-dependent peroxidases, it may be more profitable to explore the synthetic utility of non-heme peroxidases, e.g. vanadium-dependent haloperoxidases.[65]

Table 14.8 Chemocatalytic versus Biocatalytic epoxidation.

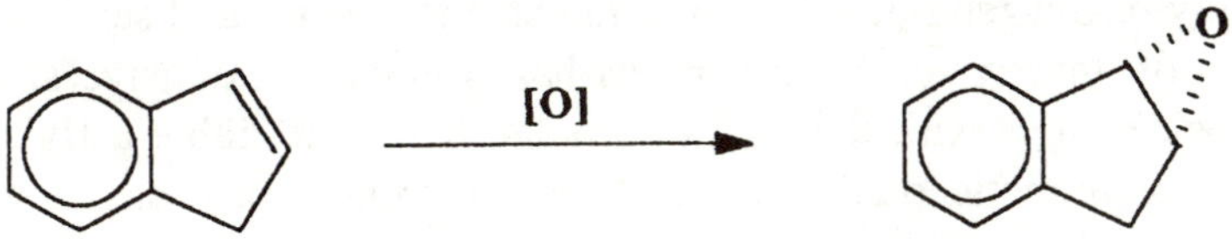

Chemocatalytic: Mn (Salen*)Cl / NaOCl
Biocatalytic: CPO / H_2O_2 (NB: product is *trans*-diol)

	Chemo catalyst	Biocatalyst (CPO)	CPO oxindole synthase
Yield %	90	85	96
ee %	88	97	–
S/C/ (mol/mol)	400	2.200	900.000
S/C/ (g/g)[a]	60	6	2.700
Catalyst costs[b] ($/kg product)	13	ca. 8.000	20

[a]M.W. of Mn (Salen*)Cl = 635; M.W. of CPO = 42.000.
[b]Assuming a bulk price of $1000/kg for Mn (Salen*)Cl and $50.000/kg for CPO.

One reaction of heme peroxidases does appear to have commercial utility: polymerization of phenols. For example, horseradish peroxidase (HRP), in the presence of H_2O_2, catalyzes the free radical polymerization of a variety of phenols. The resulting polymers constitute alternatives for the widely used phenol-formaldehyde resins which are under pressure owing to health and environmental problems associated with their use. As HRP catalyzes only the formation of chain-initiating phenoxy radicals, polymer chain growth proceeds non-enzymatically and is limited only by the poor solubility of the growing polymer in water. In organic media, in contrast, high molecular weight polymers are formed. High molecular weight polymers of *p*-phenylphenol, for example, were produced in aqueous dioxane with high productivities.[66] The resulting polymer is being commercialized as a photographic developer resin.[67]

4. Biomimetic Oxidations

The ability of oxygenases to catalyze the chemo-, regio- and stereoselective oxidation of unactivated hydrocarbons has stimulated extensive studies of biomimetic models with the goal of emulating this desirable activity.[7] The accent in biomimetic oxidations is, hence, predominantly on conversions of interest in the commodity chemicals arena. We note, however, that the same economic criteria are applicable to biomimetic oxidations, i.e. very high volumetric and catalyst productivities are required for commodity chemicals.

4.1 Cytochrome P450 mimics

The most extensively studied mimics have involved the use of metalloporphyrins, i.e. mimics of cytochrome P450.[68–70] Such biomimetic systems suffer from the same economic disadvantage as the enzyme they are meant to emulate: for oxidation with O_2, a sacrificial reductant (e.g. $NaBH_4$, Zn or Pt/H_2) is required. In the 1970s, various authors, notably Hrycay and coworkers,[71] showed that liver microsomal cytochrome P450 catalyzes the hydroxylation of hydrocarbons using a variety of single oxygen donors (e.g. H_2O_2, RO_2H, $NaOCl_2$). This pathway, which later became known as the "peroxide shunt" (see Fig. 14.6) provided a means for circumventing the need for a sacrificial reductant.

In 1979, Groves and co-workers[72] were the first to translate the "peroxide shunt" to a model system. They described the use of iron(III) meso-tetraphenylporphyrin (TPP) chloride in combination with iodoxylbenzene as the oxygen donor for the epoxidation of olefins and the hydroxylation of alkanes. Following this seminal study, metalloporphyrins in conjunction with a variety of single oxygen donors (e.g. RO_2H, H_2O_2, RCO_3H) were extensively studied.[68–70] Although these studies provided valuable insights into the mechanism of cytochrome P450-mediated oxidations, they all suffer from the inherent disadvantage of requiring a relatively expensive primary oxidant. Moreover, as with the enzyme they emulate, the inherent low stability of the porphyrin ligand is a serious drawback. Although this problem can be alleviated by the use of second and third generation catalysts, containing more stable perhalogenated porphyrins as ligands (see later), the problem of the expensive oxidant remains. In fact, such

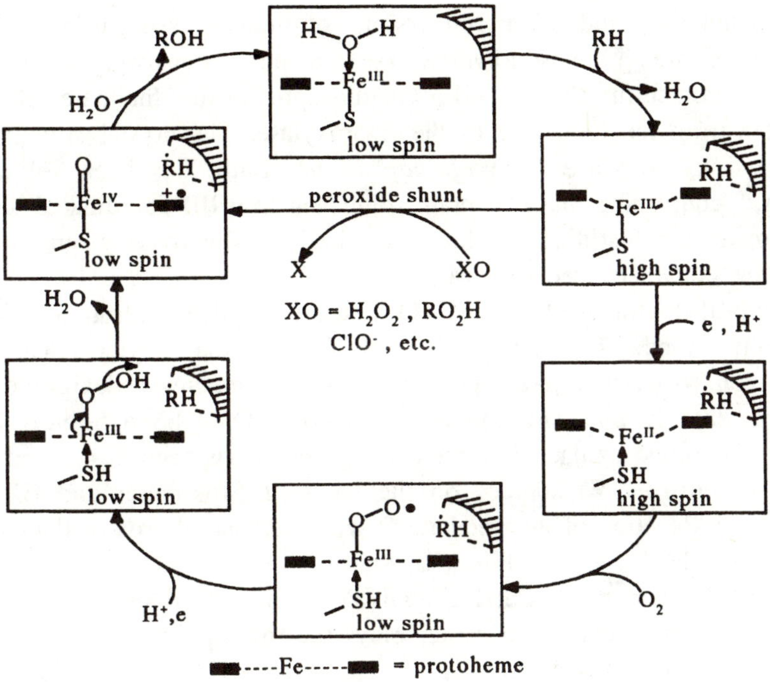

Fig. 14.6 Catalytic cycle of cytochrome P450.

systems merely transfer the need for a sacrificial reductant to an extramural activity (e.g. the production of H_2O_2 by reduction of O_2). By the same token, in the much studied co-oxidation of alkanes and olefins with aldehydes,[73] the aldehyde plays the role of the sacrificial reductant. In the context of commodity chemicals, this is likewise not a viable proposition.

4.2 *MMO mimics*

Recently, attention has shifted to the study of mimics of the non-heme monooxygenase, MMO.[12–14,74,75] MMO is a promiscuous enzyme which catalyzes the hydroxylation of a wide variety of alkanes, from methane to adamantane[12–15] and other conversions of interest (see earlier). MMO consists of three proteins: a hydroxylase, a reductase consisting of one FAD and one iron-sulfur cluster, and a coupling protein that facilitates electron transfer from the reductase to the hydroxylase. The crystal structure of oxidized hydroxylase from *M. capsulatus* (Bath) has been solved to 2.2 Å,[76] confirming that it contains dinuclear iron(III) coordinated to two glutamate and histidine residues, two bridging hydroxyl groups and a bridging glutamate (see Fig. 14.7).

Several features of the MMO-mediated hydroxylation are fairly well-understood. The reduced diiron(II) form of the enzyme reacts with dioxygen to give a μ-1,2-peroxodiiron(III) species[77] designated as intermediate P. This subsequently undergoes O–O bond homolysis to afford the active oxidant, intermediate Q, which has been designated as a bis(μ-oxo)diiron(IV) species on the basis of Mössbauer and EXAFS analysis.[78] Reaction of intermediate Q with methane affords methanol and returns the MMO to the diiron(III) state.

In MMO, the two oxidizing equivalents are stored in the diiron(IV) moiety, i.e. the second iron center plays the role of the porphyrin ligand in cytochrome P450 in storing the second oxidizing equivalent. The mononuclear nature of cytochrome P450 necessitates heterolytic cleavage of the O–O bond to avoid the formation of highly reactive OH radicals. Consequently, nature has developed an elaborate push-pull mechanism to facilitate O–O heterolysis. In contrast, binuclear iron enzymes can readily achieve the diiron(IV) state by homolysis of the O–O bond in the μ-peroxodiiron(III) species.

The mechanistic details of the hydroxylation step are, just as with cytochrome P450, still controversial.[12–15] Recent quantum chemical studies[79] are in favor of the oxygen rebound mechanism shown in Fig. 14.8.

Ligand design for MMO mimics is more complicated than with cytochrome P450. In the enzyme, the ligands consist largely of monodentate amino acid residues. In model systems, monodentate ligands undergo facile exchange, and hence, multidentate ligands have been designed to mimic the active site (see Chap. 5). For example, iron complexes of the tripodal ligands I and II have been shown to mimic various features of the proposed MMO catalytic cycle. Thus, reaction of an Fe(II) carboxylate

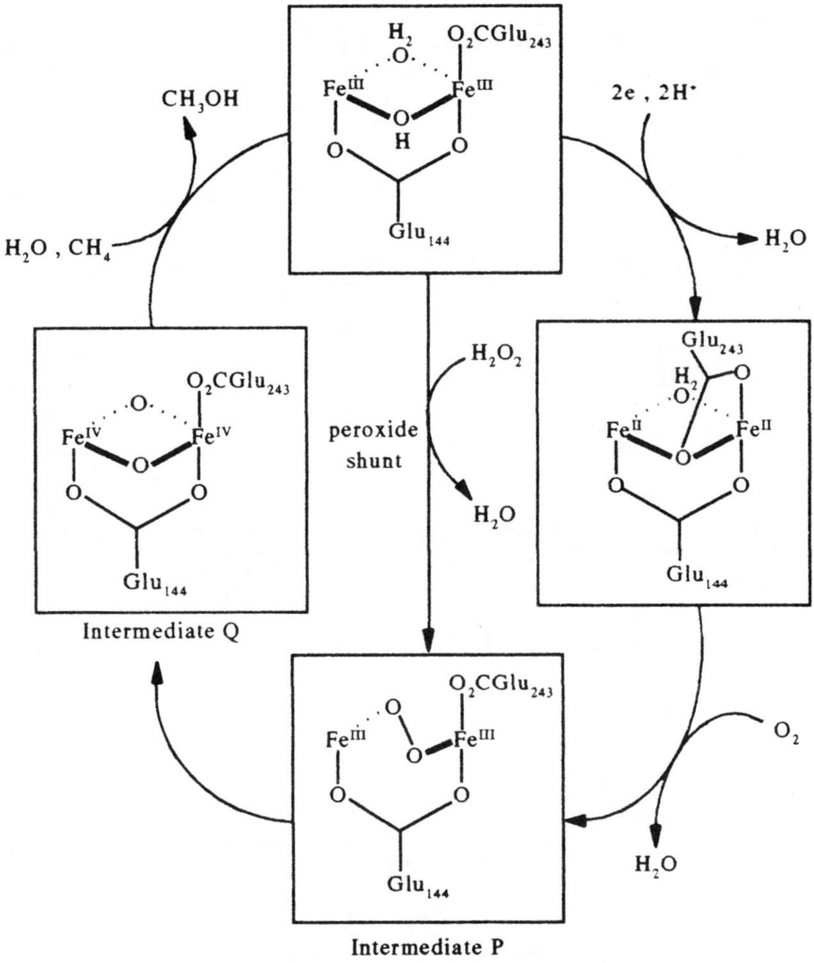

Fig. 14.7 Catalytic cycle of MMO.

complex of I (R = i-Pr) with O_2 at temperatures below –50 °C afforded a μ-1,2-peroxoiron(III) complex with bridging carboxylates,[80] and reaction of a μ-oxodiiron(III) complex of II (R = Et) with H_2O_2[75] gave a μ-dioxodiiron complex containing a formal Fe(III)/Fe(IV) oxidation state, i.e. one oxidation equivalent lower than intermediate Q in the MMO cycle.

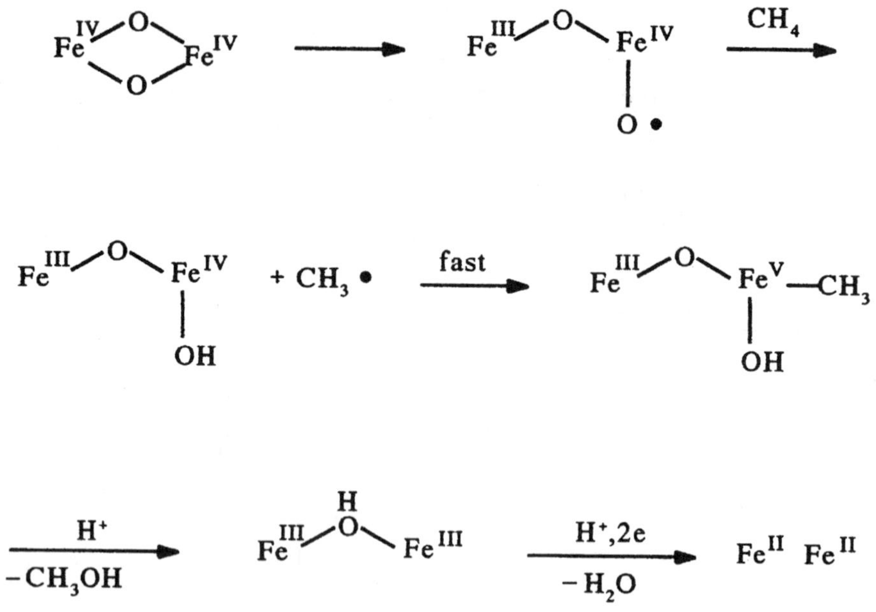

Fig. 14.8 Mechanism of C–H hydroxylation by MMO.

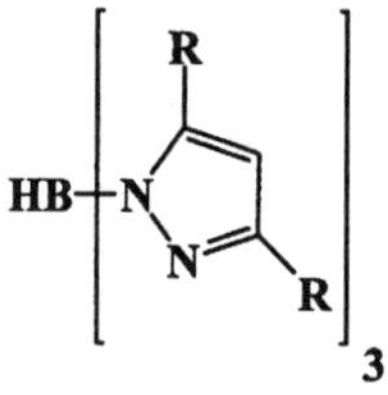

(I) (R = i-Pr)

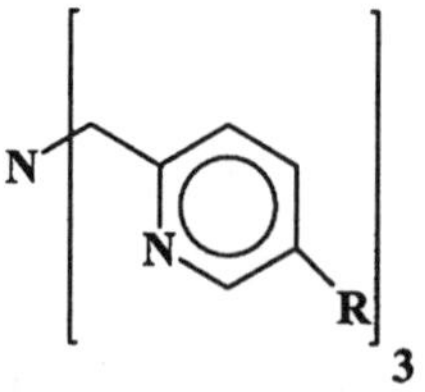

(II) (R = H, Et)

We note however, that although various μ-peroxo- and μ-dioxodiiron species have been isolated and characterized, none of these complexes have actually been shown to hydroxylate an alkane. A μ-oxodiiron complex of I (R = H) was shown to catalyze the hydroxylation of cyclohexane and adamantane with O_2 in the presence of acetic acid and Zn powder.[81] Here again, as with the cytochrome P450 models and with *in vivo* systems, a sacrificial reductant is required.

To avoid the need for a sacrificial reductant, iron(III) complexes of, for example, ligand (II) have been used by Que and co-workers[82] in conjunction with *tert*-butyl hydroperoxide (TBHP) for alkane oxidation. Thus, oxidation of cyclohexane, to a mixture of cyclohexanol and cyclohexanone, with TBHP in the presence of iron(III) complexes of II was proposed to involve initial O–O heterolysis affording a high-valent oxoiron species (Reaction 14.30) as the active oxidant (Reaction 14.31).

$$ROOH + Fe^{III} \longrightarrow Fe^{V}{=}O + ROH \qquad (14.30)$$

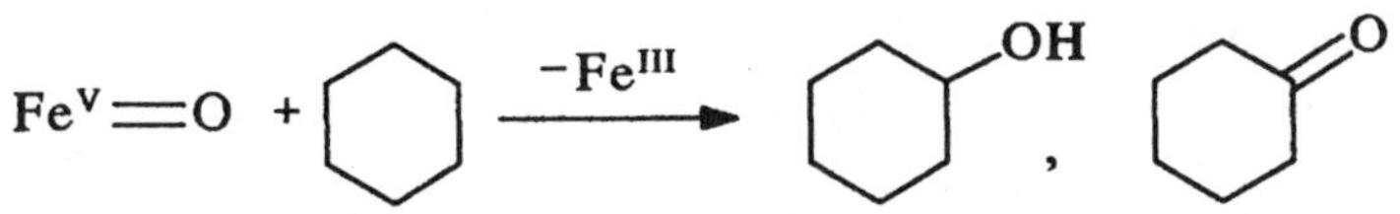

(14.31)

However, more recent studies by Ingold and co-workers (see Chap. 6)[83] have unequivocally demonstrated that such systems involve initial one-electron redox reactions of iron with RO_2H, producing *tert*-butoxy and *tert*-butylperoxy radicals (Reactions 14.32 and 14.33), which then dominate the subsequent chemistry (Reactions 14.32–14.38). As noted earlier, nature has developed an elaborate push-pull mechanism to promote O–O bond heterolysis.

$$Fe^{III} + RO_2H \longrightarrow Fe^{IV}{=}O + H^+ + RO\bullet \qquad (14.32)$$

$$Fe^{IV}{=}O + RO_2H \longrightarrow Fe^{III}{-}OH + ROO\bullet \qquad (14.33)$$

$$RO\bullet + RO_2H \longrightarrow RO_2\bullet + ROH \qquad (14.34)$$

$$2\ RO_2\bullet \longrightarrow 2\ RO\bullet + O_2 \qquad (14.35)$$

(14.36)

(14.37)

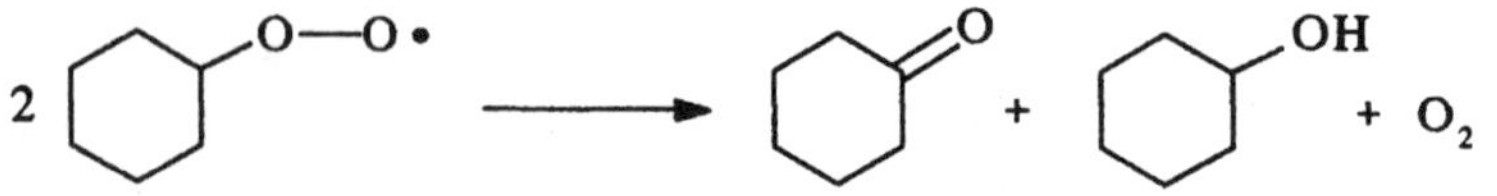 (14.38)

A convenient mechanistic probe for establishing the intermediacy of *tert*-alkoxy radicals is 1,1-dimethyl-2-phenylethyl hydroperoxide, since the resulting alkoxy radical undergoes extremely rapid ($t_{1/2} = < 5$ ns) β-scission to give a benzyl radical and acetone (Reaction 14.39). When the oxidation of cyclooctane was performed with this hydroperoxide in the presence of iron(III) complexes of (II) or "Gif-type" systems (see later), the observed oxidation products (benzaldehyde, benzyl alcohol, etc.) were all derived from the benzyl radical and the cyclooctane was virtually untouched.[83]

$$PhCH_2(CH_3)_2O\bullet \longrightarrow PhCH_2\bullet + (CH_3)_2CO \qquad (14.39)$$

The conclusion is clear: any biomimetic systems which are claimed to involve direct oxidation of hydrocarbons by putative high-valent oxoiron species must be subjected to rigorous experimental protocols to exclude the intermediacy of alkoxy radicals.

4.3 *"Gif" chemistry*

The so-called "Gif" family of systems, developed by Barton and co-workers[84] occupy a special position in the biomimetic alkane functionalization arena. The initial system reported in 1983[85] consisted of iron(II) and dioxygen in pyridine solvent in combination with zinc or iron powder and acetic acid as sources of electrons and protons, respectively. This system was shown to selectively oxidize alkanes such as cyclohexane to give the corresponding ketones directly (Reaction 14.40), rather than via further oxidation of an initially formed alcohol. Further evolution followed a similar path to the cytochrome P450 models and culminated in a system comprising picolinic acid or related ligands in combination with iron(III) and H_2O_2 (Reaction 14.41), or RO_2H as a single oxygen donor.[84] Although the "Gif" systems have several features in common with MMO (e.g. carboxylate and heterocyclic nitrogen ligands in a putative dinuclear iron complex), they appear to be fundamentally different in that they afford the ketone rather than the alcohol as the major product.

$$R_2CH_2 + 2O_2 + 4H^+ + 4e \longrightarrow R_2C{=}O + 3H_2O \quad (14.40)$$

$$R_2CH_2 + 2H_2O_2 \longrightarrow R_2C{=}O + 3H_2O \quad (14.41)$$

As noted earlier, Ingold and co-workers[83] showed that alkoxy radicals are the key hydrogen abstracting species when RO_2H is the oxidant, also with the "Gif" type system.[86]

The situation is less clear when H_2O_2 is the terminal oxidant,[87,88] and Barton distinguishes two different pathways — an Fe(II)/Fe(IV) and an Fe(III)/Fe(V) manifold — depending on the oxidation state in which the iron is introduced.[88] Both pathways are proposed to involve formation of a high-valent alkyliron species, formed via reaction of $Fe^{IV}{=}0$ or $Fe^{V}{=}O$ with the substrate, respectively. In the Fe(II)/Fe(IV) manifold, an unstable alkyliron(IV) intermediate is assumed to undergo facile homolytic cleavage to afford an alkyl radical. The putative alkyliron(V) intermediate in the Fe(III)/Fe(V) manifold, in contrast, is assumed to be more stable, and hence, affords the ketone. Formation of the alkyliron(V) species is proposed to involve reaction of a dimeric μ-peroxo-oxoiron(V) intermediate with the alkane substrate (Fig. 14.9).[88]

Alkane oxidations with Fe(II), O_2 and a sacrificial reductant are mechanistically equivalent to Fe(III) and H_2O_2, i.e. both systems are assumed to proceed via a μ-peroxoiron(III) complex. Although they afford different products, it is noteworthy that "Gif" systems afford the same C^3/C^2 oxidation ratio (3:1 per hydrogen atom) as observed with MMO *in vivo*.[89]

Irrespective of the mechanism, all of these model systems suffer from the same inherent disadvantage as the enzyme(s) they attempt to emulate: they require a sacrificial reductant or a relatively expensive oxygen donor. Is there a sacrificial reductant that is economically viable? In their original studies, Barton and colleagues used H_2S as the reductant.[85] They were inspired by the notion that primitive life

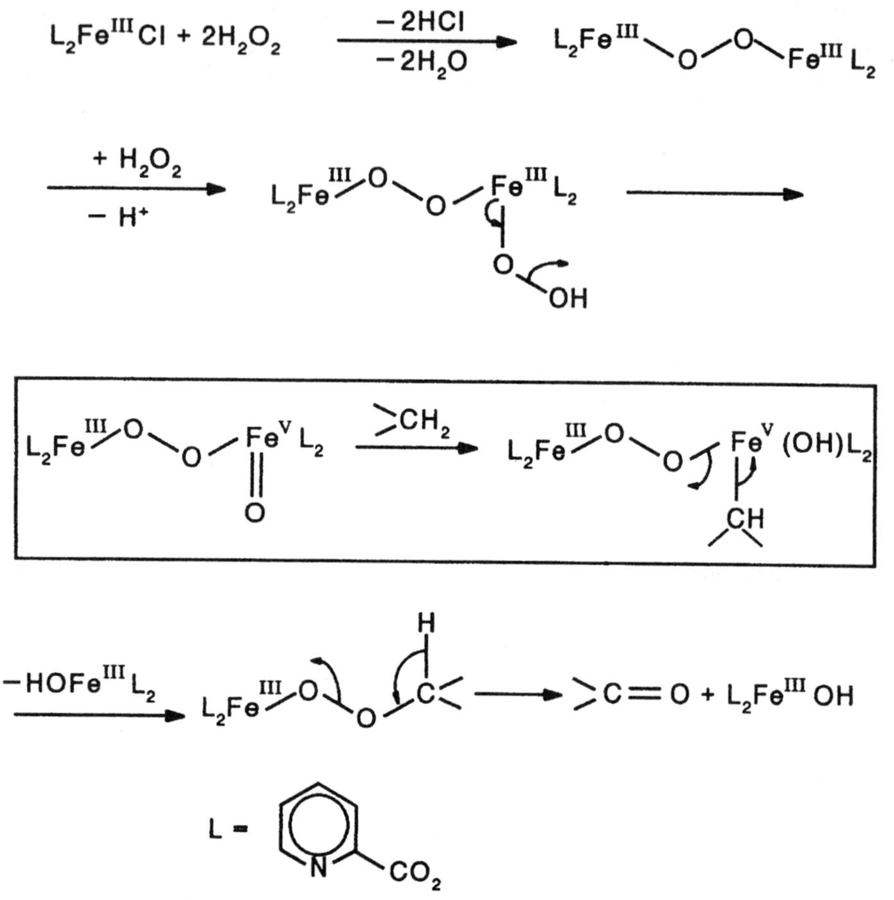

Fig. 14.9 Proposed mechanism for alkane oxidation by H_2O_2 in the Fe(III)/Fe(V) manifold.

forms may have utilized both iron and H_2S in concert with O_2 to activate hydrocarbons, and Tabushi's observation[90] that thiols, such as 2-mercaptoethanol, act as sacrificial reductants in analogous systems. More recently, Barton and co-workers[91] have improved their original system. They showed that a mixture of $FeCl_3$, picolinic acid and 4-*tert*-butylpyridine catalyzes the co-oxidation of cyclohexane and H_2S with O_2 in acetonitrile at ambient temperature and pressure reaction (Reaction 14.42).

After nine hours a cyclohexane conversion of 36% was observed, giving a 40/60 mixture, of cyclohexanone and cyclohexanol (note the different product distribution to that observed with typical "Gif" systems) and elemental sulfur. Although these results are far removed from a commercially viable process, the high selectivities at > 30% conversion are quite remarkable, and one can in principle, envisage an optimized process in which H_2S derived from an adjacent hydrodesulfurization unit is used as an economical stoichiometric reductant in the oxidation of cycloalkanes.

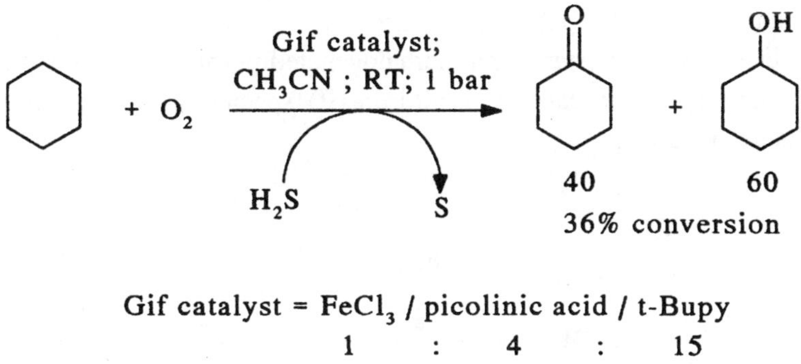

(14.42)

4.4 *The quest for suprabiotic oxidation catalysis*

As noted in the preceding sections, a fundamental shortcoming of biological and biomimetic catalysts for the aerobic oxidation of hydrocarbons is the requirement for a sacrificial reductant. There remains a great need, therefore, for suprabiotic catalysis which avoids this drawback.

In a series of publications during the period 1989–1993, Lyons and Ellis[92] showed that iron complexes of perhalogenated porphyrins (e.g. octabromotetrakins-(pentafluorophenyl)-porphyrinato iron(III) chloride, $Fe(TFPPBr_8)Cl$), are remarkably active catalysts for the hydroxylation of light alkanes by O_2 under mild conditions. For example, isobutane afforded *tert*-butanol with > 90% selectivity and turnover numbers exceeding 13000 at 25 °C. Propane required more forcing conditions (125 °C and 70 bar air) and gave a mixture of isopropanol and acetone with a turnover number of > 400.

Lyons and Ellis suggested[92] that the active oxidant was an oxoiron(IV) porphyrin produced by homolysis of a μ-peroxoiron(III) dimer. As shown in Fig. 14.10, the overall reaction constitutes an example of suprabiotic oxidation catalysis. The increased activity of perhalogenated porphyrin complexes was rationalized on the basis of the electron-withdrawing halogen substituent increasing the electrophilic character (i.e. redox potential) of the oxoiron(IV) species and, at the same time, stabilizing the porphyrin ring towards oxidative degradation. Moreover, it was suggested[92] that perhalogenated porphyrin ligands promote the dissociation of unreactive μ-oxoiron(III) species as a result of steric and electronic effects.

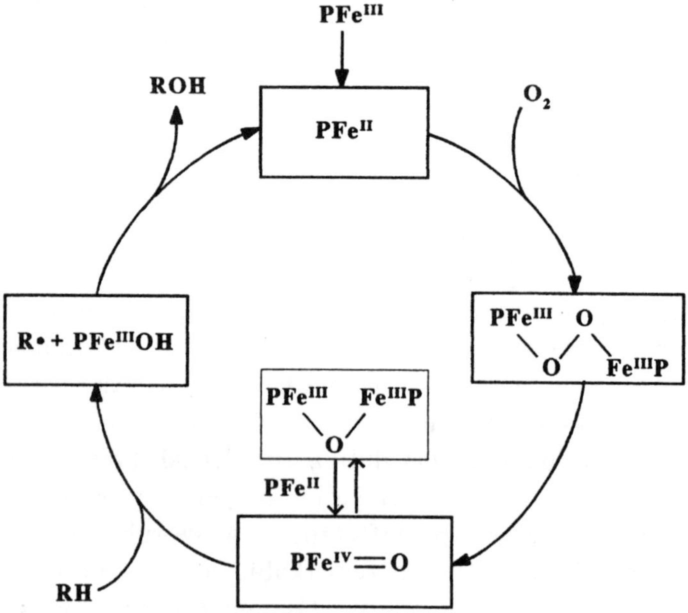

Fig. 14.10 Proposed mechanism for suprabiotic oxidation of alkanes mediated by perhalogenated iron porphyrins.

More recently, Labinger and co-workers[93] provided convincing evidence that these systems involve a classical free radical chain autoxidation pathway. Apparently, iron complexes of perhalogenated porphyrins are extraordinarily active catalysts for the homolytic decomposition of RO_2H into chain-initiating RO· and RO_2· radicals via one-electron redox reactions (Reactions 14.10 and 14.11; M^n = Fe(III)). It was suggested[93] that the main consequence of using a perhalogenated porphyrin is to accelerate the slower of the two steps, i.e. oxidation of RO_2H by Fe(III) (Reaction 14.10). Hence, this is yet another example of a C–H bond hydroxylation in which the abstracting species is an alkoxy radical, and the role of the metal catalyst is limited to promoting its generation from RO_2H.

Two possible Mars-van Krevelen pathways can be envisaged for the activation of O_2 such that both oxygen atoms are inserted into the substrate. They differ in whether a two-electron or four-electron oxidation of the metal ion is involved. With iron, one can envisage a mechanism involving a two-electron shuttle between Fe(II) and Fe(IV), as outlined in Fig. 14.10. However, such a pathway does not appear able to compete effectively with free radical autoxidation. On the other hand, *in vitro* oxidations are not limited to catalysis by elements that are abundantly available in nature. Ruthenium is of particular interest because it has a large number of readily accessible oxidation states.[94] Hence, a four-electron shuttle between Ru(II) and Ru(VI) can be envisaged, which is not likely for iron. Thus, Groves and Quinn[95] suggested the mechanism shown in Fig. 14.11, to explain the aerobic epoxidation of olefins observed with a tetrakis-mesitylporphyrinato ruthenium(II) catalyst. Ruthenium(II) reacts with O_2, affording a dioxoruthenium(VI) complex which transfers an oxygen to the olefin. The resulting oxoruthenium(IV) was proposed to disproportionate to Ru(II) and dioxoRu(VI). Similarly, Drago and co-workers[96] reported that a ruthenium(II) dimethylphenanthroline complex catalyzes the same reaction and suggested that a dioxoruthenium(VI) complex transfers both oxygens sequentially.

It should be emphasized however, that only very active olefin substrates, e.g. cyclooctene and norbornene, were used in both studies. Subsequently, Scharbert and co-workers at Hoechst showed[97] that Ru(II)TMP catalyzes

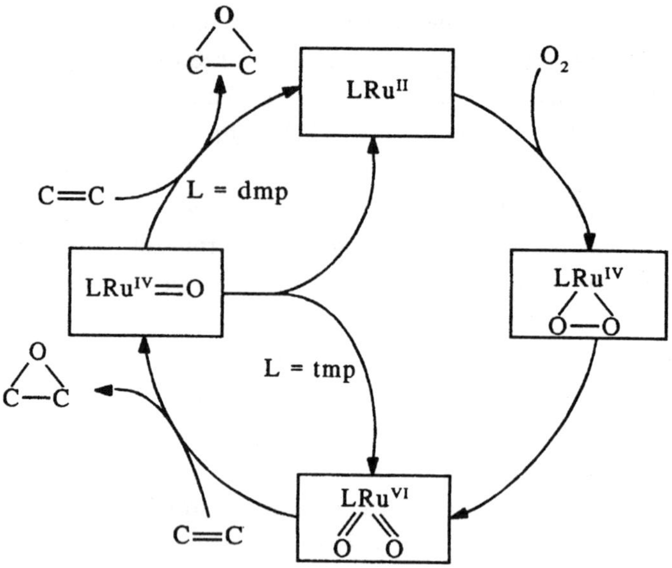

tmp = tetrakis-mesitylporphyrin
dmp = dimethylphenanthroline

Fig. 14.11 Ru-catalyzed aerobic epoxidation.

the aerobic epoxidation of propylene and 1-octene. However, the catalyst was rapidly deactivated (after ca. 40 turnovers in 24 h) due to the formation of an inactive carbonyl complex, Ru(II)(TMP)CO. The carbon monoxide was probably formed via ruthenium-catalyzed decarbonylation of the *n*-aldehyde, a by-product in the oxidation.

More recently, Neumann and co-workers[98] have reported that a ruthenium(III) polyoxometalate, $[WZnRu_2(OH)(H_2O)][ZnW_9O_{34}]_2^{11-}$, catalyzes the aerobic epoxidation of olefins and the hydroxylation of alkanes. The reactions were unaffected by radical scavengers and the observed induction period could be eliminated by prior reduction of Ru(III) to Ru(II). Molecular modeling studies demonstrated that the ruthenium is positioned in a sterically hindered pocket within the polyoxometalate framework. It was suggested that this placement of ruthenium prevents the formation of inactive μ-oxo dimer, analogous to the ruthenium porphyrin system (see above) where the sterically hindered tetramesityl porphyrin ligand performs the same function. A mechanism was proposed in which a two-electron cycle involving Ru(II)/oxoRu(IV) was favored, but a four-electron cycle involving disproportionation of the latter to dioxoRu(VI) and Ru(II) was not ruled out.

We emphasize however, that the ruthenium polyoxometalate-catalyzed epoxidations were carried out in two steps. The catalyst was first oxidized in an atmosphere of O_2 (in 1,2-dichloroethane solvent) at 80 °C. The olefin substrate was then added to the solution at room temperature, i.e. the epoxidation step is stoichiometric. High yields of epoxide (68–82%; selectivity > 99%) were observed only with very reactive olefins (cyclooctene, norbornene and 2,3-dimethyl-2-butene). 1-Octene gave only 17% yield (selectivity > 99%) and propene afforded propene oxide in 95% selectivity at an unknown conversion.

The above examples demonstrate the principle of a ruthenium-based suprabiotic approach but require considerable optimization to be economically viable. The latter example illustrates another important concept: separation of the substrate oxidation and the catalyst reoxidation steps in a Mars-van-Krevelen pathway. Since the substrate is never in contact with O_2 in this scenario, it obviates free radical autoxidation. This concept is already applied in a large scale oxidation process, namely the vanadylpyrophosphate-catalyzed gas phase oxidation of *n*-butane to maleic anhydride.[99] Such a scenario, often referred to as a single-turnover cycle, would seem to be worthy of further investigation.

4.5 *Redox molecular sieves*

A shortcoming of all the biomimetic systems discussed so far is that they mimic only the coordination of the metal in the active site. They lack the protein mantel of the enzyme which creates a hydrophobic, constrained environment conducive for selective adsorption and reaction of the substrate under quasi gas phase conditions. One approach to emulating this effect is to incorporate metal complexes within the micropores of a zeolite or a related molecular sieve by isomorphous substitution or encapsulation.[100]

Confinement of a metal (complex) in the constrained environment of a molecular sieve can endow the material with unique catalytic activity as a result of strong electrostatic interactions between acidic and basic sites on the internal surface and the substrate or reaction intermediate. Such interactions can be likened with those involving acidic carboxylic and basic amino groups of amino acid residues in the active sites of redox enzymes. Moreover, the superstructure of the molecular sieve can impose an unusual, high energy geometry at the metal site thus enhancing its redox potential. In bioinorganic chemistry, this is referred to as the entactic state[101] and it is generally believed to play a major role in determining the activity of (redox) metalloenzymes.

By an appropriate choice of pore size and hydrophobicity, one is able, just as with enzymes, to influence which molecules have ready access to the active site on the basis of their size and/or hydrophobic/hydrophilic character. Hence, tailor-made "mineral enzymes" or "zeozymes"[102] can be designed to mimic the active site of a redox enzyme more closely.

The large pore faujasites (zeolites X and Y), for example, contain supercages of 13 Å diameter intersected by channels of 7.4 Å. The supercages are large enough to accommodate metalloporphyrins or metallophthalocyanines. Hence, so-called ship-in-a-bottle catalysts can be synthesized, in which the metal complex is assembled in the supercages and is too large to exit via the micropores.[103,104] Since the complexes are physically entrapped in intrazeolite space, they can be regarded as supramolecular catalysts. Provided that the complex is stable towards dissociation and/or oxidative destruction, no leaching of the metal ion should occur.

Pioneering studies of oxidations catalyzed by faujasite-encapsulated iron phthalocyanine (FePc) were performed by Herron and co-workers.[105] The oxidation of linear and cycloalkanes with iodoxylbenzene (PhIO) were catalyzed by FePc-X and FePc-Y. Although interesting shape selectivities reminiscent of iron-dependent oxygenases were observed, activities were poor (ca. six turnovers). More recently, this chemistry was ressurrected by Jacobs and co-workers[106] who showed that FePc-Y, containing 1 FePc per 77 supercages, gave turnovers as high as 6000 in the oxidation of *n*-octane with TBHP. Turnover frequencies were rather low (12 h^{-1}), however. The use of ferrocene as the source of iron was found to be essential[106] in order to minimize the presence of free iron ions in the zeolite which leads to non-productive TBHP decomposition.

A further improvement was achieved by embedding the FePc-Y in a hydrophobic polydimethylsiloxane (PDMS) membrane.[107] The latter mimics the hydrophobic phospholipid membrane in which the cytochrome P450 resides, acting as an interface between the two immiscible phases (aqueous TBHP and alkane substrate). The catalyst exhibited a 20-fold increase in activity compared to membrane-free FePc-Y. However, reactivity orders of C–H bonds (C^3/C^2 = 3–4 in adamantane) are indicative of the intermediacy of alkoxy radicals. Hence, we conclude that, in the absence of evidence to the contrary, oxidations of alkanes with TBHP in the presence of FePc-Y involve tert-butoxy radicals as the hydrogen abstracting species, followed by a free radical autoxidation pathway. This could be confirmed by using 1,1-dimethyl-2-phenethyl hydroperoxide as a mechanstic probe for alkoxy radicals (see Sec. 14.4.2). The same applies to alkane oxidations with TBHP catalyzed by faujasite-encapsulated ruthenium perfluorophthalocyanine complexes studied by the Balkus group.[103,108] Similarly, faujasite-encapsulated iron(III) picolinate and 2-pyridinecarboxamide complexes catalyzed the oxidation of cycloalkanes with TBHP, but the normalized C^3/C^2 ratio of 3 observed with adamantane was consistent with abstraction by *tert*-butoxy radicals.[109]

In addition to the problem of the ubiquitous free radical chain autoxidation, there are several other practical problems associated with the ship-in-a-bottle catalysts. As noted earlier, the use of TBHP (or H_2O_2) is not an economically viable proposition for most commodity chemicals. Facile leaching of the metal can also be a problem in liquid phase oxidations[110] and rigorous proof of heterogeneity needs to be established. Thirdly, diffusion limitations can lead to low productivities in liquid phase reactions with ship-in-a-bottle type catalysts.

A possible solution to these problems has been provided by Panov and co-workers[110] who showed that a specially prepared FeZSM-5 molecular sieve catalyzes the selective gas phase oxidation of benzene to phenol using N_2O as the stoichiometric oxidant (Reaction 14.43). A phenol selectivity of 99% was obtained at 27% benzene conversion with a residence time of 2 secs at 350 °C.[111]

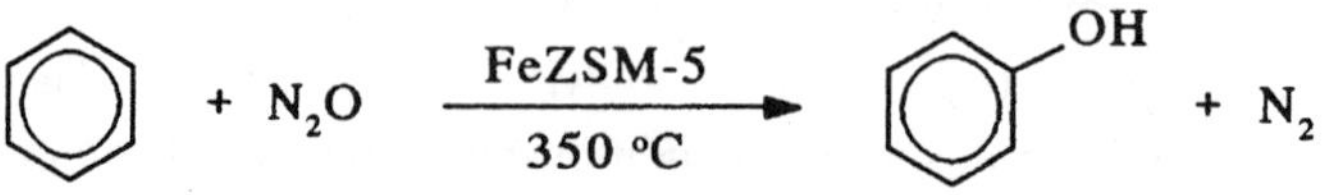

(14.43)

Monsanto developed this discovery into a commercially viable process for phenol manufacture.[112] In the commercial production of adipic acid by HNO_3 oxidation of cyclohexanone, ca. 1 mole of N_2O is produced per mole of adipic acid. As N_2O emissions contribute to atmospheric ozone depletion and global warming, adipic acid producers are stimulated to seek methods for reducing or eliminating N_2O emissions. An elegant means for achieving this is to incorporate the hydroxylation of benzene to phenol with N_2O into the overall process (as shown in Fig. 14.12).

The catalyst productivity is ca. 4 mmole, of phenol per gram of catalyst per hour, which ranks it amongst the best industrial oxidation catalysts.[112] Substituted aromatics are converted to the corresponding phenols.[111] Phenol itself undergoes hydroxylation, in 98% selectivity, to a mixture of catechol (17%) and hydroquinone (83%).[111]

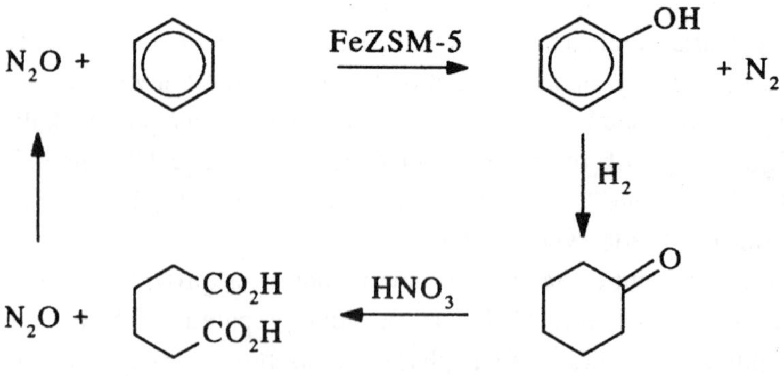

Fig. 14.12 Adipic acid manufacture incorporating N_2O oxidation of benzene to phenol.

The interesting question that remains is: what is the nature of the active oxidant in this remarkable catalyst? The FeZSM-5 (SiO_2/Al_2O_3 = 72; 0.56 wt % Fe) is prepared by hydrothermal synthesis from a gel containing $FeCl_3$. The resulting material is converted into the H-form by ammonia exchange and calcination at 550 °C. To increase the concentration of active sites, so-called α-sites,[111–113] the catalyst is heated at 900 °C in vacuum for one hour and finally treated with O_2 (0.8 Torr) at 550 °C. The Mössbauer spectra of the reduced and oxidized states of the resulting FeZSM-5 closely resembled those of the corresponding states of MMO. This was taken to be strong evidence in favor of a dinuclear oxoiron species being the active catalyst, i.e. an inorganic analogue of MMO.

Interestingly, in single turnover cycles (see earlier), the active oxidant quantitatively converts methane into methanol at ambient temperature or even at −50 °C.[113] The kinetic isotope effect (KIE), measured using the intramolecular competition of C–H and C–D bonds in CH_2D_2 was 5.5 to 1.9 in the temperature range −50 °C to 100 °C, consistent with rate-determining C–H bond cleavage analogous to MMO.[113]

We note however, that there is not a consensus on the nature of the active catalyst, and some authors even question whether the iron is essential for activity.[114] Nevertheless, irrespective of the exact nature of the active site, this catalyst constitutes an important breakthrough in biomimetic oxidation catalysis.

5. Conclusions and Future Prospects

A cursory examination of the vast body of literature on biocatalytic and biomimetic oxidations leads one to the general conclusion that they are largely aiming at different targets. Biocatalytic oxidations target the production of high value added chemicals, capitalizing on the high chemo-, regio- and enantioselectivities of these conversions. Research on biomimetic oxidation catalysis, in contrast, is largely focused on commodity chemicals, in particular the conversion of simple hydrocarbons via the "dream reactions" 14.1–14.3.

The use of biocatalytic oxidations in fine chemicals production, generally with whole microbial cells under fermentation conditions, is on the increase. The use of isolated redox enzymes, in contrast is, with a few notable exceptions, not practical even for fine chemicals production. Limited enzyme stability, which translates into high catalyst costs, is a major obstacle. Development of more stable redox enzymes, e.g. peroxidases, preferably for use in organic media, is desperately needed.

Notwithstanding the enormous effort that has been devoted to biomimetic catalysis, it has not led to much of practical use. Nevertheless, the recently developed FeZSM-5/N_2O can be regarded as a benchmark in biomimetic catalysis, and interesting leads in the area of ruthenium-based suprabiotic chemistry appear to hold promise for the future.

References

1. Sheldon, R.A. and Kochi, J.K. *Metal Catalyzed Oxidations of Organic Compounds.* Academic Press, New York, 1981; Franz, G. and Sheldon, R.A. in *Ullmann's Encyclopedia of Industrial Chemistry*, 5th Ed. VCH, Weinheim, 1991.
2. Sheldon, R.A. *J. Chem. Tech. Biotechnol.* 1997, **68:** 381; Sheldon, R.A. *Chem. Ind. (London)*, 1997, 12.
3. Holland, H.L. *Organic Synthesis with Oxidative Enzymes.* VCH, Weinheim, 1992.
4. Peterson, D.H. and Murray, H.C. *J. Am. Chem. Soc.* 1951, **75:** 5513.
5. Semonov, N.N. *Chem. Brit.* 1974, 471.
6. Mars, P. and van Krevelen, D.W. *Chem. Eng. Sci. Spec. Suppl.* 1954, **3:** 41.
7. Funabiki, T. (ed.) *Oxygenases and Model Systems.* Kluwer, Dordrecht, 1997.
8. For a possible exception see: Nguyen, H.H.T., Shiemka, A.K., Jacobs, S.J., Hales, B.J., Lidstrom, M.E. and Chan, S.I. *J. Biol. Chem.* 1994, **269:** 14995.
9. Sono, M., Roach, M.P., Coulter, E.D. and Dawson, J.H. *Chem. Rev.* 1996, **96:** 2841.
10. Ortiz de Montellano, P.R. (ed.) *Cytochrome P450: Structure, Mechanism and Biochemistry*, 2nd Ed. Plenum, New York, 1995.

11. Guengerich, F.P. (ed.) *Mammalian Cytochrome P450*, Vols. 1 and 2. CRC Press, Boca Raton, 1987.
12. Wallar, B.J. and Lipscomb, J.D. *Chem. Rev.* 1996, **96:** 2625.
13. Que, L. in *Bioinorganic Catalysis* (ed.) Reedijk, J. Marcel Dekker, New York, 1993.
14. Feig, L. and Lippard, S.J. *Chem. Rev.* 1994, **94:** 759.
15. Dalton, H. *Catal. Today* 1992, **13:** 445.
16. Van Deurzen, M.P.J., van Rantwijk, F. and Sheldon, R.A. *Tetrahedron* 1997, **53:** 13183.
17. Everse, J., Everse, K.E. and Grisham, M.B. (eds.) *Peroxidases in Chemistry and Biology*, Vols. 1 and 2. CRC Press, Boca Raton, 1991.
18. Green, J. and Dalton, H. *J. Biol. Chem.* 1989, **264:** 17698.
19. Van Beilen, J.B., Kingma, J. and Witholt, B. *Enzyme Microb. Technol.* 1984, **16:** 904.
20. Stirling, D.I. and Dalton, H. *J. Gen. Microbiol.* 1980, **116:** 277.
21. Kiener, A. *Chemtech*, September 1995, p. 31 and references cited therein.
22. Kulla, H.G., *Chimia* 1991, **45:** 81.
23. Japanese Patent 2119777 (1990), to Mitsubishi Chemical.
24. Sheldrake, G.N. in *Chirality in Industry* (eds.) Collins, A.N., Sheldrake, G.N. and Crosby, J. Wiley, New York, 1992, p. 127.
25. Furuhashi, K. in *Chirality in Industry* (eds.) Collins, A.N., Sheldrake, G.N. and Crosby, J. Wiley, New York, 1992, p. 167.
26. Leak, D.J., Aikens, P.J. and Seyed-Mahmoudian, M. *TibTech* 1992, **10:** 256.
27. Besse, P. and Veschambre, H., *Tetrahedron* 1994, **50:** 8885.
28. May, S.W. *Enz. Microb. Technol.* 1979, **1:** 15.
29. De Bont, J.A.M. *Tet. Asymm.* 1993, **4:** 1331.
30. Weijers, C.A.G.M., van Ginkel, C.G. and de Bont, J.A.M. *Enz. Microb. Technol.* 1988, **10:** 214.
31. Katopodis, A.G., Wimalasena, K., Lee, J. and May, S.W. *J. Am. Chem. Soc.* 1984, **106:** 7928.
32. De Smet, M.-J., Kingma, J., Wynberg, H. and Witholt, B. *Enz. Microb. Technol.* 1983, **5:** 352.
33. De Smet, M.-J., Witholt, B. and Wynberg, H. *J. Org. Chem.* 1981, **46:** 3128.
34. Willetts, A. *TibTech* 1997, **15:** 55.
35. Taschner, M.J., Peddada, P.C., Chen, Q.Z. and Black, D.J. in *Microbial Reagents in Organic Synthesis* (ed.) Servi, S. Kluwer, Dordrecht, 1992, p. 347.
36. Roberts, S.M. and Wan, P.W.H. *J. Mol. Catal. B: Enzymatic*, in press.

37. Donoghue, N.A. and Trudgill, P.W. *Eur. J. Biochem.* 1975, **60:** 1; Donoghue, N.A., Norris, D.B. and Trudgill, P.W. *Eur. J. Biochem.* 1976, **63:** 175.
38. Furstoss, R. and Alphand, V. *J. Org. Chem.* 1992, **57:** 1306.
39. Chen, Y.C.J., Peoples, O.P. and Walsh, C.T. *J. Bacteriol.* 1988, **170:** 781.
40. Secundo, F., Carrea, G., Riva, S., Battistel, E. and Bianchi, D. *Biotechnol. Lett.* 1993, **15:** 865; Pasta, P., Carrea, G., Gaggero, N., Grogan, G. and Willetts, A. *Biotechnol. Lett.* 1996, **18:** 1123.
41. Alphand, V., Archelas, A. and Furstoss, R. *J. Org. Chem.* 1990, **55:** 347 and *Biocatalysis* 1990, **3:** 73.
42. Taschner, M.J. and Black, D.J. *J. Am. Chem. Soc.* 1988, **110:** 6892.
43. Carrea, G., Redigolo, B., Riva, S., Colonna, S., Gaggero, N., Battistel, E. and Bianchi, D. *Tet. Asymm.* 1992, **3:** 1063; Colonna, S., Gaggero, N., Pasta, P. and Ottolina, G. *J. Chem. Soc., Chem. Commun.* 1996, 2303; Kelly, D.R., Knowles, C.J., Mahdi, J.G., Taylor, I.N. and Wright, M.A. *Tet. Asymm.* 1996, **6:** 365.
44. Khün, H., Schewe, T. and Rapoport, S.M. *Adv. Enzymol.* 1986, **58:** 273.
45. Gardner, H.W. *J. Am. Oil Chem. Soc.* 1996, **73:** 1347.
46. Whitehead, I.M., Muller, B.L. and Dean, C. *Cereal Foods World* 1995, **40:** 193; Taylor, A.J. and Mottram, D.S. *Flavour Science.* Royal Society of Chemistry, Cambridge, UK, 1996.
47. Drauz, K. and Waldmann, H. *Enzyme Catalysis in Organic Synthesis.* VCH, Weinheim, 1995, Chap. 6.
48. Weenen, H., Boog, R.L.G.M. and Apeldoorn, W. in *Bioflavour 95.* INRA, Paris, 1995, p. 375.
49. Gavagan, J.E., Fager, S.K., Seip, J.E., Payne, M.S., Anton, D.L. and DiCosimo, R. *J. Org. Chem.* 1995, **60:** 3957.
50. Eisenberg, A., Seip, J.E., Gagavan, J.E., Payne, M.S., Anton, D.L. and DiCosimo, R. *J. Mol. Catal. B: Enzymatic* 1997, **2:** 223.
51. Kremer, D.R. and Meiberg, J.B.M. *Carbohydrates in Europe* 1995, **13:** 22.
52. Klibanov, A.M., Alberti, B.N. and Marletta, M.A. *Biochem. Biophys. Res. Comm.* 1982, **108:** 804; Root, J.R., Durrwachter, J.R. and Wong, C.H. *J. Am. Chem. Soc.* 1985, **107:** 2997; Mazur, A.W. in *Enzymes in Carbohydrate Synthesis* (eds.) Bednarski, M.D. and Simon, E.S. ACS, Washington DC, 1991.
53. Burton, S.G. *Catal. Today* 1994, **22:** 459.
54. Kazandjian, R.Z. and Klibanov, A.M. *J. Am. Chem. Soc.* 1985, **107:** 5448.
55. Sun, W-Q., Payne, G.F., Moas, M.S.G.L., Chu, J.H. and Wallace, K.K. *Biotechnol. Progr.* 1992, **8:** 179.

56. Potthast, A., Rosenau, T., Chen, C.L. and Gratzl, J.S. *J. Mol. Catal. A: Chemical* 1996, **108:** 5.
57. Griffin, B.W. in Ref. 17, Vol. 2, Chap. 4.
58. Van Deurzen, M.P.J., Remkes, I.J., van Rantwijk, F. and Sheldon, R.A. *J. Mol. Catal. A: Chemical* 1997, **117:** 329.
59. Colonna, S., Gaggero, N., Casella, L., Carrea, G. and Pasta, P. *Tet. Asymm.* 1992, **3:** 95.
60. Allain, E.J., Hager, L.P., Deng, L. and Jacobsen, E.N. *J. Am. Chem. Soc.* 1993, **115:** 4415.
61. Van Deurzen, M.P.J., van Rantwijk, F. and Sheldon, R.A. *J. Mol. Catal. B: Enzymatic* 1996, **2:** 33.
62. Van Deurzen, M.P.J., Seelbach, K., van Rantwijk, F., Kragl, U. and Sheldon, R.A. *Biocat. Biotrans.* 1997, **15:** 1.
63. Seelbach, K., van Deurzen, M.P.J., van Rantwijk, F., Sheldon, R.A. and Kragl, U. *Biotechnol. Bioeng.* 1997, **55:** 283.
64. Jacobsen, E.N. in *Catalytic Asymmetric Synthesis* (ed.) Ojima, I. VCH, Weinheim, 1993, p. 159.
65. Butler, A. in *Bioinorganic Catalysis* (ed.) Reedijk, J. Marcel Dekker, New York, 1993, p. 425; Messerschmidt, A. and Wever, R. *Proc. Nat. Acad. Sci. USA* 1996, **93:** 392.
66. Dordick, J.S., Marletta, M.A. and Klibanov, A.M. *Biotechnol. Bioeng.* 1987, **13:** 964.
67. Dordick, J.S. *TibTech* 1992, **10:** 187.
68. Sheldon, R.A. (ed.) *Metalloporphyrins in Catalytic Oxidations.* Marcel Dekker, New York, 1994.
69. Montanari, F. and Casella, L. (eds.) *Metalloporphyrin Catalyzed Oxidations.* Kluwer, Dordrecht, 1994.
70. Watanabe, Y., in Ref. 7, Chap 6.
71. Hrycay, E.G., Gustafsson, J.A., Ingelman-Sundberg, M. and Ernster, L. *Biochim. Biophys. Res. Commun.* 1975, **66:** 209.
72. Groves, J.T., Nemo, T.E. and Myers, S. *J. Am. Chem. Soc.* 1979, **101:** 1032.
73. Murahashi, S.-I., Oda, Y. and Naota, T. *J. Am. Chem. Soc.* 1992, **114:** 7913; Nam, W., Kim, H.J., Kim, S.H., Ho, R.Y.N. and Valentine, J.S. *Inorg. Chem.* 1996, **35:** 1045.
74. Ito, M., Fujisawa, K., Kitajima, N. and Moro-oka, Y. in Ref. 7, Chap. 8.
75. Que, L. and Dong, Y. *Acc. Chem. Res.* 1996, **29:** 190.

76. Rosenzweig, A.C., Frederick, C.A., Lippard, S.J. and Nordlund, P. *Nature* 1993, **366:** 537.
77. Liu, K.E., Valentine, A.M., Qiu, D., Edmondson, D.E., Appelman, E.H., Spiro, T.G. and Lippard, S.J. *J. Am. Chem. Soc.* 1995, **117:** 4997.
78. Shu, L., Nesheim, J.C., Kaufmann, K., Munck, E., Lipscomb, J.D. and Que, L. *Science* 1997, **275:** 515.
79. Siegbahn, P.E.M. and Crabtree, R.H. *J. Am. Chem. Soc.* 1997, **119:** 2103.
80. Kim, K. and Lippard, S.J. *J. Am. Chem. Soc.* 1996, **118:** 4914.
81. Kitajima, N., Fukui, H., and Moro-oka, Y. *J. Chem. Soc., Chem. Commun.* 1988, 485; Kitajima, N., Ito, M., Fukui, H. and Moro-oka, Y. *J. Chem. Soc., Chem. Commun.* 1991, 102.
82. Kim, J., Kim, C., Harrison, R.G., Wilkinson, E.C. and Que, L. *J. Mol. Catal. A: Chemical* 1997, **117:** 83 and references cited therein.
83. Arends, I.W.C.E., Ingold, K.U. and Wayner, D.D.M. *J. Am. Chem. Soc.* 1995, **117:** 4710; MacFaul, P.A., Ingold, K.U., Wayner, D.D.M. and Que, L. *J. Am. Chem. Soc.* 1997, **119:** 10594; Arends, I.W.C.E., MacFaul, P.A., Snelgrove, D.W., Wayner, D.D.M. and Ingold, K.U. in *Free Radicals in Biology and Environment* (ed.) Minisci, F. Kluwer, Dordrecht, 1997, p. 79; MacFaul, P.A., Arends, I.W.C.E., Ingold, K.U. and Wayner, D.D.M. *J. Chem. Soc., Perkin Trans. II*, 1997, 135.
84. Barton, D.H.R. and Doller, D. *Acc. Chem. Res.* 1992, **25:** 504; Barton, D.H.R. and Doller, D. *Pure Appl. Chem.* 1991, **63:** 1567; Barton, D.H.R. and Taylor, D.K. *Pure Appl. Chem.* 1996, **68:** 497; Barton, D.H.R. *J. Mol. Catal. A: Chemical* 1997, **117:** 3.
85. Barton, D.H.R. Gastiger, M.J. and Motherwell, W.B. *J. Chem. Soc., Chem. Commun.* 1983, 731.
86. Snelgrove, D.W., MacFaul, P.A., Ingold, K.U. and Wayer, D.D.M. *Tetrahedron Lett.* 1996, **37:** 823.
87. Perkins, M.J. *Chem. Soc. Rev.* 1996, 229.
88. Barton, D.H.R. *Chem. Soc. Rev.* 1996, 237.
89. Kodera, M., Kano, K. and Funabiki, T. in Ref. 7, Chap. 7.
90. Tabushi, I., Nakajima, T. and Seto, K. *Tetrahedron Lett.* 1980, **21:** 2565.
91. Barton, D.H.R., Li, T. and MacKinnon, J. *J. Chem. Soc., Chem. Commun.* 1997, 557.
92. Ellis, P.E. and Lyons, J.E. *Coord. Chem. Rev.* 1990, **105:** 181; Lyons, J.E. and Ellis, P.E. in Ref. 68 and references cited therein.
93. Grinstaff, M.W., Hill, M.G., Labinger, J.A. and Gray, H.B. *Science* 1994, **264:** 1311; Labinger, J.A. *Catal. Lett.* 1994, **26:** 95; Böttcher, A., Birbaum, E.R.,

Day, M.W., Gray, H.B., Grinstaff, M.W. and Labinger, J.A. *J. Mol. Catal. A: Chemical* 1997, **117:** 229.

94. Barf, G.A. and Sheldon, R.A. *J. Mol. Catal. A: Chemical* 1995, **102:** 23.
95. Groves, J.T. and Quinn, R. *J. Am. Chem. Soc.* 1985, **107:** 5790.
96. Goldstein, A.S., Beer, R.H. and Drago, R.S. *J. Am. Chem. Soc.* 1994, **116:** 2424.
97. Scharbert, B., Zeisberger, E. and Paulus, E. *J. Organometal. Chem.* 1995, **493:** 143.
98. Neumann, R. and Dahan, M. *Nature* 1997, **388:** 353; Neumann, R., Khenkin, A.M. and Dahan, M. *Angew. Chem. Int. Ed. Engl.* **34:** 1995, 1587.
99. Coulston, G.W., Bare, S.R., Kung, H., Birkeland, K., Bethke, G.K., Harlow, R., Herron, N. and Lee, P.L. *Science* 1997, **275:** 191.
100. For a recent review see: Arends, I.W.C.E., Sheldon, R.A., Wallau, M. and Schuchardt, U. *Angew. Chem. Int. Ed. Engl.* 1997, **36:** 1144.
101. Williams, R.J.P. *J. Mol. Catal., Review Issue*, 1986, 1.
102. Parton, R.F., DeVos, D. and Jacobs, P.A. in *Zeolites and Microporous Solids. Synthesis, Structure and Reactivity* (ed.) Derouane, E.G. Kluwer, Dordrecht, 1992, p. 555.
103. Balkus, K.J. and Gabrielov, A.G. *J. Incl. Phen. Mol. Recogn. Chem.* 1995, **21:** 159.
104. DeVos, D.E., Thibault-Starzyk, F., Knops-Gerrits, P.P., Parton, R.F. and Jacobs, P.A. *Macromol. Symp.* 1994, **80:** 157.
105. Herron, N., Stucky, G.D. and Tolman, C.A. *J. Chem. Soc., Chem. Commun.* 1986, 1521.
106. Parton, R.F., Uytterhoeven, L. and Jacobs, P.A., *Stud. Surf. Sci. Catal.* 1991, **59:** 395.
107. Parton, R.F., Vankelecom, I.F.J., Casselman, M.J.A., Bezoukhanova, C.P., Uytterhoeven, J.B. and Jacobs, P.A. *Nature* 1994, **370:** 541; Parton, R.F., Vankelecom, I.F.J., Tas, D., Janssen, K.B.M., Knops-Gerrits, P.P. and Jacobs, P.A. *J. Mol. Catal. A: Chemical* 1996, **113:** 283.
108. Balkus, K.J., Eissa, M. and Lavado, R. *Stud. Surf. Sci. Catal.* 1995, **94:** 713.
109. Knops-Gerrits, P.P., L'abbé, M., Leung, W.H., van Bavel, A.M., Langouche, G., Bruynseraede, I. and Jacobs, P.A. *Stud. Surf. Sci. Catal.* 1996, **101:** 811.
110. Lempers, H.E.B. and Sheldon, R.A. *Stud. Surf. Sci. Catal.* 1996, **105:** 1061.
111. Panov, G.I., Kharitonov, A.S. and Sobolev, V.I. *Appl. Catal. A* 1993, **98:** 1 and references cited therein.
112. Uriarte, A.K., Rodkin, M.A., Gross, M.J., Kharitonov, A.S. and Panov, G.I. *Stud. Surf. Sci. Catal.* 1997, **110:** 857.

113. Dubkov, K.A., Sobolev, V.I., Talsi, E.P., Rodkin, M.A., Watkins, N.H., Shteinman, A.A. and Panov, G.I. *J. Mol. Catal. A: Chemical* 1997, **123:** 155; Panov, G.I., Sobolev, V.I., Dubkov, K.A., Parmon, V.N., Ovanesyan, N.S., Shilov, A.E. and Shteinman, A.A. *React. Kinet. Catal. Lett.* 1997, **61:** 251; Arbuznikov, A.V. and Zhidomirov, G.M. *Catal. Lett.* 1996, **40:** 17.
114. Häfele, M., Reitzmann, A., Klemm, E. and Emig, G. *Stud. Surf. Sci. Catal.* 1997, **110:** 847.

Glossary and Abbreviations

(views expressed on these pages are those of Bernard Meunier and not necessarily those of contributors of the book)

Activation of an oxidant: modification of the properties of an oxidant by coordination to a metal center (this coordination leads to modifications of the primary oxidant, for example the generation of a metal-oxo species with a single oxygen atom donor, the leaving group of the oxidant being removed in the activation step).

Aqua (not aquo): according to IUPAC rules, a water molecule coordinated to a metal center in a complex.

Autoxidation: oxidation of an organic substrate with molecular oxygen involving a free radical mechanism. The reaction of $R^{\bullet}$ with O_2 generates alkylperoxyl $ROO^{\bullet}$ and alkyloxyl $RO^{\bullet}$ radicals. This latter radical is able to abstract H-atoms from the organic substrate and contributes to the free radical chain oxidation.

Catalyst: in the context of this book, a metal complex able to increase the rate of a chemical reaction (oxidation catalysts are often difficult to recover since they are themselves submitted to oxidative degradation). In metal-catalyzed oxidations, active species are generated during the catalytic cycle after modification of the catalyst precursor by the primary oxidant.

Charges on metal complexes: the number of charges on a transition metal complex is usually expressed in arabic numerals (n+ or n–) to avoid confusion with the formal oxidation state of the metal expressed in roman numerals (III, IV, V, ...).

Compound I and II (Cpd I and II): Compound I corresponds to the peroxidase intermediate obtained after oxidation of the iron(III) resting state of the enzyme by hydrogen peroxide (for a heme-peroxidase, Cpd I is an iron(IV)-oxo entity with a radical cation on the porphyrin ligand, Cpd I is two redox equivalents above the resting state of the enzyme) and Compound II is the intermediate obtained after the one-electron reduction of Cpd I by a substrate (Cpd II is an iron(IV)-oxo entity, one redox equivalent above the resting state).

Conversion: expressed as the molar percentage of the amount of a substrate consumed in a chemical or enzymatic reaction with respect to the initial quantity of material (substrate) engaged in the reaction. To avoid the over-oxidation of products in metal-catalyzed oxidations, the primary oxidant might be the limiting factor, not the organic substrate (it should be clearly indicated to avoid confusions on data interpretation).

Distal amino acid: see proximal ligand.

Ferryl (perferryl): used for an iron(IV)-oxo species. Perferryl is used for an iron(V)-oxo species (or an iron(IV)-oxo with a radical-cation on a ligand of the iron ion, like in Compound I of peroxidases).

Heme: the iron(II) complex (ferroheme) or the iron(III) complex (ferriheme) of protoporphyrin IX (haem in British English). Hematin is usually extracted from bovin blood and the iron(III) center has a hydroxo axial ligand. Hemin is the corresponding chloroiron(III)porphyrinate complex.

Heterolytic cleavage: fragmentation of a molecule (or a ligand) in two charged entities (cation and anion) without formation of a radical species after the bond cleavage.

Homolytic cleavage: fragmentation of a molecule (or a ligand) in two entities with formation of two radicals after the bond cleavage. A particular case corresponds to the transfer of an electron into the antibonding orbital of a peroxide ROOR to produce an anion RO^- and a radical $RO^\bullet$.

HRP: horseradish peroxidase, a robust plant peroxidase used in many studies.

L: a neutral ligand.

MCPBA: *meta*-chloroperbenzoic acid.

MMPP: magnesium salt of monoperoxyphthalic acid.

Monopersulfate (or monoperoxysulfate): an inorganic peroxide available as potassium derivative of the triple salt $2KHSO_5 \cdot KHSO_4 \cdot K_2SO_4$ (current commercial names: Oxone®, Curox® or Caroat®).

NAD(P)H: reduced form of $NAD(P)^+$, the nicotinamide adenine dinucleotide (phosphate) coenzyme.

Oxene: formally an oxygen atom with six electrons "O" linked to a metal center without modification of the oxidation state of the metal (equivalent to the coordination of a carbene to a low-valent metal complex). A metal-oxene is formally a limit form of a metal-oxo species with an oxygen atom with eight electrons O^{2-} linked to a 2e-oxidized metal center. Physico-chemical data obtained on isolated metalloporphyrins having an oxygen atom ligand indicated that the iron or the manganese centers are at the oxidation state IV or V, not III.

$$\underset{\text{iron(III)-oxene}}{Fe^{III} \rightarrow \overline{\underline{O}}|} \rightarrow \underset{\text{iron(IV)-oxyl}}{Fe^{IV} - \overline{\underline{O}}^{\bullet}} \longleftrightarrow \underset{\text{iron(V)-oxo}}{Fe^{V} = \overline{\underline{O}}}$$

The intermediate mesomeric form with a single metal-oxygen bond and a radical centered on the oxygen atom is often used to explain the capacity of metal-oxo species to abstract H atoms in catalytic hydroxylations. The iron(IV)-oxyl species can be regarded as a triplet state and the iron(V)-oxo complex as a singlet state.

Oxidation number (or oxidation state): the formal oxidation state of a metal center is usually expressed in roman numerals (III, IV, V, ...) to avoid confusion with the indication of the charges on the complex itself.

Oxidative addition: Oxidation of a complex by formation of two new metal-ligand bonds (two electrons of the metal are engaged) accompanied by cleavage of a bond within the substrate reacting with the metal center.

Oxide: The bond in a metal-oxide has a strong ionic character, but a metal-oxo bond is covalent by nature; for example, CaO is calcium oxide, not calcium-oxo, and NaCl is sodium chloride, not chloro-sodium.

Oxo: a metal-oxo species contains an oxygen atom covalently linked to a metal complex with a double bond between the metal and the oxygen atom. The formal oxidation state of the metal is n + 2 with respect to the oxidation state n of the metal before the formation of the $M^{n+2} = O$ bond. Depending on the nature of the metal and the ligands, a metal-oxo can be represented as metal-oxyl entity $M^{n+1} - O^{\bullet}$, the oxidation state of the metal center being only n + 1 in this latter case (see Oxene for a schematic representation of these species).

Oxometallacyclobutane: proposed intermediate in the reaction of a metal-oxo species with an olefin via a 2 + 2 cycloaddition.

Perferryl: see ferryl.

Peroxyl: a metal-peroxyl complex contains a peroxidic ligand with a radical on the terminal oxygen atom, ex. Fe(III)-O-$O^{\bullet}$. The addition of an electron will give rise to a metal-peroxo anion, ex. Fe(III)-O-O^{-}, and the protonation of this latter species will provide a metal-hydroperoxo, ex. Fe(III)-O-OH.

PhIO: iodosylbenzene (preferred to iodosobenzene by Chemical Abstracts).

Por (or Porph): abbreviation for a porphyrin ligand.

Prosthetic group: The active site of an enzyme which is usually linked to the apoprotein (the enzymatically inactive protein without the cofactor essential for the catalytic activity). For example, heme is the prosthetic group of cytochrome P450.

Proximal ligand: the amino acid residue directly coordinated on the metal center of a heme (proximal ligands of hemoglobin and cytochrome P450 are a histidine and a cysteine, respectively). Distal amino acids are located on the other side of the heme without direct links with the metal center, but they are involved, for example, in the heterolytic cleavage of hydrogen peroxide by peroxidases or the stabilization of dioxygen coordination in oxyhemoglobin.

Radical initiator: an organic derivative like AIBN (azobisisobutyronitrile) or a transition metal salt able to generate free organic radicals in solution.

Reductive elimination: the reverse of an oxidative addition. The cleavage of two metal-ligand bonds produces a 2e-reduction of the metal center.

Salen: *N,N*-bis(salicylidene)ethylenediamino (a Schiff base ligand).

Selectivity: measured by product distribution in a chemical or enzymatic reaction. Selectivity is expressed in percentage of the molar amount of a particular product divided by the molar amount of the substrate consumed.

Single oxygen atom donor: a chemical oxidant able to transfer one oxygen atom to a metal center to generate a metal-oxo species (this latter activated complex is then able to transfer the O atom to an organic substrate).

Soret: the Soret band is the most intense absorption band in the visible spectra of porphyrin derivatives (corresponding mainly to a $\pi - \pi^*$ transition with very high ε values). This intense band was initially observed by Soret for hemoglobin (J. L. Soret, Compt. Rend. 1883, **97:** 1267).

Tetrapyrrole nomenclature: for recommendations on tetrapyrrole nomenclature, see *Pure Appl. Chem.* 1987, **59:** 779–832.

TPP: *meso*-tetraphenylporphyrinato ligand.

Turnover number: total number of catalytic cycles obtained with the same batch of catalyst ("life" of the catalyst expressed as the total number of substrate molecules transformed by one catalyst molecule).

Turnover rate: number of catalytic cycles (or "rotations") obtained by time unit (it must be indicated if the turnover rate is based on the initial rate, the first 10% of conversion, or on the half-conversion).

X: an anionic ligand (usually a halide).

Yield: The molar amount of a particular product divided by the molar amount of engaged organic substrate and expressed as a percentage. The yield is equal to the conversion multiplied by the selectivity (confusion between selectivity and yield is still too frequent in the oxidation literature). To emphasize results in catalytic oxidations, yields are often expressed with respect to the oxidant and not to the substrate used in large excess with respect to the oxidant. In the case of a ten-fold excess of substrate/oxidant, a 90% yield versus oxidant is only a 9% yield versus substrate. This is correct and may be useful for mechanistic studies, but the real challenge in catalytic oxidations is to obtain the highest possible selective transformation of all substrate molecules with the minimum of oxidant molecules required by the stoichiometry of the reaction. A "yield based on conversion" is in fact a selectivity.

Index